ENCYCLOPEDIA OF TOURISM

ENCYCLOPEDIA OF TOURISM

Chief editor
Jafar Jafari

LONDON AND NEW YORK

First published 2000
by Routledge
2 Park Square, Milton Park, Abingdon, Oxon, OX14 4RN

Simultaneously published in the USA and Canada
by Routledge
270 Madison Ave, New York NY 10016

Routledge World Reference paperback edition first published 2003

Routledge is an imprint of the Taylor & Francis Group

Transferred to Digital Printing 2005

Typeset in Sabon by Taylor & Francis Ltd

British Library Cataloguing in Publication Data
A catalogue record for this book is available from the British Library

Library of Congress Cataloging in Publication Data
Encyclopedia of tourism/ edited by Jafar Jafari
p.cm
Includes bibliographical references
ISBN 0–415–15405–7 (alk. paper)
1. Tourism–Encyclopedias. I. Jafari, Jafar/

G155 .A1 E4295 2000
338.4'791–dc21 99–052878

ISBN 0–415–15405–7 (hbk)
ISBN 0–415–30890–9 (pbk)

Contents

Editorial team vi
List of Contributors vii
Introduction xvii
Acknowledgements xxiv
Entry list xxv

Entries A–Z 1

Index 637

Editorial team

List of contributors

Ted Abernethy
St Thomas University, USA

Wesal Abu-Alam
Helwan University, Egypt

Eugeni Aguilo
Universitat de les Illes Balears, Spain

Donald Anderson
University of Calgary, Canada

Hazel Andrews
UK

Antonios Andronikou
Cyprus Tourism Organisation, Cyprus

Yorghos Apostolopoulos
Arizona State University, USA

Julio R. Aramberri
Drexel University, USA

Brian Archer
University of Surrey, UK

Simon Archer
University of Surrey, UK

G.J. Ashworth
University of Groningen, The Netherlands

Joni E. Baker
Texas A&M University, USA

Esteban Bardolet
Universitat de les Illes Balears, Spain

Rene Baretje
Centre International de Recherches et d'Etudes Touristiques, France

Lynn A. Barnett
University of Illinois, USA

Clayton W. Barrows
University of Guelph, Canada

Joachim Barth
University of Guelph, Canada

Francesc J. Batle Larente
Universitat de les Illes Balears, Spain

Tom Baum
University of Strathclyde, UK

Jay Beaman
Auctor Consulting Associates, Canada

Lionel Bécherel
International Tourism Consultancy, UK

Francois Bedard
University of Quebec, Canada

Inmaculada Benito
Universitat de les Illes Balears, Spain

Michael A. Blazey
California State University, Long Beach, USA

Kevin Boberg
New Mexico State University, USA

Jeremy Boissevain
University of Amsterdam, The Netherlands

Nev Brah
Quality Assurance Services, Australia

Deborah Breiter
New Mexico State University, USA

Robert Brotherton
Manchester Metropolitan University, UK

Frances Brown
UK

Tom Broxon
Linn-Benton Community College, USA

Edward M. Bruner
University of Illinois, Urbana-Champaign, USA

Dimitrios Buhalis
University of Westminster, UK

Robyn Bushell
University of Western Sydney, Australia

Richard Butler
University of Surrey, UK

Roger Calantone
Michigan State University, USA

Roger J. Callan
Manchester Metropolitan University, UK

Stephen Calver
Bournemouth University, UK

Barbara A. Carmichael
Wilfrid Laurier University, Canada

Mary Cawley
University College, Galway, Ireland

Georges Cazes
University of Paris I Panthéon-Sorbonne, France

Yeong-Shgang Chen
Taipei, Taiwan

Jerald W. Chesser
University of Central Florida, USA

Gary Chick
Pennsylvania State University, USA

Eliza Ching-Yick Tse
The Chinese University of Hong Kong, China

Bae-Haeng Cho
University of New England, Australia

K.S. (Kaye) Chon
University of Houston, USA

Adrian Clark
University of Westminster, UK

Christine J. Clements
University of Wisconsin-Stout, USA

Erik Cohen
Hebrew University, Israel

Fredrick M. Collison
University of Hawaii, USA

Jean Colvin
University of California, Berkeley, USA

Chris Cooper
University of Queensland, Australia

Malcolm Cooper
University of Southern Queensland, Australia

Jorge Costa
Fernando Pessoa University, Portugal

Stephen J. Craig-Smith
University of Queensland, Australia

John C. Crotts
College of Charleston, USA

Geoffrey I. Crouch
LaTrobe University, Australia

Daniel K. Crowley
Port of Spain, Trinidad and Tobago

Russell R. Currie
Lakehead University, Canada

Louis J. D'Amore
International Institute for Peace, Canada

Peter D'Souza
University of Wisconsin-Stout, USA

Graham M.S. Dann
University of Luton, UK

Carolyn M. Daugherty
Northern Arizona University, USA

Rob Davidson
University of Westminster, UK

Charles R. de Burlo
University of Vermont, USA

Alain Decrop
University of Namur, Belgium

Keith Dewar
Massey University, New Zealand

Kadir Hajj Din
Universiti Ultara Malaysia, Malaysia

Peiyi Ding
University of New England, Australia

Ngaire Douglas
Southern Cross University, Australia

Ross K. Dowling
Edith Cowan University, Australia

John Eade
Univesity of Surrey Roehampton, UK

William R. Eadington
University of Nevada, Reno, USA

Gavin Eccles
University of Surrey, UK

Charlotte M. Echtner
James Cook University, Australia

David L. Edgell, Snr.
International and Ethnic Marketing Service, USA

Taylor Ellis
University of Central Florida, USA

Kisangani N.F. Emizet
Kansas State University, USA

Cynthia Enloe
Clark University, USA

Carlos Ernesto
Secretaría de Tourismo, Argentina

Margaret Erstad
UK

Nigel G. Evans
University of Northumbria at Newcastle, UK

Gordon Ewing
McGill University, Canada

Michael Fagence
University of Queensland, Australia

Tracy Farrell
Virginia Tech, USA

Eduardo Fayos-Solà
World Tourism Organization, Spain

Daniel R. Fesenmaier
University of Illinois, Urbana-Champaign, USA

D. Manuel Figuerola Polomo
Escuelo Official de Turismo, Spain

John Fletcher
Bournemouth University, UK

Douglas C. Frechtling
George Washington University, USA

Bernard Fried
University of Nevada, Las Vegas, USA

Martin Friel
University of Buckingham, UK

Maria Fuenmayor
Texas A&M University, USA

Jan Armstrong Gamradt
University of New Mexico, USA

William C. Gartner
University of Minnesota, USA

Donald Getz
University of Calgary, Canada

Irene Gil Saura
Universitad de Valencia, Spain

Sharon S. Giroux
University of Wisconsin-Stout, USA

Juergen Gnoth
University of Otago, New Zealand

Frank M. Go
Erasmus University Rotterdam, The Netherlands

Charles R. Goeldner
University of Colorado, USA

Ramon Cèsar Gòmez Viveros
Servicio Nacional de Turismo, Chile

Martina Gonzales Gallarza
Universidad Politecnica de Valencia, Spain

Steven Goss-Turner
University of Brighton, UK

Peter Grabowski
University of Luton, UK

Nelson H.H. Graburn
University of California, Berkeley, USA

Anne Graham
University of Westminster, UK

Marcus Grant
Environmental Stewardship, UK

Deborah Grieve
University of Westminster, UK

Zheng Gu
University of Nevada, Las Vegas, USA

Clare A. Gunn
Texas A&M University, USA

Carlos Ernesto Gutierrez
Secretario de Turismo, Argentina

C. Michael Hall
University of Otago, New Zealand

Monica Hanefors
Göteborg University, Sweden

Dennis Hardy
Middlesex University, UK

David Harrison
University of North London, UK

Duncan Hartshorne
University of South Australia, Australia

Atsuko Hashimoto
Brock University, Canada

Souad Hassoun
Institut Supérieur International du Tourisme, Morocco

Jan Vidar Haukeland
Institute of Transport Economics, Norway

Hana Havlová
Czech Tourist Authority, Czech Republic

Donald E. Hawkins
George Washington University, USA

Michael Haywood
University of Guelph, Canada

Ernie Heath
University of Pretoria, South Africa

Frederick M. Helleiner
Trent University, Canada

Thriné Hely
University of Brighton, UK

Richard Hill
Fachhochschule Heilbronn, Germany

Tomislav Hitrec
Institute of Tourism, Croatia

Anne-Mette Hjalager
Aarhus School of Business, Denmark

David Hogg
David Hogg Ltd, Australia

Peter Holden
Ecumenical Coalition on Third World Tourism, Thailand

Keith Hollinshead
University of Luton, UK

J. Christopher Holloway
University of the West of England, UK

Willem J. Homan
Western Michigan University, USA

William Cannon Hunter
Texas A&M University, USA

Orhan Icoz
Dokuz Eylul University, Turkey

Hadyn Ingram
University of Surrey, UK

Joe Inguanez
University of Malta, Malta

Antoni Jackowski
Jagellonian University, Poland

Jens Kristian Steen Jacobsen
Institute of Transport Economics, Norway

Jafar Jafari
University of Wisconsin-Stout, USA

Csilla Jandala
Hungarian-Austrian Consultancy, Hungary

Myriam Jansen-Verbeke
Catholic University of Leuven, Belgium

Jiann-Min Jeng
University of Illinois, Urbana-Champaign, USA

John M. Jenkins
University of Newcastle, Australia

Jens Friis Jensen
Turisnes Uddannelsessekretariat, Denmark

Nick Johns
Norwich City College, UK

Keith Johnson
University of Huddersfield, UK

Barbara Rose Johnston
Center for Political Ecology, USA

Charles S. Johnston
University of Aukland, New Zealand

Peter Jones
University of Surrey, UK

Tom Jones
University of Nevada, Las Vegas, USA

Kathryn A. Kamp
Grinnell College, USA

Michael L. Kasavana
Michigan State University, USA

Hanan Kattara
University of Alexandria, Egypt

Peter Keller
Université de Lausanne, Switzerland

Mahmoud Khan
Virginia Polytechnic Institute, USA

Maryam Khan
Virginia Polytechnic Institute, USA

Y.J. Edward Kim
James Cook University, Australia

Kyung-Hwan Kim
Kyonggi University, Korea

Brian King
Victoria University of Technology, Australia

Vivian Kinnaird
University of Sunderland, UK

David Kirk
Queen Margaret University College, UK

Auvo A. Kostiainen
University of Tampere, Finland

Tan Kovic
University of Zagreb, Croatia

Lee M. Kreul
Purdue University, USA

Joseph Kurtzman
Sports Tourism International Council, Canada

Helmut Kurz
Wirtschaftsuniversität Wien, Austria

Stephen M. Lebruto
University of Central Florida, USA

Bang Sik Lee
Kyongju University, Korea

Bong-Koo Lee
Texas A&M University, USA

Choong Ki Lee
Dongguk University, Korea

Christine L.H. Lee
Monash University, Australia

Neil Leiper
Southern Cross University, Australia

David Leslie
Glasgow Caledonian University, UK

Alan A. Lew
Northern Arizona University, USA

Sarah Li
Murdoch University, Australia

Zhang Liansheng
Oficina Nacional de Turismo de China en Espana, Spain

Giuli Liebman Parrinello
Rome, Italy

Mary A. Littrel
Iowa State University, USA

Andrew Lockwood
University of Surrey, UK

Veronica Long
University of Minnesota, USA

Linda Low
National University of Singapore, Singapore

Deborah Luhrman
World Tourism Organisation, Spain

Maria Luz Rufilanchas
Secretaria de Estado de Commercio, Spain

Dean MacCannell
University of California, Davis, USA

Gunther Maier
University of Economics, Austria

Robert Maitland
University of Westminster, UK

Roger C. Mannell
University of Waterloo, Canada

Ajay K. Manrai
University of Delaware, USA

Lalita A. Manrai
University of Delaware, USA

Yoel Mansfeld
University of Haifa, Israel

Peter Mason
Massey University, New Zealand

Harry G. Matthews
University of South Carolina, USA

Josef A. Mazanec
Vienna University of Economics and Business Administration, Austria

Ian McDonnell
University of Technology, Sydney, Australia

Eric McGuckin
Portland OR, USA

Philip McGuirk
University of Wisconsin-Stout, USA

Robert W. McIntosh
Michigan State University, USA

Bob McKercher
Hong Kong Polytechnic University, China

Gina K. McLellan
Clemson University, USA

Richard G. McNeill
Northern Arizona University, USA

Scott M. Meis
Canadian Tourism Commission, Canada

David Mercer
Monash University, Australia

Lluis Mesalles
International Hotel Consultant, Spain

Klaus J. Meyer-Arendt
University of West Florida, USA

Victor T.C. Middleton
University of Surrey, UK

Tanja Mihalič
University of Ljubljana, Slovenia

Simon Milne
Auckland University of Technology, New Zealand

Lisle S. Mitchell
University of South Carplina, USA

Sulong Mohamad
Universiti Kebangsaan Malaysia, Malaysia

Janet Henshall Momsen
University of California, Davis, USA

Philip E. Mondor
Canadian Tourism Human Resource Council, Canada

Roland S. Moore
Prevention Research Center, USA

Josephine M. Moreno
University of Rhode Island, USA

Alison Morrison
University of Strathclyde, UK

Alastair M. Morrison
Purdue University, USA

Gianna Moscardo
James Cook University, Australia

Luiz Moutinho
University of Glasgow, UK

Hansruedi Müller
Universität Bern, Switzerland

Joseph Mugler
Institute of Small Business, Austria

Laurie Murphy
James Cook University, Australia

Bvsan Murthy
Nanyang Business School, Singapore

Ahmed Naim
Morocco

Dennison Nash
University of Connecticut, USA

Mary Lee Nolan
Oregon State University, USA

Michael Nowlis
International Hotel and Restaurant Association, France

Wiendu Nuryanti
Gadjah Mata University, Indonesia

Lars Nyberg
Mid Sweden University, Sweden

Robert M. O'Halloran
University of Denver, USA

Ainsley O'Reilly
University of the West Indies, Bahamas

Tim Oakes
University of Colorado at Boulder, USA

Mark J. Okrant
Plymouth State College, USA

Michael D. Olsen
University of Central Florida, USA

Ong Lei Tin
Temasek Polytechnic, Singapore

Martin Oppermann
Griffith University, Australia

John Ozment
University of Arkansas, USA

Marja Paajanen
Helsinki School of Economics, Finland

Stephen Page
Massey University, New Zealand

Latiffah Pawanteh
Universiti Kebangsaan Malaysia, Malaysia

Robert J. Payne
Lakehead University, Canada

Douglas G. Pearce
Victoria University of Wellingtton, New Zealand

Philip L. Pearce
James Cook University, Australia

Harald Pechlaner
University of Innsbruck, Austria

Aurora Pedro
University of Valencia, Spain

Arvo Peltonen
University of Helsinki, Finland

Tania Penalosa
Virginia Polytechnic Institute, USA

Lesley Pender
University of Newcastle, UK

Serge Perrot
New World Association for Hotel and Tourism Training, France

Ying Yang Petersen
Newport Beach CA, USA

Carl Pfaffenberg
University of Tennessee, USA

Jim Pickworth
University of Guelph, Canada

John J. Pigram
University of New England, Australia

Bengt Pihlström
Ministry of Trade and Industry, Finland

Ray Pine
Hong Kong Polytechnic University, China

Abraham Pizam
University of Central Florida, USA

Robert A. Poirier
Northern Arizona University, USA

Robert B. Potter
University of London, UK

Krzysztof Przeclawski
Warsaw University, Poland

Leslie M. Reid
Texas A&M University, USA

Laurel J. Reid
Brock University, Canada

Linda K. Richter
Kansas State University, USA

Antoni Riera
Universitat de les Illes Balears, Spain

German Rijalba
Centro de Formacion en Turismo, Peru

Gisbert Rinschede
Universität Regensburg, Germany

J.R. Brent Ritchie
University of Calgary, Canada

Robin J.B. Ritchie
University of British Columbia, Canada

Mike Robinson
University of Northumbria at Newcastle, UK

Marco Antonio Robledo
Universitat de les Illes Balears, Spain

Rafael Rodriguez
Universidad Simon Bolivar, Venezuela

Mariano Rojas
Universidad de las Americas – Puebla, Mexico

Angela Roper
Oxford Brookes University, UK

Glenn F. Ross
James Cook University, Australia

Maria Luz Rufilanchas
Secretaria de Estado de Comercio, Spain

Chris Ryan
University of Waikato, New Zealand

Amparo Sancho-Perez
University of Valencia, Spain

Francisco Sastre
Universitat de les Illes Balears, Spain

Antonio Sastre Alberti
Spain

Raymond S. Schmidgall
Michigan State University, USA

Marcus Schmidt
University of Southern Denmark, Denmark

Zvi Schwartz
Ben Gurion University of Negev, Israel

A.V. Seaton
University of Luton, UK

Tom Selanniemi
Finnish University Network for Tourism Studies, Finland

Abdal-Rahman Selim
Egyptian National Federation of Tourist Chambers, Egypt

Tom Selwyn
University of North London, UK

Antoni Serra
Universitat de les Illes Balears, Spain

Alberto Sessa
Scuola Internazionale di Scienze Turistiche, Italy

Richard Sharpley
University of Northumbria at Newcastle, UK

Robin Shaw
Victoria University, Australia

Pauline J. Sheldon
University of Hawaii, USA

Natela Shengela
State Committee for Sport and Tourism, Russia

Anthony G. Sheppard
California State University, Sacramento, USA

Patti J. Shock
University of Nevada Las Vegas, USA

David G.T. Short
Lincoln University, New Zealand

David G. Simmons
Lincoln University, New Zealand

Shalini Singh
Centre for Tourism Research and Development, India

Tej Vir Singh
Centre for Tourism Research and Development, India

M.K. Sio
Kenya Utalii College, Kenya

Egon Smeral
Osterreichisches Institut fur Wirtschaftsforschung, Austria

Ginger Smith
George Washington University, USA

Russell Arthur Smith
Nanyang Technological University, Singapore

Stephen Smith
University of Waterloo, Canada

Valene L. Smith
California State University, Chico, USA

Trevor H.B. Sofield
Murdoch University, Australia

Heidi H. Sung
New York University, USA

Margaret B. Swain
University of California, Davis, USA

Semso Tankovic
University of Zagreb, Croatia

Richard Teare
Bournemouth Polytechnic, UK

David J. Telfer
Brock University, Canada

Dallen J. Timothy
Arizona State University, USA

John Towner
Newcastle upon Tyne Polytechnic, UK

Duncan Tyler
South Bank University, UK

Randall Upchurch
University of Central Florida, USA

John Urry
Lancaster University, UK

Muzaffer S. Uysal
Virginia Polytechnic Institute, USA

Hubert B. Van Hoof
Northern Arizona University, USA

Turgut Var
Texas A&M University, USA

Ursula A. Vavrik
Paris, France

François Vellas
Université des Sciences Sociales, France

David B. Vellenga
Calvin College, USA

Christine Vogt
Arizona State University, USA

Boris Vukonić
University of Zagreb, Croatia

Jeffrey M. Wachtel
Florida International University, USA

Salah E.A. Wahab
TourismPlan, Egypt

John R. Walker
United States International University, USA

Geoffrey Wall
University of Waterloo, Canada

Alf H. Walle
Mercyhurst College, USA

Ning Wang
Zhongshan University, China

Stephen R.C. Wanhill
Bournemouth University, UK

William L. Waugh, Jr.
Georgia State University, USA

Kathryn Webster
University of Brighton, UK

Betty Weiler
RMIT, Australia

Klaus Weiermaier
Institut für Tourismus und Dienstleistungs-wirtschaft, Austria

Hannes Werthner
University of Vienna, Austria

John Westlake
Bournemouth University, UK

Paul F. Wilkinson
York University, Canada

David Wilson
Queen's University of Belfast, UK

Stephen F. Witt
University of Surrey, UK, and Victoria University, Australia

Ineke Witzel
International Training, The Netherlands

Karl Wöber
Wirtschaftsuniversität Wien, Austria

John A.J. Wolfenden
University of New England, Australia

Robert A.G. Wong
Georgian College of Applied Arts and Technology, Canada

Robert H. Woods
Michigan State University, USA

Arch G. Woodside
Boston College, USA

Yu Ming Xi
China

Tetsuro Yamashita
WTO Regional Support Officer for Asia & the Pacific, Japan

Jordan Yee
Sageware, USA

Isabel Zaragosa
Mexico

Hamid Zargham
Allameh Tabataba'i University, Iran

John Zauhar
Sports Tourism International Council, Canada

Andreas Zins
Wirtschaftsuniversität Wien, Austria

Introduction

The landmarking of tourism

The purpose of this encyclopedia is to act as a quick reference source or guide to the wide range of basic definitions, concepts, themes, issues, methods, perspectives and institutions embraced by tourism in its disparate manifestations. The volume is designed to represent its multidisciplinary scope, nature, composition, and transformation; to facilitate research, education, consultancy and practical inquiries; and/or to act as a starting point for both wide-ranging and specialised explorations in this ubiquitous phenomenon, which is now the world's largest industry. This reference book, like those in its genre, as will be noted shortly, has succeeded in meeting some of its goals but remains unfulfilled in others.

The original aim at fulfilment, the beginning, is one of the most problematic aspects of any new project, especially an unprecedented one. So it was with the development and production of this *Encyclopedia of Tourism*, as it started its journey into the landscape of knowledge, with no models to choose from and no blue prints to follow: travelling uncharted terrains, crossing unbridged passes, falling into unmarked canyons, finding unattended niches, and even discovering virgin lands and meeting other fellow backpackers and explorers. Almost every move – situated between the starting point of this searching journey and its finishing line – suggests a long tale, to be merely summarised here. But first should come the exploratory context which gave rise to this academic expedition.

The scientification of tourism

As a field of study evolves toward maturity, new informed measures, which denote successive progression and signal desired transition, are introduced and pursued: towards promising goals which potentially advance the body of knowledge to new frontiers. An overview of such cumulative strides in tourism would illustrate this unfolding scientification journey in general and its coming of age. To this point, mention of many developments would fit the occasion. But only three are cited here, and all are among those with which this writer is most familiar, from their origins onward.

The scientification process, for the purpose of this introduction, may begin with the 1960s, when tourism eventually graduated from its economically-driven phase (concerned solely with real or perceived monetary returns). With this transformation, the time had come to create new pathways toward the inclusive picture of this knowledge area. It seemed reasonable that, as the vista broadened, the move would help uncover its hidden dimensions, especially its sociocultural structure which had remained buried and thus least understood earlier on. At about this juncture, a new journal saw its debut, with no commitments to the tourism industry as such and to the already-surfaced and well-polished economic pillars. The first issue of *Annals of Tourism Research: A Social Sciences Journal* appeared in 1973. With its take-off, another augmenting academic gateway was burst open, a cornerstone was installed, and suddenly a small yet directed journey into the far-stretching landscape was on its way.

To reach its scholarly purpose – of being 'ultimately dedicated to developing theoretical constructs . . . [in order] . . . to expand frontiers of knowledge in tourism' – *Annals* continued with a long and challenging expedition, moving along its mission-driven course. Since the early days of this social science journal, its multidisciplinary editors have been guiding digs and discoveries of an army of authors, transforming the initial narrow pathway into a paved highway for the long-haul journey to the heartland. Together, editors and authors have already mapped and landmarked the territory along their direct and indirect exploratory routes and niches. This team, many times stronger today, has continued its persistent digging for the main structures and functions of tourism: by experimenting with various tools, by excavating neglected sociocultural depths, by unearthing unexpected treasures, by posing new questions, and by putting together their finds to explain what constitutes this phenomenon. *Annals* people (whose twenty-five years of finds are 'inventoried' in the *Silver Anniversary Supplement* of their journal), along with the finds of other searching and digging groups, now spread across multiple plateaux, have already met some initial goals, with many touchstones in place and landmarks installed for succeeding expeditions. Significantly, all along the widespread exploration, teams have been inspired by the vastness of open fields and the refreshingly holistic insights this ever-searching quest or 'pilgrimage' inspires. These efforts, which saw their beginning mostly in the 1970s (often without any co-ordination among them) helped in slowly but surely emancipating tourism from its earlier unidirectional economic/marketing bounds, in favour of a bird's-eye survey of what lies around.

Thus, *Annals* as a single, planned departure in this direction, alongside many other efforts initiated about the same time, led and paved alternative pathways, to build and occupy new niches and to make advancements on many fronts. These now-expanded and fertile grounds soon gave birth to, among other things, a score of thematically diverse academic journals and a larger number of knowledge-based books, with most diverging from the earlier pioneers. These publications, mostly based on fieldwork, were further reinforced and especially legitimised as more universities seriously committed to education and research in this field. These collective acts of research and education, of discovery and conquest, of pushing the boundaries, of mapping and landmarking the whole landscape, were at all times guided *a fortiori* by challenging questions nurtured with the multidisciplinary carry-on baggage of tools and concepts. The population of trained guides, researchers and educators, alert to the surrounding conditions and movements, indubitably signalled that the scientification transformation was indeed on the march in many directions.

Continuation of these purposeful explorations, characterised by a sense of mission, perseverance and endurance, have encountered many high-points, inevitably leaving their own marks and signs pointing toward higher goals. Celebration of new grounds covered and summits reached came in 1988, with the founding of a multidisciplinary community of scholars in tourism research under the flagship of the International Academy for the Study of Tourism. Obviously no informed action is born or no ship embarks without a gestation period of design and building of the vessel; in this case, years of consultation among like-minded individuals to prepare for the Academy, contrasted with years of solitary contemplation for the *Annals*.

A third scholarly example that marks this evolving scientification – of forming and expanding the knowledge base, crossing already established frontiers, building infrastructural bridges for reaching new plateaux, and more – is the birth of this encyclopedia. Some of the developmental aspects of the encyclopedia should be *a propos* for this introduction. Its brief story – including exploratory processes, with fixed itineraries, detours and short-cuts, for its own landing – is again a part of the travelogue of the longer journey, offering a fitting commentary on the formation of knowledge in this field.

The encyclopedia project

Progress in research and scholarship should not be seen in isolation. Attempts in and by various fields of investigation bring to light many similarities and differences, operating as additional lenses to view and interpret what has been achieved and what lies still ahead. The search for parallels, especially in established disciplines, compellingly clarifies that each field has its own beginnings, with ideas shaped and nurtured, definitions formed and challenged, signposts marked and shifted, boundaries formed and redrawn, legitimatisation attempted and reasoned, unique issues identified and analysed, theories invented and contested, methodologies copied and adjusted, applications made and sharpened. These efforts, whether performed individually or collectively, all contribute paving bricks (some half-baked) which shape the contour of purposeful journeys (or 'pilgrimages' to many) for knowledge (or the 'truth'), leading the exploration beyond the visible peaks and valleys, to destinations mostly hidden from view.

In the case of tourism, during its short course of research and scholarship, the journey has occasionally reached new heights affording panoramic views of the expanding landscape and inviting new thoughts and perspectives. On such rare occasions, the march may be halted momentarily to write 'memoirs' or travelogues. For this discussion, two types of such expositions come to mind. One, a register or travelogue of the whole journey, recalls what preceded the beginning, takes stock of what has been crossed, claims discoveries made, names peaks and valleys passed, lists new lands mapped, and even notes sites bypassed and opportunities lost. Such an annals of knowledge stored and merits acquired can refine the hindsight into a history of the field, first broadly brushed then gradually detailed, to show the patterns and conditions shaping the flow, from one development to the next, from one phase to its successor.

Significantly from this overlooking vantage, during this special recollection moment, the windshield view can be synergized with the rear-view for the production of a different 'memoir' of the journey. This second type gathers into its covers extracts (or snapshots) from the whole expedition, by featuring an inventory of individual bricks and mortars, tools and methods, shapes and patterns, occupied niches and unfilled cavities. Such a registry of the known and tentative, practices and philosophies, can result in a handy assortment of brief texts which would constitute the contents of an *encyclopedia*: intended to enable and guide the rest of the expedition, to influence research itineraries, to suggest goals and paths, to indicate challenges and opportunities, and more. While the first 'memoir' – the history of this scientification (including its main events and players) – is yet to be fully researched and written, the encyclopedia suggested itself as an urgently needed travel companion, as a carry-on baggage on this scholarly journey. With this outlook, from this heightened plateau, the ambition to publish this reference book was conceived, gradually gestated, and slowly brought to fruition.

Seeing a need for and making a commitment to this reference source was easy from that high altitude, but its practical development in the ordinary plains below proved quite otherwise. The real challenge of producing this publication began immediately after the publishing contract was signed. Initial questions were many, including what actually constitutes a tourism encyclopedia, what should be included or excluded and why, how long each entry should be, who should write the selected entries, what format would work best, and much more. To comment on one of these concerns, for example, even if a list of 'headwords' (for eventual entries) could be magically produced, there still must be coherent procedures to generate, co-ordinate, evaluate and discipline some hundreds of contributions for inclusion. Despite these ambiguities, or in the absence of a tested plan or a blueprint in this field, this first traditional encyclopedia devoted to tourism seemed almost of itself to be determined to begin.

The initial decision, recorded in the 1995 project proposal submitted to the publisher, was that the volume be international in scope and multidisciplinary in nature and treatment, thus offering the best road map to those landscape topographies native to tourism. In many respects, *Annals of Tourism Research* served to model the development of the project at hand. From personal knowledge of expertise in the field and outside insights sought, a multidisciplinary editorial team was formed in early 1996. Their disciplinary representations, in alphabetic order, included anthropology, economics, education, environment, geography, history, hospitality, leisure, management, marketing, political science, psychology, recreation, religion, sociology and transportation. Additional fields could indeed have been added, but to expedite the project and keep it timely, this team, with the rich range of expertise among them, was deemed a representative compromise. In addition to disciplinary editors, a smaller number were appointed to oversee the development of entries relating to major tourism generating/

receiving countries, important associations and selected academic journals. The intent for the whole team was to methodically enlarge the coverage into a more comprehensive reference book, but without forgetting that the whole could not contain more than 500,000 words.

Once appointed, the editorial board members were invited to suggest headwords that they believed should be developed for inclusion. As much as possible, they were to focus on the thematic areas assigned to each. The disciplinary editors, for example, were asked especially to name those concepts, methods, tools and so on imported from their fields to this. The country, association and journal editors were also asked to recommend headword lists for their respective areas. For instance, it was left entirely to the two associate editors for countries to suggest, from their knowledge of international tourism, which actually or potentially important national level markets/destinations should be covered. All editors were asked to suggest entry lengths for their proposed headwords. Four length choices were possible: 60, 400, 1,000 or 1,600 words.

When editors' separate lists were collated, several repeated headwords had been independently suggested by a number of them. This clearly suggested that, among other things, these terms are not unique to one but common to several fields, something eventual authors had to keep in mind when preparing their entries. The collated list, with comments and notes, was circulated among the editorial team, inviting further review, and to suggest more additions, deletions, revisions and consolidations. The product of this process, a more refined master list, still once more was forwarded to the whole team, in quest of further adjustments. After this long Delphi-like procedure, a list of headwords (and entry lengths) was finally ready.

At this point, the master list was subdivided amongst the disciplines/specialities, with each tailored list assigned to its respective associate editor. The board members also received author guidelines and sample entries for each of the four aforementioned lengths. They had already been authorized to choose their own authors. Their names and institutional affiliations were gradually submitted to the chief editor. While the editorial team received comments and suggestions, including names for those headwords still without authors, the management of their respective tasks was left to them. The editors set their own strategies in motion, reaffirming assignments, sending their authors guidelines and sample entries through the publisher, guiding the development of each entry, receiving and reading submissions and requesting revisions, until the entries were ready and recommended for inclusion. This process took more than two years, filled with many frustrating moments for almost everyone, including unforeseen circumstances, avoidable and unavoidable duplications among contributions, as well as attempts to quickly seek authorial volunteers for the unforthcoming or somehow unattended entries. An account of these consuming moments (to use a mild term) does not belong to this brief story.

When the bulk of entries were submitted and the chief editor began the initial round of editing, this development phase proved as absorbing as the earlier ones. To produce a useful volume, for instance, the needs of eventual international and multidisciplinary users had to influence the editing process and shape the outcome. The work had to be closely co-ordinated with each editor, many authors had to be reached (directly or through their associate editors), and further revisions had to be made. All this in order to adjust the coverage, to clarify the discussion, to promote uniformity among the entries, to monitor the length (often favouring reduction in order to accommodate more entries), to use terms better understood by the mixed audiences, and the like. The easiest of these tasks was to insure adherence to writing guidelines and other stylistic requirements (the publisher was to finally correct those still not in conformity). The actual editing process as a whole proved to be very complicated, troubled by language problems or choice of words/terms, cross-cultural differences of authors, their disciplinary distinctions or orientations, misunderstood assignments or scopes, incomplete coverage and exaggerated 'promotional' pieces (especially in relation to the country entries), among others. Emendations during this phase made some authors and their editors less than pleased, but the book's integrity, intent and style could not be compromised. Even so, on a few occasions the rules had to be mildly flexed simply to bring the project to a closure. As entries went through this process, the revised submissions were forwarded to the publisher, in alphabetic blocks. With the transfer of the last block of entries, the work of the Chief Editor – some four years later – seemed to be over.

It was at this time that the actual production work by the publisher began, with their copy-editor also reviewing every entry – to catch missed inconsistencies, to further edit each submission for grammar and flow, to delete entry-less headwords, to send page proofs to the respective authors, to incorporate allowable changes into their entries and, *inter alia*, to finally polish all to a publishable, well-integrated whole. The publisher also produced the author and subjects indices, as well as introduced other measures to enhance this reference book's quality and usefulness. The editorial team was not involved during this phase, but the chief editor and publisher continuously worked together, their geographical separation bridged with e-mails. The full story – of some edited entries lost in the cyberspace between Wisconsin and London, of some original mildly-edited versions popping up as actual page proofs, of some chasing and locating of authors who had moved in the meantime, etc. – would make this introduction unnecessarily long.

The structure and contents

This volume contains over 1200 alphabetically arranged, signed entries. In the body of each entry, direct cross-references, noted in **bold type**, indicate that an entry under this headword appears elsewhere in the encyclopedia. Moreover, the 'see also' list at the end of some entries directs the reader to additional related headwords. The Subject Index suggests a vast range of terms under their own headwords or discussed within other entries. In this way, the index becomes a major key in more fully utilising the encyclopedia.

The number of words allocated to each headword determined its scope. Thus, shorter entries use every word to quickly cover the subject, while longer ones obviously have more room to manoeuvre. Many entries come with references. Authors themselves decided to devote the allotted words entirely to discussing the subject itself, or to pose limits in order to make room for appending useful 'further references' (with or without annotation, representing important/relevant publications on the theme). Entries exceeding their word allocations were often shortened during the editing process (with some small length deviations allowed).

When this volume is used to quickly look up a term, one may tend to place 'tourism' in front of the headword. But this, as well as 'tourist', 'travel' and 'traveller', are used only when absolutely necessary. For instance, if one wants to read about 'tourism carrying capacity', the first word is understood and thus omitted in the headword. As a matter of fact, since the whole collection is devoted to the subject of tourism, the constant use of the term in both headwords and in discussion seemed redundant. The decision in favour of reducing such occurrences was made at the outset, sometimes to the dismay of a few authors. Similarly, inclusion of 'tourism' as a headword by itself was not even in the plan, because the whole publication is attempting to discuss what tourism is, what it covers, and how it is studied. But, at last, a short yet holistic frame found its way in. Another editorial decision, to enhance uniformity, was to have almost all headword titles appear in their singular forms. Only a few, for the sake of clarity or necessity, appear in the plural.

An examination of the headword list (pages xv–xxxv), or the volume as a whole, would reveal that, despite all efforts, still a few 'obvious' entries are not included in a publication calling itself an encyclopedia. Some missing ones had actually been assigned, but not submitted, despite promises. When deadlines could be extended, they were reassigned. To move on, it was decided not to further jeopardise the whole production process by attempting to properly cross-reference late submissions or to wait for unforthcoming entries. It was hoped that 'dummy' headwords and/or the Subject Index would lead the reader to entries which partly compensate for this limitation. While exclusion of some headwords in the volume and keywords in the index is unfortunate, one category of entries was intentionally left out. In contrast to most other encyclopaedias, here it was decided to do without biographies of the leading scholars in this field. It was reasoned that it would be extremely difficult to decide whom to include and why in this early stage of the journey. Page limitations made this 'convenient' decision appealing. When tourism eventually reaches a more advanced level of maturity, future studies covering the history of this new academic field, as well as next editions of this or similar reference publications, will be in a better position to feature profiles of its pioneer architects.

Among other problems, some related entries prepared by different authors made the same or similar points. When feasible, the degree of duplication was reduced, through editing and/or cross-referencing. But on a number of occasions, few changes could be made, lest a discussion be incomplete or incomprehensive. Further, many entries were conveying particular 'flavours' closely related to their authors' disciplinary position/ geographic location. Some were altered, but a few, to keep their integrity, remained almost unchanged. Also, certain knowledge/subject areas are unevenly represented in this volume, due to how comprehensively each editor defined the task. Whilst this 'uneven' treatment may be appreciated by some readers, others would think differently. Similarly, there is a bias or greater representation of associations headquartered in North America and/or which are US-specific. This was in some ways due to their dominant or lead positions in the field, especially when some editors thought so. One could argue that many of these associations could still, despite this bias, serve as 'models' for the rest of the world.

The publisher's preference for British spelling was agreed to without hesitation. While this was amongst the simpler issues, linguistic differences were not. The specific use of a given term in one (English) language circle versus another is a case in point: campground or caravan park, catering or restaurant business, and hiring or renting a car exemplify a manifold pesky problem. So, too, differently framed concepts and themes can be difficult to comprehend outside their immediate circles. As much as possible, attempts were made to add synonymous terms or notations, to help the reader make the connection. Even such a seemingly universal term as 'industry', often used in association with tourism, is confusing in some languages, especially in Spanish-speaking countries. But because of the growing international acceptance of tourism as an industry, as spearheaded and advocated by leading organisations, the encyclopedia opted in favour of the term (reserving 'sector' for categories typically placed under it, like the hotel sector). Therefore, such complications with linguistic distinctions, disciplinary boundaries, and even geopolitical views proved a constant challenge to this very international publication produced by a very international group of editors and authors.

Voilà, the end result

This *Encyclopedia of Tourism* and the *Silver Anniversary Supplement* of *Annals* (volume 25, pp. 989–1119), while different from one another, stand as symbolic scholarly landmarks. The anniversary issue of *Annals* is a guide to twenty-five years of knowledge accumulation appearing in one single research journal. But the encyclopedia is not bounded by a given source or time. It is a fuller reference to all that applies to tourism, by covering the range of what is known, what is tentative, and all that falls in between. Together, despite this contrast, these two reference sources provide a rear-view mirror and a windshield toward the future of research and scholarship in this field. Based on these two contributions, amongst many more, today the landscape appears better mapped and understood, the research foundation in place, with a sense of purpose and direction in the air.

This young multidisciplinary field has indeed achieved much to its credit (and credibility) in a short span of time, and this 'memoir' celebrates the cumulative accomplishment. According to a Chinese proverb, 'a journey of a thousand miles begins with a single step'. As this volume demonstrates, the scientification journey has indeed progressed, but all participants in it are conscious of the long road ahead. Its successful continuation will have to include many augmenting and innovative measures in order to expand and accentuate the multidisciplinarity of tourism; to more fully recognize and articulate both its art and science properties; to cross-fertilise these qualities of scope and nature in order to engineer a research infrastructure conducive to rigor and passion, method and interpretation; to envision new summits to be reached through pathbreaking pursuits; to keep track of the shifting sands as the landscape transforms (as the scientification histories are yet to be written); and thus to systematically and cumulatively form, embody, and fortify this corpus of knowledge as the bedrock upon which future tourism scholarship sets its course, on its way to attain the legitimacy it seeks for positioning itself in higher scholarly circles.

Returning to now and the encyclopedia, this project finally reaches its users after five years of working co-operation, as briefly detailed in this introduction. It was accomplished through the efforts of a committed team of editors and authors, from many disciplines, from many countries, from a multitude of perspectives. Together, they have built a truly international and multidisciplinary monument, depicting the whole field whose bricks and building blocks of knowledge are featured as over 1200 snapshots in this volume. This collection, intended as a reference book and a landmark for the scientification journey, aims to facilitate and promote research and scholarship in this field's broad and opening future. For those in education, this publication can both inform and arm them with various multidisciplinary lenses and tools and, further, it can guide their students toward a *tour de force* in this landscape, with the hope that they take residency in their choice niches. For those active in operational organs, business headquarters of the industry, consultants, and other 'practitioners', this is a source of knowledge and answers to some of their initial questions. To the newcomer to this line of investigation, this is a treasure ready to inform and guide.

Could this project have been cast better in scope, focus, treatment and packaging? Now the user is in the best position to answer such questions. As the journey continues, the emerging retrospectives gained with the passage of time will suggest how (in)sufficiently it was prepared for the road ahead. Alas, the clock cannot be turned back and the book appears as is. Despite its limitations (some outlined above), this reference volume takes satisfaction in having partly charted and mapped the landscape and thus in aiding to the ongoing expansion of knowledge, for the journey to reach its next height affording a fuller vista of new frontiers extending beyond this dawn. As such, the project has served its time-framed and book-bound goal at this particular juncture, by making available a register of an emerging knowledge mosaic, by providing a model for other publications, including its own future editions which should be pursued in the near future. In fact this very encyclopedia has been a journey of its own, not a destination. This single-mission excursion, with its scholarly souvenirs stock-piled on board, has reached home: but the stay will have to be short. With the academic explorations in tourism so rapidly expanding worldwide, this *Encyclopedia of Tourism* will participate in further accelerating, advancing expeditions, as the scientification journey continues, with much of its corpus still lying in the horizons ahead.

Jafar Jafari
Chief editor

Acknowledgements

Writing a full and fair acknowledgement for a project as involved and time consuming as this one is not an easy task. The twenty-three associate editors are the main architects of this encyclopedia. Significantly, some of them went out of their way to set the stage, write the script and play their respective parts, including even on many instances taking care of those entries left unattended by delinquent authors. Together, the board members and their authors contributed to the formation of this structure, brick by brick. The ensemble are the ones who have shaped the corpus and soul of this reference book.

On the publisher side, it was a sheer pleasure to work with Denise Rea at Routledge. With almost daily e-mail contacts – and immediate replies from her – progress was made as efficiently as a project of this nature deserves. At the end, it took Tarquin Acevedo and Morgen Witzel to bring the project to its final production stage. During the proofreading stage, University of Wisconsin-Stout graduate students Scott Graff and Richard Frazier assisted in checking over some parts of the volume. The beginning of this project was aided by another graduate student, Seon Kim, who helped with the master headword list and checked on and sorted out submitted entries.

Special acknowledgements are due for the support and co-operation of the Universitat de les Illes Balears Tourism School in Spain, where the chief editor spent his 1997–8 sabbatical leave. This was the period during which the project was taking shape, with the University of the Balearic Islands doing all it could to facilitate its development during the said period. The home institution of the chief editor, where this project saw its beginning and its completion, is also dutifully recognised. Finally, personally he acknowledges the accommodating part that his family played in tolerating him, literally locked in his office on the campuses of University of Wisconsin-Stout or Universitat de les Illes Balears, day in and day out, and then again hopelessly glued to his laptop when at home. In addition to this support and accommodation, his wife and two children also helped during different phases of this project, including double-checking on the headwords, printing documents, faxing contributions, and searching for missing entries.

Although only the name of chief editor appears on the cover of the book (as is the tradition with Routledge), in truth this publication, as it must be clear by now, represents the labour of all editors and authors of this production and those named above (several more not listed for the sake of brevity). This is their volume. On behalf of all editors and contributors to this *Encyclopedia of Tourism* – committed to the advancement of knowledge in this field – this volume is dedicated to researchers who are 'religiously' conquering, populating, mapping and landmarking this newly formed multidisciplinary landscape of knowledge.

Jafar Jafari, Chief Editor

Entry list

Aborigine
accessibility
accommodation
accounting
acculturation
achievement
ACTA Turistica
action research
activity
activity space
adaptation
adventure tourism
advertisements, envy in
advertisements, format of
advertising
African Travel Association
Agenda 21
agrotourism
AIDS
Air Transport Association of America
aircraft, supersonic
airline distribution systems
Airline Reporting Corporation
airline
airport
alcohol
alienation
allocentric
alternative tourism
amenity, user-oriented
American Hotel and Motel Association
American Society of Travel Agents
Anatolia
animation
Annals of Tourism Research
anomie
Antarctic tourism
anthropology
anticipation
anti-tourism
appropriate tourism
appropriation
archaeology
Arctic tourism
Argentina
arrival/departure card
art
articulation, programme
Asia Pacific Journal of Tourism Research
Asia Pacific Tourism Association
asset management
Association Internationale d'Experts Scientifiques du Tourisme
Association Mondiale pour la Formation Hôtelière et Touristique
Association of Conference and Events Directors-International
Association of Travel Marketing Executives
attention
attitude
attraction
attraction, religious
attractivity
attribution theory
auditing
Australia
Austria
authenticity
automatic interaction detection
automation

back office
back-stage
Bahamas
balance of payments

bar
Barbados
bed and breakfast
behaviour
behaviour, recreation
benchmarking
benefit–cost analysis
benefits
Bermuda 1
binary structure of advertisements
biological diversity
board
borders
branding
Brazil
break
break-even point analysis
brochure
budget hotel
budgetary control
budgeting
Bureau International de Tourisme Social
Bureau International pour le Tourisme et les Echanges de la Jeunesse
burnout
bushwalking
business format
business travel
buying decisions

cafeteria
Cahier Espaces
camping
Canada
car rental
caravan park
career
Caribbean Tourism Organisation
carrying capacity
carrying capacity, recreational
cash flow
catering
catering, airline
causal model
central reservation
centrally planned economy
centre–periphery
chain hotels
change
change, sociocultural
charter, air
Chile
China
choice set
Chunnel
circuit tourism
city office
civil aviation authority
class
class of service
classification
cliché
climate
club
cluster analysis
code of ethics
codes of ethics, environmental
code sharing
cognition
cognitive dissonance
collaborative education
collective bargaining
Colombia
colonisation
commercialisation
commissary
commoditisation
communication mix
community
community approach
community attitude
community development
community planning
community recreation
comparative advantage
comparative study
compatible
compensation administration
competitive advantage
competitiveness
compliance
computer reservation system
concentration ratio
concession
concierge
conflict
conjoint analysis
connotation
conservation
consortium

conspicuous consumption
consumerism
content analysis
contestation
contingent valuation
continuum model
Contours
control system
convention and visitor bureau
convention business
Cornell Quarterly
corporate finance
corporate strategy
corporate structure
correspondence analysis
cosmopolitanism
cost
Costa Rica
Council on Hotel Restaurant and Institutional Education
country house hotel
countryside
creative tourism
crime
critics
cross-cultural education
cross-cultural management
cross-cultural study
cross-training
crowding
cruise line
Cruise Lines International Association
Cuba
cuisine
cultural conservation
cultural survival
cultural tourism
culture broker
culture shock
culture, corporate
culture, invention of
culture, organisational
culture, tourism
curative tourism
currency control
curriculum design
cycle
Cyprus
Czech Republic

database marketing
decision making
decision support system
definition
Delphi technique
demand
demand, air travel
demand, recreational
democratisation
demography
demonstration effect
Denmark
denotation
dependency theory
deregulation, airline
design and layout
destination
destination choice
destination management
destination management company
destination management organisation
developing country
development
development era
deviance
diagonal integration
diet
differentiation
direct marketing
disciplinary action
discourse
discretionary travel
discriminant analysis
disease
displacement
distance decay
distance education
distribution channel
diversionary tourist
domestic tourism
Dominican Republic
drinking

ecoethics
ecologically sustainable tourism
ecology
economic development
economics
economics, ecological
economies of scale

ecoresort
ecotourism
Ecotourism Society, The
Ecuador
Ecumenical Coalition on Third World Tourism
education
education, computer-assisted
education effectiveness
education, entrepreneurial
education, environmental
education/industry relationship
education level
education media
education method
education, multidisciplinary
education policy
education, recreation
efficiency
Egypt
electronic promotion source
emergency management
employment
employment law
enclave tourism
encounter
energy
entrepreneurship
environment
environmental aesthetics
environmental auditing
environmental compatibility
environmental engineering
environmental management, best practice
environmental management systems
environmental rehabilitation
environmental valuation
eroticism
escape
Espaces
Estudios y Perspectivas en Turismo
Estudios Turisticos
ethnic group
ethnic tourism
ethnicity
ethnocentrism
ethnography
ethnology
European Travel Commission
European Union
event management
event marketing
event
evolution
exclave
excursion
excursionist
executive development
exoticism
expatriate
expectation
expenses estimation
experience
experimental research
exploration
externalities

facilitation
facilities management
factor analysis
factory tourism
fairs
fairy tales
fantasy
farm tourism
fast food
feasibility study
feast
Federal Aviation Administration
Federation of International Youth Travel Organisation
feminism
ferry
festival
Festival Management and Event Tourism
festival, religious
film
financial control
financial objectives
Finland
Finnish University Network for Tourism Studies
FIU Hospitality Review
flow
food
food and beverage cost analysis
food-borne illness
food service, contract
forecasting
foreign exchange
foreign independent tours
foreign investment

forest recreation
formulae
Fourth World
France
franchising
freedom
freedom, perceived
frequent flyer programme
fun
future

gambling
game park reserve
game theory
gastronomy
gateway
gaze
gender
geographical information system
geography
geography, recreational
Germany
ghetto
gift
global distribution systems
globalisation
golf tourism
government
grading system
Grand Tour
gravity model
Greece
green marketing
green tourism
greenhouse effect
greenspeak
gross profit
group business market
group inclusive tour
group travel market
growth
guest
guidebook
guided tour

hajj
handicapped
handicraft
hard tourism
health
hedonism
heliocentrism
heritage
hiking
hill stations
hippie
historical tourism
history
holiday camp
holistic approach
Holy Land
home delivery
homecoming
homelessness
homesickness
homosexuality
Hong Kong
horizontal integration
hospitality
hospitality information system
Hospitality Sales and Marketing Association International
host and guest
hotel
hotel, airport
human resource development
humour
Hungary
hyperreality

iconography
identity
ideology
image
image, destination
impact
impact, economic
impact, environmental
impact assessment, environmental
imperialism
import substitution
inbound
incentive
index, trip
India
indigenous
indirect tourism
individual mass
individualism
Indonesia
industrial recreation

industrial tourism
industry
inevitability
information centre
informal economy
information source
information technology
Information Technology and Tourism
infrastructure
innovation
input–output analysis
Institute of Certified Travel Agents
Instituto de Estudios Turísticos
integrated environmental management
intellectual property
internal marketing
International Academy for the Study of Tourism
International Air Transport Association
International Association of Amusement Parks and Attractions
International Association of Convention and Visitors Bureaus
international aviation bilateral
international aviation liberalisation
international aviation organisations
international aviation rights
International Civil Aviation Organisation
International Forum of Travel and Tourism Advocates
International Geographical Union
International Hotel and Restaurant Association
International Journal of Contemporary Hospitality Management
International Journal of Hospitality Management
International Journal of Tourism Research
International Labour Organisation
international organisation
International Society of Travel and Tourism Educators
International Sociological Association
International Tourism Reports
international tourism
international understanding
internationalisation
Internet
interpretation
intervening opportunity
interview
inventory
inversion
investment
investment decision
Iran
Ireland
Israel
Italy
itinerary

Jamaica
Japan
job design analysis
joint venture
Journal of Applied Recreation Research
Journal of Gambling Studies
Journal of Hospitality and Leisure Marketing
Journal of Hospitality and Tourism Research
Journal of International Hospitality, Leisure and Tourism Management
Journal of Leisure Research
Journal of Park and Recreation Administration
Journal of Restaurant and Foodservice Marketing
Journal of Sustainable Tourism
Journal of Tourism Studies
Journal of Travel & Tourism Marketing
Journal of Travel Research
Journal of Vacation Marketing
journalism
jungle tourism

Kaabah
Kenya
keying
knowledge acquisition
Korea

labour relations
land tenure
landscape
landscape evaluation
language of tourism
Latin American Confederation of Tourist Organisations
law
leadership
leakage
learning curve
legal aspects
legislation
legislation, environmental
leisure
Leisure Sciences

Leisure Studies
leisure tourist
length of stay
licensing
life cycle
life cycle, destination
lifeseeing
lifestyle
liminality
limit of acceptable change
literary tourism
local
local organisation
localisation curve
location
location quotient
locational analysis
location quotient
longitudinal study
long range
loyalty
ludic

magazine
magic
Malaysia
Malta
management
management accounting
management contract
Managing Leisure
manpower development
map
marginality
marker
market
market analysis
market segmentation
markets, contestable
marketing
marketing audit
marketing, destination
marketing information system
marketing, international
marketing mix
marketing objective
marketing plan
marketing research
mass tourism
mass tourism, organised
master plan
media
media planning
media, recreational
meeting planner
meetings business
merchandising
merger
metaphor
methodology
Mexico
middleman
migration
minorities
model
modernity
modular programme
Morocco
motel
motion sickness
motivation
motivation, intrinsic
motive manipulation
motor coach tourism
mountaineering
multidimensional scaling
multinational firm
multiplier effect
museum
music
myth

national character
national park
National Recreation and Park Association
National Restaurant Association
National Tour Foundation
National tourism administration
nationalism
natives
nature
nature tourism
nature trail
nearest neighbour analysis
need, recreational
neo-colonialism
Nepal
Netherlands, the
neurocomputing
new product development

New Zealand
non-discretionary income
non-governmental organisation
non-profit organisation
Norway
nostalgia
novelty

occupancy rate
occupational safety
off-road vehicle
operating cost
opportunity cost
opportunity set
optimal arousal
organisation design
Organisation for Economic Cooperation and Development
orientalism
orienteering
other
outbound
outdoor recreation
outsourcing
overseas office

Pacific Asia Travel Association
Pacific Rim
Pacific Tourism Review
package tour
paid vacation
paradigm
paradise
paratransit
park
participation, recreation
partnership
pastoral care
patrimony
payroll cost analysis
peace
People on the Move
perception, environmental
perceptual mapping
performance appraisal
performance indicators
performance standard
personal selling
Peru
phenomenology
Philippines, the
photography
piety
pilgrim
pilgrimage
pilgrimage route
pilgrimage site
placelessness
planning
planning, environmental
planning, recreation
play
pleasure periphery
pleasure tourist
Poland
polar
policy
political development
political economy
political science
political socialisation
political stability
pollution management
pornography
Portugal
postcard
post-industrial
postmodernism
power
precautionary principle
preservation
prestige
pricing
principal components analysis
product
product life cycle
product planning
product positioning
productivity
professional native
professionalism
profit
profit centre
profit margin
profit sensitivity analysis
profit variance analysis
promotion, place
promotion, puns in
property management system
prostitution
protected area

prototypical tourism form
pseudo-event
psychographics
psychology
public goods
public health
public participation
public relations
purchasing
purpose
push–pull factors

qualitative research
quality
quality, environmental
quality, transportation
quantitative method
quest

race
radio
rail
ratio analysis
rebirth
reception
recreation
recreation business district
recreation centre
recreation experience
recreation manager
recreation opportunity spectrum
recruitment
recycling
region
regional organisation
register
regression
regulation, self
regulatory agency
relationship marketing
relics
religion
religious centre
remote sensing
rent
repeat tourist
representation
research and development
reservation
residential recreation
resort
resort club
resort development, integrated
resort enclave
resort hotel
resort morphology
resource-based amenity
resource evaluation
resources
responsible tourism
restaurant
return on investment
revenue forecasting
rhetoric
risk
risk analysis
risk, perceived
rite of passage
ritual
roles
romanticism
room night
route system
rules
rural recreation
rural tourism
Russia

sacred journey
safari
safety
sales force management
sales promotions
sanitation
satellite account
satisfaction
satisfaction, customer
scale of development
scenic drive
seamless service
seasonality
second home
security
segmentation, a posteriori
segmentation, a priori
self-actualisation
self-discovery
semiotics
senior tourism
service
service delivery system

service quality
sex tourism
Seychelles
shopping
shrine
sight
sightseeing
sign
significant omission
Singapore
site
site analysis
site, sacred
sites, biblical
skiing
Slovenia
small business management
smuggling
social control
social interaction
social recreation
social relations
social situation
social tourism
socialisation
Society and Leisure/Loisir et Société
sociolinguistics
sociology
soft tourism
South Africa
souvenir
souvenir, religious
spa
space allocation
Spain
spatial interaction
special interest group
sponsored event
sport, recreational
sports tourism
staged authenticity
standardisation
statistics
stereotype
strangeness
stranger
strategic business unit
strategic marketing
strategic planning
strategy formulation
structuralism
substitution
summer cottage
sun, sand, sea and sex
sunlust
supply
survey
survey, guest
sustainable development
Sweden
Switzerland
SWOT analysis
symbolism
systems theory

Taiwan
target marketing
tax
taxi
teacher education
technology
telemarketing
television
Teoros
terrorism
testimony
Thailand
thanatourism
theme park
theory
Third World
time
time budget
timeshare
tipping
tour
tour guide
tour operator
tour wholesaler
tourism
Tourism Analysis
Tourism Economics
Tourism Management
tourism organisations
Tourism Recreation Research
tourism, secular
Tourism Society
tourism system
Tourisme
tourist

tourist as child
Tourist Review, The
tourist, recreational
tourist space
tourist trap
trade show
trading area
tradition
training
transactional analysis
transit
transportation
transportation, globalisation of
transportation pricing
travel
travel advisory
travel agency
Travel and Tourism Analyst
Travel and Tourism Research Association
travel cost method
Travel Industry Association of America
travel literature
travel writer
treaty
trekking
trend
Trinet
Tunisia
Turismo em Analise
Turizam
Turkey
turnover
typology, tourist

underdevelopment
UNESCO
unfamiliarity
uniform system of accounting
United Kingdom
United States
urban recreation
urban tourism
US Travel Data Center

vacation
vacation hinterland
vacationscape
value added
values
Venezuela
vertical integration
video
vineyard
virtual reality
Visions in Leisure and Business
visiting friends and relatives
visitor
voluntary sector

walking tour
wanderlust
war
waste management
wayfinding
weekend
wilderness
wildland recreation
work ethic
World Bank
World Leisure and Recreation
World Recreation and Leisure Association
World Tourism Organization
World Travel and Tourism Council
World Wide Web

xenophobia

yield management
yield percentage
youth pilgrimage
youth tourism

zoning

Aborigine

Having reviewed research on the impacts of tourism on Aboriginal communities in **Australia**, Altman and Finlayson (1991) concluded that the demands of tourism and the provision of goods and **services** for tourism by Aboriginal populations are fundamentally incompatible. Aboriginal peoples are not integrated into the dominant Euroaustralian economy, and the business of tourism has rigorous service encounter requirements, resulting in an intensity of management performance and expectancy with which Aboriginal people tend not to be comfortable. However, Altman and Finlayson's synopsis adjudged that there are clear benefits to both the industry and to **indigenous** communities from an enhanced Aboriginal involvement in tourism, although for the foreseeable future it envisaged that Aboriginal participation would be largely indirect, informal, small scale and in need of public subvention.

Hollinshead (1996) has further explored the fit of Aboriginal communities within the tourism marketplace, yielding ten precepts to guide future research and management action. These propositions emphasise the fragility of the Aboriginal cultural and environmental 'product' for tourism, stressing that no Aboriginal involvement in tourism can meaningfully occur until the host group or community senses (see **host and guest**) that such participation will help render it secure within its traditional lands. These precepts also underscore the high value Aboriginal populations place on '**identity**', '**community**' and 'spirituality' above any desire to engage in tourism itself, however financially rewarding that call may appear. With regard to the external tourism industry, the latter study sees the immediate challenge being whether tourism can sincerely learn to project the mystical and now agonistical numenology of particular, differentiated Aboriginal inheritances, rather than insipidly portraying some homogenous and objectified phenomenology of those cultures. To that end, it will have to now learn – after Pêcheux (1982) – to disidentify its narratives (that is, compose them within what is decidedly the new/different/oppositional ideological structures of indigenous cosmology).

Such genuine attempts at disidentification will constitute a significant attempt to contest the incremental appropriation of indigenous knowledge (and culture) by mainstream industry **discourse**. That battle for reflexive perspectivity will be a prodigious dialectical battleground in **cultural tourism** during the twenty-first century, and will not be limited to indigenous Australia.

References

Altman, J.C. and Finlayson, J. (1991) 'Aborigines and tourism: an issues paper', Canberra: Centre for Aboriginal Economic Policy Research.

Hollinshead, K. (1996) 'Marketing and metaphysical realism: the disidentification of Aboriginal life and traditions through tourism', in R. Butler and T. Hinch (eds), *Tourism and Indigenous Peoples*, London: International Thomson Business Press, 308–48.

Pêcheux, M. (1982) *Language, Semantics and Ideology*, trans. H. Nagpal, London: Macmillan.

KEITH HOLLINSHEAD, UK

accessibility

Accessibility refers to the ease or difficulty of taking advantage of an opportunity. The concept is commonly used in a spatial context in regards to the physical distance between a user and a tourism facility, that distance being measured in a straight line or along a route, by travel time, or by cost measured in monetary terms, or even in the degree of fatigue that is involved in overcoming the friction of distance. **Maps** may be drawn to display visually the travel times to a **destination** using isochrones (lines of equal travel time), with the lines generally extending outwards along major **transportation** corridors. However, there may be considerable differences between accessibility as determined by such measures and what is being perceived by potential **tourists**, and they may trade off travel time for money by substituting faster but more expensive transportation modes.

At a regional level, statistical methods have been used to assess the interaction of **demand** and **supply** for a particular resource at any point to calculate the potential, or accessibility, of that point. This concept has been used widely in transportation studies and **marketing** (Smith 1983). Changes in accessibility have major implications for tourism. Thus, for example, extensions of airline networks, frequency of flights and sizes of planes have greatly increased long-haul travel and the accessibility of distant locations. However, improved accessibility does not always stimulate tourism businesses. For example, Lundgren (1983) has shown how improvements in road transportation have allowed individuals to travel farther more quickly, thereby turning overnight into day-trip destinations and undermining the market for small **accommodation** establishments in intermediate locations. More generally, accessibility may be influenced by factors other than proximity, for example price, membership status and **licensing** regulations. Thus, even though a potential user may be very close to an opportunity, use may be prevented by barriers other than distance.

See also: distance decay; gravity model

References

Lundgren, J. (1983) 'Development patterns and lessons in the Montreal Laurentians', in P. Murphy (ed.), *Tourism in Canada: Selected Issues and Options*, Victoria: University of Victoria, 95–126.

Smith, S.L.J. (1983) *Recreational Geography*, London: Harlow, 45–57.

GEOFFREY WALL, CANADA

accommodation

Accommodation is a term used to encompass the provision of bedroom facilities on a commercial basis within the hospitality/tourism industry. Primarily it is associated with the **hotel** sector, and is readily applied to properties as diverse as business and conference hotels, **resort hotels**, **motels** and **budget hotels**, guest houses and **bed and breakfast** establishments. Accommodation also embraces other forms of **hospitality** outlets, such as university halls of residence, youth hostels, residential care facilities and hospital hotel services.

The demand for accommodation has developed in response to the rapid advances in **transportation**, with both business and recreational travel expanding at a phenomenal rate throughout the twentieth century. The **demand** itself is divided into the major and distinct markets, including business, **recreation** or **leisure**. In turn, these and other markets may be further fragmented into finer segments, users with differing needs, all concerned with the determining factors of price, **location** and **quality**. This overall demand, and the subsequent provision of a range of accommodation, emanates from the economic circumstances of both the developed and developing countries.

When considering the accommodation sector's wider economic context, it is instructive to review the dynamic relationship between **economic development** and the demand for accommodation by business tourists and their companies. The initial phase in a developing economy is dominated by manufacturing. There will be some accommodation demand as a result but it will not be high volume, with business people only requiring a small number of hotel rooms for very short stays.

The second phase is normally when the **service** industries begin to assume greater importance, in response to the previous economic growth created by manufacturing industry. Service sector companies mature into multi-unit operations, as in banking, other financial services, retail and leisure. This second phase boosts demand for accommodation, as a greater number of mobile business executives travel to more sales calls, meetings, training courses and conferences. The third phase sees all demand for accommodation stabilising, with any real **growth** being marginal, perhaps from international markets. Firms are now in the process of developing their own regional networks of offices, training centres and conference facilities. Further, with the increasing use of **technology**, more and more executives are working from home and there is less need to be away from it on expensive business trips. In this mature economic scenario, competition is fierce and cost reductions, such as travel and hotel expenses, are an important means of maintaining profitability.

Recent **statistics** confirm the worldwide importance of the business market, estimating that it accounts for 38 per cent of accommodation demand, compared with 36 per cent for **tourists** and 14 per cent for conference delegates. The remainder is made up of government officials and **visiting friends and relatives**. Hotels in appropriate locations have developed their goods and services and product to capture the lucrative business market, which sustains high room rates and high **profit** margins. The executive requires quality accommodation, with private bathroom facilities, telephone and an area for working. Some hotels are developing bedrooms which double up as an 'office away from the office', including well-lit desks, laptop computers, fax and modem plug-in points, teletext television and access to a professionally managed business and communications centre. Some refurbished hotels in London and elsewhere include all the above executive facilities plus a number of apartment-style suites, with full kitchen and dining room, twenty-four-hour room and valet service, air conditioning, power showers and trouser press, express laundry, personal room safes, satellite television, CD and video with free 'library' service and an **Internet** link.

The demand for accommodation may also be of great economic value, for instance worth over £8 billion to the UK annually. However its characteristics and needs can be very different. The family market tends to have a lower average expenditure but visitors often stay for much longer than the business traveller, and sometimes at different times such as weekends and traditional holiday periods. The **pleasure tourist** is more price-sensitive, as visitors are paying from their own pocket and are sensitive to economic conditions at both the point of origin and the point of destination. Demand can also be drastically affected by civil unrest, transportation problems and **terrorism**. A recreational tourist is constrained by **vacation** time, by disposable income and by a destination's **image** and reputation for such basics as weather, language and friendliness (see also **tourist, recreational**). This form of accommodation demand has been related to the 'push' end 'pull' factors of holiday choice (see **motivation**; **push–pull factors**). The 'push' factors are those which motivate the tourist to travel to a specific **destination**, while the 'pull' factors are those which positively attract a tourist to a certain location. The quality of accommodation and its value for money can be a powerful 'pull' factor.

The accommodation needs of the pleasure and business tourists are in some ways different. Holiday-makers want a more relaxed **environment**, with the emphasis on enjoyment and entertainment. The tourist would rather have comfortable armchairs and sofas than a formal work station or desk area. In-house video and access to swimming pool and leisure activities become important needs within, and supplementary to, the bedroom, which needs to be more of a 'home from home' than an 'office away from the office'.

The increase in the demand for leisure and for business-oriented accommodation has been international in range and **competitiveness** is on a global basis. Large international hotel chains have expanded by acquisition and by **franchising** across Europe, the Americas and the Far East. Many such companies have clearly segmented their markets, providing accommodation of various types and quality levels to ensure that a range of prices and packages are available. This global expansion has accelerated the impact of technology on the **reservations** side of the valuable accommodation market. Major companies and consortia must continuously review and update their **central reservation** networks, now regularly interfaced with the airlines, car-hire firms, travel agents and tour operators. It was recently estimated that around 30 per cent of hotel rooms are now booked via central reservations networks, that another 30 per cent are made through **travel agencies**, and that only about 36 per cent of all bookings are made directly by the client. Such trends seem certain to continue, being gradually influenced by Internet bookings made from home or office (see also **information technology**).

Apart from advanced reservations systems and staff, the accommodation sector consists of several key operational departments within hotels, namely **reception**, housekeeping and accounting. From the check-in at reception, the success of the guest's **experience** is significantly determined by the friendly and efficient service given by the accommodation department's employees. Reception, including porters and the **concierge**, is a focal point of customer care and the provision of information and guest services. The housekeeping department has the responsibility of ensuring that there is strict maintenance of standards of cleanliness and presentation of bedrooms, bathrooms and corridors. At the check-out end of the accommodation experience, there needs to be accurate billing and charging of guest accounts, with the **professionalism** to be sure that the method of payment is valid and that the company or personal account is in order, with all services received included at the agreed rate. Sales **turnover** must be maximised, as the rooms division of hotels is often the most important in terms of turnover and profitability.

Further reading

Goodall, B. (1991) 'Understanding holiday choice', in C.P. Cooper (ed.), *Tourism, Recreation and Hospitality Management*, vol. 3, London: Bellhaven Press. (An introduction to why consumers stay away from home in accommodation.)

Goss-Turner, S. (1996) 'The accommodation sector' and 'Business and conference hotels', in P. Jones (ed.), *Introduction to Hospitality Operations*, London: Cassell. (Provides a overview of the accommodation sector of the UK hospitality industry.)

Kleinwort Benson Securities (1991) *Quoted Hotel Companies: The World Markets 5th Annual Review*, London: Kleinwort Benson Securities.

STEVEN GOSS-TURNER, UK

accounting

Accounting is the generic term applied to systems for the quantitative description or projection of income, wealth and their various components. Income and wealth are related in an accounting system as follows. Income is a *flow* concept, while wealth is a *stock* concept; a positive flow of income is an increase in wealth, and a negative flow of income (a loss) is a decrease in wealth. Income as a flow relates to a period of time (such as a year, a quarter or a month) and is typically depicted in a statement of income, while wealth as a stock relates to a point in time (normally the beginning or end of such a period) and is depicted in a balance sheet or statement of financial position.

Types of accounting system include national accounts (the preserve of economic statisticians rather than accountants, and sometimes termed macro accounting), and the accounts of organisations in national and local **government** and of private sector organisations including companies, such as hotels, **restaurants** and **airline** (micro-accounting). National accounts are based not on capturing and analysing data by means of book-keeping, but on statistical estimates (see **balance of payments**). The other types of accounts, namely micro accounting, employ double-entry book-keeping, but not all governmental accounts recognise the distinction between capital and revenue which is crucial to the depiction of income and wealth.

Capital expenditure results in the acquisition of long-lived assets, which are positive components of net wealth, while revenue expenditure results in expense, which is a negative component of net income. Thus, with capital expenditure, cash (an asset) is typically exchanged for some other kind of asset, either tangible such as a hotel building or intangible such as a legal property right; net wealth is unaffected. By contrast, with revenue expenditure cash is spent (for example, on paying salaries) but no asset is acquired; net wealth is reduced. Most assets are used up or otherwise lose value over time, and this loss of value is recognised in the accounting system as an expense and a reduction of net wealth.

As well as acquiring long-lived assets such as buildings or machines (sometimes called fixed assets), organisations may purchase items for resale or for use in production. The term stock (or '**inventory**' in American English) is used to refer to such items, which are included among the organisation's assets. Examples of stocks would include goods purchased for resale in **souvenir** shops, duty-free outlets and other forms of retail distribution, whereas goods purchased for use in production may include material to make up T-shirts, wood for carving and so on. In addition, the giving and receiving of trade credit gives rise to accounts receivable from trade debtors and accounts payable to trade creditors. The former are part of the organisation's assets (positive components of net wealth), while the latter are liabilities (negative components of net wealth). Amounts borrowed by the organisation as loans are another form of liability. The total of assets less liabilities (the net wealth) equals the equity of the organisation. A common problem within the tourism industry is the operation of businesses with high ratios of liabilities, which may result in business failure.

The double-entry principle requires every transaction to be analysed according to the destination (debit) and source (credit) of the flow of wealth involved, and book-keeping entries are made to reflect both debit and credit. Thus, the purchase of an item of stock, such as wine for the cellar of a restaurant, using trade credit requires a debit entry to the stock account (the destination) and a credit entry to the trade creditor's account (the source). When the creditor is paid, the trade creditor's account is debited and the cash or bank account is credited. Neither of these transactions affects the total of assets less liabilities (net wealth which corresponds to the organisation's equity). By contrast, when the restaurant staff's salaries are paid, a payroll expense account is debited and the cash or bank account is credited. In this case, the total of assets less liabilities is reduced by the amount of the decrease in cash, and correspondingly the organisation's equity is reduced by the amount of the expense.

Within a single-entry system of accounting, net income for a period (surplus or deficit) is calculated by deducting the total of assets less liabilities at the beginning of the period from the total of assets less liabilities at the end of the period. By contrast, within a double-entry accounting system, net income for a period is calculated by deducting the total of expenses from the total of gross income or revenue. Within such a system, the relationship between the balance sheet and the statement of income in double-entry terms means that a flow of gross income or revenue (credit) is reflected in an increase in assets or a decrease in liabilities (debit). Likewise, expense (debit) is reflected in a decrease in assets or an increase in liabilities (credit). Thus, an excess of revenue over expense for a period (positive net income) is reflected in an increase (credit) to the organisation's equity as well as an increase (debit) to the total of its assets less liabilities, while an excess of expense over revenue (negative net income) is reflected in a decrease (debit) to the equity as well as a decrease (credit) to the total of assets less liabilities.

For commercial organisations, positive net income is referred to as **profit**, and negative net income as loss. Non-commercial or not-for-profit organisations tend to use the terms 'surplus' and 'deficit'. In addition to the focus on flows of wealth-components which make up profit or loss, accounting is concerned with the reasons for changes in an organisation's cash resources (cash flows) during a period. These changes are analysed into various categories in a periodic cash flow statement. This is the third component, together with the balance sheet and the statement of income, of the organisation's financial statements. The cash flow of tourism-related businesses can be extremely important because of the seasonal nature of this industry (see also **seasonality**). There may be times of the year when business is highly active, but others when the business may become slow or even dormant. In order to survive, businesses subject to such seasonal swings in trade must take into account their cash flow projections throughout the year. The effects of seasonality on the ability of businesses in the tourism industry to earn enough revenue and to control their expenses so as to make an annual profit are similarly crucial.

Finally, for any but very small organisations, the accounting system consists of two interlocking subsystems: financial accounting and **management accounting**. The financial statements produced by the former are concerned with the organisation as a whole, and are produced for the benefit of those not involved in the management of the organisation such as shareholders. Production of financial accounting statements is normally annual with interim statements being produced either six-monthly or quarterly. The annual financial statements may be subject to **auditing**. The statements produced by management accounting are concerned with sub-units within the organisation, some of which (**profit centres**) may have separately identified revenue and expenses so that their contribution to overall profit (profit contribution) can be calculated, while for others (cost centres) only the expenses are separately identified. Management accounting statements are produced at frequent intervals, such as monthly, and in conjunction with **budgeting** are an important part of an organisation's financial controls. However, given the nature of the tourism industry, many establishments are quite small and, as such, are unlikely to involve shareholders outside of the operators and their families.

See also: uniform system of accounting

SIMON ARCHER, UK

acculturation

Acculturation is a concept that has been closely associated with American **anthropology**, but has also been used in **sociology**, **psychology** and elsewhere. The term gained currency in the late nineteenth century as a way of looking at a certain kind of culture change. Usually, it refers to the social processes and consequent social and psychological changes that occur when peoples of different cultures come into contact. Along with the concept of diffusion, which refers to a people's acquisition of traits from another culture, it may be contrasted with independent invention which has been employed to indicate internally generated sociocultural change. It should be distinguished from assimilation, which refers to the absorption of one group by another, and not confused with enculturation or **socialisation** which are concepts for the process whereby growing individuals acquire the culture of their native group. British anthropologists have used the concept of culture contact to embrace roughly the same phenomena as acculturation.

Although acculturation theoretically embraces all cases of culture contact and the resulting changes in all of the parties involved, like the notion of **development**, it has mostly been employed in first-hand studies of the impact of a hegemonic Western culture on native cultures, as in colonialism or neo-colonialism. Considering the state of the world during the rise of Western **imperialism** and its often profound effect on less-developed **natives** who were the principal subjects of anthropological studies, it is easy to see why terms like development or the narrower notion of acculturation came into being. Rarely does one encounter in the literature the term being used in the broader sense to indicate a two-way process. Though there have been some attempts to connect different aspects of acculturation in a systematic way – as, for example, concerning the relationship between cultural compatibilities and personal and social conflict – one would be hard put to claim that acculturation, like evolution, ever has acquired the status of a full-blown theory. It has mostly been employed in a descriptive way.

Perhaps the most authoritative discussion of the concept of acculturation is to be found in the formulation of the seminar convened by the (American) Social Science Research Council (1954), a group made up of psychologically sensitive anthropologists and sociologists. This seminar, in its 'exploratory formulation' envisaged four principal facets of the phenomenon: the evolving (autonomous) cultural systems of different peoples who have come into contact, the contact situation, the associated conjunctive (social) relationships involved, and the resulting processes of sociocultural (including psychological) change. The general causal direction, therefore, is from the first to the fourth facets, although feedback loops would not seem to be ruled out.

From the viewpoint of present-day anthropology, this formulation has a number of inadequacies that include its functionalist orientation towards autonomous, integrated, self-regulating cultural systems (which tends to de-emphasise fragmentation, contentiousness, and reproductive dynamics), its failure to attend to relevant sociocultural contexts – as in, for instance, colonialism – and its lack of consideration of the production and distribution of **power** in acculturative transactions. If these inadequacies are addressed and the necessary adjustments made, it would seem that the concept of acculturation could continue to be a powerful master scheme for viewing culture contact, not only in the contemporary world but others as well.

The investigation of tourism as a transcultural phenomenon of many forms and aspects can benefit from the adoption of such a scheme. Indeed, anthropological studies of tourism, which so far have been mostly confined to issues associated with the impact of Western tourism on the **development** of native cultures, already have made liberal use of acculturation as well as the concept of development. Further studies over the entire range of touristic contact situations are needed. Continuing work on the consequences of **tourism** for **tourists** and their agents also can be comprehended as a kind of reverse acculturation, although the job of following through such reactions back into tourism generating cultures should be seriously considered. Finally, the study of the production of tourism and tourists (with their various **motivations**), which may be conceived of as setting in motion the touristic acculturative process in the first place, merits extended attention. It would seem, therefore, that a suitably revised view of acculturation would provide a useful anthropologically oriented overview for cross-cultural investigations of various aspects of tourism in all of its forms from a variety of theoretical perspectives.

Further reading

Berry, J. W. (1979) 'Social and cultural change', in H.C. Triandis and R. Brislin (eds), *Handbook of Cross-Cultural Psychology*, vol. 5, Boston: Allyn and Bacon. (Acculturation seen from a psychological point of view.)

Lindon, R. (ed.) (1940) *Acculturation in Seven American Indian Tribes*, New York and London: Appleton-Century. (A range of examples of acculturation involving whites and Indians.)

Nash, D. (1996) *Anthropology of Tourism*, Oxford: Pergamon Press. (Chapter 5 includes a discussion of acculturation as a master scheme for the anthropological study of tourism.)

Nuñez, T. (1989) 'Touristic studies in anthropological perspective', in V. Smith (ed.), *Hosts and Guests: The Anthropology of Tourism*, 2nd edn, Philadelphia: University of Pennsylvania Press. (The study of tourism in anthropology, with some exceptions, is seen as following the same line of development as the study of acculturation.)

DENNISON NASH, USA

achievement

The motive to achieve or succeed is not limited to work settings and has been used to understand tourism **motivation** in diverse activities and places. The formative ideas in achievement motivation date back to early **psychology** writers such as McClelland and Adler, who emphasised the importance of social factors in understanding human motivation. In a tourism context, achievement has been linked to concepts of mastery and competence and has been featured in studies of moderate **risk** activities such as scuba diving and **skiing**.

Achievement is frequently related to a discussion of intrinsic and extrinsic motivation. The former refers to behaviours that individuals conduct because these **behaviours** are inherently rewarding and satisfying. For intrinsically motivated behaviours, tourists care little about the social reactions of other people. For extrinsically motivated ones, reaching the goal is vital and the success is fully dependent on the perceived or direct judgement of other people. In the tourism context, extrinsic motivation has been seen as equivalent to achievement motivation. For example, climbing a specific mountain might be either extrinsically motivated (it is only worthwhile if everyone knows the climber reaches the top) or intrinsically motivated (it is worthwhile no matter how far one climbs). There is some evidence that achievement motivation holds potential hazards for a **tourist**, since a failure to reach a publicly sanctioned goal may result in disappointment and a sense of failure. Highly intrinsically motivated individuals are less likely to be distressed if their actions do not meet their ambitions.

There are also links between achievement motivation and the perception of **time**, as people high in the need for achievement show greater signs of frustration and boredom as they perceive time to pass more slowly than it actually does. Achievement motivation measures have not been used widely in tourism research, but may hold some promise for individuals who are learning new activities and who are seeking social approval for their pursuits. Understanding the mix of motives may also help some tourism operators provide enough time for highly achievement oriented individuals to reach the goals and achieve the outcome so necessary for their **satisfaction**.

Further reading

Fielding, K., Pearce, P.L. and Hughes, K. (1992) 'Climbing Ayers Rock: relating motivation, time perception and enjoyment', *Journal of Tourism Studies* 3(2): 49–57.

Graef, R., Csikszentmihalyi, M. and Giannino, S.M. (1983) 'Measuring intrinsic motivation in everyday life', *Leisure Studies* 2: 155–68.

PHILIP L. PEARCE, AUSTRALIA

acquisition *see* merger

ACTA Turistica

Acta Turistica is a bilingual (Croatian and English) journal. It publishes articles resulting from research conducted at the University of Zagreb, as well as submissions received from outside researchers. Its articles are aimed at academics, students and practitioners. All submissions are subject to editorial approval (double-blind review). Only those not published previously may be considered. Articles written in other languages than English will also be considered. First published in 1989, the journal is published twice yearly by the University of Zagreb Faculty of Economics (ISSN 0353–5177).

RENE BARETJE, FRANCE

action research

Action research is concerned with diagnosing a problem in a specific context and attempting to solve it in that frame. It has been used extensively in social science research since the 1960s. This research method, however, has become widely known partly because of the supposed incompatibility between the apparently separate entities of action and research. As each has its own ideology and procedures, when linked, action and research appear as uneasy allies.

An apparent incompatibility between action and research is that while research requires a clear **purpose**, action is random, tentative and non-committal. However, action research can be systematic, although the techniques used are not necessarily set in advance. Both **qualitative research** and **quantitative methods** can be used, although the former are more common. The research process is reflexive, involving **planning**, acting, observing and reflecting on data to generate new research questions, before continuing with the next stage of planning, acting and so on.

Action research has been employed when a change in a **social situation** is sought, and involves an investigation of the intervention and its implications. It frequently involves different practitioner groups researching their own practices. Such groups are attempting to gain understanding of their social practices and the contexts in which the practices are conducted in order to improve these practices. Most action research is small scale and collaborative within and sometimes between practitioner groups. For some researchers, collaboration is the attraction of the research method and a key to understanding what is under investigation. However, defining the roles of those conducting this form of research is critical, but when collaborative action research works well it can be an emancipating experience.

In tourism, action research can be used by employee groups, such as **travel agency** staff, airline cabin crew (see **airline**) or **heritage interpretation** officers, to better understand their work environment in an attempt to improve it. Action research can also aid tourism instructors to improve the quality of their activities.

Action research has the advantage that it is clearly goal-driven, may provide innovative approaches to problem-solving and can be rewarding for participants. Although it may lack the rigour of true scientific research, it is viewed as a preferable alternative to more subjective and impressionistic approaches to problem-solving. The chief disadvantages are that far too much data may be collected, some of which is later rejected, the research can be very time-consuming, the direction of the research may often appear unclear and knowing when to stop is not easy.

Further reading

Carr, S. and Kemmis, S. (1986) *Becoming Critical: Education, Knowledge and Action Research*, Lewes: Falmer Press. (Contributes to the theory of action research in an educational setting.)

Kemmis, S. and McTaggart, R. (1982) *The Action Research Planner*, Geelong, Victoria: Deakin University Press. (Provides guidance on how to conduct action research.)

McNiff, J. (1988) *Action Research: Principles and Practice*, London: Macmillan. (Comments on both the theory and practice of action research.)

PETER MASON, NEW ZEALAND

activity

Activity is a general term which refers to patterns of **behaviour** or movement. In the tourism context, activities can be seen as those things done while on **vacation**. In psychology, activities are seen as specific responses to environments and as such represent the link between what individuals are seeking from an **experience** (their motives or expectations) and what the **environment** allows. Thus destinations and attractions can be seen as offering opportunities for **visitors** to engage in various activities.

There are several areas of tourism research and **management** practice which are concerned with activities. In **market segmentation** research, for example, activities that visitors seek or actually choose to participate in can be used as a basis for distinguishing between tourist groups. Activities are also an important component of the management of visitors in natural environments. The **recreation opportunity spectrum**, for instance, is a system for deciding on the locations of facilities and access which allow for certain activities to be restricted to certain areas. At the core of this system is the proposal that visitors choose activities which reflect their **motivations** or needs.

The analysis of activities has also been a research area in environmental psychology and in human **geography**. In the latter case, the most common focus has been on the spatial locations and distributions of activities. In tourism, this work can be seen in studies of tourist time budgets. In environmental psychology, researchers have concentrated on either understanding how environments influence activities or on how activities can be used to define places.

Further reading

Canter, D. (1977) *The Psychology of Place*, London: Architectural Press. (Discusses the central role of activities in understanding places.)

Moscardo, G., Morrison, A.M., Pearce, P.L., Lang, C. and O'Leary, J.T. (1996) 'Understanding vacation destination choice through travel motivation and activities', *Journal of Vacation Marketing* 2(2): 109–22. (Provides a review of the role of activities in tourism, as a link between motivation and destination choice.)

GIANNA MOSCARDO, AUSTRALIA

activity space

The activity or action space of an individual refers to the assemblage of places in which specific activities, such as sleeping, feeding and working usually occur. All these places are connected by an elaborate network of paths, along which users of these places travel according to more or less strict daily, weekly or seasonal routines. The activity spaces of individuals are akin to territories, but these overlap to varying degrees and differ according to such variables as **gender**, age, **race** and socioeconomic status. The concept has been applied frequently to studies of **leisure** activities (Elson 1976) and linked to the patterns of movement in and the images of cities (Lynch 1960).

Husbands (1986) has shown that the evaluations of the qualities of **sites** and the activity spaces of residents of **destination** areas differ markedly from those of **tourists**. Furthermore, the patterns of movement and associated perceptions of place vary greatly between visitors of different types. Pearce (1987) reports work on Lourdes which shows that **pilgrims** travelling in groups have more confined movements than those who are independent. Somewhat similarly, Lucas (1964) found that visitors to a **wilderness** area travelling by canoe had larger activity spaces and more constraining definitions of wilderness than those going by boat. Some work has been conducted, for example by Cooper (1981), on visitor search **behaviours** which helps to understand how visitors explore destinations and how the activity spaces of visitors in destination areas are formed.

References

Cooper, C.P. (1981) 'Spatial and temporal patterns of tourist behaviour', *Regional Studies* 15: 359–71.

Elson, M.J. (1976) 'Activity spaces and recreational spatial behaviour', *Town Planning Review* 47: 241–55.

Husbands, W.C. (1986) 'Leisure activity resources and activity space formation in periphery resorts: the response of tourist and resident in Barbados', *The Candadian Geographer* 30(3): 243–9.

Lucas, R.C. (1964) 'Wilderness perception and use: the example of the Boundary Waters Canoe Area', *Natural Resources Journal* 3: 394–411.

Lynch, K. (1960) *The Image of the City*, Cambridge, MA: MIT Press.

Pearce, D. (1987) *Tourism Today: A Geographical Analysis*, Longman: Harlow, 78–81.

GEOFFREY WALL, CANADA

adaptation

When culturally diverse groups of individuals come into contact, there will be inevitable **impacts** on each others' value system, social fabric, cultural structure, political and economic processes. At the social level, the outcome is usually assimilation, **acculturation**, segregation or integration. At the individual level, the result may be more transitory and assume the form of 'passing', exaggerated chauvinism, **marginality** or cultural mediation. Adaptation thus occurs at both social and individual levels.

Adaptation is often used to describe the process of psychological reaction whereby groups or individuals accept and adjust themselves to fit into a novel or unfamiliar **environment**. The original concept of adaptation emerged from evolutionary selection. Cultural **innovation** must either adapt or alter pre-existing traits within a culture, in neighbouring cultural systems or in the natural environment. The more adaptable the cultural group is, the better the chance of its survival. As some cultures continuously grow more adaptive and powerful, they will tend to dominate less privileged cultures. Such dominance may be manifested in genocide, assimilation, segregation, **imperialism** or colonialism (see **colonialisation**).

Although both domestic and international tourism involve cultural exchange, the cultural clash is more evident in the latter. According to Valene Smith's tourist typology (see **typology, tourist**), explorers can easily adapt themselves to local norms, but incipient mass tourists cannot adjust to the **lifestyles** of the destination **community**. However, tourists' levels of adaptation are not always a concern since they are expected to have little involvement in the core relations of the host culture. It is the host population that must find a way to accommodate the norms of tourists. When the destination people impose their **values** and norms on to visitors and deny the value of the tourist culture (see **culture, tourism**), this is 'exaggerated chauvinism'. This type of adaptation is evident when the host culture is in the First World. On the other hand, when the local inhabitants, particularly in the **Third World**, forego their traditional values and norms in return for a perceived superior tourist culture, this type of adaptation is called 'passing'. Yet, host adaptation is not always affected by 'passing'. Rather, it often chooses to adapt to the tourist culture in order to derive **benefits** from tourism **development**.

Further reading

Smith, V. (1989) *Hosts and Guests: The Anthropology of Tourism*, 2nd edn, Philadelphia: University of Pennsylvania Press.

Furnham, A. and Bochem, S. (1986) *Culture Shock: Psychological Reactions to Unfamiliar Environments*, London: Routledge.

ATSUKO HASHIMOTO, CANADA

adventure tourism

Adventure tourism involves trips with the specific purpose of exploring a new **experience**, often involving perceived risk or controlled danger associated with personal challenges, in a natural **environment** or exotic setting. One can think of a continuum of adventure tourism from 'soft' to 'hard' adventure. The 'harder' the choice, the greater is the physical challenge, perceived risk and the element of danger (see **motivation**; **recreation**; **typology, tourist**). The adventurer choice of potential activities range from **hiking** and **walking tours** to mountain climbing (see **mountaineering**), whitewater racing and kayaking. Today, such excursions are available for all ages worldwide, but countries such as **New Zealand**, **Nepal**, **Kenya** and Tanzania are among popular adventure destinations.

By definition, adventure tourism involves the use of a destination's natural resource base (see also **nature tourism**). The topography and natural features of some countries and regions are especially well suited for this form of tourism. In some parts of the world, such as in Nepal, there is concern about the long-term sustainability of adventure tourism. Another important issue is with respect to the **safety** of certain activities and the need for rigorous **training** for the staff of adventure **tour operators**. This has been as the result of the growing number of fatalities and serious injuries that have occurred during adventure tourism excursions.

Adventure tourism involves both a special interest **market segmentation** and a growing number of specialised tour operators who provide the equipment, experienced **tour guides** and **itineraries**. Adventure tourism is considered to be a growth market which has been accompanied by the development of an ever-increasing **inventory** of adventure tours, the increased sale of recreational equipment and the publishing of specialised **magazines** and journals related to this subject. The increasing interest in adventure tourism has also led to the staging of specialised adventure travel shows and exhibitions, and to the formation of professional associations.

See also: alternative tourism; typology, tourist

Further reading

Sung, H.H., Morrison, A.M. and O'Leary, J.T. (1997) 'Definitions of adventure travel: conceptual framework for empirical application from the providers' perspective', *Asia Pacific Journal of Tourism Research* 1(2): 47–68. (A study of how adventure operators define this market and its tourists.)

ALASTAIR M. MORRISON, USA
HEIDI H. SUNG, USA

advertisements, envy in

Like other forms of **advertising**, tourism promotion often relies on the creation of envy. Thus, potential tourists who view a glamorous couple surrounded by symbols of luxury in a foreign setting envy them to the point of wishing to exchange places with them and, in their desire, create self-envy. This technique is frequently employed in **hotel** and **resort** publicity.

See also: promotion, place

GRAHAM M.S. DANN, UK

advertisements, format of

In tourism marketing, format relates principally to the positioning, size, shape, content and structure of advertisements appearing in newspapers and brochures. Of these considerations, structure has received the most attention, focusing as it does on the appellative function of language and the general AIDA format of **advertising** (capture attention, maintain interest, create desire and get action).

GRAHAM M.S. DANN, UK

advertising

Advertising is commercial or paid communication using one or more media other than personal selling, face-to-face meetings of salespersons and customers. Media used to transmit advertising include broadcast (such as television, radio, and movie theatre commercials), print (newspaper and magazine advertisements, catalogues, information guides or brochures, and direct mail literature), electronic (computer websites and telephone marketing), signs (in-store displays, outdoor billboards), and speciality advertising (brand names/symbols on merchandise, premiums or free merchandise, sky-writing and ads on blimps). Among other objectives, advertising is used to gain customer awareness of a **product**, **service**, brand or idea; to increase customer knowledge about how to go about buying and using the product or service; to generate a positive **image**, as well as positive attitudes and preferences for buying the product or service, among customers; and to cause purchases of the same during the current time period and in the future by new and repeat customers (see also **marketing**).

Tourism advertising may be categorised into five types: image advertising only, image advertising with a linkage-advertising offer, offers focused on linkage-advertising, linkage-advertising materials, and direct-response advertising. Image advertising is advertising that attempts to build a three-way association among the behaviour being advertised, one or more benefits and target market customers. Image advertising usually includes both an illustration and ad copy to gain comprehension and acceptance of the three-way association: the image. Image advertising includes the headline and some of the copy of a magazine advertisement promoting, for example, leisure travel to Cyprus:

> 800 BC. What is Aphrodite doing in Cyprus? Loving it. On the Island of Cyprus, the Goddess of Seduction is said to have emerged from the waves at Aphrodite's rock. Her passion was shared with many others, including Cleopatra and Leonardo da Vinci. Now you can share it, too! Fly back nine millennia to a land of sunshine, Greek temples, Roman amphitheatres, and some of the best hotels and food in the Mediterranean.

The above quote is actually the contents of two full-colour photographs (published in *Condé Nast Traveler* in March 1997) included in this one-third page ad. One shows 'Aphrodite's rock', and the other a tourist in a bathing suit on the beach facing the rock. A primary objective of such an image ad is to cause automatic retrieval from memory of the three-way association ad promoting Cyprus as a leisure destination.

After being exposed to this image message, might Cyprus be retrieved automatically when next time one hears or reads the name 'Aphrodite'? Building such automatic retrievals is a major objective of image advertising. Achieving automatic retrieval of a brand's name when thinking about another object or a benefit causes the brand to be more available for thinking. Because a brand is more likely to be purchased if a customer is thinking about the brand versus brands not being thought about, increases in availability in customers' minds are often goals of advertisers (see **branding**; **product positioning**).

The image technique with a linkage-ad offer has the highest share of advertising expenditures in tourism. The Cyprus–Aphrodite–tourist image also included a linkage offer: 'Call... for a full-colour brochure or visit [our website]'. The linkage-ad is the full-colour brochure itself. The use of image advertising coupled with the linkage can be very effective because several objectives may be accomplished. The image projected in the ad helps to build awareness and positive attitude toward the product-service, and the linkage helps to build knowledge and skills to use in buying and experiencing the product-service. Small-space, classified ads are examples of advertising focused on making offers of linkage; usually such materials are free and are mailed to inquirers with no charges for postage and handling. Given the high cost per inquiry of publishing, handling and mailing linkage-advertising, as well as the free linkage offer, a key issue is whether or not the revenue return resulting from the adverting is greater than the cost. Extensive examinations of its effectiveness are performed each year to provide specific answers to this question.

The linkage-advertising materials can be categorised into two types: detailed information on how to go about buying and using the tourism product or service versus a brochure of photographs with sparse information. An example of the first category is the information guide to Prince Edward Island, Canada. Each year this guide includes a brochure that includes date and details of festivals; the unique facilities and rates per night of hotels, **motels**, **bed and breakfast** homes and campgrounds; specific attractions to see and things to do in each county and city; detailed features of all attractions and parks in this Canadian province; and detailed maps of alternative touring routes and where to stop in each of its three counties.

Direct response advertising (see **direct marketing**) are communications that include an offer to sell a product or service within the ad itself. For example, another ad, included in a one-sixth page ad in *Condé Nast Traveler* (March 1997) by Hideaways International, says: 'Enrol me for membership at $99 per year – including my bonus (Enclose check, MC, Visa, Discover or Am Ex number and expiration date).' Many advertisers with products–services that might be sold using direct response versus other types have determined that the former is less effective than using image in combination with linkage-advertising. However, the ad by Hideaways International may be an exception to this rule of thumb.

Steps in the advertising process include deciding on target customers to guide the design of the creative message, selection of the media, choice of specific media vehicles (which television programme, radio stations or newspapers); selection of an ad schedule (which months, days, hours to run the ads and with what frequencies). Possibly the biggest mistake in planning an advertising campaign is to begin with the assumption that the objective is to reach everyone. Such an objective follows from believing the false proposition that a given tourism product–service has something to offer everybody. To build a positive image, a product–service needs to be focused on a limited appeal, highly desired by a given target group of customers. Several research studies support the proposition that a specific tourism brand must first be retrieved from memory quickly when the customer thinks about a specific benefit. If the ad message does not build a unique association with such a desired benefit, the product–service is unlikely to be retrieved when the customer is thinking about buying it. Trying to appeal to everybody weakens the likelihood that the offer is uniquely satisfying for anyone (see **market segmentation**).

Tourism related advertising is done each year by many specific destinations. Cities, regions, counties, states, provinces and countries plan annual advertising campaigns. Advertising is also done each year by hotel, restaurant, transportation, and attraction, as well as other tourism-related industries. Most of these firms do both push and pull advertising. Push advertising includes commercial communications directed to reach travel agents and persons in related industries who influence tourist choices. Pull advertising includes commercial communications directed at reaching the potential tourist. Measures of advertising effectiveness are focused on the amount of additional revenue generated from the advertising campaign that would not have occurred without running the campaign. High-quality research methods are available to make the assessments, but their execution is usually expensive (see **marketing research**).

Consequently, proxy measures of net profit impact are used in most research on advertising effectiveness. One such proxy is unaided awareness, as to how much increase resulted from running the ads that would not have occurred without them. However, the most widely used measures of advertising effectiveness include cost per thousand readers or viewers exposed to the advertisement and aided recall measures. This measure equals the cost of the ad divided by the number of thousands of persons exposed to it. The aided recall measure is showing ads to persons, usually to subscribers of magazines or television viewers, after the persons had an opportunity to read or view them, and asking if they recalled seeing them. Aided recall has not been found to be an accurate predictor of the influence of advertising on purchases. However, this measure has been used for many decades. It has historical top-of-mind-awareness as a measure of advertising effectiveness.

Further reading

Ritchie, J.R.B. (1996) 'Beacons of light in an expanding universe: an assessment of the state-of-the-art in tourism marketing/marketing research', *Journal of Travel and Tourism Marketing* 5(4): 49–84.

Woodside, A.G. (1990) 'Measuring advertising effectiveness in destination marketing strategies', *Journal of Travel Research* 29(2): 3–8.

—— (1996) *Measuring the Effectiveness of Image and Linkage Advertising*, Westport, CT: Quorum Books.

ARCH G. WOODSIDE, USA

African Travel Association

The purpose of the African Travel Association (ATA) is to promote African destinations, particularly in the North American market. Established in 1975, it is an international non-profit, non-political industry association. A different African country hosts its annual congresses and **trade shows** each time. Headquartered in New York, *Africa Travel News* is one of its publications.

TURGUT VAR, USA

Agenda 21

Agenda 21 is a comprehensive program of action adopted in June 1992 by 182 governments at the United Nations conference on Environment and Development, known as the Earth Summit. It provides a blueprint for securing the sustainable future of the planet into the twenty-first century. It identifies issues that represent economic and ecological disasters and presents a strategy for transition to more **sustainable development** practices.

SALAH E.A. WAHAB, EGYPT

agrotourism

Agrotourism, sometimes spelled agritourism and also often called **farm tourism**, refers to the provision of touristic opportunities on working farms. Many farms are located in attractive rural settings which, although work environments, may also be interesting **leisure locations** to urban residents. Agrotourism implies a closer interaction with rural **landscapes** and residents than the mere viewing of scenery. **Accommodation** may be provided on farms for **visitors** in such forms as **bed and breakfast** or campsites, produce may be sold through 'pick-your-own' schemes or farm markets, blossom or other tours may be arranged depending on location. Opportunities for active recreations such as **hiking**, fishing, horseback riding and snowmobiling may be provided depending upon the resources available. In these and other ways, farmers can gain a supplementary income from tourism and, if successful, this activity may even eventually replace farming as the major income earner of the enterprise.

Agrotourism is often viewed as a means of stimulating the economies of declining rural areas. However, success depends very much on access to markets and the nature of the farm enterprise, with mixed farms raising both animals and crops in varied topography being usually more attractive than monocultures. Furthermore, it has been found that those farmers most in need of a supplementary income are often least able to benefit from agrotourism because they lack the necessary capital, **management** and **marketing** skills.

Further reading

Cox, L.J. and Fox, M. (1991) 'Agriculturally based leisure activities', *The Journal of Tourism Studies* 2: 18–27.

Davies, E.T. and Gilbert, D.C. (1992) 'A case study of the development of farm tourism in Wales', *Tourism Management* 13: 56–63.

Evans, N.J. and Ilbery, B.W. (1989) 'A conceptual framework for investigating farm-based accommodation and tourism in Britain', *Journal of Rural Studies* 5: 257–66.

Telfer, D. and Wall, G. (1996) 'Linkages between tourism and food production', *Annals of Tourism Research* 23(3): 635–53.

Weaver, D.B. and Fennell, D.A. (1997) 'The vacation farm sector in Saskatchewan: a profile of operations', *Tourism Management* 18: 357–65.

GEOFFREY WALL, CANADA

AIDS

An estimated 17 million people (according to the World Health Organisation in 1995) suffer from Acquired Immunodeficiency Syndrome (AIDS), mainly transmitted through unprotected sexual intercourse. Increased **risk** behaviour on vacation has led to various assessments of infections during **international tourism**, though no reliable global estimates of tourists' AIDS acquisition abroad are available. Some countries have adopted AIDS-related entry restrictions and some have legislated against **sex tourism**.

PETER GRABOWSKI, UK

Air Transport Association of America

The means of travel from one **destination** to another has changed drastically over the past one hundred years and continues to change at an increasing rate. **Technology**, advances in communication, ease of travel and market **demand** have increased the volume of air transportation, cargo and passengers alike. Nine out of ten commercial travellers (buses, trains, ships and planes) use air transportation within the United States, and 95 per cent do so to go abroad. As the airline industry or sector has grown and evolved, so has the influence and involvement of **special interest groups** and governmental bodies, which have varying agendas. Some airline carriers have joined forces to meet the needs and challenges of these outside influences and involvements. One such organisation is Air Transport Association of America (ATAA).

The Association was founded in 1936 with its primary mission to, according to the association, 'help stabilise a fledgling industry through government regulation of routes and rates'. As this sector has modified and changed over the years, so has the mission of ATAA, which is to support its member carriers by promoting **safety**, cost effectiveness and technological advances. The association currently has twenty-two members, as well as three non-USA air carriers.

As the demand for air travel continues to grow throughout the world, a greater emphasis will be placed on airline carriers to be more efficient and effective. Those carriers which can co-operate with one another will have a **competitive advantage**. ATAA provides its members a venue to develop co-operative efforts to fulfil the demands of governmental agencies, special interest groups and consumers. The headquarters of the ATAA is located in Washington, DC.

RUSSELL R. CURRIE, CANADA

air travel demand *see* demand, air travel

aircraft, supersonic

A supersonic aircraft is one capable of flying faster than the speed of sound. For example, within the world of commercial aviation, the Concorde carries 100 passengers and a crew of nine at a cruising speed of 2,170 kilometres per hour, or

1,350 miles per hour. By comparison, a Boeing 747 'jumbo jet' cruises at 1,015 kmh.

ANTHONY G. SHEPPARD, USA

airline deregulation *see* deregulation, airline

airline distribution systems

Airline distribution systems are the means by which airlines make their flight capacity available to customers for purchase. Airlines tend to use a variety of distribution channels to reach customers, often a mixture of both direct and indirect means. The latter methods involve the use of intermediaries such as retail travel agents (see **travel agencies**) or **tour operators**. Some airlines use direct distribution channels without any intermediary organisations. The role of the travel agent has traditionally been important in airline distribution systems, although this appears to be changing.

More direct means of distributing airline products have assumed significance also and include bookings by telephone (often a toll-free number) or online. Airline distribution systems which exploit **technology** have been evolving rapidly. **Computer reservation systems**, interactive kiosks and the **Internet** are examples of these. This is a dynamic field of activity and further developments are likely.

Some airline seats are sold as part of an air inclusive **package tour**, with **tour wholesalers** such as tour operators combining the different components of these packages. Airlines can perform the wholesaling function themselves, and many do so today. This is particularly true in countries such as the **United States** where the domestic market is deregulated. Airlines may also own their own retail travel outlets, although many do not.

The importance of travel agencies as a system used to distribute airline capacity varies among countries. Their role is to act as an intermediary between the airlines, wholesalers and consumers. Commission is paid by the airline or wholesaler for the service that they offer. Travel agents usually employ some form of computer system due to the complexities of meeting consumer needs when making bookings on their behalf. Some act as consolidators, distributing scheduled and flight-only charter seats for airlines. The latter will often have some form of affiliation with the other organisations in their chain of distribution.

Viewdata is a generic term used to describe information systems that previously dominated the leisure travel agency network in some countries. These early systems used telephone lines to transmit data among channel members. Computer reservation systems and more recently the Internet have advantages over prior communication systems, although they are more controversial since they can bypass travel agents in some instances.

Further reading

Poon, A. (1993) *Tourism, Technology and Competitive Strategies*, Oxford: CAB International. (Provides an overview of the different stages in the development of CRS and major issues surrounding them, especially Chapter 7, pp. 172–94.)

LESLEY PENDER, UK

Airline Reporting Corporation

The purpose of the Airline Reporting Corporation (ARC) is to provide a method of approving authorised agency locations for the sale of **transportation** and cost effective procedures for processing records and funds of such sales to carrier customers. The process provides a combination of fixed reporting requirements, for ease of record keeping, with variation of payment terms for competitive purpose. The Industry Agents' Handbook, provides, among other things, the basic ticketing and reporting requirements for over 47,000 ARC-approved **travel agency** locations to report and settle ticketing transactions with the carrier participants by means of ARC Standard Ticket and Area Settlement Plan. ARC is located in Arlington, Virginia in the **United States**.

TURGUT VAR, USA

airline

Airlines (air carriers) provide commercial **transportation** linking numerous origins and destinations worldwide on both domestic and international routes. Airlines are owned in either the private or public sectors. In the **United States** airlines have always been privately owned, while in most countries airlines developed under government ownership. For those with government ownership, generally only one airline of that country provides international service. Recently there has been a movement toward deregulation of the airline industry, domestically and internationally.

A number of different types of airlines provide service. One way of **classification** is by what is transported: passengers only, cargo only, or both passengers and cargo. The latter type predominates in both domestic and international transportation, especially for wide-body aircraft service. Some airlines such as Korean Airlines and Northwest Airlines carry substantial amounts of cargo in addition to passengers.

Because airlines often operate in competitive environments, certain managerial functions are significant. **Marketing** receives considerable attention due to the need to convince potential passengers to fly on a particular airline. This can be a difficult task, since the basic airline **product** is generally homogeneous. Airlines often differentiate themselves from competition through such elements as in-flight services, schedules, points served and frequent flyer programmes.

Operations is another major managerial function of an airline. In its simplest form, operations is the provision of the basic service of transporting passengers between two airports. Operations, however, is much more complex and includes other elements of service. The primary operating areas are flight operations, terminal operations, and maintenance and repair. Flight operations includes the activities of cockpit and cabin crews, along with in-flight service. Terminal operations includes services such as ticketing, check-in and baggage handling. Maintenance and repair supports the overall operations of the airline and its equipment.

A number of significant changes are taking place which affect airlines. Among the foremost is deregulation of the industry, with many government-owned airlines such as British Airways and Japan Airlines having been privatised. The increasing use of code sharing and alliances amongst airlines has changed industry competition. Managers and others adapt to these changes or risk failure of the airline.

Further reading

Guillebaud, D. and Bond, R. (1997) 'Surviving the customer', *Airline Business* March: 54ff. (Discusses some important airline strategies.)

Hanlon, J.P. (1996) *Global Airlines: Competition in a Transnational Industry*, Oxford: Butterworth-Heinemann. (An overview of the current and emerging airline industry.)

FREDRICK M. COLLISON, USA

airport hotels *see* hotel, airport

airport

Airports are a key link between airlines, **tourists** and travel destinations. While vast attention has recently been given to airline services and growing air travel demand (see **demand, air travel**), airports received little attention despite their importance. Carriers, travellers and airport authorities now realise that more attention should be paid to the subject, especially for airport congestion.

Technology has much to do with congestion, particularly with the introduction of wide-body aircraft. Airports built to accommodate narrow-body aircraft could not easily accommodate the larger wide-body jets, which forced airport planners to rethink, rebuild and refinance airport facilities. Today's aircraft have greater range and larger **carrying capacity** and require longer runways, necessitating additional land and reconfigured airport buildings.

Deregulation contributed to congestion, with many carriers adopting hub-and-spoke route structures. These **route systems** reduce costs and enhance revenue by increasing dominance at the hub airport, but expand demand for facilities, approaches, and take-off and landing slots. Increased service offerings and lower prices brought about by deregulation increased demand for travel, and such demand funnelled through hub airports worsened airport congestion.

Proposed solutions are as numerous as the problems. Some countries plan to build new facilities to deal with causes of congestion and to realise future opportunities. This strategy requires vast amounts of land and capital resources which are not always available. Airports in this position may only be able to expand existing facilities. Both of these strategies are especially evident in Asia. For many hub airports, such as traditional **gateway** airports in the **United States** and Europe, even modest physical expansion is impossible. In response, airports generally have developed allocation mechanisms for take-off and landing slots. Given the projected growth of air travel into the next century, these are stopgap measures at best.

Further reading

Doganis, R. (1992) *The Airport Business*, London: Routledge. (A basic text on airport management and operations, especially the overview of airports in Chapter One and management issues in Chapter Two.)

Golaszewski, R. (1992) 'Aviation infrastructure: a time for perestroika?', *Logistics and Transportation Review* 28(1): 75–101. (Examines the provision of airports and air traffic control in the USA, the funding of these systems, and potential changes in governance, pricing, and allocation methods.)

Saunders, L.F. and Shepperd, W.G. (1993) 'Airlines: setting constraints on hub dominance', *Logistics and Transportation Review* 29(3): 201–20. (Focuses on the relationship between hub dominance and airline pricing and suggests that limiting hub dominance might improve air system efficiency.)

KEVIN BOBERG, USA

alcohol

Alcohol is a chemical compound consisting of carbon, hydrogen and oxygen atoms. It is produced when a liquid containing sugar is fermented and/or distilled. Depending on the type of alcoholic beverage, there are different ingredients used.

Beer is usually brewed from malted barley, although corn and rye have been used to make beer. It is then mixed with yeast and flavoured with hops. The term 'beer' includes lager (a clear, light-bodied, refreshing beer), ale (fuller-bodied and more bitter than lager), stout (a dark ale with a sweet, strong, malt flavour) and pilsner (a term which means that the beer is made in the style of the famous beer brewed in Pilsner, Bohemia). The alcoholic content of beer ranges from 4 per cent to about 16 per cent by volume.

Wine is made from fermented grape juice, but other fruits have been used to produce wines. Wines can be categorised into four main groups. Still wines can be dry, semi-sweet or sweet, and may be red, white or rosé. Sparkling wines have added carbon dioxide, making them bubbly. Champagnes are fermented in the bottle with champagne yeast. Fortified wines have been enhanced in flavour and texture by the addition of brandy or other liquor. Examples include sherry and port. Aromatic wines are wines flavoured with herbs or spices (an example is vermouth).

Liquor, also known as spirit, is an alcoholic beverage made from grains or plants that have been fermented and later distilled. These beverages are noticeably higher in alcohol content, averaging 40 per cent, or 80 proof, although some liquors can have as much as 151 per cent alcohol content. The most common grains used are corn, rye and barley. Such beverages made from these ingredients include Scotch whisky, bourbon, malt liquor and gin. Other popular beverages are vodka, which is made from grains; rum, which is made from fermented sugarcane juice or molasses; cognac, which is distilled from wine and aged in wood; and tequila, which is made from the fermented and distilled sap of the agave plant.

Alcohol has been a part of tourism for centuries, being enjoyed by some as part of a social setting where adults can relax and enjoy each other's company. **Tourists** frequently partake of local beverages, and **vineyards** are now popular attractions.

See also: drinking; licensing

Further reading

Katsigris, C. and Porter, M. (1991) *The Bar and Beverage Book*, New York: Wiley. (Provides an overview of bars and beverages.)

Lipinski, R. and Lipinski, K. (1989) *Professional Guide to Alcoholic Beverages*, New York: Van Nostrand Reinhold. (Discusses storing, serving, and selling wines, beers and spirits.)

JOHN R. WALKER, USA

alienation

Alienation is the sense of estrangement from a situation, society, group or culture. It is the feeling or state of uninvolvement, of literally being an alien. In other words, it suggests the sense of not belonging.

Although use of the term can be traced back to classical philosophy, its modern usage is most closely associated with **sociology** and the work of Karl Marx (1818–83). According to Marx, the advent of the Industrial Revolution, capitalism and the ensuing division of labour and 'relations of production' between the workers and the capitalists (the owners of production systems) resulted in people becoming increasingly alienated from the society in which they lived and worked. Continuing economic and technological advance, particularly in transport and communications, has done little to reverse this trend and, for many, alienation is a defining characteristic of modern societies. That is, their people are believed to suffer from a feeling of isolation, **placelessness**, powerlessness and meaninglessness (see **anomie**).

However, tourism provides the opportunity to **escape**, albeit temporarily, from this alienated condition, and it is thus within the sphere of **motivation** that the concept of alienation is of most relevance to tourism. For example, Cohen's (1979) **phenomenology** of tourist experiences is based upon the notion that people experience varying degrees of alienation within their home environment (see **continuum models**). Those who are more alienated, whose spiritual centre lies 'out there', are more disposed to seek out tourism experiences that will embellish or add meaning to their lives than those less alienated **tourists** who seek more recreational experiences. In a similar vein, MacCannell suggests that all people experience alienation in modern societies and tourists are, in effect, modern day **pilgrims** seeking reality or **authenticity** in other times and places. In this latter sense, tourism can also be considered a form of time travel. Thus, the increasing popularity of **heritage** tourism is also linked to alienation in as much as the desire to visit historic **sites** may be motivated by the nostalgic yearning (see **nostalgia**) for a premodern (implicitly, non-alienated) era.

References

Cohen, E. (1979) 'A Phenomenology of Tourist Experiences', *Sociology* 13: 179–201.

Further reading

MacCannell, D. (1989) *The Tourist: A New Theory of the Leisure Class*, 2nd edn, New York: Schocken Books.

Sharpley, R. (1994) *Tourism, Tourists and Society*, Huntingdon: Elm Publications.

RICHARD SHARPLEY, UK

allocentric

Stanley Plog proposed that travellers could be classified along a single travel confidence dimension which resulted in three major categories; allocentrics, midcentrics and psychocentrics. Allocentrics are confident, enthusiastic, internationally oriented travellers. The scheme was popularised by being widely reported in tourism textbooks, but its continuing usefulness is questionable due to its simple unidimensional nature, lack of measurement details and its dominant North American focus.

See also: motivation

PHILIP L. PEARCE, AUSTRALIA

alternative tourism

The concept and term 'alternative tourism' was particularly fashionable during the 1970s and 1980s, when criticism of the excesses of **mass tourism** culminated (see **critics**). Affecting a wide variety of fields, alternative tourism became part of a broader movement searching for active and innovative solutions to replace a situation considered intolerable and with dangerous implications for the future. The nature of this reaction is at the same time ethical, ideological and political; the concept is clearly a rejection of existing ideas, which are deemed inadequate and obsolete.

The difficulty encountered in defining 'alternative' tourism stems from the numerous meanings of this term which is both ambitious and vague. As an examination of various **World Tourism Organization** publications would reveal, it is frequently confused with adjectives of a similar meaning, such as integrated, adapted, controlled, endogenous, responsible, authentic, equitable, convivial and participative, with the list being constantly extended depending on its semantic form. The term 'alternative' is the largest and most encompassing of these concepts, each corresponding to the search for new aims in diverse fields but converging in their desire to stop an uncontrollable **development** of tourism. This form of tourism, it is argued, comes with dubious consequences, including overpopulated and inundated **resorts** and social problems, with **tourists** confined to their golden ghettos and locals to their impoverished conditions.

The search for a substitute to the traditional model and its consequences encompasses six main areas which constitute the criteria for alternative tourism. The first concerns the inspiration which leads to a decision and realisation of the holiday. It can be considered alternative when it follows a search for originality instead of choosing present practices, which could lead to a wide range of outdoor activities, both sporting and cultural. In many cases, this search can prove to be elitist and expensive, like the luxury holidays which shun the crammed, conventional resorts; on the other hand this also appeals to young adventurous travellers with limited financial means but with a desire to stray from the beaten track of commercial tourism (see **youth tourism**).

The second criterion concerns the pioneers of alternative holidays, especially those who refuse to assimilate with ordinary tourists through a desire to be different, and those who because of lack of money cannot access the more expensive amenities and resorts. This also includes the followers of **social tourism**, the backpackers and the wealthy devotees of organised, long-distance holidays of discovery and adventure, as well as enthusiasts of ethno-tourism and **ecotourism**. In this way, alternative tourism contributes to the continuing process of **market** and **product** segmentation which characterises modern tourism.

The **destination** of the holiday and the location chosen by tourists form additional criteria for **differentiation**. This same desire to be different leads people to choose new destinations, described in tourism advertisements as exotic, pioneering, undiscovered, new, unexplored or primitive. For as long as the signposted route and the most visited places (towns, developed resorts) remain part of the **international tourism** network, they will be carefully avoided and bypassed, to the benefit of protected regions and deeply rural areas which in theory offer a more authentic experience (see **authenticity**), preserving traditions which elsewhere are in danger of disappearing forever. In general, these destinations, once deemed a far cry from the standard tour and only visited by 'alternative' tourists, are eventually discovered by mass tourism and moulded into the image of the developed resorts.

Alternative tourism is distinguished by the most original forms of welcome, echoing the principles which define it and the chosen style of the tourism development. It can be either a specific type of **accommodation** (usually, one of the various types of local housing or ecological lodgings) or the **adaptation** of a classic formula (such as family hotels, chalets, lodges and camps, as well as holiday villages and furnished houses) where appropriate. Alternative tourism can also be defined by the specialised **tour operators** who function in this sector with specific objectives. Depending on the country and product offered, this includes either non-profit organisations or **travel agencies** and tour operators whose 'alternative' philosophy can be expressed in such terms as ecological preservation of **energy** and water, the recycling of used water and waste, subsidised transport and accommodation to encourage a wide clientele, **training** and **employment** of specialised guides for environmental exploration, and the like.

Finally, the uniqueness of alternative tourism lies in the nature and quality of the relationship it has with the local **environment**. The term should be reserved for progressive developments which are planned, controlled and expanded by the local **community**, and which, in fair **partnership** with future developers, will remain respectful of natural and human ecosystems and economical in their use of essential **resources**. In this way, alternative tourism embraces the principles of lasting and **sustainable development** and its principle commercial manifestation, **ecotourism**.

Further reading

Cazes, G. (1989) 'Alternative tourism: reflections on an ambiguous concept', *Towards Appropriate Tourism: The Case of Developing Countries*, London: Peter Lang, 117–26.

Cohen, E. (1989) 'Alternative tourism: a critique', in H.L. Theuns and F.M. Ter Ver Singh (eds), *Towards Appropriate Tourism: The Case of Developing Countries*, London: Peter Lang, 127–42.

Lanfant, M.F., Allcock, J.B. and Bruner, E. (1995) *International Tourism: Identity and Change*, London: Sage.

Pearce, D. (1989) *Tourist Development*, London and New York: Longman.

Theuns, H.L. and Ter Vir Singh, F.M. (1989) *Towards Appropriate Tourism: The Case of Developing Countries*, London: Peter Lang.

GEORGES CAZES, FRANCE

amenity, user-oriented

Destinations located in proximity to population concentrations are user-oriented. Providers of tourism goods and **services** attempt to maximise the delivery of leisure amenities by positioning their facilities at accessible locations within short distances or travel times of potential **markets**. Hotels and **motels**, and **restaurants** and entertainment venues in **resorts** strive to minimise **travel** costs.

LISLE S. MITCHELL, USA

American Hotel and Motel Association

The American Hotel and Motel Association (AHMA) is a trade association representing the $72 billion lodging industry in the **United States**. It was founded in Chicago in 1910. It is a federation of hotel and motel associations with headquarters in Washington DC, with associations in all fifty states, the District of Columbia, Puerto Rico and the US Virgin Islands.

AHMA represents over 10,000 individual hotels, **motels**, **resorts**, comprising 1.4 million transient rooms. Membership is also open to professional and academic personnel. The association has over forty committees that provide **leadership** and guidance in developing programmes and activities to its members. Members receive *Register* (the official newsletter), unlimited access to the AHMA information centre, *Lodging* magazine and other services. The association organises conventions and workshops across the United States.

The American Hotel Association Directory Corporation, a subsidiary of AHMA, publishes *Lodging* and the *Directory of Hotel and Motel Companies*. Its Educational Institute, another subsidiary, supports educational and institutional research for the lodging industry. Founded in 1952 as a non-profit educational foundation, it is the largest source of quality **training** and educational materials for the industry.

MARYAM KHAN, USA

American Society of Travel Agents

An increase in **demand** for accessibility to destinations throughout the world has forced travel-related businesses to overcome or remove barriers to tourism. One such barrier is 'reliability' in service delivery. Continuity among **service** providers fosters 'assurances' that tourism experiences will take place as planned. This in turn decreases apprehension toward tourism, making it more likeable and less arduous both physically and emotionally for individuals. In response to a need by **tourists**, businesses are now working together and forming **partnerships** for the provision of tourism-related services. The American Society of Travel Agents (ASTA) is one such organisation.

The society was founded in 1931 as the American Steamship and Tourist Agents Association. ASTA's mission is to enhance the **professionalism** of its members worldwide through effective representation in business and government affairs, **education** and **training** and by identifying and responding to the needs of the travelling public. ASTA claims to be the largest travel trade organisation in the world. The society is comprised of 28,600 members in 168 countries, although the majority are USA-based members.

Its membership is divided into two categories, active and allied. The former are **travel agencies** and the other suppliers and providers of tourism **products**. The typical active member of ASTA has 58 per cent of its business in **leisure** and 42 per cent in commercial travel sales. To provide cohesiveness and reliable service from one business to another, members must operate and abide by a **code of ethics**, which in turn is monitored by the association.

ASTA provides a number of services to its members and the public at large. The society convenes several meetings and conferences per year in addition to **trade shows**. In the form of bursaries and scholarships, about $50,000 is made available to students of tourism and academic researchers each year. Support services provide members with access to legal and industry affairs information.

As the number of individuals who travel continues to grow, a greater **demand** will be placed on the tourism industry for the uniformity and reliability of related services. ASTA and its member agencies have taken steps to respond to this demand. The society updates members and the public on scam operations, and provides resources and travel tips information.

RUSSELL R. CURRIE, CANADA

amusement park *see* theme park

Anatolia

Anatolia: An International Journal of Tourism and Hospitality Research, was published in Turkish from its launch in 1990, but switched to English in 1998. Its overall mission is to provide an outlet for studies which make a significant contribution to the understanding, practice and **education** of tourism and **hospitality**. Through its update on Mediterranean tourism, *Anatolia* also aims to heighten awareness of the region as a significant player in tourism. Submissions are reviewed by a double-blind review system. The journal is published quarterly by Nazmi Kozak (ISSN 1300–4220).

RENE BARETJE, FRANCE

animation

Animation is a form of care for guests or tourists with the aim of fostering their behaviour and consequently enticing enthusiasm for certain touristic activities in order to promote specific experiences. Tourist and cultural institutions increasingly interpret animation as a form of communicative leisure work with the objective to free up tourists from their homebound or frequent passivity, to create more intensive sensitivity for new forms of experience and to even facilitate joint adventures, leading to customer satisfaction.

HARALD PECHLANER, AUSTRIA

Annals of Tourism Research

Annals of Tourism Research is a social sciences journal focusing upon the academic perspectives on tourism. While striving for a balance of theory and application, *Annals* is ultimately dedicated to developing conceptual constructs. To perform its role in the development of a theoretically integrated and methodologically enriched multidisciplinary body of knowledge, *Annals* publishes manuscripts dealing with various aspects of the tourism phenomenon. All manuscripts are refereed anonymously (double blind) by at least three reviewers from different disciplines. It regularly publishes full-length articles (abstracts in English and French), research notes and reports, comments and reviews. First published in 1973, the journal is published quarterly by Pergamon Press (ISSN 0160-7383).

RENE BARETJE, FRANCE

anomie

Anomie is a state of normlessness, powerlessness and meaninglessness that pervades society, in contrast to its effects on personality. It was identified by the French sociologist Emile Durkheim as leading in extreme cases to suicide. In tourism research, anomie has been linked to **motivation**, particularly among people whose dearth of interpersonal contacts in the home environment, such as the elderly, encourages them to look for a sense of belonging elsewhere. It may also be experienced by the relatively privileged who do not perceive the need to seek enhanced status through tourism (via 'trip dropping', mixing with the rich and famous and so on), but who instead gain satisfaction from well-tried solutions (repeat vacationers) (see **prestige**).

Anomie is a push factor since it is found in the home setting (see **push–pull factors**), although pull factors of the **destination** can be promoted as appealing to those who lack adequate communication in their monotonous lives. Hence the popularity of group pursuits (such as dancing, bingo or coach tours) which emphasise the communitarian aspects of touristic encounters, and of **activities** which are presented as timeless alternatives to the dull nine-to-five routine of the factory or office.

However, anomie should not be confused with **alienation**. The latter, attributed to Karl Marx, refers to the estrangement undergone at the individual level through relationships based on forced labour. Here, a feeling of exploitation derives from loss of creativity in the workplace, thus rendering workers alienated from the product, themselves, their species being, their colleagues and the owners of the means of production. Where the concept of strangeness is used in devising tourist typologies, it is connected with the theorising of Georg Simmel rather than that of Karl Marx (see also **typology, tourist**).

Further reading

Dann, G. (1977) 'Anomie, ego-enhancement and tourism', *Annals of Tourism Research* 4: 184–94.

Durkheim, E. (1970) *Suicide: A Study in Sociology*, trans. J. Spaulding and G. Simpson, London: Routledge and Kegan Paul.

GRAHAM M.S. DANN, UK

Antarctic tourism

All touristic **activities** in Antarctica other than those directly associated with scientific research and the normal operations of **government** bases are referred to as Antarctic tourism. This 'new' destination receives over 10,000 visitors per annum, 90 per cent of whom arrive by sea. Because of the extreme climatic conditions, Antarctic tourism is highly seasonal (see **seasonality**) and relatively expensive with **safety** and environmental issues being major **management** concerns. This **development**, similar to that of **Arctic tourism**, is now receiving serious scholarly attention.

C. MICHAEL HALL, NEW ZEALAND

anthropology

Anthropology, a broad based discipline that studies humankind in all places and times, can be subdivided into physical, archaeological (see **archaeology**), linguistic and sociocultural fields. If the focus is on anthropology in tourism, aspects of tourism **sights**, **sites**, **behaviours** and relationships may be examined and understood from the applicable purview of all these fields. However, almost all anthropological research of tourism is by sociocultural anthropologists.

A review of the literature encompasses a variety of sources on the subject, written and used by social scientists in a range of disciplines. Major works documenting a new academic interest in the anthropology of tourism were published in the 1970s. A special issue of ***Annals of Tourism Research*** published in 1983, and entries on tourism in the *Annual Review of Sociology* and *Annual Review of Anthropology*, published in 1984 and 1989 respectively, marked this emerging field. The applied and interpretative questions have led to such research themes as tourism **political economy**, social change and **semiotics**. Following general trends in anthropology since the 1970s, tourism studies have moved away from a unilineal progress-oriented perspective on tourism development as modernisation toward postmodern concepts of **globalisation** and hyperreality (see **postmodernism**).

Focus on the **tourist** has led to analyses of **motivation** as a search for structure and meaning in modern and postmodern times. There is some general agreement that a tourist can be defined as a person who uses leisure time to voluntarily visit away from home for the purpose of experiencing change. While most accept these characteristics in common, many scholars argue that there is no generic tourist. Rather, types of tourist can be understood and classified by a range of motivations linked to social and cultural factors. Theories of tourist motivation have ranged from **pilgrimage** to **play** to **imperialism**. Regional histories of tourism vary significantly from religiously inspired travel to elite educational experiences that over time may have been democratised into **mass tourism**, **domestic tourism** and **international tourism**. The consumption of culture as ideas and artefacts (see **souvenirs**), marketed as **myths** and commodities to tourists seeking **authenticity**, is a major topic of study in neo-Marxist and semiotic analyses.

The producers of these commodities and services and inhabitants of destinations, labelled variously as hosts (in opposition to guests, see **hosts and guests**) are also a major subject of investigation. These studies range from functional descriptions to designs for **development**. Tourism has often been perceived by researchers as a form of imperialism, a negative force that is inflicted by elite metropoles on poor populations. Documented tourism impacts include ghettoisation (see **ghetto**) of locals into low-status jobs (see **employment**) and denigration of the local **environment**, social system and cultural **heritage**. Prostitution for **sex tourism** is a global example. However with local control, tourism can provide positive local change in rural and urban settings. A more moderate perspective on the possible **impacts** of tourism uses concepts such as **alternative tourism** and **sustainable development**. Local producers can have positive links with tourists, intranational institutions, international agencies, transnational tourism industry and anthropologists/academicians. These applied political economy studies search for more equitable distribution of the rewards and costs of tourism. Its potential in promoting the **cultural survival** of **indigenous** peoples living in protected **regions** and **parks** is a topic of particular concern for anthropologists working in these communities.

Criteria of difference, including **gender**, sexuality, **race**, **ethnicity**, nationality, age and local versus tourist create the **stereotypes**, hierarchies and relationships basic to tourism production and consumption. Longitudinal ethnographies documenting social change in industry, institutions and communities, multidisciplinary research and the emic studies of tourism are providing data for theorists. Ethnographies that detail specific situations are being used to construct an **ethnology** of tourism for cross-cultural comparisons on a wide range of variables. As the anthropology of tourism engages a variety of theoretical paradigms including **feminism**, critical theory and political ecology, significant understandings of this phenomenon will emerge.

Social scientists in other disciplines including **geography** and **economics** began studying tourism **impacts** before anthropologists took note. There is regional variation in the practice and acceptance of the anthropology of tourism within the discipline. Anthropology departments in the United States are significantly less engaged with the topic, despite the fact that a number of the primary figures in the anthropology of tourism since the 1970s have been from this country. Anthropology of tourism programmes are well established in some European and Asia-Pacific academies.

Anthropologists have long worked with peoples situated in what has been labelled the **Third World** and **Fourth World**, but they often did not incorporate the tourism **activities** they witnessed into their research. A good deal of rumination about why this may have been points to issues of the identification of anthropologists in the field as tourists, their discomfort in recognising that tourists conduct business with 'their' exotic people (see **exoticism**), and the perception of some Western anthropologists that tourism is a frivolous subject. Despite the pervasive nature of global tourism, its environmental impact, socio-cultural impact and economic impact, not all anthropologists consider tourism a worthy topic of investigation.

What may be a very thin line between the perceived knowledge of a **ludic** tourist and a playful anthropologist is a topic that some scholars contest and others celebrate as reflections of **gaze**. Links of local knowledge systems to the global **political economy** and concepts of **authenticity** of **experience** lead to questions about whether the 'Other' is for real, and in whose terms. The anthropology of tourism needs to grapple with these questions, and other complex topics about tourism and the movement and displacement of peoples, the re-creation of local, ethnic and national **identities**, and the legitimisation of cultural forms and ideas in a world shaped by globalisation and balkanisation. Tourism embodies globalising and localising forces in the transnational **marketing** (see **multinational firm**) of specific places and peoples.

Anthropology in tourism encompasses the **markers** of sameness and difference of human populations that tourists consume. Knowledge of ethnic customs, local heritage, the production of tradition in response to **modernity**, the history of human settlement, arts, dance, theatre, **music**, **rituals**, language, environmental **adaptation**, social norms and population movement are some of the themes that draw tourists to specific sites. Local arts and handicrafts produced for souvenirs may be simplified, mass-produced simulacra, or new expressions of a cultural aesthetic. Heritage sites may be restorations of actual buildings and **landscape**, while cultural **theme parks** like Disney's Epcot Centre are constructed samplers.

Archaeological research provides information and artefacts that may become the basis of tourism attractions. Tourists visit reconstructed sites, from Mesa Verde to Williamsburg in the United States to Chichén Itzá in **Mexico**, to name just a few examples. They are drawn there often by popularised cultural anthropology and **archaeology** accounts. Museums around the world display the artefacts of past cultures obtained through archaeological research. Replica sites, such as the Lascaux cave paintings in France, are another way of presenting past cultures while protecting the actual site.

Physical anthropology research also provides information used in tourism attractions. Early hominid sites, **museum** displays and interpretative primate exhibits link humankind to their ancestors. Research **vacations**, where volunteers work with sociocultural anthropologists, physical anthropologists and archaeologists to help collect data, is another aspect of anthropology in tourism. Linguistic anthropology informs socio-linguistic analysis of language in tourism marketing, putting a focus on how cultural content in the consumer's thinking is being shaped and reflected by market forces.

The ethics of anthropology are a critical but often covert aspect of tourism studies. An anthropologist's first responsibility is to the **community** they are studying. Whether they are engaged in an applied project or theoretical research, anthropologists are mindful of the effects of their work. It may be easy to see that one needs to promote equity for impoverished tourism producers. But the host community could also include **government** bureaucrats, multinational business executives and wealthy tourists. Anthropologists' relationship to tourism profit-makers and the tourists themselves may not be as clear, but ultimately is a critical part to the whole **representation** and understanding of tourism. The anthropology of and in tourism are linked by their ethical imperative to represent people fairly and equally in all their rich diversity through time and space.

Further reading

Castañeda, Q. (1996) *In the Museum of Maya Culture: Touring Chichén Itzá*, Minneapolis, MN: University of Minnesota Press. (Provides an example of the anthropology of tourism.)

Crick, M. (1994) 'Anthropology and the study of tourism: theoretical and personal reflections' , in *Resplendent Sites, Discordant Voices: Sri Lankans and International Tourism*, Char: Harwood Academic Publishers, 1–17. (An ethnography with a cogent overview of the global political economy of tourism.)

Graburn, N. (1996) 'Tourism', in D. Levinson and M. Ember (eds), *The Encyclopedia of Cultural Anthropology*, vol. 4, New York: Henry Holt & Co, 1316–20. (An introduction, with emphasis on pilgrimage, play, historical and comparative perspectives in the study of tourism.)

Nash, D. (1996) *Anthropology of Tourism*, Oxford: Elsevier Science. (Provides perspectives on tourism as a form of development or acculturation, as a personal transition, and as a social superstructure.)

Selwyn, T. (1995) 'The anthropology of tourism: reflections on the state of the art', in A.V. Seaton (ed.), *Tourism: The State of the Art*, Chichester: Wiley, 729–36. (A overview of the field with insight into future directions.)

Smith, V. (1989) *Host and Guests: The Anthropology of Tourism*, 2nd edn, Philadelphia: University of Pennsylvania Press. (A classic collection of articles which helped to popularise the study of tourism.)

MARGARET B. SWAIN, USA

anticipation

Anticipation is the first phase of the tourism **experience**, followed by travel to the **site**, on-site stay, return travel and post-tourism recollections. As the first phase, it is an umbrella term for the heightened **attention**, information collection and tourist **attitudes** which precede the experience. Studying attitudes in the anticipation phase is valuable for **marketing** projects and **satisfaction** research.

Anticipation may vary in its time frames. For many tourists there is a two to three-month period between making bookings and being at the **destination**, while others might have much longer or shorter anticipation periods. The length of the period will affect the processes which are active during this phase. Some prepare themselves extensively for their experience, particularly when this anticipated experience involves novel cultural contexts and high risk medical areas (see **health**). These physical and conceptual preparations for tourism are invariably augmented by a spontaneous information search, making news items and other material pertaining to the destination more noticeable to the tourist and more relevant than before.

Tourists' anticipation of their holiday or **vacation** may be a mix of holistic destination **images** and specific expectations. Such expectations are a subgroup of attitudes which they will hold towards tourism **products** and **services**. It is commonly suggested that attitudes or specific expectations held in the anticipation phase need to be equalled or surpassed for the traveller to have a satisfying holiday experience. There are some problems with this approach. First, it fails to consider that the experience may also alter the more holistic destination images and attributes which could by themselves contribute substantially to satisfaction. Further, the close alignment between specific expectations and outcomes may be conceptually sound but in practice difficult to assess. For example it is uncertain exactly when expectations should be measured, since in a long anticipation phase pre-travel attitudes might change markedly as new information is absorbed. Second, the logistical difficulties of accessing tourists in their own homes in the anticipation phase may limit effective research in this area. More importantly, there is the problem of reactivity in measuring expectations. The questions asked may focus tourists' minds on issues which they had not considered and shape the very attitudes being assessed. Measuring holistic images and using motives as a level of analysis rather than expectations may be possible solutions to these dilemmas.

PHILIP L. PEARCE, AUSTRALIA

anti-tourism

Anti-tourism is a generic term for adverse criticism of **tourists** and tourism. Tourism has always had its **critics**, often members of elite groups who have themselves been frequent tourists. In the early nineteenth century, European aristocrats ridiculed middle-**class** tourists arriving at **locations** formerly patronised only by the privileged. Later, the middle classes themselves made fun of working-class trippers taking seaside excursions as the railways opened up more places to more people. The main criticisms of tourists were that they travelled in large conformist groups, had little knowledge of the areas or cultures they were visiting, were spoon-fed elementary information, were crass in their **behaviour**, and strove to ape the **lifestyles** of their betters. Inter-class friction at destinations was a common theme in serious and satirical literature well into the twentieth century.

In the twentieth century, an influential variant of anti-tourism polemic has come from writers who nostalgically lament that travel is being replaced by its banal substitute: tourism. The historical validity of the distinction between travel and tourism is debatable. Early **pilgrims**, for example, often travelled in large groups, used guidebooks for instant information and behaved badly 'en route'. Moreover, the **Grand Tour**, the supposed apotheosis of independent travel, was a conformist itinerary of set-piece stops, undertaken by young aristocrats who took tutor-guides with them to explain the sights, and who were often as likely to drink and fornicate as the 'lager louts' of contemporary **mass tourism**. The first use of the word 'tourist' in a periodical of 1800 indicates that tourists and travellers were synonymous.

Today, the rhetoric of anti-tourism has shifted from the inadequacies of the tourist to the negative sociocultural and environmental **impacts** of tourism as a social process. Anti-tourism factions include social scientists who have often been caustic about the tourism **industry** and its impacts, and who have advocated a conservationist, protection-from-tourists **attitude** to 'traditional' cultures and environments abroad (while preserving their own rights to go anywhere). For anti-tourism commentators, the tourist problem is nearly always other people, not those like themselves. Ironically, anti-tourism has created **marketing** opportunities for the tourism industry. Specialist **tour operators** now promote exclusive products to affluent markets by calling them 'travel' rather than 'tourism' packages. Other operators have developed **alternative tourism products** for those unwilling to see themselves as 'ordinary' tourists.

Further reading

Boorstin, D. (1964) *The Image: A Guide to Pseudo Events in America*, New York: Atheneum.

Mitchell, R.J. (1964) *The Spring Voyage: The Jerusalem Pilgrimage, 1458*, London: John Murray.

Wheeller, B. (1993) 'Sustaining the ego?', *Journal of Sustainable Tourism* 1(2).

A.V. SEATON, UK

appropriate tourism

Appropriate tourism emerged as a response to political and sociocultural **quests** of the past two decades, as well as to dislusionment with **mass tourism**. It is associated with sustainable and **soft tourism** forms. It emphasises small-scale **development**, recognition of needs other than those of material consumption, and **preservation** of the quality and stability of both natural **resources** and human resources.

See also: alternative tourism

YORGHOS APOSTOLOPOULOS, USA

appropriation

In tourism, appropriation usually refers to the transfer of ownership or control of the **industry** in destination areas from local to outside interests. Related to **dependency theory** and **neo-colonialism**, it also describes the process whereby local communities become economically and culturally dependent on a foreign-dominated tourism industry. As regards **heritage** or culture, appropriation is the social construction of contested reality by political discourse.

RICHARD SHARPLEY, UK

archaeology

Archaeology is the scientific study of past cultures, both prehistoric and historic. Although documents are not eschewed when available, most of the data for archaeological research consists of the material remains of past societies. While tourism *per se* has not been a major traditional focus of academic archaeological research, the extensive use of archaeological **sites** as tourism destinations has generated interest in the interaction of archaeological sites and tourism. The major areas of discussion include decision making about which archaeological **resources** should be developed for **tourists**, methods for developing sites, the effect of **visitors** on archaeological sites and ways to protect sites, and the politics of the presentation of the past. Archaeologists interested in cultural resource management, **conservation** and public archaeology in particular deal with issues central to tourism.

Archaeological sites and **museums** housing archaeological remains are integral to much of **heritage** and **cultural tourism**, but **marketing** strategies also sometimes stress adventure, education, **exoticism** or education. Most of archaeological tourism is oriented toward short visits to such sites; however, opportunities for individuals to participate directly in this research are also possible. Some archaeological projects are funded partially with money derived from fees paid by volunteer workers while others, particularly in Europe, use volunteers as labour and provide room and board.

Almost every **region** and nation has archaeological sites that can potentially be developed for tourism. The choice of sites for **development** and their subsequent visitation rates has tended to depend on both the **accessibility** of the site and whether or not it is spectacular. Many archaeological resources represent an important and interesting part of heritage, but are visually unimpressive. The reasons for developing such sites as **attractions** are both economic and political. When used as a part of a general marketing scheme, as in **Mexico**, **Greece** and the American Southwest, archaeology can attract numerous visitors to an area, increasing the use of hotels, **restaurants** and other facilities as well as generating money from entrance fees and the sales of related **souvenirs**. Money earned through entrance fees is often reinvested in maintenance and development, but most archaeological sites and museums do not obtain enough revenue on entrance fees alone to be self-sufficient.

At minimum, the development of a site for tourism entails providing access, rudimentary services such as water and toilets, and basic protection of the site's resources. The attractiveness of a local site can be enhanced by the addition of museums, guidebooks, guided walks, reconstructions and sound and lights shows. Some of these developments are controversial. For example, archaeologists generally object to sound and lights shows because they entail unnecessary destruction of sites when cables and lights are installed, and often romanticise and trivialise the past in their presentation of it. There is increasing interest in **planning** the logical development of archaeological resources, including attention to traffic flow, handicap accessibility (see **handicapped**) and the diverse desires of consumers.

The development of archaeological resources for tourism increases the speed of degradations of sites. As the intensity of visitation increases, costs to the resource also increase. Foot and vehicular traffic, human breath, exhaust, the construction of facilities to serve visitors, vandalism and the collection of artefacts or architectural remains for souvenirs all degrade archaeological resources. Furthermore, touristic development of any **region**, because it entails the construction of roads, sewage systems, airports, hotels and other elements of **infrastructure**, threatens archaeological resources even when they themselves are not the direct focus of tourism activities. Paradoxically, however, development also provides a measure of protection against overt looting and illegal excavations, since resources become more visible and more heavily guarded. Measures taken to protect archaeological sites include stabilisation of ruins, **signs** and videos instructing visitors about appropriate site etiquette, guards and restricting access. While the democratic ideal would be unlimited access, at some sites it has been necessary to limit visitors to only portions of the site or to viewing it from behind barriers. At some sites such as the cave of Lascaux, where the breath from visitors was causing irreparable damage to the cave painting, access has been totally denied. Researchers are allowed within the cave for very limited time periods, but tourists must visit a replica instead.

Because there is a constant tension between **preservation** and use, there are arguments in the field about the extent to which archaeological resources should be developed for tourism. On the one hand, all use of a site entails some degradation of the resource. On the other, a large part the goal of archaeology is understanding the past and communicating that information to others. The general public is a major consumer of archaeological knowledge, and visitation to sites and archaeological museums provides an important aspect of public education. Some archaeologists argue that there is an ethical and in some countries perhaps even a legal obligation to provide the public access to the past and information about it.

Archaeological resources also have propagandistic value and the control of interpretation is a contested arena. Governments, for example, can use archaeological displays to demonstrate a glorious past, define the position and **role** of **minorities** or classes within the society, and create notions of unity. Archaeologists are anxious to use the same sites to explain archaeological strategies for understanding the past, and to build public interest in the preservation of the past and support for archaeological activities. Other groups, such as **indigenous** peoples, have both alternative agendas and ways of interpreting the past, and these are increasingly incorporated in the presentation of the past.

Further reading

Boniface, P. (1995) *Managing Quality Cultural Tourism*, London and New York: Routledge. (See particularly chapters 3 and 4.)

Silberman, N.A. (1989) *Between Past and Present: Archaeology, Ideology, and Nationalism in the Modern Middle East*, New York: Henry Holt and Co. (An analysis of the use of archaeology as propaganda.)

Smith, G.S. and Ehrenhard, J.E. (1991) *Protecting the Past*, Boca Raton, FL: CRC Press. (Covers archaeology and public expectations, education, site disturbance and site protection programmes.)

KATHRYN A. KAMP, USA

Arctic tourism

Tourism **activities** which occur north of the Arctic Circle constitute Arctic tourism. Road access is possible in Alaska, **Canada**, **Russia** and Scandinavia. Northern Scandinavia receives over 500,000 visitors a year, with tourist numbers increasing in other parts of the continental Arctic. Cruising and **ecotourism** also exist in Greenland, northern Russia and the Canadian Arctic, and is often associated with the interests of **indigenous** peoples. Tourism development in both Arctic and Antarctic regions has been focus of recent scholarly treatments.

See also: Antarctic tourism; polar

C. MICHAEL HALL, NEW ZEALAND

Argentina

Argentina is a country of considerable natural diversity. In the northeastern and northern areas rain forests predominate, with the internationally renowned Iguazu Falls being a particular focus. Along the western border of the country are the mountainous Andes. To the south, the Patagonian Andes are a wilder country featuring wide open spaces and unspoiled habitats. Further south still is the Antarctic region with its unique fauna and flora. The Valdes Peninsula has an especially large concentrations of South Atlantic fauna. In central Argentina, the pampas offer the traditional **image** of the country, that of ranching, cattle raising and the South American equivalent of the cowboy, the gaucho. The city of Buenos Aires offers a blend of European styles and a full range of commercial and

shopping activities as well as the seat of government.

Tourism in Argentina has traditionally been of minor economic importance, but since the beginning of the 1990s this has begun to change as the rate of growth in arrivals accelerated, outperforming world averages by a considerable margin, albeit from a small base relative to the country's size and apparent potential. In 1995, for example, 4.1 million international tourists spent $4.3 billion dollars when visiting this country. **Investment** in the sector is growing rapidly. During the period 1991–4, 206 **accommodation** establishments were opened, 59 were expanded and 142 new projects were started, together offering over 277,000 hotel beds. Some 450,000 jobs are directly supported by the tourism industry. Adequate air transport access is essential for Argentina, and the government is in the process of restructuring its airports, setting up a national **airports** system and encouraging privatisation and open skies **policies**. Air services to the country have doubled in number since 1991. Tourism is being developed on the basis of the National Secretary of Tourism's Strategic Marketing Plan, which aims to attract tourists from Europe, North America and South American markets as well as the Asian countries.

CARLOS ERNESTO, ARGENTINA

arrival/departure card

Arrival and departure cards, completed by passengers at **borders** and international airports, are used by governments to control the international flow of travellers. They contain information on the passenger's name, place and date of birth, nationality, arrival or departure details and reason for **travel**. They contain useful information used by national tourism offices and the tourism industry itself.

IAN McDONNELL, AUSTRALIA

art

Technically an oxymoron or 'denial in terms,' the phrase 'tourist art' is universally used to describe all that is false, ugly, and overpriced in the worldwide marketplace for contemporary arts and crafts. It includes not only sleazy **souvenirs** bought in airports ('airport art',) but spiritless 'folk dance' shows around hotel pools and lacklustre carvings, paintings, clothing, and artefacts sold in shops, galleries and so-called duty-free zones. But the objects bought and sold through tourism represent an immense amount of economic, not to say political and cultural interchange. But not all of this is bad. Indeed, some truly remarkable and important arts were created for or first recognised in tourism settings.

From time immemorial, **tourists** have brought back trophies; sailors brought from ancient Greece the souvenirs that stimulated classical Roman art, and travellers on the Silk Road exchanged the goods of China for those of Europe, not only silk in all its variety but pasta, porcelain and gunpowder. The earliest known African art works in Europe were ivory carvings brought across the Sahara from Nigeria, followed by 'Afro-Portuguese ivories' made by African artisans brought up to Europe. Some 400 years later, African sculptures traded by sailors for drinks in Paris bars influenced Braque, Picasso, Modigliani and the Cubists. Napoleonic grave-robbers brought back not only the Rosetta Stone but also the artefacts that produced the Egyptian Revival in architecture. The myth of El Dorado was supported by the gold jewellery brought from Cocle, Quimbaya and other South American **sites**, with designs so popular that reproductions of many pieces are still sold in **museum** shops.

Much more recent developments also illustrate the innate and unpredictable power of cultural interchange, no matter how casual, unplanned, ill-intentioned or misunderstood. Geoffrey Bardon, teaching in the Australian outback in 1971, suggested that the local Aboriginal men (see **Aborigine**) decorate a bare wall on a shed using mythical subject-matter and the 'pointillist' technique they used in body and sand paintings, but replacing the traditional coloured muds with more permanent gouache watercolours. The results were so popular that smaller examples were painted on bark panels and soon on fibre board and canvas for sale to **tourists** en route to the nearby immense stone monolith and Aboriginal sacred area now called Uluru. The unique paintings soon provided a much-needed source of local income. As women developed another technique derived from body painting, galleries in Sydney, Melbourne and elsewhere displayed the brilliant results as both tourists and Euro-Australians began to collect avidly. But, much to the frustration of art historians, the Aboriginal artists consider their myth-derived subject matter as sacred and hence secret, and will only explain that their paintings are maps seen from above outlining the activities of mythical beings in the 'Dream-ing' or Creation Era.

Another successful tourist art, Haitian painting, was originated by American watercolorist DeWitt Peters in the late 1940s when he persuaded a group of young men to try their hand at drawing and painting with materials provided by him. They soon developed a straightforward descriptive style which was fun to do and likely to sell. Genre subjects include **landscapes** and market scenes, Vodun **rituals**, the RaRa Carnival and other festivals, and magic subjects such as a profusely flowering tree that suggests a kneeling human form. Wood carvings and pierced wrought iron hangings made from discarded and flattened oil drums are also now made, bringing both honour and income to this island nation.

Artists sometimes try to 'psych out' tourists, like the Congo basket-maker who noted that Belgian colonial housewives seemed to love the ceramic tea sets made by his neighbouring ceramicists. Thus he made a basketry teaset which, though not unattractive, was totally useless and nonsensical tourist art. But a colonial housewife did indeed buy the teaset as a trophy/keepsake of her adventures in the Congo.

At the heart of the issue of **authenticity** is the intention of the artist. If he is producing an object for his own people, authenticity is assured because his audience is as informed as he of correct design and execution. But when he makes an object for tourists, he has a much less well-informed audience, and can play with themes and materials and be more casual with technique (see **staged authenticity**). But even if he does his best to produce an object perfect in his culture's terms, it may somehow have lost its spirit. The thousands of Swiss cuckoo clocks richly carved with leaves and birds in fine hardwoods have lost any sense of artistry. They are merely skilful, complete, one of many. Indeed, in the interplay between artist's intent and 'the eye of the beholder', a wide range of authenticity can be found.

Other examples of fine tourist art are the imaginative hardwood tree-of-life carvings of the Makonde of Tanzania; the pottery, turquoise and silver jewellery, and weavings of the Hopi and Navajo of the US Southwest; the batiks of **Indonesia**; the tie-dyes of West Africa; the silks of **Thailand** and **India**; the arts of **China** and **Japan**, many of them in styles dating back two millennia; the woven and embroidered textiles of the Andes; the masks and other artefacts from Papua New Guinea; the stone carvings of the Canadian Inuit (Eskimo); and the storyboards and *tapa* barkcloth of many South Pacific islands.

Finally, it is important to report that sensitive and far-seeing postmodernist art historians have begun collecting Hawaiian rubber dolls that do the *hula*; Mexican papier-mache toys; miniature Eiffel Towers and Statues of Liberty; and china plates displaying the portraits of 'Chuck and Di', all in sincere appreciation of international 'pop culture' and the 'kitsch' it produces. Thus, even though much tourist art is good, when it is bad enough it becomes 'significant.'

Further reading

Graburn, N.H.H. (ed.) (1976) *Ethnic and Tourist Arts*, Berkeley, CA.

Jules-Rosette, B. (1984) *The Messages of Tourist Art*, New York: Plenum.

DANIEL K. CROWLEY, TRINIDAD AND TOBAGO

articulation, programme

Coordinating educational requirements, levels or units among varying institutions to enable the student to successfully progress through various stages in the learning process on a continuous uninterrupted basis is an example of vertical programme articulation. By integrating the curriculum, teaching and learning experiences across a programme, for example by incorporating facets of tourism, **marketing** and research methods, a student benefits from horizontal articulation.

See also: education; training

K.S. (KAYE) CHON, USA

Asia Pacific Journal of Tourism Research

The *Asia Pacific Journal of Tourism Research* publishes both empirically and theoretically based studies which advance **education**, research and professional standards in the field of tourism in the Asia Pacific region. The journal seeks to publish original full-length articles and short research notes, reports and reviews, and uses a double-blind review process. First published in 1996, it appears twice yearly and is published by the **Asia Pacific Tourism Association** (ISSN 1094–1665).

RENE BARETJE, FRANCE

Asia Pacific Tourism Association

The Asia Pacific Tourism Association was formed by academic representatives from twelve Asian–Pacific Rim countries in 1995. With headquarters in Pusan, **Korea**, the association now has 365 members representing seventeen countries. Its conferences are held annually in a member country. The ***Asia Pacific Journal of Tourism Research*** is the official organ of the association.

BANG SIK LEE, KOREA

asset management

The term 'asset (or assets) management' indicates an aspect of managing a business organisation which focuses on its assets. In general, the objective is to improve **return on investment** by optimising asset turnover, or by ensuring that the organisation does not tie up **investment** which it does not strictly require. For this purpose, assets may be divided into cash and cash equivalents, working capital and fixed assets. Cash equivalents are highly liquid securities such as treasury bills which are an alternative to bank deposits for holding liquid funds. Working capital consists mainly of accounts receivable from debtors plus stocks held for production or sale; net working capital is arrived at by deducting accounts payable to creditors.

The **management** of cash and cash equivalents is part of the treasurer's function in an organisation. However, only in large organisations will one or more people be appointed to exercise this function; in smaller organisations it is one of the functions of the chief financial executive. The latter is also responsible for determining the **policy** regarding customer credit and monitoring the execution of this policy so that the amount of trade debtors is optimal for the business. A liberal policy leads to an excessive amount of funds being tied up in debtors and an increased risk of bad debts; a tight policy means discontented customers and lost sales. Related to this is the negotiation of payment terms with suppliers, there may be a trade-off between obtaining longer supplier credit and achieving price reductions. For tour operators, a key aspect of working capital management is its timing of advance payments receivable from customers and the timing of payments to suppliers of **hotel** rooms and airline seats.

Stock or **inventory** control is an important part of asset management in manufacturing and retailing organisations. In high-volume retailing, for example, efficient **information technology**-based stock control is a key competitive feature. Fixed assets include plant and machinery, vehicles (including ships and aircraft) and property (real estate). In general, the task of asset management is to ensure that a proper level of usage of all such assets is achieved, and that surplus assets are sold off. In the hotel sector, **property management systems** are a key feature of managing property assets. The term 'asset management' is also used to refer to the activities of the managers of investment funds, whose objectives are to achieve a satisfactory combination of income and capital **growth** for the fund.

SIMON ARCHER, UK

Association Internationale d'Experts Scientifiques du Tourisme

The aim of this tourism research association, known as AIEST, is to foster friendly relations among its members, to promote scientific activity on the part of its members, to provide documentation services, to facilitate exchange of views and experience, to support the activity of tourism centres of research and education, and to develop relations among its members and institutions. Its main activities include organising an annual conference and publishing ***The Tourist Review***, its official organ, as well as conference proceedings. The Association was founded in 1951, and had 380 members from some fifty countries in 1998. Its headquarters is in St. Gallen, Switzerland.

ORHAN ICOZ, TURKEY

Association Mondiale pour la Formation Hôtelière et Touristique

The Association Mondiale pour la Formation Hôtelière et Touristique (World Association for Hospitality and Tourism Training) or AMFORHT was founded in 1969 in Nice, France, as a platform for exchange and cooperation among official international organisations (especially the **World Tourism Organization**), private industry operators and companies, and educators, in order to define, adapt, develop and promote hospitality and tourism education and training. The membership of AMFORHT (institutes, universities, educators and consultants) is worldwide. Its headquarters are located in Nice.

SERGE PERROT, FRANCE

Association of Conference and Events Directors-International

The Association of Conference and Events Directors-International (ACED-I) promotes high standards of business and ethical conduct and fosters communication and sharing of concepts and ideas among its members. It collaborates with sister **associations** in an effort to unite the profession. The association encourages research projects that enhance the effectiveness and credibility of conference and event services. ACED-I is located in Fort Collins, Colorado, United States.

TURGUT VAR, USA

Association of Travel Marketing Executives

The Association of Travel Marketing Executives (ATME), with its headquarters in New York, represents the interest of tourism **marketing** executives. Its mission is to facilitate education and **professionalism** in this field. It provides and facilitates networking opportunities, insights and information, strategic **partnership** opportunities, new realms of professional possibilities, and solutions to **marketing** challenges to its members.

ATME has over 600 national members and about thirty international members. Among other benefits, members receive subscriptions to *Travel Marketing Decisions Magazine* (quarterly) which publishes articles ranging from destination marketing to new technology. ATME organises annual conferences and monthly meetings.

MARYAM KHAN, USA

attention

Attention is a concept from **psychology** which refers to the concentration of senses, thought and/ or awareness on an object or feature of the **environment**. Directing and focusing senses and thoughts on a particular feature of the environment is a necessary prerequisite for processing information about the feature and for acting or behaving. Attention is an important concept in psychology because it determines what is considered in conscious thought. It precedes perception and **cognition**. In simpler terms, if people are to process and remember information, then one must first be sure that their attention is directed towards that information. In tourism, attention can be seen as a necessary, but not necessarily sufficient, condition for such communication exercises as the **advertising** campaigns used in destination **marketing**, **safety** information and **interpretation**.

Two types of attention have been described by psychologists: passive or reflexive, and active or volitional. Active attention refers to situations in which people consciously choose to direct their attention towards something, which requires effort. An example would be a tourist with a particular interest in a period of **history** who deliberately concentrates on reading text and displays on that topic in a **museum**. Passive or reflexive attention is a response to some characteristic of the environment and requires no conscious effort or decisions on the part of the individual. The classic example is the ability of people to recognise and focus on their own name being used in another conversation at a crowded party.

Much research has been directed towards understanding the aspects of the environment which attract human attention. An understanding of such aspects is important in the design of successful communication efforts. Several factors have been found to consistently draw people's attention. The first of these is the intensity of stimuli used. Attention is automatically given to very loud, large or brightly coloured stimuli. **Change** and movement in the environment also draw human attention, as do **conflict** and the appearance of something unexpected. Attention is also given to objects or features which are connected to something of personal interest or relevance. This is an example of the influence of **motivation** on people's perceptions and actions.

Further reading

Hayes, N. (1994) *Foundations of Psychology*, London: Routledge. (A comprehensive introduction to psychology as a discipline; pp. 49–62 provide a good introduction to the concept of attention and its use in psychology.)

Kahneman, D. (1973) *Attention and Effort*, Engelwoods Cliffs, NJ: Prentice-Hall. (A review of attention research in psychology and the presentation of a theoretical model to describe the relationship between attention and cognition.)

GIANNA MOSCARDO, AUSTRALIA

attitude

There is broad agreement that attitudes are cognitive structures for organising one's experience of the world (see **cognition**). The term has its origins in the notion of people taking a stance, position or even a certain type of posture towards objects, people and settings. The development of the term in the last fifty years has been to clarify three main components of attitudes: the knowledge of or beliefs about an object or topic; a positive or negative evaluation of that object or topic; and a direction or imperative on how to behave when the object or topic is encountered. Attitudes may be held at a broad or a fine-grained scale and they may be focused towards either activities, people or abstract concepts.

Attitudes are a prime organising construct in the field of tourist **behaviour**, and many **survey** studies of visitors and host communities are centrally concerned with assessing attitudes (see **satisfaction**). In the broad array of terms which exist to describe cognitive structures and the psychological functioning of tourists, the term 'attitude' may be seen as the pre-eminent concept. It is virtually synonymous with the term 'opinion', which it has largely replaced in the social science literature. Attitudes is a more generic term than beliefs which refers exclusively to the knowledge component of attitudes, while the expression '**values**' summarises clusters of attitudes on a particular topic. Social **representations** are an even larger framework for understanding attitudes, as it considers not just the aggregated attitudes of individuals into values but the shared views of a community towards major topics of social significance. On a more detailed scale, expectations can be defined as pre-travel or pre-experience attitudes, while satisfaction studies can be conceptualised as attitude research conducted after the trip.

Much of the research conducted by social psychologists has concentrated on understanding the structure of attitudes and how to change attitudes. Contemporary thinking about attitudes suggests that there are linked sets of attitudes on key topics (such as **health**, politics, sexuality or tourism development). These linked attitudes have been termed 'social representations'. In order to understand attitudes to a specific topic such as tourism **development**, it is often useful to understand people's network of attitudes or the social representation of the whole issue and to assess people's social **identity** or group membership, since social forces are important determinants of why people hold attitudes.

Current attitude change theories propose that there are two different ways in which people can process information to create or change attitudes: either through a central route, which is also referred to as systematic processing, or through a peripheral route, sometimes known as heuristic processing. When peripheral processing is at work, people use very little of the available information and they employ short cuts and familiar routines to guide behaviour. In central processing, people carefully examine and analyse much of the available information. Attitudes developed through systematic processing are likely to be more detailed, more consistently related to behaviour and harder to change than those adopted through a peripheral route. Peripheral processing is illustrated by tourists accepting stereotypic clichéd views of the visited hosts and holding attitudes towards them which are largely influenced by **media** and external sources rather than by a thoughtful consideration of the actual contact experiences. Research on mindfulness is also linked to this approach, with the concept of mindfulness referring to a central processing route and that of mindlessness characterising the peripheral or heuristic approach.

In assessing attitudes, it is valuable to use a number of control groups or baseline measures in field studies since the very process of researching attitudes and trying to monitor their change can influence the investigation. Attitudes are traditionally measured on five to seven-point rating scales with ranges such as 'strongly agree' to 'strongly disagree'. In evaluating attitude research, it is usually insightful to ask both the direction (evaluative) component of the attitude and the importance of holding that attitude on the topic to the individual.

There are frequent comments in the literature that attitudes do not form a close link to actual behaviour. There are several reasons why such inconsistencies might exist. It may be that the questions asked to elicit attitudes were not sufficiently specific. Additionally, if the topic is of little relevance to the individual, the respondent may express an attitude which has few immediate behavioural consequences. It is notable that negative attitudes are often more detailed and well-organised and are likely to have been subject to central or systematic processing. Much existing tourism consumer research has been cast within a loose attitudinal framework, as have studies relating to the recreational use of the **environment**. These studies could be enhanced by closer attention to contemporary advances in attitude theory and measurement. Several studies of tourists' post-return attitude changes have been reported. In these studies group package tourists were shown to change their attitudes as a consequence of the actual **experience**. When favourable holidays took place and initial pre-departure attitudes were positive, even more positive attitudes were acquired. Less satisfactory holiday experiences and moderate pre-trip state of mind resulted in some negative attitudes to host cultures.

See also: community attitude

Further reading

Chaiken, S. and Stangor, C. (1987) 'Attitudes and attitude change', *Annual Review of Psychology* 38: 575–630.

Pearce, P.L. (1982) *The Social Psychology of Tourist Behaviour*, Oxford: Pergamon.

Pearce, P.L., Moscardo, G.M. and Ross, G.F. (1996) *Tourism Community Relationships*, Oxford: Elsevier.

Ryan, C. (1995) *Assessing Tourist Satisfaction*, London: Routledge.

PHILIP L. PEARCE, AUSTRALIA

attraction

Attractions are more than just a **site** or an **event**; they are an integral part of a larger tourism system that also consists of **tourists** and **markers**. Attraction typologies vary considerably depending on whether they are being used for **marketing** or **planning** purposes. No **site**, **sight** or event is an attractions in itself. It only becomes one when a tourism system is created to designate and elevate it to the status of an attraction. Almost any object – real or intangible – may be designated as having some special quality which allows it to be elevated through **advertising** to the status of an attraction.

The only intrinsic requirement of the object is that it is associated with a **location**. This differentiates attractions from other consumable goods. Rather than bringing the goods to the consumer, the tourist must go to the attraction to experience it. Thus, the system that creates and supports an attraction must have three major components to exist: an object or event located at a site, a tourist or consumer, and a marker, an **image** that tells the tourist why the object or event is of interest. This definition has been widely accepted since MacCanell's work on the **semiotics** of tourism in the 1970s. It is around these three basic elements that the entire tourism **industry** is constructed.

The objects from which attractions are created are typically environmental and cultural **resources**. Tourism can be viewed as an industry that turns these resources into **products**. From the perspective of the tourist, they consist of objects to see, **activities** to do and experiences to remember. Most attractions, including events, have some type of nucleus that epitomises the **experience**. How this nucleus is presented is important in influencing the experience. The tourist's expectations of the attraction nucleus begin to be shaped from the very first realisation that it exists. Through various types of **media** (including word of mouth), an image of the attraction is created over time, often well before the actual trip. These sources of information are known as generating markers as they are located in the location from which the trip originates. Expectations and **anticipation** continue to be shaped as the trip is underway (through transit markers) and, in spatial terms, the closer the tourist comes to the attraction nucleus, the more prominent the messages and markers become.

Marker messages come from all types of communication, including schooling, news media and popular media. The popularity of a book or movie can create an attraction where none existed before. Professional marketing firms are often hired to create or change the image of an attraction. A failure in the marker portion of the system can often lead to **visitor** disappointment. Most major destinations consist of a collection of attraction nuclei, the most important of which are considered primary while less important ones are considered secondary and tertiary. From the tourist's perspective, the destination itself may be considered as much of an attraction as any of the separate attraction components. This is especially true at the scale of the major countries and **regions** of the world. At the same time, if a primary attraction is not included on an **itinerary**, most first-time visitors to a destination will not consider taking the **tour**.

Attractions may be either clustered in a single location or distributed in a linear or dispersed pattern. These different spatial patterns are important in the planning and **management** of attractions. Linear dispersion requires circuit touring by automobile or rail, and **accessibility** is a major concern in their planning and development. Destinations in which attractions are clustered must plan for a proper mix of activities to provide variety and stimulation, which will encourage both longer visitor stays and repeat visitation. The clustering of attraction nuclei can also be used to concentrate tourism in isolated enclaves, thereby conserving natural resources (see **conservation**) and limiting some of the negative **impacts** of **mass tourism**. Such enclaves, however, can sometimes turn into tourist ghettos.

The assessment of attractions is a common part of planning and marketing and is undertaken to understand the **competitive advantage** of one place over others. Attractions are inventoried, and their potential for **development** (or need for protection) are studied. No single agreed-upon typology of attractions exists to conduct an **inventory**, in part because most places have their own distinctive qualities. Attraction inventories have been approached in one or more of three ways. The most common approach is to group attractions into nominal categories (also referred to as formal and ideographic). Such categories include cultural artefacts and **nature**. Examples of the former include special structures (buildings, bridges, monuments), communities, **theme parks**, **cuisine** and works of **art**. Nature includes mountains and other scenery, vegetation, **climate** and nature preserves and **parks**. Depending on the place and the purpose of the attraction inventory, other types of categories are often combined with the nominal ones.

Attractions may also be classified into cognitive or perceptual categories (see **cognition**), such as **authenticity**, educational, adventurous and recreational. They can be inventoried based on their organisational or structural characteristics, including isolated or clustered, urban or rural, low or high capacity, and seasonal or year-round attraction (see **seasonality**). The cognitive approach to attraction inventories is used when the destination image is of primary interest for marketing purposes. The organisational approach is used when undertaking **community planning** and controlling the development process are the main concern.

Further reading

Gunn, C.A. (1972) *Vacationscape: Designing Tourist Regions*, Austin, TX: Bureau of Business Research, University of Texas.

Leiper, N. (1990) 'Tourist attraction systems', *Annals of Tourism Research* 17: 367–84.

Lew, Alan A. (1987) 'A framework of tourist attraction research', *Annals of Tourism Research* 14: 533–75.

MacCannell, D. (1976) *A New Theory of the Leisure Class*, New York: Schocken Books.

Walsh-Heron, J. and Stevens, T. (1990) *The Management of Visitor Attractions and Events*, Englewood Cliffs, NJ: Prentice Hall.

ALAN A. LEW, USA

attraction, religious

Religious **attractions** are places or events which are recognised by individuals or groups as worthy of devotion and visitation. As religious or sacred phenomena, they stand out from the profane and commonplace. They can be distinguished primarily by the intensity of the perceived holiness. Not all religious phenomena that are venerated or visited by devout believers or tourists have equal status or perceived holiness. So-called mystico-religious sites are perceived as the most sacred because adherents of a **religion** believe that God and man are brought into direct context through them. Homelands, the second level of sanctity, are sacred attractions because they represent the roots of individuals or peoples. These are historical sacred sites that have been assigned sanctity as a result of an event occurring there (see also **site, sacred**). **Israel** as the homeland of Judaism, and Utah together with the Rocky Mountains and Jackson County (Missouri) as the homeland of Mormonism, are typical examples of such religious **attractions** that influence the touristic activities of their believers. At the lowest level of sacred space are the historical attractions that have been assigned sanctity as a result of significant events occurring there. Mormon examples include Joseph Smith's birthplace and Kirtland (Ohio) where the first Mormon temple was built.

Religious attractions can also be objects as well certain events and activities that are both venerated and visited by believers. Of special importance among the religious objects in all religions are natural sacred attractions such as holy mountains, caves, rivers, rocks and stones, animals, plants and trees. Wayside sanctuaries like figures, symbols and chapters are to be found especially in the main Christian denominations. Temples, stupas, pagodas, shrines, synagogues, churches and mosques are the most striking religious attractions that are built because they house the deity or give shelter to the devotees. Sometimes even religiously-influenced settlements are attractions for the adherents of a religion, such as the monasteries in various religions. So-called civil religions and ideologies also have at their disposal quasi-religious attractions like historical battlegrounds, tombs, mausoleums and monuments venerated and visited by their adherents.

Religious festivals, conferences, church meetings and congresses with local, regional, national and international catchment areas are attractions for a different number of believers. Diocese meetings in the Roman Catholic Church have their **location** at the bishop's seat. Annual meetings of Christian organisations and associations as well as national and regional church meetings have variable locations. From this one can recognise that for the tourist the religious event is the religious attraction and not the variable location.

Further reading

Park, C.C. (1994) *Sacred Worlds: An Introduction to Geography and Religion*, London: Routledge.

Rinschede, G. (1992) 'Forms of religious tourism', *Annals of Tourism Research* 19: 51–67.

—— (1999) *Religionsgeographie*, Braunschweig: Westermann.

Vukonić, B. (1997) *Tourism and Religion*, Oxford: Elsevier Science.

GISBERT RINSCHEDE, GERMANY

attractivity

Attractivity is a quantitative measure of the ability of a **destination** or facility to draw tourists. It is often used in the context of gravity **models**, although analysts use the concept in other tools as well. Regardless of the specific tool, the concept is normally used as a variable in a model to explain or forecast visitor levels.

There are four basic approaches to measuring the attractiveness of a **region** : simple exogenous, complex exogenous, inferred exogenous and endogenous. Simple exogenous refers to the use of a single variable, such as the number of **hotel** rooms, to represent the attractiveness of a destination. The complex exogenous approach combines two or more descriptive variables, such as mean hours of sunshine and mean high temperatures, into a single composite index. Both of these measures require the analyst to select the relevant variable(s) and, in the case of the complex exogenous approach, to specify how the variables are to be combined.

The inferred exogenous approach takes a different tack. This method is based on the revealed preferences for destinations, either through comparing visitor levels or through a **survey** ranking alternative destinations. The endogenous approach defines attractivity as a statistically estimated coefficient in a visitor forecasting model. Ultimately, the choice of the approach to measuring attractivity is a function of the purpose of the study, the data available and the analytical abilities of the researcher.

See also: attraction

Further reading

Cesario, F.J. (1973) 'A generalized trip distribution model', *Journal of Regional Science* 13: 233–248. (Use of endogenous approach to measuring attractivity.)

Ellis, J.B. and Van Doren, C.S. (1966) 'A comparative evaluation of gravity and systems theory models for statewide recreational flows', *Journal of Regional Science* 6: 57–70. (Classic reference for trip forecasting; includes examples of complex exogenous attractivity indices.)

Ewing, G.O. and Kulka, T. (1979) 'Revealed and stated preference analysis of ski resort attractiveness', *Leisure Sciences* 2: 249–76. (As an example of inferred exogenous approach, identifies potential problems and suggests solutions in the approach.)

STEPHEN SMITH, CANADA

attribution theory

When misfortune occurs, such as when a holiday **experience** turns out badly and/or a **tourist** receives a personal affront, there is a need to explain or account for the problem. Attribution theory is directed at understanding how people account for, or explain, problematic events. The negative event prompts individuals to ask why did this happen and, further, perhaps identify who or what is responsible for this sequence of events. The answers to these questions are important to researchers, since explanations which are finally located at the door of the tourism business or organisation could have long term consequences for profitability and repeat visitation.

The rich literature on attribution theory in psychological inquiry is usually seen as beginning with the work of Heider, who noted there were three personal disposition factors (ability, power, and intention) and three situational factors (task difficulty, opportunity or luck) which must be taken into account when explaining unsatisfactory outcomes (see **satisfaction**). Other researchers such as Jones and Davis added to Heider's idea by proposing that attributions will vary according to whether or not individuals are likely to have some knowledge of the consequences of their **behaviour** and are able to perform the behaviour. Kelley noted important differences between repeated situations and individual events as well as highlighting what has been termed the fundamental attribution error. The 'error' refers to people's predisposition to make themselves look good in a social context by attributing unexpected negative outcomes to the behaviour of others, while positive ones are attributed to their own skill or influence.

In the tourism context, attribution theory has been applied to studies of tourists' complaints. It can be argued that tourists who attribute negative outcomes to individual factors (such as explaining seasickness in terms of their own constitution) will be less likely to be dissatisfied than those who attribute the negative situations to tourism businesses (such as feeling ill because of poor boat management). Attitudes towards tourism **development** may also be understood through attribution theory. For example, when a controversial development proceeds, attributions based on negative individual dispositional characteristics (such as crookedness, collusion, insensitivity to **community** perspectives) may ultimately be harmful to tourism decision making processes and thus to business interests.

Further reading

Pearce, P.L. and Moscardo, G.M. (1984) 'Making sense of tourists' complaints', *International Journal of Tourism Management* 5: 20–3.

Shaver, K.G. (1985) *The Attribution of Blame*, New York: Springer Verlag.

PHILIP L. PEARCE, AUSTRALIA

auditing

Auditing is the examination of the financial statements of an organisation, with reference to the underlying **accounting** records and other evidence, by an independent party (the auditor) in order to express an opinion on the adequacy of those financial statements. In the case of external auditing, what counts as adequacy is generally laid down by **law** or by a **regulatory agency**. In the United Kingdom and other member states of the **European Union**, the general criterion of adequacy is whether the financial statements give a 'true and fair view' of the financial position and results of the organisation. However, the term 'true and fair view' is interpreted somewhat differently in different member states. For example, in Germany there is a specific requirement that the true and fair view should be given in compliance with generally accepted accounting principles. In the United States, the criterion is fair presentation in accordance with generally accepted accounting principles. On the basis of the audit, the auditor writes a statement of opinion. The form and wording of such statements of opinion usually follows standards issued by the auditor's professional body in the light of legal requirements. Typically, the auditor's opinion states either that the financial statements satisfy the criterion of adequacy without reservation (a clean opinion), or that they do so subject to certain specified matters or with the exception of the treatment of one or more specified items (a qualified opinion), or that the auditor is unable to express an opinion (a disclaimer).

The independent party who carries out an external audit must be an accountant who holds a legally recognised qualification as an auditor. In some countries, a firm (professional partnership) of auditors may act as auditor, and the auditor's opinion is signed in the firm's name. In other countries, a qualified auditor must sign in his or her own name; in the case of a firm, this means that it is the partner in charge of the audit who signs in his or her name. The scope of an auditor's work is based on testing the proper functioning of the organisation's accounting systems using sampling techniques. This includes obtaining independent verification of bank balances and of a sample of debtor and creditor balances, and physical verification of a sample of **inventories** and fixed assets. Principles of auditing in tourism are not any different from those used in other businesses. However, because of the specific nature of tourism services, some variations have been introduced.

SIMON ARCHER, UK

Australia

The large island continent of Australia, with an area of more than 4.7 million square kilometres (almost 3 million square miles) ranks as the sixth largest country in the world. It is a large relatively compact land mass, stretching from the tropics to the southern island of Tasmania. The population of some 19 million people is concentrated around the east and southeast coast and in the southwest, and in the major cities like Sydney and Melbourne. The arid interior is sparsely populated, as is the monsoonal region in the tropical north.

The Australian **landscape** is noted for some remarkable topographic features such as the massive monolith of Uluru (Ayers Rock) in the Red Centre of the continent. The country is dotted with hundreds of **national parks**, and off the northeastern coast of Queensland is the largest and longest coral formation in the world, the Great Barrier Reef (now a marine national park and a World Heritage site). This is one of Australia's most compelling attractions, both for **domestic tourism** and **international tourism** with over thirty **resorts** established on the reef islands and coral cays. Equally remarkable is the strange and varied animal and bird life, which is significantly different from that of other continents and is thought to be the result of isolation over millions of years. Fauna characteristic of Australia, such as kangaroos and koalas, frill-neck lizards and an amazing variety of colourful and unusual birds, are an endless source of fascination for **tourists** from other parts of the world.

The original inhabitants of Australia, the Aborigines, have been displaced over the past 200 years of European settlement and are now highly urbanised and often disadvantaged. Large streams of immigrants since the Second World War, from Europe and increasingly from Asia, are changing the character of the Australian people and making Australia a multicultural nation. While this is regretted in some quarters, it does add to the appeal of the country to tourists and to its many natural and cultural attractions for tourism.

In Australia, as in comparable Western societies, tourism is a major national industry of economic, sociocultural and environmental significance. However, this position of strength was relatively slow to evolve. The great distances separating Australia from major centres of world populations and the unknown and unpublicised nature of its attractions meant that Australia was neglected for tourism **development** until late in the twentieth century. Established destinations in the northern hemisphere remained dominant as attractions, and it was not until the 1960s and 1970s that Australia was 'discovered' as a region with tourism potential.

Distance was a problem not only internationally but also within Australia, because of the size of the landmass and the skewed distribution of population centres around selected parts of the coastline. Domestic tourism was a challenge, and even today remains time-consuming and expensive; distance continues to inhibit the **growth** of isolated destinations. Despite the history of Australian settlement, characterised by immigrants from abroad (some of them involuntary convict labour), exposure of the continent to the world at large had to await improved **transportation** technology and communications and changes in social **attitudes** towards tourism and tourists.

Such advances are relative, and even the fastest jets continue to take upwards of fourteen hours to cross the Pacific from North America and as much as twenty-four hours to reach parts of Europe from Sydney and Melbourne. Improvements in communications and **information technology** have certainly helped, but awareness and attitudes in the outside world regarding Australia as a place to visit remain somewhat distorted. Typical are false perceptions of kangaroos and other native animals roaming the city streets, a countryside overrun with deadly snakes and spiders, coastal waters teeming with sharks, and urban settlements devoid of modern comforts, medical care or cultural attractions.

The real picture is in stark contrast, and the diverse and unique attractions of Australia, supported by increasingly sophisticated **infrastructure** and solidly promoted around the world, have led to some of the highest growth rates in tourism across the globe. Since the Second World War tourism in Australia has emerged as a significant economic, social and cultural activity. Rising real incomes, increased **leisure** time, growing awareness of tourism opportunities, rising mobility and improvements in transportation, and shifts in tastes have led to an upsurge in the quantity and **quality** of tourism activities demanded by the Australian population. At the same time, growing affluence in many countries – in particular, the emergence of leisured classes in neighbouring parts of East and South Asia – relaxation of **travel** restrictions and the easing of political tensions, along with declining real costs and time involved in international tourism, have seen Australia emerge as a popular **destination** for foreign tourists. The industry now generates export earnings which exceed those from traditional exports such as coal, wool, wheat and minerals.

One of the main reasons for the rapid growth in Australia tourism in the past two decades, as already noted, has been the desire of both domestic and international tourists to **experience** at first hand Australia's unique natural and cultural **environment**. This is both an opportunity and a challenge for the future development of tourism in Australia, with a need to encourage a healthy and growing **industry** in accordance with the principles of ecological sustainability.

Despite a downturn in tourism activity in the early 1990s (experienced worldwide), the number of international visitors to Australia is predicted to grow at an impressive rate into the next century, reaching around 9 million by the year 2005. This will translate into over $17 billion annually in export earnings during a period when many of Australia's staple export industries are in decline. Between 1980 and 1995, visitor arrivals increased by an average of 10 per cent a year, while tourism export earnings grew in real terms by 12 per cent annually. When coupled with domestic tourism, it is predicted that in year 2005 about 160 million tourist nights will be spent in Australia each year.

Japan remains the largest single source of international tourism to Australia, with close to one million arrivals. Despite some lingering pockets of anti-Japanese sentiment, this market is expected to grow by a healthy 7 per cent per year. Other Asian sources of tourists include **Hong Kong**, **Korea**, **Malaysia**, **Singapore**, **Thailand**, Taiwan and **China**. Visitors from North America, **New Zealand** and Europe also contribute strongly to this international volume. The staging of the 2000 Olympics in Sydney and 2001 Federation celebrations should enhance Australia's appeal as a destination.

The most favoured destinations within the country are the major cities of Sydney and Melbourne, followed by the Gold Coast near Brisbane in southeast Queensland and the Great Barrier Reef. The inland region, or 'outback' as it is known, is increasing in popularity as facilities for tourists improve, transport is upgraded and a wider variety of natural and cultural opportunities is on offer. An interesting **trend** is the involvement of more Aboriginal groups in tourism enterprises. **Indigenous** tourism is fast establishing a recognised role in Australian tourism, particularly in the northern interior. Interest in Aboriginal culture is strong, and commercial opportunities are being developed in the sale of handicrafts, provision of traditional foods ('bush tucker') and visits to Aboriginal-managed national parks such as Kakadu and Uluru.

In common with many parts of the world, a strong interest is developing in **nature tourism** or **ecotourism** in Australia. A National Ecotourism Strategy has been formulated, complemented by some initiatives at the state level. An international research centre in ecotourism has been established and a National Ecotourism Association formed. A number of international conferences have been organised, and strong moves are being made towards a national ecotourism accreditation scheme. This is important because the growing **demand** for tourism based on Australia's natural and cultural **heritage** is leading to a rapid increase in the number of such operators and establishments. This in turn places pressure on sensitive environments and communities, including indigenous groups. Tourism remains a relatively unregulated activity in Australia, but this may change as evidence grows for the need for some guidelines in the interests of **sustainable development**. Sites of cultural significance and the physical fabric of heritage areas may need at least interim protection until educational and interpretative programmes can bring about positive changes in **attitude** and **behaviour** (see **education**; **interpretation**).

The environmental consequences of tourism are now well documented in Australia, and support for more sustainable forms of it is growing among industry operatives. Several groups have drawn up environmental codes of ethics, and regulation and self-regulation of tourism development are widespread and accepted, especially in the larger scale corporate undertakings (see also **regulation, self**). **Benchmarking**, **performance indicators** and best practice environmental management are features of a more environmentally aware approach to tourism management. However, a range of incentives and sanctions may be needed to encourage the adoption of these practices among smaller scale operators.

The **future** growth of Australian tourism seems assured, with industry backing and a strong commitment by governments at national and state levels. The endorsement of a National Tourism Strategy reflects this confidence in the future of the Australian tourism product. A further encouraging development is the establishment in 1997 of a Co-operative Research Centre for Sustainable Tourism as a collaborative venture between government, research institutions and the tourism industry.

Further reading

McKnight, T. (1995) *Oceania, The Geography of Australia, New Zealand, and the Pacific Islands*, Englewood Cliffs, NJ: Prentice Hall.

Office of National Tourism (1996) *Forecast*, Canberra: Commonwealth Department of Industry, Science and Tourism.

JOHN J. PIGRAM, AUSTRALIA

Austria

Austria is a small country, with a land area of 83,845 square kilometres (ranked 113 in the world in terms of size), located in the centre of Central Europe. With a total population of 7,862,000 (1993) it has a population density of 94 inhabitants per square kilometre. Its gross national product per inhabitant in 1993 amounted to $23,510, which puts it in the upper 10 per cent of European countries.

Similar to other industrialised European countries, Austria is showing a decline in both its agricultural and industrial sectors in terms of their contribution to the national product (GNP), and has similarly seen a long-term rise of its tertiary (service) sector, which in 1995 achieved 65 per cent of GNP. Tourism and related products and services occupy a prominent role within the tertiary sector. Depending on the sectorial accounts included in and counted as tourism activity, it amounted to 7.5 per cent of the GNP in 1995 if expenditure of **international tourism** is included only, or 13 per cent if all direct and indirect tourism-related expenditures are counted. In terms of **employment**, 6.5 per cent of Austria's labour force works in tourism (1995), which puts it among the most tourism employment-intensive countries in the world. In the same year, Austria recorded 18.2 million tourist arrivals from abroad which, translated into 117.2 million overnight stays, puts Austria in terms of international tourism into fifth position following **France**, **Italy**, the **United States** and **Spain**. On a per capita basis, Austria's tourist industry produces receipts per inhabitant amounting to $2,100 (1993), the highest such figure in the world; Austria is followed by **Switzerland** as a distant runner-up with a per inhabitant contribution of $1,200.

Although Austria's tourism has had a long history dating back to the times of the Austro-Hungarian monarchy, its present position has been rather the result of post Second World War developments. Notably, the 1960s and 1970s, which coincided with a rapid postwar expansion of **Germany**'s economy, saw the development of **mass tourism** fed to a large measure by the neighbouring countries of Germany and Switzerland, which traditionally have accounted for three-quarters of all overnight stays. The Benelux countries, Great Britain and France usually account for another 20–22 per cent, which leaves a relatively small share for long-haul travel and tourism originating in the Americas and Asia. Austria's central location in the heart of the Alps and its proximity to some of the most prosperous European countries, combined with the liberalisation measures adopted with the formation of the European Common Market and **European Union** (of which Austria became a member in 1995), have been among the major determinants of Austria's postwar **growth** in tourism.

Austria's built attractions are a rich cultural **heritage** and a thousand-year **history** in which it had formerly played a far more important political and international role than it does today. Austria's large and medium-sized cities offer both the remnants of various centuries of cultural development, particularly from the periods of the baroque, rococo and fin-de-siècle, which can be seen in numerous museums and galleries and the still-existing architecture, and which is recreated through musical and theatrical events (see also **museums**).

At the same time, Austria possesses natural **attractions** in the form of scenic mountains, idyllic mountain villages and traditions associated with the alpine rural way of life. Six out of Austria's nine provinces are either entirely covered by mountains, like Tyrol, or have large mountain districts, some of which combined with adjoining lake districts. Only Austria's most eastern province, Burgenland, features flat lands which form part of the Pannonian plains. In many regions, alpine tourism operates in both the winter and summer seasons (see **seasonality**). Austria's 117 million total overnight stays in 1995 (inclusive of domestic tourists) can be broken down as follows: alpine summer tourism, 41.9 per cent; alpine winter tourism, 33.0 per cent; **spa** tourism, 15.4 per cent; and city tourism, 7 per cent.

Out of the total 117 million overnight stays, 30.1 million (30 per cent) were domestic travellers. The average length of **vacation** in 1995 was recorded as 4.3 days for domestic and 5.07 days for foreign travellers. To accommodate tourism flows of this size, Austria has 80,947 enterprises in the field of **accommodation**, with a total capacity of 1,222,577 beds (1995 figures). Half of this capacity represents, however, private non-registered inns, rest and holiday homes, specialised spa and **recreation** institutions, and mountain refuges. Officially registered commercial enterprises of the accommodation and **food** industry include 18,120 firms, with a bed capacity of 678,349 (1995), with 25.4 per cent falling into the 4/5-star category, 27.4 per cent being 3-star operations and the remaining 46.9 per cent belonging to the 2-star and 1-star categories (see **classification**). The overall occupancy rate in 1995 stood at 26.9 per cent (30.7 per cent for registered enterprises and 16.2 per cent for all the other non-registered accommodation organisations).

Tourism development and **marketing** beyond the level of individual enterprises is carried out by local, regional, provincial and national boards. By regulation, all registered enterprises are members of their respective industry sections of the provincial chambers of commerce, which in turn are represented in the Federal Chamber of Commerce, a very important lobbying institution in terms of industrial policies aimed at tourism. The importance of international tourism for Austria is probably best highlighted by two critical figures. On average, more than 11 per cent of Austria's payments for imported goods and services are covered by proceeds from international tourism, which on average also represents over 18 per cent of Austria's total foreign income from exporting goods and services as shown in the balance of payments.

Further reading

Austrian Central Statistical Office (1996) *Der Fremdenverkehr in Österreich im Jahre 1995*, Vienna: Republic of Austria, Austrian Central Statistical Office.

Ministry of Economic Affairs (1996) *Bericht über die Lage der Tourismus- und Freizeitwirtschaft in Österreich 1995*, Vienna: Ministry of Economic Affairs.

Smeral, E. (1994) *Tourismus 2005*, Vienna: Ueberreuter Verlag.

KLAUS WEIERMAIR, AUSTRIA

authenticity

A concern for authenticity is a modern value and ideal that resulted from the **experience** of inauthenticity and alienation in modern society. Such a development can be dated back to Rousseau. A touristic concern for authenticity appeared in nineteenth-century England, where travellers tried to distinguish themselves, the agents of *genuine* **travel**, from 'tourists'. However, it is only about two decades ago that authenticity was introduced by MacCannell (1973) into tourism social sciences as a research programme. Since then it has become an agenda for study, and there has been a rapid growth in literature focusing on this theme. Although the '**quest** for authenticity' is criticised as oversimplifying tourist experiences, it qualifies as a key category in the **sociology** of **motivations**.

The original usage of authenticity was in the **museum**, conveying the meaning of whether objects of **art** are what they appear to be or are claimed to be, and hence worth the price that is asked for them. It is perhaps such a museum-linked usage of authenticity which has been extended to tourism. As a result, **products** of tourism such as works of art, festivals, **rituals**, **cuisine**, dress, housing and others are usually described as authentic or inauthentic in terms of the criterion of whether they are made or enacted by locals according to custom or tradition. Bruner (1994) has transcended this narrow meaning of authenticity by identifying four meanings of authenticity based on verisimilitude, genuineness, originality and authority. Thus, 'authenticity' not only means 'original', but also refers to the authenticity of reproduction and simulation, both of which also involve authentication by authorities and powers. Selwyn (1996) divides authenticity into two separate types: authenticity as knowledge ('cool authenticity') and authenticity as feeling ('hot authenticity'). Wang (1999) further classifies authenticity in tourism into three different types: objective, constructive (or symbolic) and existential. The sociologists usually ask three kinds of questions with respect to authenticity in tourism: why tourists quest for authenticity; how authenticity is experienced, constructed or produced in tourism; and what the consequences of the search for authenticity in tourism are.

While tourism was criticised as pseudo-events which tourists enjoyed, MacCannell argued that modern **tourists** are in search of authentic experiences, which 'parallels concerns for the sacred in primitive society' (1973: 590). Tourism is thus seen as a kind of secular **pilgrimage** on a quest for authenticity which exists elsewhere and in other cultures, especially primitive or pre-modern cultures. For MacCannell, the reason why tourists seek authenticity is a structural one; it is the inauthenticity and **alienation** of **modernity** that motivate tourists to search for authentic experiences elsewhere.

How authenticity is experienced, constructed or produced in tourism is, however, controversial. Generally speaking, five approaches to this issue can be identified. First is the approach of *cognitive objectivism*. Authenticity is treated as the originals or origins, including traditional cultures and people, which are to be cognised by tourists. Paradoxically, the quest for such a kind of authenticity often ends up as contrived experiences of '**staged authenticity**' (MacCannell 1973). Later on, this approach is loosened by stratifying tourists in terms of the relationships between social **class**, travel career and varying cognitive abilities (see **cognition**) or demands for different degrees of authenticity. Second is *constructivist* approach. This appears as a revision of the objectivist approach, which is sometimes criticised as too crude a position. Authenticity is seen as a product of social or cultural construction rather than an objective attribute of reality out there, waiting to be unearthed and cognised. Authenticity is thus described as 'negotiable', emerging, hermeneutic (Cohen 1988), or involving a power struggle regarding interpretation of **heritage** or toured objects (Bruner 1994). Seen this way, the authenticity of the past or of traditions, for example, is no more than an 'invention' in terms of the need and power of the present. Third is the *semiotic* approach (see **semiotics**). The holders of this approach argue that tourists are the 'armies of semioticians' who search for signs of authenticity. According to Culler, 'to be fully satisfying the sight needs to be certified as authentic. It must have markers of authenticity attached to it. Without those markers, it could not be experienced as authentic' (1981: 137). Tourism is thus no more than a collection of *signs* of authenticity, and what tourists quest for is merely *symbolic* authenticity (see **symbolism**).

Fourth is the *critical* approach. Supporters of this approach critically examine the representation of the 'Other' in tourism **marketing** (for example, tourism brochures) and reveal a neo-colonialist **ideology** that is hidden in the Western discourse, or **image**, of authenticity (see **neo-colonialism**). The touristic quest for the authenticity of the Other is seen as no more than a *projection* of Western stereotyped, biased and neo-colonialist imagery of the noble savage, which has nothing to do with any real assessment of the **natives** in the **Third World**. Fifth is the *postmodernist* approach. This approach is characterised by abolishing the distinction between copies and originals, or between signs and reality. The modern world is seen as a hyperreality, neither real nor false. The modern world is also explained as a simulation which admits no originals, no origins, and no real referents but only endless simulation. Accordingly, postmodernists declare an end of authenticity and justify inauthenticity in tourism.

It is argued that the quest for authenticity may bring about certain social and cultural consequences. This is often called the 'dilemma of authenticity'. For some, this dilemma means that, to be experienced as authentic, a sight must be marked as one; however, what is being marked as authentic is simultaneously inauthentic because it is mediated. For others, the dilemma of authenticity is rather that the very act of quest for authenticity may destroy authenticity itself, since the only full condition of authenticity is isolation. Tourism, as an agent of cultural **commoditisation**, may thus destroy the meaning of traditional culture. However, this thesis was criticised as an overgeneralisation, for commoditisation is *not necessarily* destructive to the meaning of cultural products (Cohen 1988). As a postmodern response to the dilemma of authenticity, Cohen (1995) suggests a 'sustainable authenticity' by justifying the contrived or copied authenticity in order to use it to prevent authentic cultures from being tourismified and hence destroyed.

As such, tourism in search of authenticity is one of the modern indicators or indexes of the ambivalence of modernity. People are tourists away from home because they 'hate' something relating to modernity such as the lack of authenticity and the loss of real self. Simultaneously, however, tourists are able to get away just because of certain enabling conditions of modernity which they 'love', such as higher living standards and so on. One cannot ultimately solve the contradiction of modernity and overcome the ambivalence of modernity. Thus, to be 'away and at home' may be a persisting dialectic of the contemporary **lifestyles**, and a touristic search for authenticity may be a sociocultural responsive action with respect to the existential condition of modernity.

References

Bruner, E.M. (1994) 'Abraham Lincoln as authentic reproduction: a critique of postmodernism', *American Anthropologist* 96(2): 397–415.

Cohen, E. (1988) 'Authenticity and commoditization in tourism', *Annals of Tourism Research* 15 (3): 371–86.

—— (1995) 'Contemporary tourism – trends and challenges: sustainable authenticity or contrived post-modernity?', in R. Butler and D. Pearce (eds), *Change in Tourism: People, Places, Processes*, London: Routledge, 12–29.

Culler, J. (1981) 'Semiotics of tourism', *American Journal of Semiotics* 1(1–2): 127–40.

MacCannell, D. (1973) 'Staged authenticity: arrangements of social space in tourist settings', *American Journal of Sociology* 79(3): 589–603.

Selwyn, T. (1996) 'Introduction', in T. Selwyn (ed.), *The Tourist Image: Myths and Myth Making in Tourism*, Chichester: Wiley, 1–32.

Wang, N. (1999) 'Rethinking authenticity in tourism experiences', *Annals of Tourism Research* 26 (2): 349–70.

NING WANG, CHINA

automatic interaction detection

Automatic interaction detection (AID) is a multivariate data analysis technique designed to segment **markets** and define or optimise customer-oriented **management** strategies. AID was introduced into social research in the late 1960s as an a priori or criterion segmentation technique. It is considered to be an explanatory method to support consumer or product-related **decision making**.

It has been applied to tourism research since the late 1970s. In 1990, some methodological improvements were realised, mainly through validation of AID results using **discriminant analysis**.

Fields of application include the identification of target groups such as different **tourist** types, the explanation of consumer or tourism **behaviour**, the detection of **market** niches and gaps in the tourism **supply** or **quality** deficiencies. The most important underlying assumptions are correctly drawn master samples with sample sizes of more than 2,000 cases, and a fairly evenly distributed dependent variable. With regard to the scaling level, all types of variables may be used in the analysis, the dependent variable being treated as dichotomous (with values of zero and one). Restrictive statistical assumptions are not required; the data may contain non-linearities, correlations and interaction effects.

The AID algorithm divides the master sample sequentially into a symmetric tree diagram of non-overlapping subgroups (cells). It defines the target group/cell by showing at each step or branch the explanatory variable (or the combination of its categories) with the highest significance. Each cell is characterised by two parameters, cell size N and cell mean Y. N represents the number of cases in the cell as a percentage of all cases, while Y denotes the portion of cases where the dependent variable has the value of unity. At each partitioning step, the pair of subgroups attaining the highest explanatory power (explained variance) is computed by the formula.

The technique's major advantages lie in the transparency and the easy interpretation of its results. Advanced applications may include the use of combined dependent variables or factorised predictors.

Further reading

Assael, H. (1970) 'Segmenting markets by group purchasing behaviour: an application of the AID technique', *Journal of Marketing Research* 7: 153–8. (Highlights the fundamental ideas of the AID technique.)

Vavrik, U.A. and Mazanec, J.A. (1990) 'A priori and a posteriori travel market segmentation: tailoring automatic interaction detection and cluster analysis for tourism marketing', *Les Cahiers du Tourisme*, Sèrie C, no. 62. (These authors demonstrate how to combine AID and discriminant analysis for significance testing.)

URSULA A. L. VAVRIK, FRANCE

automation

During the past several years, nothing has increased **professionalism** or more greatly enhanced customer services within the tourism **industry** than automation. **Technology** will continue to change the way tourism companies plan, co-ordinate, control and evaluate operations. Automated information system design and implementation is one of the fastest changing aspects of the industry. Future technological developments will be more intuitive, object-oriented, global and portable. The traditional approach of cultivating an appropriate level of computer literacy in order to render a more effective utilisation of technology is rapidly changing.

As the industry continues to increase its **Internet** participation through the development of home pages and web sites, there is an increasing worldwide exposure for global distribution of **products** and **services** (see **information technology**). Although a problem in the past, secure electronic transactions are becoming more reliable given sophisticated encryption schemes for encoding and decoding proprietary data. Together, higher transmission speeds and increased reliability will create a much more feasible environment for cyber-tourism and virtual commerce.

Advanced user interfaces have become increasingly intuitive, thereby enabling clients to quickly become familiar with applications that have traditionally been in the domain of tourism specialists. Newer applications require correspondingly less formal **training**, and recent developments in computer operating systems have resulted in movement from a character user interface to a graphical user interface. This factor has been of significant importance in the design and implementation of tourism information systems (see **marketing information system**). The near future **development** of a multimedia user interface ensures that such applications will remain an integral part of the tourism landscape.

In addition to these user interfaces, there has also been a growing interest in intuitive recognition technologies. Touch recognition involves data entry through physical contact with a special surface or screen. Touchscreen terminals are capable of moving large quantities of data easily, and hence provide an effective data processing option for a variety of tourism industry applications. Some applications rely upon a hand-held terminal, while others depend upon optical character recognition. The computer's ability to decipher and process more intuitive forms of input, for example, handwritten documentation, extends automation beyond previous expectations. The ability to configure a tourism system using open architecture provides optimal flexibility for fitting the computer to operations rather than vice versa. It is anticipated that many future technological applications, such as self-service applets, will be focused directly on the client.

See also: computer reservation systems; global distribution systems; hospitality information system

Further reading

Kasavana, M.L. and Cahill, J.J. (1997) *Managing Computers in the Hospitality Industry*, EI of AH & MA: East Lansing, MI.

MICHAEL L. KASAVANA, USA

aviation bilateral *see* international aviation bilateral

aviation liberalisation *see* international aviation liberalisation

aviation rights *see* international aviation rights

B

back office

The back office, such as in a **hotel**, refers to the location of those administrative procedures not carried out in full view of the customer (see **service delivery system**). This includes managerial and financial procedures such as **accounting**, financial and managerial reports, and payroll. Work is allocated to the back office on the basis that there is no requirement for a direct contact with the guest. Developments in computing **technology** are blurring the distinctions between front and back office activities.

DAVID KIRK, UK

back-stage

The notion of back-stage was introduced by Erving Goffman as part of his dramaturgical perspective. In considering life as theatre, back-stage (where actors prepare) is contrasted with front-stage (where actors perform). MacCannell later employed the approach for his analysis of **authenticity** in tourism. Whereas tourists seek genuine back-stage experiences among **destination** people, they are often provided with contrived events and attractions by the **industry**.

GRAHAM M.S. DANN, UK

Bahamas

The Bahamas is an archipelago of about 700 islands, of which seventeen are inhabited, situated in the Atlantic east of Florida and north of Cuba and Haiti. With an estimated population of 280,000, its main economic activity is tourism, which accounts for about 70 per cent of GDP. It is a member of the Commonwealth of Nations. The capital is Nassau, a vibrant tourism and banking centre.

AINSLEY O'REILLY, BAHAMAS

balance of payments

A country's balance of payments reflects its transactions with the rest of the world, as part of its system of national or macro **accounting**. The overall balance of payments is divided into capital account and current account, and the latter is subdivided between visible and invisible items. The capital account reflects flows of cash in and out of the national economy relating to inward or outward foreign **investment** (such as the receipt of loans or grants from abroad, or the making or repayment of the same) and purchases of financial assets (such as shares or bonds) or real assets (property) abroad, or sales of such assets in the domestic economy to foreigners. The term 'hot money' is used to refer to rapid short-term movements of cash on capital account (inward foreign investment) triggered by factors such as short-term interest rates. Hot money, being likely to depart as rapidly as it comes, tends to destabilize a country's balance of payments. Governments may seek to control these flows by the use of currency controls.

The 'visible' current account (sometimes termed the balance of trade) reflects flows of cash in and out of the national economy in respect of purchases and sales (imports and exports) of visible goods. This includes flows relating to long-lived assets (such as ships and aircraft) which in micro accounting would be treated as capital expenditure. The 'invisible' current account reflects a heterogeneous set of flows of cash including those for exports and imports of services such as banking, insurance and tourism (an increasingly important component, which it would be more logical to include in the balance of trade) as well as payments and receipts of returns on **investment** (interest and dividends), and private and intergovernmental transfers other than those on capital account. Among the **service** industries, tourism gives rise to very significant invisible export and import flows. The term 'tourism balance of payments' is used to indicate the balance between these flows. For example, **France** is a country with a large positive tourism balance of payments, as its inflows (tourist expenditure by foreigners in France) greatly exceed its outflows (tourism expenditure by French people abroad).

A country's balance of payments is important for the maintenance of the value of its currency in **foreign exchange**. Continuing balance of payments deficits for a country normally lead to an imbalance between the international **supply** of that country's currency and the international **demand** for it, in the direction of an excess supply. This in turn leads to the value (or parity) of the country's currency, in terms of the currencies of other countries' currencies, tending to fall; this is known as the country's currency 'weakening'. The converse is true if the imbalance is in the direction of an excess demand; the country's currency tends to 'strengthen'. Because a weakening currency makes imports more expensive and exports cheaper or more profitable, it tends to have an inflationary effect on the country's economy; prices in the domestic economy tend to rise. Conversely, a strengthening currency tends to have a deflationary effect, with domestic prices tending to fall or at least to remain stable. The inflationary effect may be countered by increasing the country's interest rates so as to attract foreign capital while restraining domestic demand. However, one consequence of using interest rates for this purpose may be to attract 'hot money'. Similarly, the deflationary effect of a strengthening currency may be countered by reducing the country's interest rates. A hard currency is one which has tended to strengthen and is not expected to weaken. This fluctuation would have immediate influence on the number of international tourist arrivals to a given country (see **inbound**), as well as on the number of its residents going abroad (see **outbound**).

However, even strong currencies may be subject to slight inflation; they are strong because other countries' currencies are subject to higher inflation. One reason for this inflationary tendency even in countries with strong currencies is public expenditure deficits, with government receipts from **taxes** and other sources being insufficient to meet public sector and other government expenditure. Bridging this deficit leads the government to increase the money supply, which in turn leads to the imbalance of 'too much money chasing too few goods and services'. In the short term, this imbalance is corrected by some combination of prices increase in the country's economy (i.e., inflation), the shortfall of goods and services is made good by imports, leading to a deterioration in the country's balance of payments.

One strategy which may be used in developing countries in the hope of improving their balance of payments on current account is import **substitution**. However, this strategy is likely to frustrate longer term **economic development** if it ignores the principle of **comparative advantage**. Another strategy which may be adopted for similar purposes is the **development** of tourism as an invisible export. The success of the latter strategy depends on a number of factors, including the extent to which imports of goods and services required for tourism development (such as building materials and construction services) can be kept to a minimum, and (for longer term economic development) the **multiplier effect**. Such a strategy may also raise issues of sustainable tourism.

SIMON ARCHER, UK

bar

Bars have been social gathering places for centuries, places in which beverages, mostly alcoholic, are consumed (see also **alcohol**). Bars can be found in different places such as hotels, airports, cruise liners, **restaurants** or free-standing locations. They are based on a variety of themes and can act as meeting grounds between the **host and guest** populations. The famous British pub is a bar. Bars are usually licensed by the state in some way.

JOHN R. WALKER, USA

Barbados

Barbados is the easternmost Caribbean island, with a well-developed, up-market tourism industry catering for 447,000 tourist arrivals in 1996 (up from 385,000 in 1992), plus a further 510,000 cruise ship arrivals. Tourism is the main source of **foreign exchange** earnings with receipts of $646 million in 1996. The **United Kingdom** is now the main generating country, displacing the **United States**.

MARTIN FRIEL, UK

bed and breakfast

Bed and breakfast **accommodation** is generally in a house, farmhouse or pub. Frequently, it is the owner's home. Meals other than breakfast may also be offered. This type of accommodation appeals to **tourists**, especially those from overseas, as it allows guests an insight into the local way of life. They are generally less expensive than hotels, and are thus attractive to those travelling on a limited budget.

ALISON J. MORRISON, UK

behaviour

The concept of behaviour in tourism considers customers and their behaviour specifically as it relates to touristic activities. Some distinctive behaviour topics include cross-cultural interaction, **authenticity**, tourist–guide interaction and post-travel attitudes. This subsumes both the observable behaviour of tourists and their mental and psychological processes involving **decision making**, **motivations** and **cognition**. The study of tourist **satisfaction** provides a link to business and management research. Knowledge of the behaviour of tourists in space and time is valuable to assist planners and managers of **attractions** and environments. Tourist behaviour and **experience** is assessed by **survey** studies as well as observational and field research.

PHILP L. PEARCE, AUSTRALIA

behaviour, recreation

Recreation behaviour can be classified in a number of ways. For example, it can be judged 'healthy' and 'active' or 'deviant/anti-social', 'passive' and so on. A fivefold typology divides activities into appreciative-symbolic, extractive-symbolic, passive free play, sociable learning and active-expressive (Hendee *et al.* 1971).

The main independent variables affecting exhibited recreation **behaviour** at both the national and individual/family level have been widely discussed and documented in the literature. At the highly generalised national or state level, a number of studies have pointed to five significant factors affecting the total pattern: population, economic growth/prosperity, car ownership, education and **leisure** time. At the disaggregated, individual or family level, nine explanatory variables are income, education, occupation, **age** and **gender**, cultural background (including **race**, **religion** and **ethnicity**), available leisure time, personality, fashion and place of residence. The latter is particularly important. Large metropolitan areas, for example, offer a quite different range of potential recreational opportunities (both indoors and outdoors) from rural regions. Similarly, coastal locations provide opportunities that are not available in inland settings. **Climate** is also a crucial factor influencing the time during the year that is potentially available for outdoor as opposed to indoor recreation (compare, for example, climatic **seasonality** between **Sweden** and **Australia** or **Canada** and **Malaysia**). Recreation behaviour is *learned* behaviour. Many recreational pursuits, for instance, involve the learning of specific skills that may take years to acquire. These skills – in such activities as yachting, golf, rock-climbing, snow **skiing** or playing a musical instrument – are often learned at an early age in the family or school situation, and developed upon through the life cycle. Many active pursuits are closely correlated with youth. Others, like swimming, tennis, **hiking** and golf can, and are, enjoyed by both young and old people alike.

All recreation activities, either directly or indirectly, require some financial outlay. Travelling and **accommodation** are expensive items, and so too is the equipment, including boats, golf clubs, fishing rods, and so on, that is essential for participation in most recreational pursuits. Thus, family or individual income, particularly discretionary or available income, is one of the single most important variables influencing recreation behaviour.

See also: participation, recreation; recreational demand; recreational need

Further reading

Dwyer, J.F. and Hutchison, R. (1990) 'Outdoor recreation participation and preferences by black and white Chicago households', in J. Vining (ed.) *Social Science and Natural Resource Recreation Management*, Boulder, CO: Westview, 49–67.

Hendee, J.C., Gale, R.P. and Catton, W.R., Jr (1971) 'A typology of outdoor recreation activity preferences', *Journal of Environmental Education* 3: 28–34.

Rodgers, B. (1967) 'The weekend life', *New Society*, 20 July: 78–9.

DAVID MERCER, AUSTRALIA

benchmarking

Benchmarking is a process whereby a business enterprise identifies industry leaders, compares **products**, **services** and practices, then implements procedures to upgrade its performance to match or surpass its competitors. Benchmarking has ready application in tourism where examples include the **accommodation** guides published by motoring organisations from which the ratings may provide a stimulus towards improved facilities and performance for participants.

JOHN J. PIGRAM, AUSTRALIA

benefit–cost analysis

Benefit–cost analysis is a systematic method for evaluating the economic feasibility of a public **investment**, project, plan or other action. Its aim is to lead decision makers in public organisations towards the optimal use of **resources** by displaying the outcome of a variety of alternative scenarios. The questions this analysis is designed to answer is whether this project is feasible (that is, do benefits exceed costs), which of a series of competing projects should be developed, or in what order projects should be developed.

Benefit–cost analysis serves as a tool in public organisations, where decision making is based widely on economic, social and environmental grounds rather than on the economic arguments alone. Its use in public organisations is analogous to the use of capital **budgeting** in private organisations. For instance, in a tourism enterprise that planned on the construction of a new establishment, such as a **hotel** property, capital budgeting would be useful to compare the amount of capital spent to future profit expectations. The scope of the analysis is however wider, as it evaluates projects from a public or social perspective rather than just from the perspective of a single agency.

This method of analysis serves in questions in which both the public interest and economic performance must be considered. Such decision making could touch, for instance, road building or **preservation** of **wilderness** areas for **recreation**. The problems may also be comprehensive in nature and even comprise ideological facets, such as whether massive tourism investments should be allowed to penetrate into sensitive local communities.

The process of the analysis starts with the identification of the stream of costs and **benefits** that take place in the course of the project. The realisation of the project, such as building and maintenance of a road, causes some immediate costs, such as working capital, material and labour expenses. Correspondingly, the road, once ready and in use, would directly serve motorists and public transport service. In addition to these apparent benefits and costs, the analysis may take into account issues that are more complex to identify. For instance, the new road might cut down the number of traffic accidents, and these gains include savings in insurance payments. In other words, benefit–cost analysis strives for holism by taking also intangible benefits and costs into account. It may, however, be somewhat sensitive to double processing. This means that some costs and benefits can simply consist of transfer payments from one agency to another (this is often the case with taxes). Rise in land values due to economic **growth** also portrays the risk of double counting. Therefore, the analysis should follow attentively that all benefits listed have a basis on actual change in **productivity** or wealth and, correspondingly, all costs draw on utilisation of resources.

The analysis synthesises benefits and costs that occur over time. Typically, costs come first, with benefits delayed. This is the case, for instance, with a tourism **resort** where the basic investments (such as **infrastructure**, building and **recruitment**) must be taken care of before the project starts bringing revenues in entrance fees. In order to fully compare costs with benefits, all such future transactions are converted by discounting them into their present value.

All costs and benefits in this analysis associated with the proposed **development** of a project are analysed regardless of to whomsoever they accrue. Both **tourists** and the host locality must be considered, as well as other stakeholders. The most obvious costs fall on those who are responsible for the realisation of the project, and the benefits are due to those who will most profit from the project. In addition to these apparent consequences, the analysis observes effects on more distant parties including people outside the marketplace. The road project taken as an example may cause reverberations that have no immediate market price, such as changes in the scenery, **climate**, or in air quality. These positive and negative third-party effects, also called **externalities**, are usually unintentional side effects of production, consumption or other economic transactions. Pollution caused by the tourism **industry** serves as an example of a negative externality provoking environmental damage to those living near the destination.

As the process draws on a wide scope of analysis, the total benefit of the given project is measured as its total utility to its users. The total utility of a product or service usually turns out greater than the price paid for it. For instance, the recreational value of a day visit to a **national park** may well seem greater than the minor entrance fee paid at the gate. The visitors then feel that they would be ready to pay a larger entrance fee, if needed. The gap between the total utility of the product or service and the price paid for it is called consumer surplus. In benefit–cost analysis, consumer surplus is taken into examination when identifying and assessing the benefits.

Finally, the scope of benefit–cost analysis may be enlarged to touch social costs and benefits that fall not on their origin but on the whole society. For example, the costs of environmental damages caused by massive tourism development and accommodation of mass tourists may eventually fall on the whole society, as people living in the damaged areas may start to migrate to more attractive areas. This may cause unemployment in receiving areas and pressure on the social security system, as individuals in working age are paid unemployment benefits.

Further reading

Curry, S. (1989) 'Cost-benefit analysis', in S.F. Witt and L. Moutinho (eds), *Tourism Marketing and Management Handbook*, Hemel Hempstead: Prentice Hall, 83–7. (Discusses the benefit–cost analysis and introduces its use in commercial organisations.)

Mishan, E.J. (ed.) (1994) *Cost–Benefit Analysis: An Informal Introduction*, 4th edn, London and New York: Routledge. (Covers the benefit–cost analysis including the conceptual frame and empirical illustrations.)

Schmid, A.A. (1989) *Benefit–Cost Analysis: A Political Economy Approach*, Boulder, CO and London: Westview Press. (Examines the benefit–cost analysis in public decision making.)

Smith, S.L.J. (1995) 'Benefit–cost analysis', in *Tourism Analysis: A Handbook*, 2nd edn, Harlow: Longman, 284–95. (Summarises the key features of the method.)

MARJA PAAJANEN, FINLAND

benefits

The term 'benefits' refers to anything someone receives that has perceived value. Benefits of the tourism **experience** may be in the form of psychosocial benefits to the individual, such as a satisfying experience. Benefits of tourism to a **community** may be in the form of environmental, psychological, social, cultural or economic gains or **impacts**. These benefits may be realised by the entire community or distributed to individual groups or business firms.

CHRISTINE J. CLEMENTS, USA

Bermuda 1

International aviation is based on the absolute sovereignty of a nation over the airspace above its territory. This was established through the Paris Convention of 1919. Sovereignty is exercised through control over the 'freedoms of the air' or **transit** privileges which regulate air traffic across national borders. This concept was defined as a principle of air law in the Chicago Convention on International Civil Aviation in 1944, when representatives of fifty-two countries met to consider the future of international aviation. The Chicago conference reinforced the notion of national sovereign control over airspace similar to the way in which countries exercise control over territorial seas.

The Chicago conference adopted a Form of Standard Agreement for Provisional Air Routes as a model for future bilateral agreements. However, this was superseded by the air service agreement (ASA) negotiated between the UK and the USA in 1946, called Bermuda 1 (the location of the negotiations). This became the accepted model for the next thirty years, although increasing trends towards liberalisation are now reflected in more recent ASAs.

The main characteristics of Bermuda 1 are that governments, not airlines, negotiate the terms of the agreement; government authorities of each country have the right to designate carriers; the designated carriers of the two countries have discretion to set capacity; the routes and designation of carriers authorised to operate on those routes are identified in the agreement; and fares are set by the **International Air Transport Association** (IATA) and require the approval of the participating governments.

Following the Bermuda 1 format, bilateral ASAs usually have three components. The first is the Air Service Agreement. Through its articles, the agreement specifies arrangements for setting fares and capacity levels, customs duties, transfers and airport charges, and other technical details. Second is the schedule of routes. This details traffic rights and specific routes and **airports** to be served by the designated airline(s) of each country. Finally, the Memorandum of Understanding (or Exchange of Notes in Agreed Minutes) clarifies points in the principal ASA and may be confidential to preserve commercial or national interests.

Further reading

Dempsey, P. (1987) *Law and Foreign Policy in International Aviation*, New York: Transnational Publishers. (Provides the background to current developments on this subject.)

Hanlon, P. (1996) *Global Airlines: Competition in a Transnational Industry*, Oxford: Butterworth Heinemann. (Presents an overview of international airline operations.)

International Civil Aviation Organisation (1994) 'The evolution of the air transport industry',

ICAO Journal 49(7): 46–58. (Historical focus on the civil aviation industry.)

TREVOR SOFIELD, AUSTRALIA

biblical sites *see* sites, biblical

binary structure of advertisements

Structuralists maintain that the language of publicity is built around opposites (for example, happiness and misery) in order to offer solutions to life's major problems. Tourism promotion follows this pattern both pictorially and textually (see also **promotion, place**). The binary structure of **advertising** is carried over into brochures, pamphlets and guidebooks, where the alluring qualities of the **destination** are contrasted with the monotonous attributes of home.

GRAHAM M.S. DANN, UK

biological diversity

Biological diversity or biodiversity is the variety of all life forms: the different plants, animals and micro organisms, the genes they contain, and the ecosystems of which they form a part. Biodiversity is often considered at three levels: genetic, species and ecosystem. Genetic diversity is the variability of genetic material within species. It occurs within and among the populations of organisms that comprise individual species as well as among species. Species diversity refers to the number and range of different species. While there is no certainty about numbers, there could be as many as nearly 100 million species on earth. Ecosystem diversity relates to the variety of habitats, biotic communities and ecological processes.

Biological diversity is emerging as the central concept of conservation strategies worldwide. At the Earth Summit in 1992, an agreement on a Biodiversity Convention was signed to conserve the same and to ensure sustainable use of its components. Agenda 21 deals with biological diversity in detail. At the first World Conference on Sustainable Tourism in 1995, a Charter on Sustainable Tourism was adopted which applies the recommendations of the 1992 Rio Declaration and Agenda 21 to the industry. The Charter asserts that tourism **development** should be based on sustainability.

Conservation of biodiversity is fundamental to ecologically **sustainable development** in general and to sustainable tourism development in particular. Biodiversity must be treated more seriously as a global resource, to be managed and preserved. It is the source of fresh air and water, food, medicines and industrial products. It is also a source of inspiration and cultural identity and underpins recreational and tourism opportunities.

The diversity of ecosystems and the uniqueness of a **destination**'s natural resources are the basis of the tourism industry. The integration of tourism development and biodiversity conservation is an important task for sustainable development. It is the tourism industry's role and responsibility to ensure that natural **resources** are managed in a manner which protects their intrinsic values. Properly developed and managed **ecotourism** can play a part in conserving biodiversity. Ecotourism can contribute to generating funds for **protected area** management and biodiversity conservation. Ecotourism also presents an opportunity for increasing awareness of the importance of the maintenance of biodiversity to tourists.

See also: conservation; ecologically sustainable tourism; ecotourism

Further reading

Goodwin, H. and Swingland, I.R. (1996) 'Ecotourism, biodiversity and local development', *Biodiversity and Conservation* 5(3): 275–369. (An issue devoted to ecotourism and biodiversity conservation.)

BAE-HAENG CHO, AUSTRALIA

board

Tourism or tourist boards (also called tourism authorities or commissions) is a term generally reserved for national or regional tourism organisations, and is more commonly used outside of North America. Their main roles encompass policy making and administration, **marketing** and promotion, industry facilitation, and other related services. There is a current trend towards the reorganisation and the funding of tourism boards as public–private authorities.

DONALD ANDERSON, CANADA

borders

Borders, or boundaries, mark the territorial limits up to which a political entity can exercise its sovereign authority. Various scales of border exist and they can be divided into international and subnational. International boundaries separate countries, or nations, while sub-national boundaries divide lower level civil divisions such as provinces, states and counties. Sub-national boundaries often have significant implications for tourism, especially in terms of **planning**, promotion and taxation. International boundaries, however, influence tourism in more noticeable ways. Flows of tourists, their choice of destinations, planning and the physical **development** of tourism are all affected by the nature of international boundaries.

Borders influence tourism in at least four ways: they act as barriers, they determine the form and function of destinations in their vicinity, they function as destinations and they are zones of **transit**. Borders as barriers can be viewed from two perspectives: perceived and real. For some tourists, having to cross a border may add a perceived distance owing to formalities or cultural differences on each side. Many borders are physically impermeable or highly restricted, which creates a real barrier.

The spatial development of destinations is often influenced by proximity to international frontiers. The most common pattern in frontier areas is for tourism facilities to be located adjacent to border crossing points or within easy walking distance. Furthermore, some borderlands are important destinations when they include popular attractions as Niagara Falls, Victoria Falls and the International Peace Garden. Other border-related attractions include the border markers themselves, **shopping**, **gambling** and **prostitution**. Cross-border shopping is popular in frontier areas where taxes and prices are lower and when a different variety of goods is available across the border. Gambling and prostitution tend to develop in border areas where such **activities** of vice are not permitted in neighbouring countries or provinces. Borders are also viewed by many tourists merely as transit zones, not as destinations, because this is the point where they cross from one political entity into another. Welcome signs, information centres and government buildings are the dominant tourism features of the transit **landscape**.

Further reading

Timothy, D.J. (1995) 'International boundaries: new frontiers for tourism research', *Progress in Tourism and Hospitality Research* 1(2): 141–52.

—— (1995) 'Political boundaries and tourism: borders as tourist attractions', *Tourism Management* 16(7): 525–32.

Timothy, D.J. and R.W. Butler (1995) 'Cross-border shopping: a North American perspective', *Annals of Tourism Research* 22(1): 16–34.

DALLEN J. TIMOTHY, USA

branding

A brand is the name, symbol, term, design or any combination of these used to differentiate **products** or **services** from those of competitors. It is possible to give brand names to individual products or to a complete product line. Branding is the process by which the company decides what brands it should offer. Some, like Hoover and Xerox, have become so popular that they started to be used as generic terms for the product itself (vacuum cleaner and photocopier, respectively). By branding the products or services, companies are able to differentiate them so they can be brought to the attention of buyers.

A brand is seen as a guarantee of consistent **quality** and can be used to attract loyal customers. In fact, the value of a brand is based on its perception in the customer's mind. By differentiating a product or service into several brands, where each appeals to a different group of customers, companies can use branding to help **market segmentation**.

In legal terms, the protection of brands was first established by the end of the nineteenth century, to allow manufacturers to differentiate their products from those without a brand name and other competitors' products. The major growth of branding, though, happened after the US Civil War when American firms and the **advertising media** industry started to grow.

It is important for companies to manage their brand names as if they were any other asset. If brands are well-managed, they can produce revenues and considerable profits that can either be invested in the same brand again or in other company projects. If on the contrary they are poorly managed, they can drain important resources until it is decided to discontinue the brand. If managers are able to know customers' perception of their brands, they are able to successfully manage and even extend them to other products and services.

Further reading

Kotler, P., Bowen, J. and Makens, J. (1996) *Marketing for Hospitality and Tourism*, New Jersey: Prentice-Hall. (Defines branding, provides the main concepts related such as brand name, brand mark, brand decision and identifies the main characteristics of a brand.)

Slattery, P. (1991) 'Hotel branding in the 1900s', *Travel and Tourism Analyst*: 23–35. (Addresses the issue of branding and makes prediction for the hotel industry of the growth of superchains.)

RICHARD TEARE, UK
JORGE COSTA, PORTUGAL

Brazil

Unlike the former Spanish colonies of Latin America, Brazil maintained its political unity after independence in 1822 and became the largest Latin American nation state. Brazil is vast in terms of **geography**, population and natural **resources**. It has a physical extension of 3.3 million square miles, the fifth largest country in the world. The population count reached an estimated 160 million in 1997, also the world's fifth largest. The country has many natural resources, ranging from iron ore to petroleum and natural gas to copper, gold and diamonds.

Although it has been politically unified along its recent **history** and capable to manage its ethnic diversity and varied cultural **heritage**, Brazil has not yet created a modern economy. Only recently was the country able to overcome some of the former roadblocks to **economic development**. The overall rate of growth for the period 1990–5 was 4–5 per cent. However, its gross national product in 1996 was about $580 billion and the country's per capita income in the same year was $3,640, far away from the present international threshold to development. At the same time, Brazil remains one of the most unequal countries of Latin America, and its uneven social development helps to explain the many twists and turns that have affected its economic performance. In 1996, 72 per cent of the Brazilian population had not completed the primary level of education, acting as a drag on the road to growth.

Something similar has happened in the tourism field. On one side, Brazil has many of the attractions that draw **mass tourism**. It has beautiful tropical and sub-tropical beaches, year-round warm weather and some excellent **resorts**, old and new. The Amazon basin is a magnet for ecotourists and the Iguaçu waterfalls are a sight enjoyed by many. Rio de Janeiro Bay is very beautiful, and the city itself has some world-class beaches (Copacabana, Ipanema, Leblon). Every year, it hosts the well-known Carnival celebration and thousands of people, many of them tourists, participate in the parades organised by the *samba* schools, which thanks to **television** are watched by millions. On the cultural side, Bahia, the town of Jorge Amado, has all the quaint charm of colonial times, beautiful monuments and buildings, and a rich black cultural heritage.

On the other hand, the Brazilian tourism industry has been unable to turn those excellent assets into a world-class **destination**. International arrivals fell sharply at the end of the 1980s to reach just over one million in 1990, nearly halving the figures for 1986 and 1987. However, there was a revival in numbers during the 1990s. In 1995, international tourists topped those two best years with 1.99 million arrivals, and records were broken in 1996 when Brazil received 2.67 million foreign visitors.

There are several reasons for this on-and-off performance. To start with, Brazil is a long haul from the main international **markets** of North America and Europe. Also, internal prices went up sharply, especially in touristic areas. Both trends put the country beyond the pale of most vacationers from those regions. Unlike the 1980s, the majority of present international arrivals originate in the Latin American region, mostly from nearby countries such as **Argentina**, Uruguay and Paraguay. Both average stays and expenditure per capita have decreased in the last few years; in 1996 these were 13.1 days and $70.4, respectively, as compared with 14.4 and $82.2 in 1990. At the same time, there was a high number of repeat visitors to the country, as 65 per cent of international arrivals in 1996 declared that this was not their first visit to Brazil. Post-trip **satisfaction** also ran high. Nearly 90 per cent of those polled in 1996 said they expected to come back at some time in the future.

There are other reasons for the uneven performance of Brazil. World-class **hospitality** is still missing, even in Rio, and most hotels across the country do not reach international standards of comfort and **service**. At the same time, with widespread lack of literacy and schooling in the population, it is difficult for tourism workers to excel. Further, **safety**, especially in and around Rio de Janeiro, has been one of the concerns of international tourists to Brazil and its lagging levels have been cited as one of the reasons for the downturn of the early 1990s.

But international tourism is not all that there is to Brazilian tourism. Improved economic performance has brought about an upsurge in **domestic tourism** during the last few years. There are no reliable data on this area, but it is estimated that domestic trips reached 128 million in 1994–5. Altogether, the industry contributed about 8 per cent to the gross national product during the 1990s, significantly more than 5.1 per cent for the rest of Latin America. Not to be forgotten, Brazilian top economic and social groups are avid consumers of tourism services abroad. Outgoing Brazilian tourists grew 495 per cent in the period 1980–95, as compared with a 17 per cent increase in international arrivals for the same period. Their favourite destinations are the **United States** (particularly Orlando and Miami) and **Argentina**.

This poses some questions both practical and general. On one hand, it is difficult to see how the Brazilian deficit in its touristic **balance of payments** can keep on growing at the same quick pace of the last few years, having already reached $3.3 billion in 1996. During the 1990s, Brazilians spent much more in foreign countries than they received from their international guests. On the other hand, this **behaviour** is an interesting puzzle for those theories that see **international tourism** in general as a form of neo-imperialism, where the so called southern countries are always at the beck and call of their northern counterparts. All in all, but for the outgoing end of the business, tourism development in Brazil has until now been slow and uneven. This means that the country is not receiving its due share from its outstanding attractions, and that there is still considerable space for its industry to grow.

Further reading

Abresi-Embratur (1996) *A Indústria do Turismo no Brasil. Perfil & Tendencias*, Brasilia.

Embraur (1997) *Anuário Estatístico 1997*, Brasilia.

JULIO R. ARAMBERRI, USA

break

Breaks are interludes or interruptions in an activity which features some refreshing and invigorating aspects so that the participant, upon returning to the previous activity, benefits from a renewed dedication or enthusiasm. **Vacations**, holidays and excursions are frequently seen as breaks from routine ordinary activities. 'Short breaks' are holidays of a few days duration, typically a weekend. In **hospitality**, breaks are refreshment opportunities scheduled between meeting activity.

CARL PFAFFENBERG, USA

break-even point analysis

Break-even point analysis is concerned with determining the number of units that must be sold, or the level of revenues that must be achieved, to produce a situation where revenues equal expenses. This analysis can determine the amount of **profit** earned at sales levels above break-even point; the quantification of additional units needed to be sold to reach profit levels at various points above the break-even point; the effect of changes in fixed charges; the effect of changes in variable costs; and the effect of changes in selling price.

Certain assumptions are required when employing break-even point analysis. These assumptions are that fixed costs will remain constant during the period being analysed; that total variable costs will fluctuate in a linear fashion with revenues; that variable costs will remain constant on a per unit basis; that labour **productivity** will remain constant regardless of volume; that revenues will remain constant on a per unit basis are proportional to variable costs; that there will be no volume discounts for bulk purchases; that all mixed costs will be broken down into their fixed and variable components; and that the sales mix remains constant. Furthermore, it is understood that any costs that are shared between more than one department will not be eliminated if one of the departments is closed, but rather will be allocated among the remaining departments. The basic formula for break-even point analysis is fixed costs divided by contribution margin. Contribution margin is revenue(s) minus variable cost(s). The result is the number of units required to be sold to break-even. A variation on this formula is to add a sum certain to the fixed costs to provide for **profit**.

Most tourism operations sell multiple products with different contribution margins for each **product**. Therefore, an additional tool, weighted average contribution margin percentage, is needed for break-even analysis. Contribution margin percentage is the item contribution margin divided by the selling price. Weighted average contribution margin percentage is total contribution margin divided by total sales. To get break-even in sales dollars, fixed costs are divided by the contribution margin percentage for a single product or the weighted average contribution margin for multiple products. Since most products in the tourism industry have different contribution margins and percentages, a weighted average contribution margin percentage is used to arrive at the total volume of sales in dollars to reach the break-even point.

Further reading

Coltman, M.M. (1994) *Hospitality Management Accounting*, 5th edn, New York: Van Nostrand Reinhold.

Schmidgall, R. (1997) *Hospitality Industry Managerial Accounting*, 4th edn, East Lansing, MI: Educational Institute of the American Hotel and Motel Association.

STEPHEN M. LEBRUTO, USA

brochure

A brochure is a form of printed promotional material which is designed to communicate with existing or potential visitors. Although there has been limited research on the nature and role of the brochure in tourism promotion, it is recognised that promotional material, including brochures, have been used throughout history to attract tourists (see also **promotion, place**). Today, brochures are one of the most commonly used vehicles for **destination** promotion.

Bochures are generally designed with one of two basic functions in mind: to provide practical information which visitors may use in their trip **decision making** and **planning** processes, and/or to establish an **image** of the destination as a viable alternative when planning future trips. Image brochures may be further described as 'promotional' with a goal of selling a particular business or attraction or 'lure', with the goal of promoting a destination area. These functions have generally been established using four distinct strategies or modes of presentations: attractions which provide a vehicle for **tourists** to address desired goals and/or needs; tours which 'bundle' attractions and activities within a manageable framework; a map which provides basic directional/location information; and information which provides detailed functional descriptions (such as price, availability and so on) of tourism-related facilities such as hotels, **motels**, **camping**, **restaurants** and the like.

The effectiveness of tourism brochures has been examined in a number of studies. Similar to all communication materials, they must be developed to meet specific objectives. Important characteristics of target markets include **location**, special interests and experiences, the nature of the trip and demographic characteristics. With this in mind, brochures can be a very effective means with which to influence tourism **behaviour**, often generating a relatively high **return on investment**.

Further reading

Buck, R.C. (1977) 'The ubiquitous tourist brochure: explorations in its intended and unintended use', *Annals of Tourism Research* 14(4): 195–207. (Examines the format and content of tourist attraction brochures provided in Lancaster County, Pennsylvania from a sociological perspective.)

Getz, D. and Sailor, L. (1993) 'Design of destination and attraction-specific brochures', *Journal of Travel and Tourism Marketing* 4(2/3): 111–31. (Discusses ways to design effective brochures for building image and affecting destination choice.)

Wicks, B.E. and Schuett, M.A. (1993) 'Using travel brochures to target frequent travellers and "big spenders"', *Journal of Travel and Tourism Marketing* 4(2/3): 77–90. (Examines the usefulness of travel brochures and suggests that targeting is an important principle for having impact.)

DANIEL R. FESENMAIER, USA

broker *see* culture broker

budget hotel

A budget hotel is a low-tariff, branded and often purpose-built **accommodation** product, characterised by good **quality** but limited facilities operated in accordance with standardised procedures. Mostly located on major tourism routes or outside urban settings. They have minimal public space; **food**, where provided, is often served in an adjacent **restaurant**. Tariffs are uniform nationwide and rarely discounted. Occupancy levels are high.

KEITH JOHNSON, UK

budgetary control

Budgetary controls are the management systems employed by an organisation to effectively determine variances and identify those that are significant, to analyse the significant variances, to determine the source of the problems that created the variances, and to initiate action to correct the problems. Budgetary controls measure and evaluate the actual results of operation to the budgeted results. The main purposes of **budgeting** are to provide an organised estimate of future revenues and expenses, a statement of management policy expressed in accounting terms and a method of control.

In a tourism enterprise, the budget addresses many issues. For example, from an operations budget the forecasted revenues for a period of time are openly stated (see **revenue forecasting**). In addition, wage and salary increases can be projected to provide the budgeted labour for the year. If price changes are necessary, the budget will clearly indicate this. Definite goals are established for **occupancy rates** and average daily rates in advance. All revenue producing departments should provide departmental income. The budget makes that commitment and requires management to be participatory. In fact, the best budgets are prepared when management reaches down to the lower levels of the organisation for input into the process. This creates stakeholders and an ownership in the financial aspects of the business.

The budgetary control cycle begins with establishing realistic financial goals for the organisation, followed by developing an action plan that will implement these goals. The actual results are then compared with projected results and corrective action is taken where appropriate to facilitate improvement. Those items that are appropriate for management action are determined from analysing the variances and determining those that are significant.

Often an organisation will employ a technique called zero-base budgeting (ZBB). This form of **budgeting** requires all participants in the process to start at zero and justify what they need in expenses. This technique is most effective during periods of excessive spending because it prevents a department from saying that they will need whatever they spent the previous year plus an additional amount for cost increases. With ZBB, the manager is forced to examine every cost incurred. ZBB is a form of budgetary control.

Budgetary controls require that periodic reports be prepared for each level of responsibility. The most common period of time for such reports is monthly. Since financial statements are usually prepared under the uniform system of accounts, the budgetary control reports should also be prepared using this format. Schedules are required for each operating and service department. These periodic reports must be relevant and timely to have any meaning and to be of use in a control process. The relevance issue means that operating managers should only be given departmental statements for which they have control. For example, if an enterprise is allocating service department costs to operating departments, the departmental manager will not be able to make the necessary changes since the comparison and the significant variances could possibly include allocated costs, which are beyond the scope of this individual's control. If the report is not timely, the necessary corrections will not be made until further deviations from the standard have been perpetuated.

Although budgetary control reports can take many forms, the most common ones show actual results for a particular period in the first column and the budgeted amount in the second column, with a difference column expressed in dollars and a relative difference column expressed in percentages. The budget can be of little use as a control vehicle if it is not prepared in a manner consistent with the financial reporting employed by the enterprise.

Budgetary controls require that an operation establish standards and standard procedures for every activity. The budgetary control process assumes that all employees are properly trained in the established standards and procedures. Periodic measurement and comparison of actual results to the budget allow management to make the determination if the established standards of performance are being met, and if not corrective action is put into place.

As a result of budgetary controls, errors and irregularities are kept to a minimum. An error is an unintentional mistake, and an irregularity is an intentional mistake. Errors and irregularities can be made by both management and hourly employees. The most common irregularities by employees is theft of assets and the omission of transactions. Management can create equity and record fictitious transactions.

The hospitality **industry** is especially vulnerable to errors and irregularities for a variety of reasons. This industry is a cash business with many transactions, and often is the source of **employment** for low-skilled individuals and employees first entering the workforce, groups that generally earn low wages. The items used in a typical hospitality enterprise all have a street value and are used by the general population. For these reasons, it is important to maintain good budgetary controls to monitor the expected performance versus the actual performance. Budgetary controls are preventative controls, that is, controls that are implemented before a problem occurs. Since a budget is prepared in advance of the activity, it becomes a standard to compare actual results. Detective controls, on the other hand, are designed to discover problems after the fact; an example is external audit. Detective controls also monitor preventative controls to make certain that they are working.

Budgetary control includes reforecasting the operations budget. There are times when unknown factors have caused the operations budget to lose its effectiveness as a control vehicle. Reforecasting is necessary when actual results vary significantly from the budgeted results, and the analysis has determined that the reasons for this variance is that the operations budget is inaccurate. Creating a new revised operations budget allows management to use budgetary control techniques.

Further reading

Coltman, M.M. (1994) *Hospitality Management Accounting*, 5th edn, New York: Van Nostrand Reinhold.

Schmidgall, R. (1996) *Hospitality Industry Managerial Accounting*, 3rd edn, East Lansing, MI: Educational Institute of the American Hotel and Motel Association.

Uniform System of Accounts for the Lodging Industry, 9th edn, East Lansing, MI: Educational Institute of the American Hotel and Motel Association.

STEPHEN M. LEBRUTO, USA

budgeting

Budgeting is a **management** planning activity covering all phases of operations for a definite period in the **future**. It is a formal expression of the plans, objectives, and goals established by management. Budgets are presented for the concern as a whole and for each subdivision of the operation. The budget is an organised estimate of the future and a method of control.

Some of the most common types of budgets are operating budgets, capital budgets and cash budgets. The former are revenue and expense projections for both operated and service departments. The summary of all departmental budgets are referred to as a master budget. Capital budgets refer to the acquisition of fixed assets and the management of debt and equity. Cash budgets are summaries of the cash expected to be received and disbursed over a finite period of time. Budgets can be established at a specific level of activity, such as room occupancy percentage. These budgets are called fixed budgets. Often, management requests budgets prepared at different levels of activity, such as **hotel** room occupancy percentage. These types of budgets are called flexible budgets.

The benefits of budgeting are that they force management to examine alternatives prior to a course of action, and view the facts relative to the financial health of the organisation. Budgets provide a standard for comparison, allow management to prepare for the future, measure progress and assist in self-evaluation, and clearly state the objectives of the organisation. Reasons not to budget are the time and cost of the process, the fact that there are many unknown factors, the possibility of a breach in the confidentiality of financial information, and the promotion of excessive spending merely because the budget dollars are available.

Operating budgets in **hospitality** operations begin with department managers **forecasting** revenues and expenses. Labour, the largest expense in a hospitality and tourism operation, is shown separately. Both operating departments and service centres prepare budgets. Operating budgets best serve management when they are prepared in accordance with the **uniform system of accounting** for Lodging Properties. The hospitality industry is fixed asset intensive as compared to manufacturing businesses. Therefore, capital budgeting is an important major concern. Management must decide which capital projects to fund under a capital rationing system, which chooses the most appropriate projects to undertake.

The hospitality industry is seasonal, and **demand** for its services are variable. Cash budgets must be prepared to alert managers of projected shortages and excesses of cash. By knowing in advance when shortages are expected, managers can make arrangements to cover these deficits in the most cost-effective manner. When management is alerted to projected cash excesses, these funds can be invested.

Further reading

Coltman, M.M. (1994) *Hospitality Management Accounting*, 5th edn, New York: Van Nostrand Reinhold.

Schmidgall, R. (1997) *Hospitality Industry Managerial Accounting*, 4th edn, East Lansing, MI: Educational Institute of the American Hotel and Motel Association.

STEPHEN M. LEBRUTO, USA

Bureau International de Tourisme Social

Established in June 1963, with its headquarters in Brussels, Belgium, the goal of the Bureau International de Tourisme Social (BITS) is to promote the fundamental human right to rest and leisure. The majority of its ninety-plus members are national organisations whose main, if not exclusive, aim is to develop **social tourism**. The organisation's activities include study and research, coordination and mutual operational aid, and representation.

JONI E. BAKER, USA

Bureau International pour le Tourisme et les Echanges de la Jeunesse

The aim of the Bureau International pour le Tourisme et les Echanges de la Jeunesse is to create favourable conditions for the development of **youth tourism** through the promotion of international friendship and cooperation among young people in the fields of culture, science, sport and **leisure** activities. Founded in 1960, its headquarters is located in Prague, Czech Republic.

TURGUT VAR, USA

burnout

The term 'burnout' was coined by Freudenberger (1980), who defined it as a state of fatigue or frustration brought about by devotion to a cause, a way of life, or a relationship that failed to produce the expected reward. It is a state of physical and emotional depletion resulting from conditions of work. The symptoms include a feeling of emotional and physical exhaustion; a sense of **alienation**, cynicism, impatience or negativism; feelings of detachment; and resentment of work and colleagues. Burnout victims often have a wide range of physical symptoms including headaches, stomach problems, insomnia and heart diseases. In extreme cases, burnout can lead to even serious psychological illnesses such as depression or suicide.

The occurrence of burnout appears to be correlated with numerous self-reported measures of personal distress, and in particularly with difficulty in controlling work-related stress. The common element to most work-related stresses is a feeling on the part of the individual affected that no matter how hard the person works, the payoffs in terms of accomplishment, recognition, advancement or appreciation are not there.

Empirical studies by Maslach and Jackson (1986) identified three components of burnout that were found to be most relevant for human services professionals. The first of these was emotional exhaustion, a state of depleted energy which is assumed to be a response to excessive psychological and emotional demands made on an individual. The second was a feeling of low personal accomplishments, a decline in one's feelings of competence and successful achievements in one's work and life. The third was depersonalization, a term associated with a heartless response toward clients. Research on burnout suggests that human service professionals such as teachers, health services workers and hospitality/tourism employees are particularly vulnerable to burnout. Because of the stress of interpersonal contact – the pressure to meet emotional needs of others while remaining 'selfish' – service professionals appear to experience burnout more than professionals in other fields.

Most researchers agree that external factors as well as the individual himself/herself contribute to burnout. External factors that have been found to be associated with high level of burnout are role ambiguity, role conflict, inadequate training, inability to advance, monotony, work overload, little or no emotional support, and little reinforcement from clients, colleagues or administrators. To recover from burnout, it is commonly believed that one should take a **vacation** and travel away from home and work as a **tourist**.

References

Freudenberger, H.J. (1980) *Burnout: The High Cost of High Achievement*, Garden City, NY: Anchor Press.

Maslach, C. and Jackson, S.E. (1986) *Maslach*

Burnout Inventory Manual, 2nd edn, Palo Alto, CA: Consulting Psychological Press.

ABRAHAM PIZAM, USA

bushwalking

Bushwalking is a form of active **outdoor recreation** involving walking along tracks or trails in forested or treed **landscapes**. While often a secondary tourist activity, bushwalking may be a pivotal activity supporting **adventure tourism**, **nature tourism** and **ecotourism**. While most often confined to short walks, in its longer forms bushwalking may represent a special form of tourism requiring **camping** and other accommodation (see **trekking**).

DAVID G. SIMMONS, NEW ZEALAND

business format

Multi-unit organisations in tourism may expand on the basis of one or more business formats, including full ownership and operation via **acquisitions** or **mergers**, joint ventures, leasehold agreements, **management contracts**, franchises, consortia or strategic alliance formats. Each has implications for the organisational and operational tie between units and organisations. Formats chosen depend upon whether the business is **cash flow**-based or asset-based.

ANGELA ROPER, UK

business tourist *see* business travel; convention business; market segmentation; tourist

business travel

Business travel, a tourism **market** segment, is a term used to describe all non-discretionary trips which occur either explicitly for the purpose of engaging in work, or incidentally in the course of conducting work-related activities. It includes travel associated with everyday business operations as well as travel for corporate or organisation-based meetings, conventions and congresses. Incentive travel, in which firms use trips to reward top-performing employees, is also generally included under this heading.

Business travel, by its very nature, may involve last-minute booking and frequent changes or cancellations. Because the cost is generally covered by the company or organisation rather than the individual, price may play a less important role than for pleasure trips. This does not mean that businesses have little interest in managing their travel expenses. Companies, governments and other organisations are increasingly taking steps to control costs through expense claims, and by negotiation of preferred supplier arrangements with specific airlines and **hotel** chains.

Business trips are different from pleasure holidays in several other respects. The preferred destinations for the former are generally cities rather than **resorts**. As a consequence, regions with well-developed industrial or **service** economies tend to attract the highest volumes of business travel. Such trips tend to be considerably less seasonal and more concentrated in mid-week, with **transportation** occurring during mornings and evenings to maximise time at the **destination** during business hours. Thus, hotels that cater to this market face important challenges resulting from spare weekend capacity.

Because of their professional status, business travellers are generally well-educated and affluent. These characteristics make them desirable targets for suppliers of the pleasure sector, who encourage them to take **vacation** time at the destination several days before or after the meeting or conference. Persons travelling on business may also choose to bring their spouse, or may return to the meeting site for pleasure at a later date.

Further reading

Cleverdon, R. and O'Brien, K. (1988) *International Business Travel 1988*, London: Economist Intelligence Unit.

ROBIN J.B. RITCHIE, CANADA

buying decisions

The ways in which consumers choose **products** and **services** have critical implications in the planning and evaluation of marketing strategies. Buyers consider a limited number of products/brands when planning purchases, and categorise these into different sets according to perception, preference and **experience** (see also **branding**). Thus, a central strategy is to gain entry into tourists' long-term memory and their abilities to retrieve such information into active memory when product choice is affected by both intention to purchase and situational variables.

LUIZ MOUTINHO, UK

C

cafeteria

A cafeteria is a self-service **restaurant** frequented by both **tourist** and resident populations. There are two main types. The 'in-line' cafeteria consists of a single counter along which customers pass to make choices. The free-flow system, also known as the 'scramble system' or 'hollow square', has several counters each serving different meal items such as hot foods, sandwiches, salads, desserts and beverages.

PETER JONES, UK

Cahier Espaces

Cahier Espaces publishes theoretical articles and initiatives on tourism **planning** and **development**, **leisure**, culture, **transportation**, national and regional tourism organisations, and educational centres and institutions. Each issue of Cahier Espaces deals with a theme such as *Accueil and Animation*, *Gestion des Stations* and *Nouvelles Technologies*. Articles are published in French, but can be submitted in English. First published in1985, the journal appears five times per year. It is published by Editions Touristiques Européennes (ISSN 0992–3950).

RENE BARETJE, FRANCE

camping

Camping is the act of temporarily living in close proximity with the out-of-doors, often in semi-permanent or mobile structures. It may be for purely diversionary purposes connected with recreation and tourism activities, focused on educational or therapeutic goals, done individually, in family or friendship groups, or conducted in formally organised settings. Its origins are the temporary living places associated with the military, subsistence hunting-gathering and exploratory expeditions.

Places to camp, or campsites, range from a climber's cramped bivouac on a mountainside ledge, to a designated spot with no ancillary facilities along a **wilderness hiking** trail, to a fully-developed parking place for a luxury motor-home, travel trailer or recreational vehicle in a public or private park. In popular usage, especially in the southern and western parts of the **United States**, a camp can also be a small house, or cabin, to which the user returns periodically as a base for recreational activities. In other areas, such temporary living places may be referred to as cottages or stations. These terms have evolved in recent years into the general category of **vacation** homes, or **second homes**, thus removing such structures from the popular perception of camping.

Modern recreational camping dates to the last century. In North America, civilian camping for other than the purely practical purposes of

exploration and overland **migration** into undeveloped territory began in the 1870s, with individuals and groups venturing into the wilderness by foot, horseback or small boats. The growing popularity of the automobile stimulated an expansion of relatively long-distance vacation camping in North America following the First World War. Public and private campgrounds and travel trailer parks developed to serve this increasingly mobile population. However, it was not until the years following the Second World War that vacation camping became a commonly popular activity, in North America, Europe and elsewhere.

Organised camping aimed at educational and socially therapeutic goals, frequently referred to as 'the camping movement,' originated under the sponsorship of schools and religious institutions in the mid-nineteenth century. These camps, usually conducted in the summer months for children and young adults, may be associated with organised youth groups, operated by public recreational and educational agencies, or private enterprise. Specialised camps may be organised around themes such as environmental **education**, the rehabilitation of emotionally troubled or physically **handicapped** persons, work–skill training, or instruction in computer use, mathematics, sciences or **music**.

MARY LEE NOLAN, USA
TOM BROXON, USA

Canada

Canada, a country with two official languages (English and French), combines a relatively small but ethnically diverse population of 29.3 million people with the thirteenth largest economy and the second largest geographical area in the world. The combination of a generally prosperous and cosmopolitan population, geographic grandeur, cultural variation, well-developed **infrastructure**, proximity to the **United States** and ties to Europe and Asia make this country both a leading origin and **destination** for **international tourism**. Like its neighbours, Canada is more a continent than a country. The **landscape** is a collage of different environments – mountain, prairie, forest, farmland, marine and tundra – spread over 10 million square kilometres in ten provinces and three territories.

Canada's proximity to the United States is a major influence on tourism **development** and patterns. Canada is the nearest international destination for the majority of the US population. The converse is true for Canadians: most live within 250 kilometres of the US border. They form an ethnic mosaic as diverse as the land itself. More than ninety distinct cultural groups, from Indians and Inuit through descendants of French and British pioneers to recent immigrants from Asia, constitute the social fabric.

Canada's tourism products compare favourably with those of its international competition, with key strengths in products that relate closely to its scenic natural **resources**. Two out of three of Canada's globally competitive product categories rely heavily on its natural scenery: **wilderness** (for example, Baffin-Pangnirtung or Nahanni River), **resorts** (such as Mount Tremblant, Banff or Whistler) and urban centres (including Toronto, Montreal and Vancouver).

According to the **World Tourism Organization** (1996), Canada accounts for 3 per cent of international tourist arrivals and 2 per cent of receipts. In 1995, Canada earned $11 billion from sales of both goods and services to 17 million international visitors, making tourism one of the most important exports for the country. US visitors account for about 75 per cent of foreign arrivals. While the number of Americans visiting Canada has increased modestly in recent years, their market share has dropped considerably from the 1970s when they accounted for almost 90 per cent of all international arrivals. By contrast, Asian market shares, especially from **Japan**, Taiwan and South **Korea**, have been rising rapidly. Increases in European market share have not been as great as from Asia, but Europe, especially the **United Kingdom**, **France** and **Germany**, remains an important market for Canada.

Canada also accounts for approximately 3 per cent of international tourist departures and expenditures. In 1995, Canadians spent about $14 billion on purchases of both goods and services involving 18 million international visits, making tourism one of the most important imports for the country. About 80 per cent of all Canadians visit the United States, although this percentage is slowly declining in favour of **outbound** visits to other destinations. In recent years, the tourism account balance – the difference between expenditures by Canadians going abroad and receipts from foreigners – has declined. In 1992, the deficit was over $6 billion; in 1995, it was $3 billion. Moreover, this inbound/outbound gap has closed. In 1992, the gap was almost 7 million; in 1995, it was less than 1.3 million.

While international tourism is important, **domestic tourism** far exceeds it in economic significance. Approximately 85 per cent of all tourism spending in Canada comes from domestic tourism receipts. The small ratio of international to domestic is due in large part to the size of Canada and its domestic attractions, especially **outdoor recreation** opportunities. The long-term trend, however, is one of declining domestic appeal, due to increasing tourism imports as a result of intense international competition for the very active and relatively wealthy Canadian tourists, and declining inter-provincial tourism coincident with a gradual shift to multiple shorter duration vacations. Although domestic tourism is practised by the majority (80 per cent), it represents only about 50 per cent of the total annual tourism expenditure by Canadians. The relatively low expenditures are in part because most domestic trips are taken by private automobile to destinations within the tourists' own province for **visiting friends and relatives**.

Until recently, tourism had trouble convincing sceptics that it played a significant economic role in the Canadian economy. This was due to the fact that tourism is not an industry in the traditional sense. Its consumption includes both tourism and non-tourism commodities (such as **accommodation** and groceries, respectively) and cuts across various industries (such as **transportation** and retail trade). These problems have been overcome by creating a tourism **satellite account** and associated statistical indicators. In fact, Canada plays a lead role in developing new tools for tourism analysis.

In recent years, the industry has performed better than both the business sector and the Canadian economy as a whole. For example, total tourism spending in 1995 was $41.8 billion, up 7.1 per cent from 1994. Its resulting total direct effect on the economy is estimated at $17.0 billion, up 7.0 per cent over 1994. By comparison, the Canadian gross domestic product rose only 3.9 per cent in the same period. **Employment** estimates also indicate that tourism job creation has exceeded that in the overall business sector.

In another vein, tourism in Canada has a well-developed social infrastructure with many different public and private sector organisations including **government** departments, trades, industry advocacy groups, destination **marketing** organisations and other tourism bodies at all levels. Within the public sector, the federal, provincial or territorial governments share roles and responsibilities in respect to tourism. Both orders of government share involvements in overall **strategic planning** and co-ordination, marketing activities, research, data and information activities, and **reception**, awareness and **hospitality** activities. Tourism development in terms of **planning** and plant development, however, is the prime responsibility of the provinces and territories, as well as of private enterprises. Similarly, while education is the responsibility of the provinces and territories, the task for **human resource development** falls primarily on the industry itself.

Further, within the non-government sector, many of the various national trade associations and advocacy organisations have associated provincial, territorial and local regional organisations. Examples include local convention and visitor bureaus, local destination marketing organisations and regional industry or sector associations. Each is concerned with **legislation**, **policy** and coordinated action at the relevant level and focus. The two most influential organisations at the national level are the Canadian Tourism Commission (CTC) and the Tourism Industry Association of Canada (TIAC). The CTC is a national partnership between the industry and all levels of government in Canada. Its main objectives are to market the country as a destination, and to provide timely and accurate information to aid decision making. TIAC is a national advocacy organisation that promotes the interests of Canada's tourism by formulating and advocating policies and programmes to the federal government.

Despite this progress and commitment, Canada's tourism industry faces numerous challenges. Three of the most significant are changing market demographics, the impact of new technologies and increased international competition. All Canada's main markets (domestic Canadians, Americans, Europeans and Japanese) have ageing populations. Matching the different product interests of this growing senior citizen segment will require changes in many existing products and services, in addition to developing new ones. Similarly, the organisation of the supply side of tourism, particularly the product distribution processes, is changing with the introduction of new information and telecommunication technologies. Examples include the establishment of **destination management** systems by some provinces, plus the increasing use of new technologies such as the **Internet** to distribute products directly to tourists without intermediaries. Moreover, international competition continues to intensify. In response, Canada has introduced a number of new trade **facilitation** initiatives. Significant interventions include the creation of the Canadian Tourism Commission, the historic 'open-skies' agreement between Canada and the USA, the 'two-way travel' agreement signed between Canada and Japan to triple volume between the two countries by the year 2005, and new arrangements to facilitate the movement of tourists on the ground as well as at major air terminals.

Canada's tourism industry faces the twenty-first century with a new positive **attitude** of confidence and determination arising out of the emergence of a strong sense of common national vision, industry partnership and national leadership support. Recently, all major Canadian competitive indicators are turning positive. Canada is now well positioned to develop its tourism industry further and capitalise on its growth potential as a destination with open spaces, beautiful scenery, a relatively clean environment, excellent infrastructure, high standards of **health** and **safety**, sophisticated cities, friendly small communities and a diverse culture.

Further reading

Aditus (1996) *The Aditus White Paper: Facts, Issues and Trends in Travel and Tourism*, Toronto: Southam.

Buchanan, J. (1994) *The Buchanan Report on Tourism*, Ottawa: Government of Canada.

Canadian Tourism Commission (1996) *1995/96 Annual Report: A Solid Base For Partnerships*, Ottawa: Canadian Tourism Commission.

—— (1997) *Challenges for Canada's Tourism Industry*, Ottawa: Canadian Tourism Commission.

Federal–Provincial–Territorial Conference of Ministers of Tourism (1985) *Statement of Principles on Federal, Provincial Territorial Roles and Responsibilities in Respect of Tourism*, Ottawa: Minister of Regional Industrial Expansion, Canada.

Lapierre, J. and Hayes, D. (1994) 'The tourism satellite account': national income and expenditure accounts, quarterly estimates, second quarter (Catalogue No. 13–001), Ottawa: Statistics Canada.

Tourism Canada (1990) 'Federal tourism policy', Annex, Tourism on the Threshold (Catalogue No. C86–38/1990), Ottawa: Minister of Supply and Services, Canada.

World Tourism Organization (1996) *Yearbook of Tourism Statistics*, Madrid: World Tourism Organization.

SCOTT M. MEIS, CANADA
FRANCOIS BEDARD, CANADA

car rental

Car rental or hire services provide the **tourist** with the flexibility of having a substitute private car at, or en route to, their **destination**. Such services are particularly popular with business tourists (see **business travel**). Car rental is largely provided either by major international companies or smaller, more locally based, independent firms. The former is dominated by Hertz, Avis, Budget and Europcar/Budget, and the **United States** is the most important market. **Franchising** has helped this rental business to expand. Some of the large companies are owned by car manufacturers, which use the rental outlets as a means of introducing and gaining acceptance of their new models. They have fairly similar prices but offer a wide variety of cars and hiring **locations**. This convenience appeals to the business **tourists**. The smaller firms have a more limited choice of cars and pick-up and set-down places. Their lower prices tends to attract the more price-conscious but more flexible **leisure** tourists, especially within Europe. There are also a few specialised companies which provide very luxurious vehicles for the up-scale leisure or business markets.

Much of the car rental business, particularly in the United States, is generated at airports. Railway stations are also important outlet locations outside North America. Express pick-up and return services are very often provided for the business tourist. Technological innovations, such as self-service machines and computerised mileage counts, have speeded up the transaction processes. Fly-drive packages, which offer an all-inclusive price for the cost of the air travel trip to the destination, combined with the cost of car and **accommodation**, have also increased the use of rental services for leisure purposes (see **package tour**).

The large car rental companies all have links with major airlines and **hotel** chains and can be accessed through their computer reservations systems. As with other parts of the tourism industry, reward schemes are offered for frequent car users which give them benefits such as price reductions, free upgrades and other special offers on rental and other tourism products plus priority booking (see **frequent flyer programme**). Companies with large numbers of employees using car rental services regularly on business are often able to negotiate favourable corporate rates. Other **yield management** techniques, originally pioneered in the airline sector, which vary rates charged according to patterns and type of demand, have also been adopted by the car rental business.

Further reading

Loverseed, H. (1996) 'Car rental in the USA', *Travel and Tourism Analyst* 4.

Smith, P.D. (1991) *The European Car Hire Industry*, London: Economist Intelligence Unit.

UK Euromonitor (1992) *Car Rental: The International Market*, London: Euromonitor Publications Ltd.

ANNE GRAHAM, UK

caravan park

A caravan park is a facility where tourists driving recreational vehicles can stop. It commonly provides such basic services as electric power, shower, laundry and grocery. Some also offer recreational amenities, including swimming pools and tennis courts. Public acceptance of this form of tourism in the 1960s led to its growth. After the 1970s energy crises, it also became popular to rent recreational vehicles at caravan parks, in a manner similar to renting a hotel room.

CHARLES S. JOHNSTON, NEW ZEALAND

career

A career is comprised of a sequence of work-related activities and strategies that an individual experiences, perceives and acts on during a lifetime. A career may be seen as being individually perceived and experienced and as being associated with work, however defined. Moreover, it is likely to be influenced by and to exercise influence on all other life activities, including the family and social spheres. It also has a longitudinal perspective in that it cannot be observed at a single point in time. It is a process that covers the lifetime of the individual. There are a number of commonly held views that are not necessarily components of an individual's concept of a career. It does not always imply upward mobility; there are linear careers, but also other types such as contractor, pilot or **motel** owner that do not necessarily lead to progression up a hierarchy.

Careers are not associated with any particular occupation but refer to all work-related activities that extend over a relatively long period. They need not necessarily be experiences associated with one organisation. While some people may spend their entire working life in one **hotel** management company, many others are highly mobile and pursue their careers in a variety of different organisations. Furthermore, it should be noted that career success cannot be adequately measured by such attributes as rank, salary and speed of promotion. Although these factors may affect an individual's self-perception of career success, there are many other factors that might affect individual perceptions.

In a career development orientation, there is a continual matching process between the organisation and the individual. The organisation needs to recruit, train, motivate, manage and develop human resources in order to maintain its effectiveness through survival and growth. Individual workers, on the other, hand need to find work contexts that provide challenge, security and opportunities for development throughout their entire working life. A psychological contract may be said to develop, which is continually renegotiated throughout the career of the individual within the organisation. The outcomes for the organisation may be seen in terms of **productivity**, creativity and effectiveness. Individual outcomes may be measured by **satisfaction**, security, personal development and integration of work with other aspects of life. The ideal situation is one in which this matching process leads to a mutually beneficial relationship.

A variety of commentators have pointed out that each career transition step requires decisions. As such, there are many important and difficult milestones facing people during their working lifetime, including deciding on a particular career to pursue, obtaining the **education** and **training** necessary for the chosen career, selecting a starting point to fulfil career plans, developing a strategy for obtaining a specific position in an appropriate organisation, selecting work offers from among alternatives, deciding the assignments and tasks to pursue within an organisation, developing a career path (such as technical or managerial), obtaining a position in another location or with another organisation, preparing for the next position (for example, strategies for continuing education and development), continuing self-appraisal and development of one's career goals, deciding to step down or move laterally during the latter part of a career, and selecting the time for retirement and an appropriate strategy of disengagement.

According to career development counsellors, there are a large number and great diversity of decisions to be made. Like all other workers, tourism employees face many career transitions and choices, and they need to actively manage the decision making process. Such choices should be made in the context of long-range career plans, and workers are more likely to enhance vocational satisfaction if they develop such plans. One of the most important theoretical formulations in the context of career planning and development involves the notion of career anchors. As a result of education, early organisational socialisation and work experience, an individual is said to develop certain knowledge about the match between self and job.

Schein (1992) holds that the career anchor functions in the individual's work life as a way of organising experience and identifying one's area of contribution over a working life. It determines those types of activities in which an individual feels competent. Schein identified a number of areas where the categories of career anchors. In Anchor 1, the individual seeks and values opportunities to manage. There is a strong motivation to rise to positions of managerial responsibility. In Anchor 2, the individual seeks and values opportunities to exercise various technical talents and areas of competence, and is interested primarily in the technical content of the job whether the work is finance, engineering, **marketing** or some other functional area. In Anchor 3, the individual is motivated by the need to stabilise the career situation. This person will do whatever is required to maintain job security, a decent income and the potential of a good retirement program. In Anchor 4, the individual has an overarching need to build or create something that is entirely his or her own. This is self-extension through the creation of a new **product**, process, or **theory**, a company of their own or a personal fortune as an indication of achievement that seems to be the career objective of these people. In Anchor 5, the individual seeks work situations that will be maximally free of organisational constraints so as to pursue their professional competence. Freedom from constraints and the opportunity to pursue one's own life and work style appear to be a primary need.

It is argued that workers do not firmly develop career anchors until after they have been involved in a work situation for an extended period. Career anchors clearly reflect the underlying needs and motives which the person brings into adulthood, but they also reflect the person's values and, most important, discovered talents. By definition, there cannot be an anchor until there has been work experience, even though motives and values may already be present from earlier experience (Schein 1978: 171).

Ladkins and Riley (1994) have proposed a range of structural variables which form the basis of tourism career research. They point out that the literature on career theory and measurement contains few applications to tourism. Within the tourism literature, there is a good deal of exhortation as to the benefits of a career, with the universal problem of staff retention tending to dominate the labour planning literature. Ladkins and Riley argue that the central problem of tourism career sequence analysis is that both measurement and interpretation are dependent upon the classification scheme employed by the researcher. Approaches to sequence analysis follow two possible directions, those that measure what happened and what was learned in each job, and those that concentrate on the differences between jobs putting the accent on the job transitions rather than the job. For the former, the interpretation is additive, while for the latter it is developmental. The design of career path research usually has to consider both job content and transition processes. The choice of emphasis determines the form of classification scheme applied to the sequences. Ladkins and Riley go on to make the point that most career paths are influenced by interventions. Their literature review indicates an increasing interest in sequence analysis. They point out that within the tourism contexts the players are not simply education, the individual and the organisation. The labour market is also a key player. Their perspective brings the additional advantage of taking into account labour turnover which is often critical for tourism/**hospitality** industry enterprises. Thus, any understanding of career processes within this field must encompass notions such as sequence, individual diversity and business context.

References

Ladkins, A. and Riley, M. (1994) 'Hotel management careers', *Tourism Management* 15: 221–2.

Riley, M. and Ladkins, A. (1994) 'Career theory and tourism', *Progress in Tourism, Recreation and Human Management* 6: 225–37.

Schein, E.H. (1978) *Career Dynamics: Matching Individual and Organisational Needs*, Reading: Addison-Wesley.

—— (1992) 'Career anchors and job/role planning', in D.H. Montross and C.J. Skinkman (eds), *Career Development*, Springfield: Thomas.

GLENN F. ROSS, AUSTRALIA

Caribbean Tourism Organisation

The Caribbean Tourism Organisation (CTO) was established in January of 1989 after the integration of the Caribbean Tourism Association (CTA) and the Caribbean Tourism Research and Development Centre (CTRDC), founded in 1951 and 1974, respectively. The mission of CTO is to promote tourism in the Caribbean countries through **education**, research, **marketing** and institutional relations.

The Caribbean Tourism Organisation, through its mission, attempts to increase the value and volume of tourism flows to member states; to heighten awareness and understanding of Caribbean tourism in the marketplace; to increase public awareness and understanding of tourism by host countries; to develop the highest level of skill and **professionalism** in the personnel serving Caribbean in tourism and related areas; to design and develop a comprehensive tourism information system; to ensure a harmonious interaction between tourism and the social and natural **environment**; to give support to those countries not able to represent themselves; to foster close links with regional institutions and international donor agencies; to develop a capacity to assist members in defining and responding to tourism related needs; and to develop strong linkages between tourism and other economic sectors (such as agriculture, manufacturing, and services) in each host country.

The CTO's responsibilities are regionwide and encompass both public and private members. Its **government** members include English, French, Spanish and Dutch-speaking territories of the Caribbean Region. Private sector members includes companies, organisations and persons providing goods and services to the tourism industry. The CTO programme is established in a strategic plan approved for execution every two years. The plan implementation is carried out by the various divisions responsible for CTO areas of operation. These divisions are marketing, research and statistics, education and **training**, and consulting and technical assistance.

In addition to its extensive collection of reference books and research studies on Caribbean tourism, CTO organises an annual tourism conference on **sustainable development** whose goals are to model programmes that enable the **region** to sustain a feasible tourism **product** while maintaining its environmental resources. CTO is an international development agency and a member of the **World Tourism Organization** (WTO), functioning under a headquarters agreement with the government of Barbados.

Further reading

Holder, J. (1989) 'Caribbean tourism organisation', *Annals of Tourism Research* 16(4): 589–91.

MARIA FUENMAYOR, USA

carrying capacity

The term 'carrying capacity' originated in range management, where it is used to refer to the number of stock that can be supported by a unit of land. It has been used at other scales, for example, with reference to the number of people that can be sustained by the earth. When used by **recreation** and, more recently, by tourism researchers and managers, it refers to the maximum number of people that can use an area without there being an unacceptable decline in the quality of the visitor **experience** or the **environment**. This definition indicates that it is a multi-faceted concept involving both human (or social) and physical, and perhaps other, dimensions. The above definition raises questions as to what is acceptable and who should determine that. Furthermore, it would be wrong to assume that there are simple relationships between an increase in the number of users, changes in the physical environment and the **quality** of experiences which can be obtained.

Early proponents of the concept of carrying capacity implied that recreation areas had a fixed ability to sustain use. They hoped to be able to calculate the critical level of use beyond which deterioration would set in. While intuitively appealing, and one can acknowledge that the characteristics of some places make them more resilient and able to withstand higher levels of use than others, carrying capacity has remained an elusive concept, and efforts to develop methodologies for determining carrying capacities before they have been exceeded have been largely unsuccessful. There are a number of reasons for this. Land does not have an inherent capacity to support tourism. The same **location** could be used for different purposes, such as a **nature** reserve or a golf course, with associated implications for the type of experience provided and the number of people that can participate in and enjoy that experience. Thus, in the absence of clearly-specified goals and objectives, carrying capacity does not have a specific meaning which is amenable to calculation in advance of it being exceeded.

Even in the context of range management, weather has implications for **food** production and the number of stock that can be supported. The recreation and tourism contexts are much more complex, with issues of **seasonality**, different types of users and activities, and a diversity of facilities within a single location, being among the factors frustrating the undertaking of simple calculations. However, the capacity of ancillary facilities, such as the number of parking spaces or **accommodation** units, may be readily calculated and the level of provision of such supporting **infrastructure** may be employed as a check on levels of use of other more fragile facilities or areas. If people are unable to park or stay, they must of necessity go elsewhere.

Much of the early work on carrying capacity undertaken in **wilderness** areas assumed that increases in the number of users would result in greater visitor interaction and undermine conditions of solitude sought by many wilderness visitors. Further research showed that the number of visitor interactions was only one, and not necessarily the most important, factor influencing visitor experiences. When transposed to other areas, it soon became evident that there is not a simple relationship between number of users (density of use) and crowding, and that for some activities, such as festivals or concerts, the quality of experience may actually increase with larger numbers of visitors, at least to a point. Further, surveys of visitors usually indicate positive experiences, with those who have not been satisfied previously moving on elsewhere to be replaced by others who are tolerant of the new situation.

While it is not difficult to criticise the carrying capacity concept, it is not easy to come up with alternatives. While lacking in managerial specificity, the search for capacities has spawned a great deal of valuable research on the environmental impacts of tourism and recreation, and on relationships between levels and types of use and the qualities of the experiences which can be obtained. While carrying capacity is a term which is still widely used in the literature to refer to a level of use which should not be exceeded, difficulties in its application have resulted in the development of newer concepts such as **limits of acceptable change** and the **recreation opportunity spectrum** which lend themselves more readily to managerial application.

Further reading

Anderson, D.H. and Brown, P.J. (1984) 'The displacement process in recreation', *Journal of Leisure Research* 16: 61–73.

Shelby, B. and Heberlein, T.A. (1986) *Carrying Capacity in Recreation Settings*, Corvallis, OR: Oregon State University Press.

Stankey, G.H. and McCool, S.F. (1984) 'Carrying capacity in recreational settings: evolution, appraisal and application', *Leisure Sciences* 6: 453–74.

Wall, G. (1982) 'Cycles and capacity: incipient theory or conceptual contradiction?', *Tourism Management* 3: 188–92.

GEOFFREY WALL, CANADA

carrying capacity, recreational

Recreational **carrying capacity** is the threshold level of use where **impacts** exceed levels specified by evaluative standards. From an early search for a set number of users, the term has evolved into a **management** concept encompassing biophysical (ecological), sociocultural and facility assessments and capacities.

DAVID G. SIMMONS, NEW ZEALAND

cash flow

All organisations receive and disburse cash, and in **accounting** these receipts and payments are reflected by means of a statement of cash flow. This is a tool for **financial control** and **asset management**, accounts for the change in the organisation's total balances of cash and cash equivalents (i.e., its total cash flow) during a financial period by analysing and summarising its inflows and outflows under various headings. For a business organisation, five headings are typical of international usage: cash flow from operations, interest and dividends received and paid, taxation paid and refunded, purchases and sales of fixed assets, including financial fixed assets (investments) and businesses, and financing activities (raising or repaying capital in the form of equity or loans).

Cash flow from operations may be presented either by the direct or indirect method. Under the former, for example, the total amounts of cash received from tourism sales and those paid for purchases of stock and for operating expenses are shown in the statement. The indirect method is less informative; the cash flow from tourism operations is derived from the operating profit by making adjustments for non-cash items (such as depreciation of fixed assets) and for changes in working capital (stock, debtors and creditors) which absorb or release cash.

While profitability (broadly, the ability to create economic value by producing outputs with a value greater than that of the inputs used in the process) is the key to the long-term survival of a business, cash flow is crucial in the shorter term, as it is the inability to make essential payments which causes a business to fail. Examination of the five headings shows that, apart from operating cash flow (which is associated with profitability), a tourism business has various ways in which it can generate cash. For example, loans may be repaid out of the proceeds of selling fixed assets, or by reducing working capital. However, there are clearly limits to the extent to which fixed assets and working capital can be reduced if the business is to be maintained, so that in the longer run it is operating cash flow and hence profitability which matter. The crucial importance of cash flow means that the financial controls of an organisation should include cash flow planning and control.

SIMON ARCHER, UK

casino *see* gambling

catering

Catering, in the United States, is the provision of **food**, supplies and services for specific functions on particular dates and at given locations. Elsewhere, the term 'catering' may indicate the full spectrum of what is termed in the US as food service, and includes the provision of food and service in any venue. Catering in the US has developed as a distinct segment of the food service sector spanning a broad spectrum of commercial operations providing a variety of catering services. These services include three primary types: banquet hall, off-premise and mobile unit.

Banquet hall catering, termed function catering in the UK, comprises independent units and those attached to hotels, convention or conference centres, and restaurants. This type of catering can provide several million dollars in revenue for a medium-size or large hotel. Generally, every major hotel and convention centre has banquet hall space. Off-premise catering involves the taking of all food, supplies and services to a **location** designated by the client. Off-premise events vary from the simple picnic in the **park** for fifty to the formal dinner for eight in the home, to the corporate entertainment tent for 3,000 at the Olympics. Special event catering is a type of off-premise catering which requires a particularly high level of **planning** and coordination due to the high volume. Events such as the Super Bowl require 300 servers just for the VIP boxes, but for only eight hours on one day. Mobile unit catering utilises specially designed trucks for provision of **service** to work sites. The units are equipped to serve **fast food** items and snacks, and are generally supplied by a **commissary** kitchen facility.

Catering differs from other segments of the food service sector by virtue of the diverse **roles** often assumed by the caterer. Clients often expect the caterer to provide more than the food and service. A partial list of additional supplies and services which may be requested include entertainment, equipment, **transportation** of guests and valet parking at **site**. The caterer provides quality food and service in the comfort of their on-site banquet hall today, and 100 miles from their property and all other services the next day. Catering is a growth area for the food service sector. The expansion of the corporate business market and the home meal replacement market will both continue to contribute to the growth of the catering industry.

Further reading

Jones, P. (1996) 'Outdoor catering', in P. Jones (ed.), *Introduction to Hospitality Operations*, London: Cassell. (Provides a overview of this sector of the UK hospitality industry.)

JERALD W. CHESSER, USA

catering, airline

Airline catering is the provision of food service by an **airline** for in-flight service and sometimes for employee cafeteria service. An airline's flight kitchen(s) prepare meals, snacks and other food items, usually at the airline's main **airport** hub(s). The catering may also be performed for other airlines, thus serving as a revenue source. Some airlines outsource their food service to either another airline or a third-party provider.

FREDRICK M. COLLISON, USA

causal model

Causal models are mathematical representations of relationships between two variables (X and Y) based on theoretical reasoning which meet three conditions of existence: there is a relationship between X and Y; the relationship is asymmetrical such that a change in X results in a change in Y but not vice versa; and a change in X results in a change in Y regardless of the changes in other factors. Generally, it is also assumed that the change in X must occur prior to any change in Y, although some definitions of causation allow for the possibility that a change in X and Y may occur simultaneously. Having met these conditions, one may conclude that X causes Y.

Many theoreticians argue, however, that the conditions of causality often cannot be verified empirically, and thus the concept of causality reflects the working assumptions of researchers rather than descriptions of reality. It is widely accepted that while the conditions of causation cannot be strictly met in real world applications, it is possible to infer causal relationships within the system under study. That is, since it is always possible that some unknown factors may be operating, the only way one can make causal inferences is by establishing simplifying assumptions about the lack of existence of such factors.

Path analysis is a statistical tool which provides information about causal processes by decomposing the correlation between two variables into a sum of simple and compound paths. This decomposition process enables the researcher to measure the direct and indirect effects that one variable may have on another which, in turn, allows one to examine the causal process underlying the hypothesised relationships and to estimate the importance of alternative paths of influence.

Causal modelling has been widely used in tourism research. Recent examples include the examination of factors which influence touristic behaviour (such as **destination image**, destination **choice**, visitor expenditures), the factors which affect residents' attitudes towards tourism **development**, and the factors which determine visitors' perceptions of the services provided at a destination.

Further reading

Blalock, H.M. (1964) *Causal Inferences in Nonexperimental Research*, Chapel Hill, NC: The University of North Carolina Press. (Discusses causation in social science research.)

Pedhazur, E.J. (1982) *Multiple Regression in Behavioral Research: Explanation and Prediction*, 2nd edn, Forth Worth, TX: Harcourt Brace Jovanovich College Publishers. (Provides a discussion of causation, causal models and path analysis.)

Vogt, C.A. and Fesenmaier, D.R. (1995) 'Tourists and retailers' perceptions of services', *Annals of Tourism Research* 22(4): 763–80. (An example of causal modelling within a tourism setting.)

DANIEL R. FESENMAIER, USA

central reservation

Tourism firms with multiple **locations** need a central location to handle incoming reservations from consumers and from their intermediaries. A central location is needed so that their **inventory** can be accurately tracked and updated. Most central reservation offices use **computer reservation systems** (CRS) to store and process reservations, and these in turn are often linked to **global distribution systems** via data communication links.

PAULINE J. SHELDON, USA

centrally planned economy

Economic systems can produce tourism products in a number of ways. These systems can be categorised into three groups: **market** economies, centrally planned economies and mixed economies. The centrally planned economy (or command economy) operates under the rules whereby the means of production, such as the **airlines** and hotels, are deemed to be owned by the public and all forms of economic activity are controlled by the **government**. This means that the central planning authorities are responsible for assigning production goals of the tourism industry as well as the allocation of factors of production. Within such a system, the proportion of total output absorbed as **investment** as opposed to private consumption, together with the distribution of income and the range of tourism **activities** produced, becomes a problem of **social control** rather than market forces.

Therefore, in centrally planned economies the decision to produce particular levels of output of tourism products is a social and political rather than an economic decision. Unlike market economies, rewards are distributed amongst residents according to social criteria and the system has the capability to attribute all forms of factor usage to segments of the population and industries. In contrast the market economies face the problems associated with the depletion of non-priced elements of production and consumption, such as the culture and the **environment**. Within mixed economies, there is some attempt to cover the production costs of non-priced goods, such as road networks and public health, by means of government fiscal revenue. It can be argued that this introduces some social organisational control over the distribution of income and costs. More recent times seem to have indicated that market-based economies have been more successful than their centrally planned counterparts, as exemplified by the break-up of the Soviet Union. However, it would be misleading to assume that market economies do not present problems that will need to be overcome to ensure their survival.

Further reading

Allcock, J. and Przeclawski, K. (eds) *Tourism in Centrally-Planned Economies*, a special issue of *Annals of Tourism Research* 17(1).

JOHN FLETCHER, UK

centre–periphery

Centre–periphery, also termed core–periphery, relationships refer to the links between a powerful centre, such as an urban concentration of **demand**, and more distant, less powerful areas which are often suppliers of **wilderness**, rural and dispersed tourism opportunities. The Mediterranean coast and the Caribbean may be regarded as peripheral to the urban areas of western Europe and the northeast United States, respectively.

See also: pleasure periphery

GEOFFREY WALL, CANADA

chain hotels

The **hotel** sector is comprised of both independent and chain hotels. The latter now account for over 60 per cent of the total of 6.9 million hotel bedrooms worldwide. Chain hotels are made up of affiliated properties by virtue of the fact that the chain is contracted on a continuing basis to be responsible for putting in place at least one management function in the whole system. A hotel chain is thus an organisation that competes in the tourism industry, either locally, nationally, regionally or internationally, with more than one unit of similar concept or theme.

The affiliation between a hotel and a chain can be of several types, and these exhibit different degrees of control of the chain over its hotels. The degree of control depends upon the **business format**. The hotel could be wholly owned and operated by the chain, or it could be a leased, franchised, contracted or co-operative unit. A hotel chain would have greater control over its hotels if they were fully owned rather than, for example, if they were franchised.

As a member of a chain, the hotel is likely to be able to gain access to a wide range of markets, have better representation and referral within a chain and be part of sophisticated **marketing** strategies (including corporate loyalty schemes, for example). It may also have brand name recognition and bargaining power over suppliers and buyers. Access may be available to the most up-to-date systems and procedures, project design and management, technical services, more favourable capital terms and central reservation systems. The chain hotel might be able to participate in **training** and management development programmes and thus offer **quality** and consistency guarantees in its products and services. The hotel chain itself may be affiliated to a conglomerate or related organisation, thus providing additional market value and financial linkages.

A large proportion of the top hotel chains have at least 70 per cent of their rooms in hotels outside of their home countries, and it is likely that the industry will consolidate further, making chain hotels the predominant type of hotel unit worldwide.

Further reading

Slattery, P., Feehely, G. and Savage, M. (1995) *Quoted Hotel Companies: The World Markets 1995*, Kleinwort Benson Securities. (Compares the performance of quoted hotel companies worldwide and provides an examination of the investment issues related to international expansion, via low capital cost formats.)

ANGELA ROPER, UK

change

Apart from cyclical changes, such as inflation, currency fluctuations and fashion, few institutions can match the graph of (non) scientific discovery, and structural or permanent changes, including demographic and technological changes (see **demography**; **technology**). Therefore, tourism institutions need to reduce the discrepancy between environmental change and institutional response. Quantum **theory** of strategic change suggests that large institutions tend to change only marginally.

See also: change, sociocultural; lifestyle

FRANK M. GO, THE NETHERLANDS

change, sociocultural

A shift in a social group's norms and rules of **behaviour** and/or cultural **ideology** and identities over time constitutes sociocultural change, or **acculturation** when this is due to cultural contact between groups in either mutual (trade) or unequal (**colonisation**) relationships. Tourism's **impacts** range from changes in the local political economy due to **globalisation** to the changing expectations of tourists.

MARGARET B. SWAIN, USA

charter, air

Air charters usually involve **tour operators** in negotiations with airlines to secure capacity on aircraft for a finite period of **time**. Charter flights are distinctive mainly as a result of a number of charter conditions. These include flight times and levels of **service**, amongst others. Charter flights aim to maximise the utilisation of aircraft. Charters play a pronounced role in promoting group and mass forms of tourism.

See also: mass tourism

LESLEY PENDER, UK

Chile

Chile's tourism activity is facing a new set of circumstances. For decades, the tremendous distances which separated this **destination** from the most dynamic markets, and the prevailing deficiency in communications, meant that the country found **development** difficult thanks to its geographical **location**. However, Chile has been able to take advantage of the opportunities offered by modernisation. The concept of the world as a 'global village', shortened flight times as a result of new routes and larger aircraft, and the tremendous development in communications have enabled Chile to be positioned rather differently in world markets, thus breaking the negative effect of its isolation.

An extremely long country – 4,400 kilometres from north to south, flanked by the Andes mountain range and the Pacific Ocean – Chile has a wide variety of natural landscapes and climates. From a tourism point of view, this variety of **resources** means that it can offer different products, most of which are based on particularly natural attractions such as the Andes, the Atacama desert and Patagonia. **Ecotourism** and **nature tourism**, winter sports and water sports, among others, are available.

Tourism has grown considerably in Chile, and its relative weight and implications have made it an increasingly important productive activity in the national economy. Estimates regarding the place of tourism in the Chilean economy indicate that it contributes between 3 and 3.5 per cent of the gross domestic product, generates **foreign exchange** worth nearly one billion dollars per year (which places it fifth in terms of export rankings), and generates about 200,000 jobs, of which 60,000 are direct.

Some 85.7 per cent of all **tourist** arrivals are from the Americas. Of these, the neighbouring countries of **Argentina**, **Peru**, Bolivia and **Brazil** are the main markets, along with the **United States**. Some 11.6 per cent of foreign arrivals come from Europe. The remaining 2.7 per cent of foreigners come from other continents, especially from the Asia Pacific region. In 1995, Chile was the fourth most popular destination in South America behind Argentina, Uruguay and Brazil. If growth continues at the same rate as over the past five years, a total of 2.6 million foreign visitors are projected to visit the country in the year 2000, generating some $1.6 billion in foreign exchange.

Parallel to the growth of international arrivals, the country has also experienced growth in **domestic tourism**. The country's economic growth has been remarkable. The increase in real income, decrease in unemployment, increased incorporation of women into the workforce and notable increase in ownership of family cars are all conditions which have contributed to the growth of internal tourism, allowing new socioeconomic groups to be added to the traditional groups which enjoy access to tourism **activities** within the country.

Given the country's **geography** and the fact that an excellent highway connects the country from north to south, the most used method of transportation in Chile is by road. The railway only runs from Santiago to the south, while air **transportation** is only used by a small number of people and is not used extensively for tourism. The most important aspect of the country's **hotel** trade is the way in which it has improved and become increasingly more professional since 1990. The large hotel chains not only have properties in Santiago but have also diversified to other regions in the country, and their investment plans call for even more territorial diversification. In general, tourism investments in Chile are mainly related to natural attractions. Chile's most important competitive advantage lies in its natural beauty. The country features 32 national parks, 47 national reserves and 13 natural monuments, which together cover a surface area of 15 million hectares and comprise the National Network of Protected Natural Areas.

RAMON CÈSAR GÒMEZ VIVEROS, CHILE

China

With its ancient civilization having lasted several thousand years, China is bestowed with rich tourism **resources** of both natural and cultural types on its land spreading 9.6 million square metres. Tourism in China started in 1954, but the industry did not develop itself in large scale for a long period. It was in the late 1970s that tourism as an economic activity entered into a rapid period of development, accelerating that of national economy

In 1995, major generating countries with more than 100,000 tourist arrivals were: **Japan** (1,305,200), Republic of **Korea** (529,500), USA (514,900), Russian Federation (489,300), Mongolia (261,900), **Singapore** (261,500), **Malaysia** (251,800), **Philippines** (219,700), **United Kingdom** (184,900), **Thailand** (173,300), **Germany** (166,500), **Indonesia** (132,800), **Australia** (129,400), **Canada** (128,800) and **France** (118,500). Over the years, the main tourism product of China was **sightseeing** of historical and cultural attractions. In the last five years, distinct progress has been made in the development of **resort**, **ecotourism**, **sports tourism**, rural and health tourism products.

According to 1995 statistics, China recorded 46.39 million visitor arrivals, rising 6 per cent over the previous year and putting China in the fifth place among the top destination countries in the world. Foreign exchange income from international tourism amounted to $8.7 billion, representing 19 per cent growth over the previous year and placing the country in the ninth rank in the world. There were 629 million domestic tourists representing a 20 per cent increase over the previous year. Its earnings rose by 34 per cent reaching 137.6 billion yuan. Outbound tourism recorded

4.52 million departures with a 21 per cent growth rate. China's total income from international and domestic tourism was 209.8 billion yuan accounting for 3.64 per cent of gross domestic product.

In 1995, the Chinese government invested 1,027 billion yuan including $44.5 million bank loans for capital construction in 120 tourism projects. There were 47 joint-venture projects in hotel establishments with a total investment of $1.336 billion.There are 8,896 tourism enterprises in China out of which 3,826 are travel agencies and 3,720 are hotels with a capacity of 490,000 rooms. Among the hotels, 38 are classified as five-star and 106 as four-star. The capital assets of all tourism enterprises are 125.2 billion yuan by their original value with direct outputs of 209.8 billion yuan.

The tourism industry in China employs a direct workforce of 1.12 million people while indirect employment is 5 million. The country has a total of 138 tourism institutions of higher learning including those universities that have exclusive tourism departments. Further, 484 secondary technical schools have been established. Total enrolment of tourism students is 139,260 at time of writing.

China National Tourism Administration (CNTA) is the administrative body. It has various departments such as Department of Planning and Statistics, Department of Marketing and Promotion, Department of Tourism Enterprise Administration, Department of International Liaison, Department of Personnel and Education and Department of Finance. There are thirty-one tourism bureaus at provincial level under which operational divisions correspond to those of CNTA. There are also municipal tourism bureaus functioning as administrative bodies at local level.

In 1996, CNTA formulated China's Ninth Five-Year Tourism Development Plan and the Long-Term Outline Objectives to the Year 2010. The salient policy feature of the Ninth Five-Year Plan (1996–2000) is a 'government-led strategy' including: increased investment to tourism by central and local governments; joint efforts of state, local, collective and private bodies in tourism development; increasing foreign investment in tourism; further open-up of civil aviation, railway and land transportation; strengthening of tourism infrastructure; expanding tourism marketing overseas; parallel development of domestic and international tourism; development of tourism education and application of new technology; tourism awareness education of nationals through press, television, radio and schools; and implementation of state policies for environmental protection and strategy for sustainable development.

With a view to ensuring the quality of tourism services and protecting the interest of consumers, the CNTA introduced in 1994 the Deposit System of Quality Guarantee to travel agencies. In 1996, the Regulation on Travel Agency was promulgated. The China Tourism Law is now in the process of formulation. It was projected in the Ninth Five-Year Plan that by year 2000, China would receive 55 million overseas visitors with an income of $14 billion from international tourism. Domestic tourists will reach 1 billion with an income of 360 billion yuan. This will represent 5 per cent of China's GDP.

It is expected that between 2001 and 2010, China will maintain an annual growth rate of 8 per cent for international arrivals with its earnings at 15 per cent growth. Tourism will account for 8 per cent in the gross domestic product. By then, tourism will become a major pillar industry of the national economy of China.

Further reading

China National Tourism Administration (1996) *The Year Book of China Tourism Statistics, 1996.*

—— *The China Ninth Five-Year Tourism Development Plan and the Long-term Outline Objectives to the Year 2010.*

ZHANG LIANSHENG, SPAIN

choice set

Potential tourists have a vast range of destinations, tour packages (see **package tours**) and other travel attributes from which to choose. The range of possible alternatives is the choice set. Tourists are likely to spontaneously recognise or seriously consider only a small proportion of all possibilities. This is the evoked set. The term is used in psychological research on **decision making**. In this context, it may be used to refer to the range of options which is presented to a respondent.

GEOFFREY WALL, CANADA

Chunnel

The Chunnel is a railway tunnel below the English Channel, connecting Folkestone in England to Calais in France, making possible non-stop **rail** service between London and Paris. The Chunnel, one of the most important tourism **infrastructure** projects in recent years, is fifty kilometres long. Of this distance, 75 per cent is under the sea, about forty-six metres or below the Channel floor. It is composed of three separate tunnels, each over seven meters in diameter. Two tubes carry trains in opposite directions, while a third is a service tunnel to provide ventilation and access.

Passenger trains operate through the Chunnel at speeds up to 300 km per hour. Service between Waterloo station in London and Gare du Nord station in Paris via the 'Eurostar' takes approximately three hours and costs about the same as flying. Although flying takes only about two and a half hours, **taxi** rides from Heathrow and the Paris **airports** to downtown London and Paris, respectively, add another hour and a half to the trip, costing an additional $40–60 US.

The Chunnel is not available for individual automobile or passenger usage. Eurostar transports vehicles and their passengers between Folkestone and Calais via the Chunnel. The trip takes about thirty-five minutes and is competitively priced with ferries and hovercraft. Service via hovercraft also takes about thirty-five minutes, but is subject to poor weather conditions common in the English Channel, often adding delays of two to ten hours to surface crossing times. Travellers drive onto Eurostar and must stay with their vehicles; passport checks normally take place in the station before boarding. There are no lounge or snack cars, so ferry operators taking seventy-five to ninety minutes have used their bars, restaurants, sun decks and children's play areas to compete against Eurostar.

The Chunnel can also handle up to sixteen million tons of freight per year, increasing **efficiency** by eliminating transloading between land and water carriers. Freight customers receive faster, more dependable service independent of weather conditions, which affects air and water carriers. Construction of the Chunnel, the largest civil engineering project of the twentieth century costing approximately $16 billion, was started in 1987. Service began in 1994.

Further reading

Fisher, J. (1994), 'Finally, London to Paris by train', *U.S. News and World Report*, 21 November: 95–6.

JOHN OZMENT, USA

circuit tourism

Circuit tourism involves visits to more than one **destination** during a trip away from home. Circuit tourism is typified by short lengths of stay at each destination on the circuit, a pre-planned **itinerary**, and regional or national clustering of attractions (such as the Golden Triangle in **India**). Circuit tourism is problematic in terms of accurate data collection.

CHRIS COOPER, AUSTRALIA

city office

Numerous villages, small towns and cities have established organisations to encourage tourism and related **economic development**. They are usually affiliated with a chamber of commerce, a parks and **recreation** authority, an economic **development** board or some other branch of a municipal or local **government**. Responsibility for generating tourism dollars may thus only be a sideline activity. With recognition that tourism is an important industry within a **community**, however, a city tourism office may grow in stature and become a semi-autonomous agency.

The mandate of most such offices is **marketing**-related. Since marketing to broadly dispersed **leisure** and business markets is an extensive and expensive activity, the office is usually required to attract funding from a wide variety of sources such as grants, membership, revenues and taxes. Based on a marketing plan, **advertising**, promotions, special events, convention services and **public relations** are prioritised, budgeted and implemented. In some communities, a marketing plan may be an outgrowth of a tourism action plan which provides a framework to analyse **resources** and concerns. In any case, a well-designed and executed marketing plan allows businesses and organisations to profile and evaluate markets, and to generate visitor activity beyond the levels that could be generated by businesses acting independently. This necessitates collaboration among arts groups and **accommodation** businesses, among other enterprises, in the pursuit of marketing opportunities.

Because a city tourism office provides a venue whereby important **economic development** and community concerns can be identified and discussed, public involvement may be sought. In an industry that recognises the high level of tourism **competitiveness**, it is vital that considerable attention be devoted to the development and/or improvement of a community's tourism attractions, **infrastructure** and services. A city tourism office with an understanding of the broad quality of life issues can provide an effective political voice which, when bolstered with research on the significance of tourism, can champion community **growth** and well-being.

MICHAEL HAYWOOD, CANADA

civil aviation authority

These organisations (sometimes called administrations) are national governmental agencies responsible for oversight of **airline** regulation for **transportation** service to, from and within nations. They may be responsible for economic and non-economic regulation of air transportation. Economic regulation of air transportation encompasses service characteristics and fares and rates charged. Agencies are sometimes concerned only with non-economic regulation such as **safety** requirements and provision of air traffic control, as is true of the **Federal Aviation Administration** in the **United States**. Occasionally, the agency may manage and/or operate the national airline, as has been the case of the Civil Aviation Administration of **China**.

FREDRICK M. COLLISON, USA

class

Class refers to socially stratified groups, the members of which are distinguished from one another by the amount of **power**, wealth and education they possess. In social scientific analyses of tourism, class is as fundamental a variable as are age, **gender** and **ethnicity**.

The nineteenth-century theorist Karl Marx distinctively formulated classes as groups of people serving analogous roles in production. In contrast to Marx's view that positions in the social hierarchy are ultimately due to relative ownership in the means of production, Max Weber emphasised the role of **ideology** and status (from such attributes as **gender**, **religion** and **ethnicity**) as significant shapers of each person's social standing.

In the study of tourism, class striving or emulation of higher classes is demonstrated by the mass appeal of formerly elite **resorts**. As destinations become more popular and middle-class, upper-class vacationers find other ways to mark their **vacation** spots as elite, such as creating exclusive, shielded enclaves (Thurot and Thurot 1983). This process is also illustrated by the gradual move to **mass tourism** of the British middle classes in the eighteenth century on the routes of the Grand Tours established by the British nobility in earlier centuries.

As Graburn (1983) has written, patterns of tourism are differentiated by class for three major reasons: access to discretionary income, 'cultural self-confidence' (for example, lower-class member not feeling comfortable outside their cultural milieu, versus middle-class members eager to consume new experiences), and a desire to reserve the conditions of everyday life. Upper-class members enjoying a rustic cabin holiday serve as an example of this latter **ritual inversion**.

Marx's idea of class differentiation by relations to capital (productive wealth) has been revised with reference to cultural capital and commodities of symbolic value. Heritage tourism is such a commoditised **experience** (see **commoditization**), which appeals to many in the middle class as a way to accumulate cultural capital (Richards 1996).

References

Graburn N.H.H. (1983) 'The anthropology of tourism', *Annals of Tourism Research* 10(1): 9–33.

Richards, G. (1996) 'Production and consumption of European cultural tourism', *Annals of Tourism Research* 23(2): 261–83.

Thurot, J.M. and Thurot, G. (1983) 'The ideology of class and tourism: confronting the discourse of advertising', *Annals of Tourism Research* 10(1): 173–89.

Further reading

MacCannell, D. (1989) *The Tourist: A New Theory of the Leisure Class*, revised edn, New York: Schocken Books. (Updating Thorstein Veblen's 1899 work on **conspicuous consumption**, MacCannell points out a variety of ways in which class shapes touristic experience.)

ROLAND S. MOORE, USA

class of service

Carriers in a number of modes of **transportation** may offer multiple classes of **service**, although only one is sometimes found. The main types of transportation carriers that provide different classes of service are airlines, cruise lines and railroads (see **rail**).

Airlines offer between one and three classes of service. The basic type of service provided by nearly all airlines is economy class, with passengers paying a relatively low base fare (often reduced by one or more discount fare categories). This is the lowest level of the three, with seats having the smallest pitch (distance between one seat back and the next) and width; inflight and ground services are also usually at the lowest level offered. Airlines also offer one or two further options. First class, typically the highest priced, provides the foremost level of service with wide seats which offer a large pitch and recline substantially. Ground and inflight services, including **food** and beverage and flight attendant **hospitality**, are superior to that of the other classes. Business class is, in most respects, intermediate between the other two classes.

Cruise lines offer multiple classes of service differentiated by the types of cabins on the vessel. Differentiating features of a cabin include whether it is inside or outside, the floor area, furnishings and vertical location on the ship. Generally, the better the features of the cabin, the higher the fare the passenger pays. Some cruise lines may also differentiate service based on the type of meal options. Differences are also found from one **cruise line** to another, with some offering deluxe and others providing a more basic level of service.

Rail also offers a number of classes, particularly first, second and third, with each being a lower level of service, respectively. For long-distance rail service, a train may offer sleeper option (with enclosed sleeping compartments) and coach service. Another difference is that of an express versus a local train, with the former providing a much more timely service, often using faster equipment such as with the Shinkansen. In some cases, a train like the Orient Express offers luxury service throughout.

The challenge to a transportation carrier's management is to accurately match the classes of service offered with the **demand** of the relevant **market**. Another challenge is to stay on the top of ever-changing wishes and means of both pleasure and business tourists.

See also: transportation pricing

FREDRICK M. COLLISON, USA

classification

Classification is the process or system of arranging according to class. In the context of tourism, classification usually refers to the system by which particular tourist facilities or services are categorised according to certain criteria, the purpose being to provide tourists with an indication of the standard of **quality** or **service** which they might reasonably expect and, implicitly, the likely price they will have to pay. It is also a means of maintaining or raising standards across sectors of the industry.

Most commonly, classification systems are applied in the **accommodation** sector. Hotels are graded on the widely-recognised star rating system, ranging from the basic one-star rating to the five-star luxury category. However, a variety of other facilities and services, such as **restaurants** or **transportation**, are also frequently classified. Airlines, for example, may classify different levels of service and comfort as economy class, business class and first class.

Complications inherent in tourism classification systems are the frequent lack of internationally agreed parameters for classification, and multi-classification systems for particular sectors of the industry. Thus, similar star ratings for hotels in different countries do not necessarily signify comparable levels of quality and service. At the same time, it is not unusual for different organisations, including national tourism organisations, motoring associations or clubs, guidebooks and **tour operators**, to apply their own classification systems. This idiosyncratic approach often results in accommodation or **catering** establishments receiving varying and conflicting ratings. However, although an internationally standardised classification system may be desirable, it is generally accepted to be an unattainable objective, particularly because ratings are based upon a combination of both tangible features and intangible subjective assessments (such as 'quality' or 'comfort').

For research purposes, tourists themselves are also subject to classification. In statistical surveys, tourists are classified by **purpose** of travel (such as **vacation**, business, **visiting friends and relatives**), while theoretical categorisations of tourist **roles** and **behaviour** (see **typology, tourist**) are sometimes referred to as classifications of tourists.

Further reading

Holloway, J.C. (1994) *The Business of Tourism*, London: Pitman Publishing, 119–121. (A discussion of the terminology and problems inherent in the application of classification systems to the accommodation sector.)

RICHARD SHARPLEY, UK

cliché

The term 'cliché' refers to outworn stereotypical expressions (verbal) and pictorial **representations** (visual). In tourism promotion, the former are typically found in advertisements, brochures, travelogues and **postcards**, while the latter additionally feature in **videos**. Cliché operates via indexical transference (for example, a sunset signifying romance is transferred to a couple), but it is the overuse of written and iconic metaphor (such as ever-smiling **natives** connoting happiness) which dilutes its meaning (see also **promotion, place**).

GRAHAM M.S. DANN, UK

climate

Weather is the state of the atmospheric system of a place at a particular point in time, whereas climate is the result of a compilation of measurements of atmospheric variables, such as temperature and precipitation, as observed over at least a thirty-year period. Thus, climate is what is expected on the basis of past **experience**, whereas weather is what is actually experienced. Both climate and weather have far-reaching implications for tourism. For example, the latter influences whether or not people are likely to participate in a particular activity and the **quality** of the experience of those who do. However, the exact relationships between climate and tourism vary from **region** to region and activity to activity. For example, it is self-evident that seaside **resorts** and ski hills exhibit different relationships with climate.

One of the most obvious and important aspects of climate tourism relationships is **seasonality**, which is strongly but not exclusively a consequence of climate, and has considerable implications for the viability of tourism enterprises. Climate is a major influence upon the length of operating seasons in middle and high latitude destinations. This has economic consequences, for capital must be invested in fixed plant such as marinas and ski tows year-round, but may only generate an income for a limited period in the year.

There are at least four perspectives from which relationships between climate and tourism can be viewed. First, climate as setting is a part of the environmental context in which tourism occurs. For example, the warm land and sea temperatures of many tropical coastal places are an asset for tourism in beach locations, and reliable snow is an important attribute of most well-established ski areas. A second perspective views atmospheric processes as generators of change in tourism participation, although most such studies stress relationships between weather and **recreation** rather than implications of longer term changes in atmospheric processes. Third, climate is viewed as hazard and atmospheric events, such as storms, are disruptive to tourism. For example, hurricanes and avalanches cause damage to tourism plant in vulnerable locations as well as loss of life. The fourth perspective combines the second and third approaches and views climate as both a resource and a resistance, which requires managerial responses

Most climatologists believe that concentrations of so-called greenhouse gases, particularly carbon dioxide, are causing changes to the global climate. It is projected that future climates will be warmer and drier, increased evapotranspiration reducing water availability even if precipitation increases. Further, it is expected that sea levels will rise through thermal expansion of ocean water and melting of ice. Such changes could have far-reaching implications for tourism, reducing snow cover in middle latitudes and low elevations, modifying the lengths of operating seasons, inundating tourism **infrastructure** in oceanic coastal locations, and altering the ecosystems on which much tourism to natural areas is based. Although climatic information has been used at a wide variety of scales from large regions to individual facilities as an input into tourism **planning**, the potential value of climatic records for planning is not widely recognised.

Further reading

Mieczkowski, Z.T. (1985) 'The tourism climate index: a method of evaluating world climates for tourism', *The Canadian Geographer* 29(3): 220–3.

Wall, G. (1992) 'Tourism alternatives in an era of global climate change', in W. Eadington and V. Smith (eds), *Tourism Alternatives: Potentials and Problems in the Development of Tourism*, Philadelphia, PA: University of Pennsylvania Press, 194–215.

—— (1993) 'Tourism and recreation in a warmer world', in S. Glyptis (ed.), *Leisure and the Environment*, London: Belhaven, 298–306.

—— (1996) 'The implications of climate change for tourism in small islands', in L. Briguglio, B. Archer, J. Jafari and G. Wall (eds), *Sustainable Tourism in Islands and Small States: Issues and Policies*, London: Cassell, 206–16.

GEOFFREY WALL, CANADA

club

Clubs, also referred to as 'private clubs', consist of groups of people who share a common interest or bond and generally have a central meeting place. Clubs are usually exclusive in that there is a requirement for admission and a cost to join. They are often distinguished from associations on this basis. Private clubs in the United States date back to the early 1800s, while private clubs in Europe predate even these. Club Managers Association of America (CMAA) is perhaps the leading association in the field worldwide.

Clubs can and do offer many different types of services. Some clubs specialise in one particular area while others offer a whole array of products and services. For example, while some city clubs offer **food** and beverage services, lodging and athletic facilities, most limit their offerings to food and beverage (dining). Country clubs, on the other hand, may offer a full range of dining options (multiple food service outlets) as well as golf, tennis, swimming and health facilities. Other types include yacht, tennis and hunting clubs, among others. In short, no two clubs offer exactly the same services, nor are they exactly alike.

As far as ownership goes, clubs are generally classified into one of two broad types, either member-owned or privately owned. Clubs run by and for the military are also sometimes considered as a separate **classification** because of their exclusivity. Member-owned clubs are the most common type found in the United States. Members own stock or shares in the club. When the time comes that the member leaves the club, the shares are sold and/or transferred. The 'owners' make up the bulk of the membership and are the group with the most rights/privileges. These types are generally governed by a board of governors, which is comprised of elected full stock-holding members. It is this group that sets guidelines and policies. A general manager, generally from outside of the membership, is retained to manage the property and work with the board on long-range goals. Member-owned clubs are generally operated as not-for-profit organisations. Privately owned clubs are owned by companies, corporations or individuals. These are operated on a for-profit basis and managed by an individual appointed by the owner. While they rely on some member input, this is less so than in member-owned clubs. Some clubs provide access privileges known as reciprocal agreements to their members and their guests at tourism destinations worldwide.

CLAYTON W. BARROWS, CANADA

cluster analysis

Cluster analysis is a widely used family of multivariate techniques for grouping individuals, objects or behaviours into similar clusters. In tourism research, for example, cluster analysis is often used to identify **market segments** in order to improve the effectiveness of **marketing** efforts. These segments may be based on a variety of variables including demographic characteristics of the tourists (such as age, income, **gender** and **location**) and trip characteristics (such as trip length, purpose, group size and benefits). Cluster analysis has also been used to develop a typology for classifying destinations into a schema such as developed/undeveloped, accessible/inaccessible and natural/manmade (see **typology, tourist**). The flexibility of cluster analysis to accommodate a wide range of applications makes it one of the most useful tools for understanding the natural structures among observations.

There are several approaches to cluster analysis that can be classified into two general categories: hierarchical and non-hierarchical. The former uses agglomerative procedures whereby each observation or object (the individual visitor or **attraction**) starts by defining its own group, but on subsequent steps the two closest clusters are combined into a new aggregate cluster. Eventually, all observations/objects are combined into one large cluster. Non-hierarchical clustering procedures take the opposite approach whereby the observations included in the study are split into common groups. An important difference in these two approaches is that hierarchical clustering assumes an underlying hierarchial structure among objects (that is, all individuals or attractions assigned to a cluster are maintained throughout the process of clustering), whereas in non-hierarchical clustering objects have free assignment, depending upon the number and underlying structure of the observations/objects.

Interpretation and validation of the resulting clusters are important steps in cluster analysis. The interpretation stage involves developing a profile of each cluster and identifying the variables that distinguish one cluster from another. This information enables the researcher to develop substantive descriptions of each of the respective clusters. Validation in cluster analysis describes the process to assess the generalisability or stability of the clustering solutions. The most simple and direct approach to evaluating validity involves cluster analysing of two or more separate samples (or subsamples) and then comparing the results to insure correspondence. Profiling clusters using several independent variables such as demographic and behavioural descriptors also provides a means for validation and further interpretation/explanation for the identified clusters.

See also: classification; discriminant analysis; marketing; multidimensional scaling

Further reading

Green, P.E. and Carroll, J.D. (1978) *Mathematical Tools for Applied Multivariate Analysis*, New York: Academic Press. (Discusses quantitative tools used to evaluate consumer behaviour.)

Hartigan, J.A. (1975) *Clustering Algorithms*, New

York: John Wiley. (Provides a detailed description of the approaches used in cluster analysis.)
Smith, S.L.J. (1995) *Tourism Analysis: A Handbook*, 2nd edn, London: Longman. (Chapter 5, 'Segmenting the tourism market', discuses basic theory underlying market segmentation within the tourism context.)

DANIEL R. FESENMAIER, USA
JIANN-MIN JENG, USA

code of ethics

A code of ethics is a standard of acceptable performance, often in written form, that assists in the establishment and maintenance of **professionalism**. Tourism organisations, such as the **World Tourism Organization**, have implemented codes of ethics that prescribe and sanction credible conducts for members of **tourist**, resident and business communities.

See also: code of ethics, environmental

RANDALL UPCHURCH, USA

code of ethics, environmental

An environmental **code of ethics** is a list of ethical **rules** for conducting tourism **development** and/ or **management**. The aim is to ensure the adequate incorporation of environmental protection measures either during the **planning**, development or management of tourism. This concern for **conservation** and the **environment** is a mix of ethics and self-preservation. It arises from a sense of both what is right and what is necessary.

A large range of environmental codes of ethics already exist. They include codes for tourists, developers, businesses, operators and countries. For example, an environmental code of ethics drawn up for **tourists** by the Ecotourism Association of Australia includes pre-tour, actual tour and post-tour guidelines for ecotourists. The Ecotourism Society's (1993) *Ecotourism Guidelines for Nature Tour Operators* focuses on guidelines for pre-departure, guiding, monitoring and management. This publication discusses such issues as preparing tourists before departure so as to minimise their negative **impacts** while visiting sensitive environments and cultures, reducing visitor impacts on the environment by offering literature, briefings, leading by example and taking corrective actions, and being a contributor to the conservation of the regions being visited.

The New Zealand Tourist Industry Federation has a *Code of Environmental Principles for Tourism in New Zealand*. The guiding principles are to promote **ecologically sustainable tourism** development so as to ensure that the industry can continue to be based upon the natural resources of New Zealand in the long term, and to recognise that both development and conservation can be valid and complementary uses of national resources. Examples of the code's guidelines include managing existing natural and cultural areas associated with tourism development and using them in such a way that they are protected and enhanced; ensuring that an ongoing responsibility for environmental care and protection and **community** concerns is adopted; and enhancing visitors' appreciation and understanding of the natural environment through the provision of accurate interpretation and information.

See also: ecologically sustainable tourism; environmental auditing; environmental management systems; planning, environmental

References

New Zealand Tourist Industry Federation (n.d.) *Code of Environment Principles for Tourism in New Zealand*, Wellington: New Zealand Tourist Industry Federation.
The Ecotourism Association of Australia (n.d.) *Code of Practice for Ecotourism Operators*, Brisbane: The Ecotourism Association of Australia.
The Ecotourism Society (1993) *Ecotourism Guidelines For Nature Tour Operators*, Vermont: The Ecotourism Society.

ROSS K. DOWLING, AUSTRALIA

code sharing

Code sharing is a mutually beneficial agreement where two or more airlines use the same two-letter code in their **reservation** systems to facilitate joint bookings and fare calculations. For example, two airlines may sell connecting segments on their respective routes (see **route networks**) and honour a combined fare regardless of mixed codes, or one airline may sell seats under their own code on another airline's service in order to present an apparently continuous routing. The resultant joint fare is typically lower than if the separate segments had been calculated independently.

ANTHONY G. SHEPPARD, USA

cognition

Cognition is a psychological term used to refer to all the mental processes associated with gathering and using knowledge. It is sometimes also seen as a synonym for knowledge or thinking. Cognition is often contrasted in **psychology** with affect (emotion or feelings) and **behaviour** (actions), although all three are seen as being interrelated. Cognition includes the processes of perception, memory, imagining, reasoning, learning, evaluating, problem solving and **decision making**. Many areas of tourism are concerned with cognition or cognitive processes. These include research into **destination choice**, destination image and decision making, studies of resident attitudes towards tourism (see **community attitude**) and analyses of tourist **satisfaction** and **service quality**.

Cognition was an important focus for early psychology, but was neglected for several decades because of a change in theoretical emphasis. In the 1970s it made a return to the psychological literature, and became such a dominant part that many texts refer to the 1970s and 1980s as the cognitive revolution. An important part of this revolution was the use of the computer as a metaphor for how people dealt with information. This described people as rational, logical information processors. Unfortunately this approach did not stand up well to empirical testing, and it became apparent that the computer metaphor was not an accurate representation of the way people think.

A substantial area within cognitive psychology has concentrated on the way people make decisions under conditions of uncertainty. This work is clearly of relevance for those in tourism who seek to understand destination choice and other related decisions. The major conclusion of the cognitive psychology research has been that economic and rational models do not adequately explain such decision making. Rather, the research has uncovered many different types of strategies (called heuristics) which are used to limit the amount of information processing required to reach a decision. Several factors, including cultural background, previous **experience**, **motivations**, values, social group pressures and existing beliefs, have all been found to have significant influences on various cognitive processes.

Further reading

Zimbardo, P.G. (1992) *Psychology and Life*, New York: HarperCollins. (Pp. 378–421 provide an introduction to all the various processes involved in cognition and review research into heuristics and decision making.)

GIANNA MOSCARDO, AUSTRALIA

cognitive dissonance

Cognitive dissonance is a member of a larger class of theories known as consistency theories. It may be defined as a state of psychological discomfort or uneasiness that occurs when a logical inconsistency exists among cognitive elements. Leon Festinger, the originator of dissonance theory, stated that 'two elements are in a dissonance relation if, considering these two alone, the obverse of one element would follow from the other'. For example, as a **tourist** one may evaluate alternatives prior to making a choice among a set of airlines to a given **destination**. The cognitive elements might include knowledge about airlines, knowledge that some are on time more often, and knowledge about fares and number of connections. The tourist wants to be a careful decision maker and being on time is important, but **cost** is a factor. The decision is then made with worry as to whether the right airline has been selected. These ideas can be in conflict and create a dissonant state. Higher importance and cost of a **product** may also serve as a source of dissonance.

The simultaneous production and consumption of most of the tourism services adds a unique challenge to the application and use of cognitive dissonance in this field. Although consumer **behaviour** researchers and marketers in general relate the cognitive dissonance theory to post-purchase **experience**, the state of cognitive dissonance can take place throughout the different phases of tourism experience, ranging from pre-trip **planning** and **anticipation** stage and the on-site **experience** to the post-trip reflective stage. One of the aspects of the cognitive dissonance theory in relation to its practical use is dissonance reduction. The experience or feeling of dissonance is an aversive state, and people act to reduce it.

The response of tourists to their **destination** choice and experiences is a complex process which is difficult to measure and evaluate. From a managerial perspective, the recognition of the effects of dissonance is important. One strategy which destination promoters and tourism service providers may implement is to keep in touch with tourists and users after purchasing and consuming **services**. Another strategy is to create a database of users so that their needs and plans can be better accommodated in the future. Managers should employ marketing research and develop a monitoring mechanism in order to ensure that the strategy is having the desired impact.

See also: cognition; expectation; experience; internal marketing; motivation; purchasing; risk; satisfaction

Further reading

Festinger, L. (1957) *A Theory of Cognitive Dissonance*, Stanford, CA: Stanford University Press.

MUZAFFER S. UYSAL, USA

cognitive mapping *see* map; perceptual mapping; wayfinding

collaborative education

Collaborative **education** is an approach whereby educational institutions team up, both among themselves and with the business community (via guest speakers, mind/hands-on study opportunities and field trips), to develop comprehensive tourism-related teaching materials (such as textbooks) and educational programmes. It may also include the creation of shared educational experiences involving learners from different backgrounds, disciplines and tourism sectors. The goal is to combine **resources** in order to enhance the **quality** and reduce the costs of a given learning experience.

ROBERT W. McINTOSH, USA

collective bargaining

Collective bargaining is a complex process which can best be understood by participating in it. Mandatory bargaining issues include items that directly relate to the relationship between employer and employee. Included in this list are such issues as wages, hours of work, incentive pay, overtime, layoffs and recalls of employees, union security clauses, management rights clauses, grievance procedure, seniority, **safety** issues, benefits, retirement issues and others. All other issues are referred to as voluntary, or permissive bargaining issues.

Voluntary issues may be discussed if both sides want to do so. However, neither side is required to discuss them by **law**. Examples of such items include pension and benefit rights for retired personnel, supervisory compensation, supervisory discipline, performance bonds for union or management, contract ratification processes and company price or **product** issues. During all negotiation, both sides are restricted from bringing illegal or prohibited issues such as those that violate **employment** opportunity laws, closed shop agreements, featherbedding and union or agency shop clauses in right to work states to the table.

Collective bargaining negotiations occur over both economic and non-economic issues. In recent years they have become more important to both unions and management. For instance, management typically wants a management rights clause in contracts. Such statements insure management rights to control types of products and services made or delivered, how tourism supervision is carried out and what kinds of discipline are used and how they are carried out. Unions, on the other hand, typically want just cause issues outlined in contracts, to establish limitations to management rights to discipline or discharge employees. Included in the list of non-economic issues important to both union and management are items such as quality of work life issues, union security, work rules, size of work crew, types of work that can and cannot be done by various employee groups, and grievance procedures.

Different types of bargaining may occur during negotiations. Distributive bargaining occurs when management and the union are in **conflict** with one another over a major issue. When this occurs, each side will likely present a proposal in which one side is a loser and the other a winner. This type is called a zero sum proposal, because there is always one winner and one loser. Disagreement over wages paid to tourism guides provides a good example of this type of disagreement. Integrative bargaining occurs when the two sides are not necessarily in conflict over an issue. For instance, quality of work life issues often result in integrative bargaining because both sides have something to gain from the resulting decision. Instead, this type of issue calls for more collaborative bargaining.

Mediation and arbitration are both means of using third parties to help unions and management to reach agreement during collective bargaining negotiations. Differences between them are primarily in the amount of **power** given the third party. Both require the voluntary support of unions and management, except in cases of court ordered mediation or arbitration. A mediator is a third party who attempts to get both sides to reach an agreement. Mediators may make recommendations but cannot enforce agreement while arbitrators have the power to force agreements. Arbitrators generally review all of the information available from both sides and dictate an agreement which they view as near the middle ground. Arbitration has been criticised because in many cases both sides will take the most extreme positions prior to arbitration in order to gain as much as possible. This problem can be avoided by using final offer arbitration in which an arbitrator must choose one offer or the other. Unlike regular arbitration, final offer arbitration tends to bring the two sides closer together. As a result of this movement toward the centre, agreement is often reached without the use of an arbitrator.

ROBERT H. WOODS, USA

Colombia

Colombia, the only country whose name relates to Christopher Columbus, is located in the north region of South America, with coasts on both Pacific Ocean and Caribbean. The San Andres and Providencia Islands, physically opposite the Costa Rican coast, are also part of its territory. The Pan-American Highway, running all along the Pacific coast of South America, is interrupted at the Darien jungle on the north border of Colombia with Panama.

Along with climatic diversity, historic and cultural assets are the most important potentialities used to attract **international tourism**. The Spanish colonial **heritage** has left walled fortress cities like Cartagena de Indias and small treasures like Villa the Leyva, among others. Coffee, once the axis of the economy, provides an opportunity to **experience** a taste of Colombian **hospitality**, by following the 'Rutas Cafeteras' and staying at 'haciendas' all along the areas where coffee is still grown today. Folklore is rich, and handicrafts and local products are available all over the country. The main popular musical themes, the rhythms of 'La Cumbia' and 'El Ballenato', are integrated in local and national festivals and tourism products.

In recent years there has been a negative **image** attached to the name 'Colombia', affecting directly the opportunities of the country for a greater tourism **market** share. **Ecotourism** has been discouraged by the implicit **risk** posed by guerrilla groups or drug barons in some areas. However, the reality suggests differently. **Ecuador**, Panama and **Venezuela** have poor access connections and economies unable to **supply** an important flow of visitors. Moreover, they have somewhat similar attractions. A new tourism law has been approved in 1996, aiming to increase competitiveness in the long-range international tourism markets. The decision on a positive national **branding** and **image** concept is probably the most important question to be addressed in the implementation of the stated objectives of the tourism law.

Colombia has close ties with the **United States**, and presence of international **hotel** operators through management and franchise agreements is visible in many main cities. The efforts of these latter will certainly help to improve image and promotional effectiveness. International arrivals, most of whom were business visitors, declined to 1,162,300 in 1997, from 1,400,000 in 1995. At the same time, tourism income increased to $958 million from $822 million in 1995. In 1997 tourism represented 2.5 per cent of the gross national product.

LLUIS MESALLES, SPAIN

colonisation

Colonialism prompted overseas travel and helped provide the basis for modern tourism. Holiday centres and second homes were established where the **climate** was favourable, for example, the Himalayan hill stations in British India. In many colonies, rest centres, small hotels and guest houses were built at and between administrative centres, near border crossings or at ports. They were normally operated by and for Europeans, as colonised people lacked the required capital and expertise.

The earliest establishments catered for traders and settlers and were often somewhat disreputable. However, networks of small, clean and functional hotels emerged, providing administrators, business travellers and (later) more intrepid tourists with a respite from untarred roads or uncomfortable sea crossings. Guidebooks to the colonies in the nineteenth and early twentieth centuries, listing available hotels and local attractions, including hunting, were directed at settlers and tourists. During this period, upper-class tourists often obtained letters of introduction to settlers, who willingly entertained them.

Much modern tourism reflects colonial patterns and policies. Tourists from metropolitan 'centres' to less-developed countries tend to visit former colonies, where communications are already established and there are similarities in language and culture. The prominence of transnational companies in global **hotel** operations has led to accusations of 'cultural **imperialism**,' and some commentators have deplored the sociocultural effects of tourism on 'host' societies. However, much depends on the specific context and the strength of local institutions. In the South Pacific, much land in former British colonies is still communally owned and has hindered large-scale tourism development. The issue has been more straightforward in islands under French jurisdiction, where land commoditisation has been more extensive. The type and growth of tourism has been conditioned by colonialism, but its economic, social and cultural **impacts** vary considerably according to the local context.

See also: developing country; land tenure; motivation; neo-colonialism

Further reading

Finney, B.R. and Watson, K.A. (eds) (1975) *A New Kind of Sugar: Tourism in the Pacific*, Honolulu, HA: East-West Culture Learning Institute. (An early collection of readings linking tourism to colonialism.)

Harrison, D. (1995) 'Development of tourism in Swaziland,' *Annals of Tourism Research* 22(1): 135–56. (A historical study of tourism development in a former British colony.)

Nash, D. (1989) 'Tourism as a form of imperialism,' in V. Smith (ed.), *Hosts and Guests: The Anthropology of Tourism*, 2nd edn, Philadelphia, PA: University of Pennsylvania Press, 37–52. (A provocative article linking tourist systems to metropolitan needs.)

DAVID HARRISON, UK

commercialisation

Commercialisation is the effect of modern and postmodern **socialisation** on material things and received ideas. It is the process by which 'real' things/events/places/narratives become goods or services, predominantly known for their exchange value. Under commercialisation in tourism, folk **art**, rituals, ceremonies and sites lose original cultural meaning as they are commoditised (see **commoditisation**) and rendered for commercial consumption.

KEITH HOLLINSHEAD, UK

commissary

A commissary is a centralised kitchen or **food** production unit serving a number of **food** service outlets. Such centralisation achieves **economies of scale** and lowers unit costs. Originally food was transported hot, but increasingly cooked items are chilled, frozen or vacuum-packed to ensure safe handling (see **ready prepared system**). This system is found in airline catering, social **catering** and other settings in tourism.

See also: food service, contract

PETER JONES, UK

commoditisation

Commoditisation is the process of making available for purchase or barter cultural productions which include material objects, events and performances, or even people and ways of life. In the context of tourism analyses, this term has usually been used not just for items which are ordinarily for sale (such as airline tickets or camera film), but particularly for the sale of items which are not normally or originally designed for trade. Thus, commoditisation has a pejorative connotation, especially when applied to cultural **patrimony**, sacred performances, child labour, sexual services, or rare and endangered species. When such items are reproduced endlessly for money and are sold cheaply, the process may be called trinketisation, Disneyfication or McDonaldisation (see **commercialisation**).

Appadurai (1986) shows how objects and performances embedded in traditional societies are usually only available for consumption through inherited rights, kinship or religious relations, or caste or ethnic status. They may be acquired or enjoyed through gifts or **hospitality**, mutual exchange, noblesse oblige or sumptuary laws. He shows that in some cases, and increasingly through economic **acculturation**, they may move into 'commodity status' (that is, they become available for sale to anyone who has the money, regardless of status and prior relationships). Thus the social fabric is strained by commercial transactions superseding prior arrangements. Appadurai also shows that commodities can also move back from this to other status; for instance, the Elgin Marbles from the Parthenon, which were purchased in the early nineteenth century by the British from the Turkish rulers of Greece, are now treated as objects of world **heritage** which Greece cannot repurchase at any price.

The most celebrated case of the corruption brought about by commoditisation is Greenwood's (1989) account of the Spanish government's attempt to rearrange for **mass tourism** the annual performance of the Basque nationalistic celebration, the *alarde*. The proposal that the **ritual** be performed twice, in front of paying tourists sitting in specially erected stands or bleachers, so upset the Basque townspeople that they nearly refused to go through with it. Some thought they should, as actors, be paid. A contrary case occurred when an impresario suggested to the authorities in Sienna that they perform their city's annual inter-ward horse **race**, the Palio, in a number of Italian cities each summer; to which they replied 'We do not perform the Palio, we live it!'

Much as it suits the moralistic and nostalgic leanings of Western analysts who tend to see money and commerce as automatically corrupting, most detailed studies of tourism settings show that all change has complex outcomes and that this industry is rarely the sole or even the prime factor in economic **change**. Cohen (1988) has shown that **authenticity** is not antithetical to commercialism and that it may co-exist, be modified or assume new forms in modern touristic situations. It has been commonly found that commercial modifications of objects for sale or performances for viewing may stem further decline or loss of interest, may stimulate new creativity, or may even result in the invention of new cultural expressions which enhance pride of identity and soon become accepted as 'traditional'.

This is particularly true of **music** and dance forms, which have been revived and intensified in the wake of the world's recent interest in all forms of ethnic **music**, as shown in **cultural tourism** and **ethnic tourism**. Performers may have larger and more appreciative audiences than dwindling local populations. New outlets such as the ability to make and to sell recordings, the adaptation of new instruments and **technology** and even the chance for formerly marginalised musicians to tour, have strengthened as well as transformed many local folk and ethnic traditions. Further, in music and dance the performance itself is likely to engender feelings of authenticity in the performers regardless of whether the audience paid or who they are.

A number of collections of case studies have appeared which examine the influence of tourism on the production, distribution and consumption of the arts and material culture are under heavy influence in the last few decades (Appadurai 1986; Cohen 1993; Graburn 1976; Jules-Rosette 1984). A tremendous range of outcomes characterise the differential effects of commoditisation: extinction, replications, miniaturisation, gigantism, simplification, exotification, loss of functionalism, modified or new motifs, syncretism and changes in methods of manufacturing and materials. In addition, the audiences may be locals, tourists or both. Regardless of the changes, if the commercial forms are appreciated and demanded, the performers and artists will feel rewarded and stimulated. However, if creativity is not rewarded and only repetitive and/or cheap productions are bought, the creators will get bored or discouraged.

In these times of increasing appreciation of local and ethnic productions, when performances and **art** objects can reach global audiences, the commoditised output of those under the tourist **gaze** may in fact enhance self-esteem. This is especially so when these arts are upheld as representative of the best that the national society has to offer. The commodisation of cultural forms for tourism is probably a more enlightened and profitable avenue than **employment** as a menial labour force in the **service** of the institutions of the industry.

References

Appadurai, A. (ed.) (1986) *The Social Life of Things*, Cambridge: Cambridge University Press. (Includes studies of the production, exchange, commodity status, cross-cultural dissemination and consumption of material culture.)

Cohen, E. (1988) 'Authenticity and commoditization in tourism', *Annals of Tourism Research* 15(2): 371–86. (Shows that authenticity is negotiable and that commoditisation does not destroy authenticity.)

—— (ed.) (1993) 'Tourist arts', special issue of *Annals of Tourism Research* 20(1). (Analyses commoditisation of history and heritage, colonial and post-colonial folk arts and souvenirs.)

Graburn, N.H.H. (ed.) (1976) *Ethnic and Tourist Arts: Cultural Expressions from the Fourth World*, Berkeley, CA: University of California Press. (Case studies of the stimulation, production, trade and consumption of the arts and crafts of native and minority peoples around the world.)

Greenwood, D.J. (1989) 'Culture by the pound', in V. Smith (ed.), *Hosts and Guests: The Anthropology of Tourism*, 2nd edn, Philadelphia, PA: University of Pennsylvania Press.

Jules-Rosette, B. (1984) *The Messages of Tourist Art: An African Semiotic System in Comparative Perspective*, New York.

NELSON H.H. GRABURN, USA

communication mix

The communication mix is the combination of **advertising**, **personal selling**, public relations and **sales promotions** used within the same campaign. Most tourism **marketing** programmes include the use of such communication mixes because different ingredients in the mix help best in achieving related, but somewhat unique, objectives of the marketing programme. Thus, a communication mix for a given **destination**, Disney World, for example, may include **television** and **radio** commercials, newspaper and magazine ads, free **brochures** mailed to inquirers exposed to advertising offers of the free brochure, publicity in the form of news stories about family outings at Disney World. The central premise for designing a mix that includes multiple media and related tools (for example, advertising both to tourism **marketing** professionals as well to **leisure** tourists directly) is that one execution cannot accomplish all the objectives of a communication campaign.

One of the most popular communication mix strategies in tourism is the combined use of **image** and linkage-advertising in the same campaign (see **advertising**). Television and radio ads are often used to increase awareness and provide customers with knowledge that a free brochure (linkage-advertising) is available by request. Magazine ads are often used to provide 'hard copy' of the offer of the free brochure so that consumers do not have to hurry and write down a telephone number of the television monitor. The brochure or catalogue provides detailed information on what to do and buy and how to do it; increasing customers skills for buying.

See also: marketing mix; marketing plan; marketing research

ARCH G. WOODSIDE, USA

community

A community is a combination of social units and systems that afford people daily access to those broad areas of activity which are necessary in day-to-day living. They have five major functions: economic (i.e. production, distribution and consumption), **socialisation**, **social control**, social participation and mutual support. Communities are usually defined on the bases of three major attributes: a geographical area or territory, social interaction reflecting interdependencies among social units, and common norms that are a set of shared behavioural expectations which community members help to define and, in turn, are expected to follow. Ideally these attributes should coincide, as may be the case in an isolated village. But in complex modern societies in which there is great mobility, people may sometimes have more interaction and more in common with people living at a distance than with their next-door neighbours. Thus, it is possible to speak of 'community without propinquity'.

In tourism studies, although tourists emanate from communities, greater emphasis is often placed on **destination** areas and the places which tourists visit may be viewed as host communities. The attributes of a place, including its **environment** and people, may be termed a community tourism **product**. Less commonly, assemblages of visitors in destination areas may possess community attributes as in the case of seasonal, cottage or **camping** communities. Great concern is usually expressed both by residents and researchers, for maintenance of the characteristics of host communities, particularly in remote or developing areas, and for the changes which their residents, economies, landscapes and political structures undergo as tourism evolves. Residents of such communities are often encouraged to take greater control of their futures by becoming involved in **community planning** and thereby influencing decisions about tourism developments in their home areas and protecting desired community attributes.

See also: community planning

Further reading

Bowles, R.T. (1981) *Social Impact Assessment in Small Communities: An Integrative Review of Selected Literature*, Toronto: Butterworth and Co. (Reviews definitions of community and methods of examining change in such locations.)

Murphy, P. (1985) *Tourism: A Community Approach*, London: Methuen. (A comprehensive discussion of tourism planning from a community perspective.)

GEOFFREY WALL, CANADA

community approach

Tourism **development** can evoke a variety of both positive and negative perceptions among host **community** members. Many governments and large corporations focus principally upon positive economic benefits. It is generally acknowledged that for this industry to survive, residents must be favourably disposed to tourism and must perceive that they exercise some influence over its **planning** process. It is further recognised that their attitudes should be constantly monitored and problems promptly rectified.

GLENN F. ROSS, AUSTRALIA

community attitude

Tourism **development** can evoke a variety of both positive and negative perceptions among host **community** members. Tourism-related social changes may lead to rapid and widespread development, representing advances in the community, higher standards of living, and an overall sociocultural enrichment which leads to perceptions of wellbeing. Changes may also result in dependency (see **dependency theory**), involving economic **growth** which leaves an underdeveloped structure and enhances existing social inequities. In this context, relatively few members of the host community may gain substantially from the growth and development; however, the majority of the residents may not participate in or benefit economically or socially in any meaningful way from the developments. This situation can lead to feelings of resentment, bitterness and expressions of hostility toward fellow residents and visitors.

Many governments and large corporations focus principally upon positive economic benefits. But there is now increasing recognition of the potential social and environmental costs associated with tourism development and the necessity for a careful investigation of non-economic effects. In such situations, commentators have called for careful **planning** aimed at minimising tourism's negative impacts and maximising benefits for the host. It is generally acknowledged that for the industry to survive, residents must be favourably disposed to tourists and must perceive that they exercise some influence over the planning process. Moreover, it is suggested that their attitudes toward tourism impacts on community life should be constantly monitored, and problems promptly rectified.

How any community responds to the opportunities and difficulties posed by a growth in the tourism depends on a variety of factors, not the least of which is its fundamental attitudes to the industry. A major influencing factor and reactions to this business is has the level of contact between tourists and locals, which is not uniform, but spatially selective. Residents living in close proximity to the major tourism activity **location** are more likely to be most aware of the industry and **experience** the full **impact** of its disruption in their daily lives. In contrast, those who live out in the suburbs or further from the hub of the industry are less likely to be aware of tourism and are likely to come in contact with tourists and the infrastructure less frequently.

Specific identifiable resident types have been found to develop much more positive attitudes to tourism than others. Host community members with a commercial interest in tourism are more likely to be favourably disposed to the industry than other residents. Individuals owning or operating businesses, together with those who work in those operations, are more likely to have prominent and positive attitudes to tourism, and vice versa. One of the reasons many residents generally place a lower value on tourism, when compared with other groups such as business operators and local administrators, is said to involve a lack of awareness of the extent of the economic advantage of the industry which may flow on to their community. Some commentators suggest that the general public is often largely ill-informed regarding its contribution both to the local and the national economies. Moreover, this is said to demonstrate a general weakness in the industry's **public relations**, when it fails to explain its role in maintaining and increasing the standard of living for the people of the local community, whose cooperation and goodwill are seen as essential.

This wide range of community attitudes toward the industry and tourists has led to the construction of several host–guest models within the social sciences, which attempt to incorporate both the positive and negative reactions. For any tourism **site** or **region** there may be a saturation level, and if that level is exceeded the resulting costs begin to be seen to outweigh the benefits. The Irridex model (as proposed by Doxey 1975) seeks to identify and explain the cumulative effects of this development over time on social relationships, and postulates a direct link between increased community irritation, or stress, and continual growth. This model suggests that in its early stages tourists are likely to be regarded with enthusiasm by local residents. This emergent industry is perceived to bring **employment** and revenue, and the earlier tourists are perceived to be interested in and appreciative of local customs and **lifestyles**. However, as the flow of tourists gathers pace, contact between them and residents becomes less personal and more commercialised. Tourists are commonly perceived to demand facilities built specifically for them. The industry is now no longer new or a **novelty**, but is rather taken for granted in the life of the city. Many local residents are said to have a generally sympathetic attitude to it. However if development continues, this may undermine tourism's acceptance because of such factors as increased congestion, rising prices and changes to customary ways of life. A growing number of host community members may feel that their place is being greatly altered and they have not been consulted about this. The costs of accommodating tourism may now be perceived to exceed the benefits. Such annoyance can change to antagonism if the industry and its facilities are perceived to be the cause of local economic and social problems. Doxey here cites the instances of murder of wealthy white tourists in some less-developed countries.

This model has now attracted criticism in that it suggests a unidirectional sequence, where residents' attitudes and reactions will change over time and within a predictable sequence. Murphy (1985) points out that reactions are often more complex than this. The product **life cycle** concept, as suggested by Butler (1980), may be seen to be more adequate here. In this model, sales of a product proceed slowly at first and then experience a rapid rate of growth before stabilising, and finally (often) declining. Tourists are initially said to come to an area in relatively small numbers. However, as facilities are provided and the **destination** becomes better known, the number of arrivals increases. Further promotion and the growth of facilities result in the venue's popularity growing rapidly. This rate of increase, though, suffers a decline at the point where levels of **carrying capacity** are achieved. Butler suggests that at this point the attractiveness of an area may decline in comparison to other tourism areas because of overuse and the general impact of industry. The number of arrivals may also start to decline. Butler's model involves **exploration**, involvement, development, consolidation, stagnation, decline or rejuvenation stages. It is generally regarded as more flexible and less deterministic than the Irridex model. This flexibility is particularly evident at the stagnation stage, where a destination may choose either to rejuvenate or decline, depending upon the circumstances prevailing within the destination at that time.

There are at least two identifiable groups of factors that are able to influence host–guest relationships. The characteristics of tourists will have an influence that extends beyond the physical impact of increasing numbers. The tourists' length of stay and their racial and economic characteristics need to be considered as well as their numbers. Moreover, a destination's own characteristics are likely to help determine its ability to absorb the growing number of arrivals. Pearce *et al.* (1996) have pointed out that destination characteristics such as its level of **economic development**, the spatial distribution of its tourism focus in relation to other economic activities, the strength of its local culture and local political and community attitudes will effect host–guest relationships. Large metropolitan areas, with their tourists mainly concentrated in and around a core zone, are able to handle millions of arrivals, whereas small rural settlements may have trouble handling a few thousand. Thus both **tourist** and destination characteristics are likely to have a major influence upon the positive and negative impacts of tourism upon host community members.

References

Butler, R.W. (1980) 'The concept of a tourism area cycle of evolution: implications for management of resources', *Canadian Geographer* 24: 5–12.

Doxey, G.V. (1975) 'The causation theory of visitor-resident irritants, methodology and research inferences', *The Impacts of Tourism*, Sixth Annual Conference proceedings of the Travel Research Association, San Diego, CA, 195–198.

Murphy, P.E. (1985) *Tourism: A Community Approach*, New York: Methuen.

Pearce, P.L., Moscardo, G.M. and Ross, G.F. (1996)

Understanding and Managing the Tourist-Community Relationships, London: Elsevier.

GLENN F. ROSS, AUSTRALIA

community development

Community development is a process of economic and social progress based on **local** initiatives. Tourism **development** can result in **community** problems, but its **planning** and development can potentially contribute by fostering awareness of issues and opportunities, empowering citizens to make decisions, **training** residents for **leadership** positions, providing more and better community facilities and services, and facilitating stronger local institutions and feelings of interdependence.

DONALD GETZ, CANADA

community planning

Community planning refers to **planning** which takes place at a local level, usually with considerable input from local residents through **public participation**. In this way, goals for the **community** which are likely to receive widespread local support are identified and means to achieve them are determined. This orientation views tourism potential as a local resource which should be developed primarily for the benefit of the local host. It requires the redirection of planning from a purely business or narrow **economic development** approach, to a more open one which focuses upon the enhancement of resident lifestyles and opportunities. The approach involves focusing upon the ecological and human qualities of a **destination** area in addition to business considerations and care in the integration of tourism with other land uses and activities. The common good is the criterion upon which the performance of the tourism industry is judged as opposed to financial criteria alone. Thus, in essence, community planning requires that communities take control of the process, set their own goals, determine what they wish to present to the tourism market, and plan accordingly.

Further reading

Murphy, P. (1985) *Tourism: A Community Approach*, New York and London: Methuen. (A comprehensive discussion of tourism planning from a community perspective.)

GEOFFREY WALL, CANADA

community recreation

Community recreation is derived from the facilities and services of local **parks** and **recreation** agencies, aimed at meeting residents' needs. Tourists are sometimes targeted as an additional user group so as to generate revenue. Community recreation facilities are often used as venues for special events with tourist appeal, while the hosting of events can help justify new facilities. Resident–tourist conflicts can arise over accessibility and differential **pricing**.

DONALD GETZ, CANADA

comparative advantage

The term 'comparative advantage' refers to the advantage possessed by a country that is engaged in the international exchange of goods and services and is able to produce a given good or tourism **service** at a lower resource input **cost** than other countries or destinations. Comparative advantage arises because the marginal **opportunity costs** of one good or service in terms of the other(s) differ among them. The most frequently used explanation of comparative advantage in international trade is the Heckscher–Ohlin factor proportion **theory**, which is based on differences in factor endowments among countries. The Heckscher–Ohlin model ignores **demand** and can be used with caution to explain the intersectoral exchange between industrial goods and services. The provision of tourism services is relatively labour-intensive. The factors are measured in terms of flows such as capital depreciation and working hours. In the industrialised countries, the capital–labour ratio is generally higher than in developing countries. The result is that the former nations tend to specialise in the production of capital goods, while the latter ones focus on the production of (low qualified) services, such as tourism. Both country types have their own specific manner of carrying out trade according to their own specialisation pattern. Not considered are the competitive advantages of tourism destinations (special attractions or events).

The classical Ricardo model, although also not a complete explanation for the international trade in goods and tourism services, is useful for understanding the international exchange between developing and industrial countries. The exchange is basically complementary (raw material or island beaches for industrial goods). For example, a country which only has raw materials or unspoilt landscapes with certain **climate** conditions must specialise in these products in order to have goods to exchange for industrial ones or technologically advanced products not produced in that country. On the other hand, developed countries having their own natural **resources** such as climate, island beaches and mountains, cultural or special technologies, do not need to specialise to the same degree.

See also: attraction; competitive advantage; destination; event; marketing, destination; opportunity costs; sunlust; wanderlust

Further reading

Porter, M. (1990) *The Competitive Advantage of Nations*, New York: Macmillan.

Smeral, E. (1996) 'Globalisation and changes in the competitiveness of tourism destinations', in P. Keller (ed.), *Globalisation and Tourism*, St. Gallen.

EGON SMERAL, AUSTRIA

comparative study

Comparative studies involve investigations of the same phenomena in similar situations in the search for commonalities and differences among them. The term is usually used for studies which analyse the same problem in two or more places, in a cross-national context involving more than one social system or, less frequently, in the same social system at more than one point in time. Thus, for example, one might examine the impacts of the construction and operation of a number of hotels in coastal destinations in the search for commonalties and differences among them. The comparative approach is a step in the direction of generalisation in that it can permit the separation of typical occurrences from those which are unique to particular cases. Pearce (1993) has argued that the comparative approach has yet to emerge as a recognisable **methodology** in tourism research, which has relied too heavily on case studies. However, if the comparative approach is to be adopted successfully, it is important that the cases which are being investigated are not so dissimilar that there is little basis for comparison. In a practical context, comparative studies may also permit the learning experiences derived from one situation to be applied in another. For example, interventions which have been successful in one situation may be initiated in another similar situation with expectations of positive consequences.

Further reading

Pearce, D.G. (1993) 'Comparative studies in tourism research', in D.G. Pearce and R.W. Butler (eds), *Tourism Research: Critiques and Challenges*, London: Routledge, 20–35.

Wall, G. (1995) 'Tourism and heritage: the need for comparative studies', in Z.U. Ahmed (ed.), *The Business of International Tourism*, 256–272.

GEOFFREY WALL, CANADA

compatible

When tourism use or activity is generally free of short-term negative **impacts**, then this is a compatible situation. Implicit is the notion of non-exceptional tourism **management** or regulatory practices. The concept is at the bottom of a three-level hierarchy: compatible, feasible and sustainable. Feasible tourism implies medium term socio-economic benefits, while sustainable tourism implies long-term **resource** use, **conservation** and management.

See also: environmental compatibility

RUSSELL ARTHUR SMITH, SINGAPORE

compensation administration

When people hear the term 'compensation', they usually think only about wages or salaries that workers earn in return for work or performance. For instance, in the tourism industry, workers may earn monies from both employers and guests (in the form of tips or gratuities). Both are considered a part of the wages and salaries for that position. However, wages and salaries do not represent the entire compensation a worker may receive. In fact, these represent only two types of immediate compensation. There are many additional forms of immediate compensation which a tourism employee may receive; for example, merit pay or sales commissions are types of compensation. In addition, most workers also receive some form of deferred compensation which can include such items as incentives or bonuses, benefits (insurance, pensions and social security), pay for time not worked (including **vacations** or sick leave) and, in some cases, perquisites (automobiles, low-cost or free meals and rooms, travel allowances, financial **planning** or profit-sharing and so on). In the tourism industry, discounts on use of the organisation's **resources** (travel, lodging, meals and so on) are a common form of deferred compensation. In some companies, these two types of compensation are referred to as direct and indirect compensation, respectively.

While one may think of compensation programmes as relating directly to the amount or type of work done, this is rarely the case. Instead, many other factors also influence the rate of compensation in most tourism companies. Some of the influences relate to economic conditions in either the company or the **community** (for instance, the cost of living in New York is much higher than in Panama or China; therefore, tourism workers in New York would earn much more than the named two countries). Others relate to internal or external labour market conditions. Additional influencing factors include employee perceptions of pay, union influences and governmental influences. Even employee **satisfaction** and **motivation** can influence a company's compensation practices.

Compensation administration is concerned with evaluating how much each job is worth. The five most widely employed methods of conducting job evaluations are ranking, **classification**, factor comparison, point methods and skill-based pay. Compensation administration also addresses how pay is structured within an organisation. Not all jobs are worth the same amount of pay. Determining how to identify the value of different jobs is only part of the pay structure puzzle. The number of pay grades offered, how company pay compares to competition in the market and how compensation is calculated within specific pay grades are also considered. Payment calculation for each position is based on whether the organisation prefers to be the pay leader (become known for paying more than the market average on the theory that better pay will attract better employees), pay follower (on the theory that less pay will equal greater profits) and/or pay matcher.

ROBERT H. WOODS, USA

competition analysis *see* marketing audit; marketing research; marketing plan

competitive advantage

Competitive advantage refers to economic advantages realised in a competitive **market**. Traditionally it has been associated with the enterprise as the unit of analysis, but recent applications in regional economics, strategic **management** and international business have extended the concept to include its applications to industries, regions and nations. While comparative advantage is strongly associated with the initial endowment and availability of resources and their prices in different establishments, it is more connected to the evolutionary changes in the mix of **resources** employed to emphasise the dynamics of economic advantages in different market settings.

Competitive advantages are either based on superior **cost** or the **quality** position of establishments. Having a superior market position in terms of costs or being a cost leader implies the capability to produce and market goods and/or services at average market quality but below market costs and prices. Competitive advantages in terms of costs can result from **economies of scale**, from business integration across related industries, or from superior **location** or from lower resource costs. Advantages in terms of quality are associated with the existence of distinctive capabilities and their maintenance or improvement over time relative to competing firms, industries or regions. The sources of qualitative competitive advantages for tourism destinations are foremost, natural and environmental features, followed by built attractions. Included are general **infrastructure** ranging from retail services to personal and business services, and the tourism superstructure consisting of hotels, restaurants, travel and tourism organisations, special events organisations, and **recreation** and sports facilities. Attitudes of locals towards tourists are usually also cited as a competitive advantage, the extent of which depends on the uniqueness or **authenticity** of the aforementioned sources of quality advantages.

Competitive advantages in terms of cost (cost leadership) in tourism signify value for money for tourists in major tourism businesses such as lodging, **transportation**, recreation, animation and sports facilities, as well as all other related **activities** which make up the chain of tourism services. Cost **competitiveness** in turn is determined by general price levels, inflation and exchange rates for all related services, and the efficiency with which inputs (human resources, **technology**, capital, infrastructure and superstructure) are converted into tourism products.

Further reading

Gijsbrechts, E. (1993) 'Prices and pricing research in consumer marketing: some recent developments', *International Journal of Research in Marketing* 10(1): 15–151.

Porter, M. (1990) *The Competitive Advantage of Nations*, New York: The Free Press.

Scherer, F.M. (1993) *Industrial Market Structure and Economic Performance*, Boston: Houghton Mifflin.

KLAUS WEIERMAIR, AUSTRIA

competitiveness

In most of the **economics** and business strategy literature, competitiveness has been used as a measure of economic success or economic strength. Competitiveness is a relative concept: a firm, industry, **region** or tourism **destination** is more or less competitive in comparison to any other representative or comparative unit. A firm may be more competitive relative to any other firm(s) in terms of having a higher **market** share, a higher penetration of foreign markets or of displaying lower (hence more competitive) prices. If one were precise, however, the latter represents only a measure of competitiveness rather than a definition. Strict definitions must instead use the concept of **competitive advantage**, the execution of which represents competitiveness (see **comparative advantage**). Put differently, competitiveness is the materialisation or enactment of competitive advantages with respect to **quality** or prices in the market place. The positioning of firms in the market place with respect to their competitive advantages and **behaviour** is subject to competition analysis. The analytical and empirical difficulties associated with the attribution of competitiveness (as an expression of economic success) to its various determinants of the competitive advantages suggest the need to carefully define competitiveness for analytical purposes. Hotels or destinations are not just competitive; rather they are competitive in terms of specific quality attributes, **location** advantages or prices, thus suggesting a prior necessary analysis of competition and competitive advantage.

Further reading

Francis, A. and Tharakan, P.K.M. (eds.) (1989) *The Competitiveness of European Industry*, London: Routledge.

Hamel, G. and Prahalad C.K. (1994) *Competing for the Future*, Boston: Harvard Business School Press.

Kay, J. (1995) *Foundations of Corporate Success. How Business Strategies Add Value*, Oxford: Oxford University Press.

Ritchie, B.J.R. and Crouch, G.I. (1993) 'Competitiveness in international tourism: a framework for understanding and analysis', conference proceedings of AIEST-Conference in Argentina, Bern: AIEST.

KLAUS WEIERMAIR, AUSTRIA

complaint *see* cognitive dissonance; loyalty

compliance

Compliance with statutory standards is achieved primarily by inspection, backed by the sanction of criminal prosecution for breaches. For instance, officers of environmental protection authorities review whether conditions attached to environmental authorisations have been complied with. To address the problem of insufficient skills and **resources**, there has been a move in public environmental **law** towards co-operative efforts to set and meet compliance standards. **Development** and operation of nature-based tourism is thus affected through this procedure.

DUNCAN HARTSHORNE, AUSTRALIA

computer-assisted education *see* education, computer-assisted

computer reservation system

Most tourism firms process their reservations through a computerised database called Computer Reservation System (CRS), which handles details of inventory, schedules and prices among other details. CRSs run on mainframes, minicomputers or microcomputers and are usually connected through data communication links to terminals within various branches of the company. CRSs are not to be confused with **global distribution systems**, but are often electronically connected to them for bookings.

See also: automation; hospitality information system; information technology

PAULINE J. SHELDON, USA

computer technology *see* information technology

concentration ratio

The concentration ratio is a simplistic measure of the potential social **impact** of **tourists** in a **community**. It is the ratio between the number of tourists, usually measured as mean visitation per day, and the community's population. The ratio also provides a crude measure of the relative importance of tourism to a local economy; however, it does not adequately represent the absolute importance of the industry. For example, the concentration ratio for a city such as London will be relatively small (because of its large population) although the absolute size of the industry in those communities will be substantial. A related concept is the so-called 'tourist function', the ratio between the overnight **accommodation** capacity of a **destination** (usually measured in terms of beds) and population.

STEPHEN SMITH, CANADA

concession

A concession is an agreement for a food service operator to have the right, sometimes exclusively, to supply food and drink on another party's premises. The concession frequently takes the form of a seasonal or annual lease of space and/or facilities. Concessions are used in sports stadia, airports and theme parks.

PETER JONES, UK

concierge

A concierge is a **hotel** employee who offers the ultimate in personalised **service** to the **guest**. The concierge can make arrangements for theatre tickets, tours or any other specialised service. Their professional organisation is called Les Clefs d'Or, and it takes a minimum of five years work in a hotel and a rigorous test to gain admission. Up-scale **tourists** often seek the services of a concierge.

DEBORAH BREITER, USA

conflict

Differences in the goals, expectations and behaviours of participants in the tourism system result in incompatibility and even hostility among them. Thus, there may be tension between the use of parks and other protected areas for tourism and **preservation**, between users of the same area who are seeking different experiences (for example, between cross-country skiers and snowmobilers, or between boaters and anglers), and between tourists and residents of **destination** areas. Resolution of such differences are difficult to achieve and are sought in the political and **planning** domains as well as at the **site** management level through the introduction of zoning systems and other forms of regulation, as well as formal and informal compromises.

Further reading

Jackson, E.L. and Wong, R.A.G. (1982) 'Perceived conflict between urban cross-country skiers and snowmobilers in Alberta', *Journal of Leisure Research* 14: 47–62.

Jacob, G.R. and Schreyer, R. (1980) 'Conflict in outdoor recreation: a theoretical perspective', *Journal of Leisure Research* 12: 368–80.

Owens, P.L. (1977) 'Conflict as a social interaction process in environment and behaviour research', *Journal of Environmental Psychology* 5: 243–59.

GEOFFREY WALL, CANADA

conjoint analysis

Conjoint analysis is a quantitative decompositional model used to understand consumer choice behaviour. Tourism is composed of a bundle of product attributes which differ in perceived quality and importance to potential tourists. The method allows a researcher to begin with simple rank orders, such as expressed preference for a series of **vacation** packages, and to analyse these choices to determine the relative importance the respondent assigns to each attribute at a specific level ('part worth' utilities). The conjoint model for k attributes with each attribute defined at M_K levels can be formulated, according to Timmermans (1984: 203), as:

$$U_{(x)} = \sum_{k=1}^{k} \sum_{j=1}^{Mk} \alpha kj \bullet X_{kj}$$

where $U_{(x)}$ is an overall utility or preference measure; *akj* is the part worth contribution associated with the jth level of the kth attribute; and X_{kj} is the presence or absence of the jth level of the kth attribute.

Separate part worth utilities are estimated for each level of each attribute using either ordinary least squares analysis or monotonic analysis of variance. These may be added together to predict the potential attractiveness of any new product given its combination of attributes at specific levels. This model assumes salient attributes are used in the choice profiles; data collection is carefully designed; a compensatory process for attributes as respondents trade off low levels of one attribute evaluation for high levels of another; and that the combination rule for part worths is additive. For a given choice situation, conjoint analysis measures the importance of specific attributes at specific levels, assesses the overall relative value of each product attribute, predicts preferences for new products, and can serve as a basis of **market segmentation**.

References

Carmichael, B.A. (1996) 'Conjoint analysis of downhill skiers used to improve data collection for market segmentation', *Journal of Travel and Tourism Marketing* 5(3): 187–206.

Timmermans, H. (1984) 'Decompositional multi-attribute preference models in spatial choice analysis: some recent developments', *Progress in Human Geography* 8(2) :189–221.

Further reading

Carmichael, B.A. (1992) 'Using conjoint modelling

to measure tourist image and analyse ski resort choice', in P. Johnson and B. Thomas (eds), *Choice and Demand in Tourism*, New York: Mansell, 93–106. (Explains methods of data collection and analysis.)

Claxton, J.D. (1987) 'Conjoint analysis in travel research: a manager's guide', in J.R.B. Ritchie and C.R. Goeldner (eds), *Travel Tourism and Hospitality Research: A Handbook for Managers and Researchers*, New York: John Wiley, 459–69. (Discusses advantages and pitfalls of conjoint analysis in trade-off situations.)

Smith, S.L.J. (1995) *Tourism Analysis: A Handbook*, 2nd edn, Harlow: Longman, 82–91. (Explains the method and its use in restaurant choice situations.)

BARBARA A. CARMICHAEL, CANADA

connotation

Apart from their primary significations, verbal and pictorial images are often used to imply or connote additional meanings. Connotation procedures are the linguistic, symbolic and cognitive processes through which such additional meanings are generated and communicated. One of the main reasons why the understanding of connotation procedures is relevant to tourism studies is that, like representations and the objects to which they relate, tourism **sites**, together with their activities and surrounding paraphernalia, derive much of their saliency from being able to connote meanings and associations which transcend their immediate appearances. A beach, for example, may well be a 'sea shore covered with water worn pebbles', but for holidaymakers it is also a place which carries a host of connotations; for example, with childhood, freedom or romantic liaisons. Ultimately, the whole tourism industry is founded upon such structures of symbolic transference.

Psychologists in the field have been particularly interested in the ways through which the trappings of leisure and tourism are used to connote the inner states and internal dispositions of those taking part in such activities. Incorporating insights from the psychological realm, sociologists and anthropologists are more concerned with the social origins of connotation. Veblen was the first seriously to examine how leisure pursuits took part in the orchestration of status distinctions within European society (horse racing and knowledge of classical languages, to take two examples, connoting high social status). More recently, Bourdieu has carried this work forward by looking at how status differences are symbolically marked by the connotations emanating from different artistic pursuits, and Barthes's studies of the Eiffel Tower and the *Blue Guide* are recognised by many as the corner stones of semiological (see **semiotics**) discussions of tourism icons and their metaphorical associations.

Dann has developed this *genre* in his examination of some of the symbolic vehicles of the touristic imagination. He has shown how these are implicated in the cognitive construction of feelings and attitudes towards social and personal histories and biographies (steam trains resurrected for the summer tourism trade being associated with both childhood and empire, for example) and how, furthermore, the broader panoply of **iconography** enters into cognitive processes which symbolically articulate and legitimate social hierarchies and systems of **social control**. The theoretical challenge posed by the interrelated tasks of analytically evaluating the objects strewn in the tourist's path, exploring the attributes they connote and describing the connotation procedures used in the process, derives from the necessity to thread together the realms of the personal, social, psychological and sociological within a single interpretative framework.

Further reading

Barthes, R. (1983) *The Eiffel Tower*, New York: Hill and Wang.

Bourdieu, P. (1984) *Distinction: A Social Critique of the Judgement of Taste*, London: Routledge and Kegan Paul.

Dann, G. (1996) *The Language of Tourism: A Sociolinguistic Analysis*, Wallingford: CAB International.

Selwyn, T. (ed.) (1996) *The Tourist Image: Myths and Myth Making in Tourism*, Chichester: Wiley.

Veblen, T. (1970 [1899]) *The Theory of the Leisure Class*, London: Unwin Books.

TOM SELWYN, UK

conservation

Conservation is a philosophy governing the manner and timing of resource use, and may be defined as managing the resources of the **environment** – air, water, soil, mineral resources and living species including humans – so as to achieve the highest sustainable quality of life. Nature conservation is a dynamic concept which is subject to diverse understandings and interpretations, spatially and temporally, and which is supported for many different reasons (for example, ethical bases, environmental sustainability, genetic diversity, recreation, scientific research, future choices and utility, education and political reasons). Recognition of the importance of nature preservation can be seen in the recent rise of the environmental movement and, simultaneously, the development of a conservation ethic in modern society. That recognition is tangibly evident in the creation and resourcing of public and private sector agencies and interest groups; in related **legislation** and public policy; and in the establishment of resource management units such as **national parks** and **wilderness** areas, which often serve as important attractions.

Tourism and nature conservation are interdependent, and their relationship has been a lengthy one. Tourism often stimulates measures to protect or conserve nature, but, at the same time (and somewhat paradoxically), presents a significant environmental **risk**, especially because of its demands on the natural environment. These risks are intensifying as domestic and international tourism **demand** for natural areas is growing. Furthermore, the nature of that demand is such that tourists are seeking more spontaneity, independence and participation in their experiences (and thus contributing to the growth of **ecotourism**).

Researchers recognise three different relationships with respect to conservation and tourism: conflict, coexistence or symbiosis. These can exist between those promoting tourism and those advocating conservation of nature. Conflict occurs when conservationists see that tourism can have only detrimental effects on the environment. Coexistence is noted when some, though very little positive contact, occurs between the two groups. Symbiosis is reached when the relationship between tourism and conservation is organised in such a way that both derive benefit from the relationship. Conflict and coexistence are common. Symbiosis is perhaps the least represented relationship in the national and international perspective.

Tourism can cause both environmental degradation and enhancement. **Ecologically sustainable tourism** requires the conservation of nature, and thereby leads to the maintenance or substantial enhancement of natural areas and subsequent increases in visitor satisfaction. The viability of recreational and tourism resources, rather than conflicting with conservation, actually demands it; otherwise visitor satisfaction will be reduced as the inherent appeal of the resources decline.

Tourism provides an economic impetus for conservation of the environment because protected and/or scenic areas are major attractions for domestic and international tourists. This can also contribute to a wider appreciation of nature conservation by promoting and making more accessible specific sites and aspects of nature. But the role of tourism as a consistent contributor to nature conservation is often debated because, among other things, tourists trample vegetation, disturb wildlife, carry pathogens and weeds and engage in vandalism and littering, and thus do not always behave in ways which promote a symbiotic relationship between the industry and conservation. Further, tourism has fostered the intensive viewing (with resulting disturbance or damage) and export of protected and/or endangered species.

Therefore, interrelationships between tourism and nature conservation are extremely complex and dynamic, with conflict being most acute where tourism development occurs rapidly and without **strategic planning**. Unfortunately, research has developed few strong concepts or theories to guide the role and **management** of tourism in nature conservation. Many studies focus narrowly on the physical impacts of developments at a particular site, few have a longitudinal basis, and most are reactionary as research projects are initiated when impacts are identified or speculated. Often the research focus is largely limited to the effects of tourism on vegetation and to a lesser extent wildlife, with impacts on air and water quality, soils and ecosystems relatively neglected.

According to researchers, several methodological problems concerning research on tourism and the natural environment (and thus nature conservation) can be identified, including the difficulty of distinguishing between changes induced by tourism and those induced by other activities; the lack of information concerning conditions prior to the advent of tourism and, hence, the lack of a baseline against which change can be measured; the paucity of information on the numbers, types and tolerance levels of different species of flora and fauns; and the concentration of researches upon particular primary resources, such as beaches and mountains, which are ecologically sensitive. Thus, successful integration of tourism and nature conservation objectives is of increasing importance because it enhances the choices of people and helps maintain or even enhances the quality of the environment.

See also: cultural conservation

Further reading

Budowski, G. (1976) 'Tourism and conservation: conflict, coexistence, or symbiosis?', *Environmental Conservation* 3(1): 27–31. (Recognises a spectrum of possible relationships between tourism and nature conservation.)

Mathieson, A. and Wall, G. (1982) *Tourism: Economic, Physical and Social Impacts*, Harlow: Longman. (Presents a broad introduction to the diversity of tourism's impacts, presenting a mix of theoretical and practical issues.)

O'Riordan, T. (1971) *Perspective on Resource Management*, London: Pion. (Provides an analysis of resource management concepts, allocation problems and decision-making processes.)

Pearce, D.G. (1989) *Tourist Development*, 2nd edn, Harlow: Longman Scientific and Technical. (Examines development patterns and processes, contribution to growth, and future issues.)

Pigram, J.J. (1983) *Outdoor Recreation and Resource Management*, London: Croom Helm. (Recognises outdoor recreation and tourism as important forms of resource use.)

JOHN M. JENKINS, AUSTRALIA

consortium

A consortium is an organisation of individual trading units which combine for a common commercial purpose such as joint **marketing** services and purchasing. Such organisations are not limited to the tourism industry; there are consortia of grocery retailers, accountants, solicitors and builders, as well as hotels and travel agents (see **travel agency**). Hotel consortia – also termed referral systems in the United States – are a well-established feature of international operations.

There are about thirty such organisations in the United Kingdom, of which over half can be viewed as international, such as Consort hotels and Best Western hotels. The United States has twenty-five large consortia and voluntary groups (many of which are international) as well as many smaller groupings. The benefits of consortium membership are the retention of entrepreneurial freedom and (unlike a **franchising** agreement) the option of renewing annual membership.

Individual consortium members come together to combat the marketing power of corporate groups. As organisations, consortia are able to operate in ways which would be impossible for the individual member. Each unit can be part of a publicity identity as well as providing a forum for sharing expertise. Cost reduction through purchasing advantages or training programmes for staff can be foci of the organisation. Marketing services can include access to **computer reservation systems** (CRS), establishment of overseas sales offices, conference placement services and loyalty programmes. Group promotions can target particular segments, while referral of customers between members is encouraged and promoted.

The number of consortium members continues to expand, although indications are that corporate groups who were formally members are withdrawing from such systems. As the members are collections of individual operators, one brand will not reflect the characteristics of each member. Consortia are developing a family of brands, allowing a more focused appeal to customers and members. This trend is paralleled by the emergence of smaller specialised consortia, for example golf hotels, town house hotels and castles and stately homes.

See also: cultural conservation

Further reading

Harrison, L. and Johnson, K. (1992) *UK Hotel Groups Directory 1992/3*, London: Cassell. (Provides a listing of all hotel consortia and members.)

Roper, A. (1995) 'The emergence of hotel consortia as transorganisational forms', *International Journal of Contemporary Hospitality Management* 7(1): 4–9. (Discusses organisation, structure and strategy of UK hotel consortia.)

ROGER J. CALLAN, UK

conspicuous consumption

The term 'conspicuous consumption' was first used by Veblen to indicate membership of a leisured elite. Through the lavish, ostentatious, yet ultimately wasteful enjoyment of luxury goods, conspicuous consumption was seen as the basis for establishing esteem in class stratified societies. Whereas it is easy to criticise Veblen on the grounds that **class** structure and status indicators are both more complex than he maintained, and that leisure opportunities are considered an essential requirement for all in modern society, the expression has nevertheless entered everyday usage.

There are at least three ways in which tourism might be regarded as an example of conspicuous consumption. First, members of wealthy tourist-generating nations can be viewed as conspicuous consumers of the valuable resources of the developing world. The exploitation of these countries on the **pleasure periphery** has been interpreted by some commentators as a form of **neo-colonialism**, although few **tourists** would analyse their own behaviour in such a stark manner.

Second, individual holiday choices made amongst complex status hierarchies of destinations can also be understood as conspicuous consumption, especially when considered in terms of ego-enhancement motivation and trip-dropping. However, it can be demonstrated that tourism **motivation** is not just materialistic, but reflects a wide range of other psychological and cultural factors. Furthermore, postmodern tourism is said to break down traditional boundaries and replace them with a universe in which image, illusion and irony predominate (see also **postmodernism**).

Third, other potential indicators of **prestige** include **souvenirs**, gifts, photographs, **postcards** and suntan, all of which can point to the relative affluence of the holidaymaker, and, in that sense, constitute symbols of conspicuous consumption. However, this association is perhaps too simple, since there may be a multiplicity of possible meanings attached to such activities. Thus the complexities of tourist typologies, motivations, destination choice and consumer behaviour, all provide arguments against the use of general explanations of tourist behaviour such as conspicuous consumption (see **typology, tourist**).

Further reading

Littrell, M. (1990) 'Symbolic significance of textile crafts for tourists', *Annals of Tourism Research* 17: 228–45. (Describes how handicrafts purchased by tourists have at least eight different meanings attached to them.)

Urry, J. (1990) *The Tourist Gaze: Leisure and Travel in Contemporary Society*, London: Sage. (Argues that the romantic and the collective gazes constitute separate motives, and that postmodern tourists place increasing emphasis on pleasure and play.)

Veblen, T. (1970) *The Theory of the Leisure Class*, London: Unwin Books. (Contains a critical introduction by C. Wright Mills.)

DAVID WILSON, UK

consumerism

Consumerism can be defined as a social movement seeking to augment the rights and powers of buyers in relation to sellers. It can result in consumers seeking redress, restitution and remedy for dissatisfaction with products and services bought. While inevitable and enduring, consumerism is an antithetical outcome of the marketing process. Despite efforts by many companies to assure consumer satisfaction (the guiding tenet for tourism marketers) it continues to remain elusive. In the media, for example, one reads or hears about consumers who are incensed by exorbitant **transportation** costs, questionable **marketing** practices and inadequate safety and **security** precautions.

In a similar vein, with tourism activity being pursued in communities in which people live, dissatisfaction with **tourists** can be voiced by concerned citizens. Research reveals that a **community**'s constituents can become incensed at the hedonistic, utilitarian behaviour of both buyers and sellers. Satisfaction with the tourism process, therefore, can be bedevilled by both consumerism and communitarianism.

Consumer-citizen demands about the performance of tourism products and services stem from a growing concern about quality of life and the **environment**; the complexities and hazards accompanying technological sophistication of products and services; distrust of marketing promises that can not be fulfilled; attempts at rational purchase behaviour stymied by lack of information; high-pressure sales techniques; the uncaring delivery of service; and mass consumption products that are incapable of meeting the personal needs of individuals. As a consequence, consumerism has gradually affected the tourism industry, firms, and governments. Individual companies and trades have enacted codes of behaviour and ethical practice (see **code of ethics**). Better **legislation** is being established for the protection of tourists. **Safety** standards, truth in **advertising** and other legislative efforts such as consumer protection and awareness bills are becoming more evident. Some companies, and also some destinations, have created ombudsmen to listen to and resolve concerns of disgruntled tourists. By opening up the channels of communication, companies have been able to identify and change corporate practices that were perceived as deceptive, and to educate employees to the need for improved relations with customers and citizen groups.

Consumerism has broadened the understanding and application of the concept of satisfaction. A host of factors determine continuous consumer-citizen satisfaction with market offerings: **service**, warranties, accurate advertising, **pricing**, packaging, design, safety and security. As a result of the consumer movement, many organisations are becoming more responsive to tourist needs, and governments are accelerating their regulatory intervention. Marketers are beginning to provide more guidance about their products and to insist on new and higher levels of integrity.

MICHAEL HAYWOOD, CANADA

content analysis

Content analysis is a form of observational inventorying of the properties of texts. Its central idea is that any text (verbal or non-verbal) can be subjected to scientific analysis by first isolating particular categories of content (themes, images, subjects and so on) and then quantifying their absolute and relative occurrence within single or multiple communications, in order to test hypotheses about the text's properties and, in some instances, the material and psychological state of its authors and/or its intended effects on its audiences. It has been formally defined as 'a research technique for the objective, systematic, and quantitative description of the manifest content of communication' (Berelson 1952: 26).

Content analysis was developed by social scientists in America in the 1940s and 1950s and was particularly associated with the names of Kaplan, Lasswell, Leites, Nathan, de Sola Pool and Bernard Berelson. It was first used during the war as a tool for studying German propaganda through identifying its main themes and subjects, and quantifying their synchronic distribution within single messages and also their diachronic distribution in multiple communications over given periods of time. From these analyses, scientists were able to make inferences about the material conditions and psychological morale of the Germans and their intentions, including how these were changing in time. After the war, content analysis achieved wide civilian currency in the social sciences as an empirical method for studying a wide range of mass cultural phenomena including newspaper editorial practices, sociocultural aspects of teenage romance magazines, television drama themes, racial stereotyping in the media, gender relations in comics and many other kinds of mass communication.

Despite its initial acceptance as a new scientific procedure, content analysis was in essence an empirical development of rationalist, aesthetic criticism. Anyone evaluating a book, painting or musical work is implicitly engaged in content analysis to the extent that the reader is attempting to inventory the signs and their meaning within a text, and arriving at inferences about the author in terms of theme, style, intention, outlook, background and aesthetic importance. Content analysis took this basic aesthetic procedure and developed it by, first, expanding its applications from high culture texts (the traditional frame for aesthetic criticism) to those of popular culture and mass communication and, second, eliminating the unsystematic and subjective tendencies of aesthetic criticism by replacing them with analysis built on explicit statement of a priori hypotheses to be examined, rigorously defining the categories of content for scrutiny in exploring the hypotheses, and quantifying the results. The process has been well-defined by de Sola Pool (1973: 36): 'The content analyst aims at a quantitative classification of a given body of content, in terms of a system of categories devised to yield data relevant to specific hypotheses concerning that content.'

Following its early wartime utility, criticism of content analysis as a means of understanding texts came from two quarters. First, there were those within the ranks of content analysis, notably Gerbner, who argued that a text could not be fully evaluated purely on its internal characteristics without an understanding of the external sociopolitical constraints within which messages were produced and circulated. For Gerbner, industrial and market relationships and other social determinants might result in processes of textual selection, omission and juxtaposition which would not be revealed simply by inventorying existing categories of content. In tourism promotion, for example, a **brochure**'s properties cannot be simply revealed by analysing its verbal and pictorial elements (see also **promotion, place**). A fuller understanding requires awareness of external constraints impinging upon the text, including legal requirements, political and institutional forces and industry practices (for example, the size format of many brochures is dictated by the racking policies of travel agents who will not display non-standard sizes and shapes).

The other critique came from structuralists working in rationalist rather than empirical traditions, who argued that quantitative inventorying of the denotative elements of messages might be less important than interpreting the mythic dimensions of key connoted verbal and pictorial elements, even when their numerical occurrence was small. For example, in an advertisement for the Seychelles the number of times that the destination name is mentioned might be less important than exploring the mythic dimensions of important, individual phrases or images such as 'island paradise'. Moreover, structuralism was less concerned with the aggregate inventorying of elements within a text than their structural relationships and the codes of difference set up within the text by the position and juxtaposition of ideas. Even more radically, structuralists suggested that a communication's most interesting feature might be what it excludes rather than includes, an insight caught by the French writer Macheret's notion of the 'significant silences' of the text. This includes, for example, the failure of much tourism promotion to represent the social realities of the inhabitants of Third World countries packaged as dream destinations for tourists. In short, structuralists challenged the value of quantitative inventories of manifest content and advocated interpretative approaches which revealed latent and systematically excluded meanings.

Today, social scientists are most likely to use a combination of numerical counts and interpretative analysis, as well as attempting to locate the external forces and relationships constraining a text's production to arrive at evaluations of its overall effects and meanings. In tourism, content analysis has to date been mostly used by anthropologists and sociologists in studies of the ideological aspects of promotion, particularly brochures. Less commonly it has also been used to explore the relationship between **advertising** and the editorial content of newspapers, to examine the construction of destination images in newspaper and magazine articles, and to study the historical development of destinations using time-series data gathered from entries in guidebooks published over selected periods.

Content analysis, allied to interpretative analysis, offers great potential in the study of a field such as tourism which is represented in many media forms. The vast variety of tourism promotion and publicity delivered by books, press, radio, art, film and television lends itself to synchronic and diachronic study using the theoretical and methodological approaches of content analysis.

References

Berelson, B. (1952) *Content Analysis in Communication Research*, Glencoe, IL: The Free Press. (The seminal book on the theory and methodology of content analysis.)

de Sola Pool, I. (ed.) (1973) *Handbook of Communication*, Chicago: Rand McNally.

Further reading

Gerbner, G. (1958) 'On content analysis and critical research in mass communication', *Audio Visual Communication Review* 16: 9–32. (Provides a seminal statement on the importance of taking into account external factors constraining the development of messages.)

Seaton A.V. (1990) *The Occupational Ideologies of Travel Page Editors*, Sunderland: Business Education Publications. (Attempts to explore the relationship between tourism advertising and the editorial policy of travel page editors by using content analysis.)

A.V. SEATON, UK

contestation

Contestation is the social process through which interest groups handle dominance/subjugation issues in the politics of tourism, or in policy making anywhere. Special interest groups cultivate identities and niches for themselves, particularly at the issue level. Under the pluralisms of postmodernity, natural, social, cultural or economic realities are constantly redefined, and regularly decentred, detotalised and destabilised through such oppositional contestation.

KEITH HOLLINSHEAD, UK

contingent valuation

Contingent valuation is a research methodology used to estimate the economic value of a tourist amenity or resource not sold through conventional markets. It produces an economic value for a good by asking subjects to respond to a scenario describing the current amenity or resource, plausible changes to the amenity or resource to better serve the guests, and how much they are willing to pay for it. Variations include personal interviews, questionnaires, bidding games and public referenda.

See also: environmental valuation

JOHN C. CROTTS, USA

continuum model

A continuum is a structure or process that is, or should be regarded as, continuous (that is, connected in time or space). Thus, such a model is a representation of a structure or process which normally occurs uninterruptedly between two points or poles. In a general sense, the growth and **democratisation** of tourism can be viewed as a continuum through time, commencing with the emergence of tourism as an elite social activity and ending only when it has become fully democratised.

More specifically, a number of tourism theories are based upon continuum models. For instance, the development of destinations is often conceptualised as one such model, the most widely cited example being Butler's (1980) destination life cycle model, which represents the continuous linear evolution of tourism areas through identifiable stages from exploration to decline or rejuvenation.

Linked to this evolution, host communities' changing perceptions of tourism and tourists are represented by the euphoria–antagonism continuum model, a continuous process whereby host's attitudes become increasingly less favourable as tourism develops.

However, it is within the context of tourist typologies that continuum models are most commonly utilised (see **typology, tourist**). For example, in order to explain and predict tourist behaviour and potential destination choice, Plog developed his psychographic continuum model whereby tourists are classified according to characteristics and behavioural patterns ranging from psychocentrics to allocentrics. Many other typologies or classifications similarly fall within a continuum framework. These generally range from the institutionalised or charter mass tourist to the independent explorer or traveller. Importantly, these models are either implicitly or explicitly informed by Cohen's (1972) familiarity–strangerhood continuum which suggests that tourists are more or less able to escape from their 'environmental bubble' and, hence, seek either novel or familiar surroundings and experiences. The familiarity–strangerhood model is also linked to the relationship between tourists and their own society (see **alienation**; **anomie**), which in turn determines the experiences that tourists seek.

References

Butler, R. (1980) 'The concept of a tourism area cycle of evolution', *Canadian Geographer* 24: 5–12.

Cohen, E. (1972) 'Towards a sociology of international tourism', *Social Research* 39(1): 64–82.

Plog, S. (1973) 'Why destination areas rise and fall in popularity', *Cornell Quarterly* November: 13–16.

RICHARD SHARPLEY, UK

Contours

The journal *Contours* focuses on tourism and its effect on the lives of people, encourages an awareness of the role of tourism in the developing process, provides the opportunity for people affected by tourism to express their views, denounces unfair practices in tourism, stimulates research and action for more justice in tourism, and intends to empower indigenous communities to get a fair price for their exposure. *Contours* reprints articles from newspapers/newsletters and journals which are difficult to access in other regions in the world, and encourages its readers to send articles and news from their countries or regions. First published in 1982, it appears quarterly, published by the Ecumenical Coalition on Third World Tourism (ISBN 0857 491X).

RENE BARETJE, FRANCE

contract food service *see* food service, contract

control system

Hospitality and tourism operations develop control systems in order to safeguard the assets, provide accurate and reliable accounting data, measure operational efficiency and ensure adherence to managerial policies. Safeguard the assets refers to the protection of assets from loss, waste, spoilage and effective utilisation of equipment, while accurate and reliable accounting data refers to the checks and balances in an accounting system. Additionally, operational efficiency attempts to minimise costs while providing services and product. Further, adhering to management policies refers to the organisation having standard operating procedures and management ensuring that employees follow given procedures. The objectives of a control system can be implemented if the system reflects specific characteristics including leadership, structure, sound practice, personnel, segregation of duties, authorisation procedures, adequate records, procedure manuals, physical controls, internal reports and independent checks.

Leadership identifies management as responsible for internal control systems. Management communicates and enforces policies. Each organisation should have organisational charts which delineate the chain of command. Additionally, each position on the chart should have a job description detailing specific duties. Sound practice refers to management creating an environment conducive to internal control. Policies such as sound hiring procedures, a vacation policy and bonding of employees are sound practice, as is instituting a **code of ethics**. The selection and training of employees is essential to a control system to ensure employees are competent and trustworthy. Segregation of duties refers to assigning different employees to different functions, which creates checks and balances that can prevent errors and enable the detection of theft. Authorisation procedures ensure management authorise every business transaction according to set guidelines. Adequate records are necessary to record transactions and they facilitate control and create audit trails. Procedure manuals detail the requirements for each job position which leads to consistent job performance. Physical controls include safes, locks, control of document and other security devices. Budgets and internal reports provide a vehicle for management to monitor its goals. If goals are not met, management can take corrective action. Finally, independent checks on performance refers to internal auditing and external auditing on performance to ensure that control systems are functioning properly. As such, for a control system to be functional it must meet specific objectives and reflect specific characteristics in areas such as cash receipts, accounts receivable, purchasing, payroll, income control, food and revenue control and analysing budgets.

Further reading

Gellor, N. (1991) *Internal Control*, Ithaca, NY: Cornell University Press.

Schmidgall, R. (1995) *Hospitality Industry Managerial Accounting*, East Lansing, MI: Educational Institute.

BERNARD FRIED, USA

convention *see* meetings business

convention and visitor bureau

Most cities and many communities establish convention and visitor bureaus (CVBs) or destination marketing/management organisations as marketing agencies to encourage visitation, both for pleasure and business, and as advocates for tourism development initiatives.

CVBs can be structured as independent nonprofit agencies, or may be a department within government or chambers of commerce. A major mandate is to work in cooperation with hotels, convention centres, transportation companies, tour operators and other local businesses. The purpose is to generate sales leads, secure confirmed bookings and to provide services (for example, reservations, hotel commitments, tour arrangements, concierge service and on-site registration) for organisations and/or their agents planning large meetings, conventions or trade shows. Services provided are usually free since they are funded through a combination of hotel room taxes, government funding and membership dues. Not all CVBs can afford to provide the same level of services, so there is a move to subcontract or partner with private firms.

Because of intense competition among destinations for tourism dollars, CVBs have had to become masterful marketers of their destination's offerings. The overriding objective is to achieve sustained growth in the numbers of tourists, their length of stay and levels of expenditure. Strategically, the marketing mandate begins with the development of a clear, consistent image and a branding approach that is based on a well-researched understanding of various markets. Targeted advertising, describing attractions, facilities and services, is designed and delivered to generate awareness and visitation from regional, national and international tourists, tour groups and convention delegates. Information is also disseminated through trade and membership publications, call centres and, increasingly, well-designed Internet sites which are linked to tourism businesses and organisations throughout the area.

Major challenges include the championing of tourism projects, particularly those that require public funds (hence the need to encourage economic impact research to reveal distinct benefits, costs and spin-offs whether for tourism in general or for specific projects); ensuring that the 'destination experience' of tourists results in positive word of mouth and repeat visitation; obtaining community support for tourism; collaborating with the myriad of local event and tourism-related programme organisers; overcoming the difficulty in maintaining funding for CVBs that is necessitating a new 'not-for-loss' orientation; and accepting a broader responsibility to help to make communities better places to live, work and visit. To enhance professionalism, effectiveness and image of its worldwide members, the International Association of Convention and Visitor Bureaus assists through educational programming and research.

MICHAEL HAYWOOD, CANADA
DONALD ANDERSON, CANADA
ROBERT M. O'HALLORAN, USA

convention business

Convention business is a term used to describe any commercial activity that results from travel for the purpose of attending a convention, congress, exposition, **trade show** or other similar assembly, whether regional, national, continental or global in scope, organised on a regular or occasional basis. Along with corporate meetings, which refers to smaller gatherings of employees from single organisations, it comprises the segment of business tourism known as the conference and meetings market.

Because convention participants commonly meet and have room and board under the same roof, they represent a major source of revenue to hotels and meeting facilities. However, surveys indicate that accommodation generally represents only about 32 to 38 per cent of a convention delegate's spending. The remainder of the expenditures benefit the broader tourism community, reaching transportation providers, restaurants, local tourism attractions, cultural and sporting activities, and retail stores. Another benefit of this business is that it tends to be concentrated in the off-season (**seasonality**).

Conventions are also an effective means to generate repeat business and to showcase the destination to large numbers of professional tourists. Associations are the most visible convention organisers. They may be professional, union, educational, service, charitable or public organisations, and can range in scope from small regional organisations to national and international ones. Convention planners are generally employees of the firm or association sponsoring the gathering, although it is increasingly common to see third parties hired to perform this function.

Regardless of where the planning authority lies, decision makers tend to use similar criteria when selecting a convention site. These include minimum requirements in terms of meeting facilities, exhibition space and hotel rooms in the host city. Such standards exclude many destinations from bidding on events. For locations which satisfy the basic criteria, the basis of competition broadens to include other considerations such as cost, convenient air connections, **climate**, scenery, **safety** and the presence of nearby attractions.

There is growing recognition that successful marketing of a convention site extends beyond selling the destination and its related amenities. The successful delivery of the products and services promised during the sales phase is essential. While successful conferences can generate significant long-term benefits for the host city, resort or region, poorly executed ones can just as easily damage a previously positive reputation.

Further reading

Abbey, J.R. and Link, C.K. (1994) 'The convention and meetings sector: its operation and research needs,' in J.R.B. Ritchie and C.R. Goeldner (eds), *Travel Tourism and Hospitality Research: A Handbook for Manager and Researchers*, 2nd edn, New York: John Wiley & Sons, 273–84.

ROBIN J.B. RITCHIE, CANADA

core–periphery model *see* centre–periphery

Cornell Quarterly

The *Cornell Hotel and Restaurant Administration Quarterly* publishes information, research findings and opinions and advice regarding the management of hotels, restaurants and other hospitality businesses. Its contents (reviewed anonymously in-house), intended for hospitality professionals, are broad and include topics in marketing, finance, human resources, international development, tourism and general management. Its 'Educators Forum', published in the August issue, deals with issues pertaining to hospitality research and education. It was first published in 1960, and now appears six times per year, published by Elsevier (ISSN 0010–8804).

RENE BARETJE, FRANCE

corporate culture *see* culture, corporate

corporate finance

Corporate finance is the decision making process concerning the source of a tourism enterprise's funds and the allocation of its available resources in a manner consistent with measurable financial goals and objectives of the organisation. Measurable financial goals of a tourism organisation are often compared against a standard such as prior period results, budget or industry norms. Tourism organisations evaluate potential investments for the purpose of obtaining a maximisation of return measured by **return on investment** models.

Ratio analysis is a tool that is often used to communicate corporate financial information to interested parties relative to the financial goals and objectives. Ratios can be expressed as percentages, a per unit basis, turnovers and coverages. Ratios are used by tourism management in corporate finance to help determine if there is adequate cash to meet the organisation's obligations, evaluate the level of profits, compare the debt with the stockholder's investment, evaluate the inventory usage, draw relationships between the operation's earnings and the market price of the stock, and make judgements on the servicing of the accounts receivable reasonable.

There are five different groups of financial ratios used in corporate finance: liquidity ratios, which measure the tourism firm's ability to meet short term obligations; solvency ratios, which compute the extent that the tourism enterprise has been financed; activity ratios or turnover ratios, which are used to evaluate the ability of management to use the property's assets; profitability ratios, which measure management's overall effectiveness in managing the business; and operating ratios, which are used to analyse the tourism establishment's operations. Tourism corporate finance managers develop ratios from income statements, cash flow statements and balance sheet data. Income statement information may not always be the best measurement of profitability, and balance sheet data may be out of date due to the cost principle.

Further reading

Coltman, M.M. (1979) *Financial Management for the Hospitality Industry*, New York: Van Nostrand Reinhold.

Schmidgall, R. (1997) *Hospitality Industry Managerial Accounting*, 4th edn, East Lansing, MI: Educational Institute of the American Hotel and Motel Association.

STEPHEN M. LEBRUTO, USA

corporate strategy

A corporate strategy is a master game plan for managing and operating an organisation. Its primary function is the achievement of a firm's mission statement, meeting the firm's objectives and deciding what the scope or domain of the operation should be. Through a comprehensive and detailed strengths, weaknesses, opportunities, and threats analysis of both internal and external environments (see **SWOT analysis**), a firm will be able to make decisions that will enable it to create a defendable and competitive position in its operating industry. A SWOT analysis will also provide the company imperative information on how to adapt to its environment. Corporate strategies, if successfully chosen, implemented and carried out can lead to a synergistic effect. These strategies can guide a firm's allocation of resources among several business units and/or other ventures it may choose, thus leading to the creation of wealth for all its stakeholders.

Throughout the organisation, there are various levels of strategies: corporate strategies, business strategies and functional. Like business-level or functional strategies, corporate strategies have four aspects that need to be addressed: focus, specificity, responsibility and time frame. The focus of corporate strategies targets in on what type of business should a firm be in. This also entails such elements as whether or not the firm should be diversified or undiversified, and the type of products and/or services that the firm should offer. The answers to these fundamental questions derive from what the firm's SWOT analysis is comprised of. The specificity of corporate strategies tends to be very broad compared to that of a functional strategy. This level of strategy provides the blue print of the business plan in which a functional strategy can be implemented.

The responsibility of the creation of corporate strategies is that of the chief executive officer and its top management team. It is also their responsibility to ensure that these strategies are carried throughout the various levels of the organisation. It is through their strong commitment to resources, as well as fostering an organisational culture (see **culture, corporate**) and a reward system that leads to successful strategy implementation. This indeed creates a profitable future for the company. As in any level, corporate strategies have time frames. Since corporate strategies map out long-term future plans for the firm, the time frame for achieving these objectives and goals are approximately five to ten years.

Since the 1980s, the hospitality industry has been operating in a very volatile environment. Intense competition in the domestic market has made many US firms expand their operations internationally. Many compete at a global level in hopes to gain a greater market share and increased profits. There are several generic corporate growth strategies that tourism firms embark upon in order to gain both a competitive and sustainable advantage over its fierce competitors. The generic corporate growth strategies can be classified into three categories: concentration, vertical integration and diversification.

A concentration growth strategy is followed by firms whose focus is on either a single or limited number of product(s) and/or service(s). A company uses this type of strategy to grow through its current product. Three basic ways a firm can undertake a concentration strategy are through market development, product development and **horizontal integration**. Market development is the expansion into new geographic areas or new market segments in order to gain a larger market share. Some companies may add new products and/or services or may alter or add to their current products and/or services for their target markets. This strategy is product development (see also **product planning**). Horizontal integration is a strategic move for organisations which already have great knowledge of their existing operations and which actually invest in businesses similar to theirs. Firms which embark upon this strategy eliminate competitors and create a synergistic effect at the same time. McDonald's is a prime example of a company incorporating a concentration strategy.

As to the second growth strategy, some companies choose to become suppliers to their own current products and/or services in order to gain control of the distribution channel. Firms which choose this type of move are practising vertical integration. Under this category, firms which choose to serve as a supplier are involved in backward integration. Other firms may become an outlet for its own product. This type of move is forward integration. Examples of hospitality firms that follow a vertical integration strategy are Bob Evans and Papa John's International.

A third generic growth strategy is diversification which provides firms the opportunity to capitalise and manage a portfolio of **strategic business units** in different industries. There are two types of diversification strategies: concentric diversification and conglomerate diversification. The concentric diversification strategy is used by companies which acquire businesses that are different, but at the same time complementary to its current business to create a synergistic effect. Carlson Company is an example of a firm focusing on a concentric diversification strategy, currently operating hotels, restaurants, travel agencies and marketing consultant agencies. The latter strategy, conglomerate diversification is the diversification of a firm into businesses that are very much unrelated to its current operation. This type of strategy allows a firm to acquire businesses mainly for its profit potential. ITT, parent company of Sheraton, is a conglomerate following this type of strategy.

A firm chooses a corporate strategy depending upon its mission and the goals and objectives its management wish to accomplish. Every corporate strategy has its costs and benefits, but the main purpose for implementing a corporate strategy is to gain a greater return on investment for its stakeholders. This can be achieved through the co-alignment principle, which involves finding the perfect corporate strategy, environmental and **corporate structure** fit.

See also: culture, corporate

Further reading

Olsen, M.D., Tse, E. and West, J.J. (1999) *Strategic Management in the Hospitality Industry*, 2nd edn, New York: Van Nostrand Reinhold. (Deals with the concept of strategic management and strategies undergone at both domestic and international levels.)

ELIZA CHING-YICK TSE, HONG KONG

corporate structure

Corporate structures are the arrangement among people in an organisation in order to get the work done. They are the fundamental building blocks or construction of the organisation. They provide firms with an orderly system. Corporate structures define the lines of authority as well as provide channels for communication among the ranks in the corporation. They let the persons in the firm know how information will be travelled throughout the organisation. Corporate structures determine locus of control and responsibility at every level of the business.

Aspects of corporate structure include the size of the organisation, span of control and the type of hierarchy (flat or tall). The size of an organisation generally refers to the number of employees. As an organisation expands and grows in size, more levels of management are added to the firm. This results in a structure that is more formalised, complex and decentralised. Span of control determines how many levels of management an organisation will have. Another aspect of structure is the type of hierarchy an organisation will be directed under. An organisation has a tall hierarchy if it has many levels of management, while one with a flat hierarchy has few levels.

Organizations, both in tourism and other industries, can also be differentiated by the dimensions of structure in terms of formalisation, complexity and centralisation. Formalisation refers to the degree of how rules, procedures, policies and employee handbooks are used in an organisation. Complexity involves the division of labour based on the degree of knowledge and expertise in a company. Centralisation deals with the authority of decision making involving resource allocation. A firm which has solely the chief executive officer or just the top management making decisions (such as the general manager of a hotel) is said to be highly centralised. On the other hand, the authority is delegated in a decentralised firm where a wider span of managers or even employees are involved in making decisions at every level of the company.

The dimensions of structure can influence the different types of structure an organisation will select to achieve its mission and objectives. There are five structural types. The first is a simple or entrepreneurial structure that occurs when the owner or manager makes all the important decisions involving the day to day operations. All the employees report to this individual. Most 'mom and pop' or entrepreneurial tourism properties comprise of a simple structure. In the second type, a simple structure gives way to a functional structure. This is a result of a firm having to increasingly rely on expertise in the functional areas of management. Third, a divisional structure occurs in organisations that expand not only in size but also in their product/service offering as well as serving different regional markets. Fourth, if a company chooses to diversify its portfolio of businesses, then the corporate structures comprising of strategic business units would be most appropriate. Each unit represents a distinct business with its own set of budgets and, customer base and enjoys a certain degree of autonomy. Pepsi Cola and General Mills are among the examples of this type of structure; hotel chains may be regarded in the same way. Finally, a matrix structure features a dual authority in which subordinates report both functional and product/and or market executives. This type of structure allows for specialisation and ease in responding to the market more efficiently.

Organisations can choose the various dimensions of structure that it will embark upon, but generally speaking, the degree to which these dimensions are chosen enables the firm's overall structure to fall into two categories: mechanistic and organic structures. Characteristics of a mechanistic structure include high complexity, high formalisation, downward communication and little decision-making authority shared by the lower level members. In contrast, an organic structure is characterised by a low degree of formalisation and complexity. It has a comprehensive and intricate information network. This structure allows communication to move laterally and upward as well as downward throughout the organisation. There is also a high degree of participation among the employees in the decision-making process. Strategy and other forces such as the size of the firm, technology and environment affect whether a company follows a mechanistic or organic structure.

Effective corporate structures serve as mechanisms which enable tourism managers to meet their company's objectives by implementing their strategies successfully. Whether the strategy determines which type of structure to adopt or whether the structure defines what strategy one can pursue, it is pertinent for the company to align the corporate structure with its strategy. Every organisation has some form of structure. Corporate structures allow and define lines of authority. They help co-ordinate daily activities. It is also important to remember that not all corporate structures are static. Companies can be flexible enough to select an appropriate structure over the stages of the life cycle to better suit needs in a changing environment.

Further reading

Olsen, M., Tse, E. and West. J. (1999) *Strategic Management in the Hospitality Industry*, 2nd edn, New York: Van Nostrand Reinhold. (Analyses different types of corporate structure and its structural components.)

Robbins, S. (1993) *Organizational Behaviour*, Englewood Cliffs, NJ: Prentice Hall. (Discusses the importance of organisations, their concepts and applications.)

Rue, L. and Holland, P. (1989) *Strategic Management*, New York: McGraw-Hill. (Covers strategic management and how strategy and structure plays into the success of the company.)

ELIZA CHING-YICK TSE, CHINA

correspondence analysis

Correspondence analysis is an exploratory data analysis technique for the graphical display of contingency tables. As an analogue of **principal components analysis**, it is appropriate to discrete rather than to continuous variates. The input data typically is in the form of a cross-tab indicating association between the rows and the columns (that is, brand preference and gender or tourism preferences and family life cycle).

Correspondence analysis performs a scaling of rows and columns in corresponding units so that each item can be displayed graphically in the same low dimensional space. Basically, a big cell entree (frequent co-occurrence) in a cross tab will be transformed to a small physical distance between the corresponding row and column points in the derived Euclidean space (a two-dimensional map). Thus, if a market research study indicates that Disneyland is the preferred holiday destination by couples with pre-teen children, while Hawaii seems to be most popular among retired couples, the column point couples with children will be placed near the row point Disneyland, indicating strong association and farther away from row point Hawaii, symbolising weaker association (and vice versa). While **multidimensional scaling** typically requires subjects to provide similarities in information concerning items, correspondence analysis is much easier to apply on existing databases.

Historically, the methodological aspects of the analysis and its various aliases – canonical scoring, homogeneity analysis, reciprocal averaging and dual scaling – go back to the 1930s. Hence, while the technique has received considerable attention within the statistical and psychometrical literature for decades, the academic interest within marketing research only began in the mid-1980s. Still, according to Greenacre (1989), there are some disadvantages of using correspondence analysis. Since it is an exploratory data analysis, the technique is not suitable for hypothesis testing. Further, some academic concern deals with whether it is at all meaningful to interpret distance between row points and column points simultaneously. All major statistical computer packages today include programs for performing correspondence analysis.

See also: perceptual mapping

References

Greenacre, M.J. (1989) 'The Carroll-Green-Schaffer scaling in correspondence analysis: a theoretical and empirical appraisal', *Journal of Marketing Research* 26 (August): 358–65.

Further reading

Greenacre, M.J. (1993) *Correspondence Analysis in Practice*, London: Academic Press. (A nontechnical introduction by one of the leading scholars of correspondence analysis theory.)

Lebart, L., Morineau A. and Warwick, K.M. (1984) *Multivariate Descriptive Statistical Analysis: Correspondence Analysis and Related Techniques for Large Matrices*, New York: Wiley. (Provides the mathematical derivations of two-way and multiple correspondence analysis.)

MARCUS SCHMIDT, DENMARK

cosmopolitanism

The term cosmopolitan has ancient origins. While the Greeks separated the world out into those who were Greek and those who were barbarians, the Stoics argued that all human beings were citizens of the world and children of Zeus and, as such, cosmopolitans. With the crescendo of nationalism, global capitalism and the movements of money and people in the nineteenth century the term 'cosmopolitan' gained in currency in a usage which often seemed opposed to terms such as patriot. In tourism, the term sits challengingly between the realities and rhetorics of a world which is at once increasingly cosmopolitan – there are large scale permanent, semi-permanent and temporary migrations of workers, refugees and tourists – and, at the same time, increasingly nationalistic and patriotic. For both political leaders angling for the nationalist vote and tourist brochure writers, it may often seem that nationalist sites and icons have a greater power to attract the floating tourist than the more complex signs of human cosmopolitanism. In this context, as in others (see **nationalism**) tourism is seen to take part in some of the most urgent political and cultural issues of the day.

TOM SELWYN, UK

cost

The private cost of a **tourism** product is the sum of all payments made to obtain the inputs used in its production (such as **accommodation** or airlines). The full cost includes external costs, which may be economic, environmental, psychological, social or cultural, and frequently affect those not directly involved in the production or consumption of tourism and the value of free inputs (such as scenery).

ROBERT MAITLAND, UK

cost–benefit analysis *see* benefit–cost analysis

Costa Rica

The Central American nation of Costa Rica (population 3.5 million) features many attractions, with beaches on both the Caribbean and Pacific coasts, dense rainforests, volcanoes, waterfalls, and rafting and river cruising opportunities in its interior. Despite having only 19,730 square miles, Costa Rica boasts five per cent of the world's species. Due to these natural advantages, its **ecotourism** has soared in the last ten years. Recent tourism arrival rates have been growing at over 20 per cent a year, making tourism its leading industry. Costa Rica has political advantages as well. Its stable political environment, in contrast to those in the region, has contributed to this growth.

The country has also taken steps to protect its natural wonders while utilising them through tourism for economic development. Over 10 per cent of the country is covered by national parks, where development has been kept to a minimum. Outside the parks, logging and other activities threaten to erode the natural attractions of the country. In 1986 the government passed a major incentives act to encourage tourism **infrastructure** development. That began a period of tremendous growth, well beyond the government's capacity to monitor. Incentives have been cut back so that growth can become more orderly. The government owns none of the tourism infrastructure, and the single four-year term of the country's president does not encourage long-term initiatives. Still, there is recognition that some of the development, particularly at Playa Tambor resort on the Pacific Coast, has been ecologically unsound and in violation of laws.

Currently, the government is embarking on an ambitious effort to rate the hotels and resorts in terms of their social and environmental efforts. Energy-saving efforts, **recycling**, unobtrusive design and worker-friendly labour practices are factors considered. The country is taking the lead in this effort, with plans well underway to expand this practice throughout Central America. Costa Rica has taken several other measures to protect its touristic riches. Its present challenge is coping with the pace of arrival growth and the economic pressures to expand without deliberate consideration of the impact.

Further reading

Biesanz, M. and Richard, K. (1987) *The Costa Ricans*, Englewood Cliffs, NJ: Prentice Hall. (A study of the country's politics and culture.)

Head, S. and Heinzman, R. (1990) *Lessons of the Rainforest*, Sierra. (A collection of essays on the biology, ecology and economy.)

LINDA K. RICHTER, USA

Council on Hotel Restaurant and Institutional Education

The Council on Hotel Restaurant and Institutional Education (CHRIE) was founded in 1946 as a non-profit association for schools, colleges and universities offering programmes in hotel and restaurant management, food service management and culinary arts. CHRIE's mission is devoted toward promoting and facilitating exchanges of information, ideas, research, products and services related to education, training and resource development for hospitality and tourism industry. CHRIE promotes continuous development of individuals, education as the driver of quality, excellence in service, cooperation, coalitions and global networks and proactive change. It is governed by an elected board of directors and a professional staff. Currently the council has two federations and eighteen chapters around the world. Standing committees and special interest sections assist the board in planning and administering programmes and activities. Internationally, CHRIE advocates the global hospitality and tourism education through proactive professional development, research, coalitions and networks for its members and constituencies.

Membership in CHRIE is open to everyone interested in impacting the future of the hospitality and tourism industry through education and training. With more than 2,400 members from North America and worldwide, the organisation has become a network of hospitality and tourism educators. The majority of members are from industry, education and associations. Membership includes several benefits, ranging from conferences and meetings to publications. Its two quarterly journals are *Hospitality and Tourism Educator* and *Journal of Hospitality and Tourism Research*. CHRIE also publishes a *Guide to College Programs in Hospitality and Tourism*, *HOSTEUR Magazine* and *CHRIE Communiqué* and *Member Directory and Resource Guide*, which serves as a direct connection between educators and industry professionals. CHRIE's annual conference, held in the summer, provides networking opportunity for its members. The council is also involved in accreditation of programmes in hospitality field. The Accreditation Commission for Programs in the Hospitality Administration was started in 1989; since its inception, twenty-nine programmes have been accredited. The headquarters of CHRIE is located in Washington, DC.

MAHMOOD KHAN, USA

country house hotel

Country house hotels generally offer **hospitality** within the setting of a home of architectural character and/or historical interest. The location acts as a retreat, distancing guests from everyday pressures, while remaining in reach of primary centres of commerce, tourism facilities and attractions. They are often independently owned, with the product offering reflecting the unique individuality of the proprietor.

ALISON J. MORRISON, UK

countryside

The countryside became an **attraction** during the picturesque and romantic movements of the eighteenth and nineteenth centuries, as it became more accessible through a series of **transportation** revolutions. Since then, campaigns both to increase visitation and to protect the environment have proliferated. Most countries have **national parks** and other organisations dedicated to preserving rural **heritage**.

See also: agrotourism; farm tourism; rural tourism; wilderness

DAVID WILSON, UK

craft *see* handicrafts

creative tourism

Creative tourism is a specialised type of activity holiday whereby participants learn a new practical or intellectual skill. Examples include interior design, soft furnishing, special-effect painting and upholstery. Predominately a **short break**, creative tourism has grown in popularity since the 1980s as people appreciate the need for ongoing learning and use their leisure time both creatively and productively for personal development.

DEBORAH GRIEVE, UK

crime

Crime is one type of violent activity jeopardising the stability of tourism worldwide. The most documented crime is that committed against **tourists**. Two types are recognised: planned crimes (such as **terrorism**) where the perpetrator seeks to make a statement through the use of violence, and crimes of opportunity against an unknown victims in order to gain some form of gratification.

See also: security

YOEL MANSFELD, ISRAEL

critics

Critics of tourism became conspicuous with the emergence of **mass tourism**. Coming from various fields, they interpreted the phenomenon as a **pseudo-event**, a new form of **imperialism**, the loss of spiritual values and so on, even questioning tourism's benefits. In the last decade, criticism has tended to concentrate on specific negative aspects, such as destruction of the **environment**.

See also: anti-tourism

GIULI LIEBMAN PARRINELLO, ITALY

cross-cultural education

Cross-cultural education is education (college and workplace) which is designed to equip tourism employees with the necessary skills to cope effectively with a multicultural working environment. The tourism industry is a melting pot of guests and employees from diverse ethnic, cultural and social backgrounds. Sensitive preparation of staff for this diversity is a key responsibility of the educational process.

See also: cross-cultural management; cross-cultural study

Further reading

Baum, T. (1995) *Managing Human Resources in the European Tourism and Hospitality Industry: A Strategic Approach*, London: Chapman and Hall.

TOM BAUM, UK

cross-cultural management

Culture is a social mechanism that shapes and guides people's thoughts values and beliefs and ultimately controls their behaviour. It is 'the collective programming of the mind which distinguishes the members of one human group from another, the interactive aggregate of common characteristics that influence a human group's response to the environment' (Hofstede 1980).

Culture exists everywhere, at various levels of society, and everyone belongs to at least one, at the supranational level (Western and Eastern civilisations), at the national level (American, French, Japanese), at the ethnic level (Chinese and Malay in Malaysia, WASPS, Blacks and Hispanics in the USA), and so on. Culture can also be applied to other social units such as occupational group (lawyers, accountants, physicians), corporations (IBM, Shell, Disney) and even tourism sectors (**restaurants**, hotels, airlines).

One often-posed question about multinational corporations relates to the nature of the relationship between the cultures of corporations/industries and the prevailing national cultures. More specifically, the question is to what extent these corporate cultures are independent from national systems and can thus be implemented worldwide, or whether they are subjugated to national cultures and must be congruent with them.

Some researchers (Child and Tayeb 1983; Okechuku and Wai Man 1991) claim that organisational characteristics are similar, if not identical, across nations and for the most part are free from cultural dominance. Researchers identified with the 'convergence' school of thought argue that 'individuals, irrespective of culture are forced to adopt industrial attitudes such as nationalism, secularism, and mechanical time concerns in order to comply with the imperative of industrialisation'. These researchers would have probably argued that hotel organisational variance would depend more on other contingencies, such as size, technological development, geographical diversification and market segment, than on national culture.

On the other hand, those scholars who support the 'divergence' school of thought (Hofstede 1980; Laurent 1983; Lincoln *et al.* 1981) argue that organisations have always been, still are and will always be culture-bound rather than culture-free. Therefore, one should not expect to see any convergence in managerial practices, leadership styles or work attitudes across different cultures, since these are dependent on the implicit model of organisational functioning prevalent in a particular culture. Unlike their colleagues in the 'convergence' school of thought, these researchers would probably argue that despite a similarity in tasks, size and market segments, hotels operating in different cultures will differ in many of their managerial practices such as leadership styles, communication patterns and motivation techniques. These researchers believe that national and ethnic cultures are a major determinant of people's behaviour and when the culture of an industry, whether uni-national or multi-national corporation, is incongruent with the national or ethnic culture, the result would be failure.

The rapid globalisation of tourism has brought forward the question of whether its operations can be organised and managed in a uniform manner regardless of the country in which they are located. For example, can a multinational tourism corporation such as Sheraton or Disney design and organise its worldwide operations uniformly? Can it apply uniform personnel policies and procedures, and establish homogeneous managerial practices in all its operations, or do these have to vary by country and ethnic culture? In other words, what today multinational corporate managers are trying to ascertain is whether nationality and ethnic cultures have a determinant effect on organisational structure and/or behaviour.

Cross-cultural comparative analyses conducted over the last twenty years suggests that as far as organisational structure is concerned, influence of the culture is not immense. Experience shows that an 800-room hotel in Chicago is more similar in its organisational structure, division of labour, formalisation, and specialisation of tasks to an 800-room hotel in Tokyo, Bangkok or Athens, than to a 100-room hotel in Chicago. Such is the case despite the fact that the cultures of the United States, Japan, Thailand and Greece are different from each other. Indeed a casual visit to a McDonald's in New York, London, Singapore or Paris would suggest uniformity among them, including identical menu, front-of-house and back-of-house design and layout, along with uniform equipment, reporting and control mechanisms.

On the other hand, the numerous studies conducted in various industries and the handful of studies conducted in the tourism industry (Cross Cultural Hospitality Management 1997) showed that when it comes to people's behaviour in organisations, culture counts. Therefore when trying to import managerial practices such as leadership styles, communication patterns and motivation techniques from one country to another, problems arise. National and ethnic cultures are a major determinant of people's behaviour, and when the corporate culture of a hospitality company is incongruent with the national or ethnic culture, the result would be failure. Therefore, any attempt to create a culture-free or universal hospitality/tourism operation directed by managers who are 'world citizens' will not be successful.

References

Child, J. and Tayeb, M. (1983) 'Theoretical perspectives in cross-national organisational research', *International Studies of Management and Organisation* 7(3–4): 19–32.

Cross-Cultural Hospitality Management (1997) special issue of *International Journal of Hospitality Management* 16(2).

Hofstede, G. (1980) *Culture's Consequences: International Differences in Work-Related Values*, Beverly Hills, CA: Sage.

Laurent, A. (1983) 'The cultural diversity of western management conceptions', *International Studies of Management and Organisation* 8(1–2): 75–96.

Lincoln, J.R., Hanada, M. and Olson, J. (1981) 'Cultural orientations and individual reactions to organisations: a study of Japanese-owned firms', *Administrative Science Quarterly* 26: 93–115.

Okechuku, C. and Wai Man, V.Y. (1991) 'Comparison of managerial traits in Canada and Hong Kong', *Asia Pacific Journal of Management* 8(2): 223–35.

ABRAHAM PIZAM, USA

cross-cultural study

Cross-cultural studies consist of quantitative or qualitative comparative research on the behaviour and belief systems of people in different societies, most often in different countries (Dann *et al.* 1988). Cross-cultural comparisons can be universal, such an enquiry into whether peoples of all societies practice something like tourism or comparable phenomena, such as pilgrimages. Studies may be controlled by focusing on a few comparable cases, such as studies of the museum-visiting habits of people in European nations or of the management of tourism in East Asia. One can extend the notion of cross-cultural studies to various subcultures, including those of tourist behaviour by people in different regions or of classes within a complex national society.

Early tourism research was rarely cross-cultural. It often focused on the stereotypical situation of tourists from wealthier nations going 'south', to warmer regions often in marginal areas or Third World countries, especially for economic planning to increase employment and the inflow of foreign exchange. As the subject developed, cross-cultural studies of the differential impact of tourism on dissimilar kinds of host societies emerged (Smith 1976). Soon after, a new research emphasis recognised that the tourists themselves are members of varied types of societies including members of the Third World (Graburn 1983). Many special issues of *Annals of Tourism Research* have been controlled cross-cultural studies within particular areas such as Europe, Asia, the Pacific, centrally-planned economies or even Antarctica (1994), or cross-cultural comparisons of select topics such as ethnicity, tourist guides, the environment, gender and heritage.

Further reading

Dann, G.M.S., Nash, D. and Pearce, P. (eds) (1988) 'Methodological issues in tourism research,' special issue of *Annals of Tourism Research* 15(1). (A critique of methodological naiveté in research, with examples of qualitative, quantitative and cross-cultural studies.)

Graburn, N.H.H. (ed.) (1983) 'The anthropology of tourism', special issue of *Annals of Tourism Research* 10(1). (Case studies of the motivations and behaviours of tourists from several countries in a comparative framework.)

Smith, V. (ed.) (1976/1989) *Hosts and Guests: The Anthropology of Tourism*, Philadelphia, PA: University of Pennsylvania Press. (Comparative introductory chapters, followed by case studies of the impacts of tourism on minority and majority peoples in different parts of the world.)

NELSON H.H. GRABURN, USA

cross-training

Cross-training is provided outside one's normal workplace. This may occur between different departments in the same hotel, or different hotels in the same group, but it also refers to staff (often employees of a **multinational firm**) being sent to experience working in hotels in other countries or regions (for example, from China to France).

YU MING XI, HONG KONG

crowding

Crowding can occur in all scales of tourism environment, from a **hotel** or beach to a **resort** or **landscape**. It refers to a state whereby the numbers of tourists has exceeded a psychological threshold (see **carrying capacity**). It is perceived by the participants as a heightened awareness of the other tourists in the same space, and a consequentially reduced sense of pleasure.

MARCUS GRANT, UK

cruise line

The cruise line sector was born as a response to the tremendous competition commercial air travel caused to the traditional passenger liners in the 1960s. The concept of a 'floating resort' replaced that of point-to-point **transportation**. A cruise is now defined as a sea or river trip which includes a minimum of three nights spent aboard, emphasising **animation** rather than transportation. Even shorter one-night cruises on 'floating playgrounds' are now very popular in the Baltic Sea; over 15 million departures and arrivals of sea passengers were recorded in **Finland** in 1997, for example.

The cruise business has grown very rapidly, with double-digit **growth** rates in the 1970s and 1980s. Cruises are most popular in North America, accounting for 77 per cent of the world **demand** in 1995. The European **market** is catching up, with the cruise holiday being most popular in the United Kingdom, Germany, Italy and France. In the rest of the world, Japan and Australia are the dominant markets. An estimated 6 million North Americans and 706,000 British people went on cruises in 1999, representing average growth rates of 9.3 and 26.9 per cent respectively since 1995. Demand is also changing qualitatively with shorter, thematic cruises and more varied destinations. There is a socio-demographic shift from older and wealthier to younger, middle-class and married (with children) customers. Satisfaction, resulting in repeat cruising, is one of the highest in the tourism industry.

The number of ships of 1,000 gross tons (a measure of size) and above increased from 132 in

1981 to 147 in 1985 and 170 in 1990. In 1995, cruise supply consisted of 245 ships carrying 188,000 berths and thirty new vessels to be delivered by 2000 with 53,000 new berths. Although several older (and smaller) ships will be deleted from the fleet due to new regulations, scrapping or casualties, overcapacity is now a threat since the new ships are often much larger. The highly competitive environment resulting in lower fares is leading to **mergers** or joint ventures to maintain profitability and viability.

References

G.P. Wild Ltd (1996) 'State of the cruise industry and future development to the year 2000', *International Cruise and Ferry Review* Autumn/ Winter: 35–40.

ALAIN DECROP, BELGIUM

Cruise Lines International Association

The primary objective of the Cruise Lines International Association (CLIA) is to assist its over 22,000 affiliated agencies to become more successful in capitalising on the cruise market. Currently, the Association has twenty-four member lines which represent virtually 100 per cent of the cruise industry. Over 95 per cent of their business is generated by travel agents, a partnership which is stressed by CLIA. But the Association is prohibited from involvement with any operational or marketing activities of individual member lines. It does not serve as an industry ombudsman, nor is it involved with any lobbying activities. CLIA operates under by-laws established with the Federal Maritime Commission and developed with the support of ASTA, ARTA, AAA and ACTA. Unlike other organisations, CLIA maintains regular dialogue with all the major trade associations. Through joint meetings held between its Executive Committee and appropriate associations, it provides a forum for discussing important issues facing the cruise industry.

TURGUT VAR, USA

Cuba

Tourism development in Cuba has been heavily influenced by the country's changing relationships with global superpowers. Following independence from Spain (1898), neo-colonial economic links with the United States grew. Cuba produced sugar for the American market and acted as the 'winter playground' for wealthy US citizens. The tourism industry was focused on Havana and the beach resort of Varadero, emphasising 'sea, sand, sun and sex', with gaming also important. Like the Cuban economy as a whole, tourism was heavily dependent upon US **foreign investment**. By 1958, on the eve of the downfall of the Batista regime, tourist numbers had reached 350,000 and tourism ranked second to sugar as a **foreign exchange** earner.

By the early 1960s, Fidel Castro's revolutionary government had become avowedly Marxist and was building close ties with the USSR. In response, the United States declared an economic embargo against Cuba which continues to impose travel restrictions to this day. Castro's active discouragement of international tourism and promotion of domestic tourism led overseas arrivals to fall to 10,000 by 1974. In 1976 the industry was nurtured to life again with the creation of the Soviet-inspired Intur, a state body designed to attract, package and oversee international tourists. The 1980s saw further expansion with the state establishing a number of enclave beach resorts. By 1990 arrivals had almost returned to their pre-revolution peak.

The collapse of the USSR and the tightening of the US embargo led to a dramatic fall in Cuban living standards during the early 1990s. This prompted some liberalisation of the economy and **mass tourism** was placed at the centre of attempts to attract overseas capital and to generate hard currency. In 1996 over one million tourists arrived, primarily from Canada, Western Europe and South America, making tourism the leading foreign exchange earner.

The government is trying to diversify the tourism product and promote more sustainable use of Cuba's natural resources, by turning to **alternative tourism**. Cuba boasts a variety of environmental and cultural features that are of great touristic value. In many cases, areas of natural beauty lie close to towns with a rich architectural **heritage**. The government has gone to great pains to document and protect many of these resources, with a system of wildlife refuges, **national parks** and UN Biosphere Reserves. The challenge facing Cuba is how to move beyond its image as a low-priced Caribbean beach destination and make greater use of its rich and unique resources in a sustainable and appropriate manner that will meet the future needs of the Cuban people.

Further reading

Simon, F.L. (1995) 'Tourism development in transitional economies: the Cuban case', *The Columbia Journal of World Business* 30(1): 26–40.

SIMON MILNE, NEW ZEALAND
GORDON EWING, CANADA

cuisine

Cuisine can refer to a style of cooking, the manner of **food** preparation or the food prepared. The word, originally from French, refers to all of the above meanings but has the additional implication of a kitchen or place of food preparation.

Cuisine has been a major preoccupation of many civilisations and became highly developed over the millennia. China, and later Italy and France, are widely cited as having developed cuisine to its most sophisticated artistic levels through the mid-twentieth century. While other societies placed significant emphasis on the advancement of their indigenous cuisines, none have received the international recognition bestowed upon these three ethnic culinary movements. Some scholars argue that culinary customs often change between very close neighbouring regions and that the term cuisine should not be applied to vast geographic areas characterised by diverse consumption habits. Hence, they contend that while Shantung, Kweichow and Nanking each have distinctive cuisines, they are heterogeneous as a group and should not be referred to as Chinese cuisine. Various ethnic cuisines form a major **attraction** as tourists are often interested in cultural varieties in food, and cuisine can thus be used in promotion (see also **promotion, place**). Some areas of the world, like France and Thailand, are popular destinations also because of their cuisine.

In the 1920s and 1930s, the term 'haute cuisine' (literally, high cooking) was widely used throughout the Western world in reference to the elaborate and artful French cooking developed and practised during this period. Forty years later, the term 'nouvelle cuisine' (new cooking) was applied to describe the most significant turn in French culinary history. While nouvelle cuisine shocked and excited the culinary world in its most radical movement, it was also short-lived. By the early 1980s the great chefs had returned to a more conservative practice, influenced, however, by their daring experiment.

The French experience with nouvelle cuisine had a global ripple effect, spawning new cuisines from Hungary to Australia. In France there were also offshoots in 'cuisine minceur' (fine cooking), 'cuisine naturelle' (natural cooking) and 'cuisine de marché' (market cookery), all created by famous chefs. However, most gastronomes (see **gastronomy**) and scholars agree that, while there has not been a single, unified culinary movement of the nouvelle cuisine magnitude since its demise, the culinary arts of the 1990s have greatly benefited from the creativity and innovation resulting from it. Haute cuisine also features as an attraction in tourism in the form of culinary or 'wine and dine' trips and a good culinary reputation of a region or a city may have significant importance in destination choice.

MICHAEL NOWLISS, FRANCE

cultural conservation

Culture is a word that has rather different meaning in philosophy, aesthetics, literary criticism, **anthropology**, **sociology** and elsewhere. Indeed, Raymond Williams considers it to be one of the most complicated words in the English language because of its diverse usages in distinct systems of thought (1983: 87). To Kroeber and Kluckhohn, the culture of a people comprises their patterns of behaviour, and their particular achievements inclusive of their artefacts, their ideas, and their values; to others the culture of a population is the peculiar or unique way of life of that population in terms of its mores, its customs, and its explicit and implied design for living (1963: 181).

While the culture of some small-scale, tribal or isolated populations may be relatively homogenous, the culture (or rather cultures) of complex societies are never invariable, and they always change over time. Hence it is important to monitor the rates of diffusion, **acculturation**, evolution and **development** that are relatively transformative and enduring. Tourism managers/researchers working with a given population may recognise that what is significant (that is, viewable or protectable) about lands or **heritage** may change occasionally.

While preservation invokes the effort to save, restore and continue customs (particularly of built representations of culture), conservation is the effort which ensures the stability of present inheritances by maintaining them against interference of some kind in the future. Thus under conservation, unlike preservation, the inherited elements or the symbolic features of a given culture may be altered in use or sanction while they are marked, cared for and sustained for the generations to come. Where cultural conservation involves the identification/protection of a cultural area, it is important to identify associative and cultural landscapes, where the former have powerful religious, artistic or cultural associations with or without material cultural evidence. It is always critical to resolve what is being protected, for whom and for what purpose. Sometimes there may be nothing that is readily viewable by tourists; sometimes conservation purposes may clash with that site's other use values; sometimes a local population may want the whole of an identified territory rendered 'sacred' (or 'conserved'), not just a single site.

The recent history of tourism and conservation are closely connected. The conservation movement around the world is fired by the moral crusade to maintain living diversity by conserving not only the biological **wilderness** of places supposedly uncontaminated by the physical vestiges of humanity, but it seeks to maintain the maximum diversity of cultural heritage by helping retain the supposed **authenticity** of the **traditions** and customs of people, and by obviating harmful or unwanted cultural impacts. It is then those pristine places, and those untouched peoples, that constitute a very strong appeal in tourism, as increasing volumes of travellers seek to explore their vision of the **other**.

In tourism research, Marie-Franhoise Lanfant (1995) considers that there is currently quite a poor understanding of the complex relationships that surround cultural conservation and of the related processes involving the displacement of the local, the disruption of systems of reference and the endowment of heritage with new interpretations. To Lanfant, the local cannot be meaningfully understood independent of the global, and a new truly international sociology of tourism is needed to probe the processes of mirroring, reflexivity and transitivity that arise in matters of culture survival, conservation and identity. Too frequently, to Lanfant, efforts to measure or monitor cultural change and cultural conservation have been dominated by the study of social and/or cultural impacts, which have positioned tourism only as an exogenous force. To Lanfant, tourism has been repeatedly and inadequately envisaged as a force of social change which arrives unidirectionally to destroy local societal integrity and identity within given cultural areas: to her, while tourism has itself been assumed to be the acting vehicle of change or of conservation, the rightful status of local actors in deliberately inviting tourism or exploiting it for their own ends has been grossly under assessed.

There is now a fascinating medley of non-indigenous bodies like Tourism Concern (based in London) and Cultural Survival (based in Cambridge, Massachusetts) which project strong support for **indigenous** peoples in their efforts to conserve their life-ways and stand up to the claimed encroachments of tourism and related industries. Not all of such organisations have been welcomed by removed populations in isolated/undeveloped parts of the world, however, for the ideological stances they adopt are quite varied.

Hollinshead (1996) has attempted to clarify this ideological nightmare in and around cultural conservation by adapting Bodley's applied anthropological continuum of certain political orientations of such caring/activist organisations towards tourism. There are three principal perspectives on or over primal peoples. First, *primitive-environmentalist outlooks* tend to be the deepest philosophical vista on indigenous self determination, and regard tribal cultures as a superior adaptation, seeking to stop economic developments that threaten such peoples. This outlook promotes the provision of cultural/environmental sanctuaries, and strongly advocates conservation. Tourism is seen as a powerful and damaging agent of Western/non-indigenous infiltration. Second, *liberal-political outlooks* tend to regard indigenous people as economically and politically oppressed, and seek to help them defend their cultural integrity through liberation and self-determination. This outlook promotes political mobilisation and mutual consciousness-raising about the plight of the indigenous people. It holds that tourism proposals should only be encouraged where they have received extremely careful scrutiny, and where there is substantive local community involvement and local rights-of-veto. Third, *conservative-humanist outlooks* tend to regard progress as inevitable, and seek to help the local population make the most of their new opportunities. This outlook promotes eventual integration of the indigenous population into the wider national system, via humanitarian assistance programmes, provided that they can help 'preserve' ethnic identity and pride. Tourism is still seen to be not only a potentially disruptive force but also capable of yielding many rewards for the primal population. These three stances on cultural conservation/development are rather idealised, and each has its army of detractors.

References

Hollinshead, K. (1996) 'Marketing and metaphysical realism: the disidentification of Aboriginal life and traditions through tourism', in R. Butler and T. Hinch (eds), *Tourism and Indigenous Peoples*, London: International Thomson Business Press, 308–48.

Kroeber, A.L. and Kluckhohn, C. (1963) *Culture: A Critical Review of Concepts and Definitions*, New York: Vantage Books.

Lanfant, M.-F. (1995) 'Introduction', in M.-F. Lanfant, J.B. Allcock and E.M. Bruner (eds), *International Tourism: Identity and Change*, London: Sage, 1–23.

Williams, R. (1983) *Keywords*, London: Fontana.

KEITH HOLLINSHEAD, UK

cultural survival

Cultural survival is a term indicating **Fourth World**, **indigenous** peoples' struggles to maintain their own societies and cultural identities while embedded in nation-states and global economies. **Ethnic tourism** or **cultural tourism** and **conservation** in **protected areas** are sometimes avenues toward cultural survival if tourist **commoditisation** is under the control of the peoples themselves.

MARGARET B. SWAIN, USA

cultural tourism

When Herodotus of Halicarnassus first set eyes on the Pyramids 2,700 years ago, he was a cultural tourist. His comments regarding the graffiti on the monuments and the bevy of guides available for hire indicates that he was far from the first **tourist** to visit these monuments and try to gain some understanding about the people who built them. For much of written history, what is now considered tourism was related to humankind's insatiable curiosity. The desire to travel to learn about other people and their culture has always been an essential motivator. Hunziker and Krapf expressed this well when they pointed out, 'There is no tourism without culture' (World Tourism Organization 1995: 6). It was only with the arrival of **mass tourism** a scant 150 years ago that tourism took on its modern hedonistic and less enlightened complexion.

Defining what cultural tourism constitutes is a continuing debate. As Williams points out, 'culture is one of the two or three most complicated words in the English language' (1983: 87). This complexity is compounded by prefixing it with the word tourism. As a result, there is no shortage of attempts to define this phenomena (World Tourism Organization 1995). The majority of definitions suggest learning about others and their way of life as a major element. Learning about self is a second common thread that runs through many explanations on cultural tourism. One of the most elegant of these humanistic definitions is expressed by Adams simply as 'travel for personal enrichment' (1995: 32). Kneasfsey reflect and expands on this idea of a search for knowledge (1994: 105). Further, they express the ideas of development, presentation and **interpretation** of cultural resources as an essential element of tourism. This expression of cultural tourism as an industry is an important aspect of the modern understanding of it.

In short, cultural tourism can be defined broadly as the commercialised manifestation of the human desiring to see how others live. It is based on satisfying the demand of the curious tourist to see other peoples in their 'authentic' environment and to view the physical manifestations of their lives as expressed in arts and crafts, **music**, literature, dance, food and drink, **play**, handicrafts, language and **ritual**.

References

Adams, G.D. (1995) 'Cultural tourism: the arrival of the intelligent traveler', *Museum News*, December: 32–5.

Kneasfsey, M. (1994) 'The cultural tourism: patron saint of Ireland?', in U. Kockle (ed.), *Culture, Tourism and Development*, Liverpool: Liverpool University Press.

Williams, R. (1983) *Keywords: A Vocabulary of Culture and Society*, London: Fontana.

World Tourism Organization (1995) *Report of the Secretary General of the General Programme of the Work for the Period 1984–1985*, Madrid: World Tourism Organization.

KEITH DEWAR, NEW ZEALAND

culture *see* anthropology; cross-cultural study; culture, corporate; ethnography; sociology

culture broker

The term 'culture broker' refers to a person who is a middleman or a mediator between the **destination** culture and the **tourist** culture. This broker is a go-between who usually assumes the role of explaining or selling the indigenous culture to those visiting the destination. A tour guide, for example, would be a culture broker, as a person who escorts tourists to the various villages and sites along the tour **itinerary**, and who interprets or explains what they have seen. A 'culture broker', in brief, describes one culture for the benefit of the members of another (Cohen 1985).

Tour guides are not the only form of culture broker. A similar function may be performed by local intellectuals who write books or articles about the traditional culture for visitors, or by local businessmen who conduct seminars or consult for foreign companies on how to do business in a culturally alien setting. Further, an English anthropologist, for example, may write an ethnographic account of an African culture for consumption by an English or a Western audience, and in that sense the writer becomes a form of culture broker who takes on the task of explaining African culture for other English people. Journalists, artists, photographers and travel writers may serve in a similar capacity.

Tour agencies and governmental bureaus who select the sites that tourists will visit and who design the itinerary of the tour are indeed culture brokers, ones who have considerable power. They not only select which domestic attractions the tourists will visit and which sites will be excluded, but by their overall design of the tour they play a large part in constructing what the tourists will see of the host country, and hence the impression of the country that the tourists bring home with them.

A tour agency or a governmental tourist bureau may conduct a course or provide guidelines on how the local guides should conduct themselves when dealing with select groups of tourists, such as those from Japan, Italy or Germany. On these occasions the bureau is a culture broker, assuming the role of explaining the tourist culture to the local interpreters of indigenous culture. One may even think of the process as double-ended or open-ended, for no one has absolute knowledge of a culture and there is always the possibility of misunderstanding and ambiguity. Foreign tourists may be trying to figure out the local culture with the help of domestic tour guides, but the guides themselves may be trying to understand the strange ways of the foreign tourists.

See also: anthropology; entrepreneurship; ethnography

References

Cohen, E. (1985) 'Tourist guides: pathfinders, mediators, and animators', *Annals of Tourism Research* 12(1).

EDWARD M. BRUNER, USA

culture change *see* sociocultural change

culture shock

Culture shock is defined as a special kind of anxiety and stress experienced by people who enter a culture different from their own. Culture shock includes two kinds of problems: being confused, anxious and puzzled by the way others behave, and doing the same to others by behaving in one's own way. The perspective one uses for understanding events, for judging another's behaviour or for deciding one's own behaviour is largely defined by one's cultural background. This is learned from the earliest days of life, with lessons reinforced by similar patterns of behaviour observed daily at home. When one enters a new society, however, these old ways of perceiving may not work as well. As a result, some conflicts and confusion lead to culture shock.

A review of literature in tourism indicates that culture shock may derive from several causes, such as the loss of familiar cues in one's ability to interact with people of different cultural backgrounds, the language differences, the salience of the cultural differences and non-verbal communication difficulties with the local communities. Not only tourists experience the shock; the host population can also be stressed by contacts with the tourist culture (see **culture, tourism**). Culture exchange is a more acceptable model for resource maintenance, resource wealth and intergenerational equity.

See also: cross-cultural study; cultural survival; cultural tourism

Further reading

Furnham, A. (1984) 'Tourism and culture shock', *Annals of Tourism Research* 11(1): 41–57.

Pearce, P.L. (1993) 'From culture shock to culture exchange: the agenda for human resource development in cross-cultural interaction', paper presented at Global Action to Global Change: A PATA/WTO human resources for tourism conference, October, Bali, Indonesia.

Oberg, K. (1979) 'Culture shock and the problem of adjustment in new cultural environments', in Smith ll. Luce's *Toward Internationalism*, New Bury: House Publishers, 43–5.

Y.J. EDWARD KIM, AUSTRALIA

culture, corporate

Culture is defined in varying terms by different theorists. A popular definition is the one provided by Schein (1985: 9):

> A pattern of basic assumptions – invented, discovered, or developed by a given group as it learns to cope with the problems of external adaptation and integral integration – that has worked well enough to be considered valid and, therefore, to be taught to new members as the correct way to perceive, think, and feel in relation to those problems.

The terms organisational culture and corporate culture are used interchangeably in the literature to refer to the culture pervading an organisation, be it a for-profit or a non-profit entity. Corporate culture is invariably influenced by natural or regional cultures, or macro-cultures (Hampden-Turner 1994: 12). The emphasis on individualism in US corporations, the emphasis on collective decision making in Japanese firms, and the age old *burra sahib* culture of British companies in colonial days are all reflections of the influence of macro-cultures on corporate culture.

Corporate culture is not entirely monolithic or uniform throughout the organisation. Within the overall culture, there are usually subcultures. For example, marketing and accounts personnel may have different subcultures due to the differences in the nature of their work and environment. The conflicts which sometimes arise among departments can be attributed to such differences. In a similar but broader vein there is a hotel culture, which in turn takes a different hold when the corporate culture of the Hilton chain is compared to that of Sheraton. Similarly, airlines still have a different corporate culture; that of TWA and Singapore airlines are not the same.

Researchers agree that corporate culture operates at three levels of varying visibility. The first level includes audible and visible patterns of behaviour, technology and art. The second level includes the individuals' values. At the third level are the basic ideas and assumptions of individuals that affect their behaviour (Schein 1981). Since most such ideas, assumptions and values are difficult to articulate, corporate culture can mostly be inferred by studying such external manifestations as shared things, sayings, activities and feelings.

Culture has several components: beliefs, expectations and shared values, which together drive the organisation towards its goals; heroes and heroines who serve as role models and external symbols; myths and stories about key people and incidents; **rituals** and ceremonies that bond the organisation's members; and the physical arrangements of buildings, spaces, interior decor and so on (Gordon 1993: 171–5).

Can corporate cultures be classified? Several researchers have addressed this question. Among the better known of these attempts are the Deal and Kennedy model (1982) and the Harrison (1972) and Handy (1987) models. The first one shows that corporate cultures are influenced by the amount of **risk** a company has to take in its operations and the speed of feedback on the outcome of such risk taken. In this framework, firms are classified into high risk–slow feedback (for example, luxury hotels), high risk–fast feedback (such as motion picture studios), low risk–slow feedback (such as insurance providers) and low risk–fast feedback (such as **restaurants**) organisations. The Harrison and Handy models, though using varying descriptor terms, are essentially similar in conceptualisation. Both these researchers classify firms into four quadrants based on the degree (high or low) of formalisation and centralisation.

An interesting alternative view is based on the notion that 'culture's main function is to try to mediate dilemmas' (Hampden-Turner 1994: 24). Organisations constantly face dilemmas: standardisation versus customisation, external adaptation versus internal integration, and so on. Corporate cultures must change over time as organisations adapt to the ever changing environment. Yet, a cultural continuity must be maintained in order not to destabilise the organisation. This in itself is a dilemma. How well organisations avoid such two-horned dilemmas by finding synergistic solutions determines whether they succeed or not. It is the organisational leader's responsibility to build and maintain a culture that reconciles such dilemmas (Hampden-Turner 1994: 24–33).

Corporate cultures are often categorised as strong or weak, depending upon how much the employees share the organisation's core values and basic philosophies, usually laid down by either the founder or the current CEO. Well-known examples of firms with strong corporate cultures include Disney and McDonald's. Significantly, corporate cultures cannot be created overnight. They evolve over time. Building a cohesive organisational culture depends upon four conditions: developing a sense of history, creating a sense of oneness, promoting a sense of membership and increasing exchange among members. Fostering these conditions can be accomplished by elaborating on history, communications on and by heroes and others, leadership and role modelling, communicating norms and values, instituting appropriate reward systems, career management and job security, recruiting and staffing, socialisation of new staff members, training and development, member contact, participative decision making, inter-group co-ordination and personal exchange (Gross and Shichman 1987).

However, culture tends to perpetuate itself. Hiring and socialising members who fit in with existing culture, removing those who deviate from it, justifying behaviour under current cultural norms and communications biased towards current culture are some of the ways in which an organisation perpetuates its culture. If a cultural change is to be achieved, assuming that the organisation desires such a change in the first place, managers must intervene at these points.

References

Deal, T.E. and Kennedy, A.A. (1982) *Corporate Culture: The Rites and Rituals of Corporate Life*, Reading, MA: Addison-Wesley.

Gross, W. and Shichman, S. (1987) 'How to grow an organisational culture?', *Personnel*, September, 5256.

Handy, C. (1987) *The Gods of Management*, London: Souvenir Press.

Harrison, R. (1972) 'Understanding your organisation's character', *Harvard Business Review* May–June: 119–28.

Schein, E.H. (1985) *Organizational Culture and Leadership*, San Francisco: Jossey-Bass.

BVSAN MURTHY, SINGAPORE

culture, invention of

Invention of culture conveys the dynamic theoretical perspective which sees culture as constantly in process, as defined and redefined in new contexts. It is opposed to a more traditional conception of culture as fixed and stable. In tourism, the local culture presented for touristic consumption is frequently not 'timeless', as sometimes stated in brochures, but is usually recently created to satisfy the preconceptions of modern tourists.

EDWARD M. BRUNER, USA

culture, organisational

Organisational culture is the summation of the beliefs, expectations, norms and values commonly shared by all the organisational members. Used interchangeably with the term corporate culture, the term reflects the basic philosophy and core values (such as the quality, service, cleanliness and value ethos of McDonald's) of an organisation. Within an overall organisational culture, there are usually subcultures reflecting the different environments and experiences shared by groups of employees.

See also: culture, corporate

BVSAN MURTHY, SINGAPORE

culture, tourism

Tourism or tourist culture is a vague concept that describes behaviours and institutions which can be observed at tourism destinations but which are not straightforwardly parts of the cultures of either the host society or the visiting tourists. Initially a negative concept, it is defined by what it is not, rather than having common content wherever found. If a genuine culture belongs to a particular people, an independent community of a self-sustaining way of life, tourist culture is only a part-culture, a symbiotic or hybrid form incapable of sustaining itself. However, hybrid or part-cultures are becoming the global norm in this increasingly interconnected world, where there are few truly isolated populations or independent communities. Tourist culture, from this point of view, is one example typical of the emerging mix of cultures resulting from **internationalisation**, that is, the transcultural migrations of labour, capital, technology, ideology and images.

The exact nature of these new tourist cultures becomes clearer if one uses an analogy with the already well-explored variety of tourist arts, which are one aspect of this culture. It has been shown that the various hybrid art forms that result from or cater to tourism can be analysed by the intersection of two variables. One is the formal source or origin of the new cultural forms, ranging from the host's local culture to the tourists' imported culture(s), or more or less complete novelties emerging at the contact site. The second variable relates to the intended audience of the new cultural production, either the tourists or the hosts (or both). A few well-known cases can illustrate possible forms of tourist culture and arts.

The host's traditional cultural productions are often modified for touristic consumption, typically for ethnic or cultural tourism. Examples range from dances in Bali or shadow plays in Java, shortened and simplified for this purpose; Chinese

cuisine modified in **restaurants** in San Francisco, Beijing or Vancouver; sarongs in India and kimono in Japan; or even **bed and breakfast** as commercialised forms of traditional **hospitality** in many parts of the world. Some tourists take what they have learned in tastes or dress, **music** or cuisine, and seek them out back home for continued use. The consumption of the host's traditional local culture by the hosts themselves, as modified by the tourist environment, could be seen in the many art and music festivals timed to attract tourists, in religious events and **pilgrimages** modified to accommodate sightseers or non-believers, or the host's enjoyment of the kinds of ethnic restaurant mentioned here.

As such, tourism is a powerful force in exposing and modifying local and traditional culture for the '**gaze**' (which includes aural and lingual consumption). It thrives on forms of **heritage** such as **archaeology**, building and **landscape** restoration, and local festivals. But, in ethnic and **cultural tourism**, it is a conservative and nostalgic force which tends to freeze traditions in stereotypically recognisable forms (see **nostalgia**). Tourism is not yet all-powerful. Such attractions as the Palio horse race in Siena and the Changing of the Guard at Buckingham Palace would probably take place unchanged whether tourists were present or no, and the hajj, the sacred pilgrimage to Mecca, specifically excludes non-believers or sightseers.

At the opposite end of the spectrum is the production of tourist culture in the forms of behaviours and institutions which stem directly from the tourist's home culture, suitably modified for the touristic site. Here typically this culture consists of institutions such as hotels, restaurants, cafes, cleanliness systems, financial systems, shopping, daily routines, tastes in clothing, entertainment and mores, among other things. These aspects of tourist culture, typically found in **enclave tourism**, tourism **ghettos** or **resorts** in the **Third World**, stem mostly from extensions of European and North American institutional cultures and personal tastes, even while they may appropriate superficial aspects of the host culture such as local spices, decor, textiles and colour schemes.

This complex tourist culture has become 'international culture', and is not or was not the familiar home culture of many non-Western tourists such as East and Southeast Asians, who also have had to learn to use such devices as credit cards, travellers cheques, 'French' menus and genres of wine, restaurant table settings, bikini bathing suits, off-the-floor beds and sit-on toilets. However, once learned, these components of international and, ultimately Western, tourist culture take firmer roots in everyday living. Further, these imported aspects of tourist culture are novel to the host peoples. They are the forces behind the often lamented **demonstration effect** by which local behaviour and tastes are assimilated to the outside world. Though older people may revile tourist culture, the younger generations of locals may emulate it and notice that the upper and middle classes of their own society have already taken up this 'international' (Western) style.

Most interesting are novel aspects of tourist culture which relate more distinctly to either the host or guest populations. New forms of entertainment, clothing, cuisine, souvenirs or 'pidgin' speech may emerge from the cultural mix. Obviously tourist guides, guidebooks and **model** cultures (such as Ancient City near Bangkok, the Polynesian Cultural Centre near Honolulu or the various Disneylands) and other forms of **staged authenticity** would not exist without tourism. Whole tourism market areas such as Malioboro Street in Yogyakarta or the Zona Rosa in Mexico City with their betjak drivers, touts and pickpockets have grown up in response to tourism; certain kinds of **prostitution** and pornographic shows (see **pornography**) cater to tourist tastes, though local consumption is the usual origin. In a nutshell, tourist culture encompasses hybrid generative forces which preserve some cultures, while promoting cross-fertilisation and novel forms in others.

Further reading

Graburn, N.H.H. (ed.) (1976) *Ethnic and Tourist Arts: Cultural Expressions from the Fourth World*, Berkeley, CA: University of California Press. (Covers analysis and case studies of the touristic stimulation, production, trade and consumption of the arts and crafts of Native peoples around the world.)

Jafari, J. (1987) 'Tourism models: the sociocultural aspects', *Tourism Management* 8(2): 151–9. (Discusses factors which both constitute and influence the tourist culture.)

Salamone, F. (1997) 'Authenticity in tourism: the San Angel Inns', *Annals of Tourism Research* 24(2): 305–21. (Discusses the situated nature of authenticity and the pressure to conform to stereotypes.)

NELSON H.H. GRABURN, USA

curative tourism

Curative tourism includes activities and destinations which provide physical or spiritual self-improvement and well-being opportunities to tourists. Curative **resorts** offer natural resources such as mineral springs and muds, thermal baths and inviting **climate**. Their special programmes feature proper nutrition, fasting, hydrotherapy, electrotherapy, inhalation, solarium, sunshine, pampering, cosmetics and meditation.

HANAN KATTARA, EGYPT

currency control

Currency controls are a set of restrictions on the convertibility or the international transferability, or both, of a country's national currency. The purpose of these controls is to manage the country's **balance of payments**. The extreme case found in some countries (mainly those of the **Third World** or with centrally planned economies, many highly dependent on earnings from international tourism) is a national currency that is not convertible or traded in **foreign exchange** markets, and which can be transferred in and out of the country only with the authorisation of its central bank.

SIMON ARCHER, UK

curriculum design

Curricula design in tourism, similar to those in other fields, involves preparation of an entire range of studies to benefit the student toward specific objectives. A curriculum is usually organised jointly by faculty and administrators and evolves over time. Because tourism is a complex phenomenon, curricula vary greatly among institutions.

Most tourism curricula of today deal with the hospitality industry, such as **management** principles relating to various tourism businesses, including accommodations, **food** and beverage services, **car rental** companies, **travel agencies**, convention operations, commercial attractions, and transportation. Curricula sometimes distinguish between **training** and **education**. The former is directed toward specific skill objectives needed for employment. For example, trainees for food operation are given educational instruction and skill development in cooking processes, refrigeration, equipment and food storage. Frequently offered at specialised high schools, community colleges, culinary institutes and technical schools, it is highly structured and often leads to certification.

In contrast, an education curriculum is designed for students who seek a degree from a higher level institution such as a university. Although occupational objectives, such as hotel management, frequently dominate such degree programmes, basic education in the classical sense, including courses in humanities and sciences, is usually an essential part of it. Professionals consider education as broader than training.

Education usually encompasses long-range objectives for human environment rather than skills. Such hotel and restaurant schools often include a mix of education and training in their curricula. For example, in addition to courses in management, food service, housekeeping, marketing and front desk operation, the curriculum may include courses in social sciences, communication, interpersonal relations, planning and design, law and foreign languages. Those designed at the masters and doctoral levels provide curricula for future researchers and instructors of tourism. In many universities, elements of tourism curricula are also incorporated into related programs, such as those in leisure, recreation, parks and resources development. The purpose is to provide students with understandings of the interrelationships among these related fields and tourism.

Specialists in curriculum design often regard the philosophy of an educational institution as important, especially for administrative support and funding. If the philosophy and reputation are allied strongly to a liberal arts education, direct vocational training in tourism is not likely. But if the institution is strong in its relevance to operation, then a curriculum focusing on training may be appropriate and supported.

In either case, curriculum design is influenced by the pedagogy used within courses. For example, some subjects are well suited to laboratory experiment and demonstration, whereas others require lectures and study of texts. In some instances, field trips to existing establishments for observation and interviews are incorporated into the curriculum.

Often backed with tourism business advisory boards, curricula also include internships and cooperative educational programmes. Internships provide students with practical work experience lasting several weeks, a summer, a semester or even longer. A curriculum of cooperative education is a more tightly structured work experience programme that includes credit as well as pay. At issue in many instances is the lack of cooperation and integration between the several levels of education, which often blocks easy transfer of credits. For example, a junior college's curriculum in pre-tourism may not provide the prerequisites for transfer to a university of the student's choice. Counselling at an earlier stage can avoid this problem. Emerging tourism curricula would offer expanded options for specialisation in such areas as attractions (adventure tours, **museums**, convention centres); manufacturers of equipment used by travellers; policy makers, planners, and investors; and tourism trade organisations.

Further reading

Bratton, R.D., Go, F. and Ritchie, B. (eds) (1991) *New Horizons in Tourism and Hospitality Education, Training and Research*, Calgary, Al.: World Tourism Education and Research Centre, The University of Calgary.

Hobson, J., Perry, S. and Bushell, R. (1994) 'The Joint Australian National Education and Research Conference', *Journal of Travel Research* 33(2): 50–2.

Jafari, J. and Ritchie, B. (1981) 'Towards a framework of tourism education: problems and prospects', *Annals of Tourism Research* 8(1): 13–34.

McIntosh, R.W. (1983) 'A model university curriculum in tourism', *Tourism Management* 4(2): 134–7.

CLARE A. GUNN, USA

cycle

Cycles are normally repetitive patterns of **growth** or decline, frequently applied in a tourism context to destinations or countries/regions. The most common units used in measuring cycles include visitor numbers, arrivals, tourist receipts and bed-nights. Cycles in tourism typically refer to either **destination** life cycles, which may run for many decades, or to seasonal fluctuations in visitation numbers on an annual basis.

RICHARD BUTLER, UK

Cyprus

Tourism in Cyprus, a Mediterranean destination, has been a major factor in the growth of its economy over the last twenty years and is the most important sector of the economy, representing 20 per cent of GDP and providing employment to 25 per cent of the population. With 85,000 hotel beds, in 1996 Cyprus received 2 million tourists, who spent $1.8 billion. Its policy aims at the improvement of the tourism product for high-income and middle- income tourists.

ANTONIOS ANDRONIKOU, CYPRUS

Czech Republic

With Prague as its main destination, the Czech Republic features diverse cultural heritage and natural attractions. In 1998, more than 102 million foreign tourists arrived and tourism receipts were

$3.7 billion. The average number of nights spent in the country reached 3.3 in 1998. Tourism has become a significant element of national economy, amounting to 6.8 per cent of GDP and 14.1 per cent of export earnings in 1996.

HANA HAVLOVÁ, CZECH REPUBLIC

database marketing

Database **marketing** is an interactive approach and process. It uses a computerised system (including comprehensive, updated and relevant data on a given tourism **market**) and addressable communications media (such as mail, telephone and the sales force). The purpose is to develop customer-orientated programmes in a personalised and cost-effective manner that can result in stimulated **demand** and increased customer **loyalty**.

MARCO ANTONIO ROBLEDO, SPAIN

decision making

Many **models** have been proposed to simulate the decision-making process for tourism-related products. However, this is a very complicated procedure, and it increases in complexity as more people become involved. The concepts of **image**, utility maximisation, **knowledge acquisition** and others are all involved. In addition, various socioeconomic characteristics affect decision making in different ways.

WILLIAM C. GARTNER, USA

decision support system

A decision support system is the set of problem-solving **technology** containing people, knowledge, software and hardware successfully wired into the **management** process to facilitate improved **decision making** by marketing managers (Little 1979: 9–26). An executive support or information system is a special type which combines the reporting and inquiry functionality of marketing information systems with analytical modelling capabilities in a highly user-friendly form. Another type involving multiple decision makers is called group decision support system.

Analytical **models** commonly used in decision support systems are **forecasting**, simulation or optimisation models. These are either developed with a general programming language or with a statistical or mathematical standard software package. Recent developments in **information technology** offer software tools to incorporate inferential and deductive reasoning and heuristic manipulation of data (expert systems). In fast-changing environments where expert experiences are not available, **neurocomputing** and evolutionary programming technologies have also proved to be advantageous.

In tourism, the system is most commonly applied to managerial decision problems. Recent developments also focus on the optimisation of customer services which are offered by tourist information and reservation systems (see **geographical information system**; computer reservation systems). Its marketing applications have been successfully introduced for agent counselling, resource allocation problems of tourism offices, site selection analysis and **yield management** systems for hotels and **transportation** suppliers. While the system helps marketers to make decisions, it does not replace managerial judgement. An ideal system allows a manager to combine his experience and intuition with the consistent objectivity of a computer-based model.

See also: hospitality information system; investment decision; knowledge acquisition; property management systems, site analysis

References

Little, J. (1979) 'Decision support systems for marketing managers', *Journal of Marketing* 43(3): 9–26.

Further reading

Mazanec, J. (1986) 'A decision support system for optimizing advertising policy of a national tourist office: model outline and case study', *International Journal of Research in Marketing* 3(2): 63–77. (Provides a successful tourism application of a decision support system.)

Moutinho, L., Rita, P. and Curry, B. (1996) *Expert Systems in Tourism Marketing*, London and New York: Routledge. (Provides insights into knowledge-based decision support systems as applied to tourism marketing management.)

KARL WÖBER, AUSTRIA

definition

Definitions in **tourism** studies are sometimes passionately debated topics. Most concepts have multiple, sometimes conflicting, definitions, reflecting the needs or perspectives of the scholar proposing them. However, some agreement is essential for international comparisons. The **World Tourism Organization** and the United Nations have achieved consensus on core concepts such as **visitor** and **tourist**.

STEPHEN SMITH, CANADA

Delphi technique

Delphi is a qualitative **forecasting** tool used when there is no quantitative basis for developing a prediction. In tourism, the technique is used to predict the development of new **products** such as the first commercial tour into outer space, as well as the emergence of new issues to which leaders have to respond such as **growth** in tourism **marketing** on the **Internet**.

The technique involves a panel of experts who respond to two or more rounds of questionnaires. The first round invites the panel to identify possible **future** event(s), and either estimate their probability of occurring by a specified date or the most likely date by which they will occur. Results from the first round are incorporated into a second-round questionnaire. This provides each respondent the opportunity to review responses of other panellists and to either adjust his/her forecasts or to offer counter-arguments. Rounds are continued until a consensus emerges or it becomes apparent that no consensus is possible.

Challenges in conducting a Delphi forecast include the selection of an informed panel of expert representatives of the relevant tourism sector who will remain active during the project, perhaps a year or longer. Care must also be taken in constructing each questionnaire and in summarising responses from panellists in a clear, simple and coherent manner. Because a Delphi forecast is based on consensus, the researcher must ensure that no single panel member can assert his/her viewpoints through any means other than reasoned argument and evidence. This is usually done by conducting all rounds via mail, fax or e-mail.

Results of Delphi forecasts often are expressed in probabilistic terms (for example, a certain event has a 50 per cent probability of happening by a certain year). However, this precision is illusory. Forecast probabilities are subjective judgements only, and have no empirical basis. Delphi has been most successful in predicting scientific or technological breakthroughs. It is less successful in forecasting events that are functions of public **policy**, **social situation** or **marketing**, as is often the case with tourism.

Further reading

Green, H., Hunter, C. and Moore, B. (1990) 'Assessing the environmental impact of tourism development', *Tourism Management* 11: 111–20.

(Illustrates the use of Delphi; a subsequent note in the same journal issue comments on this article.)

Smith, S.L.J. (1996) *Tourism Analysis*, 2nd edn, Harlow: Longman, 143–8. (Describes the step-by-step procedures used to make a Delphi forecast.)

STEPHEN SMITH, CANADA

demand

People have needs for goods and services and they are willing to pay for them. These goods and services are provided and sold by companies forming the **supply** sector. The willingness and ability of consumers to fulfil their needs constitutes the demand for these goods and **services**. The buyers and sellers constitute the demand and supply sectors. These two meet in a **market** (whereas **public goods** are indivisibly distributed throughout the society by the state). The higher the price, the less consumers are willing to buy the product. Correspondingly, the lower the price, the less attractive the deal is from the suppliers' point of view. The interaction between buyers and sellers in the market determines the price and quantity at which the products are sold. In other words, neither the supply sector nor the demand sector alone dictates the price of the product or the quantity exchanged. The interaction will eventually lead to a market equilibrium, meaning that a balance is found between the supply of and demand for the product.

The products provided by the supply sector consist of physical goods and immaterial services. Physical goods are transported from the site of production and made available to the consumers through distribution **channels**. Services, in contrast, are intangibles, and are simultaneously produced and consumed in the market. The process of creating a product/service and making it available in the market by the supply sector entails **product planning**, **marketing**, **pricing** and other activities enhancing its sales.

In tourism, the concept of demand may be extended to cover individuals' needs for consuming and experiencing places. **Tourists** buy **souvenirs** and clothes and use services such as **accommodation** and amusement parks (see **theme park**). However, their major motive may lie in the **destination** itself, as a tourist in Paris or Rome might like to experience the local atmosphere and become part of the host **community**. The consumption of tourism places manifests itself as the tourist's subjective **experience**.

Tourism demand is the total of people participating in tourism activities, presented as the number of tourist arrivals or departures, amount of money spent or other **statistics**. Tourists have a demand for goods, services and places that are marketed as individual **products** and product packages. A **package tour** is a pre-purchased set of products including **transportation**, activities and other **services**. The tourist may buy a trip from a **tour operator** that has created an array of package tours in co-operation with airlines or other traffic contractors, hotels, **motels** or even sub-contractors.

The desire for any goods and services relates to changes in the demand and supply sectors, and their interaction. As to tourist buyers, they easily react to an increase or decrease in their personal disposable income and its distribution, **time budgets** and other personal motives (see **push/pull factors**). There may also be more general changes in the demand generating area, such as economic depression and political uncertainty. Changes among the sellers instead refer to variations in the price level compared to rival destinations, business environment, transportation connections and product assortment, among other factors. The contact between the buyers and sellers (demand and supply sectors) is sensitive to tourist promotion, attainability and marketing (see also **promotion, place**).

Reactions of demand to economic changes are expressed as elasticities. The income elasticity of demand compares the percentage change in quantity demanded in relation to an equal change in disposable income. A ratio of less than one represents an income-inelastic situation where consumers react weakly to changes in their income. A ratio of greater than one portrays an opposite situation where buyers react more easily to changes in their disposable income. Tourism has been found to be relatively income-elastic when measured by the amount of money spent on tour, but less elastic when measured by the number of tourists or tourist nights. This implies that people are unwilling to refrain from travelling but, rather, travel with smaller budgets. Correspondingly, price elasticity of demand exposes the percentage change in tourism demand in relation to an equal change in the product price.

Tourism demand may be categorised according to the willingness and ability of tourists to participate in tourism activities. Effective demand (also called actual or aggregate demand) consists of people who actually take part in tourism activities (that is, the buyers collectively). The size of this group of tourists is the easiest to measure as they, having physically moved to the site of activities, are traceable. Latent demand (also called suppressed demand) refers to would-be buyers. This type consists of the deferred and potential demand. Deferred demand refers to those who have the will to participate in tourism activities but cannot, as they lack supporting knowledge or an access to tourism facilities, or both. The passivity of this type is due to the inefficiency of the supply sector to provide and market their products. Potential demand consists of those who have the will to participate but lack supporting social and economic circumstances. In other words, they do not have enough money or other resources to accomplish their tourism plans. Their participation would require an improvement in their socioeconomic environment. Finally, people showing no demand have neither the will nor possibilities to participate in tourism activities.

Other types of tourists may be identified to fit the given task and field of study. For example, a geographer focuses on the spatial distribution of demand, an economist examines the volume of demand and its reactions to economic changes, and a psychologist observes **motivations** behind journeys.

Tourism may be an end in itself or an unavoidable side effect of some other activity. Most forms of **recreation** and holiday tourism fall into the first category, as people buy tours to selected destinations, spend the chosen period of time there, and return. **Business travel**, on the other hand, often falls to the second category, as the trip itself is not the primary goal but rather a means to accomplish some other aim, such as success in business negotiations that take place at the destination. In this case, tourism constitutes an input required in a production or learning process.

Further reading

Crouch, G. I. (1994a) 'The study of international tourism demand: a survey of practice', *Journal of Travel Research* Spring: 41–55. (Presents an overview of empirical studies on international tourism demand over three decades.)

—— (1994) 'The study of international tourism demand: a review of findings', *Journal of Travel Research* Summer: 12–57. (Reviews findings of empirical studies on international tourism demand over three decades.)

Faulkner, B. and Valerio, P. (1995) 'An integrative approach to tourism demand forecasting', *Tourism Management* 16(1): 29–37. (Presents a view to the dialogue between tourism demand analysts and decision makers based on the Australian experience.)

MARJA PAAJANEN, FINLAND

demand, air travel

The most important lesson to learn is that there are multiple demands from multiple customer segments for multiple **airline** services. The complexity of airline markets and perishability of airline service makes demand **management** a key to customer service and airline profitability.

Pricing affects the demand for airline services. One important demand dimension is elasticity, the responsiveness of demand to price changes. Availability of substitutes, travel purpose, passenger preferences, prices, market size, and complementary and competitive service prices are important demand elasticity determinants.

Much of the cost of available seat kilometres (ASK) is inescapable, with some differences between **class** of service. Since this ratio cannot be inventoried, the marginal cost of filling an otherwise empty seat is very small. The key to profitability is minimising unsold seats and thus maximising the revenue contribution. Both require a clear understanding and control of airline service demand.

Airline demand management is known as **yield management** (yield equals unit revenue per seat kilometre). Airlines define market segments by level and elasticity of demand. Consumers with relatively high, inelastic demands (such as **business travel**) do not require a discount to induce them to fly. Segments with a lower level, more highly elastic demand (such as leisure tourists) require discounts to induce them to travel.

The classification of business and leisure travellers into highly inelastic and elastic segments is now questioned. Corporate travel budgets are shrinking, video conferencing technology is improving, and the **Internet** is spreading rapidly. With fewer dollars to spend and available substitutes increasing, corporate travel is more elastic today. Leisure tourists, with increasing income and education, are beginning to demonstrate that **vacations** may be perceived as a necessity rather than as a luxury; thus, this demand is becoming less elastic. Those who forecast air travel demand face an unending challenge (see **forecasting**).

Further reading

Goeldner, C.R. (1997) 'The 1997 travel outlook', *Journal of Travel Research* 35(3): 61–5. (Describes projected changes in travel industry sectors, including demand trends.)

Kinnock, N. (1996) 'The liberalisation of the European aviation industry', *European Business Journal* 8(4): 8–13. (Examines competition changes as a major demand determinant.)

Stephensen, F.J. and Bender, A.R. (1996) 'Watershed: the future of U.S. business air travel', *Transportation Journal* 35(3): 14–32. (Discusses changing patterns in business travel demand.)

Wells, A.T. (1994) *Air Transportation: A Management Perspective*, 3rd edn, Belmont, CA: Wadsworth. (Chapter 10 contains a discussion of airline pricing and demand.)

KEVIN B. BOBERG, USA

demand, recreational

There are two broad categories of recreational **demand**. Manifest demand refers to exhibited recreational **behaviour** unconstrained by limiting factors. Data about this can be gained from such sources as **national park** visitation statistics, sales of recreational equipment, questionnaire surveys (either site-based or residence-based), or structured observation. Latent demand has to do with the potential demand that may exist in a given population or area to visit a particular **site** or enjoy a particular recreational activity, but which remains unsatisfied because of one or more limiting factors. Latent demand can only be identified through detailed social science surveys. Constraints can be of a personal nature (too poor, young or old, physically **handicapped**, ethnic-related or gender-related), or they can be caused by a lack of suitable facilities or sites (facility-deferred demand) or by inadequate transport or publicity (linkage-deferred demand). For example, a person may wish to visit a certain kind of tourism site or pursue a specific recreational activity but does not do so because they are unaware that the opportunity exists.

Manifest demand reflects **supply**. If natural or human-made recreational opportunities are available and accessible, they will be used. However, it is also the case that if they become too heavily visited or there is a deterioration in environmental quality (see **quality, environmental**), then demand can fall or there will be a significant change in the **market** segment. Conversely, if opportunities are not physically available they cannot be used; the demand will be non-existent or latent. Technological capability now exists to create virtually any kind of recreational opportunity anywhere and thus generate demand. Lakes with artificially produced surfing waves can be constructed in the desert, artificial rock climbing walls can be built inside the buildings, new access routes can be made and so on. Exhibited demand often varies dramatically on a seasonal basis in line with school and public holidays and weather variations. This means that expensive **infrastructure** in the form of hotels, ski tows, caravan parks and the like is often idle for much of the year, but is stretched to capacity at other times of the year. In many countries and regions, deliberate attempts are presently being made to even out these visitation discrepancies by a variety of measures like diversifying the tourism **product**, extending the length of the tourism season and staggering school holidays on a regional basis.

See also: behaviour, recreation; recreational need; recreational participation

Further reading

Jackson, E.L. and Burton, T. L. (eds) (1989) *Understanding Leisure and Recreation: Mapping the Past, Charting the Future*, State College, PA: Venture Publishing.

Smith, S. (1983) *Recreation Geography*, London: Longman.

DAVID MERCER, AUSTRALIA

democratisation

Democratisation is the process whereby the opportunity to participate in tourism, which until the 1850s was a socially selective activity, has become available to the majority of people in the developed societies who wish to do so. Usually linked to technological developments in **transportation** systems, it also refers to the continually increasing range of destinations, **vacations** and styles of travel available to the mass market.

See also: mass tourism; social tourism

RICHARD SHARPLEY, UK

demography

As a branch of the social sciences concerned with studying the structure and change of human populations, demography examines birth, death and **migration** rates and **lifestyle** characteristics, among others, and employs these data to define social and economic dynamics. Demography draws upon the methodologies and theoretical perspectives of **economics**, **sociology**, **statistics**, **geography**, **anthropology** and other fields.

Demography views tourism as a form of temporary migration. Demographers seek to understand such issues as the **impact** of tourism upon society and the economy, historic patterns, the consequence of **changes** in society (such as ageing population or economic growth) upon tourism tendencies, the effects of increased life expectancy and better health conditions upon tourism patterns and activities, and the effects of political **policy** (for example, protection versus multiple use of **parks**) upon pleasure tourism. Demographic data are also useful in addressing the unsolved issue of identifying underlying **motivations** for travel. An effective **methodology** for identifying the characteristics of non-visitors to a **destination**, **attraction** or **accommodation** also remains a concern of tourism researchers.

A useful tool of researchers, demography is employed primarily in **demand** forecasting for **marketing** and promotion, and in impact assessment to facilitate destination **development** and **management** (see also **promotion, place**). Numerous factors are traditionally considered in conducting demographic analysis of a **market**. Background data include geographic area of residence, age, sex, household size, **race**, household income (disposable income), education and occupation. These may be expanded to include such lifestyle characteristics as tendency to travel, **leisure** preferences, media habits, and ownership of houses, automobiles and sports equipment. However, trip characteristics (for example, purpose of travel, activities, length of stay or expenditures) are not components of tourism (or visitor) demographics.

Primary and secondary data sources are utilised in developing visitor background data. These may be collected on an individual basis or in conjunction with another destination, attraction or **event**. Primary data collection requires skills in writing **surveys** and designing distribution methodologies. Problems associated with data collection include obtaining statistically reliable response rates and inadvertent exclusion of groups such as foreign language speakers or economically disadvantaged persons. Two survey formats normally are utilised: personal **interviews** and self-administered questionnaires. A number of researchers and agencies subscribe to secondary sources of demographic data, such as the **US Travel Data Centre**, Statistics Canada and the **World Tourism Organization**.

See also: marketing research; tourist

Further reading

The Implications of America's Changing Demographics and Attitudes on the U.S. Travel Industry (1989), Washington DC: Discover America Implementation Task Force.

MARK J. OKRANT, USA

demonstration effect

The arrival of **tourists** in growing numbers at a **destination** will have many social, cultural and economic effects on host communities. Within those communities, some may wish to imitate the **behaviour** of tourists. Such imitation is known as a demonstration effect. It implies that the new behaviours are seen as desirable by at least some sectors of the host community.

The term demonstration effect is closely linked to **acculturation**, which can be defined as cultural **change** initiated by the interaction of two or more cultural systems. However, imitation is but one aspect of acculturation. The attendance of local youths at discos in a developing world holiday **resort** may be an example of a demonstration effect where Western pop music and dance is enjoyed, but the change in behaviour where local people withdraw some features of their cultural life from public view is part of a wider process of acculturation.

An understanding of demonstration effects requires consideration of why some behaviours are imitated and not others, who imitates them, how they are learned and to what extent tourism is a cause of imitation. Research has indicated that there are a number of determining factors, including the strength of the host culture and its flexibility and responsiveness to new influences, the homogeneity of that culture and its acceptance by all members of the host community, the gap between the cultures of the host and those of the tourists, the contact situation, the type of **social interactions** that result from contact situations, the respective economic power relationships between **host and guest**, the **motivation** of both **tourist** and host, the level of exposure of the former to other sources of influence besides tourism, and the role of tourism entrepreneurs and their staff.

Early work by anthropologists in tourism concentrated upon the impacts of tourism on developing countries' societies and viewed the relationship as being one of dominance–subordination. In the first edition of Smith's book, *Hosts and Guests* (1977), many case studies related change to the impact of tourism alone. However, in the second edition (Smith 1989) there is a much greater recognition of change emanating from non-tourist sources. Indeed, Smith indicates that changing practices among Inuit peoples in Alaska that she first attributed to tourism may in fact have been due more to welfare policies and the growth of extractive industries (1989: 75–7).

Other research has sought to more carefully delineate the original practice and separate the behaviour from the reason for that behaviour. For example, in Sri Lanka there was growing evidence that **pilgrims** were using transistor radios as a source of entertainment, and this was initially thought to be due to tourists' influence. It was argued that the solemnity of the **pilgrimage** was being undermined. Subsequent analysis showed that, like Chaucer's pilgrims, there had long been a tradition of frivolity. Thus the culture of the Sri Lankan pilgrimage utilised a new means to achieve old ends of **fun**.

Another growing realisation since the original work undertaken on tourism–host impacts is that host communities are not always homogeneous; indeed, one of the functions of cultural organisation is to find ways of providing legitimate means of conflict solution. Intergenerational differences are not uncommon, and thus tourist **development** may be used by the young to more clearly demarcate differences so as to obtain their own ends. Of particular significance is the demonstration effect caused by female Western tourists in some societies. These females can develop careers and economic independence, and the sight of such female tourists can reinforce other Western processes of corporatisation and education in creating changing expectations by females of the host community.

Another factor that determines the nature of the demonstration effect is the number of tourists. However, opinions differ as to how this might cause demonstration effects. With the development of **mass tourism** in a destination has come the concept of the tourist resort. Such **resorts** are often self-contained, and tourists are bussed into hotels and find much that they require within the complex. The local culture may be invited in to provide a show, an entertainment where the hosts' culture is presented as being exotic. While such events have been criticised as being **pseudo-events** which commoditise (see **commoditisation**) and simplify, such complexes reduce social contact between local people and tourists to highly scripted occasions.

Additionally, from the perspective of local people, the tourist is seen as alien, and thus as they are so different, tourists' actions have no relevance for their daily lives. Arguably, the opportunity for demonstration of different values is greater when the contact is made between independent hosts and guests, for the social relationships may progress beyond the stereotypical **role** of tourist and service/experience provider. The type of social interaction is also important in that until research in the 1990s, a general assumption was that the demonstration effect was from the tourist to the host. However, for some tourists motivated by **quest**, pilgrimage or experiential meaning, the opportunity exist for hosts to influence tourists and to demonstrate attractive alternatives.

Much of the literature about demonstration effects is concerned with tourist–host relationships where the tourist is from a developed country while the latter comes from a **developing country**. Cultural gaps are thus assumed to be large. It can be noted that economically marginal areas exist within developed countries, and that often within those countries tourism becomes important because these are often rural sites possessing scenic touristic values. There are undoubted economic implications that are well observed; for example, house prices increase when **second homes** are purchased. Some writers have written that tourism brings urbanisation to rural areas with its demands for **shopping**, **accommodation**, entertainment and more. Under such circumstances, the demonstration effects caused by tourism are incorporated within more complex patterns of social **class** and income differences. Nonetheless, the advent of tourism makes differences more obvious and immediate.

See also: social relations; staged authenticity; stereotypes

References

Smith, V. (1989) *Hosts and Guests: The Anthropology of Tourism*, 2nd edn, Philadelphia: University of Pennsylvania Press. (Discusses host–guest relationships.)

Further reading

Lever, A. (1987) 'Spanish tourism migrant: the case of Lloret de Mar', *Annals of Tourism Research* 14(4): 449–70. (Discusses the impacts of tourism upon young female migrant workers in a mass destination.)

Nash, D. (1996) 'Tourism as acculturation or development', in D. Nash, *Anthropology of Tourism*, Oxford: Pergamon, 19–38. (Examines demonstration effects within the context of acculturation.)

Pfaffenberger, B. (1983) 'Serious pilgrims and frivolous tourists: the chimera of tourism in the pilgrimages of Sri Lanka', *Annals of Tourism Research* 10(1): 57–74. (Covers changing behaviours on pilgrimages viewed by tourists.)

CHRIS RYAN, NEW ZEALAND

Denmark

Tourism in Denmark is considered the third most important business sector in economic terms. Between 1988 and 1994, Denmark experienced a growth rate in the number of bed-nights which was exceptional in Europe. This was mainly due to a strengthened co-ordinated **marketing** effort between the **industry** and the Danish Tourist Board, and a national tourism **policy** which actively supported co-operation and product development.

With 7,500 kilometres of coastline covered with beaches, a clean environmental profile and a varied cultural **landscape**, Denmark has a natural potential for tourism. The **infrastructure** is highly developed with bridges and ferries connecting the different parts of the country, and the short distances give easy access to a great variety of activities and attractions (the two best-known of which are probably Tivoli and Legoland).

The tourism industry in Denmark is dominated by small and medium-sized companies, and the main products are holiday houses, the sea and nature. Campsites have been classified since 1988, and hotels now have their own system based on international standards. The service level in terms of information is high, with tourist information offices in almost every town. The constant development of new products has resulted in activities such as fishing, surfing, cycling, golf and a number of cultural activities. The demand side of **market** is dominated by countries in close proximity, but Danish consumers themselves constitute 40 per cent of the market.

There are a number of weaknesses in Danish tourism, such as a short season, lack of non-seasonal attractions/activities and a dependency on relatively few markets. The national tourism policy involves an attempt to solve these problems through the development and promotion of new products to new markets. Furthermore, the national tourism policy involves a number of general aspects such as the creation of continued growth in terms of economy and employment, based on the principle of sustainability. Development of the traffic infrastructure in terms of international **ferry** and **rail** connections as well as **airports** and bridges is on the political agenda, together with an increased focus on co-operation between the industry and the Ministry of Commerce and Trade institutionalised in the Danish Centre of Tourism Development and the Danish Tourist Board. The Green Key, an organisation undertaking the certification of eco-friendly hotels and hostels, is an initiative in the protection of the **environment** (see also **protected area**).

Denmark is divided into nine areas/destinations, which are organised in regional tourism development companies. Thus, the continued strengthening of the positive relationship between the regions and the different agents facilitates the implementation of the national tourism strategies. The local strategy is to further the development of tourism through the facilitation of co-operation across regional boundaries.

In light of international competition, Denmark faces several strategic problems and challenges. The low level of profit, high **taxes** and lack of **innovation** are major obstacles to the continued development of its tourism. Other challenges are to obtain equal status for tourism with other industries in Denmark, to continue the focus on the environmental aspects of tourism and to enhance the level of competence.

JENS FRIIS JENSEN, DENMARK

denotation

Denotation is the process by which the **attention** of an observer is drawn to the meaning of terms. Appearing together in a single image, cameras, T-shirts and luggage probably denote **tourist**, while London's Buckingham Palace denotes royalty. Such denoted objects almost certainly connote additional second-order meanings, the former with, for example, gullibility, and the latter with social **class** and privilege.

TOM SELWYN, UK

dependency theory

Dependency theory, also referred to as core–periphery theory, is based upon the inter-relationship between development and underdevelopment. The term 'dependency' can be used to signify economic, social or political dependency, but in reality these different types of dependency are all inter-related. Economic dependency often leads to social and political dependency, with the latter two may be more difficult to establish in an objective manner. Tourism **development** may exhibit the symptoms of dependency theory when it results in the enrichment of developed metropolitan areas at the expense of poorer, underdeveloped regions. International examples often cited tend to be small island states in close proximity to large industrialised economies (such as Caribbean tourism destinations). However, dependency theory has also been cited for the larger landmasses of Afro-

Asia and Latin America. One could argue in most of these cases that the weakness of the internal political and economic structures made it likely that they would become dependent upon other nations during the late nineteenth and early twentieth centuries.

With the advent of mass tourism, even those countries that were not sought out for their mineral deposits, slave labour, strategic military locations or agricultural **climates** were inevitably going to be brought into the dependency chain. But dependency theory may also relate to **domestic tourism** where sparsely populated peripheral regions find that their economic structures have been distorted by the demands derived by tourism activity. Many issues surrounding tourism developments, such as **second homes**, often involve dependency. For instance, the land and general price inflation that may be associated with tourism development may make it impossible for local residents to purchase their own homes or to survive in the area while engaged in more traditional industries. This then makes the local economy dependent upon tourism for its survival. The underlying thesis of dependency theory is that the so-called developed areas are subject to sustained indigenous growth, whereas underdeveloped areas exhibit reliance upon external factors and, as such, derive their growth from the developed or 'metropolitan' areas: hence the term dependency.

This theory has been criticised on a number of grounds, including the suggestion that the **infrastructure** of the dependent countries have been improved as a part of the derived development. Tourism provides a number of stimuli, from foreign **investment** to the increases in domestic demand, that can all assist **indigenous** development.

JOHN FLETCHER, UK

deregulation, airline

A **market** is considered to be reasonably competitive if relatively there are a large number of buyers and sellers, **product** and **services** offerings are homogeneous, buyers and sellers have 'perfect' information, and exit and entry are easy. No industry has so challenged theories of competition and regulation than the **airline** industry. Seven decades after initial promotion, six decades after initial regulation and over two decades since deregulation, airline market competitiveness remains problematic for **policy** makers and decision makers alike.

Governments actively promoted airlines during the industry's infancy. To nurture the fledgling industry, governments funnelled money to aircraft manufacturers and operators. Airmail contracts were a favoured means of promoting growth of the industry throughout the 1920s. This unbridled promotion led to instability and airlines came and went. While perhaps commercially acceptable, the **safety** record associated with instability was not. To meet national public interest and defence needs, the industry was regulated beginning in the 1930s.

Airline regulation and promotion followed a classic model. Entry and exit were restricted. Regulated trusts were tolerated and encouraged through joint discussion and publication of tariffs. Promotions and subsidies were freely offered through a variety of means. During the 1970s, some in the **United States** and other nations began to question the economic rationale for, and benefits of, regulation. Alfred Kahn became an eloquent and outspoken advocate of deregulation in this country. Other nations soon followed.

US domestic air cargo service was deregulated in 1977 and passenger airline service in 1978. The first attempt to deregulate international airline services was launched by the US in 1980. The **European Union** followed suit in the late 1980s, committing to deregulation in 1992. Other nations – notably in Asia, the Pacific, the Caribbean and Latin America – have joined and others may soon join the fray (see **international aviation liberalisation**).

Further reading

Button, K.J. (1996) 'Aviation deregulation in the European Union: do actors learn in the regulation game?', *Contemporary Economic Policy* 14(1): 70–81. (Examines how European policymakers and airlines have learned from US deregulation.)

Meyer, J.R. and Strong, J.S. (1992) 'From closed set to open set deregulation: an assessment of the U.S. airline industry', *Logistics and Transportation*

Review 28(1): 1–22. (Discusses the change to intense competition that occurred after deregulation.)

KEVIN BOBERG, USA

design and layout

The term design and layout refers to the creative arrangement and integration of structures and **landscapes** to meet specified functional and aesthetic objectives. This process in tourism settings calls for the sensitive location and fitting of facilities and functional support elements to specific topographic and landscape features of the **site** in harmony with the surrounding **environment**, user preferences, managerial concerns and institutional requirements.

JOHN J. PIGRAM, AUSTRALIA

destination

Destination, as distinct from origin or **market**, refers to the place where **tourists** intend to spend their time away from home. This geographical unit visited by tourists may be a self-contained centre, a village or a town or a city, a region or an island or a country. Furthermore, a destination may be a single **location**, a set of multi-destinations as part of a tour, or even a moving destination such as a cruise.

The tourism phenomenon is geographically complex, and its different **products** are sought and supplied at different stages from the origin to the destination. Spatial and characteristic diversity among destinations has become so great that it is not easy to classify them. However, several models seek to explain the tourism system relevant to the destination. Origin–linkage–destination models may serve to explain the basic feature of the generating and receiving function of origins and destinations. In any country or region, there are likely to be a number of origins and destinations, with most places performing both functions. As well as sending tourists to some (multiple) destinations, a particular place may also receive visitors from those same and other places. Likewise, the routes and linkages may carry tourists from one place to the other and back again or to some third place.

Structural **models** emphasise the relationships between origins and destinations particularly in **Third World** tourism in core–periphery terms. In these models, the market is concentrated upwards through the local, regional and national hierarchy with international transfer occurring between national urban centres either as origins or destinations. Dispersal within the peripheral destination is more restricted, with the tourists moving from their arrival point to some resort enclave. Movement may occur between **resorts**, but only limited travel to other areas occurs. Finally, evolutionary models stress dynamic, change and evolving situations, either the evolution of international movements or the **development** of destinations. These models range from **pleasure periphery**, to **psychographic** positions of destinations, to the **life cycle** of a destination, with emphasis on the structural evolution of destinations through time and space.

The demand for tourism to a particular destination may be a function of push factors in the origin and pull factors in destination areas. While tourism push factors can be regarded as a person's predisposition to tourism which is determined by the sociodemographic and psychographic characteristics, pull factors relate to the relative attractiveness of destinations (see **push–pull factors**). The tourists' destination choices can be influenced by examining several elements which make up the **attractivity** of a place. The success of any geographical unit as a destination is primarily determined by several factors. First, attractions are those elements that draw a tourist to a particular destination and which serve as the main motivation or primary pull factors. They may comprise natural attractions (scenery, **climate** or beaches), built attractions (historic sites, resorts or **theme parks**), cultural attractions (**museums** or art galleries) and social attractions (meeting the residents of destination and experiencing their way of life).

Second, amenities at the destination can be viewed as the elements within the destination or linked to it which make it possible for tourists to stay there and to enjoy and participate in the attractions. They include basic **infrastructure**, **accommodation**, **transportation**, **catering** services, entertainment, **shopping** facilities and visitor information at the destination. Amenities do not usually in themselves attract tourists, but the lack of amenities might cause tourists to avoid a particular destination because these provide the basic facilities which are regarded as contributing to the quality of the destination. Amenities harmonise and enhance the destination attractions. Third, **accessibility** can be referred as the relative ease or difficulty with which tourists can reach the destinations of their choice. Destination access is mainly a matter of transport infrastructure such as airports, harbours, roads and railways. Ease of access from the origin to the destination, and ease of movement within and among destinations, are also important factors for the viability of regions.

Fourth, **images** can be regarded as the ideas and beliefs which tourists hold about the destinations. Numerous studies have revealed that a destination possesses an image and the choice is influenced by the tourists' images of alternative destinations, whether these images are true or not. Therefore, images are another focal point of destination **marketing**. Finally, one of the key factors in destination selection appears to be the price, the sum of what it costs for travel, accommodation and participation in a range of selected services when there. The price varies by choice of accommodation, **seasonality** and the distance to a destination.

The terms 'destination' and 'product' are often confused. The former is often considered to be a tourism product in itself, but some scholars argue that the destination is simply a geographical unit within which any number of differing products can be purchased and experienced. The product can be seen as the total tourism **experience** which comprises a combination of all the service elements, which the tourist consumes from the time they leave home to the time of return. Thus the total product is much larger than the destination itself, and usually embodies several tourism products. Hence, a destination may be viewed as part of the products of the tourism industry.

See also: destination choice; image; life cycle, destination; marketing, destination

Further reading

Burkart, A.J. and Medlik, S. (1981) *Tourism: Past, Present and Future*, 2nd edn, London: Heinemann.

French, C.N., Craig-Smith, S.J. and Collier, A. (1995) *Principles of Tourism*, Melbourne: Longman.

Pearce, D.G. (1995) *Tourism Today: A Geographical Analysis*, 2nd edn, Harlow: Longman Scientific & Technical.

Travis, A.S. (1989) 'Tourism destination area development (from theory into practice)', in *Tourism Marketing and Management Handbook*, New York: Prentice Hall, 487–98.

BAE-HAENG CHO, AUSTRALIA

destination choice

Destination choice is the main focus of consumer **decision making** research in tourism, aimed at improving the understanding of how tourists choose a **destination** from the many available to them. Individual's **images** or perceptions, at least partially derived from their **attitudes** toward a destination's perceived tourism attributes, have been linked to destination preference and selection. As a result, most attempts to understand and predict destination choice include measurement of the importance of various destination attributes (for example, **climate**, natural and cultural attractions, **infrastructure**) and the extent to which various alternatives are perceived to possess those attributes.

Destination choice **models** incorporate the key conceptual elements of traveller related input variables (including **motivation**), external stimuli and inputs (including destination **marketing**) and **choice** set structure and development. Issues related to decision alternatives include the attributes used to evaluate choices and the decision **rules** used to combine information from each alternative and attribute in order to make a final destination selection. Choice set theory is based on the notion that, due to information processing limitations, selections are made only after the consumer has constructed a set of acceptable alternatives. Research in tourism has identified several destination choice sets, including awareness, awareness-unavailable, unaware, early consideration, inert, inept, action, inaction and late consideration.

There is general agreement on the structure of the overall destination choice process as including some or all of the following: perceptions (belief formation) of destination attributes in the awareness set through passive information catching; a decision to undertake a pleasure trip (problem recognition/formulation); evolution of an evoked set from the awareness set of destinations (search for alternatives); perceptions (belief formation) of the destination attributes of each alternative in the evoked set through active solicitation of information (evaluation of alternatives); selection of a destination(s); and post-purchase evaluation.

Further reading

Crompton, J.L. (1992) 'Structure of vacation destination choice sets', *Annals of Tourism Research* 19(3): 420–33. (Describes and operationally defines destination choice sets.)

Moscardo, G., Morrison, A.M., Pearce, P.L., Lang, C.T. and O'Leary, J.T. (1996) 'Understanding vacation destination choice through travel motivation and activities', *Journal of Vacation Marketing* 2(2): 109–122. (Presents a model of destination choice incorporating travel motivation.)

Um, S. and Crompton, J.L. (1990) 'Attitude determinants in tourism destination choice', *Annals of Tourism Research* 17(3): 432–48. (Develops a two-stage approach to travel destination choice.)

LAURIE MURPHY, AUSTRALIA

destination image *see* image, destination

destination information system *see* decision support system; geographical information system; marketing, destination

destination life cycle *see* life cycle, destination

destination management

Destination management is the integrated process of managing any of the three tourism **destination** types (urban, resort and rural). It covers four key elements: the destination offering (visitor **experience**, destination image and attractiveness); the visitor mix (market research); **marketing** communications (awareness and promotion); and organisational responsibility (**leadership** and **partnership**). This process is typically performed by a convention and visitors bureau, a state/provincial/regional tourism office and/or a national tourist organisation.

DONALD ANDERSON, CANADA

destination management company

Destination management companies provide on-site coordination services for visiting groups. Much of their business arises from the meetings, conventions, exhibitions and **incentive** tourism segments. They liase closely with corporate and association **meeting planners**, incentive travel planners, **tour operators** and wholesalers to arrange locally oriented events and logistics such as **transportation**, banquets, and other functions and activities. They provide seamless organisation and one-stop shopping with local expertise.

GEOFFREY I. CROUCH, AUSTRALIA

destination management organisation

The term destination management organisation (DMO), also known as a destination **marketing** organisation, refers to either a convention and visitors bureau, a state/provincial/regional tourism office or a national tourist organisation/administration. These organisations are the entities mandated to undertake the process of tourism

destination management. DMOs have become the principal organisations responsible for leading, co-ordinating, stimulating and monitoring tourism **development** and marketing for a destination area. A **destination management company** is different from a DMO.

DONALD ANDERSON, CANADA

destination marketing *see* marketing, destination

developing country

Perspectives on **development** continue to be polarised. Whereas modernisation theorists generally refer to 'developing' societies, and neo-Marxists to 'underdeveloped' societies, the term 'less developed' carries no implication of movement from a less to a more desirable state. Less developed countries (LDCs) have relatively low gross national products, a limited resource base, high levels of unemployment and underemployment, a reliance on the export of one or two primary products, and a structurally dependent position in the international economy. Many are former colonies with extremes of inequality, often with substantial ethnic heterogeneity and culturally distinct social segments. Some of these features are now considered tourism attractions and LDCs, which currently receive about 25 per cent of all international tourist arrivals, are increasingly promoted as **destination** areas.

Debates on the costs and **benefits** of tourism are ideologically charged. Advocates of tourism claim it is labour-intensive and a major contributor of **foreign exchange**. It thus reduces reliance on primary export crops and, through tourism **employment** and income **multiplier effects**, provides much-needed employment and increases national income directly, indirectly and by improving overall economic buoyancy. Tourism may also attract foreign **investment** and help inculcate new skills in the labour force, making optimum use of such natural resources as sun, sea and sand, and promoting 'exotic' **indigenous** cultures as attractions, thus gaining **comparative advantage** from otherwise profitless aspects of nature and culture. Such development is allegedly environmentally sustainable because tourism is an **industry** without chimneys. Further, by exposing LDC hosts to guests from more developed societies, tourism helps incorporate the former into the global economy.

Critics of tourism as a **development** strategy assert that the economic benefits are less than claimed and that social and cultural **impact** are overwhelmingly negative. They point to **leakages** from the economy arising from imports of products used in the tourism industry and to the dominant role of the **multinational firm** in the provision of hospitality and tour operations, and argue that the employment provided is relatively unskilled, with better-paid managerial jobs filled by expatriates. In addition, the values diffused to members of the host population, especially the young, are considered examples of undesirable demonstration effects which pollute local cultures. Critics also note numerous forms of environmental pollution resulting from **mass tourism**. Some even claim that tourism is another form of **colonisation** or **imperialism**. In so far as they support any kind of tourism, they advocate alternative forms of tourism based on small-scale, participatory ventures that are socially, culturally and environmentally sustainable.

Many LDCs, especially islands and small states, have become reliant on **international tourism** but, apart from oil-producing states, it is the wealthier ones among them, along with newly-industrialising countries, that attract most international arrivals. This is partly because they had already achieved stability and a basic standard of **infrastructure** before tourism started – holiday makers are not normally attracted by civil disorder, abject poverty, poor hygiene and major health risks – and partly because tourism does bring economic benefits. Even where income multipliers are relatively low, the foreign currency and jobs obtained from high-spending **tourists** generally exceed what is obtained from backpackers or home-stay tourists, where leakages are less and income multipliers higher. Economic benefits may be unequally distributed, but this reflects state **policy** and is not an inherent disadvantage of international tourism. In many LDCs, tourism has certainly led to growth in the **informal economy**, including jobs which border on illegality, prompting critics to highlight increased crime, **deviance** and socially unacceptable behaviour, especially **prostitution**, which has been exacerbated by some forms of **sex tourism**.

The social and cultural consequences of tourism are undoubtedly problematic. The more international tourism is promoted in a destination area, the more care is required in adapting existing social institutions, developing new ones and co-operating with outside organisations engaged in promoting or opposing tourism. Along with other influences from developed societies, for example, the **globalisation** of industrial production and consumption and the spread of television and other mass media, tourism can affect politics and family life, create new economic opportunities for the young and for women, and place traditional practices under threat. Political leaders are faced with the challenge of deciding how far to encourage tourism, from which they may derive economic benefit, and risk seeing the basis of their own **power** threatened. At the cultural level, as well as the much-debated demonstration effects, **arts** and crafts in many LDCs have been influenced by tourism **demand**, and while some deplore the loss of **authenticity** and the production of airport art, others have praised the adaptability of local dancers, carvers or sculptors and the boost given by tourism to local production. However, international tourism is here to stay, and while small-scale **alternative tourism** can exist alongside mass tourism, it cannot replace it. Policy makers and planners in LDCs need to choose the most appropriate tourism course to follow and will be assisted most by careful and competent research into the costs and benefits of tourism, which vary from one society to another. Ideally, supported by all with the long-term interests of sustainable tourism at heart, they will prioritise the wider economic, social and cultural interests of their populations, but their failure to do so should not necessarily be blamed on the tourism industry (see also **sustainable developments**. Opposition to many aspects of tourism, like its promotion, is internationally organised and injustices are quickly publicised.

See also: economic development

Further reading

Harrison, D. (1992) *Tourism and the Less Developed Countries*, Chichester: Wiley. (Discusses influence of tourism in LDCs generally and in parts of Asia, Africa, Latin America and the Caribbean.)

Lea, J. (1988) *Tourism and Development in the Third World*, London: Routledge. (Briefly summarises problematic aspects of tourism in LDCs.)

Price, M.F. (ed.) (1996) *People and Tourism in Fragile Environments*, Chichester: Wiley. (Includes studies of the way small communities in LDCs and developed countries cope with tourism development.)

Smith, V. (ed.) (1989) *Hosts and Guests: The Anthropology of Tourism*, 2nd edn, Philadelphia, PA: University of Pennsylvania Press. (Presents a well-established series of anthropological studies of tourism.)

DAVID HARRISON, UK

development

Although development is now a relatively specialised topic of study, its concerns are not new. The founders of the social sciences were the precursors of more recent and more specialised studies of development. Saint-Simon, Comte, Marshall, Marx, Durkheim and later Weber, for example, saw great material benefits in industrialisation and the division of labour, but had numerous reservations about their effects on the social organisation of society. Although their remedies for the defects differed, all believed that the new social sciences would help alleviate the problems accompanying industrialisation. If the nineteenth century saw the application of social science to social problems, during the twentieth century there was a greater politicisation of some of these solutions, most notably those deriving from the writing of Marx. The Russian revolution of 1917 fostered the belief that Western capitalism was not the only possible trajectory for economic, social and political change, and the Second World War, which necessitated planning on a large scale in all participant societies, and the rapidly accelerating process of decolonisation that followed it, further prompted governments and their critics to actively intervene in the economic and social life of nations. By the 1950s, development was becoming a special topic for study.

Generally regarded as an increase in material prosperity and measured in the growth of gross national products, development preoccupied governments of developing countries, the big powers on opposite sides of the Cold War, and a burgeoning number of academic experts and aid advisers. However, the nature of development was increasingly contested. The tensions of the 1920s and 1930s were mirrored in the theoretical debates of the 1960s, and definitions of development and how it was to be achieved fell broadly and often simplistically into two main camps, with pro-capitalist, bourgeois, modernisation theory (MT) pitted against anti-capitalist, neo-Marxist underdevelopment theory (UDT). There were other voices: Barrington Moore showed that the poor had always paid the price of modernisation, irrespective of the dominant political ideology, and Dudley Seers was prominent among those emphasising the social and cultural aspects of development, especially the need for self-reliance and social justice. Such writers are not easily categorised. However, for MT, led by such theorists as Parsons, Hoselitz and Rostow and including classical Marxists, internal constraints blocked progress in developing societies. These could be countered by emulating Western economic policies and social norms, dispensing with elements of tradition that prevented modernisation and, if necessary, importing capital, technology, skills and values from the West.

While MT was being developed, an alternative view was being formulated. Based on the work of Paul Baran, Latin American dependency theory and the world systems perspective of such writers as Wallerstein and Amin, and popularised by Andre Gunder Frank, UDT denied virtually every tenet of MT. It considered underdevelopment a consequence of colonialism and structural inequality in the world trading system. Close links with the West prevented development from occurring, led to internal inequality and inappropriate production, especially by transnational firms, and perpetuated economic, social and cultural dependence. Nevertheless, both theories saw development as a process of autonomous structural economic change, with major social, cultural and political ramifications, that vastly increased the material prosperity of developed countries and which could occur in 'developing' or 'underdeveloped' societies, if they followed the appropriate policies. For MT, which favoured capitalism, these involved closer links with the West. By contrast, UDT rejected capitalism and sought autonomous economic growth through import substitution and self-reliance, and socialism, either in one country or through co-operation with other like-minded states.

Such theoretical debates had practical and political implications. Often despite a stated hostility to theory, activists at more empirical levels of development (in aid agencies and planning, for example) inevitably entertained their own views on development processes. Depending on the perspective held, they sought change agents and elites to lead the diffusion process or progressive individuals and vanguard parties to direct the revolution. Indeed, since the Second World War, entire societies have served as temporary role models for one side or the other. Puerto Rico, Kenya, Brazil and Mexico have been among those admired by advocates of MT, and Cuba, Eritrea, Tanzania and China have figured in the UDT pantheon. However, by the end of the 1980s the arguments between UDT and MT had exhausted most of their vitriol and much of their relevance. It had become evident that both internal and external factors affected less-developed countries, and blinkered concentration on one or the other was and had been simply unrealistic. In addition, no role model seemed to be of lasting value. Furthermore, the collapse of the Eastern bloc seemed to suggest that organised socialism had failed, at least in Europe, while the rise of such Asian 'tigers' as Singapore, Malaysia, South Korea and Taiwan, with their combination of egalitarian and capitalist attributes, confounded all who hankered after simplistic formulae for producing development.

The questions raised by these theories remain important, but their contributions have been incorporated into **globalisation** theory, which focuses on the economic, social, cultural, political and environmental ramifications of a world in which national barriers sometimes seem redundant. As a consequence, although development studies is still a specialist subject, its concerns are again central to mainstream social science. However, the old dilemmas remain. There may be considerable agreement on economic and other changes occurring in a society, or on the nature of modernity or the role of tourism, but whether they are perceived as progressive depends on the observer's perspective. Ultimately, all debates about development are about progress, and this inevitably involves value judgements. Tourism, as an agent of change, has frequently been proposed as a means of development. Although there has only been a tenuous link between the mainstream tourism and development literatures, the underlying issues are common to both.

See also: acculturation; community development; economic development

Further reading

Harrison, D. (1988) *The Sociology of Modernisation and Development*, London: Routledge. (An introduction to the major theories of development and underdevelopment.)

de Kadt, E. (1979) *Tourism: Passport to Development? Perspective on the Social and Cultural Effects of Tourism in Developing Countries*, New York: Oxford University Press. (A collection of articles on tourism in less developed societies.)

Sklair, L. (ed.) (1994) *Capitalism and Development*, London: Routledge. (Discusses the nature of capitalist development and some major economic sectors in less developed and newly-industrialising societies.)

Waters, G. (1995) *Globalization*, London: Routledge. (Reviews of recent perspectives on economic, political and cultural globalisation.)

DAVID HARRISON, UK

development era

The development of tourism is divided into three main eras. That of 'craftsmanship' is characterised by spontaneous **development**, and by scarce skills and technological applications. The era of 'Fordism' is characterised by the search of profit, through **standardisation** and mass production. Finally, the 'new tourism era' has seen tourism transformed primarily by segmentation of **demand**, flexibility of **supply** and diagonal integration (see also **market segmentation**).

AMPARO SANCHO-PEREZ, SPAIN

deviance

Deviance can simply mean **behaviour** which differs from the norm. However, the term is often used in a context where deviant behaviour is viewed as being undesirable. Hence, when applying the term to tourism, two perspectives can be identified. First, as for most people, holiday taking is a period of behaviour which differs from that of their daily lives, with tourism itself being a form of deviant behaviour. Second, the actual behaviour shown by some **tourists** may be termed deviant if it involves activities such as drug usage, excessive **drinking** of **alcohol** or using services of **prostitution**. However, it should be noted that such behaviour may not necessarily be criminal.

Definitions of deviancy and what constitutes deviant behaviour require a consideration of what are the norms in a given context, who or what determines them, who abides by them and why. To argue that deviance is defined by the norm, that the legitimate posits the illegitimate is only a partial answer. A sociological definition is that deviance is a structured rejection of dominant norms, a revolt against current patterns of power. Additionally deviance may be associated with concepts like **marginality** and **liminality**. Both these terms relate to the place apparently subordinate peoples might occupy in society, and the terms have been applied to **indigenous** people, gangs, prostitutes, females in general and tourists. The distinctions between deviance, marginality and liminality thus overlap. From one perspective, liminal and marginal people are those found as being at the margin of the dominant groups within society. However, writers have emphasised processes rather than social places, thereby defining marginality as a construct of the interaction, the interface between that which is the dominant set of norms and that which is subordinate, while liminality is the state of eluding classification, the slipping through the network of classifications that normally locate states and positions in cultural space.

All three concepts relate to tourism. First, it can be argued that the study of that which is deviant is a mirror image of the dominant in that it reveals more clearly concepts, values and norms that are often taken for granted and thus not verbalised. The challenge of the deviant forces an articulation of the hitherto unanalysed. Hence the behaviour of tourists represents an alternative to normal **life-styles**. While on holiday, the clock time of the organised schedule gives way to the time of the whim; adults can escape into **play** in ways often denied to them. Tourists also seek out some types of marginal people. In this way, in the linkages with native peoples, male and female prostitutes, tourism involves a search for the 'alternative'. Tourist behaviours are thus critical examinations of dominant societal processes.

See also: alienation; crime; culture, tourism; eroticism; hedonism

Further reading

Rojek, C. (1993) *Ways of Escape: Modern Transformations in Leisure and Travel*, Basingstoke: Macmillan. (Describes the history of modern tourism as emerging from processes of social control and legitimate carnival.)

Ryan, C. and Kinder, R. (1996) 'The deviant tourist and the crimogenic place – the case of the tourist and the New Zealand prostitute', in A. Pizam and Y. Mansfeld (eds), *Tourism, Crime and International Security Issues*, Chichester: Wiley, 23–36. (Discusses definition of deviant tourist in context of sex tourism.)

—— (1996) 'Sex, tourism, and sex tourism: fulfilling similar needs?', *Tourism Management* 17(7): 507–18. (Questions whether deviancy is an appropriate label to be used in sex tourism.)

CHRIS RYAN, NEW ZEALAND

diagonal integration

Diagonal integration is a process whereby firms use information technologies to logically combine services for best productivity and greatest profitability (Poon 1993). It links such services as financial services, insurance and travel in a way that important synergies are obtained from this integration. Since the tourism **industry** is increasingly driven by information and consumers, firms can diagonally integrate to control the more lucrative areas of value creation by producing a range of **services** to be consumed simultaneously at regular intervals (such as travel plus insurance plus holiday plus personal banking).

References

Poon, A. (1993) *Tourism, Technology and Competitive Strategies*, Wallingford: CAB International.

FRANCESC J. BATLE LARENTE, SPAIN

diet

Diet is a generic term encompassing the range and variety of food which an individual consumes over a period of time, determined by a number of factors. A 'healthy' diet will include a wide variety of commodities in order to provide all essential nutrients to both protect against illness and to promote **growth** and repair of the body. There are now many small and large **resorts** which offer special diet and exercise programmes in order to capitalise on this new **lifestyle** tourism market.

KATHRYN WEBSTER, UK

differentiation

Differentiation occurs under **modernity** when activities performed within some functional realm split up as industrial society becomes increasingly specialised and heterogeneous. In tourism, differentiation tends not so much to be a theory of social change, but an indicator of the distinctiveness of individuals/groups from others. Under the de-differentiated tourism of **postmodernism**, such hierarchical rankings (of wealth/status) often dissolve.

KEITH HOLLINSHEAD, UK

direct marketing

Direct marketing is a measurable **marketing** technique that involves communicating with a pre-identified audience, using one or more **media** in order to achieve a specific **market** response within a defined time frame. The overall aim of direct marketing is to achieve lower net promotional costs per measurable response than is possible with other methods.

Sometimes referred to as 'direct response marketing' or 'database marketing', direct marketing encompasses any marketing approach that involves direct one-on-one relationships between suppliers and targeted customers with the ultimate objectives of generating sales. These approaches include direct mail, telephone sales (telemarketing) and/or personal selling. Such techniques can be used individually or integrated with other promotional methods to achieve overall **marketing objectives**.

A basic tenet of direct marketing is to efficiently link tourism service suppliers and vendors with specifically targeted trade and/or consumer markets. This goal is achieved by developing relationships with current customers to sustain repeat business, inform clients of new products and secure referrals. Direct marketing creates awareness to inform referrals of available services/products and generates new leads through contact with names on proprietary and/or personally developed lists that match or augment current customer profiles.

Predicated on developing client profiles using a database, the significance of direct marketing is that it can be used by any tourism/hospitality enterprise, irrespective of size, target market or type of products/services offered. Requisites of direct marketing involve collecting detailed customer information, market **planning** and testing. The technique offers absolute control over customer contact and fulfilment, revenue and cost accountability that is attributable to specific campaigns, precise **market segmentation** that matches customer profiles, quick and immediate testing of markets and tourism products/services, special offers and/or programmes and finally, cost-effective management of promotion budgets. With these elements in mind, any travel organisation has the potential to successfully exploit direct marketing.

See also: market analysis; marketing information systems; marketing mix; marketing plan

Further reading

Burke, J.F. and Lindblom, L.A. (1989) 'Strategies for evaluating direct response tourism marketing', *Journal of Travel Research* 28(2): 33–7. (Discusses the applicability of direct response tourism programmes for destination marketing organisations.)

Fairlie, R. (1993) *Direct Marketing and Direct Mail*, 2nd edn, London: Kogan Page. (Outlines the concepts/principles involved, including database creation and applications.)

Hughes, A.M. (1994) *Strategic Database Marketing*, Chicago, IL: Probus. (Explains how to build a database and use it profitably with a focus on accountability.)

Talarzyk, W. and Widing II, R.E. 'Direct marketing and online consumer information services (OLCISs): implications and challenges', *Journal of Direct Marketing* 8(Autumn): 6–17. (Provides a comprehensive literature review of direct marketing as well as examination of trends and technologies.)

LAUREL J. REID, CANADA

disciplinary action

The goal of a discipline programme is to promote positive employee behaviours. To effectively lay the groundwork for a discipline system that promotes positive behaviours, managers first clearly establish the rules and then communicate how those rules should be carried out through the use of effective manuals, **training**, job descriptions, performance standards, posted notices and employee handbooks.

Managers must choose between two substantially different types of discipline systems. Normally referred to as the traditional approach to discipline, the first type emphasises the administration of discipline after an employee fails to follow organisational norms and standards. Obviously, since the behaviour precedes disciplinary action, this system is reactive in nature. For instance, an **airline** employee may be disciplined for arriving late to work. The other type, normally referred to as preventive discipline, is proactive in the sense that it attempts to establish a means of directing employee behaviours.

The traditional approach includes two types of popular discipline systems: the hot stove approach and progressive discipline. The hot stove approach represents what many would think of as conventional management wisdom: if one touches a hot stove, he or she gets burned. From an organisational perspective this mean when a **rule** is broken the employee is disciplined. The advantage of this approach is that it clearly establishes rules. However, the disadvantage is that it punishes a good employee as severely as it does a poor one. For example, assume that two travel agents were late for work. One has been consistently late and has other problems at work. The other is a model employee. The hot stove approach requires that both be disciplined in the same way, even though one employee is clearly doing a better job than the other. Like the hot stove approach, progressive discipline also relies on clear and complete definition of behaviours that will be punished and the type of punishment that will be meted out for each infraction. A progressive discipline programme might include, for instance, a rule that an employee who is tardy to work once will receive an oral warning, an employee who is tardy twice will receive a written warning, an employee who is tardy three times will be suspended and so on. This step-by-step approach to discipline is very popular, probably because it intuitively sounds like a good system. In addition, this type of approach generally proves that **management** has given the employee every opportunity to perform. This evidence might be valuable later.

For example, in the United States this information is often used in court cases to prove that an employee should have been released from his/her job. Without this, many former employees have won large court settlements because of what is know as 'unlawful discharge'. Most progressive discipline programmes include four steps: oral reminders, written reminders, paid decision-making leave and discharge. In each stage of this process except discharge, the emphasis is on positively encouraging good behaviour.

ROBERT H. WOODS, USA

discount pricing *see* pricing

discourse

A discourse is the statements or talk constituting a social language about a subject. That talk constructs the message in certain 'lights', collectively constraining other ways of seeing and knowing. In tourism, those within a state agency, travel trade **association** or particular population may (through their speech and writing) consciously or unconsciously privilege certain representations of history, **nature** and so on, denying rival explanations.

KEITH HOLLINSHEAD, UK

discretionary travel

Discretionary travel comprises all non-compulsory forms of tourism. It is engaged in after fulfilling obligatory trips related to institutional and social responsibilities, such as work and family. Occurring in **leisure** time, the journeys are perceived as pleasurable. The trips can be measured objectively through **time budget** studies or subjectively through perception of the degree of **freedom** of choice.

DAVID J. TELFER, CANADA

discriminant analysis

Discriminant analysis is a multivariate statistical technique used to estimate the relationship between a single categorical dependent variable (such as tourist versus non-tourist) and a set of metric (interval or ratio level) independent variables. The primary objective of this analysis is to understand differences among the groups (as defined by the dependent variable), and to predict the likelihood that an observation will belong to a particular group. Discriminant analysis can be used to investigate differences in tourism patterns (the independent variable) among groups as defined by trip purpose (for example, business versus pleasure) or **seasonality** (the dependent variable). Its predictive function might be used to classify potential customers to differing marketing programmes depending upon known social, economic and geographic characteristics.

Discriminant analysis involves deriving a linear combination of the independent variables that maximise the differences among groups. Two key assumptions that underlie the estimation process are multivariate normality of the independent variables and equal variance among independent variables for each group defined by the dependent variable. Parameter estimates are very sensitive to multicolinearity among the independent variables (this is the degree to which two independent variables are correlated), sample size and the share of observations among groups. It is generally recommended that a ratio of twenty observations is needed for each predictor variable and that each group (depending upon number) includes at least 10 per cent of all observations.

Application of discriminant analysis usually follows a three-step process: estimation, validation and interpretation. The researcher should use theory combined with the specific goal(s) of the study as the basis for selecting the dependent and independent variables. The validation process includes testing for the generalisability of the results and the classification accuracy of the model. Interpretation is accomplished by identifying significant independent variables, ordering them by degree of influence regardless of sign, calculating gaps between group means for each significant variable and then assessing these results within the context of the **theory** which underlies the study.

See also: classification; cluster analysis; marketing research

Further reading

Morrison, D. (1976) *Multivariate Statistical Methods*, New York: McGraw-Hill. (Provides a foundation for application of this technique in tourism research.)

Zimmer, Z., Brayley, R. and Searle M. (1995) 'Whether to go and where to go: identification of important influences on seniors decisions to travel', *Journal of Travel Research* 33(3): 3–10. (Discusses an example of discriminant analysis in assessing the factors affecting travel choices.)

DANIEL R. FESENMAIER, USA
JIANN-MIN JENG, USA

disease

Disease conveys lack of **health** in body or mind. From ancient times, it has been both a reason for and consequence of **travel**. The need for health advice, preventive measures and treatment have led to the establishment of medical specialism, tropical and travel medicine, and even special clinics. Diseases typically acquired when travelling include bowel infections, sexually transmitted diseases such as **AIDS** and vector-borne illnesses like malaria.

PETER GRABOWSKI, UK

disembarkation card *see* arrival/departure card

displacement

Displacement is a process resulting in something being replaced. One group of **tourists** may replace another or some of their **activities** may just be replaced. Local people, built or natural **resources** (such as wildlife) may be replaced or values/habits altered. Displacement may be confused with **substitution**, and signals lack of socially or **ecologically sustainable tourism** development.

JAY BEAMAN, CANADA

distance decay

There are costs involved in overcoming the friction of distance. Thus, other things being equal, there is more interaction between close than distant places. This applies to the availability of information on places, whether gained by personal experience or other sources such as **advertising**, as well as the volume of tourist arrivals. Thus, at a destination, there are likely to be a relatively large number of people who have travelled a short distance with declining numbers of visitors as origins become more distant. The same relationship exists in reverse, with the numbers from places of origin, such as a city, tending to diminish with increasing distance from the origin. Since there is a relationship between the time available and the distance travelled, day trip, weekend (**second home**) and broader vacation zones may be identified around cities. The relationship between distance and volume of travel is not linear, with often fewer than expected observations at very short distances, individuals being required to leave their accustomed environment to be considered as **tourists**, and momentum coupled with a sense of exploration encouraging those travelling a considerable distance to go further. The concept of distance decay underpins the **gravity model**.

See also: accessibility

Further reading

Greer, T. and Wall, G. (1979) 'Recreational hinterlands: a theoretical and empirical analysis', in G. Wall (ed.), *Recreational Land Use in Southern Ontario*, Department of Geography Publication Series No. 14, pp. 227-45, Waterloo, Ont.: University of Waterloo.

Wolfe, R.I. (1972) 'The inertia model', *Journal of Leisure Research* 4: 73–6.

GEOFFREY WALL, CANADA

distance education

Distance learning is a learning environment whereby the student is separated by time or space from the instructor and where the principals do not have direct face to face interaction defines a distance **education** experience. Communication may be facilitated by mail, telephone, **radio**, **television**, **video** cassettes, computer-mediated instruction (see **education, computer-assisted**) and/or **Internet**-based instruction, individually or in any number of combinations to achieve an optimum learning experience. The types of media employed in distance learning might be categorised as single media, such as print, radio or television; multimedia, which includes any number of combinations of media; an interactive media, like email; or interactive online real participation, such as teleconferencing, video-conferencing or computer-conferencing.

Students may have limited financial resources, physical disabilities, time constraints, specialised fields of study or educational demands that cannot be met in the traditional within school system. Several North American universities such as the University of Houston and the University of Delaware offer both graduate and undergraduate degrees in tourism and hospitality entirely based on a distance education mode. Demographic factors, achievement motivation interests and grade point average play an integral part in the students' interest in distance education programmes. Out-of-school instruction can enhance students' understanding of the subject material and improve test performance levels by allowing them to work at their own pace and repeat any points that are not clear. The flexibility of learning pace and time makes distance education appeal largely to non-traditional, working students. Most of those who are enrolled in the Master of Hospitality Management degree programme via the distance education mode at the University of Houston are employed full-time in tourism and hospitality.

This learning process can become very cost-effective when one considers **economies of scale** gained through repetitions of the delivery process, regardless of the media. Consequently, distance education significantly benefits the tourism and hospitality students in developing countries. With the advance of computer technology and the use of the Internet as an interactive learning station, students from throughout the world will be able to access information, discuss the latest theories and interact with the leading authorities in the industry at a fraction of the cost of traditional methods.

Further reading

Clark, R.E. (1983) 'Reconsidering research on learning from media', *Review of Educational Resources* 53(4): 445–59.

Iverson, K. (1996) 'Exploring student interest in hospitality distance education', *Hospitality Research Journal* 20(2): 31–43.

Kaufman, D. (1989) 'Third generation course design in distance education', in R. Sweet (ed.), *Post-Secondary Distance Education in Canada: Policies, Practices, and Priorities*, Athabasca: Athabasca University/Canadian Society for Studies in Education.

Pelton, J. (1990) 'Technology and education: friend or foe?', in M. Croft *et al.*, *Distance Education: Development and Access*, Caracas: International Council for Distance Education/Universidad Nacional Abierta.

K.S. (KAYE) CHON, USA

distribution channel

The concept of distribution channels is not limited to the distribution of physical goods. Although the principles are the same, the channel distribution for tourism differs significantly from those used for manufactured goods. This difference stems from the nature of tourism **services** and their production system and consumption patterns. Tourism services require simultaneous production and consumption, meaning the **product** is not normally 'moved' to the consumer. Further, the product is often sold in conjunction with another one, such as **airline** tickets. Because of the perishability of most of the tourism products, many traditional channels may not work. While eliminating some of the functions and problems of **transportation** and warehousing, a distribution channel in tourism should consider such reasons in reaching and catering to the market.

A tourism distribution channel may be defined as a total system of linkages between actual and potential **tourists** and the suppliers. The structure of the distribution system may be either direct (from the producer to the seller) or indirect (the sale to the consumer through an intermediary). As the definition implies, the challenge is how to get the customer to the consumption **site** (the retailer), that is, to make it convenient and accessible. This unique feature raises the need for a different kind of distribution system in tourism.

Suppliers of tourism services and destinations may use several different methods to distribute their goods and **services**. These include their own channels (partially or wholly owned), selling through **management** and **marketing** contracts, **franchising**, hiring sales representatives and using various intermediaries. The intermediaries in a tourism channel of distribution consist of three main categories: tour packagers, retail travel agents and speciality channels. Included in the latter are **incentive** travel firms, **meeting planners** and convention planners, hotel representatives, association executive marketing organisations, corporate travel offices and others.

New developments in direct access to global distribution systems make tourism arrangements instant and more accessible. For example, agents can make the flight arrangements, get a rental car, book a hotel room and buy a ticket to a show without ever using the telephone. In addition, there is enough evidence to suggest that direct selling, away from the location of production and consumption, is on the increase since consumers have more access to the world information system of the **Internet** and other available information databases.

There are three main channel strategies used by marketing professionals and destination promoters to stimulate demand: the pull strategy, the push strategy and joint promotional efforts or cross marketing. In the pull strategy, the goal is to entice the consumer to buy the product. Certain inducements are offered to make the potential tourist more interested or seek the appropriate distribution channel for the product in question. For example, this includes frequent flyer programmes or incentives for repeat visitors to a given **resort**. The push strategy, on the other hand, acts in the opposite way. The goal is to get the intermediary to sell the product to the consumer. For example, **tour**

operators and **travel agencies** that work with resorts and hotels, or convention and meeting planners may be offered commission to increase bookings for a given time period. Certain incentives, such as complimentary rooms or free tickets to destinations, are among commonly used strategies. Joint promotional efforts and alliances are useful and commonly employed strategies in expanding the market base through intermediaries. For instance, hotels, **restaurants** and attractions can utilise coupons in order to bundle products and bridge the gap among offerings, or hotels working with airlines can arrange specific marketing packages to mutual destinations. The role of **tourism organisations** (regional or local) should not be underestimated in helping the components of supply and offerings reach the appropriate distribution channels in the marketplace.

Large tourism companies can become their own suppliers of products (vertical integration). This type of distribution needs a large amount of capital to be successful. Such efforts are usually carried out within the constraints of companies' resources and ability. Examples are Thomas Cook, American Express and the Carlson Companies. Vertical integration allows businesses to have and exert control over the entire channel of distribution through retail outlet ownership and organisation of the distribution channel. For example, both the Carlson Travel Group and the Marketing Group feed reservations to the hotels and the cruise ships. All customers of the company are encouraged to stay in Radisson Hotels at all of the relevant destinations.

Major steps in the promotion and development of destinations have been linked with advancements in physical linkages and the level of **accessibility**, the system that creates the structural linkage between the market place (origin) and the destination site. This linkage between the market and destination is also embodied in information systems relating to both computerised reservation services (CRS) and communication systems in the automation of tourism financial services.

The means that are used to facilitate the flow of information between the consumer and suppliers of tourism goods and services also need special marketing and management considerations. For example, update information packages should be developed, clearly describing and illustrating the nature of the product offerings. Bochures should be made specific to destinations. It is the responsibility of destinations to provide intermediaries with updated and accurate information.

See also: destination management company; hospitality information system; information technology; management contract; marketing information system; marketing, destination; reservation; route system; tour wholesaler; travel advisory

Further reading

Bitner, M.J. and Booms B.H. (1982) 'Trends in travel and tourism marketing: the changing structure of distribution channels', *Journal of Travel Research* 21(Spring): 39–44.

Lewis, R., Chambers R.E. and Chacko H.E. (1995) *Marketing Leadership in Hospitality*, 2nd edn, New York: Van Nostrand Reinhold. (See Chapter 16 for channels of distribution.)

Middleton, T.C.V. (1988) *Marketing in Travel and Tourism*, Oxford: Heinemann Professional Publishing. (See Chapter 18 on distribution channels in travel and tourism.)

Uysal, M. and Fesenmaer, D. (eds) (1994) *Communication and Channel Systems in Tourism Marketing*, New York: Haworth Press.

MUZAFFER S. UYSAL, USA

diversionary tourist

Diversionary tourist is an intermediate category between recreational and experiential in Cohen's **phenomenology** of tourism experiences. One whose 'spiritual centre' lies neither elsewhere nor firmly rooted at home, the diversionary tourist thus seeks neither the 'centre out there' nor simple **recreation** but, rather, tourism that offers temporary **escape** or diversion from the normal home environment and experience.

See also: alienation; motivation; typology, tourist

RICHARD SHARPLEY, UK

domestic tourism

Domestic tourism involves people visiting destinations within their own country's boundaries. It is recognised as one of the three major categories of tourism, the others being **inbound** tourism (international visitors travelling to a country other than their own) and **outbound** tourism (residents of a country travelling to other countries). Domestic tourism represents the lifeblood of the tourism industry. For example, in the United States, it accounts for 99 per cent, in Australia 94 per cent and in Britain 80 per cent of tourism volume. However, it usually accounts for a proportionately smaller percentage of tourism revenue generated.

Domestic tourism is often regarded as being less significant to national economies than inbound tourism, for it does not generate foreign currency. Instead, it redistributes domestic currency spatially within the boundaries of a country. From a regional perspective, however, its net **benefits** can be the same as inbound tourism as it provides an important source of new money flowing into a region. The additional expenditure by domestic tourists has the same type of impact as inbound tourism on the country as a whole, generating business opportunities, **employment**, expenditure and **tax** revenue for state to local governments. Indeed, in many regional centres, the **industry** is totally reliant on domestic tourism.

Defining the size of the domestic tourism market and the extent of its activity is problematic. A number of researchers have commented that this sector is measured poorly and infrequently. The reasons are manifold, and begin with the challenge of defining who is and who is not considered as a domestic tourist. Most countries have adopted an arbitrary definition based on minimum distance and time criteria (for example a minimum trip of 50 kilometres one way with a minimum stay of one night), although they readily admit the limitations of such a definition. The statistical problem of describing the sector is exacerbated by the fact that it is impossible to capture all domestic tourists because of the diverse spatial nature of this activity and the fact that many do not use commercial **accommodation** facilities. In addition, certain activities, such as the use of holiday homes, defy easy inclusion or exclusion from any arbitrary definition.

A number of features distinguish domestic from inbound tourism. Domestic tourism and the resulting benefits are more widely dispersed within a country than those of inbound tourism. Whereas inbound tourism is focussed in and around international gateways and major icons, domestic tourism activity is more diffuse. Nationally or regionally significant attractions may serve to act as **demand** generators for domestic tourists, while their appeal to the inbound market may be limited.

From an industry perspective, the financial barriers for businesses wishing to enter the domestic sector are significantly lower than they are for the inbound sector. As a consequence, the domestic sector is typified by smaller, independently owned and operated accommodation facilities and attractions. By the same token, the travel trade plays a relatively smaller role in domestic tourism, as these are independent tourists. The automobile is their preferred mode of **transportation**. Likewise, domestic tourists show a proclivity to stay with friends and relatives, or prefer moderately rated facilities over international hotel properties. Domestic tourists also tend to be frequent repeat visitors to favoured destinations.

The volume of domestic tourism is influenced by a number of factors. The size of the country, the distribution of the population within that country and its proximity to other countries influence the array of tourism choices available to residents. A large country with a widely dispersed population that has few neighbours, like the United States, generates proportionately more domestic tourism than a small country with many neighbours, like Belgium.

As with all other forms of tourism, the propensity to participate in the domestic sector is influenced by the demographic profile of the population and the general level of affluence of the country (see **demography**). The more affluent the country is, the greater the likelihood that its residents will participate in tourism. Domestic tourism is influenced further by the richness of a country's tourism resources and the appeal of its primary attractions. It is further influenced by the level of national pride evident in a country that may provide the stimulus to 'discover' or connect to the essential character of one's home country. Even some forms of domestic tourism have been likened to **pilgrimages**, where residents come to pay homage to the historical or cultural symbols that have defined their collective conscience.

The number and size of geopolitical divisions within a country will have an influence on the amount of domestic tourism that occurs, especially if each of the regions has its own distinctive identity. Many Americans, Canadians and Australians, for example, are as interested in visiting their internal states or provinces as they are in going abroad. Finally, barriers to exiting a country may restrict outbound tourism, diverting people to domestic tourism. Such barriers can come in the form of high travel costs, high departure taxes, restrictions placed on the authorisation of travel visas, an increase in value of foreign exchange, or the generating country's unfavourable international reputation.

Further reading

Jafari, J. (1986) 'On domestic tourism', *Annals of Tourism Research* 13: 491–6. (Offers an overview of domestic tourism issues.)

Morrison, A.M., Hsieh, S. and O'Leary, J.T. (1994) 'Segmenting the Australian domestic travel market by holiday activity participation', *The Journal of Tourism Studies* 5(1): 39–56. (Identifies six distinct domestic market segments with unique sets of activity participation and other characteristics.)

Sindiga, I. (1996) 'Domestic tourism in Kenya', *Annals of Tourism Research* 23(1): 19–31. (Examines emergence of domestic tourism in a developing country.)

BOB McKERCHER, CHINA

Dominican Republic

The Dominican Republic shares with Haiti the Island of la Hispaniola, the second largest in the Caribbean, where Christopher Colombus first set foot in America. Its 1,600 kilometres of coast, green hills and tropical **climate** are promoted as a complete holiday **destination**. In the 1970s, the **World Bank** promoted the Puerto Plata tourism project on the North Coast. After its success, private investors, mostly European, have increased the number of rooms to more than 34,000, with tourist arrivals reaching 2,000,000 in 1996.

LLUIS MESALLES, SPAIN

drinking

The consumption of alcoholic beverages is common in various tourism settings. A drink, a glass of wine or a pint of beer is often regarded as a symbol of **leisure** or breaking away from the everyday routine. Holidays in destinations with few regulations regarding **alcohol** consumption may result in excessive drinking by **tourists**, especially those coming from more regulated environments or countries.

TOM SELANNIEMI, FINLAND

ecoethics

Ecoethics is the study of the value of the physical and biological **environment**, or the principles of environmental conduct governing an individual/group in any interaction with the biosphere. These are moral principles or values dealing with **behaviour** which have touristic environmental **impact**, implications or significance.

See also: code of ethics; environment

Further reading

Western Australian Tourism Commission (1989) *The Ecoethics of Tourism Development*, Perth: Western Australian Tourism Commission.

ROSS K. DOWLING, AUSTRALIA

ecological economics *see* economics, ecological

ecologically sustainable tourism

The system of **development** known as ecologically sustainable tourism fosters the use, **conservation** and enhancement of a **community**'s resources. It favours those actions which promote ecological processes and the total quality of life, for this and future generations. This form of development is underpinned by the philosophy that the **environment** must be conserved if the industry is to be viable in the long term.

See also: sustainable development

ROSS K. DOWLING, AUSTRALIA

ecology

Ecology is the study of the relationships among living organisms and among organisms and their **environment**. It is especially related to animal and plant communities, their energy flows and interactions with their surroundings. During recent years, the relationship between ecology and tourism has received much attention.

See also: ecotourism; environment; impact; sustainable development

ROSS K. DOWLING, AUSTRALIA

economic development

Economic development is concerned with a variety of aspects relating to the quality of life, and should not be confused with simple economic growth. The former includes references to health and welfare, whereas the latter is focused upon the more narrow aspect of the rate of **change** of gross national product or gross domestic product. A variety of definitions has been put forward to explain economic development, and most of them include some form of *conditional* economic growth. The nature of the conditions generally relates to the education levels of the population (including literacy levels), the welfare of the poorer segments of the population, and the proportion of gross national product that is attributable to agriculture. The main thrust of the argument is that economic development should result in a more educated population with less income inequalities, where other industries besides agriculture are responsible for generating income and **employment** opportunities, and that there must be some **indigenous** self-sustained technological change. The term **development** almost always includes some reference to self-sustained growth. This makes it strange to discuss **sustainable development**, because if the development is not sustainable then it tends to be short-term growth and thereby fall outside the definition of development.

Nafziger (1984: 38) provided one of the best-quoted definitions of economic development:

> Economic growth is an increase in a country's per capita output. Economic development is economic growth leading to an improvement in the economic welfare of the poorest segment of the population, a decrease in agriculture's share of output, an increase in the educational level of the labour force, and indigenous technological change.

With respect to tourism's contribution to economic development, it can be seen that there is scope for tourism to assist, if not provide a catalyst for economic development. A variety of theories have been put forward over the past century to explain economic development. With the exception of the English classical theory, which offered little prospect for any industry, all of the major theories provided a role for tourism as a means to development. This role is either in the form of providing additional stimulus to demand or as a way of overcoming the natural inertia of economies.

References

Nafziger, E.W. (1984) *The Economics of Developing Countries*, Belmont, CA: Wadsworth.

JOHN FLETCHER, UK

economic impact *see* impact, economic

economic leakage *see* leakage

economic multiplier *see* multiplier effect

economics

Economics began as a subject that examined the welfare of a nation and/or that of the individuals of a nation. It was initially regarded with some scepticism by traditional academics. In order to enhance its standing with the academic community, economics evolved into what is known as macroeconomics and microeconomics. It is now commonly regarded as being the study of the allocation of scarce resources. Central to many aspects of economic concern is the concept of opportunity cost. That is, each time a resource is allocated to a particular function, be it at the national or consumer level of activity, then other uses of the same resource are being foregone and this carries with it an inherent cost. Economics is considered to be a social science that employs scientific techniques in order to understand the economic behaviour and choices of people, businesses and governments. Furthermore, economics may be subdivided into those aspects relating to **market** systems which ostensibly allow the price mechanism to determine resource allocation and centrally planned economies which are based upon some predetermined notion of resource allocation. Economic theories and methodologies can be used to explain, quantify and predict tourism-related activities.

Economics may be categorised into seven areas. First, macroeconomics is concerned with aggregate economics such as the economics of national governments, including income, employment, **multiplier effects**, **investment**, **tax**, foreign trade, **foreign exchange**, **economic development** and economic **growth**. The study of the effects of tourism expenditure upon national economies accounts for the largest single aspect of research in relation to tourism. Numerous studies have been undertaken to determine how much income, **employment**, government revenue and foreign exchange flows are generated by tourism spending. These studies focus upon the economic linkages, **leakages** and multiplier effects associated with an economy. Of particular concern at the macroeconomic level is the degree of dependence that an economy may have on tourism activity (see **dependency theory**).

Second, microeconomics focuses upon the activities of businesses or individuals within a national, regional or local economy. Here the study of the allocation of resources can be examined with a view to estimating profitability for businesses, optimum production levels and economic responses to different market structures. The **theory** of the firm and the study of monopolistic markets, perfect markets and imperfect markets all carry with them implications for tourism business behaviour and the price elasticity of demand. Microeconomics is also concerned with matters such as production functions, which describes how businesses produce their output from sets of resources. The existence of economies of large-scale production is evident in many of the industries involved in the delivery of the tourism product such as airlines, cruise lines and hotels. At the consumer level, economic decisions can be examined in order establish behavioural responses to price and income level changes and the study of consumer choice and welfare.

Third, regional economics relates to the study of subnational economic systems, such as counties or states. It is similar in structure and framework to macroeconomics, but its study is burdened by the absence of sets of local 'national accounts'. This lack of secondary data tends to require economists to utilise alternative **models** to test their hypotheses. The lack of local secondary data often results in the need for extensive fieldwork and primary data collection. At the macroeconomic level, tourism activity is often judged by its performance as a means of generating foreign exchange, whereas at the regional level it is often seen as a vehicle for redressing economic imbalances within a national economy, by creating income and **employment** opportunities in areas where they would not otherwise exist. Because of the spatial dimension relating to regional economics, the topic of location theory and its implications for business and **resort** development can be included under this heading. The tendency for tourism **development** to occur in resorts or clusters in many destinations makes regional economics a particularly pertinent area of study.

Fourth, development economics is the study of the performance of the system in achieving accepted economic and welfare goals. Given the fact that the economic benefits associated with tourism development are one of its most significant stimulators, it is an important aspect of the economics of tourism. This industry can be chosen as a catalyst for general **economic development** for a variety of reasons. Unlike many primary goods, the international price of the tourism product is not determined by international commodity **markets**. Even though tourism may often be regarded as being highly competitive and price sensitive, there is still some room for product **differentiation** and price determination. Furthermore, because tourism is a multi-sector product, it involves a wide spectrum of industries and the development of these sectors, and their role in achieving general economic development is an area of study.

Fifth, international economics is largely concerned with the flow of trade and services between economies and the competitive and comparative advantages that may be associated with such trade. **International tourism** expenditure may be seen as an export of goods and services by the receiving destination. On the other hand, **domestic tourism** may, under certain circumstances, be seen as an import substitution activity if, in so doing, it dissuades consumers from spending money in destinations located outside the domestic economy.

Sixth, industrial economics is concerned with such matters as agglomeration theory and the concentration of activities within an industry. The study of macroeconomics relating to tourism issues is well established as an area of research and has been central to much of the impact analysis studies that have taken place. From a microeconomic point of view, there are several areas that have attracted international attention, such as the existence of economies of large-scale production, demand theory and price elasticity of tourism destinations, the effectiveness of fiscal incentives as a means to securing **investment** in tourism, and so on.

Seventh, public sector economics encompasses the fiscal and expenditure activities of central and local governments. Given the nature of tourism and its dependency upon many elements that are outside of the pricing mechanism, the study of public sector involvement and finance is particularly relevant.

The above points, although by no means exhaustive, provides an insight into the many different aspects of economics relevant to the study of tourism and its development. Different destinations will have different economic priorities with respect to tourism activity and its development. For instance, some economies may have an abundance of labour and consequently seek to implement those forms of production that are labour-intensive, thereby exploiting comparative advantage in that area. Other destinations may find that their comparative advantage may be related to capital or particular physical resources. All of these aspects will bring with them different implications for the application of economic theories and the development of tourism. The major driving force underlying tourism development is predominantly economic, whereas considerations relating to **environment** and social considerations are often secondary in importance and seen as constraints upon development rather than goals.

There is often a tendency to consider economics as being confined purely to impact, development and market failure issues. However, economics can also be used to assist in quantifying both the social and environmental consequences of tourism and its development. Indeed, more recent research utilises economic methodologies to determine the direct and indirect environmental impact associated with tourist spending.

See also: economics, ecological; informal economy

Further reading

Bull, A. (1995) *The Economics of Travel and Tourism*, 2nd edn, Australia: Longman.

Ioannides, D. and Debbage, K. (1998) *The Economic Geography of the Tourist Industry*, London: Routledge.

Lundberg, D., Stavenga, M. and Krishnamoorthy, M. (1995) *Tourism Economics*, New York: Wiley.

Sinclair, M.T. and Stabler, M. (1997) *The Economics of Tourism*, London: Routledge.

JOHN FLETCHER, UK

economics, ecological

Ecological **economics** is a field of study that has developed strongly during the final two decades of the twentieth century. An international society which publishes the *Ecological Economics* journal describes the field as transdisciplinary, one that spans and builds upon many others. A key purpose of ecological economics is to investigate and better understand notions of sustainability (see **sustainable development**) and equity as they relate to all levels of human endeavour.

An increasing international awareness that global life support systems are endangered has led to the realisation that decisions made on the basis of short-term criteria can produce disastrous long-term results, at the local, regional or global scales. Further, there has been growing recognition that the conventional approaches and recommendations of **economics** and ecology have fallen short in their ability to address complex environmental problems in a systematic way. These factors have been significant in the emergence of ecological economics as a field of study.

Ecological economics is distinct from both ecology and economics, although it may embody many concepts from each. However, it would be erroneous to presume that it is simply a more 'green' economics than environmental economics or resource economics. Indeed, practitioners may draw from diverse areas such as **sociology**, **management**, **law**, agriculture, **political science**, **geography**, biology, public health, **accounting**, urban **planning**, engineering (see **environmental engineering**) and theology (see **religion**). Ecological economics is by nature pluralistic and encourages the use of a systems approach (see **systems theory**) in understanding the dynamics of whole ecosystems inclusive of humans and their diverse activities.

An example of ecological economics being applied to tourism is found in van den Bergh (1995) on work undertaken in an island region in

Greece. Tourism has been a rapidly growing activity in this region, but many environmental pressures have been generated as a result of this. In order to better understand the complexities of the problem, the author developed a simulation **model** that allowed investigation of how various approaches to management of the environment might affect tourism, **employment** and environmental outcomes.

References

van den Bergh, J.C.J.M. (1995) 'Dynamics analysis of economic development and natural environment on the Greek Sporades Islands', in J.W. Milon and J.F. Shogren (eds), *Integrating Economic and Ecological Indicators*, Westport, CT: Praeger. (Contains many articles which are relevant to the theme of ecological economics.)

Further reading

Costanza, R. (ed.) (1991) *Ecological Economics: The Science and Management of Sustainability*, New York: Columbia University Press. (Contains thirty-two articles that capture the essence of early thinking by modern ecological economists.)

JOHN A.J. WOLFENDEN, AUSTRALIA

economies of scale

The concept of economies of scale refers to productivity increases (or diminishing average costs) derived in certain cases from equal proportional increases in all the production inputs (Samuelson and Nordhaus 1993). Within a given technological **environment**, the economies of scale can be pecuniary, real, and external. First, increases in the production level allow for lesser monetary costs in productive inputs, financing, **transportation** or distribution. There are well-established cases of pecuniary economies of scale in tourism distribution systems and in **hospitality**. Second, because of the limits to divisibility in tourism production and the economies of specialisation, there is also a real physical reduction in the inputs used per unit of output and thus in the average **cost**. Large tourism attractions, such as **theme parks** or marinas, are good examples of real economies of scale. Third, with technological changes, there is also the possibility of external economies of scale. In this case, the general increase of production of the industry as a whole allows for greater specialisation of individual productive units and lower average costs; there is even the possibility of negative slope supply curves.

In the business **paradigm** of the Fordian age of tourism, characterised by mass production and consumption of such services, the economies of scale were the basis of profitability and **competitiveness** in enterprises. All pecuniary, real and external economies of scale were used, mimicking industrial policy strategies. However, this source of profitability has encountered shortcomings with the supersegmentation of **demand** characteristic of the new business paradigm in tourism and efficiency problems derived from excess overheads and bureaucracy in large productive units. Additionally, the use of economies of scale in tourism has produced considerable negative social and environmental **impacts** in destinations, even to the point of drastically reducing their **market** competitiveness. Today, economies of scale are still important; large concerns such as **airlines** or hospitality giants use them in many areas of their business processes, and even small companies integrated in tourism clusters benefit from external economies of scale. However, the sustainable profitability of companies and destinations nowadays demands other methods, responding better to the rapidly changing environment of tourism markets.

Further reading

Bull, A. (1994) *The Economics of Travel and Tourism*, London: Pitman.

Fayos-Solà, E. (1996) 'La nueva política turística', in *Arquitectura y Turismo*, Barcelona: Universidad Politécnica de Catalunya.

Graselli, P. (1992) *Economia e política del turismo*, Milan: Centro Italiano di Studi Superiori sul Turismo, Assissi, Franco Angeli.

Poon, A. (1993) *Tourism Technology and Competitive Strategies*, CA International.

Samuelson, P.A. and Nordhaus, W.D. (1993) *Economics*, New York: McGraw-Hill.

EDUARDO FAYOS-SOLÀ, SPAIN
AURORA PEDRO, SPAIN

ecoresort

An ecoresort is a self-contained, upmarket, nature-based **accommodation** facility. It is characterised by environmentally sensitive design, **development** and **management** which minimises its adverse **impact** on the **environment**, particularly in the areas of **energy** and **waste management**, water conservation and purchasing. An ecoresort acts as a window to the natural world and as a vehicle for environmental learning and understanding.

ROSS K. DOWLING, AUSTRALIA

ecosystem *see* environment; planning, environmental

ecotourism

The term 'ecotourism' is usually attributed to Ceballos-Lascurain, who defined it as 'tourism that consists in travelling to relatively undisturbed or uncontaminated natural areas with the specific objective of studying, admiring, and enjoying the scenery and its wild plants and animals, as well as any existing cultural manifestations (both past and present) found in these areas.' However, there is no widely accepted definition. For some, it is little more than a **marketing** concept attached to almost any tourism **product** to attract those sympathetic to environmental causes. For others, true ecotourism must contribute directly to the maintenance and enhancement of **parks** and **protected areas**, the well-being of resident communities and environmental **education**, in addition to those attributes identified by Ceballos-Lascurain which, in the absence of the latter, may be viewed as **nature tourism**. In the former case, the term has little meaning whereas in the latter case the definition is so restrictive that few cases which meet the criteria can be found, and the description is one to which operators and **destination** areas may aspire rather than being a clear notion of a widely-available tourism product. Usually regarded as a form of **alternative tourism**, ecotourism overlaps with other types such as **adventure tourism** and **safaris**. A distinction is sometimes made between hard ecotourism (in which minimal facilities are provided and there is close interaction with the **environment**) and soft options (see **soft tourism**) which involve the use of considerable support facilities.

Ecotourism is often viewed as being a new form of tourism. Although the term is less than two decades old, there are many precedents and visiting natural areas has a long history. Often cited as being one of the fastest growing forms of tourism, definitional problems have frustrated the collection of data to substantiate this claim. In fact, while many tourists enjoy and participate in environmental experiences, true ecotourists and ecotourism businesses constitute only a small proportion of the industry. Ecotourism is increasingly being advocated as an environmentally benign means of stimulating development and, at the same time, preserving natural areas and their wild inhabitants in peripheral locations. As such, it is viewed as an alternative to **mass tourism** and a potential contributor to **sustainable development**, although achievement of the latter will depend very much on the form which ecotourism takes.

Most ecotourists are residents of the developed world with above average incomes and educations. While not confined to such locations, ecotourism opportunities in the developing world have attracted the greatest attention in the literature. This has led to charges of ego-tourism and eco-imperialism as a wealthy elite from developed countries with degraded environments advocates the **preservation** of relatively natural areas in the economically poor but ecologically rich less-developed world.

See also: agrotourism; ecologically sustainable tourism; rural tourism

Further reading

Boo, E. (1990) *Ecotourism: The Potentials and Pitfalls*, Washington, DC: World Wildlife Fund, 2 vols.
Lindberg, K. and Hawkins, D. (eds) (1993) *Ecotourism: A Guide for Planners and Managers*, North Bennington, VT: The Ecotourism Society.
Wall, G. (1997) 'Is ecotourism sustainable?', *Environmental Management* 21(4): 483–91.
Weaver, D.B. (1998) *Ecotourism in the Less Developed World*, Wallingford: CAB International.

GEOFFREY WALL, CANADA

Ecotourism Society, The

The Ecotourism Society (TES), launched in 1990, is an international non-profit organisation dedicated to finding the resources and building the expertise to make tourism a viable tool for **conservation** and sustainability. TES is run by a Board of Directors.

The philosophy of the Society is 'uniting conservation and travel worldwide', and its long-term objectives are to establish **education** and **training** programmes, to provide information services, to determine standards and criteria for the field, to build an international network of institutions and professionals, and to research and develop state-of-the-art models in the field of **ecotourism**.

Major activities of the Society are geared to meet its long-term objectives. A series of international ecolodge forums and field seminars have been held to formulate an international set of guidelines for ecolodge development and operation. Workshops have been held at the George Washington University since 1991 in order to review the characteristics of the ecotourism market, current approaches in government-level ecotourism **planning**, and planning and **management** tools for local communities. TES also organised the International Ecotourism Partners meeting in 1994 to explore how the Society can form a **partnership** with autonomous local ecotourism organisations, distribute its publications, and develop joint projects and membership initiatives. To provide information, TES completed and distributed its first International Membership Directory in 1995. The Directory fosters an exchange of interaction among members. The Society also provides information to tourism communities seeking to make their operations environmentally sound, to monitor ecotourism development and to support conservation. Its headquarters are in North Bennington, Vermont, USA.

BONG-KOO LEE, USA

Ecuador

Ecuador offers a diverse range of natural and historical attractions. In 1996, 420,000 **tourists** visited Ecuador (with 47,000 going to the Galapagos Islands). Income from tourism amounted to $281 million (1.7 per cent of GNP), an increase of 7.5 per cent from 1995. The Ministry of Tourism and CETUR, the Ecuadorian Tourism Corporation, plan **policy**, provide **education** and **training**, implement international **marketing** and enforce regulations for the private sector. Promotion of **nature tourism** is a high priority.

JEAN COLVIN, USA

Ecumenical Coalition on Third World Tourism

In 1982, churches from throughout the world formed the Ecumenical Coalition on Third World Tourism (ECTWT) to assist **Third World** people in responding to the **impact** of tourism. The Coalition publishes *Contours* (a quarterly journal) and other publications such as proceedings of conferences it hosts. While not initiating actions, it researches documents and publicises social justice issues and supports those working to secure the rights of those affected by tourism.

PETER HOLDEN, THAILAND

education

The increasing size and sophistication of the tourism industry has brought increasing pressure for a substantial upgrading of the professionalism and on-the-job capabilities of all its employees. This pressure is evident at all levels, from the basic front-line staff positions to the most senior leaders. The recent growth in tourism education and training is an attempt to respond to these pressures, and to do so in a way that makes this career a competitive option in the employment market. While there is a clear recognition that education and training is a lifelong learning process, a major challenge currently facing tourism is to attract and train the many young people required for entry level positions. In response to this challenge, a broad range of education and training programmes have been established.

Technical institutions, both vocational and technical, have formed the traditional backbone of training for the industry. They are typically two to three years in duration, and lead to a diploma in various specialized areas. Examples of these include bartending, food and beverage preparation, hotel sales and marketing, and motel operations. The major strength of their graduates is their ability 'to get the job done' very soon after entering the workforce. Even so, in certain areas (notably food preparation) a period of apprenticeship may be required.

At the undergraduate level, the hotel school model is the best-established approach. European schools provided early frameworks for hospitality education. Such programmes continue to be well-respected. More recently, university undergraduate programmes that focus on education for hotel management have emerged and have become increasingly popular. Perhaps the most internationally recognised programme of this type is located at Cornell University in the United States. Other well-regarded programmes, based on somewhat similar models, are at the University of Nevada at Las Vegas (USA), the University of Guelph (Canada), the Hague Hotel School (the Netherlands), and the University of Surrey (UK). Their most distinguishing characteristics have been their emphasis on preparing individuals to manage hotel and resort properties. Their graduates have been well received by the sector for many years. At the same time, some believe that although these graduates could be excellent operational managers, they may have difficulty serving in a strategic management role.

A general management with a tourism focus model of programming seeks to broaden the educational experience of students while still providing a strong industry orientation. The structure and content of programmes based on this model may be characterised as follows: its core emphasises general management education, but also includes the liberal arts, languages and mathematics as programme requirements. Rather than having students concentrate on more advanced courses in a particular functional area of business (as in a traditional management programme), this programme is structured to enable students to understand tourism by taking a number of courses related to its subsectors. In order to obtain some of the operational knowledge and skills provided by hotel schools, tourism programmes frequently include a number of practical work terms as an integral part of the learning process.

Another model, the liberal arts with a tourism focus, includes a broad range of programme types, including discipline-based programmes having a tourism component or emphasis. The recreation or leisure studies programmes frequently include a significant tourism component, and tend to be multidisciplinary in nature. These and others usually draw heavily on the social sciences (such as geography, economics, psychology and sociology), although they may also involve input from such fields as physical education and computer science. One important characteristic of these programmes is that they generally have a more academic, as opposed to a trade, orientation. Further, these multidisciplinary majors are somewhat similar to the previous model. They are distinguished by the fact that they tend to have a much stronger industry orientation. As a result, they often include some management-related courses, a greater emphasis on language training, and increasingly, courses that relate to tourism planning. While the liberal arts programmes are frequently criticised as being too theoretical or not sufficiently oriented for the needs of the industry, they possess some very real strength. They provide students with a much broader understanding of the societal dimensions of tourism and its impacts. In addition, they offer students a high degree of flexibility in selecting courses of particular interest to them. They also tend to provide them with knowledge that, while not focused on tourism, is highly relevant to success in the field.

The hybrid model of tourism/hospitality education represents a more recent programme approach that takes into account the lessons learned from programmes designed in an earlier era. This learning has at least two important dimensions. First, it provides new insights into the structure and content of programmes, a well as their development and delivery processes. This learning has also suggested new organisational arrangements that may be more appropriate for the housing of tourism/hospitality education programmes. The hybrid model espouses an approach in which two types of complementary programmes are generally required to meet the total tourism/hospitality education needs of a region. The first type of programme required is the traditional hotel school (above) that is designed to prepare managers to effectively operate hotel and resort properties. The second required programme is one designed to train managers and future leaders for the tourism industry, defined in its broadest sense. Such a programme must impart managerial skills for a range of tourism positions, as well as provide the general education that will enable graduates to continue to learn and to grow as individuals as they pass through various stages of their careers. An effective implementation of this model is a two-year technical institute diploma programme that is fully integrated with a subsequent two-year university level management programme leading to a degree in hotel and restaurant management. The University of Calgary, Canada, has been a pioneer in the implementation of such a hybrid programme.

In addition to training individuals to meet the operational needs of tourism, there is a parallel need to educate industry leaders and managers. Traditionally, this 'education' has been provided through on-the-job training. However, the size, complexity, and sophistication of large-scale enterprises – and the need for much more strategic thinking – has given birth to graduate level programmes. These are generally of two types. The interdisciplinary M.Sc. programmes seek to emphasise the multidisciplinary nature of the tourism phenomenon. They typically contain heavy social science and environmental studies components, with a complementary emphasis on economic policy and the basics of management. A second approach that is commonly taken is to adapt an existing MBA programme to include a tourism concentration. Here the core courses provide the student with a well-rounded grasp of management principles and their application. The tourism concentration provides an opportunity for students to immerse themselves in applications of management principles to specific areas (such as tourism marketing); and various other courses (such as tourism policy, tourism and the environment, destination management, tourism and the community, and tourist behaviour and management). While the demand for graduate level programming is not yet heavy, one can expect that this will change significantly as tourism is increasingly forced to interface with other industries. This will demand a more sophisticated understanding of all aspects of the political, legal, social, cultural and technological complexities that both drive and limit all businesses.

While the traditional education system focuses on the training of those who are not yet employed, there is a parallel need to upgrade the skills of those currently in the industry, many of whom have not had the opportunity, or the privilege, of an extensive formal education. These individuals fall into two main categories, each having its distinctive training needs. Front-line staff requires skill-training programmes designed to ensure they know what must be done, and how best to do it. Such programmes are commonly termed standards and certification programmes. Initially, they rigorously define the skill sets required to perform each front-line job effectively. They subsequently teach these skills, and then test the ability of the individual to perform the required tasks. Those demonstrating the required ability are then certified in their trade.

Industry managers having little formal education (as well as those who do) can benefit enormously from well-structured executive development programmes. One of the most comprehensive is that offered by the International Association of Convention and Visitor Bureaus. This programme, which leads to the Certified Destination Management Executive designation contains a number of core courses dealing with the fundamentals of tourism destination management, and a number of elective courses that focus on areas of specialisation in Destination Management. The distinguishing characters of this programming is its emphasis on training senior managers to view the destination as a complex whole, and to make decisions that integrate both competitiveness and sustainability concerns.

Both professional and academic experts in tourism have identified the quality of educators and the quality of the educational material they use as being among the most important determinants in the success of a tourism education system. They have further identified a number of issues effecting the success of efforts to prepare educators. Structural issues involve a range of contextual factors that create somewhat unique problems. These include the late arrival of tourism as a field of education and training, a lack of industry consensus on the need for education, the diverse nature of tourism/hospitality education and training, the multiple educational demands of a rapidly growing industry, the lack of institutional structures to support tourism education, a shortage of positions of tourism educators, and the lack of advanced level programmes to properly train tourism educators.

In addition, there are a range of professional issues that must be taken into account by individuals interested in pursuing a career in tourism education. The most significant of these are the lack of clear career path for tourism educators, and the conflict between the demand for a strong academic training as well as practical experience. These combined, create powerful forces that impact heavily upon both current and future educators. In addition, the need to develop a specialised disciplinary expertise, while achieving a broad interdisciplinary understanding of tourism, must be addressed by the committed educator.

All of the foregoing, combined with pressure to gain international experience, while at the same time demonstrating a strong local commitment, creates strong pressures on the young instructor/scholar. These pressures, added to a lack of well-developed supporting teaching materials, creates a serious challenge for administrators who seek to support the educational goals and efforts of future faculty members. A critical principle underlying efforts to educate future educators is the need to ensure that their training is sensitive to the needs of the industry. At the same time it is essential to stress that academic programming in tourism should not be totally driven by industry. In addition to their pragmatic responsibilities, the educators of tomorrow must be prepared to provide intellectual and conceptual leadership to a rapidly evolving industry.

See also: cross-cultural education; cross-training; curriculum design; education, entrepreneurial; executive development; human resources development; learning curve; modular programme

J.R. BRENT RITCHIE, CANADA

education, computer-assisted

Computer applications in education include (1) computer-assisted instruction which use this as a self-contained teaching machine to present discrete lessons through drill and practice, tutorials, simulations and games, and problem-solving; (2) computer-managed instruction which helps to organise instruction and track student progress; (3) computer-mediated communications which use software applications such as electronic mail, computer conferencing, groupware and electronic bulletin boards; and (4) computer-based multimedia which integrate various voice, video, and computer technologies into a single, easily accessible delivery system, including voice recognition and language translation features. A major advantage of using computers to facilitate learning includes an enhanced capacity for self-paced learning on an individualised basis with immediate feedback. A limitation of this system is the cost of software and the need to motivate and train educators to develop instructional packages and networks in the face of rapidly changing **information technology** advances. Widespread computer illiteracy still exists, in addition to those who even do not have access to computers or computer networks.

The explosive expansion of the **Internet** has emphasised its remarkable ability to link colleges, universities, schools, companies and private citizens with learning systems and resources. The Internet is a revolutionising force for improving the

efficiency and effectiveness of education and **training** through a seamless interface between desktop computers and digital **television** with Internet appliances linked to advanced multimedia information systems through wired (phone and cable) and wireless (**radio**, cellular and satellite) communication technologies.

Information-technology-based products are being designed to bring communication capabilities wherever needed and to facilitate traditional education or **distance education** through advances such as (a) distributed object technology, like Java, which allows developers to create miniature applications – known as 'applets' – transmitted over the **World Wide Web** and which are temporarily used the client end for conducting specific tasks with full platform independence; (b) the increasing effectiveness of automated search engines, which help users to more effectively and efficiently retrieve customised information, including Lycos (www.lycos.com) and Excite (www.excite.com); (c) new tools which are rapidly expanding multimedia capabilities, so as to help networked systems move large video images easily without the jerky and unpolished look of video transmitted through telephone copper wires; and (d) intelligent agents who act as personal assistants to perform tasks such as programs that deliver web pages off-line, including Freeloader (www.freeloader.com), Pointcast, inc. (www.pointcast.com) and First Floor Software (www.firstfloor.com). New advances in computer-assisted education in tourism and other fields will emphasise ubiquitous computing which focuses on a human 'high touch' delivery system by designing the computer technology to vanish in the background.

DONALD E. HAWKINS, USA

education effectiveness

The role of tourism **education** is to train and develop the knowledge, skill, mind and character of students through a process of formal schooling and teaching. This is usually carved out at an institution of learning, whereby a systematic study of methods, theories and concepts is applied. To operate effectively, all educational institutions have a goal. For most that offer a tourism studies programme, including **hospitality** and **catering** or **restaurant** management schools, the goal may be defined as providing the industry with a reliable stream of human resources, who bring with them a diversity of skills, knowledge and attitudes.

Education and work are two sides of the same coin, in that both should result in making a positive contribution to the creation of wealth and quality of life. The system is designed to utilise its organisational resources to transform individuals, through the educational process, into 'graduates' who are able to contribute to both tourism and the society. As the system and educators use the resources allocated to them, they strive to be both effective and efficient (see **efficiency**).

Effectiveness refers to the degree to which educators and institutions are capable of attaining educational objectives. To measure the effectiveness, educational institutions ask whether their output meets the demands of the tourism workplace. Institutions must define the skills, knowledge and attitudes needed for **employment**, determine acceptable levels of proficiency, and identify effective ways to assess proficiency. They must then develop a strategy to disseminate tourism and related programmes, such as hospitality education. Tourism education, to be effective, should meet the following four criteria: (1) it should occur on a systematic basis, including educational objectives, curriculum development, teaching techniques and assessment; (2) it should be concerned with developing the skills and knowledge and changing outlooks and attitudes of the students; (3) it should contribute to the improvement of the performance in the present tourism sector practice and provide the student with the desire for 'lifelong' learning; and (4) it should provide a 'bridge' between **theory** and practice, due to the applied nature of tourism.

Educational efficiency is the degree to which organisational resources contribute to the productivity of an educational system. It is measured by the proportion of organisational resources used during the educational process. Due to fiscal constraints and creeping budget cuts in most Western countries, tourism programmes are increasingly under scrutiny with regard to their **training** and research outputs. Thus, tourism education administrators concentrate on efficiency in order to respond to inquiries concerning potential inefficiency. There are four areas in which efficiencies may be pursued: the faculty, the curriculum, research, and student and industry-related matters. Such resources can be combined, applied and transformed into the output of the educational process, that is, the graduates of a tourism education programme.

Several studies into the education and training needs of the tourism workforce in Europe, the United States and East Asia reveal that the workforce is inadequately prepared to meet the needs of business and **industry**. Principal causes for this mismatch include inconsistency between the needs of the industry practice and the educational programmes available to meet them; a failure to define specific institutional and educational objectives; a failure to take into account the career expectations of students; the lack of well-qualified tourism academics, with both specialised knowledge of the field of tourism and the theoretical underpinning; and the lack of ongoing contact between educational institutions and the industry.

While there are significant differences between the three tourism education and training levels (technical, supervisory and management) and aims of tourism education ranging from business and catering/hospitality to environmental design and planning, most programmes have a common element in that they need to educate for workplace know-how and job performance. Both of the latter two attributes involve a complex interplay between two elements: competencies and a foundation.

There are five competencies (the allocation of resources, working with others, the acquisition and use of information, the understanding of complex interrelationship, and ability to work with a variety of technologies) and a three-part foundation (skills, knowledge and personal qualities or attitudes). They are at the heart of job performance. These eight requirements together are the essential preparation for all tourism students attending technical programmes, supervisory training and university programmes. For purposes of effective understanding, the cognitive sciences (see **cognition**) suggest that competencies and the foundation should be taught in an integrated manner that reflects the workplace contexts in which they are applied. To perform effectively, students should be able to demonstrate competency in the five areas. The three-part foundation is integral to each of the five competencies. Skills enable workers to conduct processes. Knowledge provides the theoretical underpinning for problem-solving. One's values form the framework of principles or accepted standards that shape the desire to achieve, succeed or lend significance to a particular pursuit.

In short, educational effectiveness and efficiency can be considered satisfactory when tourism **development** contributes to the prosperity of the society, involving actions ranging from the coordination of **recreation** activities and the organisation of regulations for responsible development to conducting research to initiate **marketing** campaigns which are integrated into place marketing strategies. Hence, effective and efficient education is a primary factor in destination countries wishing to include tourism as an essential element of the social, economic and cultural development of their respective societies. The global economy, of which the tourism industry is an integral part, will demand better trained professionals, and tensions in society necessitate people's ability to cooperate, responsibly and reliably. It is for these reasons that it is important to have the funding to operate formal advanced and technical institutions which impart the knowledge, attitudes and skill needed by their graduates to take a professional approach to resolve issues. In tourism education, due to its complex scope and nature, there is no single method to enhance the effectiveness of a programme. Neither a prescription nor a single act will transform the incredibly diverse higher education system into a mechanism routinely producing graduates with all the skills, knowledge and attitudes which the industry desires. However, tourism education is a shared responsibility: for it to be effective, the executives representing specific industry sectors should play their part.

See also: articulation, programme; education/industry relationship; human resources development; professionalism

Further reading

EIESP (1991) *Education for Careers in European Travel*

and Tourism, London: European Institute of Education and Social Policy.

Go, F.M. (1994) 'Emerging issues in tourism education', in W. Theobald (ed.), *Global Tourism: The Next Decade*, Oxford: Butterworth-Heinemann.

Go, F.M., Monachello, M. and Baum, T. (1996). *Human Resource Management in the Hospitality Industry*, New York: Wiley.

Ritchie, J.R.B., 'Tourism and hospitality education - frameworks for advanced level and integrated regional programs' in *Requirements of Higher Level Education in Tourism: 40th Annual Congress AIEST Proceedings*, St Gallen: Association for the Scientific Experts in Tourism.

FRANK M. GO, THE NETHERLANDS

education, entrepreneurial

The focus of entrepreneurial education is the development of individual ability to successfully initiate and manage new ventures. The effective cultivation of **entrepreneurship** is particularly important in tourism, where the success of any **destination** is determined by the provision of numerous attractions, **services** and amenities. This provides diverse opportunities for new venture development, ranging from small-scale family-run businesses to larger corporate enterprises. By not encouraging and building entrepreneurial capability, a significant portion of tourism potential may not be tapped.

Three components of a successful entrepreneurial development programme are important: identifying and selecting students, using appropriate programme content and teaching style and incorporating follow-up procedures. As to the first component, careful screening and selection of students is fundamental to the success of an entrepreneurial development programme. This can be better understood by breaking entrepreneurship into several capabilities. They include a combination of **innovation**, **risk** taking and managerial skills (McMullan and Long 1990). The first two components, innovation and risk taking, describe ingrained personality traits that are difficult to learn or change. However, managerial skill can be developed and enhanced through **education**. Individuals who lack the first two are not likely to be entrepreneurial, even with the appropriate **training**. Therefore, programmes screen students based on personality traits. Characteristics such as the need for achievement, capacity for risk-taking, originality, a positive self-concept, problem-solving ability and perseverance are good indicators of entrepreneurial potential. These attributes are best assessed through a series of **interviews** rather than by more rigid forms of measurement, such as written tests. Another factor which can be taken into consideration in choosing students is previous experience with entrepreneurship. An individual who has already operated an entrepreneurial venture, or been exposed to one through working for relatives or friends, has a better understanding of what is involved in the entrepreneurial process.

For the second component, traditional business education addresses **management** issues through the study of standardised functional management areas such as **marketing**, finance, **accounting**, information systems and organisational behaviour. On the other hand, entrepreneurial training should be approached chronologically rather than cross-sectionally. The dynamics of business development rather than the functional areas are emphasised. The entrepreneur is provided with the specific skills and techniques necessary to guide the new venture through the earliest stages of development, including opportunity identification, feasibility analysis, initial financing, product design and **market** development. As a result, entrepreneurship education is unique in terms of curriculum design and delivery when compared to the more traditional forms of business management education.

The subject areas generally covered by entrepreneurial programmes are **motivation** and behaviour training, opportunity assessment, venture development, strategic **decision making** and marketing skills. Specific topics include identifying **target markets**, sales techniques, understanding credit, financial planning, **inventory** control, production issues and methods for evaluating success. One of the most important components in an effective entrepreneurship development programme is the requirement that students collect and analyse data and defend the feasibility of the start-up and operation of their own business (Loucks 1988). Therefore, most programmes include projects which involve the completion of a feasibility analysis, business plan and financial proposal.

The level of detail required for each topic covered varies depending on the scale of the proposed entrepreneurial ventures, the sophistication of the market, the degree of competition and the background of the students. For example, in the developing areas of the world these programmes may have to deal with small-scale ventures in fairly unsophisticated markets and with students with little or no background in business management. For tourism education, the core entrepreneurial development programme described here needs to be customised to the industry and often to a particular cultural setting. For example, a tourism entrepreneurial development programme in Nepal needs to include not only the traditional core components of entrepreneurial education but also modules that address the unique issues of the industry in that country, such as the environmental and cultural **impact** of tourism to Mount Everest. All illustrations, examples and cases included in the course should be customised to the **industry** and to the specific national setting.

With regards to teaching style, entrepreneurial development programmes encourage the development of independent self-reliant individuals. Therefore, approaches such as case studies, real-life projects and experiential exercises are preferred to the traditional lecture style format. Useful techniques include management games, field trips to existing businesses, roleplay situations and mentoring by successful entrepreneurs. In terms of teaching staff, experienced business management educators should be assisted by practising entrepreneurs, who can bring real-life experiences to the classroom. This balanced teaching approach is facilitated through such techniques as guest lecturers, team teaching and entrepreneurs in residence.

With regard to follow-up, the final component, ongoing counselling and advisory services should be readily available to the entrepreneurial graduate. In some programmes, the instructor and/or an entrepreneurial mentor remain involved with the graduate throughout the implementation of the entrepreneur's business idea. This provides an important resource to the entrepreneur throughout the initiation and growth of the venture, and significantly increases the success rate of programme graduates.

See also: education method

References

Loucks, K.E. (1988) *Training Entrepreneurs for Small Business Creation: Lessons from Experience*, Management Development Series No. 26, Geneva: International Labour Office. (A practical manual for the development of entrepreneurial education programmes.)

McMullan, W.E. and Long, W.A. (1990) *Developing New Ventures: The Entrepreneurial Option*, San Diego, CA: Harcourt Brace Jovanovich. (A general reference text on entrepreneurism and entrepreneurial education.)

Further reading

Echtner, C.M. (1995) 'Entrepreneurial training in developing countries', *Annals of Tourism Research* 22(1): 119–34. (An overview of entrepreneurial training in tourism with emphasis on the issues faced by developing nations.)

CHARLOTTE M. ECHTNER, AUSTRALIA

education, environmental

Environmental **education** is concerned with developing people who are knowledgeable about the physical, social and economic environment of which they are a part, so that they become concerned about environmental problems and motivated to act responsibly in enhancing the **quality** of their **environment** as well as their lives. Tourism to natural areas attracts people who wish to interact with the environment and, in varying degrees, develop their knowledge, awareness and appreciation of it. The visit itself is, therefore, educational. As well, **ecotourism** operators provide an appropriate level of environmental and cultural interpretation, usually through the **employment** of qualified **tour guides** and the provision of information both prior to and during the trip.

The level and type of education will depend on the interests and expectations of **tourists** and the skills and commitment of the operators. It will include a broad range of opportunities through **interpretation**, interactive approaches and the use of various **media**. This educational or interpretative component may acknowledge the natural and cultural values of a **destination**, and could also address issues such as resource **management** and the role and **attitude** of the host **community**. Appropriate **behaviour** contributes to high quality tourism and brings the greatest possible benefit to all participants, tourists, the host population and the **industry** itself, without causing intolerable ecological and social damage. **Sustainable tourism** can be achieved partly by educating tourists about the consequences of inappropriate **behaviour** (see also **ecologically sustainable tourism**; **sustainable development**).

Ecotourism education can influence tourist and industry behaviour and assist the long-term sustainability of tourism activity in natural areas. The development and delivery of education and interpretative services for ecotourism requires well-trained operators and guides who can deliver accurate and informative materials and presentations. Thus tour leaders and operators need to have a good knowledge of ecology, cultural issues, **conservation** and minimal impact practices, and be skilful in interpretation. Multifaceted approaches should also be included to appeal to all visitors, regardless of educational background or prior knowledge of the environment. Easily accessible, accurate and stimulating information (pre-travel and during the sojourn) is also necessary to encourage responsible behaviour and an on-going interest in tourism, the environment and conservation after the visit.

Another key component of environmental education in relation to tourism is the use of interactive learning models, such as roleplays and other participative techniques. These lead to the greatest information retention, especially if combined with supporting discussions and pictures or films. In addition, the messages conveyed by television and documentary films to portray the essential relationship between tourism and its natural setting should be based on ecological sustainability and be appropriate for the environments involved. The visual media can be a very powerful and effective way of conveying best practice and increasing environmental understanding.

The success of the learning process can be measured in behavioural change. Management goals will be assisted through more thoughtful and informed use of natural areas by visitors and operators. This process for tourism is targeted (in a co-ordinated way and as an integrated system) at visitors, both at home and at their **destination**, at the tourist industry at local, national and international levels, at the host community and at the decision makers in a range of government departments, professional **associations** and non-government organisations. In this way, environmental education can contribute to the achievement of sustainable forms of tourism which the visitors, the tourist and the host community perceive to be beneficial in both the short term and the long term. As a general principle, programmes for tourists must be enjoyable as well as informative. Their **development** should be so designed so as to encourage a sense of ownership among the participants and involve them in the process.

The recommendations presented at an International Union for the Conservation of Nature (IUCN) Conference on Educating for Sustainable Tourism (held in Slovenia in 1992) summarise the various areas where environmental education can be effective in tourism. At the school level, knowledge and understanding of the potential **benefits** and costs of tourism should be an integral part of the formal and non-formal curriculum. Teachers responsible for this aspect will need professional training. An introduction to tourism and its implications should be covered. At the professional/vocation level, training courses aimed at those seeking or having employment in this industry should include treatment of sustainable tourism and the processes which support it. The training of those managing the national estate or likely to receive visitors (such as planners, foresters, farmers and rangers) should also include treatment of sustainable tourism. Those working in land management, especially in popular tourism areas, have a responsibility to be involved in education programmes themselves. This should include how to communicate the **theory** and practice of sustainable tourism.

Furthermore, at the tour operation level, the companies should learn how to subject their activities and those of their suppliers to examination using accepted methods such as environmental impact assessment, environmental audit and life cycle analysis. At the tourist level, tour operators and travel agents should provide information in an appropriate form on sustainable tourism. Specific interests, such as mountaineering, sailing and gliding clubs should provide additional information and advice in their literature. Finally, at the host **community** level, assistance should be offered to develop ways of communicating the special qualities of the locality and culture to the visitors in their area. The community also should be assisted to identify the aspects of their culture and their environment which they value and want to keep, and to learn how to introduce development plans that do not threaten what they value. Both local and national governments have a responsibility to encourage sustainable tourism amongst their citizens. This is a complex task involving all members of society. A key role is planning and coordinating the education and training needed.

See also: codes of ethics, environmental; environmental compatability; planning, environmental; precautionary principle

Further reading

Australian Commonwealth Department of Tourism (1994) *National Ecotourism Strategy*, Canberra: ACT.

International Union for the Conservation of Nature, Commission on Education and Communication (1989) *Educating for Sustainable Tourism*, Slovenia: University of Ljubljana.

ROSS K. DOWLING, AUSTRALIA

education/industry relationship

The tourism education workplace relationship represents the various points of interaction between the diverse public and private organisations and enterprises involved on the tourism **supply** side and those universities, colleges and **training** centres providing **education** in support of the industry. This relationship can be one of mutual benefit. Industry exposure can allow college students to apply their skills and knowledge in a real workplace environment, and this in turn enhances the quality and vocational relevance of the qualification which students obtain from educational institutions. Tourism companies, especially in seasonal operations (see **seasonality**), benefit from trained and enthusiastic labour during peak periods. The industry can also benefit from utilising the expertise of educational providers in the areas of research, training and development. Companies can provide excellent opportunities for faculty to update or hone their skills and expertise, thus enhancing the educational experience of students. The opportunities to develop mutually beneficial **models** of cooperation are almost endless.

There is, however, considerable criticism from both sides about the role taken by the other in this relationship. The industry frequently criticises education for lack of realism in its provision and for 'ivory tower' remoteness from the 'reality' of the profession. Education, in turn, is commonly critical of the industry for its failure to provide working opportunities beyond the most menial for students on internship and for failing to fully recognise graduation qualifications.

Mechanisms to improve the relationship are widely recognised as important. Many educational providers involve key industry players in a consultative role which may address programme content, structure and focus. Some colleges have established strategic alliances with specific industry players in order to reap the benefits of close and, generally, closed cooperation. UK national initiatives, for example, include the work of the Hospitality Partnership and the National Liaison Group, which aim to bring both sides of the relationship together to overcome differences and enhance mutual understanding. The role of the **Council on Hotel, Restaurant and Institutional Education** serves a similar function.

The relationship has the potential for mutual advantage but at the same time there are a number of inherent problems which require recognition.

Tourism is not a homogeneous industry, and therefore it does not have common educational requirements. Educational programmes likewise show considerable diversity in most countries. Still further, there is always the potential for disagreement between priorities based on educational values in the broadest sense on the one hand, and the practical and business needs of the tourism industry on the other.

See also: education policy; educational effectiveness; education method; professionalism

TOM BAUM, UK

education level

The existing types of tourism education programmes are diverse, differing greatly in duration, content and the context within which they are delivered. However, they can in general be grouped into three levels: vocational **training**, professional **education** and **executive development**. This entry focuses on the first two levels.

Adequate vocational training is critical in order to effectively deliver the **products** and **services** required by the tourism industry. The main objective is to teach skills that can be applied to a specific position, often at the front-line or supervisory level. The content of such programmes tends to be highly practical, dealing with developing certain skills and applying them to specific on-the-job problems (Cooper and Westlake 1989). Examples include training for **hotel** employees (front desk, housekeeping), **food** and beverage personnel (chef, bartender, waiter/waitress) and tourism services (**tour guide**, travel counsellor, **tour operator**).

Various tourism enterprises often participate directly in vocational training by conducting 'in-house' programmes for employees. In addition, specialised certificate and diploma programmes are offered either through public sector institutions (such as technical institutes or colleges) or private schools. The content of a vocational programme will vary depending upon the particular occupation for which individuals are being trained. However, all vocational programmes should include two components: acquisition of appropriate skills (both technical and attitudinal) and opportunities for the practical application of these skills.

For the first component, it is important to note the inclusion of both technical and attitudinal training. The former provides the techniques and standards, whereas attitudinal training addresses the issues of quality of service and the potential problems caused by the cultural differences between **host and guest**. The later is a very vital part of vocational programmes because actual 'hands on' experience serves to enhance the technical and attitudinal skills learned in the classroom. Such experience may be accomplished through site simulation (such as a student-operated or hotel-operated **restaurant** situated on campus) and/or cooperative programmes that include actual placement in tourism positions.

In order to provide vocational training at various levels, a tiered diploma or certificate programme is most effective. Goeltom (1988) suggests a four-tiered system: diploma I provides training for entry level personnel (such as front desk clerk); diploma II offers training for supervisory level personnel (example, front office supervisor); diploma III provides training for middle managerial level personnel (such as rooms manager); and diploma IV offers training for managerial level personnel (such as hotel manager). At the first level, short intensive programmes are most appropriate in order to respond quickly to changing manpower demands. At the other three levels, more flexible training programmes are needed to accommodate individuals that are already employed. For example, customised programmes consisting of specially designed training packages/modules should be made available to those who are fully employed and cannot enrol for the regular programme. At these three higher levels, students should be taught 'on the job' training techniques so that, once back in industry, they can instruct entry-level employees. By hiring such graduates, even smaller property members can provide in-house vocational training programmes for lower level line personnel.

In contrast to vocational training, professional education is generally typified by being more academic in nature and generally occurs in a university setting. Theoretical concepts and **models** are provided, and the student's ability to interpret, evaluate and analyse is developed

(Cooper and Westlake 1989). The ultimate goal is to offer a broad understanding of the tourism phenomenon and its unique issues and challenges as studied from various disciplinary perspectives, including **geography**, **economics**, **sociology**, **anthropology** and **management**. At the undergraduate level, such an approach to tourism education supplies the qualified manpower needed at the strategic level in both the public and private sectors. At the postgraduate and Ph.D. levels, it produces qualified researchers, professors and consultants for tourism studies and the industry. There is an increasingly urgent need for individuals possessing this more **holistic approach** of tourism issues. The accelerating growth of tourism has, in many cases, resulted in inept planning and severe social, economic and environmental **impacts** on host communities. Professionals, prepared with the knowledge and skills to understand tourism in its totality, are needed to manage and monitor the phenomenon.

Jenkins (1980) suggests that although the emphasis will vary among institutions and disciplines, a university-level tourism education programme should include three broad areas of instruction: the development and presentation of an analytical framework for interpreting **international tourism**, a consideration of the models of tourism at the national level, and the teaching of appropriate methodologies and tourism evaluation techniques. In a similar vein, Hawkins and Hunt (1988) outline four fundamental goals for university-level tourism **curriculum design**: a holistic understanding of the field including economic, social, cultural, environment, political, technological and physical aspects; a programme based on theoretical models of tourism which are dynamic, comprehensive, easily understood and unifying, providing a foundation around which the student can organise and synthesise knowledge; the production of graduates with a broad base of knowledge, skills and awareness; and the placement of graduates in **leadership** and managerial roles in the tourism industry.

Increasingly, undergraduate tourism programmes, especially those with a focus on administration and planning, offer the opportunity for work placement terms. These 'co-op' programmes allow the student to apply the concepts, models and approaches learned in the classroom to real-life managerial situations. Work terms are generally provided not only during the summer vacation but also during several semesters, increasing the length of the undergraduate programme (a normal four-year programme can become a five-year co-op programme). While most existing tourism education programmes are offered from within specific disciplines, there is also a need to develop interdisciplinary tourism programmes. As outlined by Ritchie (1995), such 'hybrid programmes' would include courses in the social sciences, languages, mathematics, **environment**, business and tourism studies. Recently, there has been a move toward such programmes in several institutions, indicating that tourism studies may eventually evolve into a distinct discipline at the university level.

See also: articulation, programme; Finnish University Network for Tourism Studies; human resources development; knowledge acquisition

References

Cooper, C. and Westlake, J. (1989) 'Tourism teaching into the 1990's', *Tourism Management*, March: 69–73.

Goeltom, D.R. (1988) 'Tourism education in Indonesia', paper given at Teaching Tourism into the 1990s: An International Conference for Tourism Educators, University of Surrey, Guildford, UK.

Hawkins, D.E. and Hunt, J.D. (1988) 'Travel and tourism professional education', *Hospitality & Tourism Educator* 1(1): 8–14.

Jenkins, C.L. (1980) 'Education for tourism policy makers in developing countries', *International Journal of Tourism Management*, December: 238–42.

Ritchie, J.R.B. (1995), 'Design and development of the tourism/hospitality management curriculum', *Tourism Recreation Research* 20(2): 7–13.

CHARLOTTE M. ECHTNER, AUSTRALIA

education media

Tourism education relies on many different **media** for transmitting information to stimulate and enhance the learning process. The traditional lecture format, involving teacher–student interaction, remains important. This is increasingly complemented by computer-assisted education (see **education, computer-assisted**). Experiential learning employs the medium of real-world settings to provide the student with information and knowledge directly related to its context of application. Project-based learning allows students to utilise various forms of the scientific method to gain insights into tourism. **Distance education** formats have traditionally involved print media, but increasingly utilise electronic technology to bring the teacher and learners together.

See also: education policy; educational effectiveness; education method

J.R. BRENT RITCHIE, CANADA

education method

Educational methods have been evolving with time and **technology**. The earliest known systematic method of imparting knowledge was the oral method, whereby information and knowledge was transferred through word-of-mouth from one generation to another. Then came the age of written communication, when knowledge and skills were delivered in writing. This process was and is still considered to have a lasting impact on human memory. Nevertheless the former method maintained its importance and value, and this became popular as learning by 'rote'. Both these methods prevailed for centuries and continued to form the basis of formal educational methods. The three Rs seemed to be quintessential to personality development.

The evolution of knowledge systems gave rise to multiple disciplines. In order to facilitate a proper understanding of sciences and arts, innovative methods were adopted and were fine-tuned with experience and technology. Experience thus gained showed that involvement of the learner was perhaps the most effective and efficient educational device (see **education effectiveness**). In fulfilment of this objective various methods have been identified, each characterised by their distinctive set of utilities and limitations. These may be broadly classified under two major headings: content presentation and interactive methods. The former ranges from the rather formal to informal styles. The most prevalent and best known are lecture (semi-formal and informative), demonstration (performance-based), field trip (projectional, observation-based), panels/symposium/forum (opinion and experience talks) and one-on-one training (participative and interactive).

As to the second approach, there are approximately nine well-established and commonly practised interactive methods. These include group discussion (motivational and interesting), role play (skill rehearsal), brainstorming (idea generation, increases awareness), case study (analytical, problem solving), project (assignment-based, report-oriented), learner's presentation (prepared/impromptu speeches), games and simulations (contrived/reality exercises), in-basket (management-oriented problem solving), and computer-based instruction (stimulating software usage). The other interactive methods induce the learners to think, contribute and grow. Instructors often experiment with these methodologies singularly or in combination. Choice of method(s) is made on the basis of the learning objective(s) and level of learning, the learner's level of knowledge/skill/interest and also the availability of resources. Today, formal education must complement professional education and hence the methods adopted for one remain more or less similar for the other. Professional educational/training is a complex task involving multidisciplinary knowledge, thus demanding more than one approach for a fuller understanding of the philosophies, principles and practices. The same is true of education and/or **training**.

Tourism as a phenomenon is complex and embraces all spheres of technical know-how and human expertise. Education in this sphere of activity is a daunting task, that has become a major challenge to its quality existence. Owing to its all-pervading diversity, tourism's' subject matter is treated from various aspects. For an understanding of the educational methodologies, tourism courses can be classed under three major established approaches. First, the disciplinary approach is academic in nature and includes teaching, learning and research. Second, the vocational approach trains and aids in the development of working skills for tourism and its sectors. Third, the entrepreneurial approach assists in developing entrepreneurs, largely for the local or regional tourism businesses. Of late, tourism researchers and practitioners have been placing more emphasis on educating the tourist. This is perhaps with the view to influencing their frivolous **behaviour** and enhancing enrichment of **experience**, through **interpretation** of touristic **products** and information dissemination on or off **site**. Such an approach can probably be called behavioural, as it calls for a more humane tourism. This objective teaches the tourist the 'art of travelling' and can be conveniently incorporated into a nation's public education system at any or all levels. Tourism education and research are interdependent, and mutual benefits are derived from expertise in both. Thus, all of these approaches draw heavily from each other.

Since tourism education is objective-oriented and performance-driven, and since the content of knowledge is interdisciplinary, a **holistic approach** to the subject matter is rather difficult. For this reason, McIntosh suggests nine approaches to its study: (1) the institutional approach, which assists in compiling quantitative and qualitative databases; (2) the product approach, which studies tourism products and services; (3) the historical approach, which analyses the genetics of products and services; (4) the managerial approach, which has business orientations; (5) the economic approach, which analyses the economic implications of tourism; (6) the sociological approach, which studies host–guest interactions; (7) the geographical approach, which investigates the spatio-temporal aspects of tourism; (8) the interdisciplinary approach, which provides a coordinated or systematic understanding; and (9) the ecological approach, which investigates relationships in the biocultural systems.

Methodologies of tourism education are continuously evolving, and no single method can be prescribed as the ideal one. So far, all institutions devoted to tourism, both in the industry and in academia, are seriously concerned about the **quality** issues. This problem directly pertains to manpower **development** which in turn is a concern of pedagogical practices. Research is being conducted towards identifying finer and more effective methods of education/training personnel who will be facing newer and complex problems in the tourism industry in the future. Awareness of these problems will be provided by the information-based technology that has overtaken all other channels of communication and is being increasingly utilised in education as well. In moving from traditional methods to more sophisticated ones, tourism education methods must aim towards serving the needs of students, employees, employers, governments and communities. In order to achieve quality in education, suitable methods have to be crafted for an effective **partnership** between industry and tourism education systems. Future methods need to include, among other things, a system of checks and measures, proper communication skills, participation channels for the instructor and, above all, self-evaluation techniques emphasising professional ethics.

See also: education/industry relationship; education effectiveness; education policy; educational media

Further reading

Cooper, C., Shepherd, R. and Westlake, J. (1994) *Tourism and Hospitality Education*, Guildford: University of Surrey.

Echtner, C.M. (1995) 'Tourism education in developing nations: a three-pronged approach', *Tourism Recreation Research* 20(2): 32–41.

Jafari, J. and Ritchie, J.R.B. (1981) 'Towards a framework of tourism education', *Annals of Tourism Research* 8(1): 13–34.

McIntosh, R.W. and Goeldner, C.R. (1995) *Tourism: Principles, Practices and Philosophies*, New York: Wiley.

Tourism Recreation Research (1995) special issue on tourism education and human resource development, 20(2).

TEJ VIR SINGH, INDIA
SHALINI SINGH, INDIA

education, multidisciplinary

Education programmes in many applied fields depend on more than one academic discipline. In tourism education, this is the usual **policy**. Its aims are to help students gain knowledge of tourism's many facets and a broad general education, by drawing on the multiple perspectives provided by a range of disciplines. Despite these benefits, multidisciplinary education if focused on an applied field can be problematical. A remedy to such a situation does not replace the multidisciplinary approach but adds something to its centre. Multidisciplinarity refers to the practice of bringing several academic disciplines to focus on a study. A discipline can be thought of as a body of knowledge, for study focused on a particular field, which has been organised to some extent. This process includes assumptions, definitions, jargon, and methods for research and teaching.

Early research on tourism, some ninety years ago, was done by economists. Later, academics from other disciplines identified tourism issues to which they could relate. When educational programmes began, curriculum planners formed a well-founded belief that students would gain wider knowledge by exposure to several disciplines. No single discipline can provide more than partial understanding of the complexity of tourism and its issues. All university education has (or develops) foundations in research, now observable in a range of disciplines used to investigate tourism. Opinions differ as to what constitutes a genuine discipline and certain inclusions are likely to be criticised.

Several entries in this encyclopedia represent disciplines and research areas relevant to tourism. These include **accounting**, **anthropology**, **archaeology**, **demography**, ecology, **economics**, **ethnography**, **game theory**, **geography**, **history**, **leisure**, **management**, **marketing**, **myth**, **planning**, **political science**, **psychology**, **recreation**, **semiotics**, **sociolinguistics**, **sociology**, **statistics** and **systems theory**, to name but a few. Another list can be prepared by noting the disciplines alongside names of editors of tourism journals. For example, an examination of the 1996 list of editors of ***Annals of Tourism Research*** revealed the following mix of disciplines: economics (13 editors), geography (11), anthropology (7), sociology (9), business and management (5), political science (4), leisure and recreation (4), psychology (2) and one for each of another ten categories. This pattern of disciplines was reflected in an analysis of Ph.D. research. Jafari and Asher (1988) reviewed 157 doctoral theses on tourism presented in American universities over a thirty-year period. Their analysis found fifteen prominent disciplines. The most common dissertations were in economics (40), then anthropology (25), geography (24), recreation (23) and business (11).

Important developments in multidisciplinary education for tourism occurred between 1981 and 1996. Earlier, tourism education and research were developing 'largely independent of each other' (Jafari and Ritchie 1981: 28). Fifteen years later, there are definite signs of closer links. One is an increase in the number of individual academics who are active as researchers and educators. Another sign is that many educators are now making more use of research on tourism in their syllabus design and teaching, by bringing in articles from tourism research journals and by referring to the growing body of knowledge in this field.

Universities are structured by fields, so scholars cluster in departments where they develop particular disciplines. When a new applied field of education is required, normally it is accommodated, for a while at least, within established departments with their established disciplines. In that context, four strategies exist for organising multidisciplinary education in applied fields. Each has advantages and disadvantages, some of which seem intrinsic in multidisciplinarity (Bodewes 1981; Jafari and Ritchie 1981; Leiper 1981).

First, if few students are enrolled, insufficient for sustaining special classes, they can be directed to classes in scattered departments, taking courses in economics, psychology, management and more, along with students taking those subjects for reasons other than application to tourism. One problem here is that nothing is provided to help students relate each discipline to tourism. Another is that nothing is provided to help link up the disciplines in order to form an integrated understanding of the applied field, tourism. A second strategy has instructors from independent departments conducting special classes for tourism students. Thus many tourism courses are visited by professors from statistics, geography, marketing and the like, each relating her or his discipline to aspects of tourism. But not all of them have the time or inclination to develop expertise in tourism since it is a secondary, incidental and part-time subject. Thus, the relating to tourism problem is not adequately addressed, and the complex problem of linking fragmented disciplines remains neglected. Students are often unaware that it is a specific problem.

A third strategy is a variation on the second. A small specialist tourism unit or subdepartment is formed. Its members teach core subjects in tourism studies, and arrange multidisciplinary inputs from other departments. For most of their classes, students are taught by discipline specialists from scattered departments. This strategy has the same potential problems, moderated slightly, as the second. This third strategy is clearly described in a diagram devised by Jafari, presented in various publications (Jafari and Ritchie 1981). Sixteen disciplines or departments are shown in the diagram around the rim of a wheel. The hub is termed 'Centre for Tourism Studies'. The diagram implies that specialists from these many fields (indicative, not definitive) visit periodically. Most of the education comes from the rim (for another strategy, see **Finnish University Network for Tourism Studies**).

A fourth strategy is, arguably, ideal for students (Jafari and Ritchie 1981: 24) but is not normally feasible because of typical policies in universities and colleges. In this ideal, a substantial department (or school) of tourism is formed, comprising academics specialising in tourism with expertise in various disciplines. All or most subjects in the course are delivered by a multidisciplinary team. Until recently, very few universities have followed this fourth strategy, for a number of reasons (Bodewes 1981; Jafari and Ritchie 1981). In the 1980s and 1990s, a trend occurred towards its adoption, although it remains unusual. The trend reflects academic history. Over time, new departments or schools are formed to concentrate on new specialisations, while some older departments, deemed no longer relevant to society, are closed.

All four strategies involve multidisciplinary approaches. Thus, they create the benefits and some of the problems accompanying applied multidisciplinary education. A persistent problem is the risk of excessive discipline subjectivity, the 'blinkered approach', which leads students to think they should be developing their knowledge of, for example, sociology while their main need is to understand tourism. Such a situation can divert the purpose of applied education and reverse the proper strategic direction of the flow of knowledge (Leiper 1981: 71–2). Another persistent problem is the risk that contributions from particular disciplines will be overemphasised, diluted or distorted, rendering a true synthesis impossible.

A common problem across the first three strategies is that multidisciplinary education on an applied field tends to be fragmented, and fragments of knowledge are not conducive to understanding. Given the lack of a method for integrating knowledge from various disciplines into a cohesive whole, many students fail to achieve an adequate understanding of tourism. In other words, 'the multidisciplinary miracle has to take place... within the swallowing and digesting student' (Bodewes 1981: 43).

That common problem where the first three strategies are followed can be regarded as a condition of proliferating variety in the educational system. A remedy is found in **systems theory**. Reducing a system's complexity requires, as a first step, systemic models. These provide students and teachers with a set of shared frameworks for study. These models should have structural and dynamic features ('frameworks' and 'clockworks', in systems jargon) and must be sufficiently broad and general to allow multidisciplinary approaches to continue, but also need specificity, to indicate the key topics and systemic linkages in a comprehensive and cohesive study of tourism. They are models of whole **tourism systems**.

Such models can be placed at the hub of the wheel in Jafari's diagram for multidisciplinary education. There they become foundation concepts in courses in tourism, treated as a developing discipline in its own right. The hub then functions, not as a passive dumping ground for disciplinary specialists from the wheel's rim, but as a vital point of common reference for those specialists. The fourth strategy seems most likely to encourage this process, but the third strategy is also suitable.

Tourism education is most likely to remain multidisciplinary, largely following the third and fourth strategies. Ideally, progress will be made in the development of a distinct discipline. However, this must not mean that multidisciplinary education is diminished. The benefits of multidisciplinarity outweigh the problems, especially since a basic problem can be progressively overcome with the ongoing development of a distinct discipline to stand in the middle.

References

Bodewes, T. (1981) 'Development of advanced tourism studies in Holland', *Annals of Tourism Research* 8: 35–51.

Jafari, J. and Asher, D. (1988) 'Tourism as the subject of doctoral dissertations', *Annals of Tourism Research* 15: 407–29.

Jafari, J. and Ritchie. J.R.B. (1981) 'Towards a framework for tourism education: problems and prospects', *Annals of Tourism Research* 8: 13–34.

Leiper, N. (1981) 'Towards a cohesive curriculum in tourism: the case for a distinct discipline', *Annals of Tourism Research* 8: 69–84.

NEIL LEIPER, AUSTRALIA

education policy

The primary role of tourism education policy is to provide a framework and a series of systematic guidelines for the development and implementation of an overall education and training strategy. Such a strategy defines the basic philosophy of the education system, the structure and content of major programmes, and criteria for measuring their effectiveness. As an end result, a comprehensive policy also defines a portfolio of programmes designed to serve the needs of students and the industry over both the short and long term. This portfolio should provide an academically and professionally sound hierarchy of programmes that links all levels of the education system (primary, secondary, post-secondary, graduate and postgraduate). The goal is to support the educational needs of all those pursuing different **career** paths in this field. Within each level in turn, the policy provides more detailed guidance for designing and delivering the specific programme appropriate to that level. For example, there are a number of distinctly different models or policy frameworks for tourism/ **hospitality** education at the undergraduate level.

At this level, the hotel school model (HSM) is the most well-established approach. Early frameworks for hospitality education were provided by European hotel schools. Such programmes continue to be well-respected. More recently, university undergraduate programmes that focus on education for hotel **management** have emerged and have become increasingly popular. Perhaps the most internationally recognised programme of this type is at Cornell University School of Hotel Administration in the USA. Other well-regarded programsme, based on somewhat similar models, are at the University of Nevada at Las Vegas (USA), the University of Guelph (Canada), the Hague Hotel School (the Netherlands) and the University of Surrey (UK). The most distinguishing characteristics of these programmes have been their emphasis on preparing individuals to manage hotel and resort properties. Their graduates have been well received by the hotel sector for many years. At the same time, some believe that although hotel school graduates can be excellent operational managers, they may have difficulty serving in a strategic management role.

In terms of structure and content, hotel school programmes generally contain two main types of courses. First, there are those related to operational aspects of the hotel property, such as food and beverage management, front desk operations, hotel facility operation and maintenance, and legal aspects of hotel operations. The second type of programmes are related to the functional dimensions of management required for the successful operation of a hotel or hotel chain. Examples include sales and marketing management, financial management, human resource management, hospitality accounting and information systems management. Whatever limitations they may have, the fact remains that a well-designed and well-delivered 'hotel school' programme still receives considerable support from the hotel sector.

A general tourism management model (GTMM) seeks to broaden the educational experience of students while still providing a strong industry orientation. Its structure and content feature several characteristics. First, the core of the programme emphasises general management education, but also includes the liberal arts, languages and mathematics as programme requirements. Second, rather than having students concentrate on more advanced courses in a particular functional area of business (as in a traditional management programme), a GTMM is structured to enable students to understand the tourism sector by taking a number of courses related to the tourism/ hospitality industry. Some programmes may also provide the option of focusing on one specific subsector such as event management, **transportation**, meetings and conventions, or even hotel management. Third, in order to obtain some of the operational knowledge and skills provided by hotel schools, GTMMs frequently include a number of practical work terms as an integral part of the learning process. As a result this education model may require five instead of four years to complete.

Generally speaking, GTMM programmes are located in schools of business or similar academic units. A liberal arts institution with a tourism focus policy framework includes a broad range of programme types. For example, the discipline-based programmes have a tourism component or a tourism emphasis. Such programmes commonly involve the fields of **geography**, **economics**, **anthropology** or **sociology**. The recreation or leisure studies programmes frequently include a significant tourism component, and tend to be multidisciplinary in nature. They usually draw heavily on the social sciences (for example, geography, economics, psychology, sociology), although they may also involve input from such fields as physical education and computer science. One important characteristic of these programmes is that they generally have an academic, as opposed to an industry, orientation. The multidisiplinary majors in tourism studies are somewhat similar to the previous category. They are distinguished by the fact that they tend to have a much stronger industry orientation. As a result, they often include some management-related courses, a greater emphasis on language training and, increasingly, courses which relate to tourism planning (see **Finish Network University of Tourism Studies**).

While liberal arts programmes in tourism are frequently criticised as being too theoretical or not sufficiently oriented to the needs of the industry, they possess some very real strengths. They provide students with a much broader understanding of the societal dimensions of tourism and its impacts. In addition, they offer students a high degree of flexibility in selecting courses of particular interest to them. They also tend to provide students with knowledge that, while not focused on tourism, is highly relevant to success in the field: geography and languages are examples. Further, such programmes are highly adaptable and have demonstrated an ability to incorporate new topics and issues (such as environmental studies) more readily than the structured programmes.

The hybrid model of tourism/hospitality education model represents a more recent policy approach. This takes into account the lessons learned from programmes designed in an earlier era. This learning has at least two important dimensions. First, it provides new insights into the structure and content of programmes and the processes of programme development and delivery. Second, this learning has suggested new organisational arrangements that may be more appropriate for the housing of tourism/hospitality education programmes.

The hybrid model espouses a policy in which two types of programmes are used to meet the total tourism/hospitality education needs of a region. The first type of programme required is the traditional HSM designed to prepare managers to effectively operate hotel and resort properties. However, in line with the trends of leading programmes, increasing efforts are being made to ensure that the graduates from such programs recognise their role within the broader tourism industry. The second programme is designed to train managers and future leaders for the tourism industry, defined in its broadest sense. Such a programme must impart managerial skills for a range of tourism positions, as well as provide the general education that will enable graduates to continue to learn and to grow professionally as they pass through various stages of their careers.

Hybrid models have five important features: (1) they involve a four to five-year programme of studies which includes up to twenty-four months of practical work experience as an integral part of the programme; (2) they provide a strong management orientation balanced with an emphasis on those social science disciplines that contribute to a broader understanding of tourism and its impacts; (3) they emphasise a balanced perspective concerning tourism development, environmental protection and quality of the visitor experience; (4) they recognise the need for the development of language capabilities as well as dimensions of cultural diversity; and (5) they place a strong emphasis on the development of close working relationships with tourism organisations in both the public and private sectors.

Because of the high risk of incoherency and inefficiency, a comprehensive tourism education policy framework must also include a formal means of policy and programme evaluation. Such evaluation is normally based on a number of performance indicators. Examples are measures of student satisfaction with the overall programme and its detailed content; industry satisfaction with graduates and their ability to perform effectively; industry retention (the extent to which students educated within tourism programes actually pursue careers within the industry for different periods of time); graduate placement, or the degree to which graduates from tourism education programmes are successful in finding employment; cost-effectiveness, in reference to the relative cost of educating students in tourism programmes as compared with other areas; programme competitiveness, or how well graduates from programmes for a particular destination perform compared with those from other destination programs as measured by standardised test scores and other measures of student knowledge and performance; salary levels as compared with other destination and other industry sectors; job and career satisfaction indicator over an extended period of time; industry performance measured by the ability of the destination to compete effectively in national and international tourism markets; and finally, destination well-being, or how destination residents look to tourism to enhance their economic, social and environmental position. Altogether, because hospitality/tourism education is the major factor in enhancing the well-being of destination residents over the long term, those responsible for tourism education policy must constantly keep in mind that the fundamental purpose of these programmes is to support the development of the destination, and ultimately the well-being of its residents.

See also: articulation, programme; education workplace; educational effectiveness; education method; professionalism

J.R. BRENT RITCHIE, CANADA

education, recreation

Recreation education can be viewed from two primary perspectives: **education** for **recreation** (that is, educating the consumers, or those who participate in recreational activities) and education of recreation professionals (or producers). In recent years there has been a move towards increasing education levels for all members of the society, so that people of all ages will understand the importance of **leisure** in their lives. Advocates of recreation education promote a lifelong process where leisure values are formally taught from the earliest years of primary school as part of the standard curriculum, and informally far into adulthood.

In addition to increasing understanding of the role of leisure in people's lives, recreation education aims to increase participation in a more diverse variety of activities. Education for recreation takes on a number of forms. Standard curriculum courses or classes are common in primary and secondary schools as well as in colleges and universities. Other methods include leisure counselling services, workshops and camps. Programmes which focus on recreation activities tend to follow a certain process. First, participants are introduced to unfamiliar recreation activities. Second, they acquire a knowledge of what an activity involves, or how to do it. Third, they develop at least a minimum skill level, and fourth, they have opportunities to participate in the activities they have studied.

A lack of education for recreation in many cases hinders participation in certain activities. Education is viewed as a means of increasing enjoyment by improving participants' proficiency in a wide variety of activities. This is usually viewed as one way of improving a person's quality of life. In terms of education for recreation professionals, several types of recreation **training** exist. Outdoor, therapeutic and commercial recreation are the most common forms of recreation for which professional education is sought. All types of education require basic skills on the part of the professional providers. These skills include the history and philosophy of the field, **leadership** and supervision, first aid and safety, communication and interpersonal skills, **management** techniques, evaluation skills and programme development skills.

In addition to these basic skills, each type of recreation has its own set of specific education needs. **Outdoor recreation** refers to leisure activities that depend mostly on the natural **environment**. This type of recreation, for example, requires educated professionals who understand the proper leisure use of the natural environment, as well as the possible environmental **impact** of outdoor activities. Outdoor professionals also need to be familiar with natural **resources** and **interpretation**. This will increase the enjoyment levels of participants and build awareness among them that can assist in mitigating negative environmental impacts.

To be effective, professionals in the area of therapeutic recreation, which is leisure activities for people with special needs, must understand various illnesses and their effects, medical terminology, anatomy, geriatrics and gerontology, and kinesiology. They are expected to be familiar with a wide range of activities suitable for people with special needs, and able to provide leadership and possess supervisory skills. In addition to formal education, therapeutic recreation professionals often require certification, which is a professional **licensing** process based on education and practical **experience**.

Commercial recreation activities include money-making ventures such as swimming pools, golf courses and skating rinks. Experts in this area demonstrate high levels of knowledge and proficiency in the programme areas. They are also skilled in maintenance and management of specialised facilities. Formal education usually also includes sales techniques, programme **planning**, labour **cost** control, **marketing** and **accounting**.

See also: recreation manager; recreational behaviour; recreational carrying capacity; recreational planning

Further reading

Crossley, J.C. and Jamieson, L.M. (1988) *Introduction to Commercial and Entrepreneurial Recreation*, Champaign, IL: Sagamore Publishers. (Provides a good overview of the professional training requirements for commercial recreation.)

Ford, P.M. (1981) *Principles and Practices of Outdoor/Environmental Education*, New York: Wiley. (Includes an informative section on education in, for and about outdoor activities.)

Russell, R.V. (1996) *Pastimes: The Context of Contemporary Leisure*, Dubuque, IA: Brown and Benchmark. (Provides an informative overview of therapeutic and other forms of recreation education.)

DALLEN J. TIMOTHY, USA

efficiency

The concept of efficiency refers to lack of wastage or optimal use of productive resources resulting in a maximum of output – or satisfaction – within the constraint of a given technology and input types (Samuelson and Nordhaus 1993). The achievement of efficiency is obviously dependent on the proper coordination of economic activity to go as far as possible in consumer **satisfaction** in the framework of a certain know-how and input utilisation level. As von Hayek pointed out, this coordination is subject to the available information and knowledge, which themselves have a cost of production, transmission and processing. Therefore, information and knowledge must be considered also as restrictions on the achievement of efficiency. Additional restrictions on efficiency come from the inertia of organisations and any other administrative barriers to change. It is useful to distinguish between technical efficiency (or the ratio between physical quantities of outputs and required inputs as specified in production functions) and economic efficiency (or the ratio between the value of outputs and required inputs).

In tourism, the idea of efficiency has been associated with **cost** reduction policies in the Fordian age of tourism (mass production and consumption of standardised tourism services), but only more recently has it come to play a role within the concept of **competitiveness**, central in contemporary policy. Short-term price competition has been prevalent for many years in the **industry**, but there is a better understanding now of the crucial role of quality – and consumer satisfaction and fidelisation – in the competitiveness of tourism enterprises and destinations. In this respect, it is interesting to realise that the methodology of total quality management (TQM) has implications of efficiency beyond those of consumer satisfaction. **Quality** in tourism services (or meeting the tourists' expectations) has an obvious effect in securing the clientele of a destination. But it should also be clear that satisfying the 'internal customers' of a tourism production process will eliminate wastage and may result in a 'zero defects' or efficient process. Thus, tourism quality means also efficiency in the existing production process.

However, the concept of tourism efficiency has further implications since the methodology of TQM does not go as far as comparing the relative efficiency of all processes possible. In accepting that tourism business processes have to meet the expectations of a large enough set of consumers – now or within a foreseeable time horizon – it is possible to analyse the comparative efficiency of alternative business processes fulfilling the condition of client satisfaction so as to select the optimum process. This methodology, also referred to as business process re-engineering, deals frontally with the questions of efficiency beyond the limited 'zero defects approach' which characterises TQM.

Finally, it is important to recognise that in the business **paradigm** of the new era of tourism (characterised by segmentation of **demand**, flexibility of business processes and the search for profitability beyond **economies of scale**) success in the markets may depend on strategic positioning as a prerequisite, with 'ordinary' competitiveness (quality plus efficiency) assuring only short-term survival of the firm or the destination. The positioning of the firm/**resort** within a value network of suppliers, customers, competitors and complementors becomes all-important, and the methodologies of quality and efficiency simply assist in the rationality of competitiveness.

References

Samuelson, P.A. and Nordhaus, W.D. (1993) *Economics*, New York: McGraw-Hill.

Further reading

Bull, A. (1994) *The Economics of Travel and Tourism*, London: Pitman.

Castells, M. (1989) *The Informational City*, Oxford: Blackwell.

Fayos-Solà, E. (1996) 'Tourism policy: a midsummer night's vision', *Tourism Management*, September.

Graselli, P. (1992) *Economia e política del turismo*, Milan: Centro Italiano di Studi Superiori sul Turismo, Assisi, Franco Angeli.

EDUARDO FAYOS-SOLÀ, SPAIN
AURORA PEDRO, SPAIN

ego-enhancement *see* motivation; prestige

Egypt

Egypt, one of the oldest countries in the world, is labelled as the storehouse of **history** and the land of eternity. It has a central location between the countries of Africa, Asia and Europe. It possesses a unique cultural and archaeological wealth that dates back to prehistoric periods and includes the immortal pharaonic monuments. It has a coastline on both the Mediterranean and the Red Sea that extends for more than 2,300 kilometres. The River Nile, longest river in the world (4,600 kilometres) runs from south to north in Egypt for 1,200 kilometres, creating fertile land and making possible sedentary life for more than 60 million Egyptians.

Egypt's capital city, Cairo, is a metropolis of more than 14 million inhabitants. The country has more than seventy-five other cities and towns, supplemented by more than 5,000 villages. It has some fifteen universities and more than 100 institutes of higher education with a student body of more than 1.5 million and about 50,000 teaching staff. According to the UN economic parameters, Egypt is a **developing country** with a per capita income of $850. Its main sources of foreign currency are receipts from Egyptian workers abroad, tourism, oil and the Suez Canal.

Egypt is the leading **destination** in the Middle East. The number of tourist arrivals reached 3.9 million in 1996 (accumulating 23.5 million tourist nights), with an average length of stay of six nights. This constitutes 25 per cent of the total tourist traffic to the Middle East region and 0.66 per cent of all international traffic. Receipts from tourism reached $2.9 billion in 1995, which means a tourist per capita income of $48 compared with $725 for **Spain**, $482 for Italy, $466 for **France**, $223 for the **United States** and $16 for **China**.

Since the nineteenth century, when organised tourism began in Egypt, the great pharaonic Christian and Islamic civilisations were the main **attractions** for overseas visitors from the West. The beaten track included five focal points: Cairo, Menia, Luxor, Aswan and Abu Simbel. However, other important archaeological sites exist at Beni-Sueif, Fayoum, Asiut, Sohag, Kena, Sinai, New Valley and Siwa. Some thirty years ago, Egypt started to develop its beaches and since then has succeeded in becoming a world destination for sea, sun and scuba-diving tourism. Arab tourists, about 40 per cent of Egyptian tourist traffic, visit Egypt to enjoy its diversified way of life, moderate **climate**, developed beaches and busy nightlife. Egypt's tourism accommodation capacity includes 75,000 rooms, with over 135,000 beds. More than 25,000 rooms are due to start operation in the next three years. Nile cruisers now number 210, with about 12,000 cabins.

Egypt is presently engaged in a wide-ranging privatisation process. The private sector has swiftly matured and become responsible for 80 per cent of the five-year **development** plans. Over 20,000 companies are registered and dealing through the stock exchange, and these joint stock companies are supplemented by several thousand limited liability companies and partnerships.

The national strategy for tourism **supply** development aims at the institution of an environment where the tourist sector can accomplish its optimal **sustainable development** potential on sound regulatory, technical, social, economic, financial and environmental bases. The strategy stresses the need to diversify the tourist **product** and offer the visitors the opportunity to combine culture and **leisure** activities. To do so, the strategy includes the carrying out of a **survey** of the sector and the preparation of a priority action plan.

The main objectives of this plan are to offer a vision for future tourism **development** based on analysis of international **markets** and **demand** in relation to national assets and **resources**; to identify priority development areas based on examination of potentially developable sites; to define development requirements, including **infrastructure** and environmental and **investment** needs; and to refine implementation priorities and plans.

Since 1967, the Ministry of Tourism is the main authority dealing with tourism in Egypt, succeeding the State Tourism Administration. The Ministry is presently overstaffed, with a purely mechanistic organisation. It is in need of a thorough reshuffle and the introduction of a semi-systems approach to its organisation in order to become more efficient in responding to the newly emerging changes in tourism. In cognisance to the existing deficiencies, the Ministry is being streamlined to improve **efficiency** and strengthen its technical expertise in order to cope with the structural changes and technical developments imposed by a growing fiercer competition in **planning**, **marketing** and support of a growing strong private sector. One of the major steps in streamlining was the creation of the Tourist Development Authority in 1991. This agency is becoming an active instrument in tourism development drawing on private expertise more than depending on full-time appointed personnel.

The Ministry oversees four semi-autonomous governmental organisations: the Egyptian General Authority for the Promotion of Tourism, which is responsible for promoting tourism in both domestic and international markets; the Public Authority for Conference Centre, which manages the Cairo Conference Centre, the Tourism Academy, which monitors all training institutions in the tourism sector; and the Tourism Development Authority.

The private tourism sector is represented by the Egyptian Federation of Tourist Chambers and its four units, the Chamber of Hotel Establishments, Travel Agencies, Tourist Establishments and Handicraft Industries. In the various tourism governing bodies, regional organisations for promotion have been created since 1957 and have been operational ever since. Their **impact** upon the tourism sector have been practically negligible, although few are relatively more active than others. These all suggest that the tourism institutional setting in Egypt should be re-examined in order to provide for the necessary changes in organisation, staffing motivation, operational rules and monitoring its impact upon tourism.

SALAH E.A. WAHAB, EGYPT

electronic promotion source

Electronic sources information includes information technologies (hardware, software, netware) which tourism organisations use to publicise and distribute their products to consumers and intermediaries. Examples of such media are the **Internet**, CD-ROMs, kiosks and interactive **television** for consumers, as well as computer reservation systems (CRS) and **global distribution systems** for intermediaries. They are increasingly replacing conventional **marketing** methods by facilitating interactive communication and better **market segmentation**.

DIMITRIOS BUHALIS, UK

emergency management

A growing concern in the tourism industry is the risk of major disasters. However, the industry has been slow to adopt emergency management programmes to prevent or at least to minimise the impact of disasters on lives and property, as well as to respond to and recover from disasters. Managing environmental hazards is often in **conflict** with **economic development** interests, and typically requires limitations on the use of private property. Tourism leaders frequently oppose land-use regulations, building codes and other programmes aimed at reducing the level of **risk**.

Tourism and **hospitality** firms often ignore environmental hazards and may even make them worse. **Resorts** are frequently located so close to coastlines, mountains, rivers and other scenic features that there is increased risk from hurricanes or typhoons, flooding, landslides and avalanches, snow storms, wildfire and other natural hazards. Poor construction and maintenance increase the risks of fire, structural collapse and damage from wind and flood. Inadequate **safety training** for staff and insufficient medical resources to handle major emergencies also increase the risk to **tourists** and residents. Resort areas are particularly vulnerable because tourists are generally ill-prepared for emergencies and unfamiliar with proper safety precautions. Communities may be prepared to respond to emergencies involving their own residents, but unprepared to manage disasters involving hundreds or thousands of tourists. They are also susceptible to food poisoning (see **food-borne illness**), contagious diseases and transportation accidents.

Recent disasters and near disasters involving cruise ships are encouraging countries to increase safety regulation, including closer monitoring of food service facilities, lifeboat and other safety training, the qualifications and numbers of on-board medical personnel, and crew qualifications on vessels of foreign registry. Elderly **tourists** in particular may be at significantly higher **health** risk and require more medical care than may be readily available on board. The UN-sponsored International Decade for Natural Disaster Reduction effort in the 1990s is encouraging governments to address known environmental hazards and to provide technical assistance and financial support to developing nations. Tourism facilities are a particular concern in that effort.

See also: risk analysis

Further reading

Drabek, T. (1994) *Disaster Evacuation and the Tourist Industry*, Boulder, CO: Institute of Behavioural

Science, University of Colorado. (Explores the perspective and behaviour of tourism executives on hazards and suggests that the potential for a major disaster involving tourists is growing.)

—— (1995) 'Disaster planning and response by tourist business executives,' *Cornell Hotel and Restaurant Administration Quarterly* 36: 86–96. (Argues that tourism executives resist planning for potential emergencies and are ill-prepared for a major disaster.)

WILLIAM L. WAUGH, USA

employment

Employment in tourism may be defined in **supply** terms according to the sectors which service tourist needs and the expenditure they incur. The **industry** is a range of sectors or businesses which receive revenues from **tourists** in varying degrees; hence, jobs may be only partially dependent on or supported by these revenues. Standard Industrial Classifications (SICs) operated by governments very rarely list all the sectors, and measurement of employment in tourism is difficult to undertake; no Standard Industrial Classification of this industry or tourism employment exists.

From the **demand** side perspective, employment is related to tourism expenditures in total, per day and per unit, and is supported by transactions which lead to direct, indirect and induced activities (see **multiplier effect**). A comparison of direct employment created in each sector with resulting total employment in all other industries leads to a construction of ratios and backward supply linkages to be identified. **Input–output analysis** helps to identify the precise nature of inter-industry and tourism linkages and the employment generation which results.

Perspectives on tourism employment from an industry position focus on sector, level of skills, task and specialism or function (such as hotels, unit manager or **food** and beverage). With the introduction of national vocational qualifications in countries such as the **United Kingdom**, **Canada** and **Australia**, the levels of task and the competence or the ability to do these tasks have been revalued. National qualifications, or what were called trade standards, have been developed and these have been investigated by the vocational **training** agency of the European Union. Professional profiles, occupations, **career** paths and jobs have been outlined and the **education** and training required in order to fulfil the tasks involved have been outlined. Recent **innovations** in the workplace have been the development of multiskills, teamwork and the empowerment of employees. The delayering of some levels of employees particularly in the **management** field have led to a less tiered or hierarchical structure.

In most economies, tourism employment operates in a **market** environment and there is competition for labour with other sectors. Pressures on labour supply arise from demographic changes (see **demography**) such as declining birth rates and negative **images** of working in the service sector, such as the disadvantages of unsocial hours in tourism work. In more advanced economies, problems of labour retention and recruitment plus rising labour costs has led to the reduction of labour needs through the use of new **technology** and **standardisation** of the product and service offered. Tourism still includes labour-intensive sectors, and in developing economies it creates valuable employment opportunities for lower skill workers; however, the seasonal nature of tourism (see **seasonality**) and the **migration** of workers to tourism locations are still cited as negatives.

JOHN WESTLAKE, UK

employment law

Employment could be defined as the act of hiring, implying a request and a contract for compensation. It does not necessarily import an engagement or rendering services for another. A person may as well be 'employed' about his own business as in the transaction of the same for a principal (*Black's Law Dictionary*, 4th edition, p. 618). An employer is one who employs the services of another; one for whom employees work and who pays their wages or salaries. While such notions are almost universal, legal systems prevailing in various countries differ in details according to the politico-economic system prevailing, the legislative **policy** and the relative bargaining **power** of trade unions versus the **government**.

Employment **laws** usually contain provisions regulating relationships between employers and employees, both individually and collectively. Some of the major concerns in employment laws are the employee's job security versus **rules** of hiring and firing, the scope and limits of employer's entitlement to retain the employee if he/she quits before the term ends, and **employment** agreements and structure with express and implied terms. On these issues and others, countries differ in their employment laws particularly insofar as parties, reciprocal rights and obligations. The freedom of contract in some of these details is sometimes limited by public directives in a given country. Moreover, trade unions, powers of negotiation and exerting pressure on governments and employers is usually influenced by the prevailing political and economic climate as well as by the long-established traditions in this context. In some democratic and free economy countries like the United States, trade unions are quite strong, while in some socialist countries such as Cuba or North Korea, they are mere tools in the hands of the governing socialist party.

Sometimes it is unclear whether a restrictive provision in a labour agreement is intended simply to increase employee welfare or to create monopoly in the employers' product markets. Moreover, any minimum wage law reinforces the effect of unionisation on the wage rate by limiting the competition of non-union labour. With the dominance of private enterprise in business, employer–employee relationships become more sophisticated and intensified. More detailed provision appears in employment agreements in terms of issues such as compensation and reimbursement of expenses, duration of the agreement, non-competition clauses, medical expenses entitlements, death and disability clauses, and also including employer's rights to employees work products, restrictive and proprietary covenants, trade secrets protection, compensation of employees for invention, insurance programmes, and more.

Because tourism is a labour-intensive industry in almost all destinations, all the preceding concepts and rules prevail in a clearcut fashion. The same legal provisions governing employment in various productive and service sectors define employer–employee relationships in tourism. However, racial, sex, age and cognate forms of discrimination in employment which are a part of a larger issue (the causes of discrimination, and the consequences of anti-discrimination policies) could largely apply in tourism because of its intractable, sensitive and sometimes seasonal nature (see **seasonality**).

In the **hospitality** industry, which is the largest and most labour-intensive tourism sector, permanent, limited duration and temporary employment exist. In areas where **seasonality** of tourist traffic prevails, hotels and **catering** or **restaurant** establishments suffer from high employee turnover, particularly in developing countries. Moreover, in countries with noticeable tourism growth trends, unless there is an abundant supply of trained personnel, new hotels and other **accommodation** establishments will attract employees from other well-established lodging and catering facilities through increased salaries and higher incentives. This would lead to a general deterioration in the service quality if regulatory measures are not laid down. These measures should be primarily based on enhanced applied in-plant and institutional training to guarantee a constant flow of well-trained personnel.

As the tourism **industry** heavily relies on new technology for its efficient operation, employees are required to adapt to new techniques within their contractual terms. New challenges and opportunities in this industry as it enters the twenty-first century will require special attention to employment laws at national and international levels of operation, with multinational agencies playing an increasing role.

SALAH E.A. WAHAB, EGYPT

enclave tourism

Enclave tourism is a form of self-contained **resort** development which is geographically isolated or insulated from the surrounding **indigenous** population. Such enclaves typically are coastal, offer amenities such as tennis, golf, scuba diving and horseback riding, and often contain landing strips for jets or smaller airplanes. Tourists who stay in such enclaves have no need to leave the complex as food, drink, and entertainment are provided by the resort. At all-inclusive resorts, such as Club Med, almost all daily needs and amenities are included in the overall price. Many enclave resorts in Third World destinations are associated with '4-S tourism' (**sun, sand, sea and sex**) and are exemplified by the Gulf and Western-owned Casa de Campo resort in La Romana, **Dominican Republic**, which is regarded as one of the prototypes of enclave tourism in the Caribbean.

Enclave tourism has been studied from several perspectives, notably its role in **economic development** and social **impacts**. Resort enclaves in developing countries have traditionally been developed and owned by **multinational firms**, and thus the local economic **benefits** have been low. Whereas foreign corporations have the capital to develop tourism facilities as well as the ability to bring in tourists by tapping into established **marketing** linkages, profits are subsequently taken out of the host country and there is little 'trickle down' into the local economy. In economic terms, enclave tourism has been described in terms of both vertical integration practices and also **dependency theory**.

Enclave tourism allows virtually no host–guest interactions (see **host and guest**) and this has led to resentment by both the local population and the broader national **community**. Except for low-level **resort** staff and wealthier residents, the local population is generally banned from the resort complex premises. Enclave guests are segregated from the local culture and are especially shielded from the local informal sector which includes vendors, hustlers, drug dealers and prostitutes. At the local level, such segregation not only creates a wall between host and guest but also precludes economic benefits from filtering into the community. At the national level, such segregation is often regarded by the native population as a form of **neo-colonialism**. In a typological analysis of seaside resorts, an 'interactive' enclave resort category was recognised, in which tourists did experience a limited amount of interaction with the local population and levels of local resentment were lower than those surrounding totally self-contained complexes.

Further reading

Britton, S.G. (1982) 'The political economy of tourism in the Third World', *Annals of Tourism Research* 9(3): 331–38.

Debbage, K.G. (1990) 'Oligopoly and the resort cycle in the Bahamas', *Annals of Tourism Research* 17: 513–27.

Meyer-Arendt, K.J., Sambrook, R.A. and Kermath, B.M. (1992) 'Seaside resorts in the Dominican Republic: a typology', *Journal of Geography* 91: 219–25.

KLAUS J. MEYER-ARENDT, USA

encounter

Encounters occur where two or more different individuals, groups or populations meet. In tourism, encounters are often distinct cultural 'meetings' typified by a limited width of contact and transience of **host and guest** interactivity. They are frequently characterised by the extreme incongruity of the tourist's wealth, the poverty of his/her awareness of local society, and the exotic objectification and commodified consumption involved (see **commercialisation**; **exoticism**).

KEITH HOLLINSHEAD, UK

energy

Energy drives the basic ecological systems that constitute the earth's biosphere, and it is the harnessing of animate energy (through living organisms) and the utilisation of inanimate energy (from non-living matter) that has made possible the progress of human society. As concern grows over the long-term sustainability of reserves, attention is being focused on renewable sources of energy and energy conservation. Tourism as a conspicuous user of energy is also expected to address these possibilities.

Renewable sources of energy include direct solar energy (radiation from the sun), indirect solar energy (wind, wave, hydro, biomass), geothermal energy and gravitational (tidal) energy. Apart from direct solar energy and possibly limited use of wind energy, the other sources of renewable energy are not feasible at the level of tourism **development**. Solar water heating is the most common and increasing use of a renewable source of energy in tourism operations. Active solar collectors are already competitive for pool heating, although construction and operating costs are high and, in most cases, a back-up system is required. Passive solar systems are frequently considered for residential space heating and can readily be applied to tourism establishments. In passive systems, windows serve as collectors and internal walls and structures as heat storage, perhaps with fans for improved circulation. Passive solar heating is suited to modern integrated energy-efficient design, particularly in tourism operations requiring only low temperature space heating.

Of greater significance and more widespread application in the tourism industry are energy **efficiency** and energy conservation. Cost-effective measures which can be considered include shifting from electricity to gas for heating and cooking, replacing electric hot water systems with solar hot water (gas-boosted), use of energy-efficient lighting (long life and fluorescent globes) and appliances, design and construction of energy-efficient buildings (exposure to sun/passive solar systems), consideration of alternative forms of transport, and **education** of staff, **tourists** and **management** in energy conservation.

Measures to improve energy efficiency in tourism operations are cost-effective and contribute to long-term sustainability of the enterprise. They also may be well received in a more discerning and environmentally aware clientele. At the same time, the transition to energy-efficient modes of operation can be constrained by real or perceived barriers. These include upfront costs, inadequate appreciation and communication of costs of regular energy sources, and lack of incentives.

See also: environment; environmental compatability; sustainable development

JOHN J. PIGRAM, AUSTRALIA

entreprenurial education *see* education, entrepreneurial

entrepreneurship

Entrepreneurship is an imperative activity that initiates the start-up of a profitable business. Through this activity, the goal is to maintain and continually develop a business which provides a profitable long-term **future** for the company. A person who undertakes these activities is called an entrepreneur. Entrepreneurship has become a dominate force in the US economy, especially in the **hospitality** sector. In fact, most hospitality businesses are comprised of the 'mom and pop' organisations characterised by the entrepreneur who seeks opportunities in the **market**, takes advantage of it and creates a niche. Moreover, an entrepreneur is proactive, willing to take **risks** and adaptable to the changes in the environment.

Due to different characteristics that entrepreneurs may possess, not all are alike. For example, entrepreneurial typologies prevalent in the **restaurant** sector include the creative, opportunistic and humanistic characteristics. Each describes a certain type. A humanistic entrepreneur, for instance, practices a participative style of **management**. However, this type may lack business skills. An opportunistic entrepreneur seeks out occasions which create wealth. This type of entrepreneur is more functional in a firm which is continually changing and adapts to certain environments. A creative entrepreneur is an individual who is visionary and often starts a business from scratch based on original ideas. It is this type of entrepreneur who develops the sound concept of business. An understanding of different characteristics enables the entrepreneur to match the appropriate strategies to achieve **competitive advantage** and sustain long-term growth.

Entrepreneurship is generally associated with the introductory phase of an industry's **life cycle**, but the entrepreneurial spirit is equally critical in the maturity phase. These types of entrepreneurship, individually and together, are present in all sectors of the tourism industry. Many – whether in **hotel /tour operator** or airline business – have led their businesses to national and/or international fame. As the industry matures, fostering corporate entrepreneurship is needed for innovative strategies which would strengthen the position of tourism in the new century (see **innovation**).

See also: culture broker; education, entrepreneurial; middleman

Further reading

Low, M. and MacMillan (1988) 'Entrepreneurship: past research and future challenges', *Journal of Management* 14(2).

Naman, J. and Slevin, D. (1993) 'Entrepreneurship and the concept of fit: a model and empirical tests', *Strategic Management Journal* 14: 137–53.

Olsen, M., Tse, E. and West, J. (1999) *Strategic Management in the Hospitality Industry*, New York: Van Nostrand Reinhold.

ELIZA CHING-YICK TSE, CHINA

environment

The term 'environment' can be defined as the complex of external conditions surrounding an object, an organism or a **community**; and that specific set of measurable phenomena existing during a specified period of time at a specific location. Yet the term is essentially relative because, by definition, all environments are contexts of something. An object cannot exist in isolation from its surroundings. Likewise, an environment has no meaning except in relation to the object or organism with which it is interwoven. This reasoning applies both to human and non-human organisms and social structures, including communities, organisations and segments of society. In this relational or transactional view, the environment becomes of material relevance through the interplay between some environed feature and its surroundings; an interaction which is not static but rather is dynamic and evolving.

All human activities have environmental implications, the significance of which depends to a large extent on the technological level and **lifestyle** of the society and culture involved. Also, repercussions extend far beyond the mere natural environment to encompass socioeconomic and cultural elements as well as physical phenomena. In developed countries, the effective environment in which people operate progressively has come to comprise created elements and be dominated by human-oriented problems. Inevitably, human–environment relationships have become more complex and the task of environmental management more urgent. However, unfortunately, much of the concern voiced for the environment has been expressed in negative terms, emphasising the limiting characteristics of natural systems and the potential environmental degradation involved in resource **development**. Yet, the overall goals of environmental action and resource development are increasingly coming to coincide, and such considerations are now seen as integral and vital components of a development process designed to improve the quality of life for present and future generations.

Any discussion of tourism in an environmental vacuum is meaningless because attributes of the latter form the essential framework for travel. A wide range of settings and attractions can appeal to visitors, and the resource base for tourism encompasses features, conditions and processes in both the physical and cultural environment, a diversity which is important because it provides the very necessary element of choice to the system. All attractions are to some extent environmentally based, linked either directly or by association with a specific site or **location** and each appealing within the context of that setting. This does not necessarily imply that the **quality** of magnetism is somehow inherent or that only natural features qualify. The many artificial entertainment complexes often created from an unremarkable resource base (such as Disneyland) disprove any such supposition. However, for many tourism settings, identification with place and with local influences, physical and cultural, is fundamental.

At the same time, every **destination** is partly a created environment. Even apparently natural or unique features require some embellishment, if only signposting, to cater for visitor use. It follows that tourism environments are an amalgam of resources and facilities, and such complementary support is vital to their effective functioning. The environmental setting gives the facilities meaning and purpose, yet the relationship is very much two-way. A tourism resource of average or even mediocre quality may be enhanced by imaginative harmonious amenities. Conversely, the obvious magnetism of an outstanding natural feature can be impaired by inappropriate or poorly designed facilities. The range, quality and location of ancillary structures and services thus contribute significantly to the **image** and appeal of a destination.

Clearly, the many ramifications of tourism give much scope for interaction with the environment. The degree of **impact** is related to factors such as the intensity of tourism site-use, the resilience of the ecosystem, the time perspective of the developers and the transformational character of touristic developments. In addressing the impact of tourism, there is a tendency to view the options in terms of opposing alternatives (see **alternative tourism**): protecting the environment *for* tourism, and protecting the environment *from* tourism. However, these objectives need not be mutually exclusive. Whereas most concern has been expressed regarding adverse effects, change does not necessarily equate with degradation. There appear to be several modes of expression of the impact of tourism; the net effect may well be environmental enhancement, or the interaction may be neutral.

Tourism development can contribute to substantial improvement of recreational resources and thus add to visitor enjoyment. Tourism can lead, for example, to an enhanced **transportation** system through advances in vehicle and routeway design which allow greater opportunity for pleasurable and meaningful participation in this worldwide phenomenon. Enhanced understanding of the resource base is another positive result brought about by the application of revolutionary communication **media** techniques to interpret and articulate the environment to visitors. Beneficial modifications, or adaptations to **climate** in the form of recreational structures, clothing and equipment, have been developed in response to the stimulus from tourism. Better managed habitats for fish and wildlife and control of pests and undesirable species have become possible through the economic support and motivation of increased use.

On another plane, increasing cultural consciousness has stimulated restoration of historic **sites** and antiquities. Particularly in Europe, **heritage** features such as cathedrals, castles and artefacts of past eras could not be kept intact if their existence and **preservation** had not become the ongoing concern of a great audience of **tourists**, resulting both in substantial financial contributions from them and generous state support. Design of contemporary tourism complexes also appears to be benefiting from the demands of a more discerning tourist population. Whereas there remain many examples of unfortunate additions to the touristic **landscape**, environmental modification for today's tourist increasingly is marked by quality architecture, design and engineering. Higher standards of **safety**, **sanitation** and maintenance also help to reduce the potential for pollution. These advances demonstrate that tourism need not destroy natural and cultural **values**, and in fact can contribute to an aesthetically pleasing landscape. The environment also includes human resources, and tourism may affect customs, crafts, attitudes, traditional values and the general way of life.

Although it can be conceded that tourism has much (perhaps unrealised) potential for environmental enhancement, negative **impacts** do occur from the predatory effects of seasonal **migrations** of visitors. The most obvious repercussions are likely to be in natural areas, but the built environment may also be impaired and the social fabric of host communities can be widely disrupted. Pollution, both direct and indirect and in all its forms, is a conspicuous manifestation of the detrimental effect of tourism. However, erosion of the resource base is probably a more serious environmental aspect. This can range from incidental wear and tear of fauna, flora, and structures, to vandalism and deliberate destruction or removal of features which constitute the appeal of a setting. This erosive process can be accelerated by use of technological innovations and by inferior design and inappropriate style in the construction of tourism facilities. In parts of the old world (for example, Hurghada in **Egypt**) hotels and **resorts** can only be described as an architectural affront to the natural or historical sites where they are located. It is such circumstances which give some substance to the assertion that the creation of ugliness could be one of tourism's 'greatest contributions' to the environment.

In many resort areas, environments must serve not only conflicting touristic uses but also the resident community, many of whom take a proprietorial attitude towards their surroundings.

Congestion and overtaxing of **infrastructure** and basic services can generate dissension between transients and the domestic populations, which come to resent the intrusion of tourism. Municipal services and facilities, access to recreational sites, and personal and social life are all features of the sociocultural environment which can be seasonally curtailed by the presence of vacationers (see **seasonality**). In some cases social interaction between residents and tourists is minimal, with large numbers of visitors opting for an exclusive environment which requires minimum cultural adjustment on their part. In developing countries, aspects of tourism may have long-term disruptive effects on the **lifestyles** and **employment** patterns of host communities.

With so many variations on the theme, it is difficult to generalise on the relationship between tourism and the environment. The relative importance of each factor varies with the location and situation, and negative effects need to be balanced against positive impacts. It is important to note that tourism and environment are not merely interrelated – they are functionally interdependent – and tourism and protection of the environment are more alike than contradictory. The demands of tourism, instead of conflicting with conservation of the environment, actually require it; otherwise the very appeal which lures the visitor will be eroded, and with reduced **satisfaction** will go any chance of sustained viability. Given an appropriate commitment to **planning**, design and **management**, tourism can become an active positive agent in the process of environmental enhancement.

See also: codes of ethics, environmental; environmental aesthetics; environmental auditing; environmental compatability; environmental engineering; impact assessment, environmental; perception environmental; planning, environmental; quality, environmental; sustainable development

JOHN J. PIGRAM, AUSTRALIA

environmental aesthetics

The effect of an **environment** on the senses is known as environmental aesthetics. Such qualities of tourism facilities are, like beauty, very much in the eye of the beholder, and most observers can readily distinguish between pleasing and offensive settings (see also **quality**). Attention to environmental aesthetics in the **design and layout** process both contributes to satisfying sensory experiences and strengthens the functional **efficiency** of tourism developments.

JOHN J. PIGRAM, AUSTRALIA

environmental auditing

Environmental auditing can be defined as a process comprising of a systematic, documented, regular and objective evaluation of the environmental performance of any aspect of an organisation. This may include structure, **management**, equipment, facilities and **products** with the aim of protecting the **environment** by facilitating management control of environmental practice, assessing **compliance** with environmental policies and any regulatory requirements, and minimising the negative environmental **impact**. The term 'auditing' generally refers to a methodological examination involving analyses, tests and confirmation of a facility's procedures and practices with the goal of verifying whether they comply with legal requirements and internal **policies** and accepted practices. In its common application, environmental auditing is the process whereby the operations of an organisation are monitored to determine whether they are in **compliance** with regulatory requirements and environmental policies and standards. Therefore, the essential purpose of environmental auditing is to ensure compliance with **environmental management systems**, in particular that commitments made are implemented, that environmental standards are met, and that relevant procedures are in place and are being followed.

The concept of environmental auditing is still relatively new and is a developing technique. In the context of tourism, environmental auditing as an effective management tool can provide feedback about overall environmental performance and specific problems of an organisation, and a ready means for self-regulation of its environmental performance (see also **regulation, self**).

Environmental auditing, whether required by regulation or legislation or undertaken as a self-regulatory initiative, can be a useful management tool in helping to achieve **sustainable tourism** development. As global demands on space and resources grow with increased population, technological change and greater mobility and awareness, pressure will emerge for the industry to implement appropriate approaches for monitoring and evaluating its environmental performance, as is increasingly the norm with other sectors of the economy. The challenge facing tourism development is to endorse an effective self-monitoring and evaluation process for environmental auditing before mandatory compliance measures are imposed by regulatory action from outside the industry.

See also: environmental compatability; legislation, environmental; planning, environmental; quality, environmental; sustainable development

Further reading

Buckley, R. (1991) *Perspectives in Environmental Management*, Berlin and New York: Springer Verlag.

Ding, P. and Pigram, J. (1995) 'Environmental audits: an emerging concept in sustainable tourism development', *Journal of Tourism Studies* (6)2: 2–10.

Goodall, B. (1995) 'Environmental auditing: a tool for assessing the environmental performance of tourism firms', *The Geographical Journal* 161(1): 29–37.

PEIYI DING, AUSTRALIA

environmental codes of ethics *see* codes of ethics, environmental

environmental compatibility

Environmentally compatible tourism describes a situation whereby the industry and the environment are able to exist in harmony so that the former does not detract from or harm the latter and vice versa. Increasing environmental awareness and **conservation** activities around the globe have contributed to efforts to form this philosophy. In order to reduce the conflicts and enhance the relationship between the two, environmental impact statements are now required in many countries as part of the approval and monitoring and evaluation processes for tourism projects, particularly if a **development** is large or located in or adjacent to sensitive sites such as protected areas, rainforests, coastlines or estuaries. Further developments in environmentally compatible tourism approaches can be seen in the implementation of **environmental auditing** processes in the public and private sectors.

The key to achieving such goals for a **ecologically sustainable tourism** industry is recognition of the need for environmentally sensitive **policy** making, **planning** and development. The integration of tourism and the environment is being carried out at different levels in a number of places and for a variety of reasons, with various mechanisms being utilised. Strategies and related activities range in size from small to large-scale projects and include various economic, nature **conservation**, cultural, social **heritage**, spatial/regional and political **benefits** and costs. On a broader national and global scale, approaches to integrating tourism and environmental objectives are being developed and promoted by national and international tourism agencies, and to a limited extent by multinational corporations.

Further reading

Dowling, R.K. (1992) 'Tourism and environmental integration: the journey from idealism to realism', in A.W. Seaton (ed.), *Tourism: State of the Art*, Chichester: Wiley. (Provides an overview of the history of the relationship between tourism and the environment and recognises the link between environmentally compatible tourism and sustainable development approaches.)

Farrell, B.H. and McLellan, R.W. (1987) 'Tourism and physical environmental research', *Annals of Tourism Research* 18(1): 41–56. (Presnts a broad overview of research in the physical environment with respect to tourism.)

Pigram, J.J. (1980) 'Environmental implications of tourism development', *Annals of Tourism Research* 7(4): 554–83. (An overview of tourism and environmental issues, giving explicit recognition to the ways in which tourism can maintain or enhance the environment.)

The Journal of Sustainable Tourism provides many accounts of environmentally compatible and related tourism approaches.

JOHN M. JENKINS, AUSTRALIA

environmental education *see* education, environmental

environmental engineering

Environmental engineering can be considered from two points of view. In one sense, this is concerned with the application of engineering principles and practices to the **sustainable development** of **resources** and the **management** of environmental quality. This approach aims to bring an engineering perspective to the management of the **environment**. It might be seen as the reaction of practitioners to the alternative of seeking resolution of environmental issues through political means, negotiation, social impact analysis and the like. However, some would be critical of what could be termed strictly structural solutions for environmental problems.

In another and more general sense, environmental engineering is seen as not about 'engineering the environment' for the better, but about overlaying such a perspective to all engineering practice. In this sense, the concept has much in common with environmental planning (see **planning, environmental**) in that it imposes an environmental discipline and orientation on the profession of engineering. Whereas such a refinement in approach might be seen as desirable, it is more realistic to assume that the focus of environmental engineering will remain on the creation of innovative (engineering) solutions to existing or potential environmental problems. Such solutions must not merely 'work' in the technical sense, they must satisfy legislative requirements, be cost-effective and be acceptable to the **community**. A key focus is cleaner production processes, supported by minimisation of solid, liquid and gaseous wastes associated with modern, diversified manufacturing industry. A second concern is the consideration of biophysical and sociocultural environmental issues in the design, operation and management of industrial processes.

The relevance of environmental engineering to tourism, in both a proactive and a reactive sense, is clear. Environmental engineers consider the environment at the design stage of tourism **development**. They can then act to reduce the effects of touristic activities on the surrounding natural and human setting and thus mitigate or eliminate potential environmental concerns. Tourism projects which incorporate the skills and perspective of environmental engineering recognise the need to create a functionally effective development, but one which will operate with respect for environmental and social sensitivities, and in keeping with the expectations of a more demanding regulatory regime and a more environmentally aware society. Particular aspects of tourism development which benefit from the input of environmental engineering are in selection of materials, design, management of wastes, air and water quality, noise reduction and the conservation of **energy** and water.

See also: codes of ethics, environmental; environmental compatability; environmental management systems

JOHN J. PIGRAM, AUSTRALIA

environmental impact *see* impact, environmental

environmental legislation *see* legislation, environmental

environmental management, best practice

The term 'best practice' has had a mixed reception in the world of business. To some, it remains merely jargon or a buzzword, popular with management consultants. To others, best practice represents the essential direction which firms need to take to become and remain competitive internationally. The term originated in business organisation theory and in the world of transnational corporate **planning** and **management**. It has found its strongest expression in the manufacturing industry, but is also being promoted in the services and in natural resources management, including forestry, rangeland management and the tourism industry.

Best practice readily translates and extends into 'best practice environmental management' as a means of achieving sustainable **growth** in a competitive world. It calls for radically different organisational structures and attitudes designed to bring about continuous improvement in a firm's environmental performance. Environmental excellence is fostered by enlightened management practices which incorporate new, cleaner technologies, and an emphasis on resource **conservation**, **recycling**, reuse and recovery, in progress towards sustainability.

Since the 1980s, the tourism **industry** has shown preparedness to apply the principles of best practice environmental management to its activities. Similar to manufacturing, the scale of operations appears to have a decided influence on the type and extent of initiatives undertaken. On the **international tourism** scene, a growing number of large **hotel** corporations have implemented a variety of effective environmental measures. The initiative was taken up in Britain by the Prince of Wales Business Leaders Forum, which, together with leading hotel chains, produced a joint operations manual. The International Hotels Environmental Initiative followed and in 1992, produced a revised manual. This manual, *Environmental Management for Hotels: The Industry Guide to Best Practice*, is seen as a voluntary code of conduct and offers a useful reference and blueprint for upgrading environmental procedures in areas such as **waste management**, **energy** consumption, noise and congestion, purchasing policy and staff **training**. In a more environmentally conscious world, tourism faces increasingly stringent conditions on **development**, reflecting a concern for sustainability and the long-term viability of the **resources** on which it depends. The challenge for the industry is to justify its claims on resources and the **environment**, with a commitment to their sustainable management.

See also: codes of ethics, environmental; environmental compatibility; planning, environmental; precautionary principle

Further reading

Green Hotelier, Magazine of the International Hotels Environment Initiative, London: IHEI.

JOHN J. PIGRAM, AUSTRALIA

environmental management systems

An environmental management system (EMS) provides a framework for organisations to manage environmental impacts of their operations. It involves application of contemporary concepts to management of business aspects associated with such significant **impacts**.

The EMS concept came into prominence following the emergence of ecologically **sustainable development** (ESD) principles in the late 1980s. The latter aims to promote the **economic development** initiatives consistent with **conservation** which ensure equity in natural resources between current and future generations.

In 1990, the European Commission became one of the first major **policy**-making bodies to promote ESD principles when it developed the Eco-Management and Audit Scheme (EMAS). The latter provided EMS specifications for industries operating in the **European Union**. This was proposed as a voluntary programme to encourage industries to develop management systems which include plans for continual improvement in environmental performance, processes for meeting regulatory requirements, auditing of systems, public disclosure of environmental performance statements, and independent verification of management systems and environmental performance statements. In response to **sustainable development** initiatives, the first EMS standard was published in the United Kingdom.

During the 1992 Earth Summit in Rio de Janeiro, the Business Council for Sustainable Development emphasised the need for business and industry to develop tools for measuring environmental performance and environmental management. Subsequently, the International Standardisation Organisation (ISO) embarked on development of an international standard for an EMS model. This work culminated in the ratification of ISO 14001, Environmental Management Systems: Specification with Guidance for Use, in September 1996. ISO 14001 provides a framework to proactively manage operations which may have an adverse impact on the environment through implementation of an EMS.

An EMS adopts contemporary business management concepts by using a framework to manage environmental concerns which involves **policy** commitments, defining objectives, targets and programmes, monitoring performance and conducting reviews. The tourism **industry**, with its documented impacts worldwide and with its continued dependence on the **environment**, will have to come into terms with guidelines and principles represented in the environmental management systems framework.

See also: codes of ethics, environmental; environment; environmental auditing; environmental compatability; environmental engineering; planning, environmental; impact assessment, environmental; precautionary principle

NAV BRAH, AUSTRALIA

environmental perception *see* perception, environmental

environmental planning *see* planning, environmental

environmental quality *see* quality, environmental

environmental rehabilitation

Environmental rehabilitation is a subset of environmental protection. The rehabilitation of environmental damage and the proper allocation of the costs of this within protection frameworks have been formally incorporated in **legislation** in many countries in order to ensure that businesses and private individuals behave in a manner that avoids, rather than merely controls, their potential for future environmental damage. Tourism has a vital interest in environmental protection and rehabilitation, as it relies to a significant extent for much of its appeal on the health of the natural **environment**.

With respect to existing environmental damage from tourism **development**, mechanisms have to be found that ensure accountability for this is pursued, but not at the expense of actual rehabilitation. The history of rehabilitation is littered with examples of clean-up programs being weakened by legal argument over accountability and liability. The Superfund programme in the United States is a prime example of this (75 per cent of the budget was spent on legal fees and consultant studies, and only 54 of 1200 priority sites were cleaned up in thirteen years). Nevertheless, the Superfund **legislation** has radically changed the way industry in the United States handles and discharges waste, and is beginning to have extraterritorial effects. For example, American hotel companies worldwide are now incorporating higher environmental standards in their local tourism operations through an unwillingness to risk legal action against their US-domiciled activities.

The tourism **industry** has experienced concern internationally over the physical **impact** of its activities in recent years as host nations struggle to come to grips with the often unregulated development that has been experienced in the past. Ranging from the effects of overwhelming pressure on existing **infrastructure** to the degradation of natural sites through excessive **tourist** numbers, such concerns have helped in the development of operational guidelines for the rehabilitation of environmental damage from tourism. These are in turn enabling consistent decisions to be made on the extent, feasibility, desirability and technical standards that should apply to the rehabilitation of damage, and this can only assist local acceptance of the desirability of tourism development in the future.

See also: environmental auditing; pollution management; quality, environmental

Further reading

Malone, P. (1991) 'Pollution prevention through waste minimisation: converging the interests of industry, government and the public, environment and planning', *Law Journal* 267. (Describes the move away from end of pipe waste minimisation regulatory approaches.)

MALCOLM COOPER, AUSTRALIA

environmental standards *see* environmental auditing

environmental stewardship *see* environmental management, best practice

environmental valuation

Environmental and natural resource economics has developed a whole group of methods designed to place value on environmental attributes and which could be applied to the valuation of non-priced tourism resources. The main methods that have been developed can be classified by differentiating those that produce monetary valuation directly (by asking respondents their willingness to pay for an improvement or to avoid a degradation) from those that provide the valuation indirectly (by using prices from a related market which does exist). **Contingent valuation** is an example of the former, while the travel cost and hedonic pricing are representative of the latter.

Contingent valuation is a method that uses direct questioning of individuals to show a potential market and generate estimations of compensating variation for a change in individual welfare with respect to the **supply** level of environmental goods. This method enables economic values to be estimated for a wide range of commodities not traded in markets, such as **national parks**, forestry characteristics, wildlife, saltwater beaches, recreational fishing, air visibility, water quality, **ecotourism** destinations and tourism congestion.

The travel cost method, which can claim to be the oldest of the non-market valuation techniques, is predominantly used in **outdoor recreation** modelling, with fishing, hunting, boating, bird watching, **camping** and natural area visits among the most popular applications. The method takes advantage of the fact that each individual who visits a natural area faces a trip cost, entry fees, onsite expenditures and outlay on capital equipment necessary to enjoy the services that the host areas provide. The responses of individuals to variations in the implicit price of the visit are the basis for estimating the value of a recreational service.

Hedonic pricing identifies environment service flows as elements of a vector of characteristics describing a marketed good (typically housing), to find a relationship between the level of environmental services and the prices of marketed goods. The method has been used to value such things as noise levels around an **airport**, effects of the proximity of environmental facilities on residential property price, amenity values of woodland, price competitiveness of **package tours** and environmental impact of tourism.

See also: ecologically sustainable tourism; impact assessment, environmental

Further reading

Mitchell, R.C and Carson, R.T. (1989) *Using Surveys to Value Public Goods: The Contingent Valuation Method*, Washington, DC: Resources for the Future.

Sinclair M.T. and Stabler, M. (1997) *The Economics of Tourism*, London: Routledge.

Smith, V.K. (1996) *Estimating Economic Values for Nature: Methods for Non-Market Valuation*, Cheltenham: Edward Elgar.

ANTONI RIERA, SPAIN

eroticism

This term captures the sexual excitement which permeates much of tourism. At the heart of both eroticism and tourism lies a craving for the beautiful (a characteristic ascribed by Plato to Eros, the Greek god of sexual love). Both share associations with 'beginnings' – tourism with personal recreations, eroticism (classically) with cosmogony – and both are also implicated in the transgression of boundaries, particularly those between the private and public domains.

See also: advertising; escape; fantasy; pornography; sex tourism

TOM SELWYN, UK

escape

It is widely accepted that one of the reasons why people become **tourists** is that they want temporarily to 'escape'. Two issues necessarily follow: what is it that they desire to escape from, and what do they wish to escape to?

At the most general level, part of the answer to both these questions lies in the etymology of holiday. Modern 'holy days' are periods set apart when time bound, hierarchically organised and bureaucratically controlled work routines, governed by principles of rationality, are suspended. This tedium is what contemporary holidaymakers, like celebrants of holy days in earlier times, escape from. Furthermore, holidays are, with certain exceptions (when appropriated by the state, for example), typically celebrated in an atmosphere of relatively open and egalitarian relationships unconstrained by time or formal social regulations (Turner 1969), and it is these characteristics towards which people would like to escape.

Valid and important as these generalisations are, contemporary sociologists and anthropologists of tourism have sought to place the **motivation** to escape within more specific and recent social, cultural and politico-economic contours. Two features of modernity and postmodernity have attracted particular attention: institutional fragmentation and **consumerism** (see also **postmodernism**). The former, together with its social and psychological consequences, has been variously examined. The evocative idea of the modern's (or postmodern's) 'homeless mind' (Berger *et al.* 1973) is frequently used as a starting point. It is used by MacCannell (1976) to suggest that the cognitive rootlessness of today's tourist may find temporary relief in the psychological and sociological warmth of touristic activities.

Complementing this train of thought, Lash and Urry (1987) have argued that the passage from **modernity** to postmodernity has been shaped by a general movement from 'organised' to 'disorganised' capitalism, and that one of the accompanying features of the latter is a lessening of intellectual and cultural 'certainties'. Underlying these observations is the assumption that the psychological disturbance and cognitive dislocation which has accompanied these movements has opened up a void from which the escape promised by the tourism brochures appears compellingly attractive. Some commentators have gone on to argue that part of the reason for such attraction derives from the general sense of **nostalgia** for imagined times and places which resonate with the supposed stability of castles and stately homes, the apparent scientific and religious verities embodied in cathedrals and universities, and the imagined natural and cultural harmonies of the agricultural and fishing regions of, for example, southern Europe. Underlying such fantasies is the promise of escape from the risks and disonances of postmodernity as daily experienced in the suburban hinterlands of the cold north.

As far as consumerism is concerned, it can also be shown that the promise of escape from the apparent wholesale '**commoditisation**' associated with the consumerism of the age, is another powerful motive behind modern tourism. The argument here is more complex, as in some senses tourism itself is a form of consumption. Nevertheless, the possibility of escaping from credit card and mortgage repayment worries into the illusion of a cashless world may well be one of the foundation stones of such types of tourism arrangement as the 'all-inclusive' **package tours**. Moreover, the yearning for supposedly authentic traditions (see **authenticity**), the fondness for places such as islands and historic cities, and the feelings of togetherness and community associated with mass tourist destinations all, in different ways, indicate a fascination in the tourist mind with pre-consumerist economies and societies.

Following these considerations, it may be generalised that the ultimate destination to which tourists wish to escape is part utopia, part **paradise**. One of the most enduring images of tourism utopia derives from the **myth** of Shangri La (Hutt 1996), an imaginary and remote Tibetan valley ruled by wise lamas in which competition and greed were absent, where government was benign, and where people lived long and fulfilled lives. The idea of Shangri La originated in the 1930s, but is used nowadays in tourism publicity to denote supposedly utopian destinations (see **denotation**). Deriving from the old Persian *pairidaeza* – garden or pleasure park, echoed by the Hebrew *pardes* (orchard) – ideas of paradise have ancient roots, most strikingly illustrated in the biblical Garden of Eden with its material self-sufficiency and human innocence. More modern notions of paradise derive from various sources, including (perhaps most famously) the voyages of Bougainville, James Cook and other Western explorers to Tahiti. Later elaborated in the paintings of Gauguin, Tahiti as paradise came to be associated with plenty, fertility, beauty and sexual availability; and these are precisely the characteristics which inform much of the promotional literature associated with contemporary tourism.

Another sense of escape concerns getting free from detention or control and gaining liberty by flight. There can be few more persuasive definitions of the tourist: one who escapes from the day-to-day realities of poverty and shortage, conflict, bad government and constraints of time to a paradise or utopia from which these have been banished, and in whose place (so the **fantasy** continues) are plenty, social harmony, benign forms of authority and unlimited time. Arguably, however, the defining characteristic of the tourism paradise is that it promises the possibility of the recovery of more complete senses of the self and society than the fractured **landscapes** of the postmodern condition normally allow.

References

Berger, P., Berger, B. and Kellner, H. (1973) *The Homeless Mind: Modernisation and Consciousness*, Harmondsworth: Penguin.

Hutt, M. (1996) 'Looking for Shangri-La : From Hilton to Lamichhane', in T. Selwyn (ed.) *The Tourist Image: Myths and Myth Making in Tourism*, Chichester: Wiley.

Lash, S. and Urry, J. (1987) *The End of Organised Capitalism*, Cambridge: Polity Press.

MacCannell, D. (1976) *The Tourist: A New Theory of the Leisure Class*, New York: Schocken.

Turner, V. (1969) *The Ritual Process*, Harmondsworth: Penguin.

TOM SELWYN, UK

escorted tour *see* guided tour

Espaces

Espaces publishes articles on **planning** and **development**, **leisure** and **recreation**, and **transportation** as far as incoming tourism is concerned. This journal is aimed at the **tourist** as well as the **industry**, and at national and regional organisations and educational centres. Apart from its regular sections, each issue deals with specific topics determined in advance by an editorial committee. The coverage is typically related to France and articles appear in French (but submission can be in English). First published in 1970, the journal appears twelve times a year. It is published by Editions Touristiques Européennes (ISSN 0336–1446).

RENE BARETJE, FRANCE

Estudios y Perspectivas en Turismo

Estudios y Perspectivas en Turismo is a Latin American journal devoted primarily to research papers focusing upon the academic perspective on tourism. It publishes articles dealing with the various aspects of the tourism phenomenon. The articles are published in Spanish, with abstracts in both Spanish and English. Submissions may be in

English, French, German and Portuguese; the editorial office arranges the translation of those eventually accepted for publication. All full-length articles are reviewed anonymously (double blind). In addition, the journal publishes conference reports, book reviews and commentaries. First appearing in 1992, the journal is published quarterly by Centro de Investigaciones y Estudios Turisticos (ISSN 0327–4841).

RENE BARETJE, FRANCE

Estudios Turisticos

Estudios Turisticos, an interdisciplinary journal, is devoted to original investigations on the Spanish tourism industry, including specific studies dealing with theoretical and methodological articles and empirical reports concerning **planning**, lodging, food service, **quality** management, finance, teaching and research, **statistics**, **management**, **marketing** and special interest **markets**. One of the key objectives of *Estudios Turisticos* is to provide a high level forum for communication between academics and practitioners concerned with Spain's attractiveness and competitiveness in the Mediterranean basin. Submissions, in Spanish only, are subject to a double-blind review process. Abstracts of articles appear in English and Spanish. First published in 1964, the journal appears quarterly. It is published by the Instituto de Estudios Turisticos (ISSN 0423–5-37).

RENE BARETJE, FRANCE

ethnic group

The concept ethnic group often refers to the biological features of a particular group. However, one may also speak of ethnic groups as having cultural, linguistic, religious, national or physical differences that are in contrast with others. An ethnic group may be the majority or a minority of people in a region (see also **minorities**). Although Frauke Kraas-Schneider (1992) suggests that it is difficult to offer an exact definition of this entity, it is still important for people in a particular group to realise that they do indeed belong to one. This concept is in harmony with the United Nations definition of minorities dating back to the year 1950.

How are ethnic groups formed? Migratory movements are seen as the most common basis for such new formations. Migrations may be legal or illegal in nature. In recent times, forced migrations and refugee movements have been the basis for the ethnic group formation. Even the amalgamation of **races** in a particular region can be seen as the source for a new minority. Political relations may define the actual status of a 'mixed people'.

Schism has been considered particularly important in connection with religious formation leading to the increased coherence of a group. Hostile attitudes from other groups may enhance internal coherence. Change in the status of a spoken language may also enhance internal coherence. An ethnic group may be consolidated by the activities of other stronger groups. Thus, individual native American tribes have been lumped by the white majority into a single group known as 'Indians'. In some cases, political developments have affected the status of many ethnic and minority groups, and so do border changes or domestic and international disputes (for example, German minority groups, Soviet Nationalities Policy and recent problems in former Yugoslavia).

An analysis of the self-consciousness of a minority group may explain its history from 'category' to 'group' to '**community**'. As a group within a larger population, the minority may be designated as only a 'category' or a part of that larger entity, such as the Swedish Finns. The category may develop into a real 'group' – usually after pressure from the majority group – or evolve into a 'community'. After successfully promoting the interests of the minority, it becomes institutionalised within a larger group.

An ethnic or minority group can become the target of the tourism industry, luring especially those who are interested in **cultural tourism**. Reducing an ethnic group to a touristic **attraction** has been criticised in the literature. On the other hand, members of a minority group may take initiatives to turn their community into a tourism **destination**. Whether this is done through internal means or external forces, the development should still be sustainable if negative consequences of tourism are to be minimised.

References

Kraas-Schneider, F. (1992) *Bevölkerungsgruppen und Minoritäten. Handbuch der ethnischen, sprachlichen und religiösen Bevölkerungsgruppen der Welt*, Stuttgart: Franz Steiner. (Charts various populations worldwide.)

Further reading

Abram, S., Waldron, D. and Macleod, D.V.L. (1996) *Tourists and Tourism: Identifying with People and Places*, Oxford: Berg. (Articles discuss the relationship between tourism and ethnic questions.)

Smith, A.D. (ed.) (1992) 'Ethnicity and nationalism', *International Studies in Sociology and Social Anthropology*, 60. (Articles discuss relationships between nationalism and ethnicity.)

Sollors, W. (ed.) (1996) *Theories of Ethnicity: A Classical Reader*, Basingstoke: Macmillan. (Interpretations on theory and concepts of ethnicity.)

AUVO A. KOSTIAINEN, FINLAND

ethnic tourism

Ethnic tourism may be defined as a form of tourism in which the prime **motivation** of the **tourist** involves a desire to **experience** and interact with exotic ethnic peoples. While this definition highlights the perspective of the tourist, a comprehensive approach to ethnic tourism necessarily includes in addition to tourists the local suppliers of this exotic experience, as well as the brokers (see **culture broker**) who facilitate the interaction between tourists and these local suppliers. Taking into account these different groups leads to the conclusion that ethnic tourism is not simply a particular form of interaction between 'hosts and guests', but is more fundamentally a complex process of ethnic relations, with significant implications for changing expressions of ethnic identity among locals.

The distinction between ethnic tourism and **cultural tourism** can be rather blurry, but there are two key issues which should be considered in separating them. First, the former tends to be more narrowly focused on a particular group of people whose **exoticism** is clearly marked as the prime **attraction** for the tourist. Second, ethnic tourism more fundamentally involves placing **local** people themselves 'on stage' for the tourist to view, rather than simply serving as background players facilitating the experience. Rather than viewing historical monuments, natural wonders or even a local 'cultural milieu', the ethnic tourist comes specifically to view other people whose ways of life differ greatly from that of back home. Thus, ethnic tourism most importantly depends upon the relationship between tourist and native, an encounter which is generally brokered by some third party as it becomes incorporated into the broader tourism industry (see also **natives**).

There are two aspects of this relationship which are very important from the tourist's perspective. First, the need for an exotic encounter necessitates that the relationship between tourist and native be one which bridges a vast socioeconomic divide. Ethnic tourists typically come from highly urbanised and industrial places; local suppliers, on the other hand, tend to be **'Fourth World' minorities** who occupy a marginal economic, political, cultural and geographical position within the countries in which they are found. Often, ethnic tourists consider themselves elite travellers who shun the mass **resorts** whose affordability depends on the income gap between developing and developed worlds. Yet the encounter between ethnic tourist and native perhaps even more fundamentally depends upon such a divide, since this visitor is primarily looking for exotic difference, even if one chooses to travel frugally on a limited budget in order to mask the income gap as much as possible. Second, the success of the relationship between the host and guest often depends on maintaining **authenticity** in the eyes of the former. That is, the relationship should appear to the tourist to be unmediated and spontaneous.

Paradoxically, such an experience is impossible to develop as a 'tourist **attraction** '. Indeed, many ethnic tourists would be loath to be referred to as 'tourists'. Nevertheless, the ethnic tourist impulse does lead to standardised attractions in particular locales (for example, among 'hill tribes' in South-east Asia and **China** or Navajos in New Mexico), and for these, **middlemen** or brokers are needed to facilitate interaction between tourists and natives. Brokers occupy a crucial space where industrialised modernity meets its imagined counterpart; the authentic exotic native. Brokers are instrumental in packaging and marketing local ethnic groups in such a fashion so as to appeal to ethnic tourists. They are typically members of the dominant ethnic group within their country (for example, Han Chinese in China, Thais in **Thailand** or *mestizos* in **Mexico**), or may be ambitious members of the minority group who have emigrated from their villages to their country's metropolitan centres. In many cases they may be representatives of the state. They may simply serve as **tour guides**, but more typically they arrange performances and **activities** for tourists. Many are remarkably savvy to the tourist's desire for the exotic and authentic, and may arrange activities to appear as spontaneous as possible (for example, a visit to a local **festival**, an unannounced visit or overnight stay in a village household).

It is important, however, to realise that native ethnic groups themselves actively collaborate in the ethnic tourist experience. They do this usually with considerable state encouragement in hopes of increasing local incomes. Indeed, ethnic tourism is often promoted by the state as a catalyst for economic integration and 'modernisation' among remote subsistence-oriented populations who as yet contribute little to state revenues. Aside from whatever economic **benefits** they may derive from ethnic tourism (which are largely minimal), however, ethnic groups may actively modify their **behaviour**, dress, methods of production and customary practices in order to facilitate or otherwise cope with the tourist experience. While it has been common for scholars to view these ethnic groups purely as objectified victims of the tourism industry, they in fact actively strive to maintain subjectivity over their interactions with tourists in many ways. Thus, ethnic tourism may generate reconstructed senses of **identity**, place and tradition among local groups, may lead to creative new expressions in art, and may help local groups resist long-standing attitudes of discrimination among the dominant population.

Because ethnic tourism is capable of generating such outcomes among host groups, it should more accurately be considered a process of ethnic relations. Recent studies have indeed argued with increasing consistency that ethnic tourism (along with other forms of tourism in general) be conceptualised not as an external force impacting a local ethnic or cultural group, but rather as an important component of that local group itself. It is in such a direction that research in ethnic tourism is currently moving. Important questions remain, however. The world's ethnic tourists are increasingly non-Western, and this raises questions about the assumptions of tourist **motivations**. Chinese ethnic tourists, for example, appear to be much less concerned with authenticity than their Western counterparts; they may also be less concerned with exoticism than with finding evidence of national continuity which transcends ethnic differences. More fundamentally, research on ethnic tourism has revealed the difficulty of separating tourism from other processes of change operating at global, national and local scales. Ethnic tourism can also render problematic accepted notions of ethnic and cultural boundaries, group identity formation and a host of other issues which are only now establishing themselves on the tourism research frontier.

See also: ethnicity

Further reading

Harron, S. and Weiler, B. (1992) 'Ethnic tourism', in C.M. Hall and B. Weiler (eds) *Special Interest Tourism*, London: Belhaven Press, 83–94. (A general review of scholarly work on ethnic tourism.)

Keyes, C.F. and van den Berghe, P. (eds) (1984) 'Tourism and re-created ethnicity', *Annals of Tourism Research* 11. (Sociological and anthropological foci on the relationship between ethnic tourism and the maintenance, transformation and re-creation of ethnic boundaries.)

Michaud, J. (1993) 'Tourism as catalyst of economic and political change; the case of highland minorities in Ladakh (India) and northern Thailand', *Internationales Asienforum* 24(1–2): 21–43. (A case study of ethnic tourism as a state sponsored strategy of economic development and modernisation.)

Picard, M. and Wood, R. (eds) (1997) *Tourism, Ethnicity, and the State in Asian and Pacific Societies*, Honolulu: University of Hawaii Press. (Papers dealing with ethnic and cultural tourism development as mediated by the state as a strategy of economic development and nation-building.)

van den Berghe, P. (1992) 'Tourism and the ethnic division of labour', *Annals of Tourism Research* 19: 234–49. (Discusses ethnic division of labour developed around tourism in San Cristobal, Mexico, and the impact of tourism on ethnic relations there.)

TIM OAKES, USA

ethnicity

Ethnicity is a social process of group identification which depends on both active self-definition by members and larger social forces which structure that group's interactions with others. Early scholarly uses of the term tended to equate ethnicity with particular cultural attributes of shared descent, and assumed ethnic groups to be defined by fixed linguistic and cultural boundaries. This primordial **interpretation** of ethnicity was increasingly challenged by scholars such as Leach and Barth, who were finding that **ethnic group** boundaries did not necessarily coincide with cultural or linguistic differences. Rather, it became clear that ethnicity was in part determined by structural forces in society which serve to differentiate groups politically and economically. In these terms, ethnicity was maintained less by shared descent than by social relations, in which distinguishing one's group from others was important. This interpretation was also inspired by the fact that forces of modernisation and **nationalism** – which the primordialist approach assumed would render ethnic differences increasingly meaningless – were not resulting in the disappearance of ethnic groups. If anything, ethnicity has become more important as nations become more integrated economically and politically. Thus, it has become necessary to think of ethnicity as a social process, in which structural determinants play a role in group differentiation, but where individuals also have some degree of control over manipulating the cultural attributes of their identities. In fact, individuals may claim more than one ethnicity simultaneously. They may also manipulate their ethnicity as an adaptive strategy in coping with social relations. Yet, it is clear that within the process of self-definition, their belief in primordial traits of shared descent remains an important factor in the maintenance of ethnic identities.

Understanding ethnicity as a social process is important for interpreting the relationship between tourism and ethnic identity. Tourism is an increasingly important force influencing the process of ethnic self-definition. In some cases, locals who become ethnic attractions have been found to consciously manipulate their ethnicity in order to better meet the expectations of tourists who come seeking the uncanny and exotic. Tourism thus becomes a powerful social force in determining the reconstruction (some might say **commoditisation**) of ethnic cultural attributes. Yet, it would be premature to assume that such changes represent an inauthentic version of ethnicity (see **staged authenticity**), for as a social process, it remains a dynamic and ever-changing feature of individual and group identity.

See also: ethnic tourism; identity

Further reading

Hitchcock, M. (1999) 'Tourism and ethnicity: situational perspectives', *International Journal of Tourism Research* 1: 17–32. (Explores situational adaptations to tourism among ethnic groups.)

Keyes, C. (1976) 'Towards a new formulation of the concept of ethnic group', *Ethnicity* 3: 202–13. (A seminal piece which seeks to reconcile structuralist and primordialist interpretations of ethnicity.)

MacCannell, D. (1984) 'Reconstructed ethnicity: tourism and cultural identity in Third World communities', *Annals of Tourism Research* 11: 375–91. (Provides an introduction to the concept of ethnicity, as well as an argument about the way ethnicity changes in association with tourism development.)

Okamura, J. (1981) 'Situational ethnicity', *Ethnic and Racial Studies* 4(4): 452–65. (Argues that ethnicity is both cognitively and structurally determined.)

TIM OAKES, USA

ethnocentrism

Ethnocentrism is awareness based on inflexible and suffusing ingroup/outgroup differentiations. It routinely involves stereotyped, laudatory imagery (and compliant attitudes) towards ingroups, and generalised, condemnatory imagery (and antagonistic attitudes) towards outgroups. Ethnocentrists usually have superior/authoritarian outlooks at cultural or ethnic encounters: they privilege ingroups as dominant. In tourism, ethnocentric **tourists** and ethnocentric travel trade employees find difficulty in viewing the world from '**other** '/ 'alien' perspectives.

KEITH HOLLINSHEAD, UK

ethnography

In the most general sense, ethnography refers to the firsthand study of the cultures of particular peoples. The beginnings of such study can be traced to early accounts of travellers such as missionaries and explorers whose work can be found in libraries. Today, the term is most frequently associated with **anthropology** where the tradition of fieldwork is particularly important, though it also is used in other social sciences and even the humanities.

A persistent problem with ethnographic study has been the validity of its descriptions. Earlier, when there might have been no more than one account of a people, there were few problems in this regard. But today, with often conflicting understandings of the same culture, the issue of validity is not so easily resolved. Because the subjects of ethnographic inquiry are human and often **strangers**, special **interpretation** problems emerge. Accordingly, procedures such as extended participant observation and astute use of informants are necessary. Current ethnographic practice also increasingly recognises that because they are a part of the world they study, ethnographers must pay attention to themselves and their social position in constructing their view of others.

Ethnographic work has contributed to the understanding of tourism, most often in studies of developing countries. But this grassroots approach to human behaviour also can be used on other aspects of tourism such as studying **motor coach tourism**, caravan parks, **museums**, corporate cultures, **cultural tourism**, **ethnic tourism** or **mass tourism**. Though they already have contributed a good deal to the understanding of various touristic cultures, ethnographies of tourism have so far tended to be of an impressionistic-descriptive character. Many more studies, carried out with greater methodological and theoretical sophistication are needed to provide the accurate, intimate understanding that is the mark of good ethnography.

Further reading

Dann, G., Nash, D. and Pearce, P. (eds) (1988) 'Methodological issues in tourism research', *Annals of Tourism Research* 15(1). (A special issue of the journal devoted to tourism research methodology.)

Hammersley, M. and Atkinson, P. (1993) *Ethnography*, 2nd edn, London: Routledge. (Provides a basic understanding of the subject.)

Smith, V. (ed.) (1989) *Hosts and Guests: The Anthropology of Tourism*, 2nd edn, Philadelphia, PA: University of Pennsylvania Press. (Mostly ethnographic studies of tourism development in different parts of the world.)

DENNISON NASH, USA

ethnology

Ethnology is the theoretically-oriented comparative study of cultural patterns which recur, vary and **change** in different regions. Ethnology relies upon **ethnography** as its primary data source. Cultural relativism, a non-condemnatory assessment of other societies, is a hallmark of ethnology not accepted by all tourists. However, many **museum** and 'living history' exhibits that draw tourists are based upon ethnology.

ROLAND S. MOORE, USA

European Travel Commission

The European Travel Commission (ETC) is composed of twenty-six member national organisations. One of its major activities is to conduct joint all-Europe promotional activities (see **advertising**). It is also involved in governmental tourism statistics and national and international tourism research. Its headquarters is located in Brussels, Belgium.

ORHAN ICOZ, TURKEY

European Union

Founded in 1957 with the signing of the Treaty of Rome by six European nations (Belgium, **France**, **Germany**, **Italy**, Luxembourg and the **Netherlands**), the European Economic Community (EEC) was subsequently enlarged in 1973 to include **Denmark**, **Ireland** and the **United Kingdom**; 1981 saw the admission of **Greece**, 1986 the admission of **Portugal** and **Spain**, and 1995 the admission of **Austria**, **Finland** and **Sweden**). The Treaty of the European Union, signed in Maastricht (Netherlands) in 1992, changed the name to European Union and established its new goals of monetary and economic union and strengthened intergovernmental cooperation.

At present, government **policy** making and implementation in the European Union is exerted by five institutions: the European Parliament, elected by voters in all member countries, but with no legislatory powers; the Council, representing member states; the Commission, holding executive and legislatory **powers**; the Court of Justice; and the Court of Audit, in charge of financial control. In addition there are several consultative bodies, representing social, economic or regional interests. The Union makes use of several policy-setting legislative instruments; some are of a compulsory character such as the Reglaments, Directives and Decisions, while others constitute voluntary frameworks, such as the Recommendations and Communications. All these instruments may be strictly normative or of an operative nature (action plans, operative programmes and so on).

Although the Treaty of Maastricht mentions tourism for the first time in the *acquis communautaire*, recognising that tourism should be a matter for European Union action (article 3), the Commission has failed to approve its inclusion as an area of direct concern for European Policy. Thus, tourism policy initiatives can be enacted by the European Union only in a subsidiary way, when the individual policies of member states are insufficient to achieve the commonly intended objectives. This situation is in striking contrast to the importance of tourism in Europe as recognised by the Commission itself: 'tourism represents on average a 5.5% of the Union GDP, and more than doubles that figure in some member States ... '. With regard to employment, 'It is estimated that 9 million people work in tourism in the Union ... and that direct **employment** in tourism constitutes 6% of total employment.' (Commission of the European Union 1995: 5, 7).

Taking the figures for the whole of Europe, **tourist** arrivals from abroad totalled 360 million in 1997 (58 per cent of the world figure). International tourist nights in hotels and similar establishments were 627 million (also 58.8 of the world's total) and in other establishments 1,454 million (49.9 of the world's figure). The number of trips abroad was 330 million in 1996, equivalent to 63.7 per cent of the world's total of 518 million. **International tourism** receipts, excluding international transport, were 218 billion $US in 1997, or 49.2 per cent of the world's total receipts. However positive these figures appear to be, amounting in most cases to 50 per cent or more of the world tourism **market**, they have been declining rather steadily, at a rate close to 0.5 per cent of world market share per year. Of course this decline is due to rapid **growth** elsewhere, most notably in the Asia-Pacific region, but it is nevertheless a worrying **trend** for one of the largest industries in Europe.

Therefore, it is surprising that the European Commission, so active in other sectorial economic policies (for example, the Common Agricultural Policy), has not set an explicit framework for the **competitiveness** of its tourism industry. Although billions of ecus in the EU budgets have been spent in the creation and promotion of tourism **products** in Europe and abroad (including EU cooperation programmes with developing countries overseas), this has been done through the structural funds and programmes addressing other issues (see also **promotion, place**). Most EU directives have had profound effects on tourism, but as a secondary, and in many cases unwanted, effect. Although a directorate general (DG XXIII) has been in charge of tourism issues since 1989, its programmes (such as the European Year of Tourism of 1990, the European Action Plan on Tourism 1992–5, or the non-born Phyloxenia project in 1996–7) have turned out to be failures because of very small budgets (for example, 16 million ecu was budgeted for the whole of the four-year Action Plan on Tourism), unsophisticated design of the programmes and actions, lack of co-ordination with other directorates within the Commission and with the national tourism organisations of member countries, and even inadequate management (as in the European Year of Tourism and the Action Plan).

In this context, the European private stakeholders in tourism have been understandably hesitant between the desire to have a proper supranational framework fostering the competitiveness of the industry, and the fear that more intervention by Brussels could lead to unwanted and inefficient conditions. All in all, explicit tourism policy in Europe remains in the hands of national, regional and local governments that, in most cases, work in close cooperation with the private sector. Intervention by EU institutions takes place as a result of policies only indirectly addressed to the tourism industry.

References

Commission of the European Union (1995) *The Role of the Union in Tourism*, Green Paper of the Commission, Brussels: Commission of the European Union.

Further reading

Commission of the European Union, Eurostat, DGXXIII (1997) *Tourism in Europe. Key Figures 1995–1996*, Brussels: Commission of the European Union.

World Tourism Organization (1998) *Yearbook of Tourism Statistics*, Madrid: World Tourism Organization.

EDUARDO FAYOS-SOLÀ, SPAIN
MARI LUZ RUFILANCHAS, SPAIN

event

Each day witnesses countless temporary occurrences; most are private and unrecorded, while many are unplanned news events. Others are staged for purposes of gaining publicity and, in the words of Boorstin (1961), might be called 'pseudo events' because they are neither spontaneous nor authentic. Of particular interest in the context of tourism are planned events, either one-time or periodic, which feature prominently in place promotion strategies and destination marketing (see **marketing, destination**; **promotion, place**). The sense of rarity and urgency associated with one-time events provides a major part of their appeal. However, periodic events also have a unique ambiance formed from their fixed duration, special setting, programme, **management** and those in attendance.

The major categories of planned events include cultural celebrations (such as festivals, carnivals, religious ceremonies, parades, heritage commemorations), art and entertainment (concerts and other performances, exhibits, award ceremonies), business and trade (**fairs** and sales; consumer and **trade shows**, expositions, meetings and conferences, publicity stunts, fund-raisers), sport competitions (professional and amateur), education and scientific (seminars and workshops, clinics, congresses, interpretative events), recreational (games and sports for fun, amusements), and political and state occasions (inaugurations, investitures, VIP visits, rallies).

Within the public domain is an ever-increasing number and variety of festivals and special events which enrich lives, give meaning to communities and enable a sharing between hosts and guests. Festivals, with roots in ancient religious celebrations and feasts, can attribute part of their enormous popularity to the fact that every community and interest group can create one. The essence of a festival is its public orientation and the feeling of shared **values** that it engenders.

The term 'special event' is frequently used, but 'specialness' resists precise definition. Context makes some events special to their organisers or guests, so that a special event is a one-time or infrequently occurring event outside the normal programme or activities of the sponsoring or organising body. To the customer or guest, a special event is an opportunity for a **leisure**, social or cultural **experience** outside the normal range of choices or beyond everyday experience. Many elements can be managed to contribute to the perception of 'specialness', including the fostering of traditions, uniqueness and quality. Festivals and other events can also be used to meet a variety of goals, including economic and community **development**, thereby increasing their acceptance.

Hallmark event is another frequently used term. A hallmark is a symbol of quality or **authenticity** which distinguishes goods or pertains to a distinctive feature. An event, therefore, can aspire to be the hallmark of its **destination**, facility or organisation. Some events have achieved such a high level of awareness and reputation that their image and that of their community have become inseparable. Quality, authenticity and distinctiveness are the key contributing factors. From the perspective of place promotion, the image of popular hallmark events becomes indelibly linked to that of the host communities.

Some events are so substantial that they attract the adjective 'mega'. This term means one million in the metric system, but can also refer to large-scale economic **impacts** or **attraction** of worldwide publicity. To a small community, however, any event can have major implications, so that 'mega' is a relative concept. Certainly World's Fairs and the Olympics are big enough to earn the prefix, yet an event might never attract large numbers but still generate enormous exposure through media coverage. These media events have gained in popularity, related to the power of television coverage to reach global or targeted audiences. Mega events can thus be defined as events which yield extraordinarily high levels of tourism, media coverage, prestige or economic impact for the host community or destination.

Event tourism is an important option for destinations and place marketers, as well as a vital concern for event managers who want to develop their audience. It has two dimensions: (a) the systematic planning, development and marketing of events as tourist attractions, catalysts for other developments, **image** builders, and animators of attractions and destination areas, and (b) a market segment consisting of those people who travel to attend events, or who can be motivated to attend events while away from home (see **market segmentation**).

Event tourism has almost global application, due to the importance of events in all cultures and to the desire of marketing agencies to exploit niche markets. Emphasis on **sustainable development** will also ensure that events figure prominently in cultural and **ecotourism** programmes. As events do not necessarily require infrastructural improvements, they can incorporate environmentally sound operations practices and can contribute to cultural and environmental **conservation**. Event tourism strategies are numerous: creating and promoting periodic hallmark events as core destination attractions, around which theming, image-building and packaging are created; holding occasional mega-events to attract tourists and publicity or as catalysts for **infrastructure** development; bidding on events, such as sport competitions and art exhibits, which bring with them unique appeal and publicity; theme years which feature existing or new events; fostering variety and quality among community events; adding value to resort and facility visits through event programming (**animation**); and managing the news (events can be created to reposition a destination or correct an undesirable **image**). Popularity of research on such themes as events, festivals, mega-events and more has led to the formation of scholarly groups such as International Festivals and Events Association, and of academic journals such as *Event Management*.

See also: sponsored event

Reference

Boorstin, D. (1961) *The Image: A Guide to Pseudo-Events in America*, New York: Harper and Row.

event management

To fulfil their various roles within destination **planning** and place promotion, events must be managed as high-quality products with a strong tourism orientation (see also **promotion, place**). This is a major challenge, given that most festivals and special events are managed by non-profit organisations in which volunteers dominate.

One-time events employ project management techniques to produce an event on a prescribed date, often with years of planning and **development**. They face special challenges, including protests or political interference, fast-tracking through regulatory channels, shifting priorities and uncertain resource commitments, staff turn-over or burn-out, and the temptation to make **quality** compromises when time and money run low.

Event programmes are composed of a number of generic 'elements of style', and managers have unlimited scope in combining them to achieve uniqueness. Celebration is the essence of festivals. Spectacle, including entertainment, consists of those components which please the eye or are larger than life displays. Commerce is a feature of many events, including exhibits and merchandising. **Hospitality** refers to both the reception and service quality experienced by guests and the opportunity for sponsors to host their clients, staff and associates. Games is a broad element involving competitions, **gambling**, **humour** and surprise. Educational components are often important, and cultural **authenticity** must be considered.

A **marketing audit** can be performed on events to determine their attractiveness and readiness to host tourists. Some of the key elements are sufficient in appeal and quality to attract and satisfy tourists, a theme and setting which conveys an attractive image, a targeted communications campaign with identified tourist segments, a programme which provides generic benefits for all visitors and targeted benefits for special tourism interests, provision of special services need by tourists (for example, **accessibility**, reception of tour groups, additional information, languages, reserved seating), packaging for tourists, relationships with the tour and hospitality sectors, cooperative **marketing** involving destination marketing agencies and other events (see also **marketing, destination**), sponsorship which extends the reach of the event's appeal and communications, and site and **community** capacity to accommodate substantial numbers of visitors. Recent research has given an added attention to the theme of event **management**.

Further reading

Catherwood, D. and Van Kirk, R. (1992) *The Complete Guide to Special Event Management*, New York: Wiley.

Getz, D. (1997) *Event Management and Event Tourism*, New York: Cognizant Communication Corp.

Goldblatt, J. (1990) *Special Events: The Art and Science of Celebration*, New York: Van Nostrand Reinhold.

Graham, S., Goldblatt, J. and Delpy, L. (1995) *The Ultimate Guide to Sport Event Management and Marketing*, Chicago: Irwin.

Hall, C.M. (1992) *Hallmark Tourist Events: Impacts, Management and Planning*, London: Belhaven.

DONALD GETZ, CANADA

event marketing

The sponsorship or production of events as a **marketing** tool, specifically to connect with target audiences, build relationships, sell merchandise and achieve positive publicity, includes some of what is known as event marketing. In return for money, goods or **services**, events provide sponsors with specific benefits such as visibility, sales promotions, onsite exhibitions and **hospitality** venues.

Explosive **growth** in event sponsorship occurred during and following the Los Angeles Summer Olympic Games of 1984. The events sector has since been revolutionised, giving rise to larger budgets, more attention to marketing, merchandising and **media** coverage, and the forging of corporate **partnerships**. Sponsors often augment the reach of event promotions and provide valued technical expertise. Tourism and event development organisations actively promote destinations through events, and this process is largely dependent on the creation of media-oriented events with international sponsors.

One consequence of this **trend** has been a rise in 'ambush marketing', in which corporations seek to gain advantages through surrounding promotions without actually sponsoring the event. Confusion can also result when event sponsors differ from the sponsors of participants, such as athletes, and from media advertisers. Another trend had been for corporations to seek equity in events, or even to create their own events. This could threaten relationships with some events and result in less sponsorship money. Similarly, moves in some jurisdictions to ban **advertising** and sponsorship, especially from **alcohol** or tobacco companies, is also viewed as a threat to the events sector.

To be sustainable in the long term, there must be a congruence of goals and styles among sponsors and event organisers, with clear benefits to each party, customers and other participants. In some cases these partnerships are leading to environmental and educational programmes at events, to heightened **community** involvement by corporations, and to a resurgence of the arts. Event marketing has had the effect of elevating event tourism into prominence, not just as a special-interest market but as a major tool in destination image-making.

Further reading

Getz, D. (1977) *Event Management and Event Tourism*, New York: Cognizant Communication Corp.

Graham, S., Goldblatt, J. and Delpy, L. (1995) *The Ultimate Guide to Sport Event Management and Marketing*, Chicago: Irwin.

International Event Groups Inc. (1995) *IEG's Complete Guide to Sponsorship*, Chicago.

International Festivals Association (n.d.) *IFA's Official Guide to Sponsorship*, Port Angeles.

Schreiber, A. and Lenson, B. (1994) *Lifestyle and Event Marketing: Building the New Customer Partnership*, New York: McGraw Hill.

DONALD GETZ, CANADA

Further reading

Getz, D. (1997) *Event Management and Event Tourism*, New York: Cognizant Communication Corp.

Goldblatt, J. (1990) *Special Events: The Art and Science of Celebration*, New York: Van Nostrand Reinhold.

Graham, S., Goldblatt, J. and Delpy, L. (1995) *The Ultimate Guide to Sport Event Management and Marketing*, Chicago: Irwin.

Hall, C.M. (1992) *Hallmark Tourist Events: Impacts, Management and Planning*, London: Belhaven.

Schreiber, A. and Lenson B. (1994) *Lifestyle and Event Marketing: Building the New Customer Partnership*, New York: McGraw-Hill.

DONALD GETZ, CANADA

evolution

Since the 1950s, tourism has become one of the most important activities, coinciding with a period of **peace** and economic prosperity. The number of international arrivals and receipts showed a strong **growth** until the 1970s, slowing in the next decades. The economic problems and the conflicts of the 1990s (see **war**) caused a recession, though **future** prospects are optimistic.

AURORA PEDRO, SPAIN

exclave

Exclaves, or enclaves, are small parts of one country entirely or almost entirely surrounded by a neighbouring country. Hence they are accessible from the home country only by passing through foreign territory. Several exclaves in Europe and North America have prosperous tourism and are almost entirely dependent on this industry. Examples include Campione (Italy), Llivia (Spain) and Point Roberts (Washington, USA).

DALLEN J. TIMOTHY, USA

excursion

An excursion is a short pleasure trip or a side trip. **Tour operators** use the term to describe **sightseeing** programmes during a **tour**. Optional excursions are purchased extra by **tourists** during the trip. Transport operators offer excursion fares, special discounted tickets (usually with restrictions). Airlines offer low tariff fares (Advanced Purchase Excursion, or APEX) which must be booked a certain time in advance.

LIONEL BECHEREL, UK

excursionist

A traveller on a brief recreational trip, typically not involving an overnight stay away from home. Excursionists often travel in groups, sometimes at reduced rates. They resemble **tourists** but, because of their brief stays, their use of tourism facilities is more limited. Today, excursions between neighbouring countries are a common and significant phenomenon, and are often listed separately in tourism statistics.

ERIK COHEN, ISRAEL

executive development

Executive development is also known as professional development, advanced **management** or, more generally, continuing education, in which tourism and hospitality educational and **training** programmes are prepared and delivered. This, typically done for middle and senior managers, is offered by universities, colleges, polytechnics, related education/research institutes, trade or professional, private advisory firms or governments. It is normally a form of collaborative education between educational institutions and the tourism industry (see **education/industry relationship**).

Over the past twenty years, the importance and popularity of tourism executive development programming have substantially increased, especially in **hospitality** management. This new awareness of the need to be more knowledgeable and better skilled stems from increased competition, new **technology**, changing business conditions and strategies, and the need for greater productivity in service organisations. Tourism practitioners view executive development as part of 'life-long learning', increased **professionalism** and **career** enhancement. Tourism organisations see it as an investment in their people, culminating in greater **competitive advantage** and improved corporate performance. The anticipated outcomes of executive development are complementary, with both individuals and organisations developing strong core competencies and more effective **leadership** acumen. Other potential benefits include staying current with leading-edge **industry** thinking and applications, gaining a new appreciation of global opportunities and threats, and acknowledging new **trends** and critical success factors.

The focus of most tourism executive programmes is on current and **future** understanding of the strategic implications of key business issues and challenges, general management concepts and practices, critical organisational, sectoral and functional knowledge and technical skills, leadership and business visions, and team and teamwork development. Programme design and delivery vary, but normally feature a limited participant enrolment, highly qualified course moderators/instructors, and course workbooks and texts. Executive programmes emphasise interactive class sessions that combine techniques such as pre-assigned readings, case studies, workshop exercises, idea exchanges, computer labs, videos, guest speakers and small group discussions. Courses can also be fully or partially delivered by **distance education** (correspondence or electronic mode). Most executive development programmes recognise successful completion through certificates of merit and/or equivalent university credits. There are other programmes which follow a rigorous certification process leading to a professional designation. In the tourism and hospitality field, executive development programmes are primarily prepared by sectoral, functional or destination specialisation. Some of the more well-known programmes internationally are offered by Cornell University (hotel and restaurant management), University of Hawaii (tourism), University of Calgary (tourism destination management), University of Surrey (hospitality and catering) and George Washington University (tourism sustainable development).

See also: collaborative education; career; education method; professionalism; training

Further reading

Bolt, J. (1989) *Executive Development: A Strategy for Corporate Competitiveness*, New York: Harper and Row.

Moulton, H. and Fickel, A. (1993) *Executive Development: Preparing for the 21st Century*, New York: Oxford University Press.

DONALD ANDERSON, CANADA

exoticism

Exoticism refers to the characteristics attributed to peoples, plants, animals and **landscapes** outside the familiar reference system of the speaker. Such phenomena may be viewed as exciting or fearsome, or with ambivalence. In the world of **tourists**, the exotic features of foreign places are frequently played up as attractive or exciting **images** in **advertising** and on site **markers**. The arts and craft productions of **indigenous** peoples are often subject to **commoditisation** for **ethnic tourism**. Unusual landscapes and exotic fauna and flora are frequently marked off and preserved as game park reserves or **wilderness** for touristic consumption.

Such labelling is ethnocentric because strangeness is always relative to the **experience** of the speaker or writer. Because tourism is overwhelmingly a phenomenon of the industrialised West and East Asia, and of the middle and upper classes of all countries, it is institutions of their creation, such as **travel agencies**, tourism bureaus, **transportation** and **hotel** companies, which invent and publicise these labels. The appellation exotic is usually applied to non-Western urban or non-modern rural ways of life, to tropical or arctic fauna and flora, and to minority, native and aboriginal peoples (see also **natives**). Consequently, unchecked tourism to such fragile places may cause ecological or social changes which harm the flora, fauna or peoples so that they are no longer attractions, or tourism may cause exotic peoples to act out their lives so as to conform to the **fantasy** expectations of the tourists. Some target peoples, such as the Balinese and the Torajans of Indonesia, accept tourist demands for exotic performances and have modified their material and **ritual** behaviour without losing all the important meanings. Other peoples, such as the Inupiaq of Alaska, the Aborigines of Australia and many of the Pueblo Indians of the American Southwest, have rejected tourist curiosity by forbidding entry, prohibiting **photography**, or charging entry fees to parts or all of their living spaces.

Further reading

Cohen, E. (1989) 'Primitive and remote: hill tribe trekking in Thailand', *Annals of Tourism Research* 16(1): 30–61. (Shows how the discourse of advertising in Northern Thailand exoticises the hilltribes in creating fantasies of naturalness and naivete appealing to Western tourists.)

Graburn, N.H.H. (ed.) (1976) *Ethnic and Tourist Arts: Cultural Expressions from the Fourth World*, Berkeley, CA: University of California Press. (Discusses the processes by which the arts and crafts of ethnic and minority peoples may become extinct, revived, trinketised, transformed or appropriated for tourist consumption.)

NELSON H.H. GRABURN, USA

expatriate

Expatriates are staff, often managers, who are not nationals of the country where they are working, employed for specialist operational abilities or knowledge of the employing organisation (typically in **multinational firms**). They commonly receive higher salaries, longer leave and other benefits than **local** employees, which, along with cultural differences, can create **conflict**. Typically expatriate tourism employees work in developing countries or destinations.

See also: cross-cultural management; culture change; culture shock; globalisation

RAY PINE, CHINA

expectation

People have certain expectations from their trips, especially those which are pleasure oriented (see **pleasure tourist**). These are influenced by promotion (advertisements, brochures), prior experience and even **myths** (see also **promotion, place**). This is further conditioned by the desire for a certain level of **service** or **quality** from a given tourism product/business or **destination** at large. As these expectations are related to the non-ordinary aspects of life, satisfying them is a challenge in tourism.

TOM SELANNIEMI, FINLAND

expenses estimation

Expenses estimation is the projection of the outflow of assets or increase in liabilities occurring in the production and provision of goods and services. In the budget process, it is the step undertaken immediately after **forecasting** revenues. In order to properly estimate expenses, expected **cost** increases for supplies, **food**, labour and other expenses need to be known. Expenses are either fixed, variable or mixed. Therefore, the development of their estimations cannot be based solely on historical relationships between revenues and expenses for those other than the variable ones. Once revenues have been established, the expenses that are variable can be estimated based on the historical relationships to sales, provided the revenue was developed in a manner to pass along cost increases to the consumer.

Expenses estimation for fixed costs is accomplished by identifying the precise costs that are fixed, and applying the appropriate dollar amount to each one of them. Fixed costs will remain constant over the relevant range of activity, and thus changes in revenue projections will not have an effect on the estimates of fixed costs. The most difficult category of expenses to estimate is mixed costs. Mixed costs have a fixed component and a variable component. Therefore, relationships to sales will not necessarily hold true, nor will the cost remain the same at different levels of activity. An exercise can be undertaken to break these costs into their fixed and variable components through regression analysis. Once this exercise is completed, expenses estimation for these types of costs can be computed as explained above for variable costs and fixed costs.

There are other ways to study expenses estimation. One of these methods is to base this on standard amounts. Since revenues are estimated by using some form of logic, such as number of **hotel** rooms sold, room attendant wages could be estimated by applying a **productivity** standard quantifying the number of rooms an attendant can change in a day. This information would generate the number of room attendants and number of hours worked or needed to generate the sales expected. With this information and information about wage rates and increases, the wage expense for attendants could be estimated. For departments without revenue, expenses estimation is often based on past experience and known cost changes. This **methodology** can be applied to many other job efforts in a full service hotel and some other tourism businesses.

Further reading

Schmidgall, R. (1997) *Hospitality Industry Managerial Accounting*, 4th edn, East Lansing, MI: Educational Institute of the American Hotel and Motel Association.

STEPHEN M. LEBRUTO, USA

experience

Experience is the inner state of the individual, brought about by something which is personally encountered, undergone or lived through. Tourist experiences are such states engendered in the course of a journey, especially a **sightseeing** tour or a **vacation**. The principal sociopsychological problem in the study of tourist experiences is their distinctive quality and their relation to the experiences of everyday life.

Travel was in the past a frightening rather than enjoyable experience, though some individuals, like explorers and adventurers, were exhilarated by its dangers and challenges. With the general improvement of **security** and the development of modern means of **transportation**, travel increasingly became a more comfortable and pleasurable experience. With the routinisation and stress of modern life, the disappearance of traditional **lifestyles** in industrialised societies and the despoliation of their environment, travel and **vacations** became for many modern people an opportunity for interesting, enjoyable and often memorable experiences, which often retrospectively became the highlights of their lives. Indeed, some tourist experiences resemble Maslow's 'peak experiences', an intense personal sense of loss of identity in the grandeur, awe or ecstasy of the moment, as in the presence of a 'breathtaking' natural sight. Such experiences have sometimes been called 'nature mysticism'. Most **tourists**, however, do not ordinarily reach such heights, but enjoy their journey in a spirit of quiet, often playful **animation**. However, as tourism becomes increasingly routinised and attractions and destinations are commoditised, the difference between everyday and tourist experiences decreases: tourism tends to become just another consumer activity, forfeiting much of its distinctiveness.

In the sociological discourse of tourism, the problem of the nature of the touristic experiences of moderns focused upon the question of **authenticity**: do tourists seek to experience the authenticity of other places and other times (as MacCannell would ask), or are they satisfied with merely contrived, inauthentic offerings, staged (see **staged authenticity**) for them by the tourism business? However, the concept of 'authenticity' is a social construction. One individual may conceive and experience a place, performance or object as 'authentic', while another may not. A typology of tourism experiences can be constructed on the basis of the centrality of the tourist's **quest** for the 'authentic'. Thus, Cohen's 'existential' tourists seek direct immersion in the 'authentic' life and culture of the hosts; 'experiential' tourists seek primarily a vicarious observation of the authentic life of others; while 'recreational' tourists tend to stay satisfied with the make-believe presentations of authenticity, which they playfully accept as real. It indeed appears that the playful enjoyment of 'surfaces' has superseded the 'quest' for 'authenticity' as the cultural model of the touristic experience of 'postmodern' tourists (see **postmodernism**).

See also: liminality; ludic; play; rites of passage; ritual

Further reading

Cohen, E. (1979) 'A phenomenology of touristic experiences', *Sociology* 13: 179–201.

MacCannell, D. (1973) 'Staged authenticity: arrangements of social space in tourist settings', *American Journal of Sociology* 79(3): 589–603.

Maslow, A.H. (1978) *Towards a Psychology of Being*, 2nd edn, Princeton, NJ: Van Nostrand.

ERIK COHEN, ISRAEL

experimental research

Experimentation is a deductive approach to research. It entails the intervention of the researcher, usually to introduce the independent variable, with a high degree of control over the dependent variable. There are broadly three such methodologies: experimental, quasi-experimental and **action research**. All three may have a role to play in understanding tourism. True experiments are commonly thought of as being laboratory based, due to the high level of control that good experimentation requires. However, laboratory experimentation has the weaknesses in that population validity may be low and 'ecological validity' is weak. To overcome these weaknesses, it is possible to conduct quasi-experiments outside the laboratory in real-life situations, although doing so presents different problems. Such experimentation, however, requires the study of large groups if the many variations and ambiguities involved in human behaviour are to be controlled. Thus such large-scale experiments tend to be expensive to set up and take more time. Action research, an alternative form of experimentation, involves the researcher participating in and facilitating change within the study context.

To ensure findings that are valid, objective and reapplicable, experimental research design entails four basic steps. First, the dependent variable (the phenomenon whose variance the research aims to explain) needs to be identified. Second, the independent variables (those factors that may cause or explain the changes in the dependent variable) need to be postulated. Third, these variables need to be operationalised so that the impact controlled changes to independent variables have on the dependent variable can be observed. Four, every effort should be made to ensure that extraneous variables (other factors that might cause variation in the dependent variable) are neutralised or at least controlled. Therefore, experimental research presents six challenges, three relating to the definition of these variables and three to observation and measurement of them.

Although challenging to design, experimentation has a number of advantages that make it a suitable research **methodology** for investigating tourism phenomena. It has a high degree of validity, and industry professionals as well as researchers are able to understand it. This approach has been used to investigate queue **behaviour** (such as in **fast food** restaurants and in **theme parks**), **technology** interfaces (vending machines and automated bank telling machines), work practices (ticketing and **hotel** housekeeping) and physical **design and layout** (such as cafeteria design and airports).

PETER JONES, UK

exploration

Exploration is the act of travelling to an unfamiliar territory in order to investigate or to search out novel phenomena and experiences. The sense of **adventure tourism** and danger from heading out into the uncharted and the unknown has captured the popular imagination and generated a romantic and heroic image of the explorer. Throughout history, tales of travel and discovery have formed the foundations of scientific fact and fiction. Early travel diaries, **maps** and the spoken word have helped to shape the basis of the understanding of a multiplicity of disciplines. Whether it be for commercial, religious, scientific or imperialistic reasons, names such as Herodotus, Chëng Ho, Ibn Battuta, Marco Polo, Columbus, Magellan, Cook, Lewis and Clark and Livingstone, along with sponsoring institutions such as the Royal Geographic Society, have all added to a greater understanding of the world.

Improvements in **technology** and communication have made exploring easier. **Tourists**, in part, see tourism as the exploration of the unknown and, like the great explorers of yesteryear, modern-day tourists return home with stories, photos and **souvenirs** documenting their exploits, an analogous attempt by some to increase their social status. Cohen's (1972) typology of tourists ranges from the organised mass tourist to the drifter (see **typology, tourist**). Along the continuum, there is a gravitation towards greater understanding and integration with the **destination** environment. Explorers usually plan their own trips and try to avoid touristic attractions. Although they have a desire to mix with members of the local community, they still seek the protection of the 'environmental bubble' and are not fully integrated into the indigenous society. Pearce (1982) connects an element of psychical risk to explorers. Smith's (1989) typology ranges from explorers to charter tourists. By considering the number of tourists and their relation to the **environment**, she argues that there are only a small number of true explorers left, as there are very few areas of the world which remain undiscovered. The current trend in **ecotourism** continues to push back the remaining frontiers for tourists. Smith likens the true explorers to anthropologists who seek out and investigate the culture of the host environment.

References

Cohen, E. (1972) 'Toward a sociology of international tourism', *Social Research* 39(1): 164–82.

Pearce, P. (1982) *The Social Psychology of Tourist Behaviour*, Oxford: Pergamon.

Smith, V. (ed.) (1989) *Hosts and Guests: the Anthro-*

pology of Tourism, 2nd edn, Philadelphia, PA: University of Pennsylvania Press.

DAVID J. TELFER, CANADA

exposition *see* trade show

externalities

Costs and benefits imposed on a third party as a result of another action without the interference of the **market** mechanism are known as externalities. For example, water polluted by a fish-processing plant may endanger local water-based tourism activities. Externalities are also called third-party or spill-over effects. An externality in its extreme positive form is a public good, such as tourism **destination** promotion (see also **promotion, place**).

MARJA PAAJANEN, FINLAND

F

facilitation

The concept of facilitation has grown from its narrow limits of frontier formalities and customs procedures, to denote a free and safe movement of **tourists**. Before 1914, tourists did not have to carry elaborate passports or obtain visas to go from one country to another. During the First World War, restrictions on travel were imposed as a wartime measure. However, travel became relatively free of restrictions in interwar years.

After the Second World War, freedom to travel was given a prominent place in the activities of the United Nations and its specialised organs and tourism organisations. Early in the 1950s the United Nations adopted the following instruments to facilitate travel: a convention concerning customs facilities for tourism and a protocol regarding the importation of publicity documents and materials; a customs convention on the temporary importation of private road vehicles (June 1954); a customs convention on the temporary importation for private use of aircrafts and pleasure boats (May 1956); and sn international convention on customs facilities applicable to tourists adopted by the Customs Cooperation Council and amended in 1986 to raise the aggregate value of goods of a non-commercial nature imported free of duty by non-resident travellers and returning residents. The United Nations Conference on International Travel and Tourism, held in Rome in 1963, made detailed recommendations with regard to travel facilitation. The UN declared 1967 as the International Tourism Year, and many facilitation measures for travel were made permanent by several states thereafter.

During the 1980s a number of world instruments for the facilitation of travel, visits and stays were formulated by the **World Tourism Organization** (WTO) and were endorsed by the UN. The Manila Declaration on World Tourism in 1980 invited states to consider the significance of abolishing visa requirements and removal of all restrictions detrimental to tourism. The WTO formulated the tourist Bill of Rights and Tourism Code adopted by the United Nations on 19 December 1980. The Hague declaration on tourism was adopted by the Inter-Parliamentary Conference convened in April 1989.

The WTO facilitation and tourism safety committees concluded a number of basic documents in the following fields: **security** and protection of tourists, tourism staff and operators of facilities and host countries; tourist insurance for individuals and establishments; consumer protection measures offering a draft provisions for creating an ombudsman for tourists; drug abuse and illicit trafficking control in tourism sector; protecting and facilitating travel of handicapped people; and **health** standards for tourists and **sanitation** requirements for facilities. The substance of these instruments and standards have been put into effect by different states and implemented on s bilateral and multilateral basis.

The General Agreement on Trade in Services (GATS), signed by 125 states in 1994, sets in place a system that will lead to the gradual elimination of barriers to tourism growth. More trade in both goods and services means more movement of people and more international business. GATS tries to put into action the principle that free market forces are the best means of providing consumers with the best possible products at the best possible prices. However, many countries still apply cautious and even restrictive procedures governing passports and visas, customs and exchange controls, and health protective measures. Problems such as **terrorism**, smuggling of narcotic drugs, false refugees and the **AIDS** pandemic have provided many states with the pretext for reinforcing rather than easing travel restrictions. However, clear progress in tourism facilitation has been realised in some specific geographic areas such as the European Community, North America (Free Trade Area NAFTA) and East Asia and the Pacific region. The dramatic changes in the political and economic environment, in East European countries, the Middle East and South Africa have eased restrictions on inbound and outbound tourism movements.

ABDEL RAHMAN SELIM, EGYPT

facilities management

The term facilities management has two meanings. The technical meaning relates to the **management** of building, plant and equipment to ensure they are maintained in good condition. The other meaning refers to the contracting out of this function to a contractor, who will manage the facilities on behalf of the owner/operator. In relation to tourism and its sectors, the two main elements of facilities management are routine cleaning, usually carried out on a daily or weekly basis, and routine maintenance, based either on usage or **cycle** time. Maintenance by usage refers to routinely checking the equipment after a specific number of activities, as with automobile servicing which is carried out every so many thousand kilometres. Cycle time maintenance is carried out at regularly established intervals, such as every three or six months. For instance, **airport** fire extinguishers should be checked every six months.

In hotels, the cleaning function is usually carried out by the housekeeping department, although **food** production areas may have their own cleaning staff, especially in relation to equipment and ware washing. In the United States, maintenance is usually carried out by the hotel 'engineer'. This is not common practice in Europe, with maintenance either carried out by less senior and less qualified staff, or contracted out to specialists. In most countries there are statutory regulations governing the hygiene of food preparation and public areas, the safety of equipment and the **safety** of the built environment. Hence, to safeguard against criminal prosecution, as well as civil liability, operators usually keep records of their cleaning and maintenance schedules, in order to demonstrate they have fulfilled their duty of care. There has been a growth in the contracting out of facilities management to specialist firms. This originated in office properties, but has spread into various other tourism sectors such as hotels, **resorts** and **restaurants**.

TOM JONES, USA

factor analysis

Factor analysis, rooted in **psychology** and mathematics, is a very popular analytical and data reduction technique. Typically, the method is used for exploratory purposes where the researcher seeks to disentangle interdependencies among a set of variables in order to determine the underlying structure of a theoretical construct. However, factor analysis is often used to reduce data structure, thereby achieving a parsimonious representation of the data. For example, tourism researchers recently used factor analysis to examine the underlying dimensions of twenty travel-related characteristics of individuals from five countries. Other researchers used factor analysis to develop a market **planning** index which can be used to measure the effectiveness of marketing efforts for **hotel** and other **hospitality**-related businesses.

Factor analysis is based on the geometrical representation derived from principle component analysis. However, these two techniques are different. Principal component analysis provides a unique structure and maximum approximation of observations. Interpretation of this results is usually difficult and sometimes meaningless. On the other hand, factor analysis provides for the parsimonious structure of observations which will substantially reproduce the empirical data in a meaningful way. Factor analysis uses a number of axis rotation methods including quartimax, quartimin, varimax, orthomax and oblimax to achieve meaningful solutions.

Axis rotation can be achieved either subjectively or objectively and interrelationships among extracted factors can be either orthogonal (independent) and oblique (correlated). The most commonly used rotation method is varimax. Varimax rotation maintains the orthogonality among the factors and provides the best fit for a given factor structure. Data input for factor analysis is a correlation or proximity matrix. Although the measurement level of the correlation data has no specific requirement, the robustness of estimates and parsimony of a factor **model** depend on at least an interval scale and near multinormal distribution.

There are many ways to conduct factor analysis depending on how the data is organised (for example, a three-dimensional cube which includes entities or cases, characteristics and occasions define the edges). Applied research commonly uses factor analysis to explore the interrelationships of characteristics of entities in a cross-sectional fashion (R-factor analysis). Other modes of analysis include Q-, P-, O-, S- and T-factor analysis. Q-factor analysis, based on the transposing of a data matrix in the R-factor analysis, examines the interrelationships among entities from defined characteristics. O-factor and P-factor analysis look for the temporal structure of time series data, whereas S-factor and T-factor analysis focus on the behavioural variation over various occasions.

See also: multidimensional scaling; motivation; perceptual mapping; principal component analysis

Further reading

Kim, J.O. and Mueller, C.W. (1978) *Introduction to Factor Analysis: What It is and How To Do It*, Sage University Paper Series on Quantitative Applications in the Social Sciences, 07–001, Beverly Hills, CA and London: Sage Publications. (An introduction book on factor analysis.)

Phillips, P.A. and Moutinho, l. (1998) 'The market planning index: a tool for measuring strategic marketing effectiveness', *Journal of Travel and Tourism Marketing* 7(3): 41–60. (Provides an example of the application of factor analysis within the context of tourism marketing.)

Pizam, A. and Sussman, S. (1995) 'Does nationality affect tourist behaviour?', *Annals of Tourism Research* 22(4): 901–17. (Provides an example of the application of factor analysis within the context of tourism research.)

Rummel, R.J. (1975). *Applied Factor Analysis*, 2nd edn, Evanston, IL: Northwestern University Press. (Provides a discussion on the history, development and variations of factor analysis.)

JIANN-MIN JENG, USA
DANIEL R. FESENMAIER, USA

factory tourism

Involves organised visits to factories to see things being manufactured and processes at work. Breweries and distilleries, together with manufacturers of clothes, pottery and glass, are amongst the most popular factory visits. As wet weather attractions, usually in **urban tourism** centres, facilities can include **food** service, interpretative displays and shops where first quality goods and 'seconds' are sold.

MIKE ROBINSON, UK

fairs

Fairs are events, usually periodic, held for the exchange of goods and **services**. Rooted in the history of trade, in many cultures fairs have become more than **markets**; they encompass education, amusements, competitions and festive **behaviour**. Agricultural fairs or 'shows' focus on rural traditions; trade fairs concentrate on inter-business dealings. A World's Fair (exposition) brings nations together for comparison and celebrations, thus generating more **tourist** arrivals.

See also: festival

DONALD GETZ, CANADA

fairy tales

Since tourism is analogous to fiction, and because of its magical qualities (see **magic**), it can exploit linkages with fairy tales. Thus, touristic children can be attracted to **sites** of the imaginary, to temporarily resolve some of life's problems. Hence also the make-believe titles of many travellers' tales and **tour operators**, and the appeal of **museums** and **theme parks** featuring fairy tale characters.

GRAHAM M.S. DANN, UK

familiarity *see* motivation; novelty

fantasy

The desire to engage in **activities** that are normatively proscribed by ordinary role expectations, tourist fantasy permits the realisation of dreams through travel to extraordinary places. It promises freedom from constraint by removing the **tourist** from the conventional **attitudes** and behaviour of the home **environment** to a setting where there are apparently no restrictions on conduct.

Generally speaking, the more exotic the **destination**, the more pronounced are the cultural differences between **host and guest**. However, the more the tourist remains anonymous and detached from peer pressure, the greater are the perceived chances of dissolving the differences of strangerhood (see **strangeness**) and of replacing reality by fantasy. This creation of illusion occurs primarily at the pre-trip stage of a **vacation** through various promotional devices which feed off increasingly extravagant expectations heightened by the media and **advertising**. **Images** of sun-filled **hedonism** – a world that knows no 'no' – appeal to the **tourist as child** and to limitless **satisfaction** of desire without parental figures spoiling the pleasure.

Since there are as many fantasies as there are facets of daily life, the tourism – or fantasy – industry tries to incorporate several into a single holiday. One such illusion is a naming fantasy. Here the tawdriness of Station Road and Blackhall Street is overturned by the lure of such romantic alternatives as Coconut Creek, Tamarind Cove and Discovery Bay. Another is a colour fantasy, where the drab grey of the metropolis is replaced by the brilliant hues of a tropical **paradise**. Similarly, the monotonous sounds of a modern industrial town can be substituted with a passionate flamenco or the bouzouki **music** of a taverna dance. Additionally, there are educational, religious and political fantasies, all based on the temporary reversal of current situations and all conjuring up the notion of excess *ad libitum*. Perhaps none is so enticing as the sexual fantasy (see **sex tourism**), since this is the life domain that is most surrounded by taboo. Consequently, holiday brochures and **videos** regularly feature beckoning dusky maidens who seemingly have no moral inhibitions.

See also: inversion; liminality; play

Further reading.

Dann, G. (1976) 'The holiday was simply fantastic', *Revue de Tourisme* 3: 19–23.

GRAHAM M.S. DANN, UK

farm tourism

Farm tourism is distinguished by the specific milieu where it occurs. It is also known as **agrotourism**. It is a portion of general tourism where environmental tourism and **cultural tourism** overlap. Farm tourism is where **tourists** reside and sometime participate in the working activities of farms and ranches. It has increased in popularity as farming families search for enterprises which will supplement or replace their traditional economic activities. In addition, farm tourism is a tool for the **conservation** of traditional **landscapes**.

Tourists seek farm tourism for the outdoor component and use it as a setting for outdoor activities. Others, mainly urbanites, seek to temporarily **experience** a rural **lifestyle** which often forms part of the cultural **image** of the region, or functions as a nostalgic **escape** to the 'good old days'. In the diversification of agricultural enterprises into activities such as farm tourism, some authors (Garcia-Ramon *et al.* 1995: 268) have pointed out their varying **impacts** on the labour force. They found that women in particular play an important role in the new **hospitality** functions of farms, as much of the work falls on those who traditionally have domestic duties.

References

Garcia-Ramon, M.D., Cavoves, G. and Valdovinos, N. (1995) 'Farm tourism, gender and the environment in Spain', *Annals of Tourism Research* 22: 267–82.

VALENE SMITH, USA
VERONICA LONG, USA

fast food

Fast food restaurants offer limited menus of such items as hamburgers, french fries, hot dogs, chicken sandwiches, tacos, burritos, gyros, pizzas, pancakes, various finger foods and other choices for the convenience of customers wanting a quick meal. Orders are placed at a counter often under a brightly-lit menu featuring colour photographs of what is featured on the menu, receive their **food** within a few minutes of ordering, and have the option of consuming it at or away from the **restaurant**. Some common characteristics of fast food outlets are a clean environment, simple yet pleasant decor, consistent product **quality**, automated systems and quick service. A large number of fast food properties are part of such large franchise systems as McDonald's, Pizza Hut, Kentucky Fried Chicken (KFC), International Pancake House, Subway, Taco Bell, Quick and Georgie Pie.

This restaurant subsector consists of more than 180,000 locations in the United States, generating sales of more than $100 billion in 1997. They posted an annual average sales growth of 5.4 per cent annually between 1990 and 1997. Wraps and pitas, espresso and specialty coffee, and bagels are the items cited most by operators as gaining popularity. Value meals, which include a combination of food and beverage items at a discount, are also on the rise. Introducing at least one new item every year was a goal expressed by operators. The average check per person at fast food restaurants was $3.33 in 1997, up 3.4 per cent over 1996. During 1990–6, check averages grew an average of 2.1 per cent per year. The restaurant industry sales as a whole are projected to increase 4.6 per cent in 1999 to reach $354 billion, according to a 1998 report published in *Restaurants USA*. The 1998 sales, according to National Restaurant Association, increased by about 5.1 per cent, to $105.7 billion.

The average American eats out at least four times a week in a restaurant, of which two times are in fast food properties. According to the Association's 1998 customer survey, 76 per cent of the customers felt that the value they received for the paid price met or exceeded their expectations. Today, **tourists**, along with residents, in a growing number, frequent fast food restaurants everywhere. In 1998 McDonalds alone had 24,500 stores in 115 countries – with twenty-seven of them located at major airports – where hamburgers and fries are iconised, internationalised and sought after worldwide.

Further reading

National Restaurant Association (1998) *Limited Service Restaurant Trends*, Washington DC: NRA.

Walker, J.R. (1996) 'The restaurant business', in J.R. Walker, *Introduction to Hospitality*, Engelwood Cliffs, NJ: Prentice Hall, 170–3.

PETER D'SOUZA, USA
PHIL McGUIRK, USA

feasibility study

A feasibility study is an in-depth analysis of the financial prospects for a project or property **development**. It is not intended, nor should it be construed, to be a document that is prepared to prove the profitability of a new venture. Although the document is called an economic feasibility study, the report contains much more than financial data, generally limited to statements of cash flows. These are projected commonly for a period of five years because it takes a new property about that much time to achieve a stabilised operating environment. As the timeline expands, the credibility of the financial data is weakened due to forces such as inflation, changes in competition and alterations in **marketing plans** and efforts.

The purpose of preparing a feasibility study is to determine whether or not to proceed with the project. The logic is that the cost of preparing the study is a minor expense if it avoids a series of ongoing negative cash flow situations. A lender generally will require an investor to submit a feasibility study along with other documents associated with the loan application. The lender will also generally insist that it be conducted by a group independent of the owner/investor. This situation allows for a great deal of **conflict** in that the individual(s) commissioning the study has a vested interest in receiving a positive report so that the project can continue and receive financing. A much more logical arrangement would be to have the lending institute arrange for the study, billing the cost back to the applicant.

Preparers of feasibility studies can be public **accounting** firms, real estate management companies, management consulting firms, contract management companies and independent individuals. In the selection process of a firm, the owner/investor (client) should consider the credentials and reputation of the specific individuals in the firm assigned to project. The commissioned report is the property of the owner/investor and is proprietary information. The document, projected and prepared on behalf of the client, is not to be shared with other organisations, lending institutions or potential competition by the group commissioned to perform the analysis. The workpapers and supporting schedules, however, remain the property of the preparer of the study, with no obligation to provide them to the client or to share their content with the owner/investor. In the course of completing the assignment for the client, the consultant may obtain temporary custody of certain original books and records relating to the client's business. Upon termination of the assignment, or upon request from the client, these documents need to be returned to the client, even if the fee has not been paid.

Feasibility studies for tourism projects are usually similar in format. They are used to achieve assistance in obtaining financing and to help in obtaining contracts, franchise agreements (see **franchising**) and leases. When the project is completed and operating they often are also consulted for use as operating and marketing plans, and guides for operating and capital budgets (see **budgeting**). If it is determined during the process that the project is not economically feasible, the consultant will stop work immediately and inform the client of his/her determination. All work is concluded and the negative report is prepared as a letter, which is sent to the client.

The basic format of a hotel feasibility study contains several sections. A letter of transmittal is simply correspondence from the consultant to the owner/developer detailing the contents the report. It is intended to alert the recipient to what to expect, and the magnitude of the project. This is followed by a table of contents. An executive summary reports the highlights of the findings and other significant information about the report. Its intended audience is higher management. The characteristics of the market area include general descriptions of the geographic quadrant, state, city and specific location. These comments acquaint the reader with the geographic area, how it evolved, its strengths and its weaknesses. The actual area is described and supported by specific research of economic indicators such as population, effective buying income, **employment**, retail sales, office space trends, **transportation**, and tourism and conventions (see **meetings business**). An evaluation of the specific area includes an examination of both the primary and the secondary competition. It is in this section that the competition is rated.

The competition and **demand** analysis uses the information gained in the prior sections to determine the projected occupancy and average daily rate of the new project. It also takes into account other projects that may be primary or secondary competition as well as considering facilities that currently are classified as secondary competition that could evolve into primary competition. The recommended services and facilities section is reserved for comment on the proposed project from an operational and competitive perspective. This section includes the comments of the consultant relative to the individual property. It may also include recommendations as to the concept, number and type of **hotel** rooms, **food** and beverage facilities, banquet and meeting space allocations, and other services offered. The financial section includes estimates of construction cost, debt service calculations, replacement reserve methodologies, return on equity computations, and other select financial ratios and information as well as cash flow projections. Finally, the conclusion section inserted summarises the study and presents a positive closing statement.

Further reading

Angelo, R. (1985) *Understanding Feasibility Studies: A Practical Guide*, East Lansing, MI: The Educational Institute of the American Hotel and Motel Association.

Beals, P. and Troy, D. (1982) 'Hotel feasibility analysis, part I', *The Cornell Hotel and Restaurant Administration Quarterly* 23: 10–17.

Rushmore, S. (1988). *How to Perform an Economic Feasibility Study of a Proposed Hotel/Motel*, Chicago: American Society of Real Estate Counselors.

—— (1990) *Hotel Investments: A Guide for Lenders and Owners*, Boston: Warren, Gorham and Lamont.

Troy, D. and Beals, P. (1982) 'Hotel feasibility analysis, part II', *The Cornell Hotel and Restaurant Administration Quarterly* 23: 58–65.

STEPHEN M. LEBRUTO, USA

feast

Feasts include sumptuous or elaborate meals, often with entertainment. Religious feast days are communal celebrations and have often become public holidays or festivals. Feasts commemorate a significant **event**, mark the passage of seasons, honour a deity, or allow participants to commune ritually with gods or ancestors (see **ritual**). They might also display status through **conspicuous consumption**. Food festivals are a modern version. These all potentially act as tourism attractions.

See also: festival

DONALD GETZ, CANADA

Federal Aviation Administration

An agency of the US government charged with oversight of air **transportation**. Responsibilities include air traffic control, certifying airlines, aircraft, flight crew and airports, enforcement of regulations and development of improved operating standards. The agency became part of the US Department of Transportation when the latter executive department was formed in 1967.

ANTHONY G. SHEPPARD, USA

Federation of International Youth Travel Organisation

The purpose of the Federation of International Youth Travel Organisation (FIYTO) is to promote youth mobility and to broaden the horizons of young people through travel, language acquisition, family living, cultural and social tourism and other opportunities for personal growth. As one of the largest trade organisations in the youth and student travel sector, it is the meeting point and advocate for rapidly growing **youth tourism**. It has 245 members in fifty-four countries. It offers identity cards to its youth members to facilitate access to tourism facilities and culture and leisure activities. Its major publication is *Youth Travel International*. The headquarters of the organisation is located in Copenhagen, Denmark.

TURGUT VAR, USA

feminism

Feminism covers a number of philosophical positions which question gender inequalities and exploitation in tourism-related work opportunities, leisure and representation, as well as the masculinist discourse of tourism research. Activist organsisations and academics that monitor tourism prostitution or analyse sexist images in advertising are examples of feminist praxis in tourism, with the goal of changing local practices in a global industry.

VIVIAN KINNAIRD, UK

ferry

Ferry service is offered on short sea routes as a means of urban **transportation** (for example, the Hong Kong Star or Staten Island ferry), as an extension to road/**rail** services (as in the English Channel), or as an essential transport link (for example, to the Greek islands) for people and vehicles. It is mostly provided by conventional roll-on roll-off vessels, but sometimes by faster hovercraft, hydrofoil or twin-hulled catamaran services, and is used by residents and **tourists** alike.

ANNE GRAHAM, UK

festival

All cultures celebrate, and the things, persons or themes they value provide reasons for festivals. Falassi (1987: 2) described them as 'a sacred or profane time of celebration, marked by special observances'. Pieper (1973: 32) believed that only a religious celebration could be a true festival, and that 'through it the celebrant becomes aware of, and may enter, the greater reality which gives a wider perspective on the world of everyday work'.

While traditional festivals often retain religious or sacred meanings, contemporary ones are primarily secular, or profane. Although many traditional themes have been retained, countless new ones have been established for purposes ranging from **economic development** to building community identity and pride. But the notion of contrast with everyday life remains a valid defining construct. Festivals and other forms of special events provide more than amusement; the atmosphere might be one of gaiety, joyfulness, playfulness or liberation from normal constraints.

Festivals are not **fairs**, which have their origin and significance in the exchange of goods and services and are closely linked to periodic markets. Over time, however, many fairs encompassed educational exhibitions, competitions, amusements and festival-like celebrations. Contemporary planners of many types of event frequently attempt to create a festive or celebratory atmosphere so as to appeal to a broader audience, attract more attention or fabricate a tradition. Hence, mega-events like the Superbowl or world's fairs become festival-like.

Many festivals, regardless of theme and programming, are actually celebrations of the community itself. Falassi (1987: 2) noted:

> Both the social function and the symbolic meaning of the festival are closely related to a series of overt values that the community recognizes as essential to its ideology and worldview, to its social identity, its historical continuity, and to its physical survival, which is ultimately what festival celebrates.

Festivals have a multitude of potential meanings and encompass a number of paradoxes (Lavender 1991). They can be read as 'texts' which are stories told by members of a culture about themselves; as performances, or social dramas full of conflict and power statements; as communications about social ties in the society; as art forms; as deliberate inversions or role reversals, mocking but simultaneously reinforcing social norms. As explained by Manning (1983: 4), celebration is performance: 'it is, or entails, the dramatic presentation of cultural symbols'. Celebration is public, with no social exclusion, is entertainment for the fun of it, and is participatory, actively involving the celebrant who takes time out of ordinary routine, and '... does so openly, consciously and with the general aim of aesthetic, sensual and social gratification'.

Tourists often seek out festival and other events to gain an authentic cultural experience. By sharing with residents in an important local event, the visitor can capture some of the flavour of local **lifestyles** and traditions. But when festivals and other special events are consciously developed and promoted as attractions, there is the risk that **commercialisation** will detract from celebration; that entertainment or spectacle will replace the inherent meanings of the celebrations. In other words, tourism might destroy cultural **authenticity**, the very thing many contemporary tourists appear to be seeking. The dilemma, however, is that the benefits to be realised from tourism also offer the means to create or expand festivals, restore and cultivate traditions, and foster community spirit and sharing.

Given that the essence of authenticity is its cultural meaning, host communities must determine what is meaningful to them. In this sense, authenticity is not so much the ritual, games, spectacle or celebration itself as it is the degree to which these components have been manufactured, modified or exploited just for tourists, the media or financial success. In other words, has the event any cultural meaning for the host community and the participants, or is it merely a commodity to be sold? Do the hosts and performers think of the event as having importance in their lives, or are they cynically involved in a tourist rip-off? Will people come to accept, over time, that the invented festival is an important part of their cultural life? Many valuable studies have addressed these and other related topics, add a new depth to this socio-economic theme.

See also: event management

References

Falassi, A. (ed.) (1987) *Time Out of Time: Essays on the Festival*, Albuquerque, NM: University of New Mexico Press.

Lavender, R. (1991) 'Community festivals, paradox, and the manipulation of uncertainty', *Play and Culture* 4: 153–68.

Manning, F. (ed.) (1983) *The Celebration of Society: Perspectives on Contemporary Cultural Performance*, Bowling Green, KY: Popular Press.

Pieper, J. (1973) *In Tune With The World: A Theory of Festivity*, Chicago: Franciscan Herald Press (translated by R. and C. Wilson, from the 1963 German original).

Further reading

Getz, D. (1997) *Event Management and Event Tourism*, New York: Cognizant Communication Corp.

DONALD GETZ, CANADA

Festival Management and Event Tourism

Festival Management and Event Tourism: An International Journal is designed to meet the needs of an evolving profession. The business of special events has grown to enormous proportions and now represents a significant contribution to tourism **industry**. Festivals and special events have become and will continue to be an important medium for non-profit and public organisations to carry out their missions, and will act as a catalyst for **community** development. This refereed journal aims to meet the needs of these related groups, specifically in chronicling research progress, documenting **management** and **marketing** developments and fostering interdisciplinary work about this phenomenon. Submissions are expected to contain original material and must not have been published previously. First appearing in 1994, the journal is published quarterly by Cognizant Communication Corporation (ISSN 1065–2701).

RENE BARETJE, FRANCE

festival, religious

Festivals with religious associations are termed religious festivals. Various events have been fostered in many milieus, religions and cults which today have a religious content rather than character. These events attract large numbers of tourists more by the uniqueness of the event itself and the local colour provided by the **music**, song, rituals and costume than by their religious content. Many of these ceremonies and festivals on such occasions do not take place in places of **pilgrimage** and serve as actual or potential tourism attractions. At the end of each year in the Christian world, numerous celebrations are held especially during the Christmas season. Events of the same kind are held at Easter, Pentecost and Corpus Christi. There are numerous dates throughout the year when certain saints' days are celebrated, and the most popular feast days, which include festivals, are those in honour of the Virgin. The largest number of such ceremonies and festivals take place in Mediterranean countries, which have recently used them abundantly in their tourist promotion. Religious festivals attract a large influx of tourists form near and far origins.

Further reading

Goldberg, A. (1983) 'Identity and experience in Haitian voodoo shows', *Annals of Tourism Research* 10(4): 479–95.

Greenwood, D.J. (1989) 'Culture by the pound: an anthropological perspective on tourism as cultural commoditization', in V. Smith (ed.), *Hosts and Guests: The Anthropology of Tourism*, 2nd edn, Philadelphia: Pennsylvania University Press.

Nolan, M.L. and Nolan, S. (1992) 'Religious sites as tourism attractions', *Annals of Tourism Research* 19(1): 68–78.

Vukonić, B. (1996) *Tourism and Religion*, Oxford: Pergamon Press.

BORIS VUKONIĆ, CROATIA

film

Louis and Auguste Lumière's invention of the cinématographe in 1895 marks the beginning of cinema. The Lumières premiered the first commercial film to a paying audience, *Workers Leaving the Lumière Factory*, on 18 December 1895. From the beginning, there were intimations of film's potential role in the tourism industry. At the 1900 Universal Exposition in Paris, Louis Lumière's projection of fifteen films for a twenty-five-minute programme on a 25 × 15 meter screen to an audience of 25,000 was a notable **attraction**. The 1903 St Louis Exposition showcased *Hales Tours and Scenes of the World*, an attraction which showed at free-standing venues from 1903 to 1909. Audiences seated in train coach replicas watched famous scenes from around the world linked together with footage of oncoming rail tracks.

Film's greatest impacts on tourism today lie not within the area in which it first gained audience favour – its ability to capture scenes from reality, as in the 'actuality' films of Lumière and others – but rather in its ability to render as 'real' the imaginary. Film today as a multidimensional force has proven its ability in creating **destination** awareness, transforming ordinary places into attractions, and as an organising principle for themed parks.

The Australian film *Crocodile Dundee* illustrates the medium's ability to generate **destination** awareness. Following its release in 1986, Australia achieved the greatest increase in **tourist** arrivals for any developed country in 1987. The Northern Territories of Australia were inundated with American tourists anxious to visit the Kadadu National Park and other locales shown in the film. More than thirty years after the 1964 release of the film *The Sound of Music*, visitors to Salzburg in **Austria** can still take a 3–4 hour tour called the 'Sound of Music Tour' and visit Schloss Frohnburg, a seventeenth-century country house used as the home of the von Trapp family. Austrian consul-generals have repeatedly credited this American film with having done more for promoting Austria to foreign tourists than any other form of publicity. In San Francisco, tourists buses stop at the City Hall not only because it is one of the finest examples of French Renaissance architecture in America, but also for its associations with the *Dirty Harry* character portrayed by Clint Eastwood and the 1985 James Bond film *A View to a Kill.*

The London Film Commission has indicated that 'tourism has increased by over 20 per cent a year following the success of a major film'. Acknowledging this, convention and visitors bureaus routinely have staff specifically tasked with accommodating film scouts searching for filming locations. The guidebook *Shot On This Site* directs movie-goers to locales used in their favourite films. National tourism organisations or authorities also produce their own films for use in media campaigns and as sales and training tools.

Film studios are attractions in their own right. Tourists to Bombay, the centre for India's Hindi movie industry, can take special tours during the production of a movie. In Taipei, Taiwan, there are tours of the Central Movie Studios where sets of ancient Chinese houses and streets are still used in television productions. Hollywood movie studios such as Universal have long used their film and television studios as a tourism attraction. Disney, MGM Studios, Warner Brothers and Universal have capitalised on the interest in the film by using characters and scenes from films and aspects of film making as an organising theme for their **theme parks** in California, Florida, Australia's Gold Coast, Paris, Düsseldorf and Tokyo.

The Cannes Film Festival, fifty years old in 1997, was originally founded with the intent of fostering tourism. Now mainly a film industry and press event, French officials estimate that **business travel**, led by the Festival, injects about $400 million into the local economy. In 1957, the year of the America's first film **festival**, the San Francisco International Film Festival, only about a dozen film festivals existed worldwide. In 1997, over 450 film festivals existed around the world.

Travel guide publishers such as Fodors, Lonely Planet and Rand McNally are seizing the opportunities presented by travel films by extending their own activities to include video cassettes. New distribution channels, cable television's The Travel Channel and new technologies such as digital delivery of films via CD-ROM and the **Internet** provide new and promising roles for film as travelogues and as effective promotional media for tourism worldwide.

Further reading

Garland, J. (1993) 'Magic Magnets: Destinations as Locales for Movie Making', in *PATA' 94. Conference Record. Pacific Asia Travel Association. 42nd Annual Conference. Honolulu, Hawaii. May 9–13, 1993*, San Francisco: Pacific Asia Travel Association.

Gordon, W.A. (1995) *Shot On This Site: A Traveller's Guide to the Places and Locations used to Film Famous Movies and TV Shows*, Secaucus, NJ: Carol Publishing Group.

Karney, R. (ed.) (1997) *Chronicle of the Cinema*, London: DK Publishing.

Nowell-Smith, G. (ed.) (1996) *The Oxford History of World Cinema*, Oxford: Oxford University Press.

Riley, R., Baker, D. and Van Doren, C. (1998) 'Movie induced tourism', *Annals of Tourism Research* 24(4): 919–35.

Sklar, R. (1993) *Film: An International History of the Medium*, Harry N. Abrams.

JORDAN YEE, USA

financial control

Financial controls are a set of procedures (a financial control system) for setting **financial objectives** or targets (see **budgeting**), monitoring financial outcomes against those objectives by means of **accounting**, and analysing the differences (variances) with a view to finding explanations for them. In terms of control theory or cybernetics, budgeting or financial planning represents feed forward control, while accounting statements and variance analysis represent feedback control and are provided by the organisation's management accounting. Variances between targets and outcomes may be used as financial **performance indicators**.

Financial controls are only part of an organisation's **planning** and control systems, being directly concerned only with financial targets and outcomes (such as hotel sales revenue, **profit margins**, **cost** levels and returns on capital) rather than other key success variables such as product or service **quality**, staff turnover and market share (see **marketing**). In general, achievement of financial targets will be associated with meeting other key targets. However, in the short term tourism and other managers may be able to achieve a financial target such as profit or return on capital by reducing expenses in ways that adversely affect **quality**. A well-designed financial control system will make it hard for managers to resort to this type of short-termism, for example by monitoring the expenses that are critical to quality against budgeted amounts. Financial controls may be applied at various organisational levels within the tourism industry. At the overall or corporate level, top **management** will be concerned that the organisation as a whole is performing financially in line with management's own published forecasts for such matters as earnings (or net profit) per share.

See also: corporate finance

SIMON ARCHER, UK

financial objectives

Financial objectives are the establishment of measurable financial goals which follow the mission statement. Used by operating managers, financial objectives are measured by gross operating **profit**, which is the income of an enterprise before fixed charges. Operating management does not control fixed charges such as rent, property taxes, insurance, interest, depreciation and amortisation, gains or losses on the sale of property, and income taxes, as these are ownership issues.

The comparison of an enterprise's net profits to the amount invested by the owners is referred to as return on equity (or **return on investment**) and is the measurable financial objective for owners. Mathematically, it is computed by dividing the net income by the average **investment** during the period in which the net income was earned. Often, the return on investment is a stated financial objective of the investor and is used by **management** to drive the selling price of the **product** in a 'bottom-up' approach (or Hubbart formula) to pricing. The first step in using this approach to establish the selling price of a tourism product, such as a room in a full service **hotel**, is to provide for income taxes and to add in the fixed charges (management fees, rent, property taxes, insurance, interest, depreciation and amortisation, and gains or losses on the sale of the hotel). Undistributed operating expenses, or the cost of the service departments, are considered next. This involves an estimate of the hotel payroll and related expenses of service departments such as administrative and general, data processing, human resources, **transportation**, **marketing**, property operation and maintenance, and energy costs. These costs are added to the income before tax and fixed charges. This result is the total operating profit that must be generated from the revenue producing departments.

Income from each department (other than rooms) is subtracted from the total operating profit that needs to be generated from the revenue producing departments. If an operating department is projected to have a loss, this is added. The rooms department payroll and related expenses are added to derive the required total rooms revenue. The average rooms selling price is determined by dividing the total required rooms revenue by the projected total number of units to be sold. This calculation meets the hotel investor's financial objective.

Further reading

Schmidgall, R. (1997) *Hospitality Industry Managerial Accounting*, 4th edn, East Lansing, MI: Educational Institute of the American Hotel and Motel Association.

STEPHEN M. LEBRUTO, USA

Finland

Tourism in Finland is based on natural attractions featuring lakes (187,888 in number), forests and a well-kept **environment**. Further, the **exoticism** of Lapland, north of the Arctic Circle, and the belief that Santa Claus has his home there, generates **demand** from other countries. Finland is geographically well situated in terms of air routes from North America and Asia to Europe, and the integration of Europe has strengthened its position as a gateway to **Russia**, especially to St Petersburg and the Baltic states.

The main tourism **markets** of Finland are **Sweden**, **Germany** and Russia, but growth from Asian markets has been considerable. **Domestic tourism**, however, accounts for over 70 per cent of total overnight stays. The main tourism season is summer, with **nature tourism** and cultural festivals as well as more specific **attractions** such as architecture-based **sightseeing**. However, winter tourism with cross-country **skiing** and Christmas visits has showed the strongest growth in recent years. The tourism balance is negative, with Finns spending more abroad than do foreigners in Finland. Finns are among the top international tourists in proportion to the population. In 1997, about five million took 5.2 million trips abroad (as well as 28.3 million domestic trips).

During recent years, the Finnish government has put more emphasis on service industries and a new tourism strategy (adopted in 1996) aims at improving the competitiveness of the **industry** while at the same time protecting the unique values of nature and environment. Foreign investment is also welcome and actively promoted. Tourism products offering experiences based on nature and culture are developed with the goal of positioning Finland as a **destination** of **ecologically sustainable tourism**. The government is also giving the Finnish Tourist Board more resources for **marketing** abroad.

The tourism industry is well developed with potential for growth, as the **accommodation** occupation rate is low. Air, rail and road transport **infrastructure** is extensive. Modern convention facilities have contributed to Finland's position as the thirteenth largest convention **destination** in the world. However, the geographical position of Finland, at a considerable distance from the large population centres of Europe, means that good transport connections are crucial for the development of tourism. The **accessibility** of the country could be weakened by the abolition of tax-free sales in the European Union in 1999, as ferry traffic is very dependent on these sales. Finland's membership in the **European Union** has however opened up the possibility of using EU funds for **investments** in tourism. The enlargement of the EU and the political will to integrate Russia into European cooperation are giving the Baltic Sea region a higher profile. The increasing tourism cooperation in the region will **benefit** Finland.

BENGT PIHLSTRÖM, FINLAND

Finnish University Network for Tourism Studies

The Finnish University Network for Tourism Studies (FUNTS) is an operational cooperative structure among the Finnish universities formed for the advancement of tourism research and **education**. The structure was developed in order to upgrade the academic standing of tourism studies and to gain synergetic advantages (best expertise, innovativeness, cost **efficiency**, wider spatial scope) – or a dynamic multidisciplinary tourism programme – for the whole country. This Finnish network represents a unique education **model** which can be implemented elsewhere, especially in small and/or developing countries.

Institutionalisation of academic tourism studies emerged in Finland in the late 1980s. Because limited human resources are available for an academic thrust committed to tourism, eight Finnish universities joined force to create the network located in Savo province but serving the whole country. The number of participating universities increased to fourteen in 1997. The coordination unit of FUNTS is located in Savonlinna campus of the University of Joensuu.

Annual enrolment is fifty students, who come from all the fourteen member universities. Every student stays with her/his own major (e.g.. business **management**, **anthropology**, **sociology**, **economics**, German literature, **geography**, visual arts and design) at her/his own university. The network programme covers a variety of themes and **methodologies** of tourism studies organised into discipline-oriented and problem-oriented courses. Students can take up to sixty credits ('study weeks') arranged at three levels of requirements. The programme is modular and includes distance learning (mainly telematic) through intensive courses and fieldwork. In the classroom and fieldwork settings the students – again from different academic backgrounds and coming from different regions throughout the country – meet in educational situations where advantages of creative group dynamics are gained. Since the professors also have different academic backgrounds and skills, the multidisciplinary scope and nature of the programme is reinforced.

At present, some thirty Finnish and five to ten international professors are contributing to this nationwide tourism education network. The programme was funded by the Ministry of Education from the European sources for 'less favoured regions' up to 1999, after which the sixteen participating universities have jointly supported and financed the network.

ARVO PELTONEN, FINLAND

FIU Hospitality Review

FIU Hospitality Review publishes essays on a broad range of views dealing with all facets of **hospitality** and tourism. In general, the journal pays special attention to those operational issues and **management** topics which more closely relate to **hotel** and **restaurant** businesses. First appearing in 1983, it is published twice yearly by Florida International University School of Hotel Management (ISSN 0739–7011).

RENE BARETJE, FRANCE

flow

Flow is a theoretical construct from leisure (rather than tourism) studies, principally associated with psychologist Mihalyi Csikszentmihalyi. Flow is that dynamic state of optimal arousal (and loss of time/ self-consciousness) a recreator experiences when available skills are commensurate with the demands posed by an immediate leisure pursuit. It is intense when the recreator undertakes goal-oriented tasks offering instant feedback. Other times, 'boredom' is entertained when skill exceeds challenge, or 'anxiety' when challenge exceeds skill. Since deep flow experiences are transitory, the construct is of reduced significance, as tourism activities typically involve longer time, travel, and compositional horizons.

KEITH HOLLINSHEAD, UK

food

The human body requires nourishment, which is provided by food (see **diet**). Foods are primarily a combination of protein, carbohydrates and lipids. The failure to supply the body with sufficient quantities of food or types of food can result in disease and death. Food is also a cultural phenomenon (see **cuisine**). It is dictated by **geography**, **climate** and social **class**.

The need for food in human development is not restricted to survival. The role of food in religions around the world indicates its importance as a factor in cultural development. Unleavened bread symbolises the body of Christ in the Christian faith. The flesh of animals with cloven hooves is taboo in the Jewish faith. The daily fast, to abstain from eating, is observed during Ramadan by those of the Islamic faith.

Cultural and social occasions often centre around food. The joining of man and woman in marriage and the day a person was born are celebrated with cake. Good luck for the New Year is sought in the southern United States by the consumption of black-eyed peas, a legume, on New Year's Day. Sporting events in the USA are associated with hot dogs, beer and hamburgers. Family gatherings often culminate in special meals. Offering a visitor food or drink is a universal form of welcome. In some countries, gifts of food are taken to the home where a family member has died as an expression of sympathy and respect.

Food is the global language. The foods consumed from one region to another vary widely and provide a glimpse into the various cultures. To understand a culture, an individual must experience its food. Tourists visiting a region **experience** the culture through the goods they consume, as well as through the people and attractions. The food of a region is often a primary attraction. In most areas of the world, effort has been made to create heightened flavours, textures, eye appeal and aromas through food combinations. This has resulted in many unique blends of food and culture, such as those in San Francisco, New Orleans, Paris and Singapore, which are sought by tourists.

Further reading

Tannahill, R. (1989) *Food in History*, New York: Stein and Day.

JERALD W. CHESSER, USA

food and beverage cost analysis

Food and beverage cost analysis is the examination and analysis of the cost of sales of these two operating departments in various tourism sectors. Under the uniform system of accounts for the lodging industry, the food and beverage departments have a cost of sales section. The cost of goods sold is computed by taking the beginning inventory plus inventory purchases, which equals the cost of goods available for sale. From the cost of goods available for sale, the ending **inventory** is subtracted, which equals the cost of goods consumed. An adjustment is then made for food or beverage in and out transfers. The latter can be to/from other departments within the hotel or to/ from other properties. If there were any steward sales, the value of these transactions are also subtracted. Steward sales occur when an operation sells its inventory to individuals. The value of employee and complimentary meals and the sale of used grease are also subtracted from the cost of goods consumed. After these adjustments have been made, the final result is the cost of goods sold (food or beverage).

A concern in computing the cost of goods sold is the method chosen to value the inventory. There are at least five ways that inventory can be valued, which will affect the cost of goods sold. The first-in first-out method requires that the first products received will be the first products used, which means that the remaining products at the end of the **accounting** period will be the most recent ones purchased. The modified version of this method uses the most recent purchase price to value all of the inventory.

In contrast, the-last in first-out method states that the oldest products are the ones that remain in inventory. The specific identification method values the remaining inventory by assigning the actual cost of the actual unit that remains in inventory at the end of the period. The weighted average method calculates the ending inventory based on the weighted average of the purchases during the period. Food and beverage analysis is the comparison of the cost of food or beverage sold to the net food or beverage revenue or the comparison of the contribution margin of the food or beverage department (food or beverage revenue minus cost of food or beverage sold) to prior periods or a budget.

Further reading

Schmidgall, R. (1997) *Hospitality Industry Managerial Accounting*, 4th edn, East Lansing, MI: Educational Institute of the American Hotel and Motel Association.

Uniform System of Accounts for the Lodging Industry, 9th edn, East Lansing, MI: Educational Institute of the American Hotel and Motel Association.

STEPHEN M. LEBRUTO, USA

food-borne illness

Illness from **food** is caused by the ingestion of foodstuffs which have been directly or indirectly contaminated with harmful bacteria, such as salmonella (see **sanitation**). Such illnesses frequently cause stomach pains, diarrhoea and/or vomiting. **Tourists** avoid such illnesses by drinking bottled water, eating hot foods, and avoiding salads, local fruits, cold dishes and shellfish unless provided by reputable sources.

KATHRYN WEBSTER, UK

food service, contract

A **management contract** is an arrangement under which operational control is vested by contract in a separate business which performs the necessary management functions for a fee. In the contract food service business, this fee may be a straight fee, one linked to the turnover of the business, one in which purchasing discounts are returned to the client organisation to offset against the fee, or a combination of these. Foodservice is typically contracted out for employee feeding in offices and factories, and increasingly in schools, hospitals and some tourism establishments.

See also: commissary

PETER JONES, UK

forecasting

Planning under conditions of uncertainty is both difficult and necessary, and it creates a substantial need for forecasts. Accurate forecasts of **international tourism demand** are particularly important on account of the perishable nature of the **product** (unfilled **airline** seats, empty **hotel** rooms, and unused hire cars result in lost revenue which cannot be recouped). They are essential for efficient planning by those businesses connected with the **international tourism** market, as well as of great interest to governments in origin and **destination** countries and to national tourism organisations. Forecasting techniques may be split into **qualitative research** and **quantitative methods**, and the latter may be further subdivided into causal and non-causal methods (see **causal model**).

Qualitative forecasting has received relatively little attention in the tourism literature. The most popular qualitative method is the **Delphi technique**, which is generally used to forecast long-term environmental and technological **trends** rather than the level of tourism **demand**. The focus of this entry is on quantitative forecasting methods. Causal forecasting techniques examine the quantitative relationship between international tourism demand and its determinants. The **models** are estimated (usually by multiple regression analysis) using historical data, and future values of international tourism demand are obtained by using forecasts of the demand determinants in conjunction with the estimated relationship. The most popular causal tourism demand forecasting method is econometrics, but gravity models have also been used. The factors influencing tourism demand vary according to purpose of visit, but as the majority of visits are holiday, the main factors which determine the demand for international holiday tourism are considered.

The level of foreign tourism from a generating **market** to a **destination** country is likely to be influenced by origin population (positive impact), origin income (positive **impact**), holiday price (negative impact), substitute prices (positive impact), qualitative effects (positive or negative impact), trend (positive or negative impact), destination marketing expenditure (positive impact) and a lagged dependent variable (positive impact). As far as substitute prices are concerned, potential tourists compare the price of a foreign holiday with both the price of a domestic choice and with the prices of other foreign options in reaching their decision. The impact of qualitative effects on international tourism demand can be allowed for through dummy variables. 'One-off' events such as war, terrorist attacks (see **terrorism**) and natural disasters are likely to have a negative effect on international tourism demand, whereas mega-events such as the Olympic Games and Expo are likely to have a positive effect. A lagged dependent variable can be justified on the grounds of habit persistence/'word of mouth' recommendation. It can also be justified in terms of **supply** constraints, such as shortages of **accommodation**, passenger **transportation** capacity and trained staff; if a partial adjustment mechanism is postulated to allow for rigidities in supply, this results in the presence of a lagged dependent variable as a demand determinant.

Non-causal quantitative techniques assume that tourism demand may be forecast without reference to the factors which determine the level of demand. Univariate time-series methods determine future values for tourism demand through a process of identifying a relationship for past values of the variable. It is rarely possible to justify univariate time-series models on the basis of **theory**. The reasons for their use are essentially pragmatic: they often generate acceptable forecasts at low cost. Furthermore, they may be used where causal models are inappropriate on account of lack of data or incomplete knowledge regarding the causal structure. Univariate time-series methods which are popular for forecasting international tourism demand are exponential smoothing, trend curve analysis, decomposition methods and univariate ARIMA (autoregressive integrated moving average) models.

The idea underlying exponential smoothing models is that by allowing some weight to be given to the current forecast value of tourism demand (in addition to the actual value) when forecasting for the next period, sudden changes in demand (which may be short-lived) are in effect 'smoothed out'. The single exponential smoothing model is only applicable to stationary series, whereas Brown's double exponential smoothing model is applicable to series containing a **trend**, and the Holt-Winters double exponential smoothing method is specifically designed for series exhibiting seasonality in addition to a trend. Trend curve analysis generally involves the use of regression analysis to find a curve of best fit through time-series data, which is then projected forward into the future. Using transformations, trend curve analysis can be employed to produce forecasts from data showing a range of patterns (for example, straight line progressions, exponential growth, or patterns that display a gradual approach to a saturation level).

Decomposition methods start from the premise that the observed values of a time series are usually the result of several influences (secular trend, seasonal variations, irregular or random changes and possibly cyclical fluctuations), and attempt to isolate and measure those parts of the time series that are attributable to each of these. The univariate ARIMA forecasting method is a highly sophisticated technique and is rather more difficult to apply than the other univariate time-series methods considered. The autoregressive component implies that tourism demand depends on its own past values, and the moving average component implies that tourism demand depends on previous values of the error term. ARIMA models are very flexible, and can represent many types of stationary and non-stationary series. Furthermore, they often provide relatively accurate forecasts.

Selection of an appropriate technique for forecasting international tourism demand depends upon the requirements of the forecaster. Empirical research has shown that no single forecasting method performs consistently best across different tourism demand situations, and that more complex techniques do not necessarily generate more accurate forecasts than simple methods.

Further reading

Frechtling, D.C. (1996) *Practical Tourism Forecasting*, Oxford: Butterworth-Heinemann. (Provides an introduction to various methods available for forecasting tourism demand.)

Song, H. and Witt, S.F. (2000) *Tourism Demand Modelling and Forecasting: Modern Econometric Approaches*, Oxford: Elsevier. (Provides an introduction to modern sophisticated econometric forecasting methods with tourism examples.)

Witt, S.F. and Witt, C.A. (1992) *Modeling and Forecasting Demand in Tourism*, London and San Diego: Academic Press. (Examines the accuracy of various quantitative forecasting methods in the context of international tourism demand.)

—— (1995) 'Forecasting tourism demand: A review of empirical research', *International Journal of Forecasting* 11(3): 447–75. (Reviews the main methods used to forecast tourism demand which are reported in published empirical studies, together with the empirical findings.)

STEPHEN F. WITT, UK and AUSTRALIA

foreign exchange

Foreign exchange is the exchange of one country's currency for that of another country. The exchange rate between two currencies is the price of a unit of the one currency in terms of the other currency at which the exchange takes place (for example, 10 French francs to the British pound, 1.8 deutschmarks to the US dollar and so on). The spot exchange rate is the rate for immediate delivery. There are also forward exchange rates for the more commonly used currencies, which may be used for hedging. Exchange rates may be either fixed or floating (variable). A currency union is a group of countries which have agreed to fix (or peg) the exchange rates among their currencies, perhaps with a narrow band of permitted variation. Many currencies are fully convertible into other ones in the foreign exchange markets, but some are not. Exchange controls are a form of **currency control** which restrict the transferability of a national currency into or out of the country concerned. Because exchange rate variations affect the relative prices of goods and services in different countries, they are an important factor in international trade, including **international tourism**, as well as the sudden popularity of certain destinations at the expense of others when their currencies fluctuate substantially.

By convention, exchange rates are normally given as a number greater than one (for example, US dollars 2.50 to the British pound, rather than 0.4 UK pounds sterling to the US dollar). The forward market is used for hedging. For example, a UK tour operator will need to pay a large sum in Austrian schillings in three months' time for a number of rooms in Austrian hotels booked as part of holiday packages sold to UK customers. The exchange rate between the UK pound and the Austrian schilling is liable to vary during the three months, and the tour operator may wish to avoid the risk of an adverse movement in the exchange rate (a fall of the pound against the schilling) which would increase the amount of pounds needed to pay for the rooms. The tour operator could enter into a forward exchange contact to buy the schillings in three months' time, at the current three-month forward exchange rate. By hedging in this way, the tour operator would avoid the risk of a loss due to an adverse movement of the schilling against the pound, but would also forgo the possibility of a gain if the movement were in the opposite direction. Refraining from hedging is known as maintaining an open position.

In long-run equilibrium, exchange rates are affected by purchasing power parity and by interest rate parity, which are linked by the International Fisher Effect. The first states that if there is a discrepancy in the purchasing powers of the currencies of two countries, such that the prices of goods and services in Country A are higher than those in Country B when compared using the exchange rate between the two currencies, then the inhabitants of Country A will tend to purchase goods and services as far as possible in Country B. This will affect A's **balance of payments** negatively, and B's positively. Consequently, the spot exchange rate between the two countries' currencies will tend to shift so as to remove the discrepancy in their purchasing powers. This tendency will be reflected in the forward exchange rates.

The two parity models are similar, but in the case of interest rate parity it is the price of money (that is, interest rates) and **investment** flows that are involved. Differences in interest rates between countries would, other things being equal, induce investors to shift their funds from the country with the lower interest rates to that with the higher ones. One reason why this may not happen is the International Fisher Effect; in other words, the country with the higher nominal interest rates typically also has a higher expected inflation rate, so that after correcting for expected inflation, the real interest rates do not differ. Continuing the previous example, a fall in the exchange rate of Country A's currency would increase the prices of A's imports and decrease the prices of its exports. This, in turn, might lead to inflation in Country A. Investors in interest-bearing assets in Country A would expect a higher nominal interest rate so as to be compensated for the inflation (so as to receive the same real rate).

Hence, there is a relationship between the spot and forward exchange rates of the two currencies on the one hand, and the difference between the interest rates applicable to the two currencies on the other hand. This relationship (interest rate parity) is such that the financial outcome of an inhabitant of Country A hedging an amount payable in Country B's currency in (for instance) six months should be much the same as using the money markets to borrow the amount required at A's six-month interest rate, converting it into Country B's currency in the spot market and placing it on deposit at B's six-month interest rate until required to make the payment. If the financial outcomes of using the money market and using the forward exchange market were significantly different, then market operators known as arbitrageurs could make riskless profits by exploiting the discrepancy; and by doing so they would cause interest rates and forward exchange rates to move towards the levels required to produce interest rate parity.

These relationships assume that the currencies involved are fully convertible and not subject to exchange and other currency controls. Because of the potentially disruptive effects of exchange rate variations in international trade, forcing businesses to choose between the costs of hedging and the risks of exchange losses, various attempts have been made to stabilise exchange rates. One such attempt was the Bretton Woods Agreement of 1944, setting up a worldwide system of fixed exchange rates in which the US dollar served as a standard to which all other convertible currencies were pegged. In turn, the US dollar was convertible into gold at a rate guaranteed by the US Federal Reserve Bank. To this extent, the Bretton Woods system resembled a revival of the old gold standard, which had prevailed until it was abandoned in the 1920s. Adjustments to the exchange rates of individual currencies (normally devaluations) were possible on an exceptional basis. For example, in 1948 the pound was devalued from $4 to $2.8, a devaluation of 30 per cent.

Apart from such major adjustments, exchange rates of currencies into the US dollar were expected to vary within a relatively narrow band. In the 1970s, the United States ceased to be able to guarantee the convertibility of its dollar into gold, and in connection with this the system of fixed exchange rates was abandoned. An era of floating rates ensued. Subsequently, rather than a worldwide system of fixed exchange rates, the idea of currency unions has gained some acceptance, either on a regional basis (as with the European Monetary System) or by having certain currencies pegged to a major currency such as the US dollar or the French franc. The European Monetary System is a currency union set up among the member states of the European Economic Community in 1979. Following the Treaty of Maastricht and the conversion of the Community into the **European Union** in 1991, the European monetary system took on the role of an intended first step towards European Monetary Union, or the introduction of a single European Union currency (the Euro) in 1999. The basis of this was a system of bands within which the weighted average exchange rate of each participating currency into the other participating currencies is permitted to vary. Since its inception, Denmark and the United Kingdom left this monetary system, and Italy and Spain devalued their currencies within the European monetary system. Since the euro's introduction in 1999, the exchange rates of the participating currencies have been fixed in terms of the euro.

See also: currency control

SIMON ARCHER, UK

foreign independent tours

Organised by tourists themselves, foreign independent tours or FITs are high **quality** and low **impact** international **travel** to less known and relatively undisturbed destinations, where learning about cultures and appreciating the natural environment are often prime **motivations** (see also **nature**). The self-reliant tourists are often characterised by a desire to interact with the host **community**.

DEBORAH GRIEVE, UK

foreign investment

Foreign capital channelled into a country for tourism **development** is referred to as foreign **investment**. A host country can use different channels to secure foreign investments, including direct investments by entrepreneurs and joint ventures, as well as borrowing from governments, financial institutes or individuals for tourism projects. A method of obtaining capital on a large scale is to issue stocks on international capital markets so as to raise foreign equity.

ZHENG GU, USA

forest recreation

Forests are major sites for tourist **recreation** activities such as **bushwalking** and **camping**. Short and long walks (such as **trekking**) can be undertaken, as can fishing and rafting where there are water resources. Forests are also **sites** for **nature** tourism, **adventure tourism** and **ecotourism**. Where naturally occurring forests have special significance for nature conservation, they may be protected as **national parks**.

DAVID G. SIMMONS, NEW ZEALAND

formulae

Formulae are special expressions with incantatory power. Desire to travel, for example, can be aroused by such imperatives as '**escape**' and 'forget', while attractions may be enhanced by the employment of mystery-laden words (e.g., 'kabash') or the replacement of the humdrum by the enchanting unfamiliar (e.g., 'morne' for 'hill'). Overuse of formulae can reduce their effectiveness, as in the well-worn bliss formulae of island paradises.

GRAHAM M.S. DANN, UK

Fourth World

Indigenous minority peoples incompletely integrated into nation-states constitute the Fourth World. They generally live in geographically peripheral regions and are politically and economically disadvantaged. Often considered 'primitive,' Fourth World people have been a focus of **anthropology** and are a growing **attraction** for **ethnic tourism** and **adventure tourism**. While their handicrafts and culture are increasingly commoditised (see **commercialisation**), the environments which sustain their ways of life are endangered.

ERIC McGUCKIN, USA

France

With 70 million international tourists in 1998, France was the leading destination in the world in terms of arrivals and third in terms of receipts. It owes its success to a diversified supply that is regularly upgraded and to steady demand from all the regions of the world (but mainly from Europe). Although France has often maintained its leading position since the end of the 1980s, the number of arrivals stagnated during the first half of the 1990s at around 60 million, while expenditure on tourism by French **tourists** abroad has increased steadily by an average of 3.2 per cent a year since 1991. However, from 1996 the number of arrivals to France rose steeply, to reach the top position in 1998. Compared with its closest competitors, **Spain** (47.7 million) and the **United States** (47.1 million), it is by far the world's greatest tourist receptor country, with 11.2 per cent of total world arrivals. The Secretariat of State for Tourism reports that **domestic tourism** is also very healthy. Accommodation **surveys** show that bed nights increased by 2.6 per cent in 1998 over 1997 (plus 6 per cent in hotels, plus 11 per cent in rented **second homes** and lodgings, and plus 19 per cent in 'gites').

According to the **World Tourism Organization**, France received $29.7 billion from tourism in 1998, an increase of 6 per cent on the previous year. The **balance of payments** surplus amounted to over $11 billion, the highest of all export sectors ahead of the aerospace and the agriculture and **food** industries. **Employment** in sectors associated with tourism activity in France accounts for more than one million jobs, 75 per cent of which are permanent. In 1997, the government's stated intention was to create 100,000 new jobs in the following three years.

France has a great diversity of **resources**. This creates comparative and specific advantages which explain its success in **international tourism**. France is the only European country with Atlantic and Mediterranean coasts, and also has two mountain ranges, the Alps and the Pyrennees. These tourism resources are well developed. Traditionally the main resources were the coasts, where beach tourism developed in the nineteenth century, and those resources based around **heritage**, **arts** and crafts which represent the origins of **urban tourism** and **cultural tourism**. Recently, mountain tourism, **green tourism**, **business travel**, conventions and congresses and **theme parks** have become established.

French tourism **development** started in the fourteenth century when the first British tourists visited the French Riviera, notably Nice and Monaco, and Pau in the Pyrennees, where the first golf course in France was built. From the onset, the economic activity of tourism was supported by strong domestic demand. During the same period (the Second Empire), tourism in the Basque coast and Biarritz also took off. Thus, its industry is characterised by the importance of its domestic market, which developed **hospitality** and **accommodation** superstructures over the whole territory and which served as a solid basis of the **growth** of international arrivals.

The industry expanded rapidly with the arrival of **mass tourism** in the 1960s and 1970s. During this period new **regions** became established, for instance the Aquitaine and in particular the Languedoc-Rousillion with the seaside **resort** of Cape d'Agde, which has the greatest **hotel** capacity of all the coastal regions in France. This qualitative and quantitative transformation of the supply resulted in an economic boom and the creation of employment in regions which, until then, had been bypassed by **economic development**. However, the phenomenon of overexploitation of the Mediterranean coastal regions has resulted since the early 1990s in a return to such neglected forms of tourism as **nature tourism**, the discovery of the **environment** and **ecotourism**.

The industry comprises 165,000 enterprises in the hotel, café, restaurant, **travel agency** and **tour operator** sectors. The supply is greatly diversified in terms of different categories of accommodation such as hotels, second homes, gites and open-air accommodation. In 1996, hotel capacity in France consisted of nearly 20,000 hotels and more than 600,000 rooms. The region around Paris (Ile de France) accounted for 21 per cent of total rooms, followed by the Rhône-Alps region (13 per cent) and by Provence-Alps-Côtes d'Azur (12 per cent). There are great concentrations of commercial accommodation capacity in the regions where France has its main tourism comparative advantages. There is little supply of high-category hotels outside the region around Paris, and this has a negative effect on France's international positioning in the world market.

If the non-hotel accommodation sector is compared with the hotel accommodation sector, it is clear that the foundation of France's tourism economy is no longer based on the traditional hotel sector. Today, the majority of accommodation capacity is second homes and lodgings. This reflects changing trends of tourism **behaviour** linked to the lengthening of holiday time. As a result, domestic and international tourism demand is increasingly oriented towards more competitive accommodation such as the renting of second homes, which are better adapted for long stays.

The main problem that French tourism will face in the **future** will lie in its ability to adapt to new conditions so as to be in the position of offering creative products to new clientele from Eastern and Central Europe and especially from Asia and South America. Although the Ministry of Tourism would like to attract 20 million additional international arrivals, the economic impact of these extra tourists may be reduced if average receipts fall and, therefore, their contribution may have little effect on the economy as a whole.

France's tourism **policy** took a different direction at the end of the 1980s when it was taken out of the public sector's responsibility and placed under new institutional structures based on an association between the public and the private sectors. Tourism development and promotion policies are now focused on the enterprises in the sector. However, this trend towards a better balance between the public and private sectors is essentially limited to national level. In fact, regional and local tourism policies are still conducted by municipal public authorities, with consultation rather than association with the private sector.

The main objectives of national policies are to ensure greater and more efficient tourism **activities** and to find ways of increasing flows of arrivals. From the beginning of the 1990s until 1996, the number of international arrivals had remained stable at around 60 million a year, which meant that France was losing market share. Moreover, the GDP expenditure allocated to support the tourism sector is substantially higher in other countries with well-established tourism. For instance, the expenditure in **Canada** is three times higher; it is nine times higher in **Spain**, twelve times higher in **Australia** and up to twenty-five times higher in **Ireland**.

Tourism arrival figures in 1998 reflect the **impact** of France hosting the XVIth Soccer World Cup. Although the **statistics** have not been substantiated, according to the then President of the International Federation of Football Associations, France was expected to have generated an extra 30 per cent in tourism from the event alone. The French Secretary of State for Tourism explained in an official press release on 21 January 1999 that the department had feared that staging the World Cup would mean a fall in tourism arrivals over the period, as Spain and **Italy** had experienced when they hosted the event. In fact, WTO statistics show that international arrivals to France increased by 4.7 per cent in 1998 over 1997, nearly double the world's average growth rate. These positive results have been helped by excellent organisational abilities and coordination by the French tourism industry in **partnership** with the Secretariat of State for Tourism.

In summary, the development of tourism in France has created and expanded large international tourism groups such as ACCOR, Club Méditerranée, and Nouvelles Frontières. However, in the context of **globalisation**, these groups are weak because they have no direct links to large international financial groups and thus are vulnerable to takeover bids in the markets. Furthermore, France has not been able to take advantage of **investments** by large American or Asian hotel chains and corporations apart from the notable exception of EuroDisney. The few investments that have been realised, for instance, by Hilton,

Sheraton and Hyatt, have been confined to the region around Paris. Only Holiday Inns has built hotels in some of the larger provincial towns. However, these chains have not created 'resorts' by the seaside as they have in United States and Asia. The future of tourism development in France relies strongly on its capacity to develop a real **internationalisation** of this industry in order to maintain its leadership position in Europe and in the world.

Further readings

Frangialli, F. (1991) *La France dans le Tourisme Mondial*, Paris: Economica.

Ministère du Tourisme (1996) *Statistiques du Tourisme: Partenaires Régionaux*, Paris: Direction du Tourisme.

Py, P. (1992), *Le Tourisme – un Phénomène Economique,* Paris: La Documentation Française.

Trigano A. (1988) *Pour une Industrie Touristique plus Compétitive,* Paris: Conseil Economique et Social.

Vellas, F. and Cauët, J.M. (1997) *Le Tourisme et les Iles,* Paris: L'Harmattan.

Vellas, F. and Bécherel, L. (1995) *International Tourism, an Economic Perspective,* London: Macmillan.

World Tourism Organization (1999) *Tourism Highlights 1999*, Madrid: WTO.

FRANÇOIS VELLAS, FRANCE
LIONEL BÉCHEREL, UK

franchising

In his popular book *Megatrends 2000,* John Naisbitt declares that 'franchising is the single most successful **marketing** concept ever.' A franchise is a contractual agreement whereby a firm allows another business to use its name, to sell and use its **products** for a fee. Franchise associations all agree the practice has several characteristics: it is a method of distributing products and **services**; it involves two parties linked contractually (the franchiser and the franchisee); the franchiser provides the product concept, trademark and patent rights, corporate identity, know-how and **training**, assistance and guidance, perhaps even providing raw materials and equipment depending on the contract, and a marketing plan or the benefits of group distribution; and the franchisee pays a fee to the franchiser for use of the name for an agreed period of time period within a specified territory, as well as for support. Usually, there is a continuing financial obligation either in the form of a regular fee, a proportion of the profits, or both.

When one thinks of tourism franchises, examples in related sectors such as **car rental** and **restaurant** chains come to mind. However, franchising is actually a widespread practice in tourism. **Hotel** chains use franchising to penetrate **markets** quickly and to expand, particularly abroad. Their international reputation guarantees franchisees a ready-made market. In addition, franchisers often sell a management contract to provide the expertise to run the hotel. For example, the Ramada hotel chain operates the Ramada Franchise System. Franchisees benefit from its *Property Management System,* which includes the installation of hardware and software and training courses. Support is provided regionally with advice from franchise service managers, face-to-face professional assistance in **marketing** and training in hotel operations. When an approved marketing campaign is launched, Ramada shares the cost of the effort with the franchisee. Furthermore, all franchisee employees are entered in the *Personal Best History* scheme, a quality award system designed to encourage customer loyalty, where innovative customer service and initiatives are recognised.

As another example, Uniglobe Travel has developed a worldwide travel agency franchise system with over 1,150 franchises in sixteen countries. Uniglobe provides tools, systems, and support such as professional training, a business consultation service from eighteen regional offices, software programmes, automation agreements with **computer reservation system** providers and a *Preferred Supplier Programme* in which a package of incentives and commissions is negotiated with principals. A *Franchise Owner Association* allows Uniglobe franchisees to meet and exchange ideas. As these examples suggest, franchising reduces risks for new business start-ups. According to the US Commerce Department, 95 per cent of franchises succeed in the United States, compared with only around 30% of new independent businesses.

LIONEL BÉCHEREL, UK

freedom

Being free to do whatever one wishes is an important **motivation** for modern **tourists**. Freedom of movement is the human right to move by choice among countries and destinations. However, it has already become a matter of discussion due to negative environmental **impacts** at destinations caused by tourism **growth**.

TANJA MIHALIČ, SLOVENIA

freedom, perceived

Freedom is a complex phenomenon that includes political, social and philosophical dimensions. An important component is the psychological **experience** or perception of acting voluntarily. To have **leisure**, **recreation** or tourism, people must perceive that they have more than one alternative for action in a situation. The perception of freedom is seen as a fundamental need and essential to **health** and well-being.

ROGER C. MANNELL, CANADA

frequent flyer programme

Airlines woke up to a brave new world in 1978. After forty years of protected **markets**, the industry had to learn anew about competition. As new carriers entered the **market** and passengers demonstrated that they largely perceive airline service as homogeneous, competition increased and prices plummeted. Therefore, airlines confronted a classic **marketing** problem, namely how to differentiate their services to the consumer and turn the basis of competition away from price. For American Airlines and others thereafter, the answer was simple: creating an airline and industry version of that venerated marketing tool, trading stamps.

Frequent flier programmes reward passengers for brand loyalty. They accumulate points for using a given airline. Once enough points are accumulated, they can be redeemed for awards such as upgrades or 'free' flights. The programmes have extended well beyond this simple beginning. The first extension was to broaden carrier participation from one carrier to an alliance of carriers. Thereafter, the programme was expanded to other industry segments. Today, such programmes extend well beyond the original industry confines. Credit card companies and long distance telephone services have formed alliances with airlines. Signing up with such a company earns bonus points, while using the card or service earns more frequent flier points. A large array of goods and services earn points in, and can be purchased through, frequent flier programmes. The **United Kingdom** and **Australia** are among the many countries with frequent flier programmes. True **globalisation** probably is not far off. As good as they seem to be, frequent flier programmes are fraught with risks for the carriers and their allies. The costs of redeeming frequent flier awards may soon exceed the value they bring to the carriers; according to some, this may already be the case. Although frequent flier programmes have proven a boon to travellers and carriers, this may not continue.

Further reading

Brancatelli, J. (1996), 'More bang for frequent-flier bucks', *Fortune*, September 30: 278ff.

Brown, M. (1992), 'Fighting for the frequent flier', *Management Today*, April: 85ff.

Browne, W.G., Toh, R.S. and Hu, M.Y. (1995), 'Frequent-flier programs: the Australian experience', *Transportation Journal* 35(2): 35–44.

KEVIN BOBERG, USA

friends and relatives *see* visiting friends and relatives

front stage *see* back stage

fun

Fun refers to the **ludic** dimension of tourism, where childlike tourists enjoy contrived 'as if' experiences. Playful activities range from dressing up as natives to taking a trip to Disney World. Fun appeals to the child in the adult and is associated with the postmodern condition (see also **postmodernism**). The **play paradigm** is prominent in **anthropology** and can be linked to tourist **motivation**.

GRAHAM M.S. DANN, UK

future

The future of the tourism **industry** from a medium and long-term point of view will depend upon the foreseeable **evolution** of its most significant magnitudes and upon the characteristics of the **growth model** envisaged by tourism for this time period. Analyses of possible **trends** suggest that this industry will maintain significant growth levels, higher than international economic rates. Although these will be lower than those in the past, they will still be above the earlier 4 per cent average annual growth rate of **international tourism**.

As a result of future worldwide economic growth and the high income elasticity of tourism **demand**, which is already on the upswing, there will be an increased tendency for travel by all population segments, especially those for whom this practice has not been economically feasible or attractive. From a qualitative point of view, the future of this industry will be characterised by a move away from the **mass tourism** model which started during the 1980s and the 1990s. This period, which marked the end of the industrial age, was characterised by the Fordian mass production model. During this time, the **tourist** bought a standardised product which lacked flexibility, had no distinguishing features, and was developed and consumed with little regards for the host cultural and natural heritage (see **standardisation**).

Future tourism models will be based on the development of new forms of **information technology** and the appearance of a new consumer. The next tourist generation will be more experienced, augmented to a new level with more spontaneity and greater desire to stand apart and away from the rest. Sociodemographic changes and shifts in **lifestyles** will have a decisive influence on future tourism trends. Environmental **conservation** will be a necessity in order to guarantee success of this industry, especially that the tourist of tomorrow will have greater desire for **nature** and all sorts of opportunities which such settings can offer. In the meantime, tourism sectors and their many businesses must get ready and adapt to the new and emerging **product** and market developments, including even adventuresome space travel, as limited as it might be. More innovative and flexible models of **management** and organisation must be developed, capable of making full use of different types of information technology and of offering tourism goods and services which are much more segmented and specialised. Such measures will allow tourism to maintain its lead industry position worldwide and to even attain a higher socioeconomic and political magnitude and status worldwide.

Further reading

Edwards, A. and Graham, A. *International Tourism Forecasts to 2010*, London: Travel and Tourism Intelligence.

Poon, A. (1993) *Tourism, Technology and Competitive Strategies*, Wallingford: CAB International. (Offers an analysis of various aspects of a new model of tourism development.)

EUGENI AGUILO, SPAIN

G

gambling

Gambling is an activity that was broadly legalised in many parts of the world during the last three decades of the twentieth century. Inspired by the phenomenal **growth** and economic success of Las Vegas, many other states, provinces and countries authorised a wide variety of forms of legal gambling – including casinos – for a variety of reasons, tourism **development** being one.

When gambling is taken from an illegal to a legal status, it taps into a strong latent **demand** in the population to spend time and money on the activity. In the **United States**, for example, lotteries were prohibited in all states prior to 1964, and casino gaming had been permitted only in Nevada prior to 1978, and then only in Nevada and New Jersey between 1978 and 1989. However, lotteries were authorised in thirty-six states and the District of Columbia between 1964 and 1996, and the number of states offering casinos increased from two to more than twenty-five by the latter half of the 1990s. A wide variety of types of casinos were legalised, including riverboat casinos, small stakes mining town casinos and Indian casinos. The size of commercial gaming in the United States, as measured by 'gross gaming revenues' – total revenues after payment of prizes, or aggregate customer losses (equivalent to expenditures) – grew from $10 billion in 1982 to $45 billion in 1995 (Christiansen 1996). Approximately half of gross gaming revenues were generated by casino style gaming, with lotteries accounting for another 30 per cent.

Other countries had similar experiences as the United States. **Australia** introduced its first legal casino in Tasmania in 1972. By 1997, casinos could be found in every major city in the country as well as in a number of established tourism centres, such as the Gold Coast and along the Great Barrier Reef. Furthermore, every Australian state besides Western Australia authorised gaming devices outside of casinos by 1997, and every state permitted lotteries. Canadian provinces established lotteries throughout the country in the 1970s. They then introduced government-owned monopoly casinos in various urban and destination **resort** areas, including Montreal, Halifax, Winnipeg, Niagara Falls and Windsor, between 1989 and 1997. Prior to that, western provinces had established so-called charitable casinos – small stakes, table games only – beginning in the late 1970s.

South Africa inherited seventeen casinos from the former 'homelands' when the Mandela presidency began in 1994. These casinos had been developed through arrangements between a single company, Sun International, and homeland rulers from 1979 to the early 1990s. The national government passed **legislation** in 1996 which would expand the number of casinos in South Africa to forty, spread throughout the country. Legislation would also permit slot route operations – gaming devices in locations outside of casinos – and the government also discussed permitting lottery. Other countries have also succumbed to the allure of gambling and the economic stimulus that might follow. Major casinos developed in **Spain** and the **United Kingdom** in the 1970s. More recently, casinos have opened in such diverse locations as **Greece**, **New Zealand**, **Poland**, the **Czech Republic**, **Peru** and various republics of the former Soviet Union. The National Lottery – claimed to be the largest in the world – was introduced into the United Kingdom in 1994.

The form of gambling that is most associated with tourism is casino gaming. Other forms – such as lotteries, wagering on racing, charitable gambling and non-casino gaming devices – cater predominantly to **local** markets and thus have little direct **impact** on tourism or tourism development. On the other hand, famous historic casino centres such as Las Vegas, Monte Carlo and Macao have attracted **tourists** from neighbouring or distant states or countries as their main source of business. Indeed, Las Vegas, which by 1996 was attracting 30 million tourists per year to its 100,000 **hotel** rooms and myriad casino and entertainment facilities, had become an ideal tourism **destination** resort, centred around casinos.

The spread of casinos in the United States and in other countries has occurred for a number of reasons. First, lotteries generally preceded permitted casinos and whetted government appetites by demonstrating the revenue-generating potential of gambling. Lotteries also softened the public's attitude regarding the risks associated with gambling. The major arguments against permitting 'hard' gambling, such as casinos, centre around three types of argument: links between gambling and criminal influences, claims of the immorality of gambling, and the social consequences of problem and pathological gambling. The first of these arguments has diminished in many countries as casino companies have become more mainstream and respectable. Furthermore, regulatory bodies have become more effective in fulfilling public policy mandates and in protecting a variety of public concerns. Morality arguments against gambling have diminished in many jurisdictions, perhaps reflecting the diminishing influence of organised religion in many societies. The issue of problem or pathological gambling remains as the most significant challenge which confronts permitted gambling, especially to casino-style gaming.

It is often the economic dimension that has been the driving impetus behind permitting most forms of gambling, especially casinos. However, the desired outcomes of **economic development** and tourism stimulation have not been universal. In order for significant economic stimulation to occur, a large proportion of custom must come from outside the region where the casinos are located. Alternatively, locally supplied casino facilities need to have a substantial impact on local residents who would otherwise leave the region in order to gamble. In general, most customers of urban casinos have not been tourists. Many new casino jurisdictions in the United States provide 'casinos of convenience' which cater predominantly to residents of the area where the casinos are located. In such cases, there has been little net economic stimulation to the area, though the casinos themselves have been substantial revenue generators. The same pattern is observable with casinos in most other countries that have recently authorised casinos.

The long-term **future** of casinos and other forms of 'hard gambling' depends to a great extent on society's acceptance of gambling as a legitimate consumer pursuit. The last half of the twentieth century has been characterised by a steady increase in the degree of acceptance of the general public in the activity of gambling, in spite of the fact that the other vices – **alcohol**, tobacco, illicit drugs – have been under increasing criticism and sanction in much of the world. Technology will increasingly bring gambling into the home (through the **Internet**, through interactive **television** wagering systems) whether or not it is legally sanctioned. If permitted gambling continues to expand in society, then the role of gaming in tourism will likely decline unless gambling – especially casino-style – becomes part of a wide range of entertainment offerings. This is a formula that has been well developed by Las Vegas, but not in many other locales. Casinos have been tourism generators primarily because of prohibitions of gambling in places where people live. If that prohibition disappears, then much of tourism-based gambling will diminish as well.

References

Christiansen, E. (1996) 'The 1995 gross annual wager,' *Gambling and Wagering Business Magazine* 17(8): 53–92.

WILLIAM R. EADINGTON, USA

game park reserve

Game park reserves are owned either privately or governmentally for the purpose of limited non-economic and economic human interaction with wildlife and their habitats found within their boundaries. Hunting for sport is sometimes permitted, but most reserves protect the habitat and animals from destruction and permit only the viewing and **photography** of the wildlife found therein.

CAROLYN M. DAUGHTERY, USA

game theory

Game **theory** is a mathematical modelling tool used in various social science applications. Unlike traditional economic **methodology**, economic and business game theory models do not ignore the influence of an individual's actions on others and vice versa. Therefore, the methodology is useful in modelling tourism and **hospitality** situations where individuals and firms operate in an interdependent, competitive environment.

In a game theory **model**, each decision maker (player) has a plan of action (strategy) and is assumed to be a utility (payoff) maximiser. Four categories – perfection, symmetry, certainty and completeness – define the type of information the players use in forming their strategy. Finding a solution to a game often involves calculating an equilibrium, that is, finding a strategy profile in which each player's strategy maximises the payoff, given that the others are using their own strategies.

Several studies in the field of game theory provide interesting insights into tourism and hospitality phenomena. For example, a game model of convention pricing shows how convention planners and the hotel 'agree' on room rates, and reveals the conditions under which the hotel is more likely to maximise profitability when using **yield management**. Franchising, a dominant form of contract among hospitality firms, is better understood thanks to another game theory model that explores the role of royalties (the revenue-sharing agreement) and monitoring (the franchiser effort to ensure the prescribed business format). Game theory might prove useful in modelling **service** quality in **restaurants** and hotels where information asymmetry causes difficulties in trade. These difficulties and the adequate remedies for them can be successfully analysed using tools provided by game theory. Other promising areas for applications include mergers and acquisitions in the lodging industry, location and product differentiation of planned attractions, **advertising** and promotion of destinations, and the optimal structure of distribution channels of tourism packages (see also **promotion, place**).

Further reading

Lal, R. (1990) 'Improving channel coordination through franchising', *Marketing Science* 9(4): 299–318.

Schwartz, Z. (1996) 'A dynamic equilibrium pricing model: a game theory approach to modelling conventions' room rate', *Tourism Economics* 2(3): 251–64.

—— (1997) 'Game theory: mathematical models provide insights into hospitality industry phenomena', *Hospitality Research Journal*, 21(1): 48–70.

Shubik, M. (1984) *Game Theory in the Social Sciences – Concepts and Solutions*, Cambridge, MA: MIT Press.

ZVI SCHWARTZ, USA

gastronomy

Gastronomy comes from the title of a fourth century BC Greek poem 'Gastronome', about **food**. In modern usage, the term refers to the art and science of fine dining. It can also indicate culinary **rituals** or consuetudes, particularly those of a sophisticated or highly developed character. **Restaurants** at destinations known for their **native** foods and **cuisine** become tourism **attractions** in their own right.

MICHAEL NOWLIS, FRANCE

gateway

A gateway is a port of entry or exit that functions as an international access point. **Border** crossing points, seaports and **airports** are the most common examples of gateways. Urban centres through which **tourists** enter or exit a country are referred to as gateway cities.

DALLEN J. TIMOTHY, USA

gaze

There is considerable debate in the **sociology** of tourism as to the social and cultural bases of tourism. One important strand has been to emphasise the role of the senses in generating different tourism practices. In particular, it is argued that the visual sense has been especially significant in organising the development of Western tourism since the end of the eighteenth century. It is from this time that one can date the origins of scenic tourism and more generally the development of an **industry** based upon the collecting of 'sights'. An enormous array of technologies, discourses and organisations have developed to expand and profit from people's desire to go away from their normal place of residence and to collect sights through the visual consumption of other places.

The visual gaze of **tourists** has enormous consequences upon how tourism should be organised (the need for 'rooms with a view'); on the places which are the object of such a gaze, which may feel endlessly under surveillance and hence may produce a **staged authenticity** away from the prying eyes of visitors; and on the industries which have developed to facilitate and augment such gazes (guidebooks, **postcards**, **photography**, **television** programmes, **souvenirs** and so on).

The gaze of tourists is socially organised and not simply individually variable. Analysis has developed as to how such socially significant ways of seeing vary historically. One important distinction has been drawn between the romantic and the collective tourist gaze. The former involves a solitudinous and lengthy gaze upon aspects of nature such as mountains, lakes, valleys, deserts and sunsets, which are treated as objects of awe and reverence. Other people are viewed as intrusive, as a solitary and contemplative gaze is sought. In response, the industry has had to discover new places for tourists to be able to gaze at in solitude and to organise the gaze of existing sites in new ways in time and space.

By contrast, in the collective gaze, people view a particular site together. It is other people who give sense and order to the gaze and the absence of other people make the site deserted and lacking appropriate atmosphere. Piers, towers and promenades have been developed on the basis of the collective gaze.

Further reading

Adler J. (1989) 'Origins of sightseeing', *Annals of Tourism Research* 16: 7–29. (Discusses the origins of sightseeing and the gaze.)

Jay, M. (1993) *Downcast Eyes*, Berkeley, CA: University of California Press. (On the significance of the visual sense in Western thought.)

Urry, J. (1992) 'The tourist gaze "revisited"', *American Behavioural Scientist* 36: 172–86. (On the relationship of the concept of the tourist gaze to the ideas of Michel Foucault.)

JOHN URRY, UK

gender

Throughout the social sciences, feminists have argued that gender is a social construction which draws on certain aspects of biological sex. From the moment babies are born, they are treated differently because of their sex. This differential treatment continues throughout their lives, from the toys girls and boys are given to play with to the jobs that are considered appropriate for women and men. As a result, the male and female sex are gendered by society. Boys and men are expected to exhibit masculine attributes, whereas girls and women should exhibit feminine characteristics. Individual boys, girls, men or women who do not conform to their respective masculine or feminine characteristics laid down by their society are often ridiculed and seen as unnatural (WGSG 1997).

For feminists, one of the most important points to make regarding the social construction of what it means to be male and female in a particular society is that most often the masculine and feminine characteristics attributed by society are defined in relation to one another. Boys are expected to be boisterous, girls compliant; men are expected to lead, women to follow. This oppositional relation is particularly important because each side of the pair is not equally valued. Most often, masculine attributes are valued more than feminine ones:

> Thus, it is not just the case that males and females are gendered differently; rather they are gendered differently and as a result valued differently. The social construction of the two genders relate in a way that works to the general advantage of men and to the general disadvantage of women.
>
> (WGSG 1997: 52–3)

Why does gender matter to tourism? As a leisured goal and the **industry** that supports it, tourism interacts with global and local constructs and conditions of gender relations. Tourism is a major component of the world **political economy**, affecting massive **migration** over national boundaries. Gender is a primary factor in the industry in terms of **employment**, consumption patterns and cultural representations of place. Gender, within the context of tourism, can thus be seen as a system of culturally constructed identities, expressed in ideologies of masculinity and femininity, interacting with socially structured relationships in divisions of labour and **leisure**, sexuality and **power** between women and men.

Recent articles on gender and tourism studies (Swain 1995; Kinnaird and Hall 1996) attest that while **feminism** is not often named, it does indeed sometimes inform tourism inquiry. In this literature we find ideas behind feminist theorising about sex, gender and sexual difference as sites of distinct identities and hierarchies. The gendering of tourism's attractions, commodities and work is an issue of great importance which raises many questions, including cultural relativity, hierarchy, subjectivity, aesthetics and human rights. Issues of positionality and representation in ideas about sex and gender are also important insights from feminism for tourism inquiry.

The literature pertaining to gender within tourism inquiry is relatively recent. However, one can begin to identify the ways in which researchers have and are working with the concept of gender. First, issues surrounding gender as a concept are about women. One good example of this genre is the focus on 'work'. There have been a number of case studies presented (see for example in Kinnaird and Hall 1994; Swain 1995; Sinclair 1997) which outline the ways in which women are working within various facets of the industry. All of these studies are location specific and attempt to highlight the efforts of women in contributing their labour to meet the objectives of the local, national and **international tourism** businesses. Time and time again, the hierarchical and unequal nature of tourism-related work is highlighted. Men tend to dominate those occupations at the top of the hierarchy, while the labour of women is often classified as semi-skilled or unskilled. Second, tourism is seen by many, particularly in areas developing or changing their tourism industry, to have profound implications on gender roles and relations. The work of men and women and their respective positions in society may change. Examples can be found in Swain's (1993) work in **Mexico** and **China**, Kousis's (1989) work in **Greece**, or Cukier *et al.*'s (1996) work in Bali, where the roles and responsibilities of women may be changed or be re-emphasised as part of the processes of tourism development. Third, the representation of a place through tourism strategies can be analysed from a gendered perspective. **Brochures** can evoke a gendering of representation and the tourism product can itself be a representation of either feminine or masculine (dominant) ideology (see for example Edensor and Kothari (1994) on the masculinisation of Stirling's heritage). Finally, feminist ideas about the body and self are emerging in the tourism literature. Soile Veijola and Eeva Jokinin's 'The body in tourism' (1994) moves in the direction of 'sexing the tourist'. They review prominent tourism theorists, exposing the exclusively masculine nature of 'the tourist's body' and experiences in these constructs: there is no gender here, just men. They also offer a discussion of the relationship between the **tourist** and the **Other** which, as object of the **gaze**, is rarely embodied or analysed from a feminist perspective.

Gender, as a concept within tourism studies, is analysed and worked with in a number of different ways. This is largely due to its association with a variety of feminist theories which enlighten gender inquiry. However, gender-aware frameworks for inquiry do offer the opportunity to think through a number of complexities that are involved within the processes of tourism **development**. From the values and activities of the transnational operator to the differential experiences of individuals participating as either hosts or guests, all parts of the tourism **experience** are influenced by the collective understanding of the social construction of gender.

References

Crukier, J., Norris, J. and Wall, G. (1996) 'The involvement of women in the tourism industry of Bali, Indonesia', *The Journal of Development Studies* 33(2): 248–70.

Kinnaird, V. and Hall, D. (eds) (1994) *Tourism: A Gender Analysis*, Chichester: John Wiley.

Kousis, M. (1989) 'Tourism and the family in a rural Cretan community', *Annals of Tourism Research* 16(3): 318–32.

Swain, M. (1993) 'Women producers of ethnic arts', *Annals of Tourism Research* 20(1): 32–51.

—— (ed.) (1995) Special Issue of *Annals of Tourism Research* 22(2).

Veijola, S. and Jokinen, E. (1994) 'The body in tourism', *Theory, Culture and Society* 11: 125–51.

VIVIAN KINNAIRD, UK

geographical information system

Geographical information systems (GIS) are computer systems with facilities for the storage, retrieval, manipulation and display of spatially-referenced, usually mapped, information. They can be used to organise, integrate and manipulate a wide variety of spatial and non-spatial information that is required to address land use **planning** issues, including those for tourism. Location serves as the common attribute for linking data that may differ substantially in their source, thematic context, digital format and geographic extent. Spatial data are those which describe the geographical or coordinate attributes of entities represented symbolically using graphic primitives such as points, lines, grid cells or irregular polygons.

The graphical display of data in map form can enhance the abilities of individuals to identify patterns in data or features, to conceptualise spatial relationships among different phenomena, and communicate ideas to others. In contrast to printed **maps**, this form of data presentation is highly flexible as map content and associated information and analytical techniques can be changed relatively easily. Further, the processes of exploring, understanding and communicating the ramifications of complex **policy** alternatives can be aided greatly through visualisation of distributions using GIS.

Although not yet widely used in tourism, the potential applications of GIS are large and diverse. Examples of their application include the identification of areas which meet criteria suggesting suitability for **ecotourism**, the availability and manipulation of data for computer-assisted travel counselling, and land use planning in **destination** areas.

Further reading

Boyd, S.W. and Butler, R.W. (1996) 'Seeing the forest through the trees: using geographical information systems to identify potential ecotourism sites in northern Ontario, Canada', in L.C. Harrison and W. Husbands (eds), *Practicing Responsible Tourism: International Case Studies in Tourism Planning: Policy, and Development*, New York: John Wiley and Sons, 380–403.

Loban, S. (1997) 'A framework for computer-assisted travel counselling', *Annals of Tourism Research* 24(4): 813–34.

GEOFFREY WALL, CANADA

geography

Geography can be viewed as the study of the earth as the home of human beings. While the habitat can be regarded as an object, geography is not usually considered to be an object-centred discipline. Rather, it is distinguished by the approaches taken by its practitioners. Four such approaches are widely recognised, and all can potentially be applied to the examination of tourism.

The first approach is human–environment interaction. Geographers adopting this approach are concerned with both the implications of the **environment** for human activity and the **impact** of human beings on their environment. These two emphases together account for the bulk of geographic tourism research. Much tourism is directed to very special places, such as historical monuments and **national parks**, and geographers have been interested in the characteristics of these places and how they have been modified by this **industry**. They have explored the **images** which people hold of such places and how these are created. Further, they have also been in the forefront of research on the impacts of tourism, whether these be of an environmental, economic or social nature.

The second geographic approach is spatial analysis. This is interested in the real distribution of phenomena, in this case aspects of tourism, and in the flows of people, **products**, information and money which unite areas. Such analyses attempt to describe, explain and predict patterns of tourism phenomena including the origins, routes and destinations of the **tourists** themselves as well as other associated features such as **restaurants** or different types of **accommodation**. The movement of people from place to place and the development of associated **transportation** networks have been major geographical concerns.

The third approach is regional synthesis. Geographers interested in this approach attempt to understand the character of places and the **landscapes** which have resulted from the interaction of natural processes and human activities, whether they be at the scale of world regions such as the Mediterranean coasts or the Alps, or as applied to smaller areas such as specific **resorts**. They attempt to delimit areas with common human or physical characteristics, such as a similar **landscape** or a similar dependence on tourism, and to understand the relationships which exist in particular places between tourism and other activities, such as agriculture, forestry or urbanization. There is a close link between geography, tourism and **anthropology**, for many studies are geographically framed and culturally informed. The delimitation of regions for **planning** or **marketing** purposes is also a geographical activity.

The fourth approach is physical processes. Physical geographers study the forces which act to modify the surface of the earth, such as water, wind and ice. Tourism, being a human phenomenon, does not generally attract the attention of these geographers. However, **climate** and weather significantly influence both the nature of resources and the **quality** of experiences available at particular times in specific places. Processes of erosion and deposition in coastal and mountain areas may have considerable implications for people both as tourists and as tourism suppliers. The applied aspects of physical geography thus merge with the human–environment perspective.

There has always been a close relationship among tourism, exploration and the description of foreign lands and their peoples. However, the first academic geographical publications specifically on tourism date from the 1930s. Early work consisted of the description of the **landscapes** of destination areas and, as geographers became increasingly aware that tourism resulted in the creation of towns with particular characteristics, blossomed into investigation of **resort** morphology and recreational business districts. This is still a prominent research area among geographers. However, interest in tourism in urban areas has expanded to include investigations of tourism in large cities, economic restructuring, urban heritage and special events. Geographers have also examined **summer cottages**, **parks** and protected areas, and **wilderness** travel and experiences in rural and remote areas (see **rural tourism**).

Geographical work on tourism was spearheaded in Europe, although American geographers, often working in government agencies, played a prominent role in the examination of the perception, use and **management** of **wilderness** areas. Much of this work was concerned with user **conflicts**, the impacts of visitors on fragile environments such as **national parks**, and the notion of **carrying capacity** as viewed from both environmental and social perspectives. Although of great relevance to tourism, much of this work was seen primarily as a contribution to the understanding of outdoor **recreation** and there has often been an unfortunate and unnecessary schism, particularly in the United States, between those working on recreation and those involved in tourism.

Geographers have displayed a strong interest in the role of tourism in **economic development** and have played a prominent role in investigations of tourism in developing countries and in the debates concerning tourism and **sustainable development**. They have also been among those adopting an evolutionary approach to the study of tourism. Thus, they have been prominent in research on the history of tourism, conducting work on phenomena such as **spas** and the **Grand Tour**. They have also been major contributors to the introduction, debate and testing of tourism cycles and, on a different time scale, **seasonality**.

As already noted, the spatial approach to geography has been associated particularly with the description, explanation and prediction of patterns of tourist movements at a wide variety of scales from global to local. At times this has required the compilation and mapping of large quantities of information on tourism networks and flows. It has also involved the examination of such patterns using the gravity **model** and its derivatives, which are used to describe and project patterns of movement between origins and destinations based upon population size and intervening distances. Such work has involved the development of methods to measure the **attraction** of destination areas and, more generally, the quality of landscape and scenery. Concepts such as **distance decay** and core–periphery relationships, in which the core is viewed as an area of **demand** and the periphery as an area of **supply**, have been applied to tourism in a search for widely applicable generalisations. Use of computer technology for storing, analysing and retrieving large data sets and associated techniques such as computer cartography, **remote sensing** and geographical information systems are finding increasing application in geographical tourism research and planning.

Many national and international organisations include sections or working groups which specialise in tourism. An example of the former is the **International Geographical Union** Study Group on the Geography of Sustainable Tourism: Development and Protection of Cultural and Natural Heritage, and examples of the latter are the Recreation, Tourism and Sport Specialty Group of the Association of American Geographers, and the Parks, Recreation and Tourism Study Group of the Canadian Association of Geographers. All of these groups produce newsletters. Many university geography departments offer courses on tourism and encourage students to specialise in it in their graduate degrees, so that many dissertations have been written by geographers on tourism.

Today, geography is a highly pluralistic discipline with no dominant perspective or philosophical approach, and its practitioners adopt a wide variety of research methods. Geographers have tended to be eclectic in their research and teaching leading to charges that the discipline as a whole, and its tourism scholars especially lack cohesion and focus. They have dealt with many aspects of tourism and, although often fragmented, the literature has an underlying sense of unity when viewed from a spatial perspective and provides a substantial base for the construction of a geography of tourism. Geographers have become skilled at synthesising numerous causal factors as an aid to understanding the complexity of tourism phenomena, particularly its consequences for special environments in general and destination areas. This has permitted many geographers to become involved in impact assessment and to participate as consultants and practitioners in the evaluation of many tourism issues. Therefore, their help is often sought in planning exercises involving tourism. The proclivity to synthesise large quantities of diverse information derived from many disciplinary perspectives has also permitted them to be among the more prolific producers of general tourism texts.

See also: borders; migration; scale of development

Further reading

Mitchell, L.S. (1979) 'The geography of tourism', *Annals of Tourism Research* 6(3): 235–44. (Provides a concise introduction to the study of tourism

employing a framework based on supply, demand and linkages between them.)
Mitchell, L.S. and Murphy, P.E. (1991) 'Geography and tourism', *Annals of Tourism Research* 18(1): 57–70. (An overview of the links between tourism and geography and the accomplishments of geographers with respect to tourism.)
Murphy, P. (1980) 'Tourism management in host communities', *Canadian Geographer* 24(1(: 1–80. (This special issue is devoted to papers on tourism written by geographers.)
Pearce, D.G. (1987) *Tourism Today: A Geographical Analysis*, New York: Wiley. (Describes tourism at various scales from international to local and the measurement of spatial variations in tourism.)
Shaw, G. and Williams, A.M. (1994) *Critical Issues in Tourism: A Geographical Perspective*, Oxford: Blackwell. (Gives considerable insight into tourism as an industry, tourism employment and the role of tourism in economic restructuring.)

GEOFFREY WALL, CANADA

geography, recreational

Recreational geography is a major research speciality of **geography**. Its roots can be found in geographic publications of the 1920s; however, its flowering occurred during the Great Depression with the publication of an article on the topic of land use. During its early years, tourism as well as **recreation** was investigated, and sports geography was incorporated into the subject in the 1960s (see also **sports, recreational**).

Recreational geography is defined as the methodological study of the passive and dynamic configurations and procedures of **leisure** activities that are non-basic in nature and found in proximity to users. This **definition** distinguishes the study of recreation from tourism. The latter differs from recreation with regard to economic base theory, distance, and/or travel time to destinations and the need for tourists to secure **accommodation** and **food**.

Quantitatively, recreational geography has followed a normal S-shaped growth curve. Increases in published research and papers presented at professional meetings was incremental from 1930 to 1959. During the 1960s there was a quantum leap in formal research, and rapid acceleration persisted through the 1970s. Increases continued in the 1980s and 1990s, but at a slower pace. This evolutionary process reflects the maturation of recreational geography as a research speciality. The application of research findings to practical problems such as statewide recreation plans is evidence that non-geographers value the contributions of recreational geographers. Acceptance of manuscripts by multidisciplinary journals and recognition of recreational geography by the **International Geographical Union**, the Association of American Geographers and the Canadian Association of Geographers is evidence of the quality of the subdiscipline.

Recreational geography is a pluralistic subdiscipline which examines a multitude of topics, systematic and regional as well as pragmatic and theoretical. Research efforts have concentrated on user oriented (that is, urban) recreation with emphases on perception, participation, **planning**, **market** areas (such as hinterlands), market access, market segments, **resort** morphology, individual **activity space**, **recreation business districts** and travel gradients. Additional areas of interest are **site analysis**, **carrying capacity**, open space, **second homes** and a sense of place.

The nature of recreation geography is characterised by five factors. First, research is not concentrated on a few topics but is diverse. Second, there is no research **paradigm** to influence the direction of investigations. Third, examination of research stresses the unique case rather than the general situation. Fourth, studies that replicate and/or verify original findings are not common. Fifth, the complexity of recreation experiences and the lack of data bases explain ideographic tendencies in the literature.

See also: recreation experience

LISLE S. MITCHELL, USA

Germany

The reunification of Germany in 1990 gave a new base for tourism information and **policy**. Although there has been a rapid assimilation in the type and scope of tourism activity in the former East and West, data can conceal important differences.

Out of a population of 63 million aged 14 and over, some three-quarters now take one or more main holiday trips (of at least five days long) a year. Although this travel intensity seems to have levelled off, a significant factor since reunification has been a rapid rise in the number of second and third main holidays. In addition to these main holidays, Germans undertake some 60 million short breaks per annum. Ten years after reunification, there is little difference between eastern and western Germans with regard to the percentage who take one or more main holidays per year, with roughly one-third being to domestic destinations (or 20 million holidays) and two-thirds to foreign destinations. Total expenditure on tourism has risen consistently since the 1960s and was estimated to be $780 per person per holiday in 1998. In the same year, tourism employed 2.6 million people (more than double the number of any major manufacturing industry). In addition to providing some of the world's most active tourists, Germany offers world class congress or convention facilities, **trade shows** (in sectors such as toys, books, tourism, cars and home electronics) and festivals (such as 'Fasching', Christmas markets, beer and wine).

With increasing disposable income and foreign purchasing power, Germans have steadily turned their backs on their homeland as a destination. Less than one-fifth of young singles and couples holiday in Germany (the figure rises to one-half for senior citizens) and less than one-third of all first main holidays are spent in Germany (the figure rises to 50 per cent for third main holidays). When Germans do holiday in their own country, the private car is by far the most popular means of **transportation**, resulting in huge and predictable traffic jams at peak periods throughout the year, despite one of the best motorway networks in Europe and the German zeal for environmental awareness. The most popular destination for **domestic tourism** is Bavaria, with a quarter of all domestic main holidays. Most areas have suffered falling numbers of domestic tourists, but Mecklenburg-Pre-Pommerania has bucked the trend with a sizeable increase; it is also the only area in the former East Germany in which West Germans holiday in any significant numbers.

The annual exodus to the sun and a predilection for winter sports have established themselves as major aspects of German tourism **behaviour**, and since reunification residents of the former East Germany have quickly reached western levels of foreign travel. The tourism balance of trade deficit has increased consistently since the 1960s and is expected to reach $27 billion in 1999. Germans prefer holidays abroad because they see Germany as having poor weather, being poor value for money compared with cheap foreign destinations, and being unexciting and not offering sufficient change from daily routine and location. Their most popular destinations are **Spain** (9 million holidays in 1998), **Italy** (5.9 million), **Austria** (4.4 million), **France** (2.7 million), **Turkey** (2.3 million), **Greece** (2.2 million), the **Netherlands** (1.5 million), **Denmark** (1.2 million each) and **Switzerland** (1.1 million). The most popular long-haul destinations are the **United States** (1.8 million), **Tunisia** (0.8 million), **Canada** (0.42 million), **Dominican Republic** (0.4 million), **Egypt** (0.39 million) and **Thailand** (0.38 million).

Germany has seen its share of world tourism drop from fifth place in 1985 to being no longer in the top ten in 1996. The main countries of origin are the Netherlands, the United States, the United Kingdom, **Sweden** and **Japan**. The most frequented destinations for incoming tourists are cities rich in tradition such as Munich (especially for American, British, Japanese and Austrian visitors), Berlin (American, French, Dutch, Swedish and Italian visitors) and Heidelberg (Americans, British and Japanese). Although Germany can justly lay claim to its image of a **romantic** country with abundant **landscape**, architectural **heritage** and culture, the German people still have some way to go in establishing an international image as service-friendly, considerate hosts.

There has been a consistent **trend** since the 1960s towards booking through agencies. Although Germany has traditionally had a large number of small family-run hotels, small-scale **tour operators** and independent **travel agencies**, fiercer competition, investment in **technology** and a trend to last-minute booking are squeezing profit margins and causing some high-profile bankruptcies. Four of the five biggest tour operators in Europe are German, and these 'big four' – TUI, NUR, LTU and DER – are steadily increasing their market share.

Until reunification, government attitudes in West and East Germany diverged: the former allowed a relatively free market economy, whereas the latter pursued a **policy** of directing and restricting tourism and leisure. Since reunification the Western *laissez-faire* approach has prevailed with a minimum of **government** interference. In the federal government, the interests of the tourism sector are scattered throughout almost every ministry with a consequent shortfall in overall coordination. At federal level the government merely attempts to set out basic parameters which will promote and protect tourism, although reunification provided a stimulus to massive investment in the former East Germany, much of which was consciously aimed at improving **infrastructures** with tourism in mind. At state (Land) level, each government holds responsibility for regional economic measures to develop the tourism infrastructures, health and spa resorts (which are municipal operations), **social tourism**, and support for tourism **marketing** and research projects. At local level, each municipality conceives and pursues its own **policy** through a local municipal tourism association. The trade works closely with government and local authorities on all three levels, but the diversity of interests has often led to disparate and even conflicting policies. The main trade **associations** are the DTV (German Travel Association), the DBV (German Spa Association), the DEHOGA (German Hotel and Restaurant Association) and the DRV (German Travel Agencies Association), and they have formed the DZT Standing Committee to coordinate policy with the German NTO. The DZT has been responsible for marketing Germany abroad, but from 1997 it has also been responsible for coordinating **internal marketing**. The lion's share of its funding comes from a federal government subvention, but there are plans to reduce this progressively from $22.5 million in 1996 to $15 million in 2000. Hard times lie ahead.

Further reading

British-American Tobacco/Freizeit-Forschungs institut (1997) *7. Gesamtdeutsche Tourismusanalyse* Hamburg, BAT. (An analysis of tourism motivation and intentions.)

Forschungsgemeinschaft Urlaub + Reisen, *Urlaub Reisen*, Hamburg: Gruner & Jahr. (Reports providing the basis for statistical information.)

Freyer, W. (1995) *Tourismus*, Munich: Oldenbourg. (A systematic reference work on tourism with constant references to the German market.)

Hill, R. (1993) 'Tourism in Germany', in W. Pompl and P. Lavery (eds), *Tourism in Europe*, Wallingford: CAB International, 219–41. (An analysis of structures and trends in German tourism.)

RICHARD HILL, GERMANY

ghetto

Traditionally referring to the area of an Italian or other city where Jewish people lived and to which they were confined, the term ghetto is now more commonly used to describe any area or quarter of a city inhabited by particular racial or other identifiable groups. It also implies that the area or group in question is to some extent segregated from the rest of the city or **community**. Therefore, a tourism ghetto may be defined generally as that part of a **resort** or **destination** primarily set aside or utilised for tourism and where there is little contact between **tourists** and **local** people other than those who work within the tourism **industry**.

More specifically, either purpose-built destinations in tourism development areas (such as Cancun in Mexico) or all-inclusive club-style **resorts**, such as the Club Méditerranée or Sandals holiday complexes, are frequently referred to as tourism ghettos. In both cases, the resorts are usually segregated or built at a distance from existing communities, there is often little or no opportunity (and, for many tourists, little need or desire) for contact between visitors and local people and, in effect, the resorts are little more than annexes of tourism-generating countries attached to destination areas. As a result, tourism ghettos are frequently criticised for alienating tourists from local people and culture, thus minimising the chance for meaningful encounters and exchange (see **alienation**). At the same time, they also bring little economic **benefit** to local communities as visitors have few opportunities to spend money on local goods and services; whereas resort income is maximised, the majority of tourist spending occurring within the (often foreign-owned) resort complex.

Conversely, tourism ghettos may be considered an effective means of reducing the negative **impacts** of tourism. The damage from tourism's **development** is restricted to relatively small areas of frequently non-productive land and, by segregating tourists from the local population, the chances of confrontations and the longer-term process of **acculturation** are reduced, thereby protecting the traditional culture and **lifestyle** of host communities. In practice, this is the approach adopted by some countries such as **Tunisia**, whereas other countries pursue a policy of dispersing tourism development.

See also: resort enclave

Further reading

Gunn, C. (1994) *Tourism Planning: Basics, Concepts, Cases*, 3rd edn, London: Taylor & Francis.

Pearce, D. (1989) *Tourist Development*, 2nd edn, Harlow: Longman.

RICHARD SHARPLEY, UK

gift

Although many items are acquired as holiday **souvenirs**, others are purchased as gifts for relatives and friends back home. Some **tourists** are known for buying presents; for example, the Japanese, as in their country the exchange of gifts is an important component of the social fabric. Visitors to **Third World** destinations sometimes bring inexpensive articles to distribute to local children.

EDWARD M. BRUNER, USA

global distribution systems

Global distribution systems (GDS) are computer databases used by retail travel agents to research and book all types of tourism **products**. They were originally created by the airlines but have grown to be the major electronic distribution system for the entire tourism **industry**. The GDS are Sabre, Apollo/Galileo, SystemOne/Amadeus, Worldspan and Abacus. The central site of each contains many large mainframes connected to thousands of terminals through some of the largest data communication networks in the civil world.

PAULINE J. SHELDON, USA

globalisation

Globalisation is a term under which prevailing **models** of social, economic, and political organisation have popularly been collected, although the term itself has only seen widespread use since the 1980s. Within the context of Western society, there have been three waves of globalisation: **colonisation**, **development** (in the sense of capitalist development), and what is frequently termed neoliberal capitalism. In respect to this third wave, the process of globalisation is characterised by such factors as free trade zones, technological dependency, restructuring of science and technology, privatisation, debt–equity swaps, **mergers** and **acquisitions**. It affects virtually all aspects of economic activity that are related to free-market economies: communications, **transportation**, agriculture, manufacturing and more. Further, it is a social process that affects every component of each economic activity – reorganisation of the labour process and **management**, production, distribution channels, **marketing**, exchange, **technology**, information, telecommunications, **education** and **training** – moving all of these components away from local control into a global and/or regional scope, from small scale to large scale, and from centralised to decentralised structures.

A central feature of globalisation is that many contemporary processes or problems (including tourism) cannot be adequately studied or solved at the level of the nation-state. This is because one of the main characteristics of globalisation is the centralisation and concentration of both capital and power in the private **multinational firm**, particularly firms which are located in metropolitan as opposed to peripheral countries This is not to ignore the nation-state, but rather to focus on a conception of a global system based on transnational practices that have impacts on several interacting spheres: environmental, economic, political and cultural–ideological. Multinational firms are the major source of these practices because of their enormous scale. In fact, the gigantic size of these firms is illustrated by the fact that the annual sales of many of these firms are larger than the gross national product of many countries. For example, in 1991, only about sixty countries had GNPs of more than US$10 billion, while 139 multinational firms had annual sales in excess of US$10 billion.

In one sense, tourism has long been 'global' or international in that **tourists**, as the **demand** side of the system, have for many centuries visited other countries: medieval **pilgrimages**, grand tours of Europe in the eighteenth century, the first **package tours** of Europe in the mid-nineteenth century organised by Thomas Cook, and ship-based travel in the early twentieth century to remote destinations. As tourism demand increased, however, with growing prosperity in the developed nations following the Second World War and more particularly with the advent of jet airplanes in the early 1960s, the boom in mass tourism both led to and was facilitated by increased involvement by multinational firms in the **supply** side of the tourism system.

Modern **international tourism** is therefore dominated in corporate terms by TNFs, including airlines, hotels, **tour wholesalers**, **tour operators**, travel agents and **car rental** companies. These firms are characterised by high levels of vertical and horizontal integration. Vertical integration might, for example, involve an international airline offering both regularly-scheduled flights and tour charters, operating a chain of travel agencies, and owning hotels and a car rental company. Horizontal integration might, for instance, involve a hotel company owning a range of classes of hotels, from budget **motels** to luxury business-class hotels to all-inclusive **resorts**.

A second feature of globalisation with respect to tourism has been the liberalisation of economic **policies**, such as the loosening of **foreign exchange** controls in many countries. Much of the increase in tourism from Japan since the 1970s, for example, is related to the government's policy change that allowed its nationals to convert sufficient yen into foreign currency to allow for travel abroad. A related third feature is a strong dependency on external market forces in terms of both demand and supply. Overall, international tourism is dominated in both supply and demand by developed countries. That is, most international tourists originate in and travel to developed countries, while tourism to developing countries is dominated by tourists from developed countries. Economic policies and the state of the economy in the developing countries, therefore, have great **impacts** on destinations. For example, the United States embargo on travel to Cuba has seriously slowed the development of Cuban tourism; similarly, variation in tourist arrivals in the Caribbean in the 1980s was strongly correlated with the economic recession in the United States. On the supply side, the development of a major new tourism sector in one destination may create an intervening opportunity that seriously decreases arrivals in another country; an example was the rapid expansion of tourism in the Dominican Republic in the 1980s and the resultant impacts on other Caribbean destinations.

One result of the globalisation of tourism is that host or **destination** nations see much of the economic benefits of tourism lost to them through high economic **leakage** and low economic multipliers (see **mulitiplier effect**), to the advantage of the nations in which the majority of the tourists originate and of multinational firms. While most international tourists are generated from and received by developed countries, the impact of the globalisation of tourism is most notable in developing countries because of the impacts of these firms on what Sklair (1994: 168–9) has termed the six criteria for development effects: linkages (imports and exports), foreign currency earning, upgrading of local personnel, technology transfer, conditions of work (wages, job security, hours, workplace facilities) and environmental effects. Moreover, because of the large amounts of capital required to develop a major tourism sector and of the frequent absence of available local capital, developing countries are usually forced to allow TNFs to dominate their tourism sectors. Thinking of tourism in terms of globalisation changes the shift in focus from short-term problems (such as waste disposal) to more fundamental issues (such as corporate control) and goals (such as development that improves the quality of life of local people).

See also: internationalisation; transportation, globalisation of

References

Sklair, L. (1994) 'Capitalism and development in global perspective,' in L. Sklair (ed.) *Capitalism and Development*, London and New York: Routledge, 165–85.

PAUL F. WILKINSON, CANADA

global village *see* globalisation

golf tourism

Golf has universal appeal to both residents and tourists alike. Golf originated in Scotland in the fifteenth century and was played on an open field with rough edges. In 1754, the Royal and Ancient Golf Club was established in St. Andrew's. In 1867, the first women's golf **club** was formed, introducing golf as a female recreation to the world. The first national golf championship, held in Scotland in 1860, was limited to professionals. In 1861 it was open to the world as the British Open. The first British Ladies Amateur Championship was held in 1893, at the Lytham and St Anne's Club.

Approximately 20 million people worldwide play the game. One can golf in a beautiful country surroundings, on courses twenty-four hours a day with a never-setting sun, or without a blade of grass on desert layouts in simple dunes with oiled greens. Many golf **tourists** stay in **resorts** which have specific courses and professional training in the game. Competition takes place on private, semi-private and public courses. Private courses cater to members and their guests. Semi-private establishments service members and green fee-paying outside customers. Public grounds are open to all enthusiasts. In most golf resorts a wide range of facilities is provided, such as lodging, pools, **spas**, **restaurants**, practice ranges, video swing analysers and simulators. For travelling golfers and spectators seeking golf enjoyment, most countries have tourism **boards** offering golf information. Further, there are well-organised regional, national and international tournaments for professionals and amateurs, sanctioned by the various golf associations. Major golf events draw large numbers of tourists, notably the Masters Tournament, Ryder Cup Series and worldwide Professional Golf Association tournaments.

As the greatest obstacles to golf are time, money and space, the pitch-and-putt facility, initiated in Ireland several decades ago, is gaining popularity. Distances vary from 20 yards to 80 yards with nine holes; a loft iron and putter are used. This shortened terrain and time saving hold much for golf tourists. The 'golf attaché', a computerised golf course in a briefcase, is also gaining in interest for travellers.

Further reading

Campbell, M. (1991) *The Encyclopedia of Golf*, London: Dorling Kindersley. (Covers golf's history, rules, site, personalities and traditions.)

JOSEPH KURTZMAN, CANADA
JOHN ZAUHAR, CANADA

government

There probably has always been a close connection between tourism and government. In fact, some of the earliest **tourists** were heads of state and other political dignitaries who went great distances to visit other leaders or holy sites. Much early tourism included diplomats representing their heads of state as they delivered official messages or negotiated issues between their ruler and others. As commerce grew and developed, tourism became characterised by **business travel** and travel by persons with the financial means to enjoy their **leisure** or to expand their knowledge. In the modern world, the concept of **mass tourism** recognises that this practice, both at home and internationally, has become affordable to an ever-increasing number of the world's population. In this development of mass tourism, government has been a key participant. Some governments, like Saudi Arabia, have tried deliberately to impede tourism in the country, while most others have actively promoted its **growth**. Government policies often reflect internal political debate over tourism's **impacts**, external relations with particular regimes, and the correct economic and political uses of this **industry**.

Investigation of tourism could include the study of government and how it impacts tourism activity. At the same time, a study of government could include some analysis of how tourism has affected the operation of government and how its policies and regulations have been influenced by the needs or the presence of tourism. The nature or **quality** of tourism is both an input to government decisions and a result of its decisions. When government interacts with tourism as a human activity, it does so on behalf of the society or the state. Government is the arrangement of institutions and processes that exercises power in an authoritative and legitimate fashion.

Early studies concentrated mainly on the description of **powers** contained in constitutions as a way of discerning what governments could or could not do. Constitutional law became the basis of most analyses of politics and of government. Especially since the Second World War, the study of government has expanded and evolved to cover the much broader realm of all political behaviour that affects a society, including world society. The so-called behavioural revolution in social science resulted in new generations of students of government who sought to understand politics and government through observation and analysis of political data and behaviour. They have viewed government as a complex system made up of numerous subsystems. When these subsystems mesh together and work properly, they produce an orderly society characterised by a minimum of conflict among its citizens. Inputs to the political system, in the form of demands and supports, are converted to outputs in the form of new policies or regulations. As tourism and government interface, the former as an industry provides both support to and demand on the political system. Government output affects this industry in a variety of ways and can determine what its **future** in a particular political system will be.

How governments interact with tourism depends mainly upon the nature of the political system in question, upon how powers are distributed among various agencies and by how that power is derived, whether from popular elections or some less democratic process. In a federal system, for example, governmental functions are distributed among several levels of authority such as provinces, states, counties or cities, in addition to the federal level. In a capitalistic economy there is more emphasis on interaction with the private sector to promote and regulate tourism, while in socialistic systems (see **centrally planned economy**) government will likely play a more regulatory **role** and directly sponsor more tourism ventures.

In a democratic political system one could claim that government represents the people, since through open elections officials are chosen and are held responsible to the voters. In more closed political systems, government also represents the state. Even if it appears not to have the consent of the voters or citizens, a totalitarian government is nevertheless exercising the power of the state in an authoritative manner. Whether that authority is legitimate and whether that government is legitimate is subject to interpretation. Yet, as long as there is no effective challenge to or overthrow of the existing government, it carries the banner of legitimacy. For persons who must deal with that government, in the case of tourism development or promotion, its legitimacy as the official representative of the state is assumed.

While international law recognises nation-states as having sovereignty, other types or levels, such as provinces or dependent countries, may have limited authority over certain legal matters but are subordinate to the superior nation-state. Just as governmental functions affecting tourism are distributed by the type and level of organisation, so also are there different levels of authority over governmental matters important to tourism. For example, in a sovereign nation-state which has subordinate divisions, a decision about which foreign **airlines** should be allowed to operate into its territory would be made at the national level of government. The level of foreign ownership in general would be a national decision, since no sub-national division of the government would have the authority to determine that policy.

However, in those same countries the nature of the requirements for a motor vehicle operating licence issued to a foreign **tourist** would likely be determined by the state or provincial level of government. In further example, a policy decision to permit the construction of a large **hotel** on a particular piece of land in a given city may involve primarily local government bodies. The land would require proper **zoning** classification by **local** authorities, and construction permits would be issued by local or provincial governments. Hence, the appropriate level of government to become involved in a tourism issue will depend on a number of factors, and in some cases several levels of government may become involved. In countries where there are no local or provincial governments, all issues of tourism policy would likely be handled by the national government. Such political systems are usually described as unitary governments.

There are numerous tourism-based issues which may require direct action by government(s) either in a facilitative sense or in a regulatory role. To illustrate the close connections existing between tourism as a human activity and government, which is also a human activity, for example, one could think of international travel policies and regulations, including air transport policies and international agreements; customs and immigration policies; currency restrictions and **foreign exchange** values; travel restriction on citizens because of **safety** concerns or lack of diplomatic protection; environmental issues arising from tourism development (see **environment**); land use policies; infrastructural pressures or demands from tourism development (see **infrastructure**); **crime** prevention and law enforcement needs resulting from tourism; taxation policies including **taxes** or levies unique to tourism; the development of historic sites or major cultural **attractions** which facilitate tourism; and the **preservation** of local, regional and national **identities** and culture where the volume of tourism is high relative to the local population and society. All of these are issues which can be drastically affected by government actions.

Government policies on such issues can stimulate or retard the development of both national and **international tourism**. As national governments join and support international organisations pertinent to tourism, the free movement of persons among countries is facilitated. Likewise, such movement can be restricted, as illustrated in the prohibition by some countries of travel by its citizens to specific nations. Strict immigration and customs enforcement or tight currency controls can make certain international tourism difficult. Even government intervention in major labour disputes affecting **domestic tourism** or international travel can have a significant impact on the tourism industry across national boundaries.

The quality and volume of tourism activity in any society is greatly dependent not only upon government tourism policies or the lack of policies, but also upon that government's relations with tourist generating nations and its own citizens' tourism proclivities. The above examples illustrate the range of issues on which governments may act to affect the industry. Since in most polities government is responsible for the overall maintenance of order, what it does in all sectors of society will affect tourism as well.

See also: code of ethics; demand, air travel; democratisation; environmental legislation; globalisation; ideology; imperialism; mass tourism; political development; postmodernism; regulatory agency

Further reading

Easton, D. (1965) *A Framework for Political Analysis*, Englewood Cliffs, NJ: Prentice Hall. (Applies systems theory and analysis to the study of politics and government.)

Madrigal, R. (1995) 'Residents' perceptions and the role of government', *Annals of Tourism Research* 22(1): 86–102. (Empirical research on what residents think governments, role should be in tourism development.)

Matthews, H. (1978) *International Tourism: A Social*

and Political Analysis, Cambridge: Schenkman Publishing Company. (Focuses on the Caribbean region and the political issues of tourism development.)

Richter, L. (1883) 'Tourism Politics and Political Science: A Case of Not So Benign Neglect', *Annals of Tourism Research* 10(3): 313–36. (A critique of why political scientists have not applied their research to tourism as readily as some other social scientists.)

HARRY G. MATTHEWS, USA

grading system

Grading systems are a qualitative assessment of the facilities and services provided by hotels, guest-houses, inns, self-catering establishments, caravan parks and other forms of short-let **accommodation**. Grading is a generic term which may have different interpretations. It is sometimes confused with registration of accommodation which is the recording of establishments on a register possibly requiring inspection or compliance with particular legislation. Similarly, **classification** is the stock of accommodation subdivided into categories. Each category consists of specified facilities such as the proportion of private bathrooms, minimum size of accommodation. Each country may classify differently having a number of categories covering self-catering accommodation, guesthouses, hotels and motels or different classifications for different types of accommodation. Classification does not imply a qualitative element, only that specified facilities and services are provided, with the overall understanding that the establishment is clean and well maintained. The term grading or quality grading is a qualitative assessment of the facilities described under classification.

Grading systems are operated by national tourism organisations such as English Tourist Board, motoring organisations such as Automobile Association and commercial organisations such as Michelin. These oganisations vary in the approach to grading which they adopt and the criteria which they use. This presents a confused picture for the consumer who cannot make informed comparable judgements about the symbols and grades employed. The approaches include classification such as Northern Ireland Tourist Board, classification and grading such as Automobile Association (which classify with stars and indicate a quality grade with a percentage score); a combination of classification and quality grading (where the differences are not discernible to the consumer, such as Michelin).

Initiatives are under way to united grading systems for the benefit of the customer. In the United Kingdom it is planned to harmonise the schemes of the tourism boards and the motoring organisations. In Europe the Tourism Unit of the **European Union** is examining the possibility of harmonising European grading systems.

Further reading

Callan, R.J. (1993) 'An appraisal of UK hotel quality grading schemes', *International Journal of Contemporary Hospitality Management* 5(5): 10–18 (A comparison of UK grading schemes.)

—— (1994) 'European hotel classification: book or burden?' *Hospitality* 146(October/November): 14–17. (Report highlights of European grading harmonisation.)

—— (1995) 'Hotel classification and grading schemes, a paradigm of utilisation and user characteristics', *International Journal of Hospitality Management* 14(3/4): 271–83. (Customer usage of grading schemes and their perceived importance.)

ROGER J. CALLAN, UK

Grand Tour

The Grand Tour was a circuit of Europe undertaken by the wealthy, for reasons which included culture, **education**, **health** and pleasure. Principally centred on France, Italy and Germany, some tours also extended to Spain, Portugal and Greece. The dominant nationality on the Grand Tour was the British, but there were also significant numbers of French, German and Russian participants. Grand Touring developed in the sixteenth century, reached its zenith in the eighteenth century and survived in modified form into the nineteenth century. A tour could last from anywhere between two to eight years.

The Grand Tour is an example of long-term continuity in tourism **history**. In some sense, it was a re-emergent form of cultural tourism from the ancient world where wealthy Romans, for instance, would visit the glories of Greece and Egypt. Furthermore, the route patterns and facilities of the Grand Tour built upon the **infrastructure** provided for **pilgrims** and other travellers since the Middle Ages.

The majority of tourists on the Grand Tour were the young male offspring of the affluent. In Britain, they came from the ranks of the landed classes, whose estates and other sources of wealth provided the time and money for extensive travel. The educated middle classes would provide the tutors to the young tourists, while the working classes provided their accompanying servants. By the early nineteenth century, the middle classes were increasingly becoming Grand Tourists themselves, sometimes accompanied by their families. At this stage the tour had reduced in length, often lasting for no more than a few months.

The underlying causes of the Grand Tour can be traced to the changing cultural relations between different parts of Europe which evolved from the sixteenth century onwards. Classical antiquity and the Renaissance formed the centre of the culture of educated elites in Britain, France and Germany, and this stimulated the desire to visit Italy in particular. Allied to this were developments in philosophy and science, which emphasised the importance of travel in the **quest** for knowledge. To these educational and cultural motives must also be added that of health, and the appeal of the warm south for curing real or imagined ailments.

Distinct spatial and temporal patterns of travel were created. Some of these patterns remained little changed, while others were modified by taste and travel conditions. A dominant route pattern was that which ran through France via Paris, to the Rhône Valley, thence crossing the Alps to Turin or going by sea from Marseilles to Italy. Within Italy, an enduring pattern connected Florence, Rome, Naples and Venice, combining visits to antiquities and the Renaissance heritage with the delights of cosmopolitan culture. North of the Alps, there was a circuit of the courts of Germany or a return journey via the Rhine. Route patterns could be modified by changing aesthetic tastes. The picturesque and romantic movements of the later eighteenth century encouraged more tourists to include scenic tours of Switzerland and the medieval towns of central Italy in their **itineraries**. There were few major changes in travel conditions during the period of the Grand Tour, and it is unlikely that the coming of the railways was significant for ending its significance. The railway network was still in its infancy in Europe when the tour declined as a major institution.

Tourists from Britain departed for the Grand Tour either in April–May or late August–September. This was either at the end of the fashionable season or timed to cross the Alps before the winter. Journeys were also timed to be in Rome at Christmas and Easter and Venice for Ascensiontide. Travel in Italy was generally during the winter months, avoiding the summer heat in Rome and Naples. Thus, distinct visitor seasons were created in these centres (see **seasonality**). In terms of length of stay, Rome, Venice, Florence, Naples and Paris generally dominated the time spent in particular places, with visits of several months being common.

An embryonic tourism industry supported the needs of the visitors. In many instances, of course, the tourists simply formed part of the general clientele for transport and **accommodation**, but in some centres their numbers and wealth created particular responses. For example, the Piazza di Spagna area of Rome formed the main centre for visitor apartments in that city, while Florence gained a reputation for the quality of its tourism accommodation by the later eighteenth century. **Tour guides** to attractions could be found in many centres, as could a **souvenir** industry. By the 1820s, a number of entrepreneurs (see **culture broker**) were providing inclusive packages of transport, lodgings and food for visitors to Switzerland. Information and advice could be obtained from the diplomatic and banking network around the continent.

By the middle of the nineteenth century, the Grand Tour had largely faded. The elite moved on to new areas such as Greece and the Near East or to exclusive **resorts** within Europe. Yet, the cultural centres of the continent continue to draw visitors through modern equivalents of this notable episode in tourism's history.

See also: guidebooks; history; travel literature; youth tourism

Further reading

Black, J. (1992) *The British Abroad: The Grand Tour in the Eighteenth Century*, New York: St Martins. (Presents a comprehensive survey of the Grand Tour.)

Mead, W.E. (1912) *The Grand Tour in the Eighteenth Century*, New York: Houghton Mifflin. (Offers a detailed study of the Grand Tour.)

Towner, J. (1985) 'The grand tour: a key phase in the history of tourism', *Annals of Tourism Research* 12(3): 297–333. (Provides an analysis of the tourists, tour patterns and tourism industry of the time.)

JOHN TOWNER, UK

gravity model

The gravity **model** forecasts trips between an origin and a **destination** (see also **forecasting**). As the name implies, it is based on an analogy to Newton's law of gravitation, where:

$$T_{ij} = \frac{GP_iA_j}{D^d_{ij}}$$

T_{ij} = number of trips from i to j;
G = a statistically-estimated constant;
P = population of origin, i;
A = attractiveness of destination, j;
D = distance between i and j, and;
d = a coefficient reflecting the effect of distance on travel.

In other words, the volume of travel between an origin and a destination is directly related to the population of the origin and the attractiveness of the destination, and inversely related to the distance between them.

Gravity models typically are calibrated for a single origin and multiple destinations; **forecasting** visitation for multiple origins normally requires a separate gravity model for each. Like most other **quantitative methods** used as forecasting tools, gravity models require data related to past travel patterns. Once the model is calibrated with historical data, a forecast may be made by inserting expected future values for the origin's population and/or destinations' attractiveness, and then solving.

The gravity model is a flexible and useful starting point for the development of other travel models. For example, researchers have replaced the population variable by other measures of the tendency of an origin to generate outbound tourism, such as automobile ownership. Numerous measures of attractiveness are also utilised. The distance variable is also subject to experimentation, with various combinations of physical distance, travel time, travel costs, and the number of intervening opportunities used to measure the separation between an origin and a given destination.

In its original form, the gravity model is unconstrained; that is, there is no intrinsic upper limit on the number of trips the model can predict. In practice, limits do exist and researchers have modified the gravity model to predict relative travel flows (**market** shares) as well as the absolute number of trips. Although the gravity model began as an analogy, Niedercorn and Bechdoldt (1966) have demonstrated that the model is a theoretically sound solution for the problem of maximising individual **satisfaction** from travel subject to time and budget constraints.

See also: distance decay

References

Niedercorn, J.H. and Bechdoldt, B.V. (1966) 'An economic derivation of the "gravity law" of spatial interaction', *Journal of Regional Science* 9: 273–82.

Further reading

Smith, S.L.J. (1996) *Tourism Analysis*, 2nd edn, Harlow: Longman, 131–40. (Describes how to calibrate a gravity model and reviews assumptions, problems, and modifications.)

STEPHEN L. J. SMITH, CANADA

Greece

Greece has a long tradition in tourism, mainly due to its **history** and ancient civilisation which regarded foreigners as sacred. Its tourism **product** is composed of an amalgam of natural, cultural and **heritage** attractions, including some 15,000 miles of coastline, 2,500 islands, diverse fauna and flora, mild winters and warm summers, 25,000 registered and protected monuments and archaeological sites, numerous **museums** and about 500 traditional settlements.

As a land of rich economic, religious and intellectual activity for more than three and a half millennia, geographically spread on an archipelago located at the southeastern corner of Europe on a crossroad to Africa and Asia, Greece inevitably has stimulated tourism activity since the beginning of documented humanity. In modern years, organised tourism commenced after the Second World War, and expanded rapidly after the mid-1970s. Not surprisingly, tourism grew especially on island destinations and regions with historical monuments.

Amenities are offered by a plethora of small enterprises. A total of 7,500 official hotels offer half a million beds, while about 1 million unregistered beds are provided. Moreover, 11,000 beds in cruise liners and 12,000 in yachts, as well as 83,000 camping spaces, accommodate all types of **demand**. About 20,000 **restaurants** and countless **catering** and entertainment establishments are on offer. Some 7,000 **travel agencies**, 1,500 coach rental firms and some 4,000 car rental agencies are estimated to operate throughout the country. The tourism product is distributed mainly by European **tour operators** who organise **package tours**. There are about 800 tour operators in Greece. Accessibility is facilitated through thirty-two airports, most of which can receive direct international flights. Olympic Airways and other private airlines provide an extensive network, while a complex network of sea, road and **rail transportation** facilitates passenger transportation.

Since the early 1950s Greece has enjoyed a **growth** in arrivals. In 1950, over 33,000 foreign tourists arrived; this figure rose to 11.3 million in 1994. About 75 per cent of visits are concentrated in the May to September period. **Germany** and the **United Kingdom** contribute almost 25 per cent of the total volume each, while **Italy**, Scandinavia, the **Netherlands** and **Austria** are the other main markets. In 1993, about 48.5 million bed nights were recorded, 75 per cent of which were by international tourists. More than 75 per cent of tourists arrive by air and 58 per cent by charter flights, indicating that they are on package holidays. About 90 per cent were holiday makers, while only 7 per cent travelled for business and 3 per cent for other reasons. The average **length of stay** is fourteen days, while the average expenditure per capita in 1993 was $354. Research on tourist segmentation, **motivation**, attitudes and **satisfaction** is inadequate (see also **market segmentation**).

The tourism **policy** is overseen by the Greek National Tourism Organisation and the Ministry of Tourism. The two organisations share the responsibilities for **planning**, implementation and promotion at both national and regional levels, as well as coordinating the public and private sectors. Public **investment** in commercial facilities was utilised in the past to stimulate both tourism demand and the confidence of the private sector to invest. A wide range of **incentives** have also been utilised since the early 1980s to improve the existing facilities and stimulate the **development** of high quality amenities. The national organisation operates twenty-one regional and twenty-five overseas offices (in eighteen countries) mainly responsible for promotion and marketing. There are no regional tourism boards, although counties and local agencies often establish working groups for the promotion and coordination of the local industry.

The Ministry of Development and the National Tourism Organisation are responsible for national policy. Unfortunately, there is a lack of a comprehensive **master plan**. Hence policy is usually based on generic mid-term policy objectives: to increase arrivals and **foreign exchange** income, to enhance **competitiveness**, to improve **services** in the industry, to reduce **seasonality**, to attract high spenders and **alternative tourism**, to construct up-market facilities such as luxury hotels, golf courses, congress centres, casinos and marinas, to develop **infrastructure** with emphasis on transportation, to support **social tourism** for low-income domestic tourists, to redistribute tourism **demand** and **supply** to achieve an even spread, and to train tourism employees. Occasionally, incentives for tourism development, responses to lobbying from **associations** and **regions**, training programmes and general **legislation** direct the Greek tourism industry towards the described policy directions.

The tourism industry is a vital stimulator of the Greek economy, contributing to the **balance of payments**, **employment**, gross national product, national income, consumption of output by other sectors and regional development. The Greek National Tourism Organisation estimates that in 1992 the foreign currency from tourism expenditure was $5.8 billion, contributing almost 8 per cent of the GDP and covering 41.7 per cent of the international trade deficit. During 1965–1980, the tourism output multiplier in Greece was estimated to be 1.52. Tourism is also instrumental for regional development in peripheral, insular, and problematic regions which lack opportunities for other types of development. **Domestic tourism** facilitates a certain degree of redistribution of wealth between metropolitan and peripheral regions.

Tourism has introduced a range of sociocultural **impacts** mainly due to the lack of effective **planning** and **management**. The **commoditisation** of **history**, cultural traditions and everyday manifestations affects the renowned Greek **hospitality**, commercialises human relationships with tourists and reduces the ties of solidarity of societies. Tourism professionals overwork during the summer and inevitably neglect their social, religious and cultural obligations. Family relationships and values are also under great transition and perhaps threat. The demonstration effects are enhanced not only by tourism but also by mass media, as well as by Greek students and professionals living abroad. An increase in **crime**, as well as consumption of **alcohol** and drugs, are also attributed to tourism. Tourists' **behaviour** is often offensive and unacceptable for the host population, but is tolerated due to the economic benefits.

Although tourism has contributed to the regeneration of urban and rural regions, several **resorts** have been exploited due to inadequate planning and lack of reinvestment in their sustainability. The **environment** has suffered from haphazard, uncontrolled buildings, as well as pollution of the sea, while the flora and fauna are being affected by waste disposal. Coast pollution, water shortages, inadequate sewage treatment and waste disposal, traffic congestion, noise pollution, overbuilding and aesthetic degradation are some of the impacts experienced in a number of resorts. The habitats of several endangered species are also being destroyed. Urgent coordination with all actors in the tourism industry is essential, while regulation is required to set objective and measurable limits and targets in order to preserve environmental resources. Financial difficulties in the industry only exacerbate these impacts, as the private and public sectors feel unable to reinvest in the **conservation** of resources.

Tourism in Greece has great potential due to the rich variety and quality resources which generate a healthy level of demand. A long-term strategy is required in order to enable the country to optimise the tourism impacts and develop its competitiveness in the global marketplace. The strategy would improve the prosperity of the host population, optimise the tourism impacts, maximise tourist satisfaction and maximise the profitability for enterprises. Planning and rational management of resources will be instrumental for the future of Greek tourism in the new millennium.

Further reading

Buhalis, D. (1998) *Tourism in Greece: Strategic Analysis and Challenges for the New Millennium*, Aix-en-Provence: International Centre for Research and Studies in Tourism. (An overview of tourism management and planning in Greece and its impacts.)

——(1999) 'Tourism in the Greek Islands: the issues of peripherality, competitiveness and development', *International Journal of Tourism Research* 1(5): 341–59. (Tourism marketing and management for the Greek islands.)

Cooper, C. and Buhalis, D. (1992) 'Strategic management and marketing of small and medium sized tourism enterprises in the Greek Aegean islands', in R. Teare, D. Adams and S. Messenger (eds), *Managing Projects in Hospitality Organisations*, London: Cassell, 101–25. (Provides a strategic analysis for the tourism supply of the Aegean Islands in Greece and proposes a strategy for tourism planning.)

Konsolas, N. and Zacharatos, G. (1993) 'Regionalization of tourism activity in Greece: problems and policies', in H. Briassoulis and J. Van der Straaten (eds), *Tourism and Environment: Regional Economic and Policy Issues*, Dordrecht: Kluwer Academic Publishers, 57–65. (Illustrates a critical analysis of tourism planning in Greece and demonstrates a range of structural problems.)

DIMITRIOS BUHALIS, UK

green marketing

Green **marketing** is an approach that highlights environmentally friendly practices adopted by tourism destinations and/or operations. It aims at environmentally discerning customers who make a conscious buying decision. Usually it stresses specific environmentally sound practices, while at the same time giving less prominence to other business practices that might not perform to the same eco-standards.

See also: ecoresort; ecotourism; soft tourism

MARTIN OPPERMANN, AUSTRALIA

green tourism

A form of **alternative tourism**, green tourism is generally related to **rural tourism**. It denotes a **nature tourism** seen to be environmentally compatible and with little or no ecological **impact** on the **destination** area. Like many other alternative tourism terms of the 1980s, it did not gain a wide following and was rapidly succeeded by **ecotourism**.

MARTIN OPPERMANN, AUSTRALIA

greenhouse effect

The greenhouse effect is a natural process whereby incoming short-wave radiation from the sun passes through the earth's atmosphere and is re-radiated at longer wavelengths which are intercepted by the so-called greenhouse gases. The major greenhouse gases are carbon dioxide, methane, nitrous oxide and halocarbons. Increasing concentrations of these gases are giving rise to global warming. This warming has been projected to be approximately 3.5 degrees Celsius globally for a doubling, since pre-industrial revolution times, of carbon dioxide or its equivalent in other greenhouse gases, with the magnitude of temperature increases being greatest in high latitudes, particularly in autumn and winter. At current trends, this situation will occur within the next fifty years. Less confidence can be placed in assessments of future precipitation, but there may be reduced moisture availability in inland areas and sea levels may rise, chiefly because of the thermal expansion of ocean waters and because of glacial melting. There may also be an increase in the frequency of extreme events such as tropical storms.

Should such climatic changes occur, there could be far-reaching consequences for tourism as a result of changes in the natural **resource** base and modifications in season lengths which influence the economic viability of tourism businesses. Coastal locations and those based on winter activities, such as ski **resorts**, appear to be most vulnerable. However, both natural and human systems are adapted to a considerable degree to perturbations in climate and weather, and climate change may result in opportunities as well as problems for tourism enterprises.

See also: climate

Further reading

Wall, G. (1992) 'Tourism alternatives in an era of global climate change', in W. Eadington and V. Smith (eds), *Tourism Alternatives: Potentials and Problems in the Development of Tourism*, Philadelphia: University of Pennsylvania Press, 194–215.

—— (1993) 'Tourism in a warmer world', in S. Glyptis (ed.) *Leisure and the Environment*, London: Belhaven, 293–306.

—— (1996) 'The implications of climate change for tourism in small islands', in L. Briguglio, B. Archer, J. Jafari and G. Wall (eds), *Sustainable*

Tourism in Islands and Small States: Issues and Policies, London: Cassell, 206–16.

GEOFFREY WALL, CANADA

greenspeak

Greenspeak is that **register** of the language of tourism promotion which targets the environmental tourist (see also **promotion, place**). Via the operator's name, logos, slogans, eco-explicit messages and the green themes of **nature, nostalgia** and nirvana, this multi-layered **discourse** overtly displays concern for **sustainable development**. Covertly, however, contradictions abound, since the unrestricted universe promised to the individual is a limited world prone to invasion by the masses.

GRAHAM M.S DANN, UK

gross profit

In general **accounting** terms gross **profit** is the difference between net sales and the cost of goods sold. It may be interpreted as that portion of net sales that is available to cover operating expenses and to produce a profit. It does not measure overall profitability.

In simple format, the traditional (manufacturing industry) form of income statement has gross profit as a middle level of profit from which operating expenses and fixed charges are deducted to determine net income. The contribution made by gross profit must be large enough to also cover all of these expenses for the business to earn net profit. However, the tourism **industry** has not found the traditional income statement and concept of a single level of gross profit adequate. The emphasis in the industry is on the control of labour and inventory rather than fixed assets. Therefore, the focus of profit reporting emphasises the development of information on revenues, use of inventory and labour and other resources at the appropriate level of **management**. Its statements utilise at least three levels of profitability: hospitality income statement, departmental revenues, less direct departmental expenses, profit from operated departments, less operating expenses, gross operating profit, less fixed charges, net profit.

The first two levels are both concepts of gross profit. By defining gross profitability at two levels, the focus is on managerial responsibility for revenues and costs at each level. The first level (profit from operated departments) is defined as revenue, less direct **inventory** and labour costs controlled by department heads: the individual gross profit of each department. The second level (gross operating profit) is defined as profit after all operating expenses have been deducted but before fixed charges (rent, taxes, insurance, depreciation and interest). Most **hotel** general managers, for example, do not exercise control over fixed charges as they are incurred or imposed by the owners and governmental **tax** authorities. However, a general manager is held responsible for a level of profit resulting from the revenues and expenses over which control has been given (most revenues and all operating expenses).

Further reading

Coleman, M. (1991) *Hospitality Management Accounting*, New York: Van Nostrand Reinhold. (Covers gross profit concepts applied to pricing.)

LEE M. KREUL, USA

group business market

The business tourism market can be divided into two subsegments: the individual business market, composed of individuals travelling alone or in small numbers for work-related purposes, and the group business market, which involves larger numbers of people travelling together to attend events related to their common business purpose or wider interests. The latter is commonly subsegmented into the corporate and association markets. While the corporate market may be best described in terms of the purpose of the event (annual general meetings, training seminars, product launches, shareholders' meetings, sales strategy planning meetings, **incentive** travel and so on), the association market is usually subdivided according to the characteristics of the association or its geographical coverage (scientific, religious, political, professional, regional, national, international and more).

The group business market is largely composed of events which are discretionary, in the sense that their venue is not fixed in advance. This distinguishes it from the individual business market, which is almost entirely non-discretionary in nature, the **destination** or venue being determined by the location of the client to be visited, the contract to be signed, the problem to be solved by the business traveller, and so on. Due to its discretionary nature, this market is the focus of considerable competition between venues, destinations, and other suppliers of services. A major problem facing these suppliers is the identification of decision-makers in corporate organisations and associations.

Since relatively few companies have a dedicated conference management department, identifying key corporate buyers is a major and continuous challenge to those wishing to market their facilities and services to them. The member of staff responsible to plan meetings for their company may have any one of a whole host of different functions within that company, from the managing director or marketing manager to travel manager or personal assistant/secretary. Regarding the other principal characteristics of the corporate group business market, delegate numbers are typically under 100, and hotels are usually the favoured type of venue.

By contrast, in the 'not-for-profit' association group business market, the decision on where to hold a conference is commonly taken by a committee elected by the association's membership. Destinations often put forward bids to tender for major conferences, with or without the assistance of their convention and visitor bureaus. Lead times for such events are generally much longer than for corporate events, due in part to delegate numbers being higher and thus involving far more planning. Purpose-built or converted conference centres are much more likely to be the venues for these events than events in the corporate sector. A final major distinction between the two sectors is the far greater likelihood of partners attending events organised by associations.

ROB DAVIDSON, UK

group inclusive tour

Planned, organised and marketed by commercial travel organisers, group inclusive tours (GIT) minimise participants' exposure to the perceived **risk** of outbound tourism. Travel is en masse, usually by coach or plane. GITs can be single-centred, particularly to **mass tourism** destinations; duo-centred, notably to two cities or one city and one beach **destination**; or multi-destination, with **itineraries** following on a well-defined route.

DEBORAH GRIEVE, UK

group tour *see* guided tour

group travel market

The term group travel market refers to situations where the **product /service /experience** is promoted and sold to a group of persons (such as a ski club), rather than to individuals. This group may be from either the business or pleasure sector. The actual tourism experience may be realised as a group or as an individual.

J.R. BRENT RITCHIE, CANADA

growth

Growth in tourism most often refers to an expansion of tourist numbers or an increase in tourism receipts. Growth may be measured in a wide variety of ways including **employment** and income generated, land area devoted to tourism uses, and arrivals. Lack of reliability and consistency in data makes accuracy and comparability of trends difficult.

RICHARD BUTLER, UK

guest

Guests are outsiders temporarily admitted to a social setting through invitation by a host. **Hospi-**

tality to guests is based on **rules** and **rituals** which constitute a form of social exchange (Blau 1965) between the parties, through which some fundamental contradictions in the stranger–local relationship are mediated. A **stranger** is always potentially dangerous because he or she is unknown, or is the bearer of a magico-religious menace. The rituals of hospitality neutralise that menace, while admitting the stranger and endowing them with the temporary status of guest in the host setting. This status is essentially ambivalent since the stranger has no entitlements in that setting: this acceptance and treatment are wholly dependent on the good will and generosity of the host. The status of guest also imposes some restrictions and obligations on the incumbent. While privileged, the guest's **role** is also socially, spatially and temporarily limited: he or she is precluded from playing many ordinary roles in the host setting, or from entering some, especially intimate, spaces. Most important, the guest ought not to overstay the extended hospitality.

Moreover, like the **gift** (Mauss 1994), the acceptance of hospitality offered carries some obligations of reciprocity. Indeed, although the host's generosity bestows the guest with honour, it is also mingled with interest: the former expects reciprocation from the latter, especially an invitation for a return visit. Perpetual intergroup relations based on social exchange are dependent on such reciprocity. Once reciprocity is impaired, either because return invitations are not extended or cannot be consummated for practical reasons, the **host and guest** relationship tends to break down. It can also be transformed into one based on economic exchange, namely on the immediate payment for hospitality services rendered, rather than on an obligation of future reciprocity on the part of the guest.

Tourism can be conceived as a form of commercialised hospitality, combining aspects of both social and economic exchange. The guest–host relationship is both historically and sociologically the **paradigm** for the relationship between tourists and the institutions of hospitality. In the early stages of tourism **development**, this relationship is still pronounced: tourists are seen as 'paying guests'. With the emergence of **mass tourism** and the development of the tourism system, relationships become more formalised and more strictly based on economic exchange. But the professional competence of the tourism service personnel is expressed precisely in their ability to make the tourist feel as a guest and to see the service as an act of generosity, although it is based on calculations of profitability.

References

Blau, P. (1965) *Exchange and Power in Social Life*, New York: Wiley.

Mauss, M. (1994) *The Gift*, London: Cohen & West.

Further reading

Pitt-Rivers, I. (1968) 'The stranger, the guest and the hostile host', in J.G. Peristiany (ed.), *Contributions to Mediterranean Sociology*, Paris: Mouton, 13–30.

ERIK COHEN, ISRAEL

guest survey *see* survey, guest

guide *see* culture broker; tour guide

guidebook

There is no clear definition of a guidebook. A wide range of books may serve as aids to **tourists**, and so forms of guidebooks can occur within the whole field of **travel literature**. However, a useful **classification** is to see guidebooks as lying more within the objective, informative sphere of production, as distinguished from the impressionistic, personal world of the travel book. A characteristic of mainstream guidebooks over the last 150 years would be their impersonal, systematic approach to providing information and guidance. In this form, the genre usually has a combination of **itineraries**, an inventory of places and objects considered of interest, together with practical information on **transportation**, **accommodation** and costs.

The plethora of guidebooks today have their immediate origins in the output of the great publishing houses of the last century such as Murray and Baedeker, but these in turn were derived from much earlier historical practices. De Beer (1952) suggests four main traditions shaped the evolution of the genre: **geography** and **history** books, itineraries and road books, travellers' narratives and guides to individual towns and cities. Works of **geography** and **history** can be traced back to classical times with books such as Pausanias's *Guidebook of Greece*, and this tradition re-emerged in Renaissance Europe with major accounts like Alberti's *Descrittione di tutta Italia*, published in 1550, which combined history and description. This general type of material had little influence on the format of guidebooks, but had the important ingredients of impersonality and comprehensive coverage.

A more direct influence on the guidebook came from itineraries and road books. Lists of places along particular routes occurred in the Roman Empire and were provided later for **pilgrimage** routes to Rome, Jerusalem and other centres. The increase of travel and trade in Europe provided a growing secular market for itineraries. Charles Estienne compiled a major collection of routes for France in 1552, *La Guide des Chemins de France*, and similar works later appeared for Germany and elsewhere. Details of **fairs**, markets and currencies were added to route information such as James Wadworth's 1641 *European Mercury*, which outlined routes and venues for trade on the continent. This tradition thus contributed practical information to the evolution of the guidebook.

Travellers' narratives form the third major component of guidebook development. In fact, early narratives often had sufficient factual information to act as forms of guides. Early Grand Tourists (see **Grand Tour**) frequently used the published or unpublished journals of previous visitors to help them on their journeys. The drawback was that narratives lacked a systematic and detailed approach for purely practical purposes. Nevertheless, the popularity of travel narratives in the later seventeenth and eighteenth centuries demonstrated the **demand** for travel literature and what forms of description were appreciated.

Towns of particular importance had guidebooks devoted to aspects of them from an early period. As a centre for pilgrimage, Rome had the *Mirabilia* from the twelfth century onwards, listing the **sights** of the city together with an itinerary. Lists of churches and **relics** were provided in the *Indulgentiae*. From the sixteenth century, Rome's religious attractions were supplemented by details of its antiquities. Pompilio Totti's 1638 *Ritralto di Roma Moderna* was fundamentally a secular guide. By the early eighteenth century, many European cities had guidebooks devoted to them, a tradition which obviously continues to this day. The number of guidebooks to countries also grew considerably. Italy was especially important with early works including Francois Schott's 1600 *Itinerarium Italiae* and J.H. von Pflaumern's 1625 *Mercurius Italicus*. France's first guidebook appeared in 1615, the *Itinerarium Galliae*.

Major advances in guidebooks to countries came with the enormous **growth** of travel in Europe in the eighteenth century. Tourists required both basic information and coverage of the **art**, architecture and scenic wonders encountered while abroad. Notable examples include Thomas Nugent's *Grand Tour* (1749 and subsequent editions) and Mariana Starke's 1802 *Travels in Italy*. The latter utilised a star-ranking system for sights, a method much employed by the Michelin *Green Guide* today. These eighteenth-century guidebooks firmly established the market for such productions and combined itinerary with details of transport, accommodation and costs together with the description of places. They became increasingly specialised and their practical information required a system for updating editions. Thus, by 1800, precursors of the modern guidebook were in place, but they were still inadequate as basic up-to-date practical tools. It was these deficiencies which led to the main advances of the nineteenth century.

Two publishers may be said to have had a decisive influence on the form and growth of guidebooks in the nineteenth century: John Murray and Karl Baedeker. Murray's 1836 *Handbook for Travellers on the Continent* stemmed from personal dissatisfaction with the guidebooks he encountered on his own travels, which he found unsystematic and insufficiently detailed. A significant development was his use of other writers for his guidebook series, most notably Richard Ford for the 1845 *Handbook for Travellers in Spain*. Another distinctive feature of Murray was the commitment to revising material on a regular basis in new editions. The 1838 *Handbook for Travellers in Switzerland* went through nineteen editions before the last one in 1904. This strong relationship between publisher and guidebook was adopted by Karl Baedeker. Modelled on Murray's system, Baedeker produced a series of guidebooks on areas of Europe from 1839 and their increasing dominance of the market was reflected by the start of English language editions from 1861. At its zenith, just prior to 1914, Baedeker was producing seventy-eight titles covering thirty countries. A well-organised bureaucracy of editors and agents was one factor in their success. Such was the size of the market, however, that other major guidebook publishers could also flourish, including Muirhead, Michelin, Joanne and Thomas Cook. It is from these firms that the multiplicity of modern guidebooks is descended.

References

De Beer, E.S. (1952) 'The development of the guide-book until the early nineteenth century', *Journal of the British Archaeological Association*, 3rd series, 15: 35–46. (An authoritative study of the origins of guidebooks in Europe.)

Further reading

Adams, P.G. (1983) *Travel Literature and the Evolution of the Novel*, Lexington, KY: University of Kentucky. (Chapter 2 contains a detailed discussion of early guidebooks.)

Buzard, J. (1993) *The Beaten Track*, Oxford: Clarendon. (Pages 65–77 have an excellent section on Murray and Baedeker guides.)

JOHN TOWNER, UK

guided tour

One approach to the organisation of tourism is to structure the **sightseeing** component of the trip with a guided tour. In essence, guided tours consist of one individual orienting the group and explaining and interpreting the setting. They can vary from a short structured presentation by a **tour guide** in a particular setting for a limited time period, or they may involve extended contact between a guide and a tourist group, lasting days and sometimes weeks.

Guided tours have been a part of the history of tourism since the days of the **Grand Tour**. They offer significant **employment** opportunities within the global tourism industry, they can facilitate the careful management of fragile attractions and events, and they can solve a number of tourist problems in areas such as language, orientation, **security**, cultural interaction and understanding of the **site** or activity. The **training** of guides is well advanced in many countries, with long-standing educational systems and accreditation schemes for those operating in the United Kingdom, Europe and China. There are many guides in Africa, North America, Australia and New Zealand who work in the **national parks** systems. Guide training may include topics such as language skills, environmental or resource knowledge, **health** and **safety** concerns, and social communication skills. Guides are typically both educators and entertainers, though in the longer guided tours they will be frequently called upon to resolve conflicts and provide advice on touristic choices and activities.

Research on guided tours has concentrated on the personality and style of the guide and the stress of the role. Additionally, the guided tour has been conceptualised as a special kind of social situation and the **roles**, **rules**, goals, setting and **behaviours** of the participants in these dynamic social groups have been studied. One of the most common styles is the guided bus tour, but guides also conduct walking tours, boat trips and even snorkelling trips. As a style of tourist activity, guided tours have been popular when tourists have little previous travel experience and when the host culture presents substantial cross-cultural communication problems for the visitor. Much of the early travel to Europe by North Americans was in guided tours, and more recently Japanese and Asia Pacific tourists have been strongly identified as archetypal guided tour participants.

See also: middleman; package tour

PHILIP L. PEARCE, AUSTRALIA

hajj

The **pilgrimage** to Makkah (Mecca) performed by Muslims is called the *hajj*. It is one of the five fundamental Islamic principles. The first is the Islamic testimony that 'Only Allah deserves to be worshipped, Muhammed being His Prophet'; the second is the prescribed Islamic Prayer *(solat)*; the third contains prescriptions about fasting during the month of Ramadan; the fourth describes the Islamic tax (*zakat*); and the fifth prescribes the *hajj*, the pilgrimage. This prescribed journey (according to the lunar calendar) is obligatory for every Muslim at least once in a lifetime, if he or she can afford it. The *hajj* brings millions of pilgrims (definitionally, **tourists**) each year to Saudi Arabia, where this and other holy places are located, and the pilgrims both on *hajj* and *umrah* (done not on the prescribed annual date) form the major part of foreign visitors to this country, which is only now slowly opening its doors to other types of tourist

SEMSO TANKOVIC, CROATIA

hallmark event *see* event; festival

handicapped

The term 'handicapped' refers to a visible or invisible physical or mental condition that substantially limits at least one major life activity of an individual. A **tourist** with such a disability may require changed or different accommodations in dining, lodging and other tourism facilities. **Training** and **education** may be needed to remove attitudinal and/or communication barriers when hosting persons with disabilities.

SHARON S. GIROUX, USA

handicraft

Handicrafts are objects produced using hand processes that are applied to a variety of materials, including clay, fibre, glass, metal, skin, stone and wood. Simple **technology** may augment the work of the hands; **indigenous** aesthetics for colour, motif and design are prominent. Throughout the world, handicraft production is often household-centred and village-based. Whole villages may devote their energies to a common product category such as pottery, wood carving or rug weaving. Handicraft product categories include pottery, textiles, jewellery, metal work, hide items, stained glass, carvings, basketry, furniture and housewares. Handicrafts can serve utilitarian, decorative or **ritual** purposes. In many societies, handicrafts are deeply enmeshed with ethnic, local, regional or national **heritage**. The grounding of handicrafts in cultural heritage provides a platform for **commoditisation** as tourism-oriented objects, leading to a host of labels such as airport art, ethnic art, **souvenirs**, and tourist art or crafts. Production of handicrafts is linked to global **marketing** or **globalisation**, **development**,

gender, **ethnicity**, **employment** and **tradition**.

The term handicraft is often interchanged with craft and handcraft. While handicraft tends to be applied in developing countries, the terms craft and handcraft are generally employed in scholarly literature generated in industrialised countries. Individuals who produce handicrafts may refer to themselves as artists, artisans or craftspersons. The growth of handicrafts as a tourism commodity has engendered positive and pejorative labels that focus on the intended consumers, **tourists**. In addition to terms listed in the opening paragraph, other tourism labels for handicrafts include art by metamorphosis, art of **acculturation**, art of transformation, boundary art, commercial art, ethnoart, ethnological art, folk art and transitional arts. Distinctions between **art** and craft have been made in early scholarly work; however, the boundaries are blurred between the two terms. In particular, a difference between art and craft objects may not exist for many producers. However, some contemporary scholars do delineate ethnic art or craft products from tourist arts. The former category describes objects produced with a local, regional or indigenous orientation, while tourist art production is externally oriented towards a tourist or export consumer.

The study of handicrafts and their linkage to tourism draws from a range of disciplines that includes **anthropology**, art and design, business, cultural studies, environmental studies, international development, **marketing**, **museum** studies, **sociology**, textiles and apparel, and women's studies. A seminal volume edited by Nelson Graburn is credited as one of the first collections of scholarly studies that examined handicrafts and their linkage to tourism. Since the 1970s, academicians have addressed handicraft issues that focus on the various participants in the marketing system of producers, intermediaries and tourist consumers. Both applied and theoretical applications have been explored. Topics concerning **authenticity**, ethnicity, **gender**, tradition, handicraft typologies and linkages between handicraft consumption and tourism styles have provided lively debate revealing the complexity surrounding handicraft production worldwide.

Handicraft typologies describe the stages by which products that have long been produced and used for local consumption are transformed to commodities intended for outside markets, including tourists. As handicraft products are transformed for the latter, change is often evident for raw materials, mechanisation, form, ornamentation, coloration, function, quantity and **quality** of production, and symbolic content. Local artisans associate maintenance of their handicraft traditions with retention of long-used production processes and technologies, despite their application to new tourism-related products. In contrast, outsiders, often unaware of indigenous production patterns, associate loss of handicraft traditions when they observe change in products for tourist purposes.

The intersection of handicraft production with tourism is a vital junction. Fuelled by increasing levels of tourism activity and expenditures worldwide, handicraft production provides a means of income for many marginalised peoples, including large numbers of women and ethnic minority groups. Handicraft production is often noted as a means of income production that supplements agricultural production, fits with rural and agriculture **lifestyles**, is promoted by government and **non-governmental organisations** for development, **foreign exchange** and maintenance of indigenous traditions, and provides handmade objects desired by consumers of industrialised nations. For many of the world's craftspeople, their local markets have been replaced by machine-produced products. Developing handicraft products that appeal to tourists provides one avenue by which artisans may maintain their craft for income generation.

One-third of tourists' expenditures are devoted to retail and **shopping**-related activities. Handicrafts comprise a major category of their souvenir purchases. They define handicraft authenticity in terms of raw materials, production processes, workmanship, aesthetic qualities, indigenous use, and historical and cultural integrity. Through purchasing and using handicrafts, tourists experience indigenous **lifestyles**, expand their worldview, differentiate the self from or integrate with others, express creativity and **experience** aesthetic pleasure. Tourism retailers, in attempting to attract customers, negotiate the traditionality of handicrafts as they blend and juxtapose product features for a tourism **market**.

Further reading

Cohen, E. (1992) 'Tourist arts', in C.P. Cooper and A. Lockwood (eds), *Progress in Tourism, Recreation and Hospitality Management*, London: Belhaven Press, vol. 4: 3–32. (A review of literature of handicraft objects produced for the tourism market.)

—— (ed.) (1993) special issue of *Annals of Tourism Research* 20(1). (Eleven articles are devoted exclusively to tourist arts.)

Graburn, N. (ed.) (1976) *Ethnic and Tourist Arts: Cultural Expressions from The Fourth World*, Berkeley, CA: University of California Press. (Contains an early scholarly work on ethnic art and linkages to commoditisation and tourism.)

Littrell, M. A. (1990) 'Symbolic significance of textile crafts for tourists', *Annals of Tourism Research* 17: 228–45. (Discusses a content analysis of international tourists' perceptions of inherent symbolism in textile craft objects purchased for self-consumption.)

Swain, M.B. (1993) 'Women producers of ethnic art', *Annals of Tourism Research* 20: 32–51. (Explores gender issues and the production of ethnic arts in global capitalism environment.)

JOSEPHINE M. MORENO, USA
MARY A. LITTRELL, USA

hard tourism

Hard tourism refers to **development** primarily focused on quick economic returns. It is characterised by large numbers of **tourists** seeking replication of their own culture in institutionalised settings, with little authentic cultural or environmental interaction with the host. The term is often used as a synonym for **mass tourism**, and an antonym to **soft tourism**.

ROSS K. DOWLING, AUSTRALIA

health

In 1976 the World Health Organisation defined health as 'a complete state of physical, mental and social well-being, not just the absence of **disease** or infirmity.' In spite of this long-standing definition, health in relation to tourism has been researched largely from the perspective of 'tourist illness'. This fails to recognise the full extent of tourism and health issues by limiting attention to both a negative (illness) orientation and visitor-oriented focus. Some research has explored the concept involving destinations with perceived health benefits, such as **spas**, and forms of special interest tourism in which the tourist is actively seeking improved health. Limited work has also focused on the positive concept of the **value** of tourism *per se*, to enhance good health and **quality** of life for both **host and guest**.

Health and tourism require a more rigorous connection to contemporary approaches to an expanding new ecological public health. Adoption of this **paradigm** broadens the framework for discussion of **ecologically sustainable tourism**. It augments the emphasis on economic considerations to a much broader range of variables, such as quality of life of the host **community**, the well-being of **tourists**, and interactions of human health and the natural **environment**. The sustainable tourism debate has recognised the importance of natural and social environments as underpinning economic viability of the industry, but the debate has largely neglected health as integral component.

Most research listed under tourism and health where the latter is equated with illness originates in either of two disciplinary backgrounds. The first group, medical investigators interested in the epidemiology of tourism-related disease, rarely differentiate between **tourists** and others. There has developed a specialised field of travel medicine, emporiatrics, with its own enterprises, professional associations, journals and regular conferences. A second and broader group of social scientists have demonstrated their interests in the relationship of tourism-related illness to **motivation** and **experience**. These researchers generally have little background in public or community health. Each group places their primary focus on illness and its related problems which includes the incidence and aetiology of tourism-related disease, disruption of travel and consequent impacts on consumer enjoyment, the legal responsibilities of tour operators and agents, provision of pre-travel advice on health matters, adequate primary health care at destinations and a wide range of preventative measures.

Travel-related illness encompasses diseases, injury, **recreation** and tourism-related hazards. Diseases are related to **food**, water or air, or can be insect (vector) and animal-borne or communicable through human contact (including sexually transmitted diseases). Injury includes transport-related accidents, occupational safety hazards and animal bites. Recreational hazards are sunburn and other skin conditions, water-related and sport-related accidents, or illnesses caused by recreational drugs, alcohol and sex. Travel-related hazards include air and sea sickness, exhaustion, altitude, stress and personal security incidents. As well as considering the type of illness, many factors relate to the destination, the tourists and the activity they undertake. Factors of destination which impact on health are stage of development, medical services, general **sanitation**, geographical location and **climate**. The tourist's age, gender, health status, special conditions, duration, tourism style and recreational activity are all important.

Each grouping has high, medium and low risk categories with specific advice, treatment and associated requirements. Illness prevention/control deals specifically with travel advice, insurance, immunisation, prophylaxis, epidemiology, notification and quarantine procedures. Travel health counselling deals primarily with three major areas of vaccine: preventable diseases, vector-borne diseases such as malaria, and **food** and water-borne diseases, especially diarrhoea. Infectious diseases account for a very small fraction of deaths occurring among tourists, but represents a major cause of non-fatal illness. Cardiovascular disease, more prevalent in the elderly traveller, accounts for 50 per cent of deaths; accidents and trauma for around 25 per cent of mortality in overseas travellers. A special branch of travel medicine is the development of telemedicine which delivers specialist advice and care to the tourist in situ.

There has been a mixed response from various sectors of the industry to issues relating to health. Many are concerned that reports of illness linked to tourism will be detrimental to business. The **Pacific Asia Travel Association** and the **World Tourism Organization** have led efforts to educate and inform the industry employees. International bodies such as the World Health Organisation and national bodies have assumed the work of educating health professionals and tourists on health-related issues. There have been several 'travel safe' campaigns, provided initially to respond to concerns about **AIDS** but expanded to encompass the full range of issues.

The new public health integrates human health into ecosystem well-being. Public health has its origins in mid-nineteenth century Europe, introduced to deal with urban problems of air and water quality, food-borne infection and diseases arising out of poor hygiene and poverty. The most important solutions to these problems came from changes to social conditions rather than advances in medical technology. Environmental health, preventative medicine and social reform were key components of this notion of public health. In more modern times, despite changes in the pattern of illness from infectious disease to **lifestyle**-related illnesses, this conception of public health began to fade. By the 1970s, the emphasis shifted from 'public' to 'individual' risk factors. The link between social change and public health weakened. Less emphasis was placed on the interdependence of health and the social and physical environments. Individuals were seen as responsible for their lifestyles and health risk factors.

In the 1980s, the World Health Organisation and other authorities recognised alarming trends in health outcomes related to a reduction in health equity. The Ottawa Charter of 1986 marked another reorientation of public health away from the dominant notion of the individual, simplistic cause–effect interventions and surveillance approaches, towards a more complex environmental and social model. This ecological approach served to emphasise the interconnections between humans, their physical and social environments, and their health.

Tourism is a modern phenomenon linked to a number of positive causal factors, including a rise in the standard of living, desires to improve quality of life, increases in **recreation** and **leisure** pursuits, and the desire to reduce the stress of modern living. As part of that improvement, it is logical that successful tourism should be closely linked with health promotion. The World Commission on Environment and Development report *Our Common Future* (also known as the Brundtland Report) (1987) stressed the need to reconcile economic development with the resource endowment of the natural world and equity in human well-being. The challenge posed by the Brundtland Report was to strive for **sustainable development**, though this admirable objective has been clouded somewhat by the ambiguity of the term 'sustainable'. Health is clearly part of the rhetoric of **ecologically sustainable tourism**, but is yet to inform **policy** and **planning** adequately.

For tourism development to be environmentally responsible, and if the use of the term health in tourism research is to embrace a more ecological and broader approach, it is necessary to consider biohistory, or the study of human situations in the history of life on earth. This begins with understanding biological processes, ecological principles and the sensitivities of biological systems of which humans are a part. Many argue that the future well-being of humanity depends on satisfying both the health and well-being needs of the biosphere's ecosystem, and the health and the needs of human beings. Humans in developed countries are generally well off in terms of life expectancy, though their quality of life could arguably improve; whereas those in the developing world are far worse off. The biosphere is necessarily of global concern, due to the universal ecological impact of humans, the combined effects of an extraordinarily high material standard of living and a massive increase in world population. A key question is whether it is possible for human beings to lead healthy and enjoyable lives in a society characterised by a steady rate of resource and **energy** use. Tourism, unfortunately, usually encourages high energy usage in developing as well as developed countries in a manner that is not sustainable.

Therefore, the conceptual framework for consideration of health and tourism interactions should include wellness and illness, promotion and prevention. It should not be confined only to reporting on diseases, accidents and health risks related to travel. Wellness promotion and illness prevention considers the broad interactions of physical, mental, social, economic and environmental factors with the policy and planning frameworks which impact on stresses related to quality of life. The Brundtland report, in its broadest sense, is the strategy for sustainable development which aims to promote harmony among human beings and between humanity and nature.

General concepts relevant to health and tourism include issues of equity and access, in particular relating to the treatment of illness and injury in developing countries or remote locations; anthropocentric versus biosphere attitudes and values which underpin approaches to human health; individual well-being versus community health; specialist versus holistic (see **holistic approach**) integrated human health with environmental/ ecosystem health; and sociocultural **impacts**. Tourism-specific concepts include both tourist versus host well-being and links between wellness and quality of life of host **community** with visitor **satisfaction**, as well as issues of safety and risks.

Further reading

Brown, V.A. (ed.) (1989) *A Sustainable Healthy Future: Towards an Ecology of Health*, LaTrobe University and Commission for the Future. (Papers from a World Health Organisation workshop aiming to set a new agenda for public health policy.)

Clift, S. and Page, S.J. (eds) (1996) *Health and the International Tourist*, London: Routledge. (Deals with health problems and preventative measures from the perspective of tourists.)

World Commission on Environment and Development (1987) *Our Common Future*, New York: Oxford University Press. (Global issues of sustainable development.)

World Health Organisation (updated annually) *International Travel and Health: Vaccination Requirements and Health Advice*, Geneva: WHO. (Prescriptive information on immunisation requirements, prophylactic measures and other health advice for travellers abroad.)

ROBYN BUSHELL, AUSTRALIA

hedonism

Hedonism is the unbridled pursuit of pleasure. Instant gratification of the ego and its **anticipation** are characteristic of the **tourist as child**. Experientially, hedonism can be particularly ego-enhancing in **Third World** settings visited by First World tourists, which may explain why some all-inclusive **resorts** are so designated. Motivationally, hedonism is linked to the behavioural excess exhibited in **activities** sanctioned at home.

GRAHAM M.S. DANN, UK

heliocentrism

Heliocentrism conveys the centrality of sunshine in promotion targeted at the **tourist as child**. Pictorially, the sun is a Mother icon often depicted with bright colours and a smiling face, representing sensual happiness. Verbally, the message emphasises the human body and an erotic tan (see **eroticism**). **Sunlust** tourism dates from the 1920s. The negative publicity associating it with skin cancer is far more recent.

See also: promotion, place

GRAHAM M.S. DANN, UK

heritage

Dictionary definitions of heritage emphasise the notion of inheritance, what is or may be transmitted from ancestors. Tourism usage often classifies it as cultural or natural heritage. The World Heritage Convention (1982) also enjoins signatory nations to ensure the identification, protection, **conservation**, presentation and transmission to future generations of the world's cultural and natural heritage. Culturally, heritage may be broadly defined to encompass the **history** and ideas of a people and/or a country, **values** and beliefs, buildings and monuments, **sites** of important past events, the **arts** (literature, **music**, dance, sculpture, art), traditional events and festivals, and traditional **lifestyles**. Natural heritage refers to **landscapes** in which pristine **wilderness** (unlogged forests, undammed rivers, unfarmed mountains) predominate. Landscapes modified by human endeavour are regarded as part of the cultural estate.

It is to some degree artificial to separate heritage into cultural and natural divides because the values associated with landscapes are cultural. Furthermore, for many **indigenous** peoples their cosmology is based on the indivisibility of humankind and **nature**. Their past is linked with the present into the future and heritage is a lived **experience**. These issues underline the difficulty of defining 'heritage' because, ultimately, it can be virtually anything that anyone wants it to be. This becomes apparent when the social and political significance of heritage is considered. The social significance of heritage lies in its association with **identity**. It is fundamental in helping individuals, communities and nations define who they are, both to themselves and to outsiders. It may provide a sense of belonging in a cultural sense and in terms of place. It can be personal past or impersonal heritage.

In this context, ownership of heritage is linked to ideology, and its **symbolism** takes on strong political overtones. The selective conservation of some sites rather than others and accompanying **interpretation** and presentation may be used to sustain or demolish a particular version of history or promote certain political or social **values**. Historiography (the study of writings of history) reveals that contemporary values and circumstances always influence the interpretation of historical fact. Recent studies in the **sociology** of **development** have also brought about a reconceptualisation of such key concepts as 'culture' and '**authenticity**'. They stress that each generation redefines its heritage in response to new understandings, new experiences and new inputs from an ever-increasing range of sources, both internal and external. Interpretation may change to suit or satisfy particular needs because heritage, its ownership and its presentation will involve considerations of changing values, **power** structures and politics. Heritage is thus not a 'thing', static in time, but through continuous interpretation it may be viewed as a process.

Since the end of the Second World War there has been a global proliferation of interest in heritage, particularly but not solely in the Western world. Many writers consider that this preoccupation with the past is a response to the extent and scale of social and environmental dislocation, whereby the mass of people in post-industrial societies have become separated from the continuous traditions of the past. Modern society has turned back to the past to understand how it has arrived at the present. In rural societies, such as predominate in many **Third World** countries today, the past is not distinguished as a separate realm but is continuously reproduced in the everyday lives of the present. For modernised societies, the breach with the past has meant that history, devoid of its previous role as an active constituent in informing the present, is seen as completed and becomes a matter for **nostalgia** and curiosity. All attempts to bring back the past emphasise its separation from the present. This break cannot be bridged because, just as the pace of change has created the need for a sense of the past, it has also removed any possibility of that need being satisfied.

If the pace of change in recent times has created a demand for the past, then tourism has often supplied the **product** to meet that **demand**. Globally, the heritage industry has become a very significant tourism sector. In countries such as the **United Kingdom**, **France** and **Italy**, heritage has become the major focus of their entire tourism industry. In Asian and Middle Eastern countries, such as **China**, **Indonesia**, **Thailand**, **Egypt** and **Israel**, the monuments of past civilisations are major attractions for both near and far tourists. In Central and South America the vanished civilisations of the Aztecs, the Incas and the Mayas constitute a nucleus for tourism. On a global basis, the World Heritage Convention has also given very significant impetus to natural heritage as a major component of tourism. Thus the Serengeti Plains of Africa, the Grand Canyon of the **United States**, the Great Barrier Reef of **Australia** and many national **parks** which have received World Heritage Site Listing have become recognised not just as national symbols but as part of the global heritage.

One of the conclusions drawn by those who contend the putative death of the past and its separation from present everyday life is that in some Western societies heritage becomes little more than a commodity, a sanitised package offering a glorified version of the past for public consumption at heritage shrines. Much tourism use of heritage has been criticised as exploitative, where commercial considerations outweigh historical accuracy and where **commoditisation** results in the static portrayal and production of **staged authenticity** in village folk **museums**, in pseudo-festivals and spurious events, in trivialisation, in mass-produced handicrafts and artefacts (so-called 'airport art') and similar fabrications. Invention, **substitution**, reconstruction, replication, reproduction, simulation and permutation are accepted and play havoc with the concept of authenticity.

The counter-argument is probably best presented by the World Heritage Convention, where exacting standards attesting to the unique and outstanding cultural and natural values of a site must be verified before designation as a major component of global heritage. Any site proposed for listing must also be accompanied by a **management** plan for the site's conservation and presentation in which educational and scientific values are paramount. Heritage as touristic entertainment is not considered by the Word Heritage Committee, although it may be tourism which provides the economic underpinning for preservation and conservation of the site.

Further reading

Bonniface, P. and Fowler, P. (1993) *Heritage and Tourism in the 'Global Village'*, London: Routledge. (Survey of heritage issues in the context of tourism development and globalisation.)

Hewison, R. (1987) *The Heritage Industry*, London: Methuen. (Generally critical of the commercialisation of heritage.)

Merriman, N. (1991) *Beyond the Glass Case*, Leicester: Leicester University Press. (Examines the role of museums in heritage preservation and education.)

Tilden, J. (1977) *Interpreting Our Heritage*, 3rd edn, Chapel Hill, NC: University of North Carolina Press. (Formulates principles of interpretation and their application to heritage sites.)

Uzzell, D. (ed.) (1989) *Heritage Interpretation*, vol. 1,

The Natural and Built Environment, London: Belhaven. (An anthology of papers addressing all major aspects of heritage issues.)

World Heritage Committee (1984) *Operational Guidelines for the Implementation of the World Heritage Convention*, WHC/2 revised, Paris: UNESCO, World Heritage Committee. (Specific guidelines intended for worldwide heritage preservation.)

TREVOR SOFIELD, AUSTRALIA
SARAH LI, AUSTRALIA

hiking

Hiking means to walk a great distance, usually in rural or **wilderness** settings. It can be differentiated from walking for pleasure not only by the greater distances involved, but often by the need for specialised equipment such as hiking boots, backpacks, lightweight protective clothing and camping gear, and non-refrigerated preserved foods. Hiking is usually combined with tourism ranging from a day-trip on a **package tour** to a tailored trekking trip in remote areas.

TOM BROXON, USA

hill stations

Tourism **resorts** that are nested in the highlands or mountainous areas are located in hill stations. They were developed during the colonial times as retreats from the hot and humid lowlands of the tropical/subtropical colonies. The temperature and climatic differences still provide the major attractions of these hill stations, but largely for **domestic tourism**. Hill stations are especially prevalent in South and Southeast Asia.

MARTIN OPPERMANN, AUSTRALIA

hippie

Hippies are those who reject the conventions and values of middle-**class** society. Closely associated with the 1960s 'flower power' movement, hippies espouse peace and love, wear colourful clothes and frequently adopt communal **lifestyles**. As **tourists**, they are typically long-term adventure travellers (for example, on the 'hippie trail' to Asia) who sometimes establish communities at particular destinations, most famously in Goa and Kathmandu.

RICHARD SHARPLEY, UK

historical tourism

Aspects of the past are increasingly used in the construction of tourism products for a number of expanding and varied **markets**. Such historical tourism includes trips principally motivated by aspects of the past, but also historical excursions and **activities** undertaken during non-historically motivated holidays.

The tourism **product** may be composed of preserved, collected and interpreted historical artefacts, sites, buildings, districts and even whole towns, including memories and associations of places with historical events and personalities. The selection from the past as a quarry of possibilities and its presentation for contemporary consumption results in **heritage** which may draw upon folklore, mythology and products of the human creative literary and artistic imagination. Historical **tourists** may be motivated by a search for **identity** or **fantasy**.

Historical and **heritage** tourism are almost synonymous and have an overlapping relationship with many other forms of tourism. These include the historical aspects of cultural, **art** and **festival** tourism, many aspects of place-specific and **ethnic tourism** including flows, costume and **gastronomy**, and those trips associated with **museums**, art galleries, monuments and archaeological and historical sites. More broadly, historical tourism accounts for a large proportion of **urban tourism**.

Two controversies surrounding this form of tourism are particularly prevalent. First, the heritage consumed in historical tourism is defined and shaped by the contemporary demands for it, unlike history where the aim is the accurate description of past events. Thus, **interpretation** rather than the resource itself plays the critical role in the shaping of the product. Therefore, the **authenticity** of the consumer **experience** takes precedence over the authenticity, or even the presence, of the resource. The importance of the explanatory 'marking', the interpretation centre and the historical **theme park** may lead to accusations of **staged authenticity** or **commercialisation**. Second, the past has many other contemporary uses other than in the tourism industry. This can result in competition between different users of the past. The heritage created by **local** residents in support of their identity, or by their governments seeking legitimisation, may not be the same as that created by the tourism industry, a situation which leads to conflicts between users.

See also: archaeology; commercialisation; conservation; cultural tourism; nationalism; nostalgia

Further reading

Prentice, R. (1993) *Tourism and Heritage Attractions*, London: Routledge.

Tunbridge, J.E., and Ashworth, G.J. (1996) *Dissonant Heritage: The Management of the Past as a Resource in Conflict*, London: Wiley.

G.J. ASHWORTH, NETHERLANDS

history

History is concerned with the dimension of time and attempts to understand human processes within that context. Its emphasis on the temporal dimension provides a depth which comes from viewing societies not as static, but in a continuous process of change. Fundamentally, the discipline considers the transformation of phenomena (people, places, institutions, ideas) through time, from one state into another. History can thus contribute a vital evolutionary perspective to the study of tourism. It can help an understanding of the origins of this human activity and isolate what may be essentially short-term **trends** in its **development** from what constitute basic long-term continuities.

The philosophy and methodology of history have undergone a number of changes. It has been subject to the ideographic (uniqueness) versus nomothetic (generalisation) debate, positivist approaches to knowledge via the social sciences, as well as the concepts of Marxism, humanism and **structuralism**. The influential *Annales* school in France stressed the need for an interdisciplinary approach to history and a focus on structural analysis rather than a history of events. In more recent times, some historians have returned to a more narrative mode (arranging material in a chronologically sequential order). Today, a whole range of approaches to the past exists within the discipline, but the central concern of understanding change through time remains its distinctive contribution to knowledge.

Two main strands of historical research in the field of tourism can be identified. One comes from the work of mainstream historians, the other from social science tourism researchers who seek a temporal dimension to their subject. Poor interdisciplinary links between these strands, however, continue to check the overall development of research in this area.

Historians in the United Kingdom, France, Germany and North America have undertaken significant research into the evolution of tourism. In the former, the majority of historians tend to approach the subject through the broader arena of **leisure**, and the field of leisure history has seen considerable growth in the last fifteen to twenty years. Through these efforts, the major trends in British leisure in the nineteenth and twentieth centuries have become clearer and the underlying processes of change explored through a lively theoretical and ideological debate. This depth of analysis has perhaps not yet fully developed outside this country, where the role of urbanisation, industrialisation, class structures and technology and their relationship to leisure and tourism have all been examined and questioned. This work of mainstream historians is critical for understanding a number of processes which underlie the evolution of tourism. For instance, the notion of mass follows class is frequently cited in the tourism literature as a basic aspect of development over time. But, many historians would question this mechanistic process and suggest that class interrelationships have been far more complex, with social groups often creating their own leisure worlds rather than simply imitating others.

The other main strand of tourism history has come from social science tourism researchers seeking a temporal perspective on the subject. Thus, conceptual frameworks and models of today would gain in validity if they can also be successfully related to tourism in the past. One aspect of this area of research has been the notion of the **destination life cycle**. A particular difficulty here, however, is the adequacy of the historical data available and a tendency for *post hoc* rationalisation in order to fit the model into historical contexts. This stricture applies to a number of other tourism frameworks which can have a somewhat simplistic view of historical processes. The weak links with historians contributes to this particular difficulty.

All research requires the careful consideration of the range of source material available. With historical data, questions concerning **authenticity**, purpose and representativeness are major issues. There are fundamental problems of bias in both archive and physical evidence data which can be summarised as selective deposit and selective survival. Did only certain types of people or institutions leave the kind of data being analysed? Have other vital sources of data simply disappeared over time? These questions relate to the data available for tourism history, which can be categorised into four main areas.

The first area is statistical records. The systematic measurement of tourism only began in the twentieth century, when its increasing volume and economic significance created a need for its statistical measurement. Before then, there was no official **government** role in tourism. From the 1920s, however, national governments increasingly recorded flows of tourism and expenditure, and this output accelerated markedly after 1945. Prior to this era, historians have to make use of scattered and fragmented material such as visitor lists and census data, occasional transport company **statistics**, or inquiries into living conditions and working hours. The second area, personal documents, has provided a major source of information. Diaries, letters and journals of travellers provide a biased but sometimes the only record of certain forms of tourism. Studies of elite travel in Europe (such as the **Grand Tour**), North America and elsewhere have often depended heavily on this form of data.

Mass communication is the third area. Most literate societies produce material to inform, entertain or persuade the public. In terms of tourism history, such material includes guidebooks and route **maps**, **magazines**, journals and newspapers. Other works which may refer directly or indirectly to tourism are poems and satire, cartoons and novels. Historians of literature have utilised forms of **content analysis** to gain insights into contemporary culture and the role of travel in past societies. Other sources, the fourth area, can include material used by classical historians and archaeologists in research on the ancient world. For instance, the inscriptions and graffiti on the inn walls at Pompeii or monuments in **Egypt** have been studied, as have the papyrus letters that have survived. Archaeologists have examined the remains of holiday villas, inns and **restaurants** in ancient **Italy**. For more recent times, material can include diplomatic records or the archives of travel firms. Oral history clearly has much to offer for the reconstruction of **leisure** and tourism. In addition, historians can turn to the visual arts, such as paintings, where the portrayal of leisure practices in certain ways can illuminate their role in society. **Photography** offers similar possibilities.

Tourism history research has, so far, focused on a number of basic themes. The first is that of the ancient world, where meticulous study of sources by classical historians has revealed much about tourism in ancient Greece, Egypt and Rome. For the medieval period, the principal theme has been that of the **pilgrimage** in Europe and the **Holy Land**. Another era of extensive journeys was that of the Grand Tour. Here, there is a wealth of material in the form of traveller's diaries, journals and guidebooks as well as direct connections with contemporary art and architecture. Spa and seaside resort development in the United Kingdom, Europe and North America has attracted detailed research from historians and geographers where both individual case studies and broader histories of resort growth have made this one of the strongest areas of historical research. In North America especially, the role of **national parks** and **wilderness** areas has received much attention and the history of many parks has been thoroughly documented. Partly related to this theme has been the concern of historians and geographers with the wider topic of changing aesthetic tastes for **landscapes** and the subsequent influence on tourist perceptions and patterns.

These strongly developed research themes have, however, contributed a particular slant to tourism's history as it has developed so far. In many ways, the emphasis has been on the occasional and prestigious episode, rather than the mundane and regular, in people's lives, whether rich or poor. There has also been a tendency to study tourism as isolated events rather than relating them closely to both **lifestyle** and life cycle. Certainly in the more popular literature, the activities of the wealthy continue to dominate the picture with suggestions of a past 'golden age' of tourism before the masses. In the literature, the role of certain business entrepreneurs like Thomas Cook is seen as basic to development as is the role of certain forms of transport **technology**. An appreciation of the complexity of historical processes still remains to be incorporated from the work of historians.

Fundamentally, however, the history of tourism as developed so far is that of Western tourism. The industry is portrayed as originating within a Western heartland and spreading ever outwards across the globe while at the same time permeating downwards from the wealthy to the wider mass of the population. But very little is known about forms of tourism in past societies in **India**, **China** or elsewhere. Only by expanding historical research into other cultures and societies can a fuller appreciation of the evolution of tourism be attained.

See also: archaeology; paid vacations; travel literature

Further reading

Bailey, P. (1989) 'Leisure, culture and the historian: reviewing the first generation of leisure historiography in Britain', *Leisure Studies* 8: 107–27. (Provides an outline of the ideological and theoretical debate on leisure history between British historians.)

Butler, R.W. and Wall, G. (eds) (1985) 'The Evolution of tourism: historical and contemporary perspectives', *Annals of Tourism Research* 12(3). (A special journal issue which was one of the first attempts to add a major historical dimension to the field of tourism research.)

Towner, J. (1996) *An Historical Geography of Recreation and Tourism in the Western World, 1540–1940*, Chichester: Wiley. (Links historical and geographical approaches to the growth of tourism in the United Kingdom, Europe and North America.)

Towner, J. and Wall, G. (1991) 'History and tourism', *Annals of Tourism Research* 18(1): 71–84. (A review of historical approaches to tourism in Britain and North America.)

Walton, J.K. and McGloin, P.R. (1979), 'Holiday resorts and their visitors: some sources for the local historian', *Local Historian* 13(6): 323–31. (Provides detail on the range and nature of historical data for resort studies.)

JOHN TOWNER, UK

holiday *see* vacation*;* paid vacation

holiday camp

The term 'holiday camp' refers to a form of **vacation** provision characterised by collective **experience**. Typically, the camps comprise multiple units of **accommodation**, combined with shared dining and social facilities and with organised activities.

The first camp in the **United Kingdom** was formed in 1894, and was designed to meet the needs of young men from the industrial towns of Lancashire. Campers were accommodated in bell tents, and days were spent together enjoying 'good, clean fun'. Altruistic motives led to other camps, offering opportunities for children, miners and factory workers to spend a week or more in a healthy **environment**, usually by the sea.

A distinction can be drawn between the first generation of camps through to the early 1930s, which tended to be small (accommodating a few hundred **visitors** at most) and run as family firms or to meet specific labour or other needs, and a second generation of camps which were large (catering for up to 10,000 visitors at any one time) and highly commercialised. The latter were the creation of entrepreneurs attracted in the 1930s by the anticipated growth of holiday making, supported by **legislation** requiring employers to pay their staff for a week away from work. The peak in terms of numbers of camps and visitors is the period between the 1930s and 1960s.

Holiday camps offer an exemplar of changing tourism patterns in the twentieth century. It is no coincidence that camps were most popular at a time when working-**class** families experienced their first holidays away from home, and when it was reassuring to be organised and in the company of others. It is also no coincidence that, with a progressive emphasis on consumerism in society, holiday camps as such have passed out of vogue. Modern versions, typically renamed holiday centres or **leisure parks** and financed by corporate capital, cater for needs far removed from those of visitors to the pioneer camps.

Although holiday camps occupy a distinctive place in British tourism history, there are closely related variants in other countries. These include summer camps in the **United States**, *colonies de vacances* for children in **France**, camps along the Black Sea for trade union members in the former Soviet Union, and contemporary **models** of collective holidays for adults pioneered by Club Mediteranée and for families by Disneyworld.

Further reading

Ward, C. and Hardy, D. (1986) *Goodnight Campers! The History of the British Holiday Camp*, London: Mansell.

DENNIS HARDY, UK

holiday home *see* second home; summer cottage

holiday with pay *see* paid vacation

holistic approach

Based on the principle of holism, the holistic approach provides a comprehensive and integrated approach to the study of tourism. The basic tenets for holistic approach include all key characteristics of tourism: movement of persons, **transportation**, stay and **activities** at **destination**. Determinants of **demand** and **supply** and the associated underpinning factors should be taken into account. A holistic approach involves all elements of tourism, whether regarded as an industry or an academic field of investigation.

DAVID LESLIE, UK

Holy Land

Palestine is the Greek name for the area which Christians call the Holy Land. The word is derived from the name of the ancient inhabitants of the littoral, the Philistines, and, according to the oldest sources, it was also called the land of the Amorites or Aharu (the Western Lane). In the Old Testament it is mentioned as Canaan, and is later known as the land of the Hebrews, the land of Israel or the Promised Lane. After the Babylonian Captivity it was called simply Judaea, while the name Palestine, still used today, prevailed in Greek–Hellenistic–Roman times. Politically, it mostly corresponds to the state of **Israel**, also including the West Bank and Gaza, and extending as far as some parts of Lebanon and Syria.

Geologically and geographically, this area is an extension of Syria and is contiguous with the Sinai peninsula. The parallel mountain chains of the Lebanon and Anti-Lebanon are located in the north, the West Bank in the south, and the East Bank in the east. The valley of the Jordan extends between Lake Tiberias (Sea of Galilee, Chinnereth, Chinneroth, Gennesaret) and the Dead Sea, then continuing across the Negev to the Gulf of Aqaba in the Red Sea. The Holy Land is usually divided into Galilee (the northern part), Ephraim or Samaria (the central part) and Judaea (the southern part). With respect to its natural characteristics, the prevailing **climate** is subtropical with two marked seasons of the year (the rainy season and the dry season). Archaeological finds bear witness to the culture of the inhabitants which goes back 7,000 years BC, and numerous archaeological sites and artefacts have been found dating from the fourth millennium BC onwards.

Once wooded, fertile and well-inhabited, this area was occupied by the Hebrews around 1200 BC. Later, during the reign of King David, they united and established a powerful state around 1000 BC. Situated at the crossroads of great cultures and civilisations, this land has always attracted numerous conquerors, including the Egyptians, Assyrians, Babylonians, Greeks, Romans, Arabs, Crusaders, Mameluks, Turks, French and British. The state of Israel was created in 1948, and the frequent **wars** that have broken out since then have led only to truces, with no global peace agreement with the neighbouring countries.

The Holy Land does not bear this epithet only because the Christians believe that Jesus Christ was born, lived and died there, but is also regarded as holy by the members of the other two monotheistic religions, Judaism and Islam. This is so mainly due to the numerous holy places that remind them of their national and religious history. In this respect, special importance is attached to the holy city of Jerusalem, which has existed continuously for three thousand years. Therefore, the Holy Land has been a frequent tourism **destination** for **pilgrims** and has been described in texts giving accounts of their experiences on formerly long and hazardous journeys. Modern religious tourism relies on a well-developed and specialised local **supply** of accommodation, transport and other capacities and **services** and on a timeless and unique **image** in the **market**.

Further reading

Allcock, J. (1988) 'Tourism as a sacred journey', *Loisir et société* 1(1): 33–8. (Describes elements of tourism as a phenomenon with a number of terms belonging to religion.)

Lefeuvre, A. (1980) 'Religious tourism and pilgrimage', *On the Move* 120: 80–1. (Contains the basic characteristics of journeys with religious motives, among which pilgrimages to the Holy Land have a special importance.)

Rebic, A. (1983) *Vodic po Svetoj zemlji* (A Guidebook of the Holy Land), Zagreb: KS. (In Croatian, contains all the relevant information: religious, touristic and others for the entire region of the Holy Land.)

TOMISLAV HITREC, CROATIA

home delivery

Customers who wish to eat **restaurant** food in their own homes are able to telephone their order for home delivery. Food is usually delivered hot by courier within thirty minutes of ordering. This style of service is found in the ethnic and **fast food** segments. Many tourism businesses now accommodate this concept in their properties.

PETER JONES, UK

homecoming

The term 'homecoming' denotes a sentimental and social reuniting event of specific memberships. It usually implies 'alumni' within the educational settings. Homecoming has been extended to include reunions, be they families, clans, athletic teams, institutions, military or corporate members. These membership gatherings are always polarised by special **event** activities and programmes. Most are cyclical in nature. Though some are once in a lifetime happenings, they still contribute to the tourism business of the host **destination**.

See also: regression

JOSEPH KURTZMAN, CANADA
JOHN ZAUHAR, CANADA

homelessness

One of the paradoxes of global tourism is that, in order to **escape** from the growing sense of fragmentation and 'cognitive homelessness' associated with **postmodernism**, tourists render themselves (albeit temporarily) homeless. This situation is doubly ironic, given that local hosts of destinations are themselves increasingly displaced by advancing tourism **development** in the world's **pleasure peripheries**.

TOM SELWYN, UK

homesickness

Homesickness is a form of **culture shock** experienced in tourism caused by frustration and misunderstanding of the norms and loss of control in an unfamiliar **destination** setting. The person wishes to be back in a familiar and secure home culture. Homesickness is less common among **tourists** than sojourners, due to their short stay, little involvement with locals and tendency to seek 'differences' in host cultures.

ATSUKO HASHIMOTO, CANADA

homosexuality

Homosexual tourism has two components. The first appeals solely to homosexuals. This is illustrated by gay travel guides like *Spartacus*, *Ferrari for Men* and *Gia Pied*, or by companies like Man Around. The second appeals primarily to homosexuals but also crosses into mainstream society through gay and lesbian carnivals. San Francisco, Auckland, and Sydney have 'parades' which fit this category.

CHRIS RYAN, NEW ZEALAND

Hong Kong

Formerly a British colony, Hong Kong became a Special Administrative Region of the People's Republic of China on 1 July 1997. Hong Kong is the gateway to China for many tourists, a major stopover and an aviation hub for Northeast Asia. In recent years the gateway and stopover role has expanded with the development of a new airport, redevelopment of the main conference centre and the construction of new hotels, with much of the development being focused on the waterfront. In addition, bus and rail transport **infrastructure** has also been improved. With tourism receipts of $9.7 billion, almost doubled from 1990, the industry is a substantial contributor to the economy and accounted for approximately 8 per cent of the GDP in 1995.

In 1991 Hong Kong had over 6 million visitors. By 1995 this figure had grown to over 10 million. The majority of non-mainland China inbound tourists are from Taiwan, **Japan** and Southeast Asia. The **United States** is the major non-Asian market. Approximately half of tourism expenditure is on **shopping**, indicating its importance as an **attraction** of visiting Hong Kong. With high economic growth and increased personal disposable income, Hong Kong has become a sought-after tourism market. **Outbound** tourism grew at over 10 per cent per annum, between 1990 and 1996, with trips to destinations other than Macau and China growing by approximately 20 per cent per annum. Over 3 million residents travelled overseas in 1996.

The relationship with China is critical to the region's fortunes. The political unrest in China in 1989 had a dramatic affect on the Hong Kong economy and on the tourism trade in particular. Tourist arrivals fell by 4 per cent in 1989 to 5.5 million. The inbound situation in Hong Kong only improved as political unrest in China died down, leading to a growth in inbound to China and heavy discounting on the part of hotels, **tour operators** and airlines. Following the return to Chinese sovereignty in 1997, the maintenance and promotion of stability, **security** and **safety** and Hong Kong's free economy are the key factors influencing the **future** of its tourism industry.

Further reading

Hall, C.M. (1997) *Tourism in the Pacific Rim: Development, Impacts and Markets*, 2nd edn, South Melbourne: Addison-Wesley Longman.

Hobson, J.S.P. (1995) 'Hong Kong: the transition to 1997', *Tourism Management* 16(1): 15–20.

C. MICHAEL HALL, NEW ZEALAND

horizontal integration

Horizontal integration takes place when properties in the same sector of the tourism industry, such as the **accommodation** sector, are merged with the intent of improving **market** share and increasing competitiveness. In theory, horizontal integration leads to **economies of scale** and thus to cost savings and price reductions. This allows the company to become more competitive, nationally and/or internationally.

See also: vertical integration

FRANCISCO SASTRE, SPAIN
IMMACULADA BENITO, SPAIN

hospitality

The term 'hospitality' has emerged as the name by which lodging and food service managers would like their industry to be perceived. As well as simplifying and shortening the phrases 'lodging and food service' (as used in the United States) or 'hotel and catering' (UK terminology), it also conveys an **image** that reflects the tradition of service that goes back over many centuries to the earliest days of innkeeping. Many industry associations, academic journals and publications have adopted this term, even though it is not necessarily widely used by the general public. For instance, in the United States the **Council on Hotel Restaurant and Institutional Education** (the Hospitality and Tourism Educators Association) calls their quarterly publication the ***Journal of Hospitality and Tourism Research***; likewise in the United Kingdom, the Hotel and Catering International Management Association (the industry's professional association) changed the name of its journal to *Hospitality*. The term has also been adopted by universities and colleges which offer programmes in hospitality management.

Definition of the word 'hospitality' is relatively straightforward. A typical dictionary defines it as 'kindness in welcoming strangers or guests'. However, defining the hospitality industry is much more difficult. This is because different sources, such as government agencies, market intelligence consultants and other experts, use different criteria for defining the industry. In essence, hospitality is made up of two distinct services: the provision of overnight **accommodation** for people staying away from home (lodging), and the provision of sustenance for people eating away from home (food service). Lodging is made up of two main sectors: commercial lodging is predominantly hotels, while 'institutional lodging' comprises hospitals, universities, prisons and so on. The food service industry is also made up of a number of different sectors. These include **restaurants**, offering fine dining, speciality menus or quick service; social catering; contract food service; in-flight catering; and various types of institutional catering, such as schools and hospitals.

In 1952, Walter Bachmann wrote that a hotelier and restaurateur was both 'host and businessman. ... The host should be cheerful and happy, an artist in living who enjoys everything that is good and beautiful'. This concept of hospitality has existed for thousands of years. Ancient Greece and the Roman Empire had taverns providing food and shelter for travellers. It is estimated that the Romans built nearly 10,000 inns, roughly twenty-five miles apart, to support the rapid movement of government officials and couriers, rather like the stations set up by the Pony Express in North America in the 1850s. Other great empires, such as Persia, the Chinese Empire and Japan, also developed similar provision for travelling dignitaries and people of wealth.

In Europe, such provision declined during the Dark Ages until Charlemagne in the eighth century supported the development of rest houses for **pilgrims** travelling to holy **sites**. Later on, monasteries and mediaeval guilds began to welcome travellers of all kinds. The accommodation was often rough and ready, with simple straw mattresses on the floor in a communal sleeping area. Until the 1650s, most people drank large quantities of **alcohol** as water was often polluted and unpasteurised milk was dangerous. However, coffee, chocolate and then tea began to be imported into Europe, which led to the setting up of coffee houses, the precursor to restaurants. As wealth and travel increased, inns and taverns became more sophisticated. One of the first proper hotels was built in Nantes, France in 1788. It was at about this time that the first restaurant appeared, and by 1794 there were 500 dining places in Paris alone. They arose out of the political revolution in France, since many chefs who had been employed by the nobility suddenly found they had no nobles to feed.

Throughout the 1800s both restaurants and hotels developed, serving all strata of the population. By 1848 in New York there was everything from Sweeney's (a sixpenny eating house) through Brown's (a restaurant catering for the gentility), up to Delmonico's (the top American restaurant of its day). For immigrants to the New World, setting up in the restaurant business was a relatively easy way to get established, especially if the menu catered for the large numbers of immigrants from that country. The diversity of American **cuisine** derives from the blending together of all these different ethnic traditions. At this time, most major European cities and most eastern cities in North America had luxury hotels, such as the Palmer House in Chicago and the Ritz in London.

The link between **food**, drink and lodging continued with the growth of the railways and air travel. In the United Kingdom and other countries, the railway companies were major investors in hotels, often built as part of their concourse. The companies also included restaurant cars in their rolling stock, especially on long train journeys such as the Orient Express and the Trans-Siberian railway. Even the first commercial flights, by KLM between London and Paris in 1919, included pre-packed meals. By the 1930s, Pan American's clipper service included dining at tables like in a restaurant.

At the beginning of the twentieth century, more and more people worked some distance away from home, in factories and offices. In the United States, many new kinds of restaurant were developed to cater for such people: diners (1872), lunch counters (1873), self-service (1885), cafeterias (1890) and short-order restaurants (1905) all entered into the American vocabulary. Between 1910 and 1925 the number of restaurants in America grew by 40 per cent; New York alone had 17,000 outlets in 1925. Four years earlier, in 1921, Bill Ingram and Walter Anderson started the first hamburger chain when they opened their first White Castle restaurant, although it was not until 1954 that the term '**fast food**' was coined. Since then, the hospitality industry has grown into a global industry providing food, drink and accommodation in an increasingly wide range of settings and formats.

There is considerable variation in the accuracy of data about the hospitality industry collected in different countries. For instance, there is in practical usage no definitive definition of what constitutes a hotel, although the **World Tourism Organization** has established guidelines. In the United States, the main sources of information about the industry are two trade associations, **American Hotel and Motel Association** and **National Restaurant Association**. In the United Kingdom, the official source of industry data is the Standard Industrial Classification. The classification tends to be based around the British approach to **licensing** premises, either with regards to health and safety or the sale of alcohol. While this makes classification and data collection relatively easy, licensing arrangements may not be the best way to understand and analyse the industry. Other UK sources of data include the Hotel Training Foundation, the British Tourist Authority, Home Office, British Hospitality Association and Business Monitor.

According to World Tourism Organization figures, there were 11.8 million hotel rooms in the world in 1993, with an average growth of 2 per cent over the previous five years. However, growth is variable in different parts of the world: the highest annual growth (6.8 per cent) is in the East Asia Pacific region, whilst both the Middle East and Eastern Europe experienced a reduction in the number of rooms. Most of these hotel rooms are located in Europe, with 44 per cent of the total. North American has 32 per cent and Asia Pacific 12 per cent. Africa, the Caribbean, South Asia, Latin America and the Middle East each has less than 5 per cent. Sources of business for hotels also varied by region. On average around the globe, domestic business (48 per cent) is slightly less than foreign business (52 per cent). However in North American 85 per cent of business is domestic, while in Africa, Asia and the Middle East only 20 per cent is domestic.

Similar **growth** in global restaurant brands is also occurring. In the mid-1990s, it was estimated that the top 100 companies operated nearly 110,000 units in the United States and a further 20,000 outside the country. Seventy per cent of the sales are in so-called quick service or fast food restaurants, such as Burger King, KFC and Taco

Bell. McDonald's alone has thousands of outlets, of which the majority are in the United States.

ANDREW LOCKWOOD, UK

hospitality information system

Hospitality information systems are mechanisms that deliver processed data to **management** to facilitate the decision-making process. Much of the information needed by management exists within the enterprise. Some is required to be maintained by law, and other data exists as a result of business transactions entered into by the enterprise. Some information exists within individuals and is not available without involving that person in the decision-making process. As the operation grows, **hospitality** information systems become more structured, requiring additional data production, sorting and processing. With the increased speed and capabilities of microcomputers, and the reduced cost of hardware and software, hospitality information systems are available to assist management in many ways.

Centralised purchasing management systems are available to create purchase orders that can be immediately communicated to vendors. This system works best for multi-unit organisations. There are also products available to track inventory and to compute theoretical sales. These systems can automatically route requisitions and delivery worksheets. Systems for recording **food** and beverage sales continue to improve, offering not only sales tracking data but internal control features, such as remote printers at production stations giving authorisation to release inventory. Orders can be taken tableside by hand-held wands or touch screens, eliminating the need for servers to write down guest selections.

In the area of telecommunications, hospitality information systems have proven to help enterprises generate additional revenue through facsimile machines, modem connections, conference calling and videoconferencing. As **technology** increases, this continues to be a growth area for the industry. Hospitality information systems provide entertainment to guestrooms through on demand movies, video game and videotape rentals, and premium channel selections. **Security**, a major concern in tourism, is another area where the systems have improved delivery of guest service through key cards, security cameras and motion detectors. Energy costs are also better managed with hospitality information systems by controlling heating and air conditioning.

Front office and back office accounting procedures have been greatly enhanced by hospitality information systems. The most visible to the guest is the **property management system** which helps the check-in and check-out process and stores data on each guest. This, coupled with other sales software packages, allows management to maximise revenue through **yield management**.

See also: automation; computer reservation systems; information technology

Further reading

Coltman, M.M. (1979) *Financial Management for the Hospitality Industry*, New York: Van Nostrand Reinhold.

STEPHEN M. LEBRUTO, USA

Hospitality Sales and Marketing Association International

The Hospitality Sales and Marketing Association International (HSMAI) provides **education** and networking opportunities for its members on both chapter and international basis. The Association has members from numerous industry segments such as hotels, hotel management companies, convention and visitors bureaus, airlines, car rental agencies, cruise lines, **destination** management companies, and **advertising** and **public relations** firms. HSMAI is located in Washington DC, USA.

TURGUT VAR, USA

host and guest

Host and guest have multiple and largely similar derivations, including the Latin *hostis* and the

German *gast*: stranger, foreigner, enemy. Thanks largely to Smith's (1977) eponymously entitled collection of essays on the social and cultural implications of tourism, both terms are indelibly associated with the birth of the **anthropology** of tourism. Related to such words as hospice, 'a house of rest and entertainment for **pilgrims**, travellers, or strangers', guest house (a term in use at least since the tenth century) and **hospitality**, the act of being hospitable, host and guest denote persons concerned with the process of transforming **strangers** into familiars and enemies into friends.

The relationship between host and guest thus connotes a sense of ambivalent hostility and friendship. This is a theme which has extensive literary references. For example, one may point to Chaucer (1374), 'there is right now y-come in to towne a geste, a Greek espie' (Troylus and Criseyde), and Shakespeare (1605), 'Conduct me to mine host, we love him dearly' (Macbeth). It is also a theme which has provoked a significant line of enquiry both in mainstream **sociology** and social anthropology. In the case of the former, the work of Simmel (1950) on the stranger, and the attendant social processes provoked by the presence of a stranger, is seminal. In social anthropology, the **ritual** formulas associated with relations between host and guest, and the social and cultural elaborations to which these give rise, have been the subject of a substantial corpus of ethnographic studies (see **ethnography**), most noticeably those concerned with Mediterranean and Middle Eastern society. Pitt-Rivers's (1977) 'The Law of Hospitality', for example, which itself builds on earlier studies in the genre, is one classic example of the ethnography of the relationship between host and guest in the Mediterranean.

These observations illustrate one of the ways in which tourism studies in general, and the anthropology of tourism in particular, are founded upon a preoccupation with themes which have been established in both literature and anthropology since very early days. Two questions, closely related to each other, follow. What distinctive and new insights have tourism studies brought to the understanding of the relationship between host and guest in a modern and postmodern world in which tourism plays a distinctive role? How, in the light of social processes associated with the tourism industry, have the terms host and guest acquired new meanings and connotations?

Taking these questions together, there are at least three distinct areas of enquiry in tourism studies which are concerned directly with the nature of hosts and guests, and the relationship between them. The first of these concerns the transformation of social and cultural relations in a world in which all relations are increasingly subject to **commoditisation**. Originally conceived in terms of the changing nature of relations within traditional social settings, such as villages in southern Europe, which have become tourism destinations, debates about commoditisation have more recently become concerned with the extent to which the global marketplace has come to define relations everywhere. In such a world do the **roles** (in a destination, for example) of host and guest dissolve and become transformed into versions of buyer and seller?

The second follows directly and concerns the formulaic nature throughout the world, including the destinations of the South, of host–guest relationships in such tourism-related institutions as hotels, **restaurants** and bars. What is interesting here is the extent to which the course and character of social relationships in such settings is determined not by any sort of **indigenous** cultural tradition but by standardised 'management handbooks' composed by tourism industry specialists from regions such as North America and northern Europe. The broader question concerns the extent to which all Southern cultural texts about hospitality, and the relation between host and guest, have given way to texts composed by Northern scriptwriters associated with the industry.

The third is rather different (being informed perhaps by a slightly more optimistic tone). There is a growing collection of studies which are charting the way in which the relationship between local hosts and guests, in such diverse settings as **bed and breakfast** establishments in the English West Country and the tourism circuits of Malta, have come to be articulated by women. One emphasis here is on the way in which the economic structures and demands of the industry have given rise to changes at **local** levels in both economic and social structures which have affected **gender roles** within families and households in such a way that women have moved in many cases into central socioeconomic roles.

The changing roles of host and guest constitute a fundamental touchstone of the nature of the contemporary world, particularly as far as this world is shaped by tourism. In this sense, those studies of tourism concerned with the changing relations between host and guest appear as important instruments in the understanding of **modernity** and **postmodernism** themselves.

References

Pitt-Rivers, J. (1977) *The Fate of Shechem, or the Politics of Sex: Essays in the Anthropology of the Mediterranean*, Cambridge: Cambridge University Press.

Simmel, G. (1950) *'The Stranger', The Sociology of Georg Simmel*, New York: Free Press.

Smith, V.L. (1977) *Hosts and Guests: the Anthropology of Tourism*, Philadelphia: University of Pennsylvania Press; revised 1989.

Further reading

Black, A. (1966) 'Negotiating the tourist gaze: the example from Malta', in J. Boissevain (ed.), *Coping With Tourists: European Reactions to Mass Tourism*, Oxford: Berghahn.

Bouquet, M. (1985) *Family, Servants and Visitors: The Farm Household in Nineteenth and Twentieth Century Devon*, Norwich: Geo Books.

Zarkia, C. (1996) 'Philoxenia: receiving tourists but not guests', in J. Boissevain (ed.), *Coping With Tourists: European Reactions to Mass Tourism*, Oxford: Berghahn.

TOM SELWYN, UK

hotel

A hotel is a tourism business unit which, as its main endeavour, rents room **accommodation** to the general public for a minimum duration of one night. Frequently this activity is supported by the provision of **food** and drink and other related services. Hotels vary in the number of rooms available, the level of service provision, target **markets**, tariff charged and type of ownership and operation.

KEITH JOHNSON, UK

hotel, airport

Hotels located near major airports, meeting the hospitality demands of business and leisure **tourists**, are commonly known as airport hotels. Commercial links exist with the airport authority, with airlines, **reservations** agencies and courtesy coach operators. Market segments include aircraft personnel (pilots, cabin crew), **inbound** and **outbound** tourists, and passengers affected by delayed flights. Well-equipped conference facilities are developed for the wider business market.

See also: accommodation

STEVEN GOSS-TURNER, UK

human resource development

Human resource development (HRD) can be described as a continuous process or virtuous cycle which uses investment in human capital in order to improve productive output, enhance the **quality** of that output, provide increased benefits for those employed and contribute to an improved quality of life for those involved and their dependants. At the same time, HRD is a process which is at the core of an organisation's **investment** in its human capital and, as such, can make a significant contribution to improved performance, productivity and profitability.

HRD is a key functional aspect of human resource management (HRM) and, in its broadest sense, encompasses the complementary process of ecotourists, **training** and developing personnel for or in an organisation. HRD is a term which is of relatively recent currency and has evolved as an umbrella term to assist in overcoming social, cultural and practical difficulties which can exist in defining its component areas of **education** and **training** and development.

The HRD process may be formal in its organisation, consisting of publicly accredited education courses or training modules but also includes a wide range of informal learning situations including those which may not be actually recognised as such. Formal HRD includes the general educational investment which the state and individuals make in preparing young people from pre-school age for their role in society and the workplace. It includes both general and vocational education in schools, colleges and universities. The inclusion of reference to tourism and the development of awareness about the industry and its generic skills needs (customer care, quality) is growing internationally, in both the elementary and secondary school curriculum, and thus is the starting point for HRD in tourism. The macro socioeconomic perspective on HRD is one often neglected in HRM texts. In the context of tourism, the true vocational focus may come early in the process, as within the German dual system or may take place at undergraduate or postgraduate level in universities. Formal HRD also includes programmes in which employees of organisations participate, leading to publicly accredited trade, academic or professional qualifications. Informal HRD is that which takes place within and external to the organisation but which has no formally accredited outcomes.

HRD may be located within the organisation and include activities such as on-the-job training (OJT), in-company training and the workplace classroom. In larger tourism organisations, HRD is the responsibility of a specialist department or a specific unit within human resource management. This allows HRD to be planned and organised in a strategic manner, based on identified training and educational needs within the organisation and designed to balance individual development aspirations with the wider requirements of the company. In small organisations, a dedicated training or HRD section is less common and the function may be one that is shared along with wider HRM responsibilities. In many smaller tourism organisations, HRD is likely to be a responsibility subsumed within the role of a member of the senior management team, and may not receive the attention or priority it requires. Formal HRD within tourism's micro-organisations is frequently neglected and may be one contributing cause to the declining competitiveness of the small tourism business sector in some countries. HRD may also take place external to the organisation, through programmes in colleges and training centres which are company specific or of a more general nature.

An HRD programme operates at a variety of levels and stages within an individual's career in a company. It has an important role to play at point of entry into a company (or even prior to it) through induction training, which can relate to technical preparation for the workplace but generally also has a strong organisation cultural focus. It is a training process which has equal applicability (with different focus) for junior, entry-level positions and senior managers. Further, HRD is at the core of individual and organisational development. Companies frequently recruit on the basis of the potential contribution which an individual can make, but achieving that potential requires significant, structured investment in the skills and wider development of that individual. To optimise individual potential, the programme can be through specific courses or qualifications. However, this may take the form of an extended option of exposure to and experience in various departments of the organisation over an (extended) period of time. Individuals may also be targeted for promotion within an organisation's succession planning system, and HRD initiatives are used to prepare them for their intended responsibilities.

At an organisational level, HRD plays a key role in supporting or planning for change which, in this context, may relate to new technical demands on the workforce (for example, the introduction of new types of aircraft into the fleet). It may reflect new communication systems such as the introduction of e-mail for all staff. **Marketing** and **market** changes can necessitate organisational HRD efforts in order to prepare employees for such change. For example, this may include a focus on language skills to meet new route intentions in an **airline**, or to reflect promotional investment by a **destination** in specific new tourism markets.

The traditional model of HRD in tourism is based on the 'front-loading' of skills principle in the sense that skills development is concentrated prior to entry into the industry or during the early stages of **employment**. This model assumes that once skills are honed, the working environment remains largely unchanging so that the qualified pilot, chef or **tour guide** has the capability to undertake tasks associated with the job with little modification throughout their career. This mediaeval apprenticeship model of training has little relevance to the demands of contemporary tourism employment where the role of HRD is to equip staff for constant change in the workplace, supporting the concept of 'lifelong learning' for all those active within the economy.

It is difficult to generalise about principles of HRD in the context of the international tourism sector. The limitations are imposed as a consequence of a number of factors which include the supply-side diversity within the various sectors of tourism which necessitate differing technical and **service** skills. Supply-side operations across the **transportation**, **accommodation**, food service, **attractions** and other sectors have limited skills requirements in common, beyond generic customer handling, communications, finance and related areas. The context in which HRD takes place in these sectors also varies to a considerable degree. A further factor is diversity in business structure, size and ownership, varying from the micro family operation to the mega **multinational firm**.

Tourism, as an industry, remains dominated in most countries by small family-run businesses within which the capacity to effectively deliver HRD is often limited. While the focus in these operations tends towards the informal, such investment may be insufficient in a rapidly changing business environment and a volatile labour market. Limitations to generalisation are also the result of different traditions and philosophies of education and training between countries and regions. Tourism is generally 'locked into' the structures and philosophies of public sector education and training settings in which it operates. These approaches may not be fully sensitive to the vocational HRD needs of tourism. In a few countries, the private sector, in partnership with public institutions, has been able to participate in HRD programmes.

Finally, the varying levels and forms of public sector support for HRD within the tourism sector also cause diversity. This action has generally been justified on the grounds of the fragmented and diverse nature of supply-side business. In particular, the need to ensure minimum operating and service standards throughout a destination's supply chain is at the centre of justification for financial and practical measures in support of tourism HRD. Examples of how the public sector provides incentives to companies to invest in HRD include payroll levy schemes (Malaysia and Singapore), subsidised training provision (Ireland and Scotland) and tax exemption (a widely employed mechanism).

In some respects, tourism as a sector exhibits features of immaturity when compared to other industrial sectors in terms of its HRD policies and practices. This may be a consequence of the environmental considerations discussed here. It is also a reflection of wider weaknesses which exist in relation to the image which tourism, as an industry, has for sustained employment. Investment in HRD is constrained by labour market challenges which parts of the sector face, including high labour turnover, inability to recruit top quality entrants at all levels, and a poor career image. The issues facing HRD in tourism are further compounded by a lack of planning at a macro level. This means that there are few established links at a strategic **policy** level between the tourism sector, its development, and the provision of HRD in schools, colleges and universities. HRD has the potential to play an important role in the **sustainable development** of tourism, but unfortunately this role is widely neglected.

See also: manpower development

Further reading

Baum, T. (1995) *Human Resource Management in the European Tourism arid Hospitality Industry*, London: Chapman and Hall. (Addresses the social and historical context of HRD in tourism.)

Esichaikul, R. and Baum, T. (1998) 'The case for government involvement in human resource development: A study of the Thai hotel industry', *Tourism Management* 19(4): 359–70. (Considers the role of public sector agencies in supporting HRD in tourism enterprises.)

TOM BAUM, UK

humour

Humour refers to a sense of the ludicrous, absurdly incongruous, comic or amusing in a happening, action, situation or expression of ideas. Tourism is a rich source of humour because it brings into juxtaposition people of different cultures, interests and expectations with a consequent high potential for contrast, miscommunication and misunderstanding. Tourists in particular, because they are in an unaccustomed **environment**, may engage in inappropriate **behaviours** with bizarre consequences. Lack of familiarity between tourists and residents may lead both to fall back on **stereotypes** to guide their interactions and this increases the likelihood of occurrence of communication breakdowns. Since most people in the Western world are familiar with tourism situations, they constitute a setting to which they can easily relate and which is readily exploited by humorists in many forms of media including novels, **film**, T-shirts and cartoons. Although there is a wealth if examples of tourism humour, these have yet to be collated or examined as the subject of rigorous research. Thus, tourism humour is a widespread phenomenon which has yet to receive the academic attention it deserves.

Further reading

Coren, A. (ed.) (1987) 'Travel number', *Punch* 292, 21 January. (Contains many excellent examples of tourism humour.)

Lodge, D. (1991) *Paradise News*, Harmondsworth: Penguin. (A humorous novel.)

GEOFFREY WALL, CANADA

Hungary

Hungary played a key role in the changes that took place in the Eastern European region. The transformation from a monolithic structure to a multiparty system was carried out peacefully, according to a national consensus. The Republic of Hungary is a parliamentary democracy. The level of political openness of the country has significantly increased the interest in Hungary which in turn contributes to **development** of **investment** and tourism.

As with all the ex-socialist countries of Central and Eastern Europe, Hungary is faced by serious economic problems. The acceleration of inflation is coupled with a decrease in production, and the efforts to reduce the national debt have weighed down on the standard of living. One of the most important indicators of the economic transformation is the privatisation process, which is still under way. But as the result of this process, the level of unemployment is continually increasing. Despite these problems, Hungary has a stable economy in the region which appears in the expending of foreign investments.

Over the past decades Hungary has attracted millions of **tourists**, partly due to its advantageous location in central Europe and partly due to its special political-economic system. In 1992, 20.2 million international tourists spent $1.23 billion. In 1997 these numbers were 17.3 million and $2.6 billion, respectively.

However, as a result of the changes of the 1990s, the dynamic **growth** in tourism has been modified to some extent. Although the number of tourists exceeds by nearly four times the population of Hungary, the ratio of tourists has decreased, the use of registered **accommodations** is continuously reducing and the structure of main generating countries has been altered. In respect to domestic and international tourists, Budapest and Lake Balaton represent the main attractions. About 60 per cent of the total foreign tourism takes place in these regions, and only 40 per cent of that is distributed among the other regions of Hungary. This territorial concentration is coupled with a large seasonal fluctuation, which worsens the negative **impacts** of tourism. In order to stop this process, the country has to change its tourism development strategy, Instead of **mass tourism**, Hungary has to offer special tourism products that meet the requirement of sustainability and are in harmony with the international **trends**.

Hungary has a multi-level tourism coordination system. The National Tourism Committee is the highest decisive and consultative body in the Ministry of Economics. The decisions and directives of the committee are carried out by the Tourism Department of the ministry. The Hungarian Tourism Corporation is founded and supported by the ministry taking over the tasks concerning research, **marketing**, promotion, development, **education** and control of the Tourism Development Fund from the Tourism Department.

CSILLA JANDALA, HUNGARY

hyperreality

During recent decades, distinctions between 'the real' and 'the represented' have become problematic as one travels and experiences places. To Baudrillard, one increasingly consumes signs or copies rather than real entities themselves. When the simulation of things is particularly ubiquitous or spectacular, hyperreality results, constituting a state of manipulated **discourse** where referential reason disappears – a realm where illusion is not achievable because the real itself is no longer knowable or attainable. **Images** projected in tourism frequently habituate people to hyperreal replica worlds which are surreal and exhilarative to some, but depthless, inauthentic, and commoditised (see **commoditisation**) to others.

KEITH HOLLINSHEAD, UK

I

iconography

Illustration by visual **images** connoting ulterior reality is known as iconography. In tourism, promotion icons are mainly pictorial and are typically found in brochures, pamphlets and **maps** (see also **promotion, place**). A theme such as **nature** may be signified by several icons (such as palm trees, bougainvillaea or giraffes). Some icons, however, are polysemic: they offer more than one type of meaning (such as the ocean), thus making them more difficult to interpret.

GRAHAM M.S. DANN, UK

identity

There are many different aspects of identity relevant to tourism studies, such as ethnic (see **ethnic group**), place, national and cultural identities. While such a broad term defies a simple definition, it is clear that identity is more than an objective notion of someone's essential being. Rather, identities may be thought of as social constructions in which individuals' sense of belonging to some larger group or geographical locale is in many ways influenced by their social position *vis-à-vis* other individuals and collectivities. Recent research has increasingly insisted that identity be regarded not as a given quality one takes for granted, but as a sense of belonging which is actively manipulated by individuals according to their relations with others, relations which are often structured by larger political or economic forces. Thus, individuals may claim many different identities at the same time, and may claim different identities over the course of their lifetime. Such a voluntaristic approach to identity, however, is often criticised for downplaying identity-forming attributes which one cannot always escape, such as locale, language, race, **gender** or sexuality. Therefore, identity perhaps most accurately constitutes a process whereby these basic attributes serve as a limited set of 'raw materials' for individuals to use in constructing different senses of belonging to social groups, locales, **lifestyles**, nations and so on.

In tourism studies, it has been common to discuss identity in terms of tourism's 'impact' upon different kinds of identities. Tourism, for example, has been shown to play a role in the reconstruction of ethnic identities, in altering the place identities of locals in destinations, and even in helping to build (or in some cases, break down) a sense of national identity in countries where tourism **development** occurs. One of the greatest concerns seems to be that tourism is capable of creating 'false' identities among host populations, based on 'inauthentic' representations of place or culture (see **staged authenticity**). However, it is important to understand that identities as social constructions are dynamic and changeable according to changing social contexts. Tourism is but one of many social forces which condition one's sense of identity. Overall, though, such questions have been subject to more speculation than actual research in the tourism field, indicating an important direction for future inquiries into the relationship between tourism and identity.

See also: ethnicity; ethnology

Further reading

Ashworth, G.J. and Larkham, P.J. (eds) (1994) *Building a New Heritage: Tourism, Culture, and Identity in the New Europe*, London: Routledge. (Papers discussing ways tourism is being promoted as a catalyst to the formation of new identities in Europe to facilitate European unity.)

Bradley, D. (1983) 'Identity: the persistence of minority groups', in W. Bhrukasasri and J. McKinnon (eds) *Highanders in Thailand*, Kuala Lumpur: Oxford University Press, 46–55. (Discusses the relationship between minority identity and social processes of change in Thailand.)

Lasch, S. and Friedman, J. (eds) (1992) *Modernity and Identity*, Oxford: Basil Blackwell. (Papers dealing with changing conceptions of identity in contemporary societies.)

TIM OAKES, USA

ideology

Ideology is a highly controversial and contested concept in the social sciences. It is associated with attempts to identify the social distribution systems of ideas, to disclose their origins and the interests which lie behind them, and to specify the unequal relationships of **power** they may generate or sustain. Some analysts of ideology see tourism as an ideological system in itself, constructed, elaborated and exported within international capitalism, whose effects include the commodification and **appropriation** of subordinate cultures, cultural denote and domination: in effect, **imperialism** by peaceful means. In 'weaker' versions of ideology, tourism is seen not as a single ideological system but as a field constituted by a number of separate ideological elements circulating among **tourist** populations, which may be general (such as 'travel broadens the mind') or specific (such as 'the Seychelles is an earthly paradise').

Both approaches to ideology recognise that tourism is shaped by social forces and that motivation for recreational pursuits (or lack of it) is never purely a matter of personal choice, but a function of the power of socially derived ideologies and the 'vocabularies of motive' such ideologies tend to produce in the minds and actions of individuals (including tourists' **images** of places and cultures). Tourists may think they choose; in reality they are, to a degree, 'chosen' by ideologies circulating within their social networks. Once tourist ideologies have gained acceptance, those who embrace them tend to forget whence these ideas originated and tend to absorb them as natural 'facts of life'. Thus, for example, many people in the developed world today think regular recreational travel is a natural rather than an ideologically derived **expectation**.

Ideology has acted as a stimulus for three kinds of tourism scholarship. First, the analysis of tourism as a general system of practices and **representations** focuses on the unequal power inputs and interests which structure them, and the effects these produce in relationships (direct or mediated) between **host and guest**. Researchers in this tradition have 'deconstructed' tourism representations (**advertising**, guidebooks, travel programmes) to suggest the ideological, connoted meanings about foreign cultures and peoples which they embody, the political and economic interests responsible for them, and the social effects such portrayals may have. Second, historical analysis aims at tracing at what moment specific tourism ideologies and their associated 'vocabularies of motive' emerged and why (for example, the fashion for picturesque **sightseeing**). Third, comparative social analysis shows how ideologies vary among groups (for example, by comparing tourism tastes within one society or occupational group with those of others). Compared to its widespread and much debated occurrence in other academic fields, ideology is still a relatively rare focus in tourism analysis. Nevertheless, it offers promise in understanding tourism as a social process.

See also: dependency theory; discourse; myth; neo-colonialism; orientalism; semiotics; structuralism

Further reading

Andrews, M. (1989) *The Search for the Picturesque*, Aldershot: Scholar Press. (Traces the evolution of the ideology of the picturesque and its effect upon tourism practices.)

Selwyn, T. (ed.) (1996) *The Tourist Image: Myths and Myth Making in Tourism*, London: Wiley. (Articles examining ideological meanings in international tourism representations.)

A.V. SEATON, UK

image

Image may be defined as the perceptions, beliefs, impressions, ideas and understanding one holds of objects, people, events or places. An image is a simplified, condensed version of which the holder assumes to be reality. Held and stored images are the means humans use to organise the various stimuli received and processed on a daily basis, and help to make sense of the surroundings and the world in which one lives.

Image has been an important topic in tourism research since the 1960s, and it is believed to be the key underlying factor in **destination** site selection. People hold different images of different destination areas. These images become the main determinant for maintaining or eliminating a particular destination as a possible choice, once the list of all known alternatives is subjected to a winnowing process using more tangible considerations (such as time and money).

The image formation process is more important to tourism than to other industries. Since tourism products are an amorphous mass of experiences, produced and consumed simultaneously, with no opportunity to sample the product prior to purchase, the images someone holds act as a surrogate for **product** valuation. Because of the importance of image in the tourism decision process, marketers will spend an inordinate amount of time, effort and expense presenting particular destination images to target markets (see **target marketing**) with the hope that their choice will rise above the clutter of other destinations, all vying for a share of the tourism expenditures.

Place images are comprised of three distinctly different but hierarchically interrelated components. The first one is cognitive. The sum of beliefs and **attitudes** toward some object, place or thing leading to some internal evaluation of its attributes is the basis for the cognitive component. It is object evaluation based on fact or what is believed to be the facts surrounding the object under consideration. These facts most likely have been acquired over time from a number of sources and may or may not form a realistic image of the object. 'Perception is reality' is an often heard phrase which means simply that reality is the cognitive image one holds toward any object, place or thing.

The second component of image is effective. This component is how one values the object under consideration based on personal feelings or motives. Motives determine what one wishes to obtain from the object which then determines the **value** placed on the object (see **motivation**). The third component is conative, which is the action component of image. A decision to purchase a product or choose a destination is made after the cognitive and affective components of the objects image are evaluated and valued. The conative component is the decision stage.

Image research is probably as old as the discipline of **psychology**, but received more attention as a result of the attitude measurement work conducted during the 1930s, and in the mid-1950s with the publication of Boulding's seminal work, *The Image: Knowledge in Life and Society*. With the advance of the consumer age following the Second World War, **marketing** research began to devote more attention to product attribute measurement and by association, image. In the early 1970s the first tourism image research articles and books began to appear, and this has been a prolific area of inquiry since.

The generally accepted findings on this research theme are several, five of which may be described here. First, people residing in different geographic areas will hold different images of the same destination. Second, the further one lives from a particular place, the less likely one is to have a distinct image of the place. Third, images change slowly, and in the absence of a major event will take years to modify using conventional image modification techniques. Fourth, the smaller the destination is the more likely its image is that of the larger political entity in which it is located. Fifth, images are formed and modified continuously through various means.

It is the last finding that continues to be the basis for much academic research. Images are said to be formed by induced and organic agents. Induced agents are those generally controlled by some other source (such as destination promoters) and organic components are acquired through personal **experience**. These agents have been subjected to further classification in an attempt to produce customised image development programmes based on cost, market penetration and credibility of the message delivered. One of the agents that does not fit either the induced or organic category is also one of the most powerful with the ability to change long held images in a short period of time. It has been termed 'autonomous' and consists of information received via news media or popular culture (such as movies and documentaries). Because of the multiple image formation agents that one is exposed to over a period of time it is important to assess held images before embarking on modification campaigns.

Increased attention has been given to the methods utilised to measure existing images. Most assessment research has asked **survey** respondents to evaluate selected destination attributes, activities and so on, using a type of Likert scale anchored by bipolar adjectives (such as very impressive–very unimpressive). Free elicitation has been proposed as an alternative technique. This method limits the amount of information provided to the survey recipient with the hope of achieving a richer and more varied range of responses.

Tourism image research is still a largely unexplored area of inquiry. Because of its critical importance in the tourism decision-making process, much still needs to be done. It is a fruitful area of research for many from various disciplines including, but definitely not limited to, **economics**, marketing, psychology, **recreation**, **anthropology** and **geography**.

Further reading

Boulding, K. (1956) *The Image: Knowledge in Life and Society*, Ann Arbor, MI: University of Michigan Press.

Gartner, W.C. (1996) *Tourism Development: Principles, Processes, Policies*, New York: Van Nostrand Reinhold. (See Chapter 11.)

Gunn, C. (1972) *Vacationscape: Designing Tourist Regions*, Austin, TX: Bureau of Business Research, University of Texas.

Hunt, J.D. (1971) 'Image: a factor in tourism', unpublished doctoral dissertation, Colorado State University.

WILLIAM C. GARTNER, USA

image, destination

The destination image refers to the **attitude**, perception, beliefs and ideas one holds about a particular geographic area in the world. It is formed by the cognitive image one has about a particular **destination**. Alternatively, a destination image can be viewed as the mental picture promoters are trying to instil within a target audience. This is one of the key factors in the travel selection process.

See also: destination choice

WILLIAM C. GARTNER, USA

impact

Impacts of tourism are the changes which occur as a consequence of the **industry**. Although tourists and the people that they come into contact with in their places of origin are modified by tourism, discussion of impacts focuses predominantly on changes in **destination** areas. These changes may be extremely wide-ranging, but for convenience it is usual to divide impacts into three categories: economic, environmental and social. In reality, the categories overlap and changes in all three occur at the same time. For example, it is possible to spend money to improve the **environment**, but changes in the **employment** of household members associated with the acquisition of positions in tourism may have implications for social interrelationships within families. Furthermore, there are other changes, such as those of a political nature, which do not fit well into the threefold division. In spite of these complexities, it is commonly observed that the economic impacts are often positive and the environmental and social impacts are less desirable.

There is a wealth of studies of the impacts of tourism on a wide range of topics including employment, incomes, **taxes** and **foreign exchange**, soils, vegetation, water, wildlife and **landscape**, and **crime**, language, **music**, arts and crafts, and celebrations. Most of these investigations are case studies, and a careful review would indicate that impacts are more complex than is indicated by the preceding generalisation with both positive and negative impacts being recorded in all three major impact categories. For example, while some authors laud the benefits of employment, incomes and foreign exchange resulting from tourism development, others point to inflation, high leakages, **seasonality** and low status of associated jobs. Some document the environmental ills whereas others suggest that the industry can provide a rationale for and resources to promote environmental preservation and even enhancement. Furthermore, while it has been argued frequently that tourism destroys culture, it has also been suggested that it can revive cultures and can give added value, both intellectually and economically, to cultural expressions such as music, dance and other arts and crafts. If such contradictions are to be avoided and the case studies are to contribute more effectively to the generation of cumulative knowledge, much more attention must be given in the future to the precise contexts in which specific impacts occur.

While some of the changes associated with tourism may be commonly observed, for example those in the degree of **local** control over the industry and in resident attitudes towards the industry as it moves through the seasonal cycle, the specific consequences of this business will vary with the circumstances and this makes generalisation difficult. The particular nature of impacts will be influenced by such factors as the type and **scale of development**, the specific activities undertaken, the characteristics of the destination area, the stage of development, the involvement of culture brokers and other intermediaries, the **policy** context in which tourism occurs, and the implementation and degree of success of strategies to mitigate negative impacts.

Three approaches are commonly adopted in analyses of the impacts of tourism: after-the-fact analyses, monitoring and simulation. Most academic studies of the consequences of tourism have adopted the first approach and have documented its impacts in a place after they have occurred. The advantages of this approach are that results can be derived expeditiously, but it is often not possible to determine the number of **tourists** and the exact nature of the activities which resulted in the impact, nor is it possible to avoid the undesirable impacts since they have already occurred. Monitoring involves undertaking repeated measurements in a place from the time of the onset of tourism. This potentially allows the recording of both the agents of change (such as the number of tourists and their activities), as well as their consequences; but it is an approach which is costly in terms of both time and money and it is frequently not possible to initiate investigations before the onset of tourism to establish a base level against which changes can be compared. Simulation is an approach employed in some ecological studies which impose test plots to known stresses, such as a specific number of passes of a snowmobile, following which the consequences are recorded. This approach permits the establishment of relationships between agents of change and their impacts, but it is not amenable for use in economic and social studies.

Measurements of the various types of impacts of tourism are undertaken using very different methods and indicators. Thus, for example, economists may record money and jobs using economic **multiplier effects**, environmentalists may measure such attributes as species diversity and coliform counts, and social impacts may be examined through questionnaire **surveys**. This makes it difficult, if not impossible, to combine the results of such studies to ascertain if the **benefits** exceed the costs. However, this may not be as critical a problem as it may at first seem, for regardless of whether the development proceeds, a decision which is often as much political as it is based on thorough impact analyses, the detailed information is still needed for **management** purposes.

In many jurisdictions, the completion of an environmental impact assessment is required in order to acquire permission to undertake a substantial tourism development (see **impact assessment, environmental**). Such evaluations often incorporate economic and social consequences in addition to environmental impact assessments more narrowly defined. However, the backward-looking perspective adopted by many academic impact researchers is not readily transferred to the predictive perspective required of many practitioners, nor does it direct them clearly towards variables which may be manipulated to influence the outcomes of tourism.

Since the impacts of tourism occur at the same time as other changes affecting people and places, and the consequences of the industry often exhibit similarities with those of other agents of change, it is often difficult to separate more general changes from those which are attributable specifically to it. This is not only a difficult analytical problem, it also sometimes results in tourism being blamed for changes for which it is not responsible. Nevertheless, it is true that too much tourism can result in the destruction of the very resources on which it depends, as well as degrading tourist experiences and disrupting the **lifestyles** of local residents. Recognition of such situations has led to the development of such concepts as **carrying capacity** and **limits of acceptable change** in an attempt to guide managers in their efforts to keep negative impacts within acceptable levels.

While outside observers often bemoan the existence of impacts, some impacts are desired and tourists are actively sought by residents of destination areas as a means of modifying the life opportunities of their children and themselves. However, the impacts of tourism are felt differently by various members of society reflecting such attributes as degree of involvement in tourism, **gender**, language skills, **race**, **class** and access to power. However, other things being equal, the greater the differences between the characteristics of the tourists and the residents of the destination areas, the greater the impacts of tourism are likely to be.

Further reading

Crick, M. (1989) 'Representations of international tourism in the social sciences: sun, sex, sights, savings and servility', *Annual Review of Anthropology* 18: 307–44. (Contains many references to the social and cultural impacts of tourism.)

Hunter, C. and Green, H. (1995) *Tourism and the Environment: A Sustainable Relationship?*, London: Routledge. (Includes useful chapters on environmental impacts and their assessment.)

Knight, R.L. and Gutzwiller, K.J. (1995) *Wildlife and Recreationists: Coexistence through Management and Research*, Washington, DC: Island Press. (Features a number of papers on the impacts of tourism on wildlife.)

Wall, G. (1996) 'Rethinking impacts of tourism', *Progress in Tourism and Hospitality Research* 2(3/4): 207–15. (Examines the status of research on the impacts of tourism.)

Wall, G. and Wright, C. (1977) *The Environmental Impact of Outdoor Recreation*, Department of Geography Publication Series No. 11, Waterloo, Ont.: University of Waterloo. (Reviews the environmental impacts of recreational activities.)

GEOFFREY WALL, CANADA

impact, economic

A change in the level or nature of an activity in a productive sector of an economic system causes an economic **impact**. Changes in the level or pattern of tourism expenditure will create an **impact** on the economy where that expenditure is made. The impact can be expressed as of income, **employment**, government revenue, output or **foreign exchange** flows, and is generally measured by the use of **multiplier effect**.

JOHN FLETCHER, UK

impact, environmental

Environmental impact refers to positive as well as negative aspects of a **tourism** organisation's environmental performance. It usually covers physical, biological, economic, social or cultural aspects. Environmental impacts of tourism are often cumulative. Recognising the positive and cumulative **impacts** can be just as beneficial to improving environmental performance as correcting the negative aspects.

See also: environmental valuation

PEIYI DING, AUSTRALIA

impact assessment, environmental

Environmental impact assessment (EIA) is defined as an analytical procedure for predicting and evaluating the environmental **impact** of proposed **development** programmes and projects, terminating with a written report (environmental impact statement) to prescribe safeguards. It is also a legally defined administrative procedure to involve major interest groups in the decision-making process, inform the public and resolve potential conflicts caused by multiple uses of the community's resources.

Generally, at the planning stage of a proposed project, the significant environmental impacts can be identified and examined, and measures suggested for their prevention or mitigation. The main objective of an assessment process is to identify **risks**, minimise adverse impacts and determine environmental acceptability; to achieve environmentally sound proposals through research, management and monitoring; and to manage conflict through the provision of means for effective public participation.

The evolution of the environmental impact assessment process reflects growing public concern over such issues. The process has been improved, and has become an important means of protecting and improving environmental quality as more and more countries have adopted impact assessment by legislation. EIA is designed to follow a particular format and is required by the decision-making authority for its review as part of the project approval procedure. The EIA procedures are a very useful technique to ensure that environmental impact of proposal projects including tourism projects have been taken into consideration and preventive actions taken, thus providing the basis for making any necessary adjustments to the proposal. However, in its implementation, some weaknesses can arise. It appears that improved and effective environmental impact assessment must extend beyond impact statements to include continued monitoring and revision of possible objectives and operational procedures. In addition to its predictive role, the process must allow for ongoing impact assessment. This continuing management role is particularly important in the case of more complex projects. Under these circumstances, EIA should be improved to extend to the entire process from the inception of a proposal to **environmental auditing**.

See also: codes of ethics, environmental; environmental compatibility; environmental management systems; planning, environmental; quality, environmental

Further reading

Butler, R. (1993) 'Pre- and post-impact assessment of tourism development', *Tourism Research: Critiques and Challenges*, London and New York: Routledge, 135–55.

Ding, P. and Pigram, J. (1995) 'Environmental audits: an emerging concept in sustainable tourism development', *Journal of Tourism Studies* (6)2: 2–10.

Ecologically Sustainable Development Working Group (1991) *Final Report – Tourism*, Canberra: Commonwealth of Australia.

Inskeep, E. (1991) *Tourism Planning: An Integrated and Sustainable Development Approach*, New York: Van Nostrand Reinhold.

PEIYI DING, AUSTRALIA

imperialism

The term 'imperialism' has long been employed (often carelessly) in the social sciences and in the briefer span of tourism studies. In 1968, Hans Dalder wrote that imperialism had been widely used emotively, but rarely theoretically. That pattern has continued whether the term is employed more generally to refer to transactions between societies or nations, or more narrowly to refer to specific kinds of transactions (such as economic and cultural). All usages, however, have implied a process whereby societies extend their influence over others. In the social sciences, the concept has had important links with Marxist thought, although not necessarily in the archetypal sense to denote the final stage of capitalism where the export of capital by large countries dominates. In current usage the economic aspect remains important, but the term often refers to any kind of hegemonic relationship.

In critical studies of tourism involving transactions between developed and developing countries (where negative consequences for hosts, such as **commoditisation**, exploitation and environmental degradation, are said to increase), the concept of imperialism, explicit or not, is pervasive. Thus, social scientific criticism of Western ways and their influences has been extended to this variety of tourism. Whether used in this or a broader sense to embrace (because of the service factor), all touristic transactions, the notion of tourism as imperialism has encountered problems either because of imprecise deployment or empirical exceptions. Thus it is argued that, though it is useful in calling attention to the power dimension of touristic transactions and the negative consequences for many hosts, the concept of imperialism cannot substitute for careful, empirical analysis of **power** differentials and other aspects of touristic transactions.

See also: colonisation; dependency theory; government

Further reading

Crick, M. (1989) 'Representations of international tourism in the social sciences; sun, sex, sights, savings and servility', *Annual Review of Anthropology* 16: 307–44. (An anthropological treatment of tourism in the developing world, seen as pervaded with the notion of imperialism.)

Nash, D. (1989) 'Tourism as a form of imperialism', in V. Smith (ed.), *Hosts and Guests: The Anthropology of Tourism*, 2nd edn, Philadelphia: University of Pennsylvania Press, 37–52. (Argues that because tourism involves service, it is seen potentially as imperialistic.)

—— (1996) *Anthropology of Tourism*, Oxford: Pergamon Press. (Discusses the notion of tourism as imperialism in this broader anthropological treatment.)

DENNISON NASH, USA

import substitution

Import substitution is an economic policy which consists of the encouragement or creation of domestic industries to produce goods (or, more rarely, **services**) which would otherwise have to be imported. Such a policy has two complementary objectives: to promote the industrialisation of the domestic economy (or to mitigate unemployment), and to improve the **balance of payments**. The latter aim is to be achieved by substituting domestic for imported **value added**. These objectives are characteristic of, but not confined to, **Third World** countries. More developed countries may also have incentives to pursue import substitution, both as a means of reducing unemployment and to improve their balance of payments.

A typical example of import substitution involves the automobile industry. To reduce imports of cars, trucks and buses, developing countries first set up assembly factories and import complete unassembled kits of vehicles which are assembled by local workers. Then certain components start to be fabricated locally, and this process continues until eventually entire vehicles may be produced and assembled locally. The main challenge to be met in order for this process to succeed is that of manpower development to meet international standards of **quality** and **efficiency**. Levels of **education** are relevant here.

Import substitution policies typically run into the problem that they conflict with the basic economic notion of **comparative advantage** and the principles behind **SWOT analysis**. Hence, the domestically produced products are liable to be uncompetitive with imports in terms of price, quality or both. Thus, they can only be sold by being protected from the competition of imports. Such protectionism is problematic. Because of the lack of competition, the domestic industries concerned have inadequate incentives to improve their standards of quality and efficiency, and are deprived of opportunities for **bench-**

marking. Successful implementation of import substitution policies thus typically involves joint ventures with foreign **multinational firms**, which become partners in the original assembly factories, supply the kits for assembly and subsequently participate in the production activities by facilitating the needed knowledge acquisition and **training**. This, however, entails risks of **neo-colonialism**. An alternative to import substitution is the development of domestic industries that are strongly export oriented and can compete in world **markets**, earning foreign currency which can be used to pay for imports. With regards to tourism, an export-oriented policy, namely the **development** of tourism products (together with the necessary **infrastructure**) in order to attract **inbound** tourism, may be compared to an 'import substitution' policy of developing tourism products intended to induce residents to substitute domestic for outbound tourism.

SIMON ARCHER, UK

inbound

The **World Tourism Organization** defines inbound (or incoming) tourism as that which involves non-residents of a country travelling to that **destination**. Inbound **tour operators** (also referred to as ground operators and handling agents) make arrangements and operate tours within their country or region on behalf of organisers in foreign countries who send them their clients.

LIONEL BECHEREL, UK

incentive

Incentive travel is one of a number of **management** tools used to motivate staff to accomplish clearly defined business goals. Staff are rewarded with trips for meeting financial (sales, commissions, new business) or non-financial targets (reducing staff turnover, excellence in customer **satisfaction** and so on). Incentive travel is one of the fastest growing segments of the **business travel** and the meeting, **convention business** and exhibition sectors, with an annual growth rate of 10 to 15 per cent. Estimates suggest incentive sector may grow to $50 billion by the year 2000.

Incentive travel is designed to be a self-liquidating promotion, whereby the extra income generated from the incentive programme pays for the trips that are awarded. Effective incentive programmes have challenging but attainable goals, clearly defined quotas, a short reward period and an adequate budget to promote the concept. Further, the locations chosen are usually well-known, mid-centric, **mass tourism**-oriented destinations that are 'prestigious' and generate a sense of excitement. They must also be safe, secure and stable destinations (see **security**).

Incentive travel was introduced initially as a means of motivating sales staff to increase their productivity or to increase sales of specific product lines. It was demonstrated that tourism is often a more powerful motivator than other incentives organisations may use, including cash bonuses. Today, incentive travel programmes form an integral aspect of overall **management** strategies which provide benefits to both the individual and the total enterprise. Organisations gain from attaining business objectives and also by facilitating networking and team building opportunities among key staff, fostering a stronger corporate culture (see **culture, corporate**) and generating greater employee **loyalty**. The individual benefits are from gaining status and recognition for a job well done.

Incentive travellers form an attractive **market** segment for many destinations. Because the travel component of the trip is free, incentive tourists tend to spend more money at the destination than many other visitors. As well, because they are more willing to travel during shoulder seasons, the effective peak season for destinations may be extended (see **seasonality**).

Further reading

Sheldon, P. (1995) 'The demand for incentive travel: an empirical study', *Journal of Travel Research* 33(4): 23–8.

Shinew, K.J. and Backman, S.J. (1995) 'Incentive travel: an attractive option', *Tourism Management* 15(4): 285–93.

Witt, S.F., Gammon, S. and White, J. (1992) 'Incentive travel: overview and case study of Canada as a destination for the UK market', *Tourism Management* 13(3): 275–87.

BOB McKERCHER, HONG KONG

index, trip

The trip index (Ti) represents the ratio between the total number of nights (Dn) spent at a single **destination** to the total number of nights (Tn) spent on the trip, multiplied by 100. The index is 100 for a trip spent wholly at one destination; and 50 for a destination which receives half the nights of a trip. It provides a useful summarising statistic for **circuit tourism**.

CHRIS COOPER, AUSTRALIA

India

India is the second most populous country in the world and travel has long been a central component of its culture. **Pilgrimage** is an important element in all of India's main religions, with such centres and routes that have developed over the past three thousand years remaining important to the present day. The modern era of tourism in India started with the incorporation of most of the country within the British Empire in the eighteenth and nineteenth centuries. The British developed **transportation** routes which laid the basis for tourism in India. As in Europe and North America, railway development served as a stimulus to business and **leisure** trips. Not only did the railway system reinforce the position of pilgrimage centres in this society because of their increased **accessibility**, but it also led to the development of hillside **resorts** in the Himalayas as summer holiday destinations for the British Raj.

Despite the relatively early development of resorts, **international tourism** received little consideration as an **economic development** mechanism by the **government** until the 1990s. India's federal structure has a major influence on tourism development. Under the constitution, responsibility for tourism exists at both the state and national level. At the national level, the primary source of tourism policy is the Department of Tourism of the Ministry of Civil Aviation and Tourism, as well as the India Tourism Development Corporation. The latter is a limited company which also comes under the auspices of the Department. At the state level, each state has its own ministry responsible for tourism. Governments at all levels have historically played a major role in the **supply** of tourism facilities and **infrastructure**, with the majority of higher quality **accommodation** being state-owned. Several states have established tourism development corporations in order to encourage both foreign and domestic **investment** in tourism infrastructure, often through public–private **partnerships**.

The Tourism Corporation of Gujarat, for example, developed a plan which included several **tax** concessions for investors, such as exemption from luxury tax, sales tax, electricity duty, turnover tax and entertainment tax, and long-term loans from state institutions. Concessions at the state level have also been matched by central government fiscal incentives for tourism projects, including exemptions on income tax for 50 per cent of the profits from **foreign exchange** earnings, on the remaining 50 per cent if the amount is reinvested in new tourism projects, and on import duty on imports for hotel projects. Central government has explicitly sought to encourage regional tourism development by providing interest subsidies on term loans from eligible financial institutions for hotels in cities other than Mumbai (Bombay), Delhi, Calcutta and Chennai (Madras), with higher rates of subsidy being available for hotel development in designated tourism areas and **heritage** sites.

The provision of financial incentives for the industry by the central government in the 1990s is indicative of not only increased attention to tourism's potential for generating **employment** and foreign exchange, but also wider deregulation of the Indian economy to competition and foreign investment. In the **accommodation** sector, the federal government now allows foreign management up to 51 per cent ownership of hotels. India had relatively high rates of economic growth of over 5 per cent from 1992 to 1997. The growth of domestic **demand** is predicted to lead to a doubling of passenger transport output every seven to ten years. Railway, road and aviation sectors are thus receiving high levels of public and private **investment**. The aviation sector has been substantially deregulated. In 1994 the monopoly of Indian Airlines, Air India and Vayudoot (since merged with Indian Airlines) over scheduled air transport services was ended. By 1996 six private corporations had been granted scheduled airline status, with nineteen air taxi operators being permitted to run charter and non-scheduled services. Nationwide, twelve **airports** were earmarked for upgrading with private sector participation. The new **airlines** have also encouraged some states to consider reviving unused airfields to attract tourist traffic.

Apart from the neighbouring countries of Bangladesh and Pakistan, major **inbound** markets include **United Kingdom**, the **United States**, Europe and **Japan**, the latter increasingly focused on pilgrimage-related tourism. However, while India is perceived as an exotic **destination** in the West because of the communication of **images** of the Taj Mahal, Srinigar (Kashmir), the Himalayas and its varied culture, it has also suffered from images of poverty, poor health and **sanitation** and inferior **infrastructure**. Negative perceptions of India in Western countries have long served to stifle the development of inbound tourism in India. News coverage of natural disasters, outbreaks of **disease**, political instability and ethnic unrest have created a difficult environment in which to effectively market the country as an international destination. An outbreak of plague in 1994, just as the peak winter tourism season was beginning, severely damaged growth expectations with India only receiving 1.6 million visitors, excluding nationals from neighbouring Bangladesh and Pakistan, instead of the expected two million. As a result of the plague, the capital, New Delhi, which had hoped to reach five million visitors in 1997, deferred that target to the year 2000.

The development of a large middle-class in India has substantial implications for tourism. The **outbound** market is witnessing substantial growth in the **visiting friends and relatives** market and in leisure tourism to the United Kingdom and Southeast Asia, while **domestic tourism** remains strong. Infrastructure provision will likely remain the main barrier to tourism growth, while perception of political stability and relative freedom from natural disasters will influence the foreign market. However, the development of new markets in terms of **sports tourism** (golf and skiing), conventions and continued demand for traditional heritage and pilgrimage tourism appear to augur well for continued growth.

Further reading

Bhattacharyya, D.P. (1997) 'Mediating India: an analysis of a guidebook', *Annals of Tourism Research* 24(2): 371–89.

Kale, S.H. and Weir, K.M. (1987) 'Marketing Third World countries to the Western traveller: the case of India', *Journal of Travel Research* 25(2): 2–7.

Kaur, J. (1985) *Himalayan Pilgrimage and the New Tourism*, New Delhi: Himalayan Books.

Richter, L. (1989) *The Politics of Tourism in Asia*, Honolulu, HA: University of Hawaii Press.

C. MICHAEL HALL, NEW ZEALAND

indigenous

Indigenous means native to a particular place, as in referring to a species unique to a region or a cultural trait specific to a people. 'Indigenous peoples' is a widely used term for those who are original in their habitats, who maintain a strong sense of identity with their lands and cultures. Other current names for these groups are first or original nations or **Fourth World**. Whatever term is used, indigenous peoples themselves have the right to define its meaning, and its recognition by outsiders. Fourth World is the term which the World Council of Indigenous Peoples has used to distinguish themselves and their cultures from those of the industrialised and developing worlds. They share a common **history** of early socio-political independence, followed more recently by colonisation from foreign states.

The term is a useful means of self-identification for these peoples to represent and defend their shared visions, values, and human and *intellectual property* rights. Today, indigenous peoples are often under cultural assault as minority groups within a dominant state society. Part of this modernisation assault occurs as indigenous peoples' lands and cultural practices are consumed by ethnic and nature-based tourism. These natives are significant for tourism on both cultural and environmental grounds. It is estimated that there are 5,000 distinct indigenous groups, comprising approximately 250 million people and representing 4 per cent of the world's population. Maintaining cultural diversity is their distinct achievement. They are caretakers of a significant part of the ecosystems.

Indigenous tourism is a term used in the literature to refer to those activities which directly involve indigenous peoples. In this type of tourism, the native groups are in control of enterprises which have indigenous culture as a main **attraction**. Many have organised to become managers and beneficiaries of eco-based tourism aimed at natural resources conservation. Analysis of their participation in this industry shows that tourism can challenge, and in many instances has challenged, indigenous culture and property rights.

See also: anthropology; ethnic group; ethnic tourism

Further reading

Burger, J. (1990) *The Gaia Atlas of First Peoples: A Future for the Indigenous World*, New York: Doubleday.

Butler, R. and Hinch, T. (eds) (1996) *Tourism and Indigenous Peoples*, London: ITP.

Greaves, T. (1996) 'Indigenous peoples', in D. Levinson and M. Ember (eds), *Encyclopedia of Cultural Anthropology*, New York: Holt, vol. 2, 635–7.

Swain, M.B. (1989) 'Gender roles in indigenous tourism: Kuna Mola, Kuna Yala, and cultural survival', in V. Smith (ed.), *Hosts and Guests: The Anthropology of Tourism*, Philadelphia: University of Pennsylvania Press, 83–104.

CHARLES R. DE BURLO, USA

indirect tourism

In some circles, indirect tourism is understood as the use of tourism facilities by government agencies and non-governmental organisations, traders, businessmen, developers and missionaries, among others, whose **destination** visits are for purposes other than undertaking **recreation** activities. This indirect form is considered to differ from direct forms where the primary **motivation** is pleasure tourism (see **pleasure tourist**).

MARTIN FRIEL, UK

individual mass

The individual mass is a **tourist** who shares the perceived demographic and psychographic characteristics of **mass tourism**. These may be summarised as preference for familiar atmosphere, **accommodation**, commonplace activities and sun and fun destinations; relatively low activity levels with an emphasis on relaxation; and complete tour packaging prior to departure, although not necessarily as part of a group (see **package tour**).

MICHAEL HALL, NEW ZEALAND

individualism

Greater consumer interest in individualism, coupled with the twin demands of **freedom** and independence, has strongly contributed to the fragmentation of the mass market. The **trend** towards customisation has resulted in a more diverse tourism **market**, as people turn away from the **package tour** to participate in those tourism forms more suited to their personal interests.

FRANK M. GO, THE NETHERLANDS

Indonesia

Since it was discovered by European explorers in the sixteenth century, the islands of Indonesia – Bali, Borneo, Java, Sumatra, Sulawesi and the

Spice Islands – have occupied a special place in the Western imagination. Even today, Bali is better known to many as a **destination** than the country itself. Indonesia's archipelago of more than 17,000 islands has a richness and diversity of natural resources, including the astonishing change in the species of wildlife that occurs between Bali and Lombok islands, known as the Wallace Line. The backbone of tourism products in Indonesia besides tropical beaches is its cultural mix of Buddhism, Hinduism, Islam and Christianity that flourishes among its 200 million people. This is especially evident in Java, with its **UNESCO**-designated World Heritage Sites of the tenth-century Borobudur and Prambanan temples.

Prior to 1969, the unstable political and economic climate in Indonesia did not encourage tourism **development**. However, under the 'New Order' government (1966–98), long-term **planning** and a stable political and economic environment transformed Indonesia. At the beginning of the first five-year plan in 1969, 86,000 **tourists** visited Indonesia; by 1997 the number had grown to 5.1 million foreign tourists who spent $6.7 billion. Tourism planning was first introduced in Indonesia in the 1970s with a master plan for Bali, following which the **government** set up the Bali Tourist Development Corporation. This was a state company commissioned to build **infrastructure**, promote a stronger **image** and encourage **investment** in an enclave resort, Nusa Dua (see **enclave tourism**). Today, extensive tourism infrastructure can be found in major destinations.

By the mid-1990s, tourism had become Indonesia's third most important source of foreign revenue apart from oil. Projections suggest that by the year 2005 it will reach the top position. To accomplish this, a 'Bali Plus Strategy' is being implemented and twenty-three regional **airports** are designated as international gateways. Asia Pacific countries are Indonesia's logical future tourism market and already provide 75 per cent of all visitors. Some of the challenges facing Indonesian tourism include location constraints, regional imbalances, low volume, types of resources, similar products from competing Southeast Asian countries and an international image of Indonesia that is political rather than a promotional asset.

Further reading

Nuryanti, W. (1995) 'Building on Indonesia's fifty years of tourism', in A. Alatas Moerdiono and J. Ave (eds), *Indonesia: The First 50 Years*, Singapore: Archipelago Press, 201–4.

WIENDU NURYANTI, INDONESIA

industrial recreation

The integration of recreational activities and facilities in the workplace is referred to as industrial recreation. This may involve the social mingling of management with workers, and most typically occurs in companies advocating a relaxed work atmosphere. Awareness of societal change and special needs of groups such as shift workers has led companies to increase the number of onsite recreational facilities. Owners of office parks increasingly also design recreational facilities into the setting.

CHARLES S. JOHNSTON, NEW ZEALAND

industrial tourism

This form of tourism is concerned with visits to contemporary industrial sites to see **products** made, **services** delivered and the processes and people involved. Using guided tours, viewing galleries and purpose-built centres, visitors are provided with insights into the workplace, and companies can reap **public relations** benefits. The range of industrial **attractions** is wide and sites include factories, mining operations, power stations, dairies, wineries and theatres.

MIKE ROBINSON, UK

industry

For two hundred years, people in certain **roles** have been described as **tourists**, and in those roles have used various **services** supplied by businesses and other organisations. However, the term 'tourism industry' only emerged in the 1960s. One source for the new idea was the concept of tertiary industries, an expression coined in 1935 to designate recognition of industrialised services. An industrialised service is one subjected to similar methods of financing, producing, managing and **marketing** to those used in manufacturing since the eighteenth century.

'Tourism industry', therefore, refers broadly to collections of business firms, organisations and other resources which foster or support activities of tourists, in particular by providing services. While no consensus exists about the precise components (because the collections are diverse), a custom has developed for referring to industrial sectors, such as travel services as travel agents, **tour operators**, carriers providing **transportation**, **accommodation**, food services and **attractions** (including entertainment and recreational resources). Components of various sectors are located across three geographical realms in whole **tourism** systems. For example, travel agents do their main work pre-trip in traveller-generating regions, while carriers perform their main roles along transit routes and most accommodation is in **destination** regions.

While there is now wide agreement that tourism is industrialised, there are different views regarding the nature of that industrialisation. Three positions can be suggested. First, many researchers perceive tourism as being totally industrialised; they see this industry as representing all resources used by all **tourists**. This view leads many observers to describe it as one of the largest in the world. Second, some researchers point out that referring to 'the' tourism industry is misleading if it implies that all resources used by tourists, anywhere, are linked into one industry. These observers prefer 'tourism industries' as a general expression, keeping the singular for specific cases. A third view agrees that tourism industries exist but argues that tourism is a form of human **behaviour** which is partly but not wholly industrialised. Differences between the first and third views require further elaboration.

The first view, commonly used, regards the tourism industry as the aggregate of all businesses that directly provide goods or services to business, pleasure and leisure activities away from home environments (Smith 1988). This allows researchers to measure the industry by demand-side analyses, surveying tourists' expenditure and converting the findings into a **supply**-side notion. Since tourists consume the same wide range of services and goods as people consume in non-tourist **roles** in their routine lives at home, sector analyses of the industry span the entire consumer economy. Thus, the tourism industry as commonly conceptualised permeates all consumer-related industries.

The theory that tourism in many places is partly industrialised (Leiper 1995) stems from observing that only some services and goods used by tourists come from suppliers with strategic business relationships to tourism. The proportion varies in different places and times. Many retailers serve a mix of tourists and non-tourists (such as local residents) and do nothing distinctive to promote to or service the former. Describing a firm as a unit of a tourism industry merely because tourists are among customers might be realistic if 'industry' is simply taken to reflect economic impact, but it is unrealistic and misleading if 'industry' carries its normal meaning, an assiduous, intentional, strategic application of resources. Further, many tourists' needs are satisfied to some extent by non-economic factors, such as sunshine and incidental displays of customs, which are resources used by tourists but outside any industry in the normal sense of that term.

Measuring partial industrialisation requires surveying suppliers and other resources in order to bisect resources used by tourists into one category forming tourism industries and another category representing non-industrialised or independent tourism. This allows that many firms and organisations can be in more than one industry simultaneously. Many cases are marginal, strategically positioned on the fringes of tourism industries. Suppliers in the business of tourism normally co-operate with one another to some extent. These links convert collections of independent suppliers into synergistically cohesive industries. One link is **package tours**, typically involving cooperation among airlines, hotels and **travel agencies**. Another is computerised reservation systems, in which various tourism-related businesses participate.

See also: management; marketing, resources; systems, tourism

References

Leiper, N. (1995) *Tourism Management*, Melbourne: RMIT Publishing.

Smith, S.L.J. (1988) 'Defining tourism: a supply-side view', *Annals of Tourism Research* 15: 179–90.

Further reading

Sheldon, P.J. (1989) 'Professionalism in tourism and hospitality', *Annals of Tourism Research* 16: 492–503.

NEIL LEIPER, AUSTRALIA

inevitability

There are three reasons for considering tourism to be inevitable. First, there is continual pressure to turn 'undiscovered' places into **resorts** to accommodate expansion of the world's largest industry. Second, the processes of **postmodernism** have made everyone either a **tourist** or object of tourism. Third, **travel** can be viewed as a human imperative, starting with **migration** and culminating in a growing interest in space tourism.

FRANCES BROWN, UK

information centre

Tourism information centres provide information and **reservations** for destinations and tourism enterprises. Run by local, regional or national organisations, they aim to facilitate the visit for consumers and to assist organisations to implement their **policies**, by increasing tourists' **length of stay** and expenditure or by discouraging them from visiting environmentally sensitive areas.

DIMITRIOS BUHALIS, UK

informal economy

The informal economy is a process of income generation which is unregulated by the societal institutions so that such businesses operate without legal recognition and are neither registered and enumerated nor officially taxed. In contrast to the formal sector, these enterprises generally exhibit ease of entry, reliance on **indigenous resources**, family ownership, small scale of operation, labour intensivness, skills acquired outside of the formal school system, part-time labour, and unregulated and competitive **markets**.

The informal sector is particularly important in developing countries. Within tourism, such businesses are found in lodging (unlicensed guest houses), craft-related activities and other small enterprises, including **souvenir** vendors (who also may be called peddlers or hawkers), pedicab drivers, prostitutes, unofficial guides and small **food** stalls. In spite of the number of people involved in informal activities, few tourism plans address the needs of the informal sector or even acknowledge its existence.

Further reading

Cukier, J. and Wall, G. (1994) 'Informal tourism employment: vendors in Bali', *Tourism Management* 15: 464–7.

Kermath, B. and Thomas, R. (1992) 'Spatial dynamics of resorts: Sosua, Dominican Republic', *Annals of Tourism Research* 19(2): 173–90.

Timothy, D. and Wall, G. (1997) 'Selling to tourists: Indonesian street vendors', *Annals of Tourism Research* 24(2): 322–40.

GEOFFREY WALL, CANADA

information source

Information sources are used by tourists to plan trips, the **industry** to formulate strategic and operational plans, and scholars to investigate a research area. Tourists use information centres, brochures, **tour guides** and travel literature. Professionals utilise market research and financial reports. Scholars consult previous research results published in books and journals. National tourism boards collect **statistics** and provide the initial resource base.

DIMITRIOS BUHALIS, UK

information technology

Information **technology** refers to the application of computers to process, store, transmit and display information. The information may be data, text, graphics, voice, **images** or **videos**, and the computers may be supercomputers, mainframes, minicomputers, microcomputers, laptops or network computers. Their processing speed, size, storage capabilities, input–output devices and interconnectivity vary with each installation.

In tourism, information technology plays a very important role since the industry is so information-intensive. Every firm in the **industry** must process information about its products and services and make them available to consumers. It must also receive reservations, and process its own internal company information. Even though this is a service-based industry relying heavily on human relations, tourism firms are finding that the implementation of information technology can contribute to higher levels of service. This is occurring since employees are freed from the tedious tasks now performed by information technology.

Information technology is a combination of hardware and software. The hardware includes the computer itself, monitors and input–output devices (keyboards, mouse, touch screens, scanners, joysticks, optical bar coding, printers, fax machines, modems, digital telephones), communication hardware such as multiplexors, connecting cables (twisted copper wire, coaxial cable, fibre optics), and other methods of communication including satellite and microwaves. Software applications include both system and applications software. The former refers to operating systems (such as DOS, Windows 98, Novell Netware and UNIX), utility and communication software. Operating systems are becoming more user-friendly and less proprietary, allowing different software packages to be used on different operating systems. Application software used in tourism includes both generic business software such as spreadsheets, word processing, databases, desktop publishing and accounting software. There are, however, many more applications written specifically for tourism firms.

The **airline** sector is a heavy user of information technology and has been instrumental in many significant developments which have altered the face of information technology in general. One is its development of the airline computer reservation systems (CRS) and networks in the 1960s. This involved substantial **investment** to develop a new operating system to handle online transaction processing (called transaction processing facility) which is now used in other industries, such as the banking industry. The airline CRS are used to store information on schedules, routes, fares, fare rules, tickets, passengers, seat plans and frequent flyer databases. A few of these systems have evolved into the **global distribution systems**. Other airline applications are departure management control systems, which process information as a plane departs from the **airport**, crew and route management systems, automated maintenance systems and decision support systems to assist management with the many complex choices that must be made. Other parts of the **transportation** sector also use information technology. In particular, the road network is implementing intelligent systems to monitor traffic conditions, reduce congestion and increase safety by communicating the information to onboard computers in vehicles.

The most important information technology applications used by travel agencies are terminals to one or more of the global distribution systems. The terminals allow agents to research and book all types of tourism **products**. They may be used to run other software applications such as travel policy software, and fare auditing software. Travel agents also use back office software to process their **accounting**, commission tracking, customer information for marketing purposes and to produce reports. Government tourist offices are also using information technology to manage and market their destinations. Common applications include destination information systems, which are databases used to store comprehensive, updated information on facilities which are online to travel counsellors and major marketing offices in the destination's originating markets. Various software applications are also used to assist government offices in the collections and analysis of tourism statistics.

Many applications have been written for the **hospitality** sector. The most important is the property management system used in hotels to process guest, room, facilities and accounting information. Guest history files containing detailed information of frequent guests are an important component of the system. Chain hotels also operate central reservation systems to store, process and communicate their room rates and availability for hotels in the chain. Numerous additional applications of technology including electronic locking systems, digital phone systems, guest operated devised and energy management systems are used in the **accommodation** sector. Point-of-sale systems are used in foodservice outlets and retail stores to process transactions and inventory information.

The **Internet** is a major information tool for **tourists** and firms catering to them. Hundreds of thousands of tourism companies have home websites on the **World Wide Web** providing information on their products to the millions of Internet users. Reservations are possible through the Internet, but the percentage of users of this service is lower than those who search for information. The travel distribution channels are changing as electronic access for consumers increases.

Information is beginning to incorporate higher levels of intelligence and functioning. Expert systems and robotics are two applications of artificial intelligence that are being used in the tourism industry. For example, the **airlines** are operating expert systems for crew management, maintenance of aircraft and network design. Robotic technology is being experimented with in the foodservice sector. **Virtual reality**, which uses a variety of computer technologies to give an **experience** of a different reality, are being considered as a way of giving tourists a 'taste' of a **vacation** before they purchase it. Future computer systems in tourism are likely to incorporate more intelligence and be able to assist with higher level functions.

See also: automation; hospitality information system

Further reading

Kasavana, M. and Cahill, J. (1992) *Managing Technology in the Hospitality Industry*, Michigan: AH&MA. (Covers all applications of information technology to the hospitality industry.)

Schertler, W., Schmid, B., Tjos, A.M. and Werthner, H. (eds) (1994) *Information and Communication Technologies in Tourism*, Vienna: Springer Verlag. (Proceedings of major international conference on tourism and information technology.)

Sheldon, P. (1997) *Tourism Information Systems*, Oxford: CAB International. (Examines how information technology is being used in various sectors of the tourism industry.)

PAULINE J. SHELDON, USA

Information Technology and Tourism

Information Technology and Tourism: Application – Methodologies – Techniques is a technical sciences journal focusing at the interface of information technology and tourism. It strives for a balance of theory and application; its main objective is to develop a theoretically integrated and methodologically enriched multidisciplinary body of knowledge in this emerging field. All manuscripts are refereed anonymously (double blind) by at least three reviewers from different disciplines. It regularly publishes full-length articles (abstracts in English), research notes and reports, comments and reviews. First published in 1998, it appears quarterly, published by Cognizant Communication Press (ISSN 1098–3058).

HANNES WERTHNER, AUSTRIA

infrastructure

Infrastructure refers to systems in place which allow for the efficient functioning of a business activity (**industry**) or concentration of people (**community**). Basic infrastructure systems that serve both purposes include electrical, water, sewage, communications, government services (such as police) and **transportation**.

One of the early but still often heard arguments for increased tourism **development** is that this industry can be built on existing infrastructure.

While this may be true in some cases, it is not true universally. Often existing infrastructure is at capacity or in such poor condition that increasing use will lead to overload and system breakdown. An important task in development **planning** is to determine infrastructure capacity and expected demands before proceeding with physical development. Even in urban areas where the infrastructure, is often built with excess capacity infrastructure demands must be assessed to determine whether new development of the type proposed can be accommodated.

Infrastructure **technology** has progressed from ground-based systems to include satellite telecommunications. The relatively efficient air transportation systems taken for granted today are made possible by numerous satellites used for airplane tracking and worldwide computerised reservation systems. Advanced telecommunication systems utilising microwaves for cellular telephones, and the expansion of the **Internet** and **World Wide Web** have greatly increased the ability of people to communicate with each other and to acquire information needed for **decision making**. This type of new technology allows for the development of advanced infrastructure bypassing two or three generations of old technology. This is especially important for developing countries. Often their systems are so old and inadequate that the cost of upgrading to the most modern technology would be commensurate with, or perhaps even less expensive than, repairing existing systems. This would be especially true with microwave telecommunications systems. Given this needed infrastructure investment some destinations, and hence their businesses, would be better able to compete with those in the developed world.

Infrastructure improvements are generally very expensive undertakings and are often deemed the responsibility of **government**. Because of this, government exerts a strong influence in the development of tourism. Government frequently uses the provision of infrastructure needs as an incentive to lure new businesses to an area. Significantly, the environmental impacts of a proposed development are increasingly considered before any government commitments are made for infrastructure development or improvements. The results of the **impact assessment** may help determine the amount of public funds, if any, will be made available for infrastructure development or may be used to identify the particular type of infrastructure technology required to prevent undesirable impacts.

WILLIAM C. GARTNER, USA

innovation

Four types of innovation can be identified in relation to tourism. **Product** innovations result in goods which are either entirely new or in a new context. Process innovations embrace the development and utilisation of **technology** or expertise in novel business processes. **Management** innovations improve frameworks in order to advance organisational performance. Finally, logistics innovations enhance the external liaisons with consumers and other value-chain members.

ANNE-METTE HJALAGER, DENMARK

input–output analysis

When attempting to determine the economic **impact** of tourism spending, it is necessary to estimate the **multiplier effect** associated with the expenditure. A variety of techniques can be employed to estimate economic multiplier values, including economic base theory, Keynesian multiplier **models**, ad hoc models and input–output models. The most detailed and comprehensive methodology available for this estimation is input–output analysis. If comparisons are to be made between the different models used to estimate the economic impact of changes in final **demand**, then these can be made by drawing upon the models' structures. The base **theory** and Keynesian models both have the major deficiency associated with trying to encapsulate the varied impacts felt by each sector of production by using a single equation. That is, in these two models all sectors are treated the same. The ad hoc model, on the other hand, allows the researcher to construct a different equation to represent each of the productive sectors of the economy. Further, input–output analysis, because it includes all of the economic sectors, takes the ad hoc model's approach to its logical conclusion by representing the entire economy by a system of simultaneous equations.

Input–output analysis is a general equilibrium approach to determining an economic system, and it has a pedigree stretching back to the time of Quesnay, who produced the first transactions table in 1758. The term general equilibrium means that the model encompasses all of the productive activities of the economy under study rather than just the tourism sectors. Models that are selective in terms of the productive sectors incorporated are known as partial equilibrium models and tend to yield significantly lower economic multiplier values. Although Quesnay's table could hardly be described as an operational input–output analysis, it did focus attention upon the industrial interdependencies that exist within an economy. The construction of a model that determined multiplier values was left until Leontief developed an input–output of the United States economy using 1930s data. The major issue relating to input–output analysis is the manner in which the economy is aggregated into sectors in order to satisfy the assumptions of linear and homogeneous production functions, and to provide a level of aggregation that will be acceptable for the study of tourism's economic impact.

The technique of input–output analysis may be considered in two separate stages. First there is the construction of an input–output table, similar in structure to that developed by Quesnay. This table, generally known as a transactions table, may be seen as being analogous to a set of national accounts except for the fact that attention is drawn to the transactions that take place between industries within the economy (intermediate sales and purchases) rather than final user transactions. Although this table is not a model, it provides a wealth of information to planners and policy-makers because it highlights the economic structure of the **destination**. This table also shows the direct economic effects associated with any change in final demand. It further circumvents one of the constant problems surrounding the economics of tourism in that it allows the pattern of this spending to determine which sectors are included under the umbrella title of tourism. The table can be combined with other data to look at dependencies and possible supply bottlenecks where some industries may be working close to their full capacity.

The second stage of the analysis involves the conversion of the table into an input–output model. This action requires the normalisation of the table, by dividing the value contained in each cell by the corresponding column total. This process results in a table of coefficients where the vertical columns show the production functions of each industry and if each column is summed it will yield a total of one. This coefficients table is then subject to the Leontief inversion routine, which allows the calculation of the indirect and induced economic impacts associated with any change in final demand. The model, because it is a general equilibrium model, can be used to calculate the economic impact of any change in final demand not only in tourism, although the models tend to be built in order to calculate the economic effects associated with a specific type of final demand. Input–output analysis results in the calculation of a variety of economic multiplier vales. These include the direct, indirect and induced multipliers relating to income, **employment** and government revenue output, as well as estimating the import requirements associated with any change in final demand.

This method of analysis is not without its limitations and weaknesses. For instance, because the model is general equilibrium in nature, it requires detailed information relating to the expenditures made by businesses in all of the productive sectors of the economy, not just the tourism-related sectors. This can make the construction of the input–output table an expensive exercise both in terms of time and resources. The data requirements are extensive and generally require the implementation of specific business expenditure **surveys** in order to determine the patterns of intermediate purchases. Furthermore, a variety of assumptions are necessary in order to accept the results of input–output analyses. These include that all of the businesses aggregated under the heading of a single productive sector are producing their output in an identical manner (i.e. that production functions are homogeneous). It also requires the assumption that the production functions are linear in that any increase in demand for the output of any sector will result in an increase in demand for the factor inputs of that sector in exactly the same proportion as those used on average, to produce the current level of output.

Because input–output analysis attempts to calculate the economic effects of changes in final demand, it is intended to work on marginal coefficients. However, the cost of deriving the marginal coefficients of production functions is such that the analyses tend to rely on average coefficients. This means that some accuracy is sacrificed in an attempt to keep the costs of construction at a reasonable level. Studies that have been undertaken in order to determine the significance of this inaccuracy tend to suggest that it is an acceptable level of inaccuracy, no more than 5 per cent in most instances.

Input–output analysis offers an extremely flexible and robust framework for studying the economic effects of changes within an economy. If new sectors are introduced or existing ones are removed then the framework is such that a revised model would need to be constructed. However, this shortcoming is to be found in any model that is constructed to simulate the workings of the economy. In order to construct a model that will yield the level of detail found in input–output models and to achieve similar or greater levels of accuracy, it would be necessary to construct an econometric model. This is a system of equations whereby each of the equations reflect the production and consumption functions of the economy. These latter models are even more costly to construct in terms of time and data, and their results are less accessible to the planners and policy makers who require the information.

Input–output analyses are not limited to calculating economic impacts. They have been used for a wide variety of purposes including travel models and, more recently, environmental impact studies. The advantage of using input–output analyses to determine environmental impacts are enormous because it means that the economic impacts associated with a change in the level or pattern of final demand associated with tourism spending can be determined along with the environmental effects of that change in activity. It also means that the environmental impacts can be examined at the direct, indirect and induced levels of impact in much the same way as the economic impacts.

Further reading

Fletcher, J.E. (1989) 'Input–output analysis and tourism impact studies', *Annals of Tourism Research* 16: 514–29.

Leontief, W. (1966) *Input–Output Economics*, New York: Oxford University Press.

O'Connor, E. and Henry, E.W. (1975) *Input–Output Analysis and its Applications*, Griffin's Statistical Monographs No. 36, London: Charles Griffin & Co.

JOHN FLETCHER, UK

Institute of Certified Travel Agents

The Institute of Certified Travel Agents (ICTA) is an international non-profit organisation founded in 1964 to enhance the quality of professional tourism practice through **education**, testing and certification. Its pyramid of **professionalism** offers education and recognition at every career stage. For entry-level professionals, tests are given and basic competency is stressed. This stage is followed by the ICTA's professional certification process, which offers two levels of recognition for experienced professionals. One of these offers the highest standards of professionalism and advanced knowledge of business, industry, and global issues necessary for expert practice in this field. In addition to the educational activities, ICTA conducts research in areas specific to its field of operation. The headquarters of the institute is located in Wellesley, MA, United States.

TURGUT VAR, USA

Instituto de Estudios Turísticos

The Instituto de Estudios Turísticos (IET) in Spain was founded by a 1962 decree as Centralised Public Service. Through its existence it has been reorganised several times. Different general aims have been entrusted to the IET, including market research, advancement of knowledge in the field, studying the role of public administration, collaboration in the incentive to and implementation of professional **training** and seminars, and so on. The institute has already made several contributions, including development of the bank of estimated econometric **models**, input–output tables for the tourism economy, models of national accounts, semiological processes for the design of the tourism promotion language, tourism planning models and various databases. The library of the institute contains some 22,000 titles and bibliographic references for more than 60,000 documents. Its headquarters is located in Madrid.

D. MANUEL FIGUEROLA POLOMO, SPAIN

integrated environmental management

Integrated environmental management is a **holistic approach** to managing environmental impacts (see **impact, environmental**) which considers the range of business activities and how they interact to contribute to various environmental impacts. The process involves **management** of a range of environmental impacts arising from an organisation's activities. It is aimed at managing biophysical as well as sociocultural impacts (see **change, sociocultural**).

Management of biophysical impacts involves looking at activities which may impact on water resources and quality, soil and land degradation, flora and fauna loss, and air quality. Management of sociocultural effects considers business activities which affect aesthetic quality, such as areas of visual significance, recreational opportunities and experiences, **indigenous** culture, areas of archaeological and **heritage** significance, noise, traffic, and local, regional and national economy. Integrated environmental management aims to mitigate or minimise a range of cumulative environmental impacts. For example, management of solid waste at a tourism facility would consider raw materials procurement, reducing waste generation, and reuse potential and recycling potential before suitable disposal. The process may involve management of a range of business operations which contribute to an impact, and/or management of a range of environmental influences associated with business operations.

See also: cultural conservation; environment

NEV BRAH, AUSTRALIA

intellectual property

Intellectual property is an inclusive term which refers to various values legally legitimising ownership rights to objects or ideas invented by individuals. Legally classified as copyrights, trademarks, patents, industrial designs, slogans or royalties, these rights are held by individuals or private corporations. Intellectual property also includes rights of ownership to works of art. Legally binding rights are invested in property invented by individuals. Knowledge is assumed to be a private good requiring copyright or patent protection from theft or other incursion.

In regard to tourism, intellectual property conceivably covers a huge range of protections for products and knowledge. Computer systems as intellectual property, industry slogans, trade secrets and trademarks in **marketing** exemplify the concept. Another type of intellectual property rights is that held by **indigenous** people who are the objects or attractions of tourist interest. Traditional cultures and ways of life which are attractions require protection because they are under direct threat from tourism, which converts cultural meanings held by this knowledge into a product for sale. Indigenous peoples are now asserting their rights to group cultural heritage through **laws** governing intellectual property.

The legal mechanisms for indigenous rights currently include copyright (or patent) and use of contracts. The domains of cultural property and **heritage**, which are linked to international human rights and labour law agreements, is enabling indigenous communities to protect their group intellectual property. Examples include sacred property (**music**, **ritual**, material culture and knowledge) and cultural heritage (performances, attire and crafts) which tourism appropriates as commodities. Resources and knowledge of ecosystems which nature-based tourism may use are also identified as indigenous intellectual property.

The First International Conference on Cultural and Intellectual Property Rights of Indigenous People, convened in 1993 by the Mataatua Tribes in Whakatane, Aotearoa, New Zealand, resulted in the Mataatua Declaration on Cultural and Intellectual Property Rights of Indigenous Peoples. It states that intellectual and cultural property is a basic right of self-determination, and that indigenous peoples and their direct descendants are first beneficiaries of indigenous knowledge. However, the collective character of traditional knowledge remains difficult to implement in the context of existing intellectual property law, which assumes a need to protect individual inventors or initiators of knowledge.

Further reading

Graves, T. (ed.) (1994) *Intellectual Property Rights for Indigenous Peoples: A Source Book*, Oklahoma City: Society for Applied Anthropology.

CHARLES R. DE BURLO, USA

internal marketing

While external **marketing** focuses on consumer markets, internal marketing is inward-looking, covering the planned design of corporate exchanges with inside system components for sales-led purposes. The target groups of this marketing are the company's staff, local inhabitants or, alternatively, the individual service providers in a tourism **destination**. Its objective is to obtain motivated, obliging, customer-minded, **quality**-conscious and responsible staff or host populations.

In all branches, internal marketing has a key function because, for any **product**, the aim is to sell the product or **service** to the staff first, to make them capable of satisfying customer requirements as best possible. This marketing form is crucial in tourism because individualised **service quality** is of supreme importance, because production and consumption of goods and services coincide, because the **guest** does not ask for just one isolated service but an entire service package, and because the local population as a whole is part of this service package.

In many places in recent years, a critical view of tourism has developed among host populations. There is talk of decreasing tourism awareness, meaning the carefully considered way in which locals deal with the phenomenon of tourism perceived as a whole, with all its benefits and drawbacks. The most important reasons for this are that locals are excluded from the decision-making process and psychological carrying capacities are ignored. This decline in tourism awareness results in insufficient goodwill towards the **industry**, as becomes apparent when pro-tourism policies are voted down, and in an unfriendly **attitude** towards **tourists** or even protest action. This is what is termed the 'rebellion of the hosts'. As a result, internal marketing (and especially internal **communication mix**) has had to be intensified in destinations. Its modern instruments in destination include open tourism forums, information **brochures** and exhibitions, school briefing kits with **video** films, commercials on local **television** stations and similar channels.

See also: community; community development; marketing, destination

Further reading

Bruhn, M. (1995) *Internets Marketing–Integration der Kunden–und Mitarbeiterorientierung*, Wiesbaden: Gabler.

Ferrante, C.L. (1994) *Konflikt und Diskurs im Ferienort: wirtschaftsethische Betrachtungen*, Berner Studien zu Freizeit und Tourismus Nr. 32.

Gummesson, E. (1987) 'Using internal marketing to develop a new culture', in J.A. Czepiel, C.A. Congram, and J. Shanahan (eds), *The Service Challenge, Integration for Competitive Advantage*, Chicago: AMA Proceedings Series.

Müller, H.R. and Boess, M. (1995) *Tourismusbewußtsein: Empirische Belege und Hintergründe*, Bern: FIF-Universität Bern.

HANSRUEDI MÜLLER, SWITZERLAND

International Academy for the Study of Tourism

Founded in 1989, the International Academy for the Study of Tourism is a multidisciplinary body of scholars. Membership is not automatic; a candidate must be sponsored by one of the current members, and the nomination committee circulates the credentials of the nominees for the final selection by the current membership. The officers are elected for two-year period. Various committees serve for the operation of the organisation. The Academy, according to its bylaws, can have only seventy-five members.

A selected number of papers presented at the biennial meetings are published as free-standing books. The newsletter of the Academy appears quarterly, distributed to the members free of charge and available to outsiders by request. There are five regional editors representing the five continents, who coordinate submissions of news and short scientific notes. The Academy can be reached through the headquarters of the **World Tourism Organization** in Madrid.

TURGUT VAR, USA

International Air Transport Association

The mission of the International Air Transport Association (IATA) is to represent four groups: airlines, the general public, governments and third parties such as travel and cargo. For the airlines, IATA offers joint means – beyond the resources of any single company – for developing opportunities and solving problems. Airlines knit their individual networks into a worldwide system through the association, despite differences in language, currencies, laws and national customs. For governments, industry working standards are developed within IATA. It is thus the most effective source of airline experience and expertise upon which governments can draw. In fostering safe and efficient air transport, the association serves the stated **policy** of most of the world's governments. IATA saves much effort and expense otherwise expended in bilateral negotiations on, for example, tariffs.

For third parties, IATA serves as a collective link between them and the airlines. Passenger and cargo agents are able to make representations to the industry through the Association and derive the benefit of neutrally applied agency service standards and levels of professional skill. Equipment manufacturers and others are able to join in the airline meetings which define the way air transport goes about its business. For the general public, IATA simplifies the travel and shipping process. By helping to control airline costs, it contributes to cheaper tickets and shipping costs. Thanks to airline cooperation through IATA, individual passengers can make one telephone call to reserve a ticket, pay in one currency and then use the ticket on several airlines in several countries.

The modern IATA is part of the structure of post-Second World War aviation which grew out of the Chicago Conference of 1944 (it replaced the International Air Traffic Association of 1919). In both organisation and activity, IATA has been closely associated with the **International Civil Aviation Organisation**, the United Nation's specialised agency for civil aviation. IATA's headquarters is in Montreal, with its main executive office is in Geneva and regional offices in Amman, Brussels, Dakar, London, Nairobi, Santiago, Singapore and Washington DC. In addition, there are fifty-seven offices around the world responsible for agency services, billing and settlement plans and cargo account settlements.

See also: international aviation organisations

ORHAN ICOZ, TURKEY

International Association of Amusement Parks and Attractions

The world's largest amusement industry trade association, International Association of Amusement Parks and Attractions (IAAPA), was formed in 1918 for professionals who own, manage and operate amusement parks, water parks, family entertainment centres and attractions, as well as the manufacturers and suppliers that serve them. The Association is dedicated to the preservation and prosperity of this sector. It is made up of more than 4,500 members in eighty countries, and works through a series of volunteer committees to create services, educational products and opportunities. Membership benefits and goals include improved **efficiency**, **marketing**, **safety** and profitability, together with the professional standards in the amusement sector.

TURGUT VAR, USA

International Association of Convention and Visitors Bureaus

The International Association of Convention and Visitors Bureaus (IACVB) was founded in 1914 and represents over 423 bureaus in more than thirty-two countries. Its mission is to enhance the **professionalism**, effectiveness and image of its worldwide members and the **industry** they represent. To assist member bureaus, IACVB organises several conferences per year on topics ranging from administration to sports tourism **marketing**.

RUSSELL R. CURRIE, CANADA

international aviation bilateral

For the greater part of history, transoceanic commerce could only be carried by ship. The rules and freedoms of the sea, although old and sacrosanct, provided insufficient guidance for the governance of the **airline** industry. The technologies and eras of development simply were too far removed from one another. Shortly after the beginning of the twentieth century, however, governments were confronted with a means of **transportation** with which they were ill equipped to deal. In its earliest stages the airline industry was treated as a curiosity. Military confrontation during the two world wars dramatically shifted the debate to one of urgency. The need to rebuild airline systems after the wars once again shifted the governance issue, this time to stability. More recently, airline deregulation shifted the issue once again. Open skies are not yet the norm, but the goal can be reached.

Before aircraft proved their military worth during the Second World War, **policy** makers regarded them passively. One issue to be considered was clearly the extent to which nations had sovereignty over airspace. Some contended that airspace was international in nature and no country could exert sovereign claim. Others contended that nations had legitimate national defence interests in claiming sovereignty. The freedoms of the sea seemed to support the latter, rhetoric the former, which carried on until the Second World War. The wars themselves had a great impact on aviation regulation policy. Much of the prewar debate was settled by the start of it. Although the war was far from conclusive on the battlefield, by 1919 aircraft demonstrated their ability to shift the fortunes of war.

The Paris Convention (1919) provided the first multilateral framework for governing international airlines. Prewar debates about freedom of the skies was resolved in favour of nations' relatively unquestioned control of airspace above and immediately proximate to their geographic territories. The Havana Convention (1929) reaffirmed the sovereign airspace claims of nations. Beyond ratifying the Paris Convention, the Havana Convention further delineated the rights of nations to specify airline services to, from, within and over their airspace. These rules, originally applicable only within the Americas, were later more broadly affirmed. If any doubt remained about the potential of aircraft to reshape national fortunes, the Second World War dispelled it. Aircraft could easily penetrate the sovereignty claimed by any nation.

In 1944, the United States convened what is known as the Chicago Conference. Its purposes were to foster international civil aviation development in a safe and orderly manner and to establish international air transportation on the basis of equality of opportunity and economical operation. It was generally agreed that such a system should be developed through a framework of bilateral treaties between nations. Aspects of the Chicago Conference agreement considered to be a reaffirmation of previous principles, often are referred to as a 'Chicago Standard Form' treaty. Many current treaties are still governed by these principles, and others are influenced by deliberations undertaken, but not ratified, at the Chicago Conference. The 'Two Freedoms Agreement' considered in Chicago remains the most basic of **international aviation rights** agreements. The first freedom provides for safe passage over the sovereign territory of another nation. The second freedom provides for a non-commercial stopover in another nation. The 'Five Freedoms Agreement' likewise remains significant as a statement of international aviation rights. In addition to the first two freedoms, it called for nations to permit commercial carriage to a bilateral partner, from a bilateral partner, and to a bilateral partner and beyond to another bilateral partner.

Although the areas of agreement following Chicago were considerable, the areas of disagreement were even more numerous. Further clarity of governance issues came in 1946, but without total agreement. At the behest of the **United Kingdom** and the United States, a temporary bilateral agreement between the two nations was reached in what commonly is referred to as the Bermuda Agreement. This embraced in principle the five freedom rights of the Chicago Conference – that is, overfly, technical stopover, to, from and beyond rights – but practically committed both parties to only the first four freedoms. The typical Bermuda Agreement specified the routes served, number of carriers eligible to provide service, and other capacity constraints. Most notably, the typical Bermuda Agreement deferred fare and rate issues to development by the **International Air Transport Association**, a carrier-based discussion group.

Owing to a curious mix of economics and politics, airline deregulation became the official policy of the United States in 1977. Deregulation by whatever name means placing a maximum reliance on **market** forces rather than government regulatory schemes in setting prices and capacity. The domestic air cargo industry was deregulated that year, and the domestic airline passenger service the next year. Other domestic transportation markets were largely deregulated in 1980. In the same year, the USA declared its intention to support deregulation of international airline markets. But unlike domestic markets, no nation can unilaterally impose its policies upon international markets. For the past twenty years, the United States and others have tried to negotiate open skies agreements, pushing for the full exchange of rights and a diminished role for the **International Air Transport Association**. Notable successes have been achieved in Europe and Asia, but much remains to be accomplished. Deregulation means placing a maximum reliance on market forces rather than government regulatory schemes in setting prices and capacity. Some refer to the system as one of *caveat emptor*; but only time will tell who the buyer is.

Further reading

Dobson, A.P. (1995) *Flying in the Face of Competition*, England: Avebury Aviation.

Haider, D. (1996) 'The United States–Japan Gateway Awards Case of 1990: international competition and regulatory theory', *Public Administration Review* (January/February). (Examines the impact of deregulation and increased competition on regulatory politics in the international setting.)

Hill, L. (1997) 'Bilateral ballistics', *Air Transport World* (February): 53ff. (Discusses the replacement of traditional air services agreements with open skies agreements.)

KEVIN BOBERG, USA

international aviation liberalisation

International aviation liberalisation is a result of the airline deregulation movement that began in the United States in 1977 (see **deregulation, airline**). Emphasis was placed on the removal of economic regulatory barriers, which hindered a free market approach to domestic and international air services. Gradually this freedom spread to other regions of the world, and it continues today. As nations and economies become more globally linked and interdependent, the necessity for free market approaches to international aviation services is essential.

US airline deregulation began at a specific point in time, when the air cargo industry was deregulated. The passenger airline industry followed in 1978. Often this approach is referred to as the big bang theory, where governmental economic regulations on pricing, route additions or deletions and most requirements for new entrants are eliminated all at one point in time. **Airline** managers could immediately change prices, add new routes in profitable markets, discontinue service in unprofitable markets and form new airlines with a minimal amount of **government** interference. Soon, many new carriers had entered the marketplace. The general results were lower fares for the consumers and increased frequencies in several markets.

The logical next step was to transform international aviation services in a similar manner. However, this was a much more challenging undertaking. The vast majority of international airlines were state-owned and considered to be national flag carriers, whereas US carriers were privately owned. International aviation bilaterals govern most air services beyond the domestic arena of each nation. The two participant countries selected the airlines to serve a route, set frequencies and capacities, regulated fares, and in some cases allowed share or pool revenues.

The major objectives of the two countries were to insure profits for the national flag carriers and to limit competition from other 'foreign' carriers. Each government exhibited very protectionist attitudes. The United States found it impossible to transfer its total and immediate deregulation approach to other regions of the world, as most countries wanted to maintain their traditional bilateral approach to international aviation services.

As a result of this situation, the big bang approach to deregulation was deemed inappropriate in other parts of the world. Instead, the term liberalisation was widely used. This means a more gradual, progressive process of removing governmental economic regulations from airlines, the privatisation of formerly state-owned carriers and an opening up of international markets to carriers from other countries. The most significant international aviation liberalisation programme to date has been in the **European Union**, whose officials felt that the more gradual approach was most appropriate given the fact that there were twelve countries involved (fifteen from 1998), each of which had different approaches to liberalisation. These differences included existing international bilateral air agreements; the degree of privatisation of state air carriers; the appropriate **model** for **economic development**; change methods and techniques; and administrative procedures, as well as cultural differences. In addition, this air market is different from that of the United States, in that a very significant volume (over 40 per cent) of air service consists of charter flights rather than scheduled service.

The European airline liberalisation was to take place over three distinct phases beginning in 1987 and ending in 1997. In the first stage in 1987 bilateral capacity controls were gradually changed from a 50–50 split between the carriers of the two countries involved. A carrier in one country could capture up to 60 per cent of market share. Additionally, more than one airline per country could serve a route, a practice which is termed multiple designation. Further, it became much easier for airlines to file discounted airfares which met certain prescribed conditions. The second stage in 1990 made it much easier to file more heavily discounted airfares, allowed extensive competition in markets, often adding three or more carriers per route and authorised market capacity shares to go to 75–25 percentage splits. The third and final stage began in 1993, and again more freedom was granted in terms of capacity limits, setting fares and gaining access to new markets.

Much work still remains, as the European Commission must resolve issues such as **airport** capacity restrictions and allocating landing and take off slots to new entrants. Additionally, the anti-competitive aspects of computerised reservation systems and continued state subsidies to certain airlines, which greatly distort a free **market** approach to competition, need to be addressed. Another challenge is the transition away from bilateral agreements to multinational aviation agreements in which the European Community would negotiate on behalf of countries and/or airlines in determining international air service agreements. This would mean that individual European countries would give up some of their power to the community, a concept that has met firm resistance in several member countries. Currently, many international airlines experience considerable freedom in serving markets and setting prices. The term open skies is widely used to describe these newfound freedoms. In addition, numerous strategic alliances have been formed in which airlines from several countries link their routes and services together. More and more Asian and Latin American airlines are also becoming partners in these strategic alliances (see **globalisation**).

Further reading

Balfour, J. (1994) 'The changing role of regulation in European air transport liberalisation', *Journal of Air Transport Management* 1(1): 27–36. (Examines the reasons and pressures for liberalisation in Europe and how a gradual process may work best.)

Doganis, R. (1991) *Flying Off Course*, 2nd edn, London: HarperCollins. (Provides an overview of the economics of international airlines.)

Kinnock, N. (1996) 'The liberalisation of the European aviation industry', *European Business Journal* 8(4): 8–13. (Reviews the progress of European Union/airline liberalisation.)

Ott, J. and Neidl, R.E. (1995) *Airline Odyssey*, New York: McGraw-Hill. (Offers an analysis of the changing airline industry.)

US General Accounting Office (1993) *International Aviation*, GAO/RCED-93-64, Washington, DC. (An analysis of the ability of US airlines to compete in Europe.)

Williams, G. (1993) *The Airline Industry and the Impact of Deregulation*, Aldershot: Ashgate. (Reviews deregulation in the United States, Canada and Australia and liberalisation in Europe.)

DAVID B. VELLENGA, USA
WILLEM J. HOMAN, USA

international aviation organisations

The Chicago Conference of 1944 reached agreement on establishing a regulatory regime to manage international aviation with three main components which continue today: the **International Civil Aviation Organisation** (ICAO), the **International Air Transport Association** (IATA) and air service agreements (ASAs). ASAs are international bilateral agreements negotiated between governments which allow designated **airlines** of the two parties access to each nation's air space under the **international aviation rights** or 'Freedoms of the Air'. There are several thousand ASAs registered with ICAO. While each of these components operates separately, it is necessary to harmonise some of their work to regulate aviation activities effectively.

See also: deregulation, airline; Bermuda 1

Further reading

Bureau of Transport and Communications Economics (1994) *International Aviation: Trends and Issues*, Canberra: Australian Government Publishing Service.

International Civil Aviation Organisation (1980) *Convention on International Civil Aviation*, 6th edn, Montreal: ICAO Doc 9470.

Sochor, E. (1991) *The Politics of International Aviation*, London: Macmillan.

TREVOR SOFIELD, AUSTRALIA

international aviation rights

The 'Freedoms of the Air' define the rights of commercial airlines to operate on specified routes. The first five freedoms were formally defined at an international aviation conference in Chicago in 1944, while the last three were enumerated at other times. Any of these freedoms are available only to commercial airlines once the nations involved have agreed to those rights in the applicable international aviation bilaterals.

The Technical Freedoms includes the right of a commercial airline to overfly the territory of another country; often called the 'right of innocent passage' (the first freedom) and the right of a commercial airline to land in another country for the purposes of refuelling, aircraft repair, or an emergency (the second freedom). The Commercial Freedoms include the right of a commercial airline to carry fare-paying passengers from the home country to a second country (the third freedom). The next freedom is the right of a commercial airline to carry fare-paying passengers from another country to the home country. These last two freedoms are often called 'out and back' rights. The fifth freedom is the right of a commercial airline to carry fare-paying passengers to/from a third country as part of service between the home country and a second country. There are various forms of fifth freedom rights; an airline has full fifth freedom rights if the airline can pick up and discharge passengers on any flight sector. If the airline is not permitted to carry local traffic between the two foreign nations, this is known as limited fifth freedom rights, with the segment between the two nations known as a 'blind sector'.

There are three more freedoms. The sixth freedom is the right of a commercial airline to carry fare-paying passengers between two other countries via a flight that transits the home country. The next one is the right of a commercial airline to carry fare-paying passengers between two other countries without flying via the home country. The final freedom is the right of a commercial airline to carry fare-paying passengers between two points in another country (sometimes known as cabotage rights).

See also: deregulation, airline; international aviation liberalisation

Further reading

Gidwitz, B. (1980) *The Politics of International Air Transportation*, Lexington, MA: D.C. Heath and Co. (Discusses international aviation policy, with chapter 6 being on aviation bilaterals.)

Hanlon, P. (1996) *Global Airlines: Competition in a Transnational Industry*, Oxford: Butterworth-Heinemann. (Chapter 4 contains a discussion of the eight freedoms.)

FREDRICK M. COLLISON, USA

International Civil Aviation Organisation

The International Civil Aviation Organisation (ICAO) is a United Nations-related organisation that encourages the development and application of quality air travel. Established in 1944, this is a specialised agency of the United Nations, whose purpose was set forth at the Chicago Convention which was originally signed by fifty-two countries. The Convention agreed on certain principles and arrangements in order that civil aviation might be developed in a safe and orderly manner and that international **services** might be established on the basis of equality of opportunity and operated soundly and economically. Although responsible for developing international **rules** governing all areas of civil aviation, ICAO essentially exists to allow anyone to fly safely and comfortably in a commercial aircraft anywhere in the world.

The specific goals of ICAO include facilitation of civil air travel by reducing the obstacles placed on the passage of people and cargo across international borders by immigration, customs and public **health**; the establishment of international air transport that is based on equality of opportunity and operated both soundly and economically; the **standardisation** of each technical field within aviation; the development of technical cooperation that will assist developing countries in their opportunities for access to safe civil aviation; regional **planning** for aviation problems that cannot be dealt with on a worldwide scale; development of a standard code of international air law which covers rules governing the environment; and development of a satellite-based system concept to meet the future communications, navigation and surveillance/air traffic management needs of civil aviation.

According to the terms of its constitution, ICAO is made up of an Assembly, a Council of limited membership with various subordinate bodies and a Secretariat. The chief officers are President of the Council and General Secretary. As the governing body the Council gives continuing direction to the work of ICAO. It is in the Council that standards and recommended practices are adopted and incorporated in the existing rules. The Secretariat is divided into five divisions: the Air Navigation Bureau, the Air Transport Bureau, the Technical Cooperation Bureau, the Legal Bureau and the Bureau of Administration and Services.

ICAO works closely with other members of the UN organisation, which include the World Meteorological Organisation, International Telecommunication Union, Universal Postal Union, World Health Organisation and International Maritime Organisation. ICAO has numerous publications which fall under the following categories: conventions and related acts; agreements and arrangements; rules of procedure and administrative regulations; annexes to the convention of civil aviation; and procedures for air navigation services.

See also: international aviation organisations

TURGUT VAR, USA

International Forum of Travel and Tourism Advocates

The International Forum of Travel and Tourism Advocates (IFTTA) is an organisation providing attorneys involved in tourism, with a forum for the collection and dissemination of data on international travel laws, cases and decisions. The organisation held its first international conference in 1986 in Jerusalem. The headquarters of IFTTA is located in San Francisco, California, USA.

TURGUT VAR, USA

International Geographical Union

The International Geographical Union (IGU) has long been concerned with research in tourism and **recreation**. For example, in 1994 it appointed a Study Group on the Geography of Sustainable Tourism to carry out a variety of comparative international studies. There are ten full members and 308 corresponding members from fifty countries. It publishes a semi-annual newsletter, *TOURISTinfo*.

FREDERICK M. HELLEINER, CANADA

International Hotel and Restaurant Association

The International Hotel and Restaurant Association) was founded in 1947 for prominent hoteliers. As a non-profit organisation, it represents the world **hotel** sector by identifying and articulating common positions, supporting the resolution of key issues, defending its interests before **policy** makers and international organisations, lobbying for the recognition of the contribution that hotel and **restaurant** businesses make to worldwide social and **economic development**, providing authoritative information and objective analysis of global issues and their **impact** on hotels, organising meetings in different parts of the world, and covering international events that offer business opportunities to the sector and its commercial partners.

IH&RA represents some 300,000 establishments in over 145 countries. Its congress is the main annual event which attracts delegates from all around the world. The association endeavours to be entirely collaborative and delivers its programme through partnership with a number of international and regional organisations, including **World Tourism Organization**, Universal Federation of Travel Agents Associations, International Road Transportation Union, European Tour Operators Association, International Hotels Environment Initiative and United Nations Environment Programme. The headquarters of IH&RA is located in Paris.

MARYAM KHAN, USA

International Journal of Contemporary Hospitality Management

The editorial objective of the *International Journal of Contemporary Hospitality Management*, (published in association with the Hotel, Catering and Institutional Management Association) is to communicate the latest developments and thinking on the **management** of **hospitality** operations worldwide. This multidisciplinary refereed (double blind) journal is designed for managers working in every sector of the **hotel** and **restaurant** sectors, for educators and researchers, and for managers working in other tourism sectors. Articles based on experience and evidence – rather than philosophical speculation – are encouraged. First published in 1989, the journal appears seven times per year, published by MCB University Press (ISSN 0959–6119).

RENE BARETJE, FRANCE

International Journal of Hospitality Management

The *International Journal of Hospitality Management*, supported by the International Association of Hotel Management Schools, discusses major trends and developments in a variety of disciplines as they apply to its theme. The range of topics covered includes human resources management, consumer behaviour and marketing, business forecasting and applied economics, operational and financial management, planning and design, information processing, education and training, technological developments, and national and international legislation. In addition to refereed (double blind) papers and research notes, the journal publishes discussion papers, viewpoints, letters to the editor, a calendar of forthcoming events, book reviews, and reports on conferences. First published in 1982, it appears quarterly, published by Elsevier (ISSN 0278–4319).

RENE BARETJE, FRANCE

International Journal of Tourism Research

As an international publishing platform for research practice in tourism and hospitality, *International Journal of Tourism Research* aims to promote the standing of both subject areas. The journal publishes views and debates current themes in the field, with a focus on methodological and best practice issues as well as policy related to issues and applications. In particular, this refereed journal encourages a fusion between tourism and hospitality approaches to research. First appearing in 1995, it is published quarterly by Wiley (ISSN 1077–3059).

RENE BARETJE, FRANCE

International Labour Organisation

International Labour Organisation (ILO) was created in 1919 and became the first specialised agency of the United Nations in 1946. Its headquarters is located in Geneva, Switzerland. ILO has been concerned with humanitarian, political, economic and **peace** issues. It works both to promote economic growth in harmony with social objectives and to stress technical assistance in labour and social matters, particularly in the developing countries. Its tourism sector office has produced many studies dealing with tourism workforce development and conditions worldwide.

BONG KOO LEE, USA

international organisation

Organisations representing public and/or private sector interests at international level are known as international tourism organisations. Examples include industry associations (such as the **International Hotel and Restaurant Association**), government-derived agencies (**World Tourism Organization**) and combined public and private sector organisations (**Pacific Asia Travel Association**). The term is sometimes confined to entities attempting to be global in their scope, to differentiate them from supranational organisations.

BRIAN KING, AUSTRALIA

International Society of Travel and Tourism Educators

The International Society of Travel and Tourism Educators (ISTTE) is an organisation of educators in tourism and related fields representing professional schools, high schools, four-year colleges and graduate degree-granting universities. ISTTE is represented by such varied educational and industry sectors as business administration, **economics**, gaming management, **geography**, **hospitality** administration, **hotel** and **restaurant** management, **leisure** studies, **marketing**, meetings and conventions management, **parks** and **recreation** studies, **sociology**, tour and travel operations, tourism **planning** and development, **transportation** studies, **travel agency** training and other allied areas.

ISTTE was founded in the late 1970s in response to the diverse needs of increasingly complex tourism **education** and **training**. Continued growth in this field will require an increasing pool of trained professionals at all levels, who understand the close interrelationships among all segments of the **industry**. Through ISTTE members increase their professional knowledge by interacting with all levels of tourism educators, developing a network of peers worldwide and working with other educators to advance the professional image and understanding of tourism, both as a field of study and operation. The Society offers five categories of membership: active members, associate members, graduate student members, allied members and emeritus members. Its headquarters is located in Harper Woods, Michigan, USA.

K.S. (KAYE) CHON, USA

International Sociological Association

The goal of this non-profit scientific association, founded in 1949, is to represent sociologists everywhere – regardless of their school of thought, scientific approaches or ideological opinion – and to advance this body of knowledge. The association has over 3,600 members worldwide. Many of its research activities are through its specialised committees and working and thematic groups. One of its research committees, established in 1994, deals with sociology of **international tourism**; another deals with leisure. Objectives of the tourism committee include establishing contacts among tourism sociologists, encouraging dissemination and exchange of information, and promoting international meetings, research and publications in the field of sociology of tourism. The headquarters of the association is located in Madrid, Spain.

TURGUT VAR, USA

International Tourism Reports

International Tourism Reports presents detailed profiles of the world's tourism **destination** countries. The journal is global in its coverage and includes both established and newly emerging tourism destinations. Each issue of this periodical contains detailed studies on some four destination countries. Based on statistical data available from published and unpublished sources and on original research, these studies provide a survey of major issues, developments and forecasts affecting the tourism business in that country, arrivals and expenditures, **accommodation** and **occupancy rates**, **transportation**, **length of stay** and **seasonality**, purpose of visit, tourism **policy** and **development** prospects. First published in 1983, the journal appears quarterly, published by Travel and Tourism Intelligence (ISSN 0269–3747).

RENE BARETJE, FRANCE

international tourism

According to the World Tourism Organization (WTO), international tourism differs from **domestic tourism**. The former includes tourists who cross a country's border and spend at least one night there, but not more than twelve consecutive months. International tourism consists of **inbound** and **outbound** tourism. Inbound involves non-residents travelling within a given **destination** country and outbound residents travelling to a country other than theirs. International tourism always includes border crossing, whether by land, sea or air. It also includes some form of international arrangements in private trade (mainly in distribution systems) and in public institutions (mainly through tourism bodies).

The great interest in international tourism lies mainly in its **economics**, as it is an export trade from the point of view of the visited country (an import on the side of the country of origin). However, it also has important sociocultural consequences on both tourist generating and receiving countries. Normally, the **trends** and patterns of international tourism follow the North–South direction, thus meaning an important monetary flow from rich countries towards less developed countries.

The use of **leisure** time in international tourism is influenced by several important factors, including the proximity or distance to country borders (the case of the Benelux countries in Europe); development of modern **transportation** systems within and to different countries (excellent motorways in Germany, popularity of aviation in the United Kingdom); the **climate** of the destination (sunny seaside sports in Spain in summertime, **skiing** in the Alps in wintertime); cultural **attractions** (art **heritage** in Italy); entertainment opportunities (shopping in London, dining in Paris, **gambling** in Las Vegas); exoticism and mystery appeals (Southeast Asia and Africa); the new **lifestyle** of enjoying **second homes** in foreign countries (owned or rented in the Mediterranean Riviera); and more.

When the industrial revolution took place (1750–1850), the basis for modern European **mass tourism** was created. The steam machine invention was quickly followed with manufacturing changes and new sea and land transport improvements (trains and ships). At the same time, these developments led to the creation of new middle classes with more leisure time and disposable income, and hence provided with the means to visit other countries or to **escape** from the polluted industrial urban areas of their own. The main destinations to be marketed were the foreign sea and mountains **resorts**. This development, which started in early 1880s with the French Riviera (Nice and Cannes), continued southwards into the Spanish Costa Brava (North Catalonia) and the Balearic Islands around 1905. Actually, both 'sun and beach' destinations and 'snow' resorts gained popularity simultaneously. Thus, for example, as early as 1892 Switzerland had become the most popular destination among British tourists.

In the twentieth century, modern **technology** fostered the development and expansion of mass tourism. Increased leisure time and income, the development of economic mass transportation systems and eagerness of resorts and destinations all over the world to have a share of the booming tourism **market**, among other things, led to an unprecedented worldwide expansion of this **industry**. The private car transport led to charter bus services, the public air transport espoused charter air services, and sea cruises went almost exclusively international. These **charter** services were quickly commercialised by the northern and central European tour operators, pioneers in discovering and marketing new international resorts. Their intermediation task may be regarded as one of the influential tourism developments of the century. Many resorts of the less developed countries could not have had much tourism without the investments and efforts of European **tour operators**. This was not the case in the United States, where this role was played to a lesser extent by the American air companies flying internationally.

Tourism has now become an important form of mass consumerism, standardised, protected by legislation and traded by increasingly bigger **multinational firms** all over the world. All these developments helped to position tourism as the world's largest and most diverse economic activity. Today many countries rely on this 'dynamic' industry, as a primary source for generating revenues, employment, private sector growth and **infrastructure** development (World Tourism Organization 1997a, 1997b). This is more evident in countries where traditional economic sectors like agriculture have been shrinking.

One conclusion to be drawn from the evolution of international tourism is that it is a consequence of the increasing globalisation of the economic and social activities during the twentieth century. The early travellers abroad generally did not mind using local touristic resources and supplies during their stays and thus did not change their host habitat, whereas the modern mass tourists normally expect services similar to those at home and, knowingly or not, produce untold changes in their holiday destinations. International tourism is a slow but steady equalising process among host and guest countries. Balancing tourism's positive and negative consequences is a challenge yet to be articulated and faced by governments worldwide.

According to the **World Tourism Organization** (1997b), the pattern of international flows reveals first and foremost a heavy geographical concentration of tourism arrivals. The ten leading destinations account for almost 52 per cent of the world volume. The total international arrivals (including those related to religion and family **motivations** but excluding excursionists or same day visitors) in 1997 was about 612 million. France was the most popular country destination with 66 million arrivals (10.9 per cent of world total), the United States was second with 49 million (8 per cent), Spain was third with 43 million (7.1 per cent), Italy was fourth with 34 million (5.5 per cent), and the United Kingdom was fifth with 25 million (4.2 per cent).

To have a better idea of the economic importance of international tourism receipts for these host countries, in 1997 USA earned $75 billion (16.9 per cent of world total), Italy $30 billion (6.7 per cent), France $27.9 billion (6.3 per cent), Spain $27 billion (6.1 per cent) and the United Kingdom $20.6 billion (4.64 per cent). Comparing these percentages with those of the arrivals, there is an evidence of the level of expenditure in each country: the United States at $1,531 per person is at one end and France at $418 at the other (the latter is partly due to the fact that many European tourists just travel through France to other destinations) Total international tourism receipts in 1997 amounted to $444 billion.

Viewed differently, regionally international arrivals and receipts for 1997 in percentages of world total market share in order of ranking were Europe (59 per cent of arrivals and 49.4 per cent of receipts); the Americas (19.4 per cent and 27.1 per cent, with the United States having the lion's share); East Asia/Pacific (14.7 per cent and 18.7 per cent); Africa (3.8 per cent and 2 per cent); the Middle East (2.4 per cent and 1.9 per cent); and South Asia (0.7 per cent and 0.9 per cent). Parenthetically, in the same year the Mediterranean countries of Europe, Africa and the Middle East accounted for 187 million international arrivals (30.6 per cent of the world total), thus being the single regional destination with biggest share of total world tourism (25.5 per cent of total world receipts). The Spanish archipelago of Balearic Isles has the single biggest tourism market share (5 per cent) with 9.4 million arrivals. Similar concentration was present in the Pacific (Hawaii) and in the Atlantic (Canary Islands), all being 'sun and beach' destinations.

During recent years there has been a gradual diversification of tourism markets with the emergence of new destinations, according to the World Tourism Organization (1997c). This is more evident in the East Asia and the pacific region. For example, China jumped to the sixth position in the world with 24 million arrivals (3.9 per cent of the total). On the other hand, the Russian Federation has not yet made substantial gains in its world market share (2.5 per cent), being ranked fourteenth with 15 million arrivals. These shifts and changes at regional and global levels continue to be among the most important research themes financed and pursued by major tourism companies and tourism-dependent economies and eagerly watched by both tourism generating and receiving countries.

References

World Tourism Organization (1997a) *International Tourism: A Global Perspective*, Madrid: WTO.

—— (1997b) *Tourism Highlights*, Madrid: WTO.

—— (1997c) *Tourism Market Trends: Europe*, Madrid: WTO.

ESTEBAN BARDOLET, SPAIN

international understanding

Travel has provided people with the opportunity to encounter, appreciate, interact and learn, both formally and informally, about the cultural and ecological diversity of destinations throughout the world. Based on the 'contact model' of social **psychology** of intergroup **conflict**, which stresses that contact promotes the potential for understanding and acceptance among members of different groups, it is argued that tourism, by bringing together hosts and guests from various countries, helps to bridge psychological and cultural gaps promoting trust, respect and understanding. In a throwback to the **Grand Tour** of the past, contemporary demands for speciality educational, **heritage** and **cultural tourism** represent formal examples of **tourists** engaged in actively learning about the countries and peoples they are visiting.

The 1963 United Nations Conference on international tourism proclaimed that tourism was mind-broadening and ennobling; at the 1980 World Tourism Conference in Manila, it was declared that tourism can be a vital force for **peace**. This concept went on to gain prominence in Vancouver, Canada in October 1988 at the international conference on 'Tourism: A Vital Force for Peace', which promoted the idea that tourism results in positive changes in the attitudes of tourists towards the host culture contributing towards world peace.

While there is agreement that tourism to some extent facilitates interaction between different groups, there are few empirical studies which have investigated the level of international understanding generated and the potential for tourism to promote peace raising debate in research. Crick (1989) suggests that in some regard this line of reasoning is rhetorical, and tourism may even perpetuate and reinforce **stereotypes**. Reporting on four empirical studies investigating the contribution of tourism to better understanding between unfriendly nations, Pizam (1996) speculated that tourism by itself does not automatically contribute to positive attitude changes or to the reduction of perceived national and ethnic differences.

The **motivation** of hosts and guests, preconceived **images** and perceptions, type of trip, form and level of tourism **development**, and degree and form of interaction between hosts and guests all **impact** the level of international understanding generated. Trips which present limited or staged interaction between the two groups may reinforce stereotypes. While tourism has the potential to increase international understanding through cultural contact and educational tours, not all forms of tourism promote conditions for increased understanding between different groups.

References

Crick, M. (1989) 'Representations of international tourism in the social sciences: sun, sex, sights, savings, and servility', *Annual Review of Anthropology* 18: 307–44. (Outlines various perspectives of tourism in social sciences.)

Pizam, A. (1996) 'Does tourism promote international understanding between unfriendly nations?' in A. Pizam and Y. Mansfeld (eds), *Tourism, Crime and International Security Issues*, Chichester: Wiley, 203–13. (Examines past studies and questions if all forms of tourism promote international understanding.)

Further reading

D'Amour, L. (1988) 'Tourism: the world's peace industry', *Journal of Travel Research* 27: 35–40. (Examines the potential of tourism to act as a force for world peace.)

DAVID J. TELFER, CANADA

internationalisation

The growing phenomenon of interaction and activity occurring across international boundaries is known as internationalisation. In respect of tourism, its significance derives from the acceptance of tourism as an explicit strategy of economic **development** in **Third World** nations. Sociocultural clashes and the practices of **multinational firms** mean that internationalisation is far from leading to global homogenisation.

See also: globalisation

ROBERT B. POTTER, UK

Internet

The Internet is a public computer network that spans the entire globe. Communication on the Internet is based upon certain protocol, a well-defined communication standard in the public domain. The Internet connects many different types of computers and operates over a broad range of communication channels. Because of its open architecture, the Internet has grown in a bottom-up manner and adapted to the new needs of its users.

Internet users are typically connected to the network through 'Internet providers'. These are commercial companies that provide the infrastructure that is needed for connecting to the network, administrative arrangements with other providers that ensure the proper delivery of messages, user support and so on. A user can be connected to the respective provider in many different ways, ranging from high-capacity leased lines to dial-in via regular telephone connections. The basic communication standard of the Internet allows for many different types of services, such as electronic mail (e-mail), file transfer, remote terminal access, remote database query and the like. All these services work according to the client–server concept: a client program translates the user's activities into a proper request for information sent to a server program, typically located on another computer somewhere on the Internet. The server program acts on this request and, for example, returns a certain data file, returns the answer to a database query, or sends out a set of e-mail messages. Currently, the most important Internet services are e-mail, e-mail based discussion lists, newsgroups and the **World Wide Web** (WWW). While the first three allow for communication between two or more Internet users, the WWW provides access to an enormous wealth of electronically stored information.

In the context of tourism, the Internet is used in many different ways. Airlines, **travel agencies** and city tourism offices, among others, offer electronic versions of their brochures and catalogues. More sophisticated applications show current vacancies and allow for online booking. Steady customers are easily informed about special offers via e-mail. In many areas, users share views about destinations and recreational **sites**, talk about their **experience**, ask for suggestions and help in their travel planning, and more. Such direct exchange of views may have a strong impact upon the **image** of a destination, carrier or agent.

GUNTHER MAIER, AUSTRIA

interpretation

Interpretation is any activity which seeks to explain to people the significance of an object, a culture or a place. Its three core functions are to enhance visitor experiences, to improve visitor knowledge or understanding, and to assist in the protection or conservation of places or cultures (see **cultural conservation**). It is most commonly used to refer to **activities** such as guided tours or walks and educational presentations, and to products such as guidebooks and information leaflets. Interpretation is a core activity in settings such as **museums**, art galleries, zoos, historic areas and **national parks**.

All definitions of interpretation share three features. First, interpretation is seen as communication rather than education or simply provision of information. Interpretation should offer its audiences more than just information; it seeks to encourage their interest in learning more by themselves and it must be entertaining or offer ways to enhance experiences. Second, interpretation seeks to help people understand a place or culture and in doing so should generate support for the conservation and protection of these places and cultures. Third, interpretation can be seen as an important tourism, recreation or resource management strategy. Effective interpretation should provide people with the necessary information and awareness to encourage more sustainable or appropriate behaviours.

Interpretation can be seen as sharing some of the features of both public and environmental education programmes. However, environmental education is primarily focused on students in formal educational institutions, while interpretation audiences are usually composed of people at **leisure** in tourism and recreational settings. Public education campaigns, such as those concerned with reducing litter and encouraging water conservation, are aimed at large audiences who are often the residents of an area. Interpretation, on the other hand, is most commonly designed for **visitors** to particular places and the interpreter's audience will often include both local residents and domestic and international **tourists**.

Interpretation is an important **management** strategy or tool for encouraging **ecologically sustainable tourism**. Its first function is the enhancement of visitor experiences. This is a critical and common role for interpretation in tourism **products**. Many tourism forms, including **ecotourism** as well as cultural, ethnic and **nature tourism**, are built around the interpretation and presentation of natural and/or cultural **heritage**. Visitors seeking such experiences are often accompanied by guides knowledgeable about the places or cultures being sought and are given talks, lectures and written material to prepare them for onsite experiences. Many ethnic and **indigenous** tourism products include historical and cultural information about the groups and how to behave in cross-cultural situations. Moreover, visits to archaeological sites, historic buildings, monuments and **national parks**, among others, usually include activities such as guided walks or tours, self-guiding trails with signs or brochures, with visitor centres providing audio-visual and other displays. Many of these interpretative activities provide people with both physical and mental access to the places visited. Positive visitor experiences encourage conservation attitudes. An example of this interpretation function is the presentation of codes of conduct which encourage visitors to behave in ways which create fewer negative environmental and sociocultural **impacts**. This process can also be used to lessen pressure on heavily used or sensitive sites by providing visitors with information on alternative places and activities.

The literature on this theme has been predominantly concerned with setting out principles for the design of effective interpretative activities. The available systematic and empirically based evaluation research, though limited in scope, suggests that interactive or participatory forms can be more effective than passive, didactic approaches. For example, when participants on a **guided tour** are asked questions and encouraged to engage in different activities, the outcome is likely to be more successful than when they simply follow and listen to the **tour guide**. It can also be argued that variety is important in effective interpretation. This is reflected in the changing methods of presentation being used in **museums** and zoos. Traditionally such places offered visitors only the opportunity to look at displays and read text. But today, museums and zoos are increasingly using computer displays, audio-visual displays and hands-on activities to provide more varied experiences.

In addition to the development of guidelines for effective interpretation, the academic literature has been concerned with the issues of what should be interpreted. There has been much criticism of interpretation as presenting limited or inaccurate information. Such criticisms include claims that the information provided in the interpretation of places or events often reflects the perspectives of the politically dominant regimes. Information about (or from) minority or politically powerless groups is often neglected. There have also been debates over the extent to which interpreters should seek to present **conflict**, **war**, natural disasters and other negative events. The presentation of negative material is a challenge for interpreters given the common emphasis placed on the need for interpretation to be entertaining.

Further reading

Knudson, D.M., Cable, T.T. and Beck, L. (1995) *Interpretation of Cultural and Natural Resources*, State College, PA: Venture Publishing. (Provides definitions and guides to the development of interpretative programmes in a range of settings.)

Moscardo, G. (1996) 'Mindful visitors', *Annals of Tourism Research* 23(2): 376–97. (Sets out major principles for the design of effective interpretation.)

Tilden, F. (1976) *Interpreting Our Heritage*, 3rd edn, Chapel Hill, NC: University of North Carolina Press. (Defines interpretation and provides principles for its use and design.)

GIANNA MOSCARDO, AUSTRALIA

intervening opportunity

Destinations or attractions which are closer to tourism-generating **markets** than more distant competitive places intercept visitors, and thus diminish their likelihood of visiting the latter places. The closer places are regarded as being intervening opportunities with respect to the latter places.

See also: accessibility; distance decay; gravity model

GEOFFREY WALL, CANADA

interview

Attempting to measure the wider aspects of tourism is a complex process which many national, regional and local government agencies undertake through the use of survey methods, usually focused on **tourists** completing arrival and departure cards. While such methods provide a mechanism to know their numbers, this is simplistic and does not offer any detailed insights into the **attitudes** towards tourism, how tourists perceives their experience of a **destination** and the **images** and attitudes they form.

This is a complex population to survey because it usually involves many mobile individuals. Thus the process of statistical measurement to ensure use of representative survey methods becomes complicated. However, the only method available to adequately assess the views of tourists, decision makers and residents in relation to tourism is the interview. It comprises a method of communicating with tourists or agencies/residents affected or who may have a view on tourism. It requires an active involvement of the researcher and respondent (the person being interviewed) whereas participant observation does not involve communication with the respondent. Although questioning is more economical than observational techniques, it ultimately depends upon the willingness of the respondent to cooperate and participate in the process.

Interviews are widely used by tourism researchers to gather data, which are then subject to analysis and interpretation. There are a range of different types of interviews, including personal interviews, where an interviewer may control the conditions in which the respondent is interviewed; this usually consists of a two-way conversation between a researcher and respondent. In some contexts, where planners and agencies want to survey the resident population about attitudes to tourism, they may select a telephone interview method which uses a geographical sampling framework based on a computer-assisted telephone interview system. Ultimately, a successful interview is one where the interviewer can motivate the respondent to provide the range of data that is being sought.

Further reading

Veal, A. (1992) *Research Methods for Leisure and Tourism*, Harlow: Longman.

STEPHEN PAGE, NEW ZEALAND

inventory

An inventory is a list of items held in stock. The physical inventory refers to the actual count of all items, while the perpetual inventory is the continuous records of purchases of stock and issues from it. The challenge for tourism **management** is to hold enough inventory to avoid stock-outs whilst setting stock levels to minimise the cost of holding items. An added challenge in this **industry** is the perishability of its many **products**.

PETER JONES, UK

inversion

Inversion is defined as the turning upside down or reversal of a normal position or order. Therefore, as participation in tourism is commonly seen as a form of change or short-term **escape** from the routine or ordinary, tourism may be broadly viewed as a process of inversion. That is, tourism represents, for the **tourist**, the temporary reversal of the ordered and routine character of day-to-day life of home and work into the unstructured **freedom** of the **vacation**.

The link between tourism and inversion is also manifested in more specific contexts. It is, in particular, an important element of **motivation**, because tourism offers the opportunity to escape from one **lifestyle** to another. For some, to be a tourist is to be able to indulge in a **fantasy** life of the 'idle rich', to be king or queen for a day; conversely, others may seek to escape the trappings of wealth and **modernity** for a far more simple or traditional existence. In either case, the holiday is an inversion of everyday reality. Tourists may also be motivated by the chance to escape from the rules and constraints of their home society. Freed from time commitments and codes of dress and **behaviour**, among others, tourism allows people to participate in **ludic** behaviour. Thus, the responsible, constrained, work-like nature of normal life is inverted into the frivolous, carefree and play-like existence of the holiday, and in that sense is comparable to the temporary inversion from adulthood back to childhood.

Inversion is also central to the concept of tourism as a modern religious experience or **ritual**. Traditionally, the passage of ordinary, profane time is interrupted by festivals or events, such as Christmas or Easter, which have a religious or sacred significance and which represent brief slices of sacred time. Similarly, the annual summer holiday is a modern secular ritual which people utilise to add meaning to their lives or to 're-create' and which therefore represents the temporary shift or inversion from ordinary, profane time to sacred time.

See also: tourist as child; liminality; pilgrimage; play

Further reading

Gottlieb, A. (1982) 'Americans' vacations', *Annals of Tourism Research* 9(1): 165–87.

Graburn, N. (1989) 'Tourism: the sacred journey', in V. Smith (ed.), *Hosts and Guests: The Anthropology of Tourism*, 2nd edn, Philadelphia: University of Pennsylvania Press, 21–36.

Lett, J. (1983) 'Ludic and liminoid aspects of charter yacht tourism in the Caribbean', *Annals of Tourism Research* 10(1): 35–56.

RICHARD SHARPLEY, UK

investment

Many economic variables can be classified into either flow or stock variables. Investment is a flow concept, and it relates to the rate at which the stock of capital increases, decreases or remains unchanged. Thus investment tends to be related to the purchase of goods that have some quality of longevity associated with them. If the goods do not possess this characteristic then the purchase is simply consumption.

Investment is a term that encompasses all issues relating to the maintenance or increase in productive capacity, and can take place with respect to all aspects or factors of production. For instance, businesses may invest in capital equipment in order to replace obsolete or worn-out capital, or to increase its productive capacity. Zero investment will not maintain the capital stock and it will diminish until the rate of it can at least offset the rate of capital usage (depreciation). Businesses may also invest in land in order to expand or rearrange its activities in a more effective manner. Still, businesses may invest in human capital by **training** its workforce in an attempt to make the labour factor more productive or effective. Similarly, an individual may invest in their human capital by improving the skill or knowledge base and its value can be estimated as the cost of that acquisition of skills and/or education plus the earning opportunities foregone during that acquisition. Characteristically, textbooks tend to break investment down into *gross* and *net* investments. The former is concerned with the production of new and the improvement of capital goods already in existence. This may be contrasted with the latter which relates to gross investment less the amount of the capital stock used up in the production process (depreciation).

Investment in tourism can be examined from a number of aspects. For instance, the industry generally requires investment in infrastructure in communications, **transportation**, utilities and the like. Therefore, new port and **airport** facilities, sewage treatment and desalination plants and satellite telephone systems are all examples of capital investment infrastructure. On the other hand, new **hotel** buildings, **restaurants** and theatres are investments in superstructure. But investment can include increases in the stocks of towels or crockery, or training courses for staff.

There have been a number of theories put forward to explain the level of investment taking place within an economy. These included the accelerator theory that relates the level of investment to the rate of change in the level of output. It is not the level of output that is important in determining the level of investment, but its rate of change. Significantly replacement investment must take place if the productive capacity of a business or an economy is to be maintained; because, by its very nature, capital tends to wear out or become obsolete during the production process. It is often difficult to distinguish accurately between what is replacement and what is new or induced investment. This is because technological change occurs over time, and a piece of capital equipment that is purchased to replace a worn-out item may well include characteristics deemed new investment. Investment can only be negative at the establishment level of analysis rather than at the national. For instance a single business can dis-invest by selling off some of its capital stock. However, all businesses in an economy cannot do the same because there would not be a market for purchasing the capital stock that is for sale.

Investment in productive capacity is an essential ingredient of economic growth and thus of **economic development**. In most industries, investment is a significant factor input even in the service-based ones such as finance and tourism. The high capital component associated with many aspects of tourism is responsible for the high operating leverage, and hence the drive for volume in this industry. This is more evident in for instance the **airline** and **hotel** sectors.

See also: foreign investment; joint venture

JOHN FLETCHER, UK

investment decision

Investment decisions are choices made relative to the raising of funds by a company, and the allocation of those funds within it. This capital raising and allocation process must be consistent with corporate objectives. Some of the investment decisions made in a tourism enterprise might be choosing a piece of equipment among several alternatives, determining whether to replace old equipment with new, and deciding which fixed assets should be purchased under a capital rationing situation. Capital rationing occurs when there are limited funds for capital purposes. Therefore, competition for available dollars exists among the different departments and the different projects. This makes the use of sophisticated capital **budgeting** models very important.

Investment decisions in tourism and other industries are typically evaluated by four **models**: the accounting rate of return method, the payback method, the net present value method and the internal rate of return method. The first model considers the average annual project income, but does not look at cash flows. The proposed project is accepted if it exceeds the minimum accounting rate of return required. The payback method uses annual cash flows from the project to offset its cost. This determines when the investment is recovered. There are flaws with both of these methods, as neither considers the time value of money. In addition, the first model uses net income instead of cash flows, and the second does not consider cash flows after the payback period.

The net present value model discounts future cash flows to their present values. It is computed by subtracting the project cost from the present value of the discounted future cash flow stream. The decision is usually to accept the project if the net present value is greater than zero. If alternative projects are considered, the one with the highest value should be chosen. Finally, the internal rate of return model considers cash flows and the time value of money. It determines the rate of return earned by a proposed project based on the cash flows. It is computed by setting the sum of the present values of the future cash streams minus the project cost to zero. The discount rate is then computed, and the project is accepted if the computed rate is greater than the hurdle rate. The latter is the minimum rate that a company will consider to earn on its investments.

Further reading

Coltman, M.M. (1994) *Hospitality Management*

Accounting, 5th edn, New York: Van Nostrand Reinhold.

Schmidgall, R. (19967) *Hospitality Industry Managerial Accounting*, 4th edn, East Lansing, MI: Educational Institute of the American Hotel and Motel Association.

STEPHEN M. LEBRUTO, USA

Iran

In keeping with the Holy Qur'an, tourism in Iran is an important means of encouraging **peace** and cultural **heritage** exchange. During 1996, 500,000 international tourists visited Iran. The number is forecasted to reach 5 million by 2020. The industry has also gained an important role in national **development** plans. The Deputy Minister for Tourism in Ministry of Culture and Islamic Guidance has established a **master plan** for the development and **management** of tourism based on private sector involvement. The number of universities, **training** centres and specialised publications for tourism is growing.

HAMID ZARGHAM, IRAN

Ireland

Tourism is a growing economic tool in Ireland. The Irish Tourist Board, the origins of which date to 1939, is the main developmental, promotional and regulatory body. There are seven Regional Tourism Authorities and several licensed regulators. Major expansion of **accommodation** and recreational facilities marked the 1960s, the 1980s and the 1990s. Touring, **heritage** and activity holidays are popular in Ireland. The United Kingdom, Continental Europe and North America are the key tourist generating **markets**.

MARY CAWLEY, IRELAND

Israel

The major development process in Israel started right after the Six Days War, and resulted in an expansion of its territory and, thus, incorporation of many more cultural and **pilgrimage** attractions. Its geographical proximity to European markets, on the one hand, and its highly unique tourism products, on the other, offered Israel a major potential for **economic development** through this industry. However, this potential has not been fully exploited as a result of the ongoing turbulent **security** situation in the Middle East (Bar-On 1996; Mansfeld 1996).

Over the years, Israel has suffered from major declines in **inbound** tourism as a result of various local, regional and international security crises. In most cases, when turmoil had been replaced by a peaceful period, Israel has managed to regain tourists' faith and thus maintained the further **development** of this industry (Mansfeld 1996). Hence, in 1967 the number of tourist arrivals was around 230,000, and the number grew to 2.2 million in 1995 (Central Bureau of Statistics and Ministry of Tourism 1996). The most rapid growth of arrivals and receipts from tourism spending took place between 1991 and 1995. This period is characterised by the reconciliation process between Israel and the Arab world. The new evolving political climate has contributed to a major positive change in the way Israel has been perceived as an attractive **destination**.

To date, Israel's major generating markets are located in Europe, with **Germany**, the **United Kingdom** and **France** dominating this market. About 56 per cent of inbound tourism in 1995 was generated in Europe and 23 per cent in North America, with an average **length of stay** of 8.7 days. Tourists engaged in such activities as pilgrimage to the Jewish, Christian, Muslim and Bahai holy places, cultural and heritage tourism, **visiting friends and relatives**, **sightseeing** around the country, **urban tourism** (especially in Israel's economic capital, Tel Aviv), **health** tourism on the Dead Sea shore, **countryside** tourism, sun, sea and sand tourism along Israel's Mediterranean coast in Eilat, and **ecotourism** primarily in the Negev desert.

References

Bar-On, R.R. (1996) 'Measuring the effects of violence and the promotion following violent acts'. in A. Pizam and Y. Mansfeld (eds), *Tourism: Crime and International Security Issues*, Chichester: Wiley, 159–74.

Central Bureau of Statistics and Ministry of Tourism (1996) *Tourism and Hotel Services Statistics Quarterly* 24 (2).

Mansfeld, Y. (1996) 'Wars, tourism and the "Middle East" factor', in A. Pizam and Y. Mansfeld (eds), *Tourism: Crime and International Security Issues*, Chichester: Wiley, 265–78.

YOEL MANSFELD, ISRAEL

Italy

The word *feria* (plural *feriae*) appeared for the first time in ancient Rome. The Latin word meant **festival**, and at the same time it took on the meaning of **vacation**. For the first time in human history, this word designating an ancient, religious holiday, firmly rooted inside the city, metamorphosed into an excursion from one locality to another. The Latin festival became the first model of a 'mobile' holiday, far from one's place of residence. As time passed, it gradually became linked to vacation taken in relation to changes in the seasons.

The term **Grand Tour** appeared in the seventeenth century. It was best defined by Sir Francis Bacon in his famous essay 'Of Travel', where he states 'Travel, in the younger sort, is a part of education; in the elder, a part of experience.' The most popular **destination** was Italy, far more than **France**, as the centre of ancient Roman and even Greek civilisation, a nation that no Englishman could complete his education and culture without visiting. However, real tourism (pre-**mass tourism** in this period) should not be confounded with the Grand Tour, which was a once in a lifetime journey. The first form of mass tourism was developed in Italy because until 1860 Nice was an Italian region, belonging to the Kingdom of Piedmont and the Savoy dynasty. The Italian Riviera, as the area came to be called, became renowned as a **resort** in that period. It is still known today in English as the Riviera, in keeping with the **image** that was in vogue at the end of the eighteenth century.

There were 90,000 British visitors in Italy in 1913. By then, tourism had become an opportunity, offered Italy by nature and history, to compensate for the numerous economic obstacles that its geographic and natural configuration with areas of harsh, mountainous terrain and few raw materials have always created in its social and **economic development**. From the time of political unification, Italy found itself struggling with the difficulties inherent in its foreign trade, deriving from a poor agricultural economy with a very sizeable population. This contradiction between economic possibilities and population density gave rise to two socioeconomic phenomena: unemployment and emigration. Tourism and emigration were the two traditional export items in Italy's **balance of payments**, making it possible to offset debts incurred for the imports its economic development needed.

The economic and political uncertainty between the two world wars was the cause of the national government's first intervention in tourism, especially after the devastating effects of the Great Depression of 1929. In 1919, the Italian State Tourist Office was created. It was given a mandate to gather data, to make legislative proposals, to promote its **domestic tourism** and **international tourism** and to facilitate bank credit for hotels. Then the Commissioner's Office for Tourism was created, later renamed the General Direction for Tourism. In 1932 the national **government** set up the Provincial Committees for Tourism that later became the Provincial Tourist Offices, acting as the government's local branches in the sector. Bank credit facilitation for hotels was enacted in 1924, followed by a series of measures to promote demand, especially domestic. These included the so-called 'Fascist Saturday' (Saturday had always been a working day in Italy), yearly paid vacations, snow trains (to mountain resorts in the winter) and the so-called blue trains (to seaside resorts in the summer). The government also passed a series of facilitations for foreign tourists, such as discounted gasoline coupons and an advantageous **foreign exchange** rate for foreigners, dubbed 'tourist lire'.

Italy has the most substantial **accommodation** structure in all of Europe. In 1994 there were 34,549 hotels and similar establishments, with

944,227 rooms (1,724,333 beds). There are 2,346 collective tourism establishments with 1,223,671 beds. This huge availability of accommodation was built primarily with domestic investments and is thus entirely Italian. **Hotel** chains only started taking an interest in Italy a few years ago and their market presence, even for 4-star and 5-star hotels, is fairly weak in comparison to the number of rooms handled directly by national businesses. This feature has its positive and negative aspects. On one hand, it has allowed for the totally endogenous tourism **growth** while, on the other hand, modern **management** has been late in coming to what are basically small and medium-sized (often family-run) businesses.

Italy completed its postwar economic reconstruction in 1958, leading to a fundamental restructuring from a basically agriculture-based to an industry-centred economy. Between the two world wars, tourism was a balm to the structural deficit of its balance of payments and is still an essential item among its exports. Without tourism, Italy's **balance of payments** would have been in serious difficulty over recent decades. The trend is now stable. In 1985 international tourism receipts rose to $8.7 billion, compared with free on board exports amounting to $78.9 billion. In 1995, international tourism receipts jumped to $27.4 with respect to exports totalling $231.3 billion. **Outbound** tourism amounted to $2.3 billion in 1985 with respect to cost insurance freight imports amounting to $90.9 billion. In 1995 expenditure of Italian tourists going abroad reached $12.4 billion, in comparison to cost insurance freight imports of $204.1 billion. This confirms that tourism is Italy's main export of services and its second largest export item with respect to different goods, with a total of 11.1 per cent of exports in 1985, rising to 11.9 per cent in 1995. From the viewpoint of expenditures in comparison to imports, the total rose from 2.5 per cent in 1985 to 6.1 per cent in 1995, demonstrating the high level of industrialisation that Italy has reached and, with respect to the past, the Italians' strong desire to travel abroad.

As a result, Italy has a pre-eminent position in the **international tourism** market. It ranked fourth among the world's top country destinations from 1980 to 1996. The annual average growth rate in the same years was 2.5 per cent, the share of arrivals worldwide reached 7.7 per cent in 1980 and touched 5.5 per cent in 1996. In the same years, Italy moved from third to second rank for tourism receipts among the world's top tourism earners, with an annual average growth rate of 8.4 per cent. In the share of worldwide receipts it reached 6.9 per cent in 1996 in comparison with 7.8 per cent in 1980. In the ranking of the world's top tourist spenders in the same sixteen years, Italy moved from thirteenth to sixth place with an annual average growth rate of 14 per cent. This was the highest among the first ten spender countries. In the same period, the share of expenditures worldwide rose from 1.8 per cent to 4.1 per cent, showing a net trend in the Italian **outbound** tourism demand following the country's economic evolution. Finally, among the countries with the highest surplus in the world for international tourist balance of accounts, Italy with $15.1 billion took third place in 1995, after **Spain** and the **United States** but before France.

The changes in the economic structure over the last thirty years have determined a net change in the **trend** of tourist arrivals, which consisted primarily of foreigners until the beginning of the 1960s. In 1996, over 29 million tourists from abroad stayed in Italy, compared to over 57 million Italian tourists. This is due to the economic position Italy has achieved, ranking sixth worldwide in GDP in 1995 as well as being in sixth place in terms of worldwide exports. In addition, Italy is in second place after **Japan** in terms of current balance account surplus, and in sixth place in terms of industrial production.

However, despite these statistics Italian tourism faces a series of momentous problems in the future, problems that, for two reasons, were hidden by the trends of recent years. The first reason is the almost complete monopoly of Italy's immense cultural **heritage**, which encourages foreign tourists to visit Italy at least once. The second reason is the location of Italy near a series of important European spender countries. In addition, the lira's sharp devaluation, falling almost 30 per cent with respect to the US dollar in 1992, makes Italian tourism prices a better bargain. But problems still remain. Italy's image as a seaside **resort** has declined due to territorial saturation and ecological damage, while **cultural tourism** remains a unique feature. Despite this, tourists coming to Italy for seaside **vacations** still rank high, followed by tourists arriving for the country's cultural attractions.

A series of studies have shown that, in contrast to what happens in other European countries, Italy's high tourism prices with respect to product **quality** are a determining factor in convincing tourists not to repeat the **experience** of vacations. Italy has the highest ratio of price to quality in Europe, second only to **Sweden**, which is not eminently famous for tourism. Another negative aspect is the percentage of foreign tourism that goes to the Italian Mezzogiorno, with its enormous potential from the perspective of **climate**, culture and vacation spots; fewer than 15 per cent of foreign tourists visit this area and fewer than 20 per cent of Italian tourists do so. Tourism (which could be a fundamental economic leverage for the less developed part of the country with high unemployment) only confirms the general tendency to concentrate economic growth in the country's central and northern areas.

Italy is a country with an enormous heritage of culture, art, nature and fine food that could exploit its tourism resources to a much greater extent, both economically and socially. A serious problem that has developed over recent years is the lack of any economic **policy** for tourism. In 1972, the national **government** transferred the competence for tourism to the regions, abdicating almost all its jurisdiction in the sector; only some of the central and northern regions have made good use of this responsibility. In 1993, a national referendum abolished the Ministry of Tourism. The Italian State Tourist Office, in charge only of promotion and **advertising** abroad, has also faced many problems in assuming the responsibility of co-ordinating touristic activities with the regional governments.

Further reading

Formica, S. and Uysal, M. (1996) 'The revitalization of Italy as a tourist destination', *Tourism Management* 17(5): 323–32.

McCourt Francescone, P. (1997) 'Italy', *International Tourism Reports* 1: 4–25.

Rognant, L. (1990) *Un geo-système touristique national: l'Italie, essai systémique*, Aix-en-Provence: Centre des Hautes Etudes Touristiques.

Sessa, A. (ed.) (1992) *Il sistema turismo Italia: produzione, promozione, commercializzazione*, Rome: Quaderni dell'ANIEST.

ALBERTO SESSA, ITALY

itinerary

An itinerary is a planned pattern of travel from start to its completion. It contains details of **transportation** schedules and **accommodation**, including departure and arrival at different places, duration of journey, activities at a **destination**, and type of accommodation booked. An itinerary may be priced as a total **package tour** when provided by a **tour operator**, or each component may be individually priced if an independent tourist uses a **travel agency** or books directly with the service provider. The itinerary can include non-commercial activities.

CHRIS RYAN, NEW ZEALAND

J

Jamaica

Located 90 miles south of Cuba, Jamaica is the third largest Caribbean island, with a population of 2.5 million, a developed **infrastructure** and a striking tropical **environment**. Tourism is the main source of **foreign exchange**. Earnings from tourism rose from $516 million in 1986 to $965 million in 1995. The number of tourists visiting Jamaica was about 1 million in 1987; this figure increased to 1.75 million in 1995.

JAN ARMSTRONG GAMRADT, USA

Japan

Japan is the largest tourist-generating country in the Asia Pacific region. After restrictions on carrying foreign currency abroad were lifted in 1964, at the same time as the country was enjoying considerable **economic development**, the number of Japan's **outbound** tourists soared to ten million per year in the 1990s. The 'Ten Million Project', a government-initiated plan to double outbound tourism in five years, reached its target in 1990, one year earlier than planned. Even after the economic slowdown of the 'bubble' economy, Japan's outbound tourism is still on the rise except for short periods such as the recession, the Gulf War, the 'bubble' economy collapse and the Kobe Earthquake.

Japan has developed a mature tourism business with a comprehensive **infrastructure**. Major **airline** networks fly to most important cities in the world. The new Kansai international **airport**, opened in 1994 in Osaka, strengthened the services. An efficient network of wholesalers handles outbound traffic by packaging tours and selling them to the public. **Package tours** have reached a level of excellence in quality and services never before achieved. In respect of consumer protection, remarkable progress has been made in the registration of the travel agents and in the guarantee bond or **itinerary** booking guarantee.

Japanese tourists have been important to the **development** of tourism in Asia. Tourism is important because of its size and spending levels. The Japanese spend more than any other foreign visitors in many destinations. The industry in the region shows great interest in Japan because of its sheer volume of outbound traffic, which has boosted the tourism business and development of infrastructure of many countries. This is in sharp contrast to **inbound** tourism, which has virtually stagnated to the level of 3 million since the beginning of the 1990s, a disproportionately low figure given Japan's **resources**. The high value of the yen and ambitious promotional efforts of other competing Asian countries have helped to drive international tourists away from Japan. The government launched a new 'Welcome Plan 21' project in 1996, aiming to double the inbound traffic in the coming ten years. Although Japan will likely continue to face the problems of the strong yen and competition from Asia, with natural resources, **history**, cultures, importance in the world economy and a comprehensive infrastructure, Japan still is an attractive **destination**, and this will help maintain **demand**.

Further reading

Ministry of Transport, Japan National Tourist Organisation (1996) *Tourism in Japan 1996–97*, Tokyo.

TETSURO YAMASHITA, JAPAN

job design analysis

Job analyses are rarely used in their completed form. However, information contained in the analysis is regularly used for a variety of purposes, including the creation of job descriptions and specifications, training programmes, job evaluation and compensation planning, and the development of performance appraisals. This analysis also provides the justifications why specific abilities and skills are required for a job. For this reason, a job analysis is often the frontline defence in proving to the Equal Employment Opportunity Commission (or similar entity outside the United States) that a business necessity exists which accounts for the legal discrimination on which a company has based selection and promotion decisions. The analysis also has a direct relationship to job design. Job analysis is what work is done; job design is how the work is done.

Designing jobs is a changing process, as good tourism managers know. Unfortunately, some tend to think that once a job is designed and described in an employee manual it never changes. That is rarely the case. Customers' needs change, technology associated with the job changes, and personal characteristics of the jobholders also change. Each change may require a revised job design. A travel agent's job, for instance, is much different today than it once was because of a variety of factors, including how new equipment is used, how the the **computer reservation system** shapes their day-to-day function, and how new tourism products are packaged. Selecting which jobs to study is the first step in completing a thorough job analysis. Some companies analyse each job performed in the organisation once per year; others use a rotation system wherein each job is analysed every three years. How often this is analysed primarily depends on the degree of change associated with the position. The principle kinds of information collected in job analysis are job activities (actual work), equipment used, job context, personnel requirements, human behaviours, basic tools and other work aids needed, and performance standards required.

Several methods of collecting information are available and widely used. The simplest and least expensive method is observation. Here managers simply watch employees at work and make detailed notes of their tasks and behaviours. Another popular method is interviewing (see **interview**) the employees who do the job. There is strong justification for this method, as it is the employee who knows the work better than anybody else. The critical incident method involves capturing actual events that transpire. For example, one such critical incident might read like this: 'On June 27 Mr Jones, a bellman, observed a guest fretting over how to get to his car in a lot several hundred yards away in a strong rain. Without hesitation, Mr Jones provided the guest with his own umbrella.' Over time, a large enough number of such critical incidents can be captured to form a fairly clear picture of the actual job requirements. Some companies compile job analysis information by asking their employees to keep a diary or daily log of their activities over a specific period of time. The method is very cost effective and comprehensive and it encourages employees to think about the work that they do. However, it also requires a substantial amount of employee time in writing the events into their journals.

In many cases employees perform ineffectively; thus **productivity** is low not because of poor **training**, inadequate supervision, underdeveloped employee skills or poor work habits, but because the job is designed poorly. In addition, such cases can lead to low levels of job **satisfaction** which in turn leads to low motivation, high employee turnover, and high absenteeism rates. Four techniques – job simplification, enlargement, enrichment and rotation – are widely used in designing jobs. The first technique, sometimes referred to as time and motion analysis, involves first breaking jobs down into their smallest component and then assessing how work is done in each of these job sections. Job enlargement is the process of broadening the scope of individual simplified components by adding tasks together. Typically, the tasks added together are similar and involve the same skills and abilities. Adding similar tasks in this way is also sometimes referred to as horizontal/ expansion of a job. In this case, horizontal refers to the fact that the jobs require the same or similar skills and abilities. The problem with job enlargement is that it does not necessarily lead to satisfaction on the part of the employee.

Job rotation is often used to alleviate some of the boredom that employees see in performing the same job over and over. Under this system a hotel employee, for example, would only do this job for a specific period of time. After this time period the employee would be rotated to another hotel department with different job responsibilities. This type of system requires cross training of employees in several different areas. Job enrichment involves vertical expansion of a job by adding responsibilities that are not extremely similar to the tasks already performed. As such, enlargement, rotation and enrichment – the key considerations in job design – are the content and context of the work. Each of the approaches is used to affect these two issues in some way.

ROBERT H. WOODS, USA

joint venture

A joint venture is a form of strategic alliance or co-operative arrangement where ownership is shared and a separate enterprise formed. This may strengthen existing businesses through shared expertise, capital, removal of competition and creation of **economies of scale**. **International tourism** joint ventures between foreign organisations and local partners facilitate introduction of foreign products into local **markets**.

See also: foreign investment; multinational firm

CHRISTINE L.H. LEE, AUSTRALIA

Journal of Applied Recreation Research

The *Journal of Applied Recreation Research* is devoted to applied research on a wide array of topics concerning **recreation** and **leisure**. Of interest to both academic researchers and practitioners, the journal emphasises the practical implications of empirical and conceptual recreation and leisure research. All manuscripts, refereed anonymously by three reviewers, must be original research and must not have been submitted for simultaneous review elsewhere. Open to contributors from Canada and abroad, the journal publishes manuscripts in English and in French. It first appeared in 1996, and is published quarterly by Wilfrid Laurier University Press (ISSN 0843–9117).

RENE BARETJE, FRANCE

Journal of Gambling Studies

The *Journal of Gambling Studies* is an interdisciplinary forum for the dissemination of information on the many aspects of **gambling** behaviour, both controlled and pathological as well as a variety of problems attendant to or resultant from gambling, including alcoholism, suicide and **crime**. The journal also publishes articles dealing with public **policy**, social and economic **impact** analysis, and the operation and regulation of commercial gaming industries. Its articles are representative of a cross-section of disciplines including psychiatry, **psychology**, **sociology**, **political science**, criminology, **economics** and social work, and are of interest to the professional and layperson alike. The journal uses a double-blind refereeing process. First published in 1985, initially as the *Journal of Gambling Behaviour*, it appears quarterly, published by Kluwer Academic/Human Sciences Press (ISSN 1050–5350).

WILLIAM R. EADINGTON, USA

Journal of Hospitality and Leisure Marketing

The *Journal of Hospitality and Leisure Marketing* examines relationships between hospitality and leisure, with special attention paid to innovations in applied marketing for both academicians and industry leaders in these fields. The journal publishes contributions written from a variety of perspectives, including those of the academician, the practitioner, and the public policy maker. First published in 1993, the journal appears quarterly, published by Haworth Press (ISSN 1050–7051).

RENE BARETJE, FRANCE

Journal of Hospitality and Tourism Research

Previously known as *Hospitality Research Journal* and *Hospitality Education and Research Journal*, the *Journal of Hospitality and Tourism Research* publishes refereed articles to advance the knowledge base of the hospitality and tourism field. This includes both original, theoretically and empirically based research. Submissions are expected to exhibit excellence in scholarship and make a significant contribution to the field. The journal features sections for empirical research, conceptual articles, applied articles, book and software reviews, conference reports, viewpoints and a research conference calendar. First published in 1976, it appears quarterly, published by Sage (ISSN 1096–3480).

K.S. (KAYE) CHON, USA

Journal of International Hospitality, Leisure and Tourism Management

The *Journal of International Hospitality, Leisure and Tourism Management* presents interdisciplinary works which combine studies focusing on international areas. The goal of the journal is to highlight advances in **hospitality**, **leisure** and **tourism** administrative practice. Its aim is to publish papers appealing to both practitioners and academics, research and management leadership. The *Journal* publishes articles, research notes, news and reviews, perspectives in practice, and research-in-progress. First appearing in 1997, it is published quarterly by Haworth Hospitality Press (ISSN 1092–3128).

RENE BARETJE, FRANCE

Journal of Leisure Research

The *Journal of Leisure Research* is devoted primarily to original investigations that contribute new knowledge and understanding to the field of **leisure** studies. It emphasises empirical reports, but publishes review papers as well as theoretical and methodological articles. This refereed journal publishes three different types of studies: integrative and evaluative review papers, regular papers and short notes. First published in 1968, it appears quarterly, published by the National Recreation and Park Association (ISSN 0022–2216).

RENE BARETJE, FRANCE

Journal of Park and Recreation Administration

An official publication of the American Academy for Park and Recreation Administration, the *Journal of Park and Recreation Administration* was established to bridge the gap between research and practice for administrators, educators, consultants and researchers, and to act as a forum for the analysis of **management** and organisation of the delivery of **park**, **recreation** and **leisure** services. The *Journal* publishes studies aimed at moving theoretical management concepts forward in the field and to provide implications of theory and research for problem solving and action in park and recreation organisation. First published in 1983, it appears quarterly, published by Sagamore Publishing (ISSN 0735–1968).

RENE BARETJE, FRANCE

Journal of Restaurant and Foodservice Marketing

The *Journal of Restaurant and Foodservice Marketing*, targeted at academics in **hospitality** management programmes and institutional food service programmes, is devoted to **marketing** in the food service field. Articles published in this journal include empirical research, conceptual studies, cases and industry reviews. It provides researchers and practitioners with an overview of work being done in all areas of food service marketing. First published in 1996, it appears quarterly, published by Haworth Press (ISSN 1052–214X).

RENE BARETJE, FRANCE

Journal of Sustainable Tourism

The *Journal of Sustainable Tourism*'s aim is to reduce the friction created by the complex interactions among the tourism **industry**, **visitors**, the **environment** and host communities. It foster basic and applied research for the advancement of sustainable tourism. The journal publishes a mix of studies of interest to both academics and practitioners, provides interdisciplinary perspectives on the subject, and opts for practical proactive initiatives. First published in 1993, it appears quarterly, published by Channel View Books (ISSN 0966–9582).

RENE BARETJE, FRANCE

Journal of Tourism Studies

The *Journal of Tourism Studies* publishes articles on tourism from scholars and practitioners in a range of disciplines including **economics**, commerce, biological and physical sciences, social sciences and humanities. Material considered for publication in this refereed journal includes original research reviews, issue-oriented papers and descriptive, and analytic discussions. The journal aims to be of interest to those in **planning**, **policy** making and consultancy positions, as well as academic staff teaching tourism courses. First published in 1990, the journal appears in two issues per year, published by the James Cook University, Department of Tourism, Australia (ISSN 1035–4662).

RENE BARETJE, FRANCE

Journal of Travel & Tourism Marketing

The *Journal of Travel & Tourism Marketing* is a managerially oriented and applied journal which serves as a medium through which researchers and managers can exchange ideas and keep abreast of the latest developments in the field. A refereed (double blind) journal, it includes travel services, tourism **management** organisations, meetings and convention services, and **transportation** services. The journal publishes full-length articles, research reports, case studies, viewpoints, book reviews and conference reports. First appearing in 1992, it is published quarterly by Haworth Press (ISSN 1054–8408).

RENE BARETJE, FRANCE

Journal of Travel Research

As the official journal of the Travel and Tourism Research Association, the *Journal of Travel Research* serves as a medium through which those with research interests can exchange ideas and keep abreast of the latest developments pertaining to tourism research and **marketing**. A refereed journal (double blind), it publishes information on research and marketing, new research techniques and findings, creative and critical views, generalisations and applications of research output, and synthesis of research and marketing material. The journal favours explanations of the practical applications of any research technique or set of findings for the industry. First appearing in 1963 (under a different title), it is published quarterly by Sage (ISSN 0047–2875).

RENE BARETJE, FRANCE

Journal of Vacation Marketing

The *Journal of Vacation Marketing* provides a forum for refereed papers, case studies, briefings and reviews on the latest techniques, thinking and practice relating to **vacation**. Its objective is to provide material that is of direct relevance to the practitioner while meeting standards of intellectual rigour. The main sectors covered by this journal are **accommodation**, tourist **board**, holidays organiser, **transportation** and tourism attractions. First published in 1994, it appears quarterly, published by Henry Stewart Publications (ISSN 1356–7667).

RENE BARETJE, FRANCE

journalism

The continuous expansion of tourism since the late eighteenth century to the present almost exactly coincides with the rise of popular journalism, first in America and Europe and then elsewhere. Until the 1920s the press was the major journalistic form, but **radio**, **film** and **television** have greatly expanded the journalistic field.

Two kinds of journalism affect tourism. One is non-specialist, general news coverage. Mainstream news commonly presents coverage of places, people and events which may have a positive or negative effect on tourism. Reports of **wars**, **terrorism**, **health** scares and environmental disasters in countries or regions may inhibit travel (for example, in Northern Ireland since the late 1960s). Conversely, destinations may be boosted by favourable coverage. Since the Second World War, and particularly during the last twenty years of the twentieth century, the tourism industry itself has been a subject of mainstream news (such as hallmark events and tourism innovations). The second kind is specialist tourism feature journalism. The expansion of the industry has also stimulated journalism focused exclusively on it. By the early twentieth century there were **magazines** devoted to travel (*Outing*, launched in America and UK in the 1860s, and *The Traveller*, founded in 1900). National and regional newspapers have increasingly featured weekly or monthly travel sections, and today many publish such large weekly supplements. Since the 1970s, radio and television stations worldwide have broadcast regular travel programmes and series, and in the United States there is a travel channel (Discovery).

Given the volume of journalism and its importance as an information source for tourism decisions, the almost complete absence of research into its scope and effects is surprising. A basic research agenda into tourism journalism would need to address four main issues: how much is produced, how is it produced and what factors influence the shape it takes, how many people are exposed to it and what are their profiles, and what influence does it have in the short term (on immediate tourism decisions) and long term (on destination **images** and so on). Until these questions are systematically investigated, assessments of journalistic impact on tourism must be restricted to anecdotal instances, primarily supplied by the industry. Many national tourism organisations inventory the media coverage they have achieved in the **marketing** and **public relations** sections of their annual reports, and there is recognition that such coverage may achieve cheaper, more far-reaching effects than commercial advertising.

Further reading

Seaton, A.V. (1989) *The Occupational Influences and Ideology of Travel Writers: Freebies, Puffs, Vade Mecums or Belles Lettres?*, Newcastle: The Centre for Travel and Tourism and Business Education Publishers.

A.V. SEATON, UK

jungle tourism

Jungle tours have become a major component of **green tourism** in tropical destinations. A jungle is a subclimax tropical forest consisting of a tangled growth of lianas, trees and scrub which may form an almost impenetrable barrier to the **tourist**. It is characteristic of former clearings and of riversides where light penetration is greater than the forest interior. By contrast, true climax rainforest has little undergrowth since light penetration is poor and, contrary to popular **images**, is easily negotiated. Nevertheless, both sets of environments are usually characterised by the tourism market as jungle. Jungle tours are a relatively recent phenomenon of Western **international tourism**. Although nineteenth-century travel expeditions to South America and Southeast Asia bore some of the hallmarks of jungle tours in terms of hunting trophy animals, the majority of nature-based tourists to the tropics preferred to **safari** in savannah grasslands.

The emergence of the environmental movement in Western countries in the 1950s and 1960s led to dramatic shifts in the nature of tourist consumption in jungle areas, with **photography** and **nature** appreciation replacing game hunting as core activities. Related to this shift was the rise of mass **television** and the role that natural **history** documentaries played in creating a more positive image of the jungle. The portrayal of anthropologists studying jungle primates served to turn such activities into attractions in countries such as the Congo, **Indonesia**, Rwanda and Zimbabwe, thus providing a financial base for primate **preservation**. More broadly, the enormous biodiversity of tropical rainforests served to attract tourism and encourage the establishment of **national parks**. Consequently, jungle tourism has emerged as an economic alternative to rainforest destruction for timber, dam-building and grazing in countries as diverse as Australia, Brazil, Costa Rica, Mozambique and Thailand. One of the greatest difficulties for jungle tours is the provision of access to upper levels of the rainforest canopy where most birds and primates are located. Therefore, **accessibility** is being made available through the development of canopy walkways, as in the Wet Tropics of north Queensland, Australia.

Guided tours by **indigenous** inhabitants are increasingly popular as ethno-botanical and educational tourism become distinct speciality markets. Jungle tourism is thus a significant tool for providing an economic basis for the **conservation** of ecological and cultural diversity in tropical rainforest areas, and is receiving considerable support from Western conservation and aid agencies as part of **sustainable development** programmes.

Further reading

Wallace, G.N. and Pierce, S.M. (1996) 'An evaluation of ecotourism in Amazons, Brazil', *Annals of Tourism Research* 23(4): 843–73.

C. MICHAEL HALL, NEW ZEALAND

Kaabah

The Kaabah is a cube-shaped room in the centre of the Masjid al-Haram, in Makkah (Mecca), Saudi Arabia. It is also called God's House (Baituallah) and **destination** (Qiblah), or the direction towards which the Muslims worldwide face at prayer times. The annual **Hajj pilgrimage** to this holy site by Muslims from practically everywhere generates millions of arrivals to Makkah. Saudi Arabia has developed for this truly mega-**event** a vast network of touristic **infrastructure** and superstructure, which is underused for most parts of the year outside the Hajj.

SEMSO TANKOVIC, CROATIA

Kenya

Kenya remains one of the most popular destinations in Africa, with about 6 per cent of overseas visitors to the continent. Since its independence in 1963, tourism has played a vital role for Kenya's **economic development** as it greatly contributes to the **foreign exchange** earnings and creates wage **employment** of about 8 per cent. **Growth** was rapid during the 1970s and 1980s, reaching a record high of 863,400 foreign arrivals in 1994. Since then, the growth has recorded some slight decline.

Approximately one half of tourism **development** is located in the coastal region and focused on beach **recreation**. Wildlife viewing, particularly in the popular **parks** and reserves of Amboseli, Maasai Mara, Lake Nakuru and Samburu is a major tourism activity. Other attractions include historical and archaeological **sites**, and the diverse culture of the Kenyan people. The tourism sector has experienced tremendous growth in hotels, tours and **travel agencies**. By 1993, the number of hotel bed-nights available reached about 12 million with an average **occupancy rate** of around 60 per cent. Most international tourists arrive by air in Nairobi and Mombasa. Besides Kenya Airways, the country is served by forty international **airlines** and a dozen air **charter** companies from Europe.

The **quality** of **infrastructure** for tourism, as well as for general use, varies appreciably. Due to rapid coastal hotel development, pressure on the supply of water and electricity has recently been experienced. The road network is fairly extensive. Kenya Railways provides first-class service among major towns, particularly between Nairobi and Mombasa. Over the years, Europe has been the major generator for tourists to Kenya, accounting for 60 per cent of total arrivals. The **government**, in conjunction with the private sector, has intensified promotional activities in all major tourist-generating **markets**, notably Western Europe and North America. Since 1987, tourism has been the leading foreign exchange earner in the country, with a value of $436 million (about 33 per cent of total exports in 1993).

The **policy** of allowing inclusive tour **charter** operations is controversial in terms of the overall benefits such tourists generate to the country, when compared to the **impacts** of **mass tourism**. Kenya's general tourism development strategy over the past 30 years has been to gradually increase the number of **tourists**. However, according to the current development plan (1997–2000), the focus is on attracting up-market, higher spending tourists. It is the responsibility of the government to ensure quality service is delivered to this target market through its organs, such as the Kenya Tourist Board, Kenya Tourist Development Corporation and Kenya Utalii College. The latter trains personnel to work in tourism-related organisations, while the former assists in financing tourism-oriented enterprises. Considerable progress has been accomplished in recent years on **conservation** through the activities of the Kenya Wildlife Service. However, various pressures prevail on the natural environment and major conservation efforts are still required.

For effective promotion of tourism, Kenya operates offices in major tourist generating markets, especially Germany, the United States, the United Kingdom, France and Switzerland. In addition, the country participates in international trade fairs and exhibitions in an attempt to promote the country as a destination. The 1996 World Travel Market in London witnessed the launching of the Kenya Tourism Board. It is expected that this Board will alleviate some of the existing problems associated with **planning**, funding, coordination and promotion of tourism in Kenya. As the world becomes environmentally conscious of the need to reassess its needs, each country will have to evaluate the type of tourism that suits it best. As a matter of policy, Kenya continues to discourage the low-budget tourists who have contributed to environmental degradation in other parts of the world, focusing instead on up-market tourists who are more sensitive to nature. If quality tourism **management** policies are effected, Kenya's industry could be developed and sustained for the present and future generations without disturbing nature's delicate balance.

Further reading

Kenya Economic Survey (1995, 1996) Nairobi: Government Printers.

Kenya National Development Plan (1994–6, 1997–2001) Nairobi: Government Printers.

Kenya Wildlife Service (1990) *A Policy Framework and Development Programme, 1991–1996*, Nairobi: Government Printers.

Kibara, O.N. (1994) 'Tourism development in Kenya: the government's involvement and influence', MSc. thesis, University of Surrey.

Ouma, J.P.B.M. (1970) *Evolution of Tourism in East Africa*, Nairobi: East African Literature Bureau.

M.K. SIO, KENYA

keying

Keying is a social convention by which social 'reality' is transformed and seen as something else, such as the presentation of a fight as mere horseplay. In tourism, a peculiar, inverted variety of keying is frequently employed: the 'as if' situation in which participants are induced to playfully make believe that presented settings, activities or events are 'real', when as tourists they may be well aware that such occurrences are contrived.

See also: ludic; play

ERIK COHEN, ISRAEL

knowledge acquisition

Knowledge acquisition has a variety of meanings. It has a scientific sense as well as a practical one. From a practical perspective, knowledge is acquired through actual **experience** with a **product** or by collecting information from other sources. The tourism decision process models the steps people go through when selecting a **destination**. Embodied in this process is the concept of **image** formation. With respect to credibility of the information received, autonomous and organic causes have the most credibility, whereas induced formation agents make for low credibility with higher levels of market penetration.

Friends and relatives have historically been the most highly credible source for knowledge acquisition regarding tourism products. Advertisements, and other induced **image** formation agents, are important to create particular images for consideration, but their low levels of credibility put them at the bottom of the knowledge acquisition list. Communication technology is the new entry in the knowledge acquisition process for tourism. The **World Wide Web** has opened up avenues for disseminating information including chat rooms for discussing the merits of various tourism-related products, data profiles for countries, websites with destination **advertising** and scientific data sets for researchers.

Knowledge acquisition, from a scientific perspective, is the reason for academic research in the study of tourism. 'Knowledge builds on knowledge' is the principle behind the scientific method which is the basis for all organised inquiry. Tourism, as a rather new field, began to develop its own area of inquiry, borrowing from other more established disciplines, in the early 1970s. The scientific method has seven steps. It begins with an identification of the problem in quite specific terms usually put into the format of a null hypothesis. Once the problem has been identified, related literature is thoroughly reviewed. The third step is to establish the appropriate methods to collect data followed by the actual data collection task. Data analysis, interpretation of results and final report writing complete the process. Without adherence to the scientific method, advances in tourism research, and hence the acquisition of knowledge, would not be forthcoming.

Further reading

Gartner, W.C. (1996) *Tourism Development: Principles, Processes and Policies*, New York: Van Nostrand Reinhold. (Chapters 10–11 most closely relate to this theme.)

Smith, S.L.J. (1989) *Tourism Analysis: A Handbook*, 2nd edn, Harlow: Longman.

Woodside, A. and Ronkainen, I. (1977) 'Traveller evoked set, inept set and inert sets of vacation destinations', *Journal of Travel Research* 16(1): 14–18.

WILLIAM C. GARTNER, USA

Korea

The Korean peninsula, south of China, is partitioned into Communist North Korea and the Republic of South Korea. There is very little tourism in North Korea. In South Korea a buoyant tourism industry saw visitor arrivals grow from 1.6 million to 4 million, and tourism earnings increase from $1.5 billion to $5.4 billion in the period 1986–96.

MARTIN FRIEL, UK

L

labour relations

Management sometimes mistakenly believes that employees join unions because they want more pay. As a result, many managers in every business, including tourism, attempt to fight unionisation by offering greater pay for employees. However, pay is only one of the reasons that employees join unions. Employees join unions primarily because they believe this will help them accomplish their goals, such as economic security, comfortable working conditions, respect, control over their own work and more. From an employee's perspective, managers have control over various working means and conditions and thus those who vote to certify unions in their workplace are more likely to be voting against management, than for unions.

In the United States, the first real effort to prevent mistreatment of employees by management came about with the passage of the Wagner Act in 1935. Provisions of it gave employees the right to organise and to engage in collective bargaining. The act is still in force and plays a very significant role in union-management relations and is the basis for the responsibility of employers to bargain in good faith. The five key 'unfair labour practice' provisions of this act prohibit employers from interfering with or coercing employees from forming or joining unions; attempting to dominate or influence the operation of unions; from discriminating in hiring or tenure of employees because of their membership or activity in unions; from retaliating against employees who file unfair labour practice charges with National Labour Relations Board; and failing to bargain in good faith.

The intent in the Taft-Hartley Act of 1947 was to balance power between unions and employers. Provisions of this act prohibited closed shops; established the rights of states to enact right-to-work laws; established unfair labour practice charges that could be filed against unions for coercing employees to join unions and other actions; eliminated 'featherbedding' or charging members for services not provided by the union; established the Federal Mediation and Conciliation Service to help unions and management reach agreements; provided for civil suits against either employer or employee for failure to follow terms of agreements; gave National Labour Relations Board power to enforce cease and desist orders through court injunctions if either side engaged in unfair labour practices; and gave management the right to discuss advantages and disadvantages of unions with employees as long as they did not either threaten to punish employees who joined unions or to pay extra benefits to those who did not join. The McClelland Committee on Anti-Racketeering uncovered abuses of power, unethical conduct and corrupt practices in some unions.

The US government employees were first allowed to organise and bargain collectively by Executive Order 10988 issued by President Kennedy in 1962. Legal rights of government employees were further defined by passage of the Civil Service Reform Act of 1981which allowed federal employees' roles more consistent with union members in other fields. All these labour relations development issues and acts, in the United States and elsewhere, obviously have had direct impacts in and implications for tourism – indeed a very labour intensive global industry. In particular, its many trade associations have been involved in dealing with labour relations issues and they regularly advise their membership on best workforce and workplace practices. Further, hospitality and tourism management studies have continuously contributed to a better understanding of the subject, both theoretically and practically, leading to improvement in human relations conditions and thus increasing productivity in this industry.

ROBERT H. WOODS, USA

land tenure

Land tenure refers to the different notions of ownership of land. In traditionally oriented societies, land is often communally held (custodial rights in perpetuity) by clans or other group units, based on ancestral occupancy. In contemporary capitalist society, land is a commodity which can be bought and sold by individuals. In tourism, problems may arise when communally held land becomes the object of speculations or when, for instance, **development** in coastal areas restrains access to **local** people.

TREVOR SOFIELD, AUSTRALIA

landscape

The combination of the physical and cultural **attributes** of an area reflect the interaction of natural and cultural systems, and together these are known as landscape. Cultures have shaped their own landscapes out of the raw materials provided by the earth, and each landscape reflects the culture that created it. Much can be learned about the past and present inhabitants of an area by observing the landscape which visually chronicles the most basic strivings of humankind for shelter, **food**, clothing and entertainment. Where the mix is judged to be particularly attractive, the result may be viewed as scenery, or an amenity landscape, which may be an important resource attracting tourists. Landscapes are not fixed but evolve as the interactions between people and their **environment** change over time. While attractive landscapes are an attribute of many destinations, the **development** of tourism results in their modification. The construction of hotels, **motels**, campgrounds, attractions, signage and other tourism paraphernalia results in changes to the pre-existing landscape. Sometimes the changes are so profound that it is possible to talk of tourism landscapes in which tourism dominates the uses of the land and the appearance of the area.

See also: landscape evaluation; resort morphology

GEOFFREY WALL, CANADA

landscape evaluation

Scenery may be viewed as being a natural **resource**. In order to determine which landscapes are of high **quality** and deserve attention by planners and managers, numerous attempts have been made to develop techniques to measure scenic resources in an objective and quantitative fashion, and to delimit landscapes with high amenity value. Three general approaches are in use for this purpose. The first approach is landscape consensus studies in which experts strive for a consensus about the **attributes** of an area. The second is landscape description studies which aim to identify and measure the attributes of specific landscapes or assess the quality of specific landscapes against some standard or criteria. These are essentially a form of resource **inventory** with the objective of producing a **map** which locates the presence and nature of scenic resources. The third approach is landscape preference studies which attempt to determine which aspects of the **environment** are seen as attractive or unattractive. This may be done indirectly, by inferring preferences from evidence such as literature and art, or directly, by interviewing individuals about specific landscapes.

While the research on landscape evaluation is rich empirically, it has a tendency to document the obvious fact that individuals, at least in Western countries, enjoy places with water and trees. The work lacks a strong theoretical base. Appleton (1975) has attempted to provide this with his proposition of prospect and refuge **theory** which suggests that individuals give high value to landscapes which simultaneously provide an opportunity to see and to hide and thus reflect prospects of survival. However, this theory has not received widespread acceptance.

See also: landscape; environmental valuation

References

Appleton, J. (1975) *The Experience of Landscape*, Chichester: Wiley.

Further reading

Mitchell, B. (1979) *Geography and Resource Analysis*, London: Longman. (Chapter 6 contains a concise and perceptive overview of landscape evaluation.)

GEOFFREY WALL, CANADA

language of tourism

The language of tourism covers all forms of touristic communication at every stage of a trip. Like other languages, it has several functions linking the addresser, addressee, content and context of messages. It is also structured in a similar fashion to **advertising**, especially as regards the use of tense and its promise to effect magical transformations in the receiver. But this language differs from others regarding the anonymity of the speaker (the authorless **brochure**, for example). Frequently, communication is a monologue emphasising the greater knowledge of the sender, while its content is replete with euphoria. There is as well a tautological quality to messages, since tourists generally filter their experiences according to prior expectations, and the latter in turn have been linguistically framed. Then too, when tourists talk among themselves at the **destination** and regale friends and relatives with accounts of their travels on their return home, they tend to do so using the imagery of earlier promotional sources. Therefore, the language of tourism assumes a cyclical quality reflecting the circular nature of tourism itself.

However, perhaps the most important characteristic of this language is its ability to manipulate the **attitudes** and **behaviour** of **tourists**, both individually and collectively. The need to exercise social control over the client becomes clear with the realisation that tourism is an ever-expanding worldwide phenomenon and that, without constraint, it becomes virtually impossible to manage. Moreover, unless limits are imposed, negative sociocultural and environmental consequences will outweigh the benefits, and tourism could end up destroying itself. Such control has been manifested linguistically throughout all periods of tourism's history in travellers' tales, political speeches, lectures, guidebooks and so on. Nowhere, however, are regulatory mechanisms more necessary than in contemporary forms of **mass tourism**, where they are evident in hotels, various types of transport, **tours**, **sites** and attractions.

One way of controlling the tourist is through the use of imperatives and hortatory language. Another allied method is to place him/her in a dependent childlike state. By treating the **tourist as child**, the impression can be given that, whatever the restrictions, personal **freedom** nevertheless remains. Maternal **images**, scenes evocative of infancy and themes of sun and fun are hence routinely employed by the language of tourism in order to resocialise the clientele while maintaining an illusion of liberty. The language of tourism additionally operates through several media, which correspond to the five human senses either singly or in combination. At one time, the simplest channels of communication were the written and spoken word which, before the appearance of electronic media, were the traditional ways of transmitting touristic messages. In a postmodern age of the image, however, and particularly since the advent of the **Internet**, the emphasis has switched to multimedia presentations focusing predominantly on the visual. Here it is important to acknowledge that the science of signs – **semiotics** – can play an extremely useful role in deciphering the complex messages of the language of tourism. It should further be recognised that it has the ability to alter its discourse according to the topic at hand. Alternatively stated, the language of tourism employs many **registers**, at least one for every type of tourism in an increasingly diverse variety of offerings and niche **markets**. The manner in which an ecotourist is addressed, for example, is quite different from the way **heritage** tourism is portrayed.

Finally, the language of tourism employs a multiplicity of techniques in order to enhance the effectiveness of its communication. At the verbal level, it makes great use of comparison, keywords and **keying**, testimony, **humour**, languaging and ego-targeting. Visually, it appeals through colour, format, cliché and such **connotation** procedures as trick effects, pose, indexical transference, photogenia, aestheticism and syntax. When words with pictures are combined, techniques vary from puzzles and temporal contrast to those of collage, ousting the competition, infraction of taboo and significant omission. As such:

> the language of tourism is a structured, monological, multistrategical and controlling way of communicating between often anonymous parental senders and readily identifiable childlike receivers. Through many registers, diverse media and at all stages of a trip, the language of tourism transmits timeless, magical, euphoric and tautological messages which contain the circular expectations and experiences of tourists and tourism.
>
> (Dann 1996: 249)

References

Dann, G. (1996) *The Language of Tourism: A Sociolinguistic Perspective*, Wallingford: CAB International.

GRAHAM M.S. DANN, UK

Latin American Confederation of Tourist Organisations

The Latin American Confederation of Tourist Organisations (COTAL) was established in 1957. Its mission is to promote tourism and the activities of the **travel agencies** and **tour operators** in the Latin American countries, through **marketing** and institutional relations. To achieve its mission, the confederation organises an annual congress in various Latin American countries, addresses issues inherent in the tourism activity, and maintains successful relations with official tourism organisations throughout its region. The headquarters of COTAL is located in Buenos Aires, Argentina.

MARIA FUENMAYOR, USA

law

Law is sometimes interpreted as an expression of cultural values, and sometimes as a rationalised framework of **power**. Friedman (1975) has argued that 'the function of the legal system is to distribute and maintain an allocation of values that society feels to be right... allocation, invested with a sense of rightness, is what is commonly referred to as justice' (1975).

The four traditionally recognised schools of thought about the nature of law are natural law theory, legal positivism, legal evolutionism or historical jurisprudence, and sociological jurisprudence. These emerged in the nineteenth century. At present, social scientists approach law with a distilled and selective recombination of many of these classical legal conceptions. They use these transformed paradigms blended with the new **information technology**, methods and preoccupations that the law is called upon to regulate. This resulted in a new politico-economic setting that requires new legal solutions.

Applications would be in the study of legal institutions themselves, administration of justice, behaviour of courts, lawyers and other collaborative agencies. Economic dimensions and consequences have loomed increasingly large in the study and evaluation of legal norms. The cost of justice and the nature of access to justice have become issues of major concern to governments. The high-flown individual and **community** values that legal principles express are examined by legal economists in the light of their **efficiency** and social effect, not just their self-defined moral content and rightness.

Tourism is an all-embracing phenomenon representing for various destinations an agglomeration of sectors that generate and accumulate wealth in the society. The law is called upon to address such vast fields of tourism-related activities and businesses with all its multi-faceted relationships. The legal norms in such context form a coherent body of law, which is termed the 'tourism basic law' of the **destination**. Such codes encompass all legal rules relating to national tourist **policy** directives and various touristic strategies, in addition to the legal **rules** regulating the various tourist activities within the country concerned. Examples of such activities are **tour operator** and **travel agency** activities, **accommodation** relationships, **transportation**, entertainment and **tour guides**, as well as the legal norms regulating tourism **development** and imposing limitations on **planning**, whether environmental, social, economic, cultural or political. Many of these impose themselves on law. There are limits which arise from the nature of law as a purposive system, with its test of effectiveness residing in its fulfilment of its purpose or objectives. A second type of limitations relate to the society in which law is enforced, and whether it forms a good communication system and whether it is well received and accepted. A third category of limitations touches upon the utility of law for social transformation. A last group of limitations is inherent in the prevailing political and economic system.

See also: employment law

SALAH E.A. WAHAB, EGYPT

leadership

Leadership is generally regarded as essential in the functioning of any tourism organisation. Success or failure within any such system is very commonly attributed to the quality of leadership. A fundamental concern of any organisation is how to attract, train and keep people who will be effective leaders. A plethora of research studies demonstrates that leadership quality is related to group morale and **productivity**. A major question that emerges in regard to **management** quality revolves around what is meant by good leadership. One major perspective on this central question lies in the area of leadership **roles**.

The notion of managerial role has become a valued insight into the uncertainty of leadership. The commentator Mintzberg (1973) has identified a range of distinct roles that are typically played by managers within any organisation. Mintzberg points out that roles primarily revolve around the manager's position in the group and his or her interpersonal relationships with others. The figurehead role is the first of these. The manager performs tasks that are often of a legalistic or symbolic nature, such as introducing guest speakers or writing letters of appreciation to retiring employees. The second interpersonal role is that of liaison. The manager acting as liaison person serves as a link between one's own and other groups throughout the rest of the organisation. Mintzberg suggests that managers build and maintain an informal network of contacts with individuals in other groups, because these relationships serve as the foundation upon which rest some of the manager's other roles (such as monitor, disseminator, spokesperson and negotiator). Leadership is said to be properly classified as an interpersonal role, since it involves an exchange relationship between the leader and each of his/her subordinates. In such a manner, the manager motivates group members and assists them to work together effectively in the pursuit of a desired group or organisational goals. The leadership role may be seen to overarch all managerial activity.

Nebel (1992) and Nebel and Ghie (1993) have studied the major roles of the **hotel** general manager. They describe two basic roles. As organisational developer, the roles of the general manager extends beyond the immediate confines of the hotel. They need to monitor information about both the **community** and the competitive environment, and that requires them to become a liaison between the hotel and the outside world. Further, the information they gather from the external environment needs to be analysed and disseminated to subordinates within the hotel. To effectively function as an entrepreneur, the manager must process both externally and internally obtained information. This role develops specific plans and programmes geared to improve the operating **efficiency** and service strategy of the hotel. Any new plan or programme imposes additional time obligations on both the general manager and subordinates. Moreover, new plans and programmes often involve the allocation of financial and human resources. Therefore, in the course of developing and implementing specific strategies for the hotel, the manager plays the decisional work role of resource allocator. Finally, the general manager furthers the leader work role by continuing relationships with subordinates as an integral component of the job and by choosing which specific plans and programmes to implement. As leader, the manager must contend with the challenge of ensuring that subordinates fully accept the plans and programmes outlined for the hotel and are committed to working toward their successful implementation. This requires that the general manager pays careful attention to the work roles of monitor and disseminator of information.

References

Mintzberg, H. (1973) *The Nature of Managerial Work*, New York: Harper & Row.

Nebel, E.C. (1991) *Managing Hotels Effectively*, New York: Van Nostrand Reinhold.

Nebel, E.C. (1993) 'A conceptual framework of the hotel general managers' job', *Hospitality Research Journal* 16: 27–39.

GLENN F. ROSS, AUSTRALIA

leakage

Leakages, sometimes referred to as withdrawals or outflows, are that part of national income which is not spent on domestically produced goods or services. Much of the money earned by households, for example, accrues to the public sector in the form of **taxes** and customs duties; some is saved by the households and some spent on imported goods and services. These are the principal forms of leakage. But whereas in most cases taxes and savings remain in the economy for possible future use, imports form a permanent loss of revenue out of the country and are a deficit item in the **balance of payments**.

The magnitude of the import leakages is determined by the situation of the domestic economy, in particular its ability to produce the goods and services required both directly by **tourists** and as a secondary consequence of the **development** of tourism (see **multiplier effect**). In turn, this depends upon the size of the economy and especially the nature of the trading linkages which domestic firms are able to establish with each other. The situation is compounded if the domestic firms are too small to supply the extended market. Further, during both the construction and operating phases of hotels and other tourism establishments, the domestic economy may not have the finance, expertise or ability to cope with **demand**; thus foreign loans, engineering and management companies may be used with a consequential leakage of profits and management fees abroad. This leakage element is compounded if some local staff require to be trained abroad.

Another less obvious increase in leakages may occur if there is a change in domestic consumption patterns. As the economic **benefits** of tourism filter through the economy, national income rises and the income levels of a larger number of people rise. In consequence, their purchases increase and may change in nature from domestically produced goods to items with a higher import content. This in turn leads to further leakages. Leakages out of national economies vary from as little as 10 per cent to as much as 80 per cent, but these are generally much higher in the case of local economies. The principal reason is that local economies are much more integrated into their parent national economies than are national ones with other national economies.

BRIAN ARCHER, UK

learning curve

The relationship between the performance of a particular activity and accumulated experience in doing the same is known as the learning curve. Performance normally improves as experience increases. In a business context, whether tourism or other, this relationship usually expresses performance in terms of the cost of producing one unit of output, and experience in terms of the accumulated volume of output. Two phenomena normally provide the basis for this relationship: learning and **economies of scale**.

GEOFFREY I. CROUCH, AUSTRALIA

legal aspects

The term 'business' has no single recognised definition or legal meaning. It may mean or embrace activity or enterprise for gain, benefit, advantage or livelihood; activity of some continuity, regularity or permanence; commercial or industrial establishment or enterprise; constant or continuous or habitual **employment** or occupation; efforts of men to improve their economic conditions and satisfy their desires; enterprise in which the person engaged shows willingness to invest time and capital on future outcome; and every legitimate avocation in life by which honest support for family may be obtained (*Black's Law Dictionary*, 4th edn, 248–9).

It is agreed that any occasional, single or isolated activities do not constitute business. By the same token, the meaning of business is not dependent on whether enterprise is profitable or has prospects of being profitable. Thus, the legal forces affecting business in general, and international business in particular, are boundless and varied. The national **laws** governing various aspects of business, including contracts, agency, corporations, partnerships, bailments, sales, innkeeping and negotiable instruments, differ from one country to the other.

Moreover, legal norms of the common law system vary from legal rules of the civil law orbit. The common law includes both the 'unwritten' law based on the *stare decisis* doctrine and statutory law, which is spoken of as 'written' or enacted law. Even within the same common law system, there is a distinction between two different remedies, law and equity. The civil law orbit includes varying legal systems which might have different solutions to the same problems, depending upon the nature of the theories and philosophies underlying these legal systems. However, it has been proven that there is a common core of legal systems that transcend the varying legal norms, even with legal systems originally based upon divine revelation, such as Islamic law.

International business is affected by many thousands of laws and regulations on hundreds of subjects that normally arise in the course of international business relations among nations, **multinational firms**, international organisations and national companies of different countries. Although many nations, both developed and developing, have laws and regulations that affect the activities of international firms, there has been no successful effort to coordinate them. Examples of these laws are those addressing investments, taxation, antitrust and restrictive trade practices, tariffs, quotas and other trade obstacles, product liability (civil and criminal), labour, price and wage controls, currency exchange controls, exports and imports, business contracts, patents, trademarks, copyrights and trade secrets, intellectual and technical property, restrictions on dumping and much more.

The rules applicable to the tourism business, further, have become immense and varied. Laws governing tourism, including national **policy** (if such exists), national and **regional organisations**, **travel agencies** and **tour operators**, the **hotel** business, **vacation** villages, various modes of **transportation**, environmental controls, **planning** and **development**, **marketing**, the scope and limits of governmental control, investment incentives, **facilitation**, bills of rights and the like, are all part and parcel of the new form of business law.

Thus, business law touches everyday lives through every contractual dealing undertaken. A contract, usually in the form of a commercial bargain involving some of the exchange of goods or services for a price, is a legally binding agreement made by two or more persons, enforceable by the courts. As such, contracts may be written or oral; to be binding, they must make an offer and unqualified acceptance thereof, and include intention to create legal relations, lawful cause (a consideration in common law systems) and genuine consent (that is, absence of fraud). The terms must be legal, certain and passable of performance.

Contractual dealings, as the cornerstone of all commercial transactions, have resulted in the development of specific bodies of law within the scope of business law that continually evolve to cover newly emerging areas of human activities. Thus, all areas relating to competition, anti-trust, anti-monopoly, consumer protection, computers, **global distribution systems**, other information and communication technology (see **information technology**), new forms of ownership as condominiums and timesharing, including shifting schemes in all sectors of tourism, come under the umbrella of business law.

SALAH E.A. WAHAB, EGYPT

legislation

The term 'legislation' represents a method of deliberately creating new **law** by state institutions entrusted with the **power** to do so. Among the reasons for the growing importance of legislation during the past two centuries are the need for greater complexity of governmental activities, the need for certainty in the jurisdiction and decisional norms of officials, the belief in equality of treatment among citizens, and the desire to clarify rights and obligations of individuals in the society.

Legislation includes the written constitution as well as the various statutes and enactment of such authority in a state. While the priority attached to legislation used to vary between the civil and common law orbits, such variation is gradually losing its importance. The hierarchy of legislation proceeds downward from the constitution to the statutes to executive decrees by the president, cabinet and various ministers. While legislation is uttered by state bodies having power to create legal norms in legislative form, case law is produced by courts. Thus, the literary sources of legislation are usually separated from case law. As Kelsen's theory of law illustrates, a superior legislative norm ordinarily contains power norms which authorises officials to make subordinate legislation, and it partly offers decisional norms which guide the official in determining the content of the particular norms to be made.

Legislation is undoubtedly necessary to regulate the expanding **government** function and human relations in modern states. Tourism has become one the main societal preoccupations where legislative norms and/or administrative directives are indispensable to varying degrees in different countries, depending on their political and economic systems. In this respect, tourism's national and local organisation, **tour operators** and **travel agencies**, multiple types of **accommodation**, various aspects of **development** and promotional strategies, **transportation**, socio-economic and environmental **impacts**, sustainability and the application of **Agenda 21** are some of the many areas that require legislative and/or executive enactment.

SALAH E.A. WAHAB, EGYPT

legislation, environmental

Nearly all **legislation** – statutes passed by parliaments and other **laws** (regulations, proclamations and so on) promulgated under the same authority – has an environmental aspect. There is a spectrum or hierarchy of environmental importance according to which legislation can be ranked. This can be illustrated with reference to Australian legislation.

First, there are Acts of Parliament which have an obvious environmental consequence, but the main aim or purpose of which is towards some other, often related area. An example is bushfire control legislation, where the main aim is **safety** and the bureaucratic arrangement of the fire service or the highways and main roads Acts in various parts of Australia. The primary responsibility for most aspects of direct regulation of the **environment** rests, under the Constitution, with the States.

Typically, there is a core of legislation where the central function is either to create an administrative framework for the control of environmental issues or to aim to deal with a critical environmental issue directly. This core can be divided into legislation dealing with specific topics. **Planning** and **development** legislation creates a framework for the implementation of urban and regional planning, in particular the control of what type of building (and other development) can take place in any given area. Environmental review legislation calls for **heritage** environmental impact statements (EIS) and the less onerous public environmental reports. Protection of national heritage tends to be divided into legislation relating to natural features (**national parks** and other reserves, including **wilderness**), features created by humans (buildings and movable items), native flora and fauna and endangered species. Resource legislation covers responsibility for legislative control of mining, forestry, inshore and freshwater fishing, land and water use. Finally, pollution protection applies to all media of pollution – water, air and noise – dangerous substances and special areas (for example, ship-borne oil pollution in coastal waters). All such legislation has directly and indirectly influenced the **growth** and development of tourism almost worldwide.

See also: codes of ethics, environmental; environment; environmental auditing; environmental management systems; environmental compatibility; impact assessment, environmental; precautionary principle; planning, environmental;

DUNCAN HARTSHORNE, AUSTRALIA

leisure

Leisure, along with happiness and wisdom, were according to Aristotle the goals of human life. Of these, leisure was the most fundamental in that it was a prerequisite to the other two. In this view, leisure is not merely the perceived **freedom** of not having to be busy, but the state of 'truly disinterested interest, the achievement of understanding' (Craven 1937: 402). It is worth noting that this lofty state was maintained both by the refinement of the mind and by a large slave class, to whom leisure was only a rumour.

Classical Rome superseded classical Greece chronologically; whether their view of leisure supersedes that of the Greeks is moot. For the Romans, leisure was **time** off work. Properly managed, it renewed people for work and war. In the late stages of the Roman Empire, the filling of leisure with entertainment and spectacle to prevent boredom (and possible civil unrest) was one of the challenges that faced **government**. These two views of leisure, the Greek with an emphasis on the mind and quality of life and an undercurrent of elitism, and the Roman with an emphasis on rest and entertainment and an undercurrent of social control, characterise the two dominant views of leisure through much of history.

Through the Middle Ages, the Catholic Church viewed work as the necessary atonement for original sin. The one day of leisure that God created – the Sabbath (a religious tradition created by Judaism) – was viewed more as rest for future work rather than rest from past work. With the Protestant revolution, rebellious theologians rejected many Catholic traditions but kept and reinforced the notion that the only value of leisure was for rest and worship of a God that expected and rewarded hard work.

The work ethic, founded in fundamentalist Christian values, proved to be a useful social tool to help advance the Industrial Revolution. The social costs of the Industrial Revolution, including radical changes in social structures, transformation of family organisation and values, poverty, illness and injury, and the destruction of rural environments, created a setting in which nineteenth-century reformers such as Robert Owen and John Stuart Mill became active. While these early reformers were more interested in broad political and social issues, later authors began to address leisure specifically. These included Paul Lafargue (*La droite B la paresse* or *The Right to Be Lazy*) and Thorstein Veblen (*The Theory of the Leisure Class*). Throughout this period, leisure was viewed as synonymous with free time.

In the 1920s, scholars began to explore the role of leisure in creating one's identity and in socialising individuals. One of the more important works to address leisure during the first half of the twentieth century was that of Riesman (1950). He argued that there had been only two major social revolutions in the West in the last 500 years. The first was the Renaissance, in which the force of tradition was overthrown by a new respect for one's own conscience and intellect. The second, following the Second World War, replaced the emphasis on self by concern for comparing oneself to others. This was especially manifested in terms of the rise in **consumerism** and a concern about non-hereditary social status, both of which are most clearly seen in leisure contexts.

Sociological inquiries into the nature of leisure broadened dramatically from the 1960s onward. From the role of leisure in creating a social identity, researchers explored the relationship between leisure and family, leisure and work, leisure and community, and leisure and culture. Dumazdier (1968) was one of the leading sociologists who addressed the nature of leisure during the 1960s and 1970s. He conceived leisure as a restricted set of activities that required two social preconditions. The first precondition was that society had to be sufficiently advanced so that most social activities are personally selected by individuals, rather than imposed by tradition or ritual. The second was that the economic **activities** that provided livelihoods were essentially arbitrary or synthetic rather than natural. In other words, leisure was a function of an industrial or post-industrial society.

Further, Dumazdier felt that leisure implied four personal conditions. First, leisure activities were free of commitments to work, school or **religion**. In other words, leisure was an expression of perceived freedom. Second, leisure was 'disinterested' in the sense that activities were engaged in for intrinsic **motivation**, rather than material or social gain. Third, leisure is pleasurable, and fourth, leisure must fulfil the need for rest, entertainment and self-transcendence. Other authors advocated the 'state of mind' interpretation of leisure (as opposed to a 'free time' interpretation) even further. One of the best known and perhaps most extreme views is that of de Grazia, who wrote, 'it [leisure] is an ideal, a state of being, a condition of man, which few desire and fewer achieve' (1962: 5).

While both 'leisure as free time' and 'leisure as a state of being' conceptualisations have advantages, each also has limitations. The 'free time' view must wrestle with operational questions about what constitutes free time. For example, is time spent in childcare or in religious observances 'free time'? Further, this view largely ignores the social and psychological dimensions of leisure as a human **experience**. On the other hand, the 'state of being' view presents leisure as a phenomenon that can be so difficult to operationalise with simple empirical tools that the concept has very limited utility in policy or **management** contexts. In recent decades, social and psychological researchers have refined a number of approaches in the empirical study of leisure. One of these is studying people from many walks of life through carefully structured interviews or **time** budget diaries, and determining the circumstances in which they feel at leisure. Laboratory studies have also been conducted in which the researcher attempts to induce feelings of leisure in the experimental subjects.

Modern scientific approaches to defining leisure can be classified according to two criteria (Mannell and Kleiber 1997). The first is the type of phenomenon taken as an indicator of the occurrence of leisure. Two types of phenomena are used: objective and subjective (Neulinger 1974; Lawton 1993). Objective phenomena equate leisure with participation in certain types of activity such as doing a crossword puzzle. Leisure may also be defined in terms of the setting: tennis courts, beaches or theatres are likely to be correlated with feelings of leisure. The subjective approach interprets leisure as a mental state. This requires that researchers be able to measure satisfactions, meanings, moods and cognitions associated with leisure. While these can sometimes be inferred from the observation of **behaviour**, they are more likely to be recorded through **interviews**.

Scholarly approaches to defining leisure may also be classified according to the vantage point adopted by the researcher. In other words, regardless of whether the researcher is using objective or subjective indicators, leisure may be defined in terms of what the researcher views it to be (an external vantage point) or that of the person being studied (an internal vantage point). External definitions are based on what people in a social group agree constitutes leisure. The researcher predetermines what is leisure for the people being studied based on knowledge of the group to which the people being studied belong.

From an internal vantage point, the researcher defines leisure based on whether the behaviour, setting, or experience is described as leisure by the person being studied. In other words, the researcher allows the subject to determine whether a given activity, setting or experience is leisure. By using the subject's own definition, some researchers believe they can achieve a more accurate picture of what leisure is and what it means. Personal definitions are influenced by the values and beliefs of the group to which the individual belongs. Externally and internally based definitions of leisure can agree, but they are often at odds. These various approaches – objective versus subjective criteria and external versus internal vantage points – all offer legitimate insights. Answering questions about leisure typically requires a combination of these approaches, and tourism connects and has been studied in respect to these and other perspectives.

References

Craven, I. (1937) 'Leisure', in *Encyclopedia of the Social Sciences*, New York: Macmillan, vol. 9, 402–5.

de Grazia, S. (1962) *Of Time, Work, and Leisure*, New York: Twentieth Century Fund.

Dumazdier, J. (1968) 'Leisure', in *International Encyclopedia of the Social Sciences*, New York: Macmillan, vol. 9, 248–54.

Lawton, M.P. (1993) 'Meanings of activity', in J. Kelly (ed.), *Activity and Ageing*, Newbury, CA: Sage, 125–44.

Mannell, R.C. and Kleiber, D.A. (1997) *Social Psychology of Leisure*, State College, PA: Venture Publishing.

Neulinger, J. (1974) *Psychology of Leisure: Research Approaches to the Study of Leisure*, Springfield, IL: Charles C. Thomas.

Riesman, D. (1950) *The Lonely Crowd: A Study of the Changing American Character*, New Haven, CT: Yale University Press.

STEPHEN SMITH, CANADA
ROGER C. MANNELL, CANADA

Leisure Sciences

Leisure Sciences publishes research on **leisure**, **recreation**, **play**, and **tourism** from a social science perspective. Articles address theory, the cultural, social, psychological, economic, political, and philosophical aspects of leisure, planning for leisure environments, gerontology, travel and tourism behaviour, and research methods. Manuscripts are peer reviewed anonymously by at least three reviewers. The journal publishes full-length articles, target articles with peer commentaries, research notes, book reviews and comments. First appearing in 1979, the journal is published quarterly by Taylor & Francis (ISSN 0149–0400).

GARRY CHICK, USA

Leisure Studies

An official organ of the Leisure Studies Association, *Leisure Studies* publishes articles on all aspects of **leisure** studies and from a variety of disciplinary bases, including **sociology**, **psychology**, human **geography**, **planning**, **economics** and the like. Emphasising the social sciences, the journal covers the whole range of leisure **behaviour** in the arts, sports, cultural and informal activities, tourism, urban and rural **recreation**. In addition to full-length articles, this refereed journal also publishes research notes and reviews. All submissions should be in English. First appearing in 1982, the journal is published quarterly by E. & F.N. Spon (ISSN 0261–4367).

RENE BARETJE, FRANCE

leisure tourist

Leisure **tourists**, in contrast with business travellers, travel for pleasure and thus are not under any obligations to frequent specific destinations or facilities. They tend to be price and fashion conscious, concentrate their touristic **activities** to specific (**vacation**) times and are influenced by **marketing** and publicity. Leisure tourism is heavily influenced by living standards, discretionary income levels and **vacation** entitlements.

STEPHEN J. CRAIG-SMITH. AUSTRALIA

length of stay

Length of stay refers to the period of time which people spend in a **destination**. It may be measured in hours, days, nights or weeks. Many definitions require that visitors to a destination stay at least twenty-four hours or overnight, and less than one year, to be considered a tourist. In the former case, they may be viewed as an excursionists, and in the latter as migrants. Destination areas often look for means to extend tourists' length of stay in order to increase positive economic **impacts**.

GEOFFREY WALL, CANADA

liberalisation *see* international aviation liberalisation

licensing

Each town, county, state, province or country has their own jurisdictions over the licensing laws relative to the sale and consumption of alcoholic beverages. For example, the time of day in which the alcoholic beverages are sold varies from place to place. When compared to northern Europe and the United States, the central and southern parts of Europe have relatively liberal licensing laws and maintain policies that keep the pubs open for a large portion of the day and night. The United Kingdom, however, has been saddled with a closing time of 10:30 or 11:00 pm for many years. It is said that hundreds of years ago, a group of dock workers were so drunk that they were unable to load spices for the king's banquet that evening. The king, infuriated by this incident, proclaimed that the pubs would close in the afternoons.

The drinking age also varies according to the region. For example, in the Canadian provinces the legal drinking age is set at 18 or 19. In the United States the drinking age is 21 in most states, whereas in northern Europe the drinking age is generally 18. In some regions, licenses may include the sale of all beverages or only for the sale of beer and wine. An off-sales licence allows the merchant to sale of alcoholic beverages for off-premises consumption. A licensee has a responsibility to society to serve alcoholic beverages in a responsible manner. Owners, managers, bartenders and servers may be held liable under the law if they serve alcohol to minors or to persons who are intoxicated. In most cases, the extent of the law is very severe. Tourism from the more strict regions, like Scandinavia, is often directed to more relaxed destinations with regard to alcohol, like the Mediterranean countries.

JOHN R. WALKER, USA

life cycle

The life cycle concept was introduced into tourism studies as a variable in tourist behaviour research and as a development model for **products** and destinations (as an indication of various stages in **behaviour** and demographic variables.) Traditionally, this concept has been linked to a family life cycle, with age being the context in which preferences and activities of tourists change over time. Thus the concept has been developed as an instrument for **market segmentation**. The main assumption is a correlation between life cycle and **lifestyle** which eventually expresses itself in the preferences for destinations and specific activities. Life cycle is a construct to explain and manage development of tourism **products** and destinations and has a strong connection with **economics**, having been first introduced to help understand the evolution of business organisations.

The challenge for tourism planning and research has been to adapt the life cycle model to the dynamics in development, reflecting the internal forces and the external components of **growth**. The expectations of a growing tourism market, the need to understand the growth factors in each stage and the intention of forecasting future development based on past growth rates are the key issues in this approach. The life cycle is usually presented as an S-curve in time which evolves through stages of introduction, development, consolidation, saturation and stagnation, followed either by decline or rejuvenation.

See also: life cycle, destination; opportunity cost; product life cycle

Further reading

Butler, R.W. (1980) 'The concept of a tourism area cycle of evolution: implications for management of resources', *Canadian Geographer* 14(1): 5–12.

Getz, D. (1992) 'Tourism planning and the destination life cycle', *Annals of Tourism Research* 19: 752–70.

Haywood, K.M. (1989) 'Can the tourist area life cycle be made operational?', *Tourism Management* 7: 154–67.

Martin, B.S. and Uysal, M. (1990) 'An examination of the relationship between carrying capacity and the tourism life cycle: management and policy implications', *Journal of Environmental Management* 31(4): 327–33.

MYRIAM JANSEN-VERBEKE, BELGIUM

life cycle, destination

This term describes the temporal pattern of **development** of destinations. Introduced by Stansfield (1978), the concept of a development cycle is much older and can be found in nineteenth-century literature on the development of **resorts**. These places were described as proceeding from exclusive development through expansion and into decline as fashions changed and competition developed. A description of this process in Mediterranean destinations is given by Christaller (1963). Plog (1973) related the rise and fall of destinations to the characteristics of the **tourists**, suggesting that destinations proceeded along a spectrum related to different market segments at different stages in their development. He raised the issue of inevitability of decline as older destinations become less attractive to a decreasing market.

These ideas were taken further by Butler (1980), who proposed that destinations evolve through a distinctive cycle within which there are six identifiable stages characterised by specific attributes of the tourist and the nature and scale of development. This cycle took the form of an S-shaped curve. The first stage is exploration, in which there are few **tourists** and no development specifically for tourism. The second stage is involvement, in which primarily local residents offer facilities for tourists and regular but small numbers of visitors arrive. The development phase is characterised by extensive establishment of tourism facilities, a large number of visitors, **marketing** of the destination and the influx of external capital. The fourth phase, stagnation, displays a declining **growth** rate in development and visitor numbers, the latter reaching their peak in this period. Most of the facilities will be under the control of non-locals and the destination will be experiencing problems from competition. The destination may then go into a decline phase, which may become more rapid as there is a lack of investment and renewal and it becomes less attractive. Other possibilities include a steady state pattern or, following considerable reinvestment and/or the addition of a major new **attraction** (such as **gambling**), the destination may undergo rejuvenation.

This model has been widely and frequently applied. While variations to the general pattern have been found, mostly reflecting specific local conditions, and criticisms have been made of the unidirectional nature of the model and its implied inevitability, its general arguments remain accepted in the literature.

See also: life cycle; product life cycle

References

Butler, R.W. (1980) 'The concept of a tourist area cycle of evolution: implications for the management of resources', *The Canadian Geographer* 24(2): 5–12.

Christaller, W. (1963) 'Some considerations of tourism location in Europe: the peripheral regions', *Regional Science Association Papers* 12: 95–105.

Plog, S. (1973) 'Why destination areas rise and fall in popularity', *Cornell Hotel and Restaurant Administration Quarterly* 13–16.

Stansfield, C. (1978) 'Atlantic City and the resort cycle: background to the legalization of gambling', *Annals of Tourism Research* 5(2): 238–51.

RICHARD BUTLER, UK

lifeseeing

Lifeseeing programmes, often organised, provide tourists with the opportunity to **experience** particular aspects of the host culture. This can also include visiting tourism highlights and **attractions** of the **destination**. Local experts usually coordinate the lifeseeing portion of the programme. Home visits are also arranged. Axle Dessau, former director of the Danish National Tourist Office, is credited with articulating this form of tourism.

See also: sightseeing

CHARLES R. GOELDNER, USA

lifestyle

The concept of lifestyle is connected with the form of **behaviour** specific to a social position, while the concept of ethos seems to be first of all connected with the hierarchy of **values**. The term 'lifestyle' was used by, among others, Weber, Veblen and Adler.

The types of ethos depend on the hierarchy of values of a given person. According to the most important values, one may distinguish such key ethos as the search for truth, wisdom, conformity, alterocentrism, love, egocentrism, creativity, consumption, **freedom**, independence, consent to constraint and others. The types of lifestyles depend on the affiliation to the social class or other social group to which an individual belongs. Thus, the lifestyle typology can be based on such sociocultural variables as age, social **class** or profession. These influence lifestyles of the youth, old people, workers, peasants, intelligentsia, aristocracy, soldiers and many others.

It is difficult to say to what extent the specificity of the touristic lifestyle exists, but taking into account the still growing number of **tourists** worldwide, it is reasonable to argue that tourism is a form of social category. This represents a lifestyle which characterises tourism and tourists on a voluntary and temporary move away from home, with activities at their destinations and eventual return to their home communities. The essence of this lifestyle is captured in the realisation of such touristic goals as cognition, acquaintance with the host nature and culture, involvement in **leisure** pursuits, searching for rest and relaxation opportunities, conducting studies and making family or business contacts. More multidisciplinary research on tourism will have much to say about the tourism lifestyle, its roots, the diversity in its manifestation and its meanings at and away from home.

KRYSZTOF PRZECLAWSKI, POLAND

liminality

Liminality (from the Latin *limen*, a threshold) is the middle stage of transition in a **rite of passage**. The concept has been introduced into **anthropology** by Victor Turner (1969) following the earlier work by Arnold van Gennep. Individuals in a liminal stage are 'betwixt and between', because they pass through an ambiguous cultural realm, devoid of the structural characteristics of the preceding and following stages. The liminal stage is experienced by participants in the rite of passage as one of 'non-place' and 'no-time', resisting categorisation.

Participants in the rite are stripped of **power** and levelled to a stage of social homogeneity, during which they are inculcated with sacred knowledge. They form a transient 'antistructural', undifferentiated and egalitarian formation, which Turner called 'communitas'. Turner has coined the term 'liminoid' to designate states resembling liminality in other social phenomena, including the **pilgrimage** and tourism. Although Turner himself did not specifically study tourism, some researchers have sought to apply his conceptual framework to the analysis of touristic situations, particularly **vacations** or 'holidays'. Such situations are described as 'antistructural' reversals of everyday routine existence, during which the normal social differences among the **tourists** are temporarily suspended. **Time** is experienced as 'flow' without horological divisions, norms of attire and sexual conduct are abandoned, and a spontaneous camaraderie, resembling *communitas*, develops among the participants. The dress and conduct of the tourists marks them sharply off from the locals, particularly in tourism enclaves in developing countries (Wagner 1977).

Liminoid touristic situations are markedly **ludic**. In sharp contrast to the seriousness of liminality in rites of passage, tourists may engage in reversed **role**-playing such as being 'kings for a day' or 'primitives' (Gottlieb 1982), whether spontaneously on some remote beach, or more self-consciously within the framework of a Club Méditeranée. In **sightseeing**, a liminal flow of time is more difficult to achieve owing to constricting factors, such as timetables and **itineraries**. However, groups engaging in sightseeing may occasionally approximate the closeness of a *communitas*, despite the formal, institutionalised framework of their journey.

References

Gottlieb, A. (1982) 'Americans' Vacations', *Annals of Tourism Research* 9: 165–87.

Turner, V. (1969) *The Ritual Process*, Chicago: Aldine.

Wagner, U. (1977) 'Out of time and place: mass tourism and charter trips', *Ethnos* 42(1/2): 38–52.

Further reading

Cohen, E. (1985) 'Tourism as play', *Religion* 15: 291–304.

ERIK COHEN, ISRAEL

limit of acceptable change

As part of the early debate on the ability of ecosystems to sustain damage, the term 'limits of acceptable change' was coined to describe the level of allowable variations in the quality of the **environment** before irreversible degradation is likely to occur. Discussion and research on this concept has shifted to determining the **carrying capacity** of ecosystems and, in order to achieve this, the most appropriate indicators of environmental change. Indicators of this change are physical, chemical, biological or socioeconomic measures that can be used to assess natural **resources** and environmental quality, and reduce the number of measures that are normally required to represent a given situation to the **community**. A key question that has yet to be resolved is whether or not it is possible to develop a core set of environmental indicators that could be used by all communities. Since they are shaped by community-driven sustainability goals which themselves are influenced by local environmental, economic and social conditions, there has been considerable debate as to the preferred set.

Furthermore, the biophysical dimensions of carrying capacity are not immutable, but are subject to variations brought on by technological change, greater efficiency in consumption, recycling of resources, or increasing substitutability of non-renewable resources by renewable ones. Indeed, although the carrying capacity approach has a certain validity for defining the limits of acceptable ecological change at a regional scale, it is of little help in the context of the intra-urban built environment. Environmental management rather than **development** control is of much greater importance in that context.

The Commission for Sustainable Development is the United Nations body charged with developing sustainability indicators on a worldwide basis, and is working closely with the **World Tourism Organization** to develop such indicators for the tourism **industry**. These will be based on international discussions on the limits of acceptable change to local environments from the development of tourism, and used by governments and communities to regulate and manage the industry in the **future**.

See also: biological diversity; benchmarking; codes of ethics, environmental; ecologically sustainable tourism; environmental auditing

Further reading

American Planning Association (1996) special issue on Monitoring Change, *Journal of the American Planning Association* 62(2).

MALCOLM COOPER, AUSTRALIA

literary tourism

Literary tourism is a form of tourism in which the primary **motivation** for visiting specific locations is related to an interest in literature. This may include visiting past and present homes of authors (living and dead), real and mythical places described in literature, and locations affiliated with characters and events in literature. Regions strongly associated with an author may be marketed in that vein, such as 'Shakespeare Country'.

RICHARD BUTLER, UK

local

'Local' is a term which refers to both a scale of analysis and the characteristics which are associated with a particular neighbourhood, **commu-**

nity, **location**, place or area. The term is used with reference to impact analysis, and is regarded as the most appropriate level for the **development** of sustainable tourism **planning** and **development** strategies because of the need to integrate community responses.

C. MICHAEL HALL, NEW ZEALAND

local organisation

A local tourism organisation is a governmental or non-governmental entity, or public–private **partnership** body, which represents a specific city or **community**. It is a fourth-tier organisation in countries that have national, state/provincial/territorial and regional tourism representations, and its organisation varies by country. In the United States, this body tends to be a **convention and visitor bureau** funded primarily through a room tax. Other countries, such as the United Kingdom, more frequently place this responsibility with the local government authority such as the city or municipal council, with direct funding from the local government. Public–private sector partnerships also exist in some communities where direct government funding is combined with private sector contributions.

The roles of this local agency include **marketing**, **planning**, **development**, research and visitor servicing. Most promote their communities to **pleasure tourists**, the travel trade, and potential convention, conference and meeting groups. The organisers of convention and conferences are an especially important target, because of their large potential economic **impacts** on a host community. Local tourism organisations may be involved in the planning, development or operation of convention or conference centres that serve this market. Many operate tourism information centres within their communities.

The local tourism organisation can be a catalyst in the community in analysing and enhancing its community's **image** as a **destination**. This agency often coordinates long-term planning exercises involving visioning and the development of strategies or plans. Almost certainly, it champions tourism in the local community and lobbies other agencies for greater recognition and funding of this industry. It may conduct a research programme to determine business patterns, economic impacts of tourism and visitor profiles. As an umbrella organisation for a wide range of businesses and community interests, it must often answer to a variety of 'masters' including local politicians, the **hotel** and **restaurant** operators and other local agencies, as well as local residents. Balancing the needs and expectations of these diverse interest groups is one of the major challenges for most local tourism organisations.

See also: city office

Further reading

Bramwell, B. and Rawding, L. (1994) 'Tourism marketing organisations in industrial cities: organisations, objectives and urban governance', *Tourism Management* 15(6): 425–34. (Reports on local tourism organisations in five British industrial cities.)

Judd, D.R. (1995) 'Promoting tourism in US cities', *Tourism Management* 16(3): 175–87. (Shows steps that US cities are taking to develop and promote tourism.)

ALASTAIR M. MORRISON, USA

localisation curve

Localisation curves provide a graphical representation of the level of concentration of a **tourist** activity. The curves utilise location quotients which are ratios of, for example, bednights to geographical area. Where there is a perfect match between the two variables, the curve would be represented by the diagonal between the two axes. Increasing divergence between the curve and the diagonal displays the level of concentration.

CHRIS COOPER, AUSTRALIA

location

Location is a relatively broad term which refers to a geographic area containing one or more potential **sites** where a business can be established. Unlike manufacturing industries which can transport their goods to **markets**, service industries must be situated in convenient proximity to their customers. Thus, location decisions are more important for service businesses, highlighted by the frequently cited quote ascribed to Conrad Hilton that 'there are three things important in the hotel business: location, location and location'.

For the firm, location selection is an important long-term strategic decision that often involves substantial capital investment and commitment to the market in that area. The primary consideration when selecting a location is its potential to return a good profit on the firm's **investment**. The governments regulating a location must have a favourable disposition towards business investment, as reflected in their **rules** regarding site **development**, taxation and domestic labour. It is important that a location's political environment is stable and has an economy with predictable **growth**. Additional factors affecting international locations include foreign labour **policy**, import duties, **taxes**, repatriation of profits, and stable, favourable **foreign exchange** rates.

A location which has available a low-cost labour pool and supplies is more profitable than one where the inputs must be imported. Similarly, the availability of utilities, **transportation** and communication **infrastructures** reduces the capital investment and operating expenses necessary to do business. From a competitive viewpoint, an area must have one or more sites available for development and must not be saturated with competitors relative to the size of the market being served.

The selection of a location for tourism purposes must meet the consumer's demand for **attributes** such as **climate**, terrain, activities, culture and uniqueness. The area should also have a sufficient selection of appropriate **accommodation** and transportation service providers. Consumers must perceive the location to be safe, with good social **attitudes** towards **tourists** and adequate medical facilities available in case of accident or illness. Except where transportation itself is enjoyable (cruises, motor coach tours and so on), a location which minimises the cost and rigour of travel is preferable to a similar one which does not.

Given the complexity and strategic importance of location decisions, some firms employ consultants who specialise in location decisions made by using qualitative and/or quantitative research. The latter methods of location assessment are described in the operations management literature and texts. A relatively simple approach called the factor rating method consists of the sum of the products of factor important weights and factor scores. The largest total sum indicates the most preferred location.

See also: site analysis; trading area

JOACHIM BARTH, CANADA

location quotient

The location quotient compares a **region**'s share of some **activities** with its share of some other aggregate. For example, if a region receives 20 per cent of tourist arrivals to a nation and has 30 per cent of the nation's population, its location quotient for visitor arrivals (with population as the base) would be 20/30 or 0.67.

STEPHEN SMITH, CANADA

locational analysis

Locational analysis deals with questions about why land uses occur where they do or where certain uses should be located for maximum **efficiency**. Most such analyses can be assigned to one of two schools of thought: **transportation** cost theory or central place theory. The former stems from von Thünen's (1875) modelling of agricultural land use around a village. The core of the **model** was the difference between the costs of transporting various goods versus their price at the **market**. This work eventually became the inspiration for tourism **development** models such as those of Miossec (1977) on the effect of transportation links in the shaping of destinations.

Central place theory is associated with Christaller (1933), who conceptualised urban areas as 'central places' or concentrations of economic and political activity. This theory was a major force in economic **geography** and regional science into the 1970s. Among the contributions this work made to the **geography** of business activity has been the formalisation of the concepts of threshold populations (the minimum number of people needed to support a given **industry**), hinterlands (the area containing the threshold population) and hierarchies (the pattern of a few large cities offering many goods and services and more offering fewer). Models of tourism development from the perspective of central place theory focus on the interaction of destinations (as peripheral areas) and urban markets (central places) such as Britton (1980).

Locational theory has developed a number of concepts of general utility in tourism, four of them listed here. First, location is important in determining the success of a firm. Second, the choice of a good location involves trade-offs among transportation and production costs, **resource** and labour availability, market accessibility and land costs. Third, certain types of businesses do well if they avoid locating close to competitors, others benefit from such closeness, and still others are indifferent. Finally, population size and the number and location of competing firms limit the potential for business growth.

References

Britton, S.G. (1980) 'A conceptual model of tourism in a peripheral economy', in D.G. Pearce (ed.), *Tourism in the South Pacific*, Christchurch: Department of Geography, University of Canterbury, 1–12.

Christaller, W. (1933) *Die zentralen orte in Suddentdeutschland*, Jena: Gustav Fischer.

Miossec, J.M. (1977) 'Un modele de l'espace touristique', *L'Espace Geographique* 6(1): 41–8.

Von Thünen, J.H. (1885) *Der isolierte staat in beziehung auf landwirtshaft und nationalokonomie*, 3rd edn, Berlin: Schumacher-Zarchlin.

STEPHEN SMITH, CANADA

longitudinal study

Longitudinal studies represent an examination of phenomena over a period of time. Incorporating a temporal element may involve retrospective study (a time series of data which exists from the past to the present) or prospective study (establishing a system for data collection from the present into the future). The value of this approach in tourism research can range from increasing an understanding of **development** processes in general, to the changing **experience** of tourism for the individual. The time scale employed could vary from the length of a holiday to many years.

Longitudinal studies as a particular technique stems from the growth of developmental **psychology** in the late nineteenth century and its concern with studying changes in **behaviour** over the lifespan. In the 1930s, a series of longitudinal studies were established to study child growth and behaviour in the United States, some of which are still being conducted. Problems common to most longitudinal studies include the selection of appropriate measures at the start of the study and the maintenance of comparable data collection over a time period. For instance, assessing the longer term **impacts** of tourism on communities and environments involves baseline data being gathered and then systematically traced over time. Few tourism projects include this form of research.

A technique related to longitudinal studies has been the time series approach developed by econometrics, where a set of data is ordered in time with observations made at regular intervals in order to isolate **trends**, cycles and periodic and random events. Certain forms of statistical tourism data are amenable to this type of approach and have been used in economic studies of tourism (see also **statistics**). Longitudinal and time series approaches can also be related to wider concepts of tourism development. The resort life cycle, for instance, depends for its validity on adequate, comparable data which is available over a considerable period of time. Longitudinal studies of tourism can also, in theory, be applied at the individual level. The changing role of tourism within a person's life cycle (from youth to old age) has been little researched. This approach would require a long-term, prospective research methodology, which has been used in health studies and could be applied to the field of tourism.

See also: change; economics; impact

Further reading

Pearce, D. (1989) *Tourist Development*, 2nd edn,

London: Longman. (Specifically addresses longitudinal studies in tourism.)

Préau, P. (1983) 'Le changement social dans une touristique de montagne: Saint-Bon-Tarentaise (Savoie)', *Revue de Géographie Alpine* 71(4): 407–29, and 72(2–4): 411–37. (An example of a longitudinal study of tourism impacts.)

JOHN TOWNER, UK

long range

Long range is a time frame for tourism strategic **planning** where **demand** or **supply** conditions are predicted or proposed for a considered distance into the future. It is often related to national or regional provision of tourism-related facilities, **services** and **infrastructure** in response to a forecast of tourist demand. Usually this is integrated into national **policy**. Some tourism industry suppliers have long-range plans and forecasts.

See also: longitudinal study

RUSSELL ARTHUR SMITH, SINGAPORE

loyalty

Customer loyalty is a preoccupation of marketers in all areas, including tourism. Loyalty represents a construct central to the study of tourist **behaviour**. It is a well-known fact that keeping customers is less expensive than creating new ones. The challenge is to understand and appropriately use the factors that will determine customer loyalty. What triggers the emotional link so different to being just a regular purchaser is an important focus. Customer **satisfaction** is one thing, loyalty is another. Tourism companies must find out what drives loyalty and customer retention, which are two different issues. It is the emotional factor that appears to be so important in developing loyalty drives. Loyalty is much more than just repurchase; it is people's reaction to the brand (see **branding**), its **image** and its past performance.

Past research has offered different conceptualisations of loyalty. However, the view most commonly held today comes from the distinction between intentional and spurious loyalty. Some researchers suggest loyalty indexes should be based on a mix of attitudinal and behavioural measures. Customer loyalty should be viewed as the strength of the relationship between an individual's relative **attitude** and repeat patronage. The relationship is seen as mediated by social norms and situational factors. Cognitive, affective and conative antecedents of relative attitude are identified as contributing to loyalty, along with motivational, perceptual and behavioural consequences. Retention is seen as being preceded by loyalty, and service **quality** and/or customer satisfaction/dissatisfaction are widely seen as antecedents of customer loyalty. This can be seen as an attitudinal construct, whereas customer retention as a behavioural one.

Loyalty schemes should also be used to ensure that the right customers are targeted. Lifetime tracking of customer **behaviour** and purchase patterns can bring unexpected results. Tourism companies also need to research key drivers of defection or dormancy. Issues such as loyalty and defection are much more emotional than physical. Defection is often internally defined as the loss of a customer, but a customer may see it completely differently. It appears also that when it comes to customer loyalty, researching staff attitudes is just as important as knowing what the customers think.

Further reading

Dick, A.A. and Basu, U. (1994) 'Customer loyalty: towards an integrated conceptual framework', *Journal of the Academy of Marketing Science* 22(2): 99–113.

Dimanche, F. and Havitz, M.E. (1994) 'Consumer behaviour and tourism: review and extension of four study areas', *Journal of Travel and Tourism Marketing* 3(3): 37–57.

Pritchard, M., Howard, D.R. and Havitz, M.E. (1992) 'Loyalty measurement: a critical examination and theoretical extension', *Leisure Services* 14: 155–64.

LUIZ MOUTINHO, UK

ludic

The ludic aspects of tourism or tourist **behaviour** can be best understood through the definition of **play** presented by Johan Huizinga in his classic work *Homo Ludens*. Play is:

> considered as a free activity standing quite consciously outside 'ordinary' life as being 'not serious', but at the same time absorbing the player intensely and utterly. It is an activity connected with no material interest, and no profit can be gained by it. It proceeds within its own proper boundaries of time and space. It promotes the formation of social groupings which tend, among other things, to stress their difference from the common world by disguise or other means.
>
> (1950:13)

If tourism, especially sunlust tourism or **sun, sand, sea and sex** tourism is considered an activity where **tourists** go through a voluntary spatiotemporal transition from ordinary life to something that resembles the liminal, and engage themselves in playful and hedonistic activities in bounded ritual spaces (**resorts**), the analogy between play and tourism comes forth clearly. When tourists travel to their playgrounds, they not only cross a mental boundary between the ordinary and non-ordinary, but they also travel across space from the context of the ordinary everyday life to something that lies outside it (see **rites of passage**).

Many of those theorising on tourism and tourist behaviour (including Erik Cohen, Nelson Graburn, Jafar Jafari and James Lett) seem to propose that in modern secularised and differentiated societies tourism as play with its liminoid aspects has taken over functions that religious **rituals** with their liminal aspects had in the relatively undifferentiated tribal societies. Following this line of argument, keeping in mind that play is a crucial part of human culture, it can be understood how **mass tourism** to destinations with no traces of **authenticity** in the paradigmatic sense of the word can be rewarding, recreating, uplifting and even provides deep **experience** for tourists as they play this game outside the context of everyday life.

Further reading

Cohen, E. (1985) 'Tourism as play', *Religion* 15: 291–304. (Examines the concepts of play, rites of passage and liminality in connection to tourism.)

Huizinga, J. (1950) *Homo Ludens: A Study of the Play Element in Culture*, Boston: Beacon Press.

Lett, J.W. (1983) 'Ludic and liminoid aspects of charter yacht tourism in the Caribbean', *Annals of Tourism Research* 10(1): 35–56.

TOM SELANNIEMI, FINLAND

magazine

Magazines can offer tourism promoters the unique advantages of unlimited word space, hi-fi colour, reach of either mass or highly selective niche **markets** and, in some instances, the opportunity to run advertisements alongside specialised editorial features. Magazines occur as discrete publications but also as inserts in weekly newspapers. Destination **image** advertising is often best suited to magazines.

A.V. SEATON, UK

magic

Magic is manipulative power producing extraordinary effects for a client. Tourism promotion, through the use of special words and script, promises to transport and instantly transform potential **tourists** into persons other than themselves (see also **promotion, place**). Although the word 'magic' is sometimes employed by national tourism organisations (such as 'Magic of Spain'), more often it is used in the naming or renaming of destinations.

GRAHAM M.S. DANN, UK

Malaysia

A Southeast Asia country which has vigorously and successfully developed tourism, Malaysia is a federation of thirteen states with two regions, Peninsular Malaysia and East Malaysia. Approximately 80 per cent of the population (20.7 million in 1995) lives in Peninsular Malaysia. It is a multiracial and multireligious society. The country experiences a monsoon tropical **climate**.

Tourism **development** is undertaken within the context of federal government **policy**. The leading policy, 'Vision 2020', has the aim of creating a fully developed nation by 2020. Five-year national plans (such as 'Seventh Malaysia Plan 1996–2000') are key instruments in realising this goal, and the tourism **industry** is an important component of the plan. Tourism **policy** formulation is undertaken by the federal Ministry of Culture, Arts and Tourism. Under this authority is the Malaysian Tourism Promotion Board, which is responsible for **marketing** and research. Strong and effective **partnerships** have been forged between the public and private sectors for tourism development and operation.

International tourist **demand** has grown from 2.5 million in 1981 to 7.5 million in 1995. Most tourists go to Peninsular Malaysia (95.4 per cent in 1995). The majority of international visitors are residents of Asian countries (88.2 per cent in 1995), with **Singapore** being the major market (60.8 per cent). Malaysia has extensive marine, coastal, mountain and forest **resources**, as well as cultural heritage attractions which are related to **indigenous** and immigrant cultures. These are being successfully applied to tourism development. Tourism-related **infrastructure** is being upgraded continuously. The construction of the north–south expressway in Peninsular Malaysia and other highways is greatly enhancing touristic access.

The new Kuala Lumpur International Airport is the primary air **gateway**. In 1995 there were 76,373 **hotel** rooms, of which 82.5 per cent were in Peninsular Malaysia. Many attractions such as **theme parks**, **museums** and events have been developed.

Federal **policy** and institutions are well evolved. State-level tourism institutions are in the process of being developed in terms of structure, expertise and funding. With rapid economic **growth**, a major challenge becomes the protection of important natural and cultural **heritage** resources. For **ecologically sustainable tourism**, proactive measures will be required for **conservation** and, where necessary, regeneration. Related to this is the issue of regulation of development and operation. Stronger regulatory frameworks are needed to achieve and maintain a competitive tourism **product**. **Human resource development** is also of major importance in a sector which has experienced significant labour and expertise shortages.

RUSSELL ARTHUR SMITH, SINGAPORE

Malta

The small island country of Malta is situated at the centre of the Mediterranean Sea, 93 km to the south of Sicily and 290 km to the north of Tripoli. The Maltese archipelago consists of Malta, Ghawdex and Kemmuna, covering a total area of 320 square km. It has an ethnically homogeneous population numbering slightly above 377,000. Though almost all Maltese are bilingual, they have their own native Semitic language. The country's millennial and rich **history**, together with its temperate **climate**, has made it a popular destination for cultural and **resort** tourism. The number of annual arrivals exceeds 1.2 million.

JOE INGUANEZ, MALTA

management

Tourism management can be analysed at four levels: scope, ownership, industry sector and management function. At the first level, those who are concerned with the macro-effects of the tourism industry have analysed its consequences on the economy, the ecology, and the sociocultural milieu of the host community. Economists have developed mathematical models to estimate the direct and indirect **impact** of income injected by tourists into the national, regional, or local economies. Ecologists, geographers and regional planners have studied the negative effects of tourism on the physical environment. Regional planners have attempted to develop models of physical planning and design of tourism regions that would minimise the above negative impacts and preserve the quality of life of the local community. Further, sociologists and anthropologists have examined the real and perceived consequences of tourism on host communities as a result of host–guest interaction (see **host and guest**).

At the ownership level, the provision of goods and services for tourists away from home is normally done by both public and private enterprises. With the exclusion of some centrally planned economies where the state owns and operates tourism enterprises, public tourism organisations normally devote most of their efforts to the function of **marketing** and promotion of this industry in their region (see also **promotion, place**). Public organisations can be found at the national, regional and local levels. It is private enterprises which make the bulk of the industry and specialise in the provision of commercial services to tourists. They range from large **multinational firms** who own hundreds of properties throughout the world to family owned and operated hotels and restaurants.

At the third level, tourism is composed of several sectors, some of which have grown to industries in their own right and six of them are outlined here. One, the **accommodation** sector includes hotel and motel enterprises and is concerned with managing customer **demand**, ensuring customer **service**, protecting assets, aiming at **quality**, improving employee performance, increasing **productivity** and achieving satisfactory levels of **return on investment**. The internal organisation of this sector includes front office, food and beverage, guest services, maintenance and engineering, security, reservations, human resources management, accounting and marketing.

Second, the food service sector provides tourists with foods and drinks while away from home. But not all food service properties are restricted to tourists, with a majority of them serving local residents. Food service operations can be divided by type and ownership and the latter can be either independent or franchise. The most important aspects of the management of foodservice operations relate to the service delivery system, production planning, consumer food preference, menu planning, physical design, nutrition, purchasing, cost control and marketing.

Third, the **transportation** sector in tourism includes only those passenger services which transport tourists between their origin and **destination**, as well as those used by tourists at the destination for **sightseeing** and internal travel. Commercial transportation services can be divided into land, sea and air. The former consists of rail, motor coach (buses), rental cars, taxi and limousines. The relative magnitude of each of the above and their importance to tourism varies from country to country. Rental cars are by far the fastest growing commercial land transportation mode in most destinations. Limousines are used mostly by business tourists to and from the airports, while taxi services are used by tourists as a mode of intra-city transportation as well as coming from or going to airports, seaports and railway stations. Commercial sea transportation services are dominated by cruise lines, a mode which is currently experiencing a phenomenal growth. Major cruises operate in the Caribbean, Mediterranean, Pacific, Northern Europe and Alaska regions. Most modern ships are floating luxury resorts which provide every conceivable amenity or service. These vessels can accommodate anywhere between 100 and 4,000 passengers. The next mode, **airline** operations, can be singled out as one which has totally revolutionised the tourism industry and made travel to long-haul destinations accessible, inexpensive and comfortable. The typical airline organisation consists of the line functions of engineering and maintenance, flight operations, marketing and services. Airline staff functions consist of finance and property, information services, personnel, corporate communications, economic planning, legal and medical.

Fourth, the travel agent and tour operation sector is comprised of those intermediary businesses which sell individual and combined tourism goods and services. Travel agents sell, among other things, airline tickets and hotel rooms and are rewarded through commissions received from suppliers. To illustrate, the sector supplies information and advises on destinations, **itineraries** and facilities; sell tickets on any mode of transportation; sells insurance policies for passengers and luggage; provides clients with accommodation reservations; arranges for excursions and visits to various attractions; and assists with or supplies a range of miscellaneous services such as visas, money orders, tickets to theatres, **museums** and the like. Most travel agencies are small and employ less than ten individuals. Their crucial management aspects involve marketing and sales, delivery of service, human resources management, **accounting**, **budgeting** and office automation. Still within this sector, tour operators (tour wholesalers) combine various tourism components, goods and services, and sell these through their company, retail outlets and travel agents. The most important managerial aspects related to these tour businesses are selection of destinations, planning, preparing and marketing **vacation** tours.

Fifth, the conventional and meeting sector is concerned with meeting, convention, congress, **trade show** and exposition functions or businesses. These gatherings, which vary in size from a few dozen participants to tens of thousands, can be regional, national, or international in scope. The actors in the convention and meeting sector are **meeting planners** and convention managers, trade shows, hotels, convention and visitors bureaus, exhibitors, exhibit designers, transportation services, exposition service contractors, destination management companies and food service companies.

Sixth, theme park and attraction enterprises constitute one of the newest sectors of the tourism industry. They represent particular built locations where for a single admission fee visitors are offered a mix of recreation and entertainment opportunities. Today most such attractions are centred on a unified theme or motif such as history, future, culture, geography, fantasy and others. The most important issues in theme parks relate to human resources management, consumer behaviour, **forecasting**, new product development, and maintenance and engineering.

Finally, in addition to the scope, ownership and industry sector level, there is the managerial function level. The most important such functions in the field of tourism management are marketing, financial management, legal aspects, management information systems and human resources management. In principle, the field of marketing as applied to tourism enterprises is very similar to the marketing of services. Tourism marketers are concerned with market segmentation, selection of target markets, **branding** and branding strategies, **pricing**, market positioning, effectiveness of the promotional mix, market feasibility, **motivation**, tourist **decision making behaviour** and **satisfaction**.

Financial managers in tourism enterprises are involved in the process of making decisions by interpreting and analysing appropriate data. They are directly involved in raising capital funds, are responsible for **asset management**, and allocate funds through the preparation of capital and operations budgets. Those who are involved in managerial accounting are responsible for recording, summarising and interpreting financial data. In most hospitality businesses, recording and summarising are accomplished through the use of uniform system of accounts. This process offers a formal structure within which financial data is accumulated and organised for the reporting of financial position and operation results.

Tourism enterprises of all types are affected by the legal system in which they operate. All legal systems regulate, to one extent or another, supplier–consumer relations, commercial relations, competitive marketing relations and international relations In most countries, tourism operators have legal obligations to their guests such as receiving and accommodating them, providing safe hotel rooms and so on. To prevent the occurrence of legal problems, tourism enterprises train their employees in the legal aspects of their business.

To increase productivity, most tourism enterprises own and operate management information systems or **property management systems**. These systems are composed of computer hardware and specialised software that management uses to operate their property in an efficient manner. Information systems are used for a multitude of tasks such as reservation, guest registration, accounting, purchasing and inventory control, scheduling, energy management and manpower planning. Related to the latter, the management of tourism human resources incorporates the functions of recruitment and selection of employees, orientation; motivation, communication, **leadership**, training and development, administration of wages, salary and benefits, job analysis, job evaluation, performance evaluation, promotion, health and safety; **collective bargaining** and other related concerns.

See also: cross-cultural management; disciplinary action; emergency management; environment management systems; environmental management, best practice; event management; facilities management

Further reading

Jones, P. and Pizam, A. (eds) (1993) *The International Hospitality Industry: Organizational and Operational Issues*, New York: Wiley.

Khan, M.A. (1991) *Concepts of Foodservice Operations and Management*, 2nd edn, New York: Van Nostrand Reinhold.

Nebel, E.C. (199) *Managing Hotels Effectively*, New York: Van Nostrand Reinhold.

Teare, R. and Olsen, M. (eds) (1992) *International Hospitality Management: Strater in Practice*, New York: Wiley.

ABRAHAM PIZAM, USA

management accounting

Management accounting in the hospitality sector is the communication of accounting and financial information to internal managers within its enterprise, and the evaluation of the latter from a financial perspective. Most large operations employ a controller for this function, a management accountant familiar with all aspects of the business. Therefore, **hospitality** management accountants participate in the decision-making processes. They also develop systems of internal control designed to increase the operating **efficiency** of the business and reduce fraud.

Management accounting differs from financial accounting in that the latter generates and reports accounting information that is used internally by management and accountants, and externally by stockholders, bankers, creditors and brokers. The former links very closely with cost accounting in the analysis and interpretation of financial statements, cash flows and cost–volume–profit analysis. Cost accountants are responsible for assigning costs to goods and **services**. This activity drives selling price of an enterprise's goods and services. Hospitality management accountants address the issue of whether to take a purchase discount and compute the effective interest rate on loans, and are responsible for cash management. They supervise cashiers, cost controllers and the night audit function in hotels, and it is they who prepare appropriate financial reports and are responsible for compliance with government regulations and **tax** laws.

Hospitality management accountants can earn the Certified Hospitality Accountant Executive designation from Hospitality Financial and Technology Professionals or the Certified Management Accountant designation from the Institute of Management Accountants. These professional designations require certain educational and professional requirements and the passing of an examination. Many hospitality management accountants also are Certified Public Accountants. All of these professional designations require a hospitality management accountant to maintain integrity, avoid conflicts of interest and communicate information objectively.

Other relevant areas to the hospitality management accountant are break-even point analysis, budgetary controls, **budgeting**, **corporate finance**, expense estimation, feasibility studies, **financial objectives**, **food** and beverage cost analysis, hospitality information systems, investment decisions, **payroll cost analysis**, **profit centre** analysis, **profit sensitivity analysis**, **profit variance analysis**, **property management systems**, return on **investment** and revenue **forecasting**.

Further reading

Coltman, M.M. (1994) *Hospitality Management Accounting*, 5th edn, New York: Van Nostrand Reinhold.

Ilvento, C.L. (1996) *Profit Planning and Decision Making in the Hospitality Industry*. Dubuque: Kendall Hunt.

Needles, B.E., Anderson, H.R. and Caldwell, J.C. (1994) *Financial and Managerial Accounting*, 3rd edn, Boston: Houghton Mifflin.

Schmidgall, R. (1997) *Hospitality Industry Managerial Accounting*, 4th edn, East Lansing, MI: Educational Institute of the American Hotel and Motel Association.

STEPHEN M. LEBRUTO, USA

management contract

The management contract is a **business format** which separates ownership from operation. In the **hotel** industry it has provided the opportunity for much-needed capital to fund the **demand** for new construction in world **markets**, while creating the vehicle for such tourism management companies to expand their networks and market shares with reduced exposure to **investment** and political **risks**. The **growth** in contracts has also been driven by hotel owners' need for experienced and established operators for their own peace of mind and to satisfy investor's demands.

There is no standard definition of a hotel management contract, but it is usually defined as a formal arrangement under which the owner of a hotel or another tourism business employs the services of an operator to act as his/her agent to provide professional management, in return for a fee. The operator assumes full responsibility for the management of the business, while the ultimate legal and financial responsibilities and rights of ownership of the property, its furniture and equipment, its working capital and the benefits of its profits (or burden of its losses) remain those of the owner. The owner may be a private individual, a financial institution, a real estate company or a government. The operator is most likely to be an established hotel chain offering **marketing** strength, brand names, bargaining power, systems and procedures, project design and management, technical services, **training** and management development. However, the rate of growth in the number of independent hotel management companies is increasing; these include groups that do not have their own international brand and reservations systems and who operate hotels for a variety of owners.

The owner usually seeks an effective return, the contractor an effective earnings stream. Typically, the fee structure is in two parts: a base fee of around 3 per cent of hotel turnover and an achievement fee of around 10 per cent of gross operating profit or earnings before debt, interest and tax. Economic changes impacting upon the hotel industry, accelerated competition amongst operators worldwide and greater performance demands by owners and lenders are leading to adjustments in contract provisions and increased owner bargaining strength.

Further reading

Sangree, D.J. and Hathaway, P.P. (1996) 'Trends in hotel management contracts', *The Cornell Hotel and Restaurant Administration Quarterly*, October: 26–37. (Discusses how a 1996 survey finds a vastly different landscape for hotel management contracts compared to the 1980s.)

ANGELA ROPER, UK

Managing Leisure

To inform and stimulate discussions relevant to **leisure** management, *Managing Leisure: An International Journal* publishes articles, book reviews, and research notes and conference reports dealing with this theme. It is designed to appeal to academics, practising managers, consultants, politicians and students. Its broad scope accommodates coverage of arts, media, countryside **heritage**, **hospitality**, sport (see also **sport, recreational**), tourism, **management**, **human resource development**, facility and amenity management and the like. First appearing in 1996, the journal is published quarterly by Chapman and Hall (ISSN 1360–6719).

RENE BARETJE, FRANCE

manpower development

Manpower development is an umbrella term which is used to cover a number of activities, including **education**, **training**, staff development and **human resource development**. At both macro (national) and micro (individual firm) levels, manpower development relates to **policies** and practice designed to enhance the skill base of the tourism industry to enable it to meet the expectations of all stakeholders.

TOM BAUM, UK

map

Maps are used in tourism **advertising** at the pre-trip stage as a complement to the invitation to travel, and during the visit as a zoning device for **itineraries** and **tours**. Unlike the maps found in atlases, which are purely informational in content, the cartography of brochures, pamphlets and guidebooks is additionally a medium of **social control**. For instance, **tourists** are encouraged to sightsee designated attraction spots, while leaving out others which have been significantly omitted. Similarly, locations where there is a tour operator **hotel** presence are included, whereas rival **resorts** of equal or greater host society importance may not even feature. For example, in Japanese **domestic tourism** (where such maps have been produced for the last 350 years), not only are *meibutsu* (famous things) highlighted, but so also are items which the tourist has a duty to bring back in the form of *omiyaje* (presents).

Although the study of maps is under-researched in tourism, several characteristics are worthy of note and further investigation, including the use of colour and its attempt to mould client **expectation** (coating former colonies in pink in British publicity, for instance, may conjure up images of Empire); the apparent disregard of political realities (such as the 'Turkish question' in maps of

Cyprus and the independence of Belize in maps of Guatemala); the employment of icons to underpin tourism's child-oriented themes (for example, sun and **fun**); the tendency of specialist **tour operators** to contextualise the **destination** area in relation to nearby countries, while generalists often omit the latter since they compete for custom; disproportionate sizing of key regions, cities, towns and villages that distorts the scale of the map and their relative importance; and the appendage of charts comparing temperatures and hours of sunshine in the destination with those of the origin environment as a motivational device for travel.

With the growing significance of **nostalgia** in tourism **motivation**, maps are increasingly devoted to **history** and **heritage**. There are thus maps of Roman Britain, Viking and medieval York, Dickens's London and William Shakespeare's London, published by such organisations as Ordnance Survey, Geomex and Coutt's Heritage Print and Design.

Further reading

Dann, G. (1996) *The Language of Tourism: A Sociolinguistic Perspective*, Wallingford: CAB International. (Particularly see pages 156–9.)

Graburn, N. (1995) 'The past in the present in Japan: nostalgia and neo-traditionalism in contemporary Japanese domestic tourism', in R. Butler and D. Pearce (eds), *Change in Tourism, People, Places, Processes*, London: Routledge, 47–70.

GRAHAM M.S. DANN, UK

marginality

Marginality is the state of being on the periphery or fringe of a **location**, culture, society or situation. Early writings on cross-cultural interaction by Stonequist (1937) examined the difficulties people encounter when they **experience** crises of belonging. Those who (aspire to) have membership of two cultural groups with incompatible norms will often find themselves at the margins of both. Unless the conflict is resolved, the marginal person will vacillate between the two cultures never really fitting into either.

Interactionist work in the tourism literature on hosts and guests has documented the existence of marginal types caught between cultures. Tourism generates opportunities in both the formal and informal sectors of the economy, which can result in individuals drifting away from or abandoning their traditional way of life. Smith (1989) explores the concept of 'marginal man' with reference to the indigenous population of Alaska, where ageing marginal members of the Eskimo culture have entered the marketplace to sell their former **lifestyle**. Those who take mediating positions in the industry, such as culture brokers, performers, street/beach vendors and prostitutes, or those who model their **behaviour** and dress after tourists (see **demonstration effect**) risk abandoning local traditions and thereby alienating themselves from their own culture. Never fully succeeding at moving into the tourist environment, they are also seen as trying to distance themselves or make a profit from their own culture, thus leading to marginalisation. Smith identifies the hallmark of the marginal man as being bilingual along with personality traits of charisma and wit. As tourism reaches into more remote areas, these people will be relied upon to a greater degree.

Marginality has also been identified in the development stage of the destination **life cycle** (Butler 1980). During this stage, control of tourism passes from local companies to large-scale external organisations, thereby marginalising the local residents and businesses. In a similar fashion a **destination** can be marginalised if it falls victim to negative publicity, political instability, increased competition or the alteration of a major **transportation** route, all reducing the number of tourist arrivals.

References

Butler, R.W. (1980) 'The concept of a tourist area cycle of evolution: implications for management of resources', *Canadian Geographer* 24(1): 5–12.

Smith, V. (1989) 'Eskimo tourism: micro-models and marginal men', in V. Smith (ed.), *Hosts and Guests: the Anthropology of Tourism*, 2nd edn,

Philadelphia: University of Pennsylvania Press, 55–82.
Stonequist, E.V (1937) *The Marginal Man*, New York: Scribner.

DAVID J. TELFER, CANADA

marker

There have been several efforts to construct general theoretical models of the tourist **experience** (MacCannell 1976; Urry 1990). MacCannell links his theory of tourism to general **semiotics**, positing a correspondence between the **attraction** and the semiotic signified, and the marker as 'signifier'. In restricted usage, 'marker' refers to the signage and informational plaques found at a place visited by tourists. Examples would include the brass plaques on famous old houses, the signs on the cages at zoos which give the names of the animals in the cage, and perhaps some additional information about their **behaviour**, habitat and the like. **Tourists** do not necessarily arrive at destinations knowing everything they might know to make their experience meaningful and interesting. The quality of their experience is in part a function of the ways tourism settings are marked. Too much information can detract from the object that is marked. Too little can impoverish the experience. The aesthetics of marking itself can alter the entire experience.

MacCannell expanded the use of marker to refer to all material which makes reference to a **destination**, whether this is found on-site or off-site. The fame of an attraction (such as the Eiffel Tower) is a function of the quantity of its off-site markers. The power of world-class attractions stems from the fact that they are the subject of a huge proliferation of textual and visual **images** ranging from amateur photos in family albums to being the backdrop for novels and major motion pictures (like Mount Rushmore in Hitchcock's 'North by Northwest'). This proliferation of images is usually not commercially driven by a particular agency responsible for promoting the attraction. It happens outside the framework of focused entrepreneurial activity, having many sources, most of which seem incidental, but it is crucial to the drive behind **sightseeing** which is a collective human activity even if the sightseers are not organised by anything except the perceived fame of the places they visit. The huge proliferation of off-site markers establishes the tourists' desire to visit a major attraction, to see it for themselves. Thus, attractions which have achieved fame naturally act as magnets for commercial attractions and set up the situation in which the tourists may discover lesser sights for themselves.

The act of sightseeing is technically a 'marker–sight–marker transformation'. When they are finally in the presence of the sight, tourists exchange previous markers for actual experience; when they leave, they now possess markers of their own in the form of a memory, a story, a photograph or a **souvenir**. Thus, the act of sightseeing is a key part of the formation of collective experience and a collective conscience in societies which have grown to the point where it is no longer possible for everyone physically to get together. At least everyone can 'get together' in principle, one tour group at a time, at the most sacred attractions and have a 'common memory' thereby.

See also: advertising

References

MacCannell, D. (1996) *The Tourist: A New Theory of the Leisure Class*, New York: Schocken Books.
Urry, J. (1990) *The Tourist Gaze: Leisure and Travel in Contemporary Societies*, London: Sage.

DEAN MacCANNELL, USA

market

In an everyday sense, the term 'market' is often associated with its historical past as a place where traders meet buyers to exchange their wares. Developments in electronic communications (see **computer reservation systems**; **global distribution systems**) have meant that markets no longer need to be physical settings for trade to take place. Thus economics uses the market concept to mean a system by which goods and services are exchanged between buyers and sellers. This would include such powerful global networks as the

Internet, as well as traditional street markets in urban areas that may serve as attractions.

In market-oriented economies, the prices serve as the adjustment mechanism by rewarding sellers, rationing available supplies amongst buyers and relaying information forward from sellers, showing relative costs of production or scarcity value, and backwards from customers, showing their relative preferences by what they are prepared to pay. By this means, markets are able to deal with some of the fundamental issues in **economics**, including what should be produced, the method of production and for whose use. Where markets do not operate, as in centrally planned economies, then it is customary for the state to undertake the regulatory function of matching the **demand** and supply for goods and services, thus displacing the price mechanism. For example, before the movement away from communist regimes, the primary emphasis of Eastern European countries was on **social tourism**, which allowed the population to benefit from subsidised holidays through workers' organisations and central government provision. The success criteria were based on visitor numbers and most tourism facilities were heavily subsidised.

For markets to work smoothly, it is necessary for both buyers and sellers to be well informed of the choices open to them. In industries like tourism, which are fragmented into large numbers of consumers and producers, the transaction costs of obtaining the required information is often far too high for individuals acting on their own. This gives economic opportunities for culture brokers or intermediaries, such as **tour operators**, to package and undertake **marketing** activities in order to realise the potential demand and supply in tourism. Because in this industry the consumer is buying the product unseen, there is concern amongst governments that suppliers should not misrepresent their wares to the market, and hence most countries have **legislation** giving consumers the right of redress if their purchases were not as specified.

STEPHEN R.C. WANHILL, UK

market analysis

The systematic monitoring and analysing of **tourism** demand, in terms of the overall market in general and key **market segmentation** in particular, is known as market analysis. This process consists of measuring market volumes, market shares and market revenues, as well as profiling key elements of consumer needs. These elements include **trends**, developments and changes in tourist characteristics, **motivations** and usage of facilities.

MARTIN FRIEL, UK

market research *see* marketing; marketing research

market segmentation

Market segmentation is the prerequisite for selective market operation. It means splitting consumer (industrial) markets into homogeneous subgroups of consumers (businesses). It requires decisions of the marketing manager on the classification criteria, the size and the number of segments. Market segmentation is planned in conjunction with **product positioning**, as these two strategic issues depend on each other. There are two fundamentally different concepts of market segmentation. The first is a priori (or criterion) segmentation (see also **segmentation, a priori**; **segmentation, a posteriori**). This departs from a predetermined criterion such as purchasing volume or **loyalty**. In a priori segmentation, the marketing objective determines how to divide markets into smaller units. The second concept is a posteriori (or similarity) segmentation. This rests on the assumption that subgroups in a consumer population are homogeneous in terms of motives, **attitudes** and/or activities. This mental and behavioural homogeneity is likely to make them react to product offerings and promotional efforts in a similar manner. The most popular approach to master such a market situation is known as 'benefit segmentation'. It considers consumer groups with a markedly different pattern of benefits sought to be 'natural' segments in the market. From a behavioural science point of view, the notion of 'benefit' relates to the more general concept of **attitude**. **Benefits** desired or expected are attitudes towards particular consumption goals.

More recent segmentation strategies involve values and **lifestyles**. In particular, **vacation** style is regarded as a temporary lifestyle into which a tourist escapes from his/her everyday surroundings. A vacation style represents a cognitive and emotional state of mind as well as the accompanying behaviour. Tourism activities are also useful raw material for constructing a posteriori segments. Questions about the tourist's activities are customary in travel **surveys** (see also **survey, guest**). The concept of tourist roles also suggested for segmentation purposes is closely linked to tourism activities which are regarded to come up in symptomatic combinations. According to the marketing view of segmentation strategies, it is not the aim to detect the 'true' segment structure in the market. The analyst and the **marketing** manager continuously invent new consumer typologies and segmentation schemes. Being the first to practice such a scheme gives a company or a tourism **destination** a competitive edge.

See also: segmentation, a priori; segmentation, a posteriori

Further reading

Hsieh, S., O'Leary, J.T. and Morrison A.M. (1992) 'Segmenting the international travel market by activity', *Tourism Management* 13: 209–23.

Kamakura, W.A. and Novak, T.P. (1992) 'Value-system segmentation: exploring the meaning of LOV', *Journal of Consumer Research* 19: 119–32.

Yiannakis, A. and Gibson, H. (1992) 'Roles tourists play', *Annals of Tourism Research* 19: 287–303.

JOSEF A. MAZANEC, AUSTRIA

markets, contestable

Markets which appear to be oligopolistic, or even monopolistic, can be competitive so long as they are contestable. The mere threat of competition – as evidenced by unlimited and costless entry and exit – can cause even the apparent monopolist to behave competitively. The theory has been used to support a policy of continued **airline** deregulation, despite apparently increased levels of industry concentration.

KEVIN BOBERG, USA

marketing

In 1960 the American Marketing Association described marketing as the performance of business activities that direct the flow of goods and services from producer to consumer. During the following decades this notion became more abstract, and evolved from a business function to the business doctrine of market-oriented **management**. Later it became liberated from merely profit-oriented buyer–seller relationships. The generic concept of marketing applies to all sorts of transactions and exchange relationships between a producer or service provider and a purchaser or client. Non-profit organisations such as charities, religious communities, political parties and municipalities are also looking for members or customers; they offer or promise some benefit and get something in return. The fundamental concern of the marketing effort in all these organisations is to systematically (according to well-defined objectives and a marketing plan) influence other individuals' and groups' behaviour. The members of a target group (see **target marketing**) may fulfil any role relevant for the organisation. Thus, for an economic enterprise marketing also occurs on the labour market to recruit highly qualified personnel, or in capital markets to attract potential investors. Marketing is a social and managerial process, aimed at matching an organisation's capabilities and the wants of its clients so as to reach both parties' objectives. It rests on the premise that no organisation can survive unless it meets customer wants thereby acquiring a **competitive advantage** over its rivals. The need for marketing arises once there are alternatives giving the client or customer the freedom of choice.

The tourism **industry** was among the latest branches of the world economy to experience the shift from a seller's to a buyer's market. During the 1970s the range of its products and services gradually began to exceed **demand**. Marketing thought slowly began to penetrate managerial processes in all companies related to the industry such as **tour operators** and **travel agencies**, **theme parks**, **hospitality** businesses, **transportation** modes and the non-profit organisations responsible for promoting destinations. Given the competitive pressure on national and international levels, it became paramount for the providers of tourism services to take the initiative in influencing the **behaviour** of the consumers who play their roles as potential customers (guests, visitors). Marketing is a business function which requires a considerable amount of analysis and **planning**. The basic steps involved in a marketing plan are universally applicable. The principles are the same for all companies and organisations struggling for competitive advantage.

The marketing planning process runs through five stages: goal setting, situation review, **strategic planning**, instrumental planning and measurement of results. The **marketing objectives** are derived from more fundamental corporate ones and from an organisation's mission statement. The objectives may refer to economic criteria such as increasing the number of tourists or the sales volume, or maintaining market share, or to psychological variables such as raising the awareness level for a brand name, modifying a destination **image** or strengthening customer **loyalty**. The goal serves as the yardstick for the later measuring of results. A situation review implies a **marketing audit** and a **SWOT analysis** (covering strengths, weaknesses, opportunities and threats). If the auditing is to proceed according to theory, there must be an explicit set of assumptions regarding customer **behaviour** (see **marketing research**).

The SWOT analysis benefits from using a portfolio **model**. These models define succinct evaluation criteria to assess the market chances and the competitive position of the tourism services that a company (or organisation, tourism-receiving country or region) has to offer. In its simplest form, that of the **growth** share matrix, a portfolio model uses three assessment criteria: the market growth rate, the relative market share (own share compared to the toughest competitor's share) and the importance value of a product or service (or its contribution to the company's overall sales). More sophisticated portfolio models introduce two or more criteria to derive market attractiveness and competitive position scores. To construct such compound scores, the individual criteria must be rated by their strategic importance. Managerial judgements (soft data) are admissible in these rating exercises and subject to learning by trial and error. The most popular version of the multi-factor models is known as the industry attractiveness analysis or the directional policy matrix. The SWOT analysis is never limited to the past and current situation; it includes demand estimation and **forecasting**. Thus it is not just extrapolation of the historical development, but reflects managers' expectations depending on two or more scenarios. Alternative assumptions about a company's own strategies and competitors' reactions are inherent in these scenarios.

Strategic marketing planning involves two interrelated decision areas: **product positioning** and **market segmentation**. Positioning deals with optimising the attributes of a tourism **product** or service and with the way it is presented to the public. Segmentation refers to classifying the consumers according to the wants and expectations a service provider faces in the marketplace. The positioning and segmentation decisions cannot be made in a rational manner unless they are based on an explanatory model of consumer (tourist) behaviour.

Instrumental planning guides the implementation of strategies. The tools to execute marketing plans are called the four Ps (product, place, promotion and price) or the **marketing mix**. New product planning, product differentiation and the preservation of existing products are examples of product-related policies. The management of distribution channels, including the computerised reservation systems and the **global distribution systems**, establishes contact between tourism service providers and customers. Exploiting a service provider's own customer data (database marketing) is one of the methods of **direct marketing** with no intermediaries interfering in the producer–consumer relationship. The communications mix comprises an assortment of various sub-instruments such as media **advertising**, **public relations** or **sales promotion**. Salesforce management and **personal selling** contain a communications aspect (the message conveyed in a sales talk) and a distribution facet (the establishing of provider/customer contact). **Pricing** is the fourth instrumental area where the cost restraints of the service provider have to be brought into alignment with the customers' reservation prices (their willingness to pay up to a certain amount).

The tourism industry has to observe specific restrictions when implementing a marketing programme. Its products (such as **package tours**) consist of a number of product and service modules offered by a variety of service providers. None of them fully controls the final appearance of the product. Some of its ingredients (natural resources) may be completely predetermined and largely inaccessible to managerial action. The range of options for product planning and **quality** assurance is limited. In a resort, cooperation and harmonised action on a voluntary basis is required to compose and market a consistent bundle of products and services. As the services tend to lack autonomy and distinctiveness in their respective markets, a **branding** policy cannot be easily established and maintained. The company name or the name of a business conglomerate usually fulfils the brand function rather than the individual products or service packages. Non-profit organisations (particularly in European tourist receiving countries) are heavily engaged in the marketing of industry. They take full or partial responsibility for marketing support with typical regard to commissioning market research studies, to national, regional and local advertising, and to distribution (see **reservation**).

The measurement of results concludes the marketing **planning** cycle. The simple recording of aggregate figures (number of trips or guests, dollar sales, occupancy rates, **foreign exchange** earnings) must be complemented by thorough causal research (see **causal model**) to learn about market response. Regular guest surveys, for example, are routine sources to provide the necessary feedback information for marketing managers (see **statistics**; **survey, guest**). If the results are far off target, with no apparent explanation, management will need to have recourse to more specialised information collection. Using non-reactive methods of market research, such as observation or hiring phantom customers to report on deficiencies in the distribution system, or failures during the service encounter (mystery shopping) may reveal unexpected causes. According to the principles of **quality** management, the monitoring of customer satisfaction is a continuous process rather than a sporadic event (see **satisfaction, customer**).

Quantitative methods and advanced model building characterise the scientific approach to marketing. Consumer (tourist) behaviour research offers explanations ranging from conventional econometric and stochastic process modelling to computer simulations and neural networks (see **neurocomputing**). Many optimisation models are available for each marketing instrument, but fewer are capable of tackling double-instrument optimisation (such as price and advertising), and very few models consider the entire marketing mix. A comprehensive theory to encompass strategic (positioning plus segmentation) issues, together with the instrumental decisions of the marketing mix, is still out of sight.

See also: benchmarking; electronic promotion source; event marketing; relationship marketing

Further reading

Bagozzi, R.P. (1986) *Principles of Marketing Management*, Chicago, IL: Science Research Associates. (Gives proper credit to marketing research and methodology.)

Culture, P., Armstrong, G., Saunders J. and Wong V. (1996) *Principles of Marketing: The European Edition*, London: Prentice Hall Europe. (A new version of a marketing classic.)

Eliashberg, J. and Lilien, G.L. (eds) (1993) *Marketing: Handbooks in Operations Research and Management Science*, vol. 5, Amsterdam: North-Holland. (Reflects the state of the art in marketing science.)

McDonald, M. (1995) *Marketing Plans*, 3rd edn, Oxford: Butterworth-Heinemann. (A textbook recommended for beginners.)

Teare, R., Mazanec J.A., Crawford-Welch, S. and

Calver, S. (1994) *Marketing in Hospitality and Tourism: A Consumer Focus*, London: Cassell. (Discusses recent tools for analysis and planning and their use in managerial decision making.)

Witt, S.F. and Moutinho, L. (eds) (1995) *Tourism Marketing and Management Handbook (Student Edition)*, London: Prentice Hall. (Contains a collection of major marketing issues particularly relevant for the tourism industry.)

JOSEF A. MAZANEC, AUSTRIA

marketing audit

A marketing audit is a comprehensive, systematic, independent and periodic examination of a company's (or business unit's) marketing environment and activities. The intent is to determine problem areas and opportunities and to recommend an action plan to improve the company's **marketing** performance. This is a thorough and objective evaluation of a tourism organisation's marketing philosophy, goals, policies, tactics, practices and results. Such a comprehensive procedure can provide a valuable – and sometimes disquieting – perspective on the performance of the tourism company's marketing plans. A periodic review of marketing plans is invaluable both in identifying the tasks that the organisation does well and in highlighting its failures. Periodic review, criticism and self-analysis are crucial to the vitality of any operation. Marketing audits are especially valuable in pointing out areas in which managerial perceptions differ sharply from reality. Methods of conducting audits are almost as diverse as the tourism companies that use them. Some audits follow only informal procedures. Others involve elaborate checklists, questionnaires, profiles, tests and related research instruments, which go beyond the normal control system. They are applicable to all tourism organisations, large or small, for profit or not.

The marketing audit consists of examining six major components of the tourism company's marketing situation: marketing environment audit, marketing strategy audit, marketing organisation audit, marketing systems audit, marketing productivity audit and marketing function audit. Insiders or outsiders can perform the task. External consultants can add objectivity, impartiality, continuity and breadth of experience. The greatest potential is as part of a regular auditing process in the organisation. In this way, comparison can be made between the results of each audit so that performance trends can be monitored and used as the basis for target setting, appraisal and reward schemes linked to **management** development and **training**. A common approach is to use a checklist of diagnostic questions. This approach provides a reliable short-cut in assembling information and an insurance that within the broad span that comprises corporate **policy**, no vital issue or question is omitted. Apart from providing clear guidelines as to the actions a tourism entity might undertake to improve its position, checklists have the advantage of comprehensively identifying marketing resources.

Further reading

Brownlie, D.T. (1994) 'Marketing audit', in S.F. Witt and L. Moutinho (eds), *Tourism Marketing and Management Handbook*, 2nd edn, Hemel Hempstead: Prentice Hall, 453–9.

Wilson, A. (1984) *The Marketing Audit Checklists*, New York: McGraw-Hill.

LUIZ MOUTINHO, UK

marketing communication *see* communication mix

marketing, destination

Destination **marketing** is practised around the world at both national and local levels, or wherever organisations responsible for a defined area seek to attract and influence visitors. Traditionally concerned with the overall promotion of appealing **images** to attract **tourists**, destination marketing is increasingly concerned with segmentation and managing the balance between tourism and the **environment** (see also **market segmentation**).

At national level, destination marketing is typically carried out by national tourism organisations, mostly funded by governments with tourism industry partners such as hotels and airlines. Although it can be traced back to the 1920s, the global development of destination marketing is associated with the rapid expansion of **international tourism** and **domestic tourism** since the 1960s. In larger countries there is typically a hierarchy of destination organisations ranging from the national through several area or regional bodies, to local organisations dealing with particular towns, cities or resorts. Most destination marketing is a form of public sector intervention in a commercial **market**, intended to attract more tourists. Initially **growth**-oriented, it typically focuses on **image** creation, **advertising** and **sales promotions**. It may be aimed at any or all forms of domestic and **international tourism** but, to maximise its effectiveness, it is increasingly targeted at segments selected by market research.

As destinations develop, growing numbers of tourists and the growing maturity of the industry shift the focus of destination marketing to take account of tourism capacity and visitor management issues. Because the positive and negative **impacts** of tourism differ according to the various tourist types targeted, destination marketing becomes more partnership and product-based, recognising that the overall tourism **product** is an amalgam or composite product comprised of several elements not controlled by national tourism organisations. Destination marketing thus incorporates market research for developing destination objectives and strategies for targeting, segmentation, and monitoring purposes. It focuses on facilitation, meaning intervention in the market by a national organisation to achieve identified goals by supporting the marketing of all parties involved in the provision of tourism products. It includes, for example, the organisation of workshops and travel **trade shows**, operation of destination information and booking systems, and support for **product planning** and development. All forms of **facilitation** are justified on the grounds that if they were not carried out by the national tourism office, at least initially, they would not happen. Increasingly, they are justified by objectives of sustainability.

See also: destination; image; marketing; sales promotion; sustainable development

Further reading

Kotler, P. *et al.* (1995) *Marketing for Hospitality and Tourism*, Englewood Cliffs, NJ: Prentice-Hall. (Addresses destination marketing.)

Middleton, V.T.C. (1994) *Marketing in Travel and Tourism*, Oxford: Butterworth Heinemann. (Deals with structures and functions of NTOs.)

VICTOR T.C. MIDDLETON, UK

marketing information system

A marketing information system (MIS) is a subsystem of an organisation's management information system. It consists of people and equipment, and contains procedures to gather, sort, analyse, evaluate and distribute timely and accurate information needed for marketing **decision making**. The main basis of a tourism MIS is **marketing research** information from the organisation's environment which is stored in a database format. More advanced systems also provide features to forecast the **future** state of the economy and market **demand** for tourism products. Ideally, the marketing database is maintained at the lowest possible level of aggregation in order to provide marketing managers with the highest degree of flexibility.

By integrating and disseminating complex marketing information, an MIS contributes to the improvement of managerial performance. Therefore, an effective MIS will provide query functions for quick information retrieval on a video display terminal and deliver printed forms or reports to alert the manager when an unexpected situation, whether positive or negative, has developed. The value of reporting has been enhanced greatly by recent advances in graphical display technology. New multidimensional approaches allow more options and information to be displayed than was possible with traditional presentation techniques. Further enhancements with more sophisticated analytical **models** change an MIS into a **decision support system**. Although an MIS can be conceived for all organisations and firms involved in tourism marketing, it has proven to be especially useful for national and regional offices acting in multiple markets with a highly diversified product line. In this case, tourism marketers are encouraged to benefit from comprehensive **marketing research** information.

See also: information technology; market analysis; marketing; strategic marketing

Further reading

Kotler, P. (1994) *Marketing Management: Analysis, Planning, Implementation and Control*, London: Prentice Hall International. (A general marketing management book with strong emphasis on the description of the development and use of marketing information systems.)

Marshal, K. (1996) *Marketing Information Systems: Creating Competitive Advantage in the Information Age*, Danvers: Boyd & Fraser. (Provides an understanding of the basic components of marketing information systems and their development.)

Wöber, K. (1997) 'Marketing information and decision support on the Internet: new opportunities for national, regional and city tourist offices', in *Information and Communication Technologies in Tourism Proceedings of the International ENTER Conference in Edinburgh, Scotland 1997*, Vienna and New York: Springer Verlag. (Describes a successful implementation of a marketing information system for tourist offices.)

KARL WÖBER, AUSTRIA

marketing, international

International **marketing** is the performance of business activities that direct the flow of a company's products and services to customers and other companies across national boundaries for a profit. The only difference from domestic **marketing** is that international marketing takes place in more than one country. Several journals, such as the *International Marketing Review*, are devoted to the theme.

See also: cross-cultural studies; globalisation; international tourism

MARCUS SCHMIDT, DENMARK

marketing mix

The marketing mix is the set of tools that companies use to penetrate their target **markets**. A company must decide what level of expenditures is necessary to achieve its **marketing objectives**, and how to divide such costs among the variables of the mix. The marketing mix is usually characterised as the 'four Ps': **product**, price, place (distribution) and promotion, all applicable to and frequently used in tourism studies.

Product is the firm's tangible offer to the **market**. The term 'product' is used to refer to physical goods, services or ideas that are offered for exchange. The most fundamental level of a product or **service** is its core benefit, the basic or primary benefit that the customer is really buying. A company turns the core into a service to satisfy customers. Benefits can be added to the offering which will take the customer beyond **satisfaction** to delight, exceeding normal expectations and desires with unanticipated benefits. Price is the amount of money customers pay for a service or product corresponding to the offer's perceived value (see **pricing**). This is one of the most flexible elements of the marketing mix. Unlike product features or channel commitments, it can be changed fairly quickly. Price should be set as an intrinsic element of the market positioning strategy, not independent of the other elements of the marketing mix.

Place encompasses the various activities that the firm undertakes to make the service or product accessible and available to target customers (see **market segmentation**; **target marketing**). Marketing intermediaries are used to make a product or service available for consumption. They include merchants who buy, take title to and resell merchandise; agents who search for customers and may negotiate on behalf of the producer; and facilitators who assist in the distribution process (see **distribution channel**), but do not take title to goods or negotiate purchases or sales. Place also concerns the **location** of an **attraction** and its clustering with other service locations. Finally, promotion is all the activities the firm undertakes to communicate and propose its services or products to the target market. Promotion, or the marketing communication mix, includes **sales promotion**, **advertising**, **personal selling**, **public relations** and **direct marketing**. The general responses sought through marketing communications are cognitive, putting something into the consumer's mind; affective, changing the consumer's **attitude**; and behavioural, getting the consumer to act.

Further reading

Kotler, P. (1994) *Marketing Management: Analysis, Planning and Control*, 10th edn, Englewood Cliffs, NJ: Prentice Hall.

ROGER CALANTONE, USA

marketing objective

Marketing objectives should be realistic, quantitative and consistent goals, usually arranged hierarchically from most to least important. Targets for **market** share, sales volume, customer awareness or distribution coverage are common marketing objectives in tourism. These represent the goals of the **marketing plan**, and are based upon situation, opportunity and issue analyses.

Further reading

Kotler, P. (1994) *Marketing Management: Analysis, Planning and Control*, 10th edn, Englewood Cliffs, NJ: Prentice Hall.

ROGER CALANTONE, USA

marketing plan

The **marketing** plan is the central tool for directing and coordinating the marketing effort. It includes summaries of current tourism market situation, analysis of opportunities and issues facing the product line, financial and **marketing objectives**, approaches to be used to achieve the objectives, action programmes, a projected profit and loss statement, and plans for monitoring and control.

Further reading

Kotler, P. (1994) *Marketing Management: Analysis, Planning and Control*, 10th edn, Englewood Cliffs, NJ: Prentice Hall.

ROGER CALANTONE, USA

marketing research

Marketing research is the systematic gathering, recording, processing and analysing of data. It is undertaken to improve the quality of marketing decisions. The information provided by this research is used to identify and define marketing opportunities and problems, to refine and evaluate actions, to monitor performance, and to improve the understanding of **marketing** as a process. This process specifies the information required to address these issues, designs the method for data collection, manages and implements the data collection procedures, interprets the results, and communicates the findings and their implications.

There are numerous specific uses for this research in the **marketing mix**. In the product areas, it can be used in concept development and testing, product development (see **product planning**), introduction and testing, market testing, product discontinuation decisions and product line decisions. Marketing research is used in determining the price elasticity of demand, or how sensitive **demand** for a product is to changes in price (see **pricing**). Its uses for marketing research in distribution decisions include location selection, allocation of floor space and layout. Among its uses in **sales promotion** are in determining the effectiveness of promotional effort and in the allocation of sales effort.

The research helps marketing strategy calibrate with an environment made up of many uncontrollable and changing variables: behavioural, cultural and demographic variables, economic and business variables, **technology**, and political and legal variables. It can provide information about changes in the **environment**, shifts in competitive offerings, alterations in the customer base and reactions to new or modified products or services. Marketing research plays an important role in achieving satisfactory profits by promoting understanding of how to satisfy customers' needs and wants.

The process begins with a definition of the research problem or objective from which one develops specific questions to be addressed. One or more hypotheses, or possible answers to each question, are developed. From this, the necessary information to be gathered is specified. The next step in the process is to chose a method for collecting the needed information, which can be collected from primary (customised for this project) or secondary (already existing) data sources. The sampling design must then be set up to ensure that the sample will represent the population under study. The data collection process comes next, followed by the analysis, interpretation and presentation of the data in a report.

The scientific method plays an important role in marketing research. Its use gives a higher level of confidence in the results of the research. It emphasises the objectivity of the investigator, the accuracy of measurement, and the continuing and exhaustive nature of investigation. However, several areas of marketing research present difficulties in applying the scientific method. The first area concerns the complexity of the subject. Much marketing research focuses on the consumer and his or her **motivations**, **attitudes** and beliefs. The **attributes** and their effects are often hard to measure and interpret. Another area of difficulty concerns the process of measurement. The process that is used may influence the result. Humans tend to behave differently when they realise that they are being measured. In addition, sometimes it is not easy to establish causal relationships (see **causal model**) through marketing research. Accurate prediction may also be difficult, especially given that opinions and attitudes may change. However, emphasis on the scientific method improves the quality of the research output, which in turn can improve the quality of **decision making**.

Data can be collected from primary or secondary sources. Primary data is collected to address a specific research objective. The three major data collection techniques from primary data are qualitative **interview** techniques, **survey** interviewing methods and experimental designs (see **experimental research**). Qualitative interview techniques include direct observation, in-depth interviews and focus groups; these methods focus on a small number of respondents. Survey methods use structured questionnaires and attempt to gather data from a large number of respondents (see **survey, guest**). Primary data collection requires specialised expertise and can be expensive and time consuming. Secondary data uses what is already available, collected for a purpose other than the present research problem. The secondary sources include existing company information, databanks from governments and other organisations, and syndicated sources that collect data for use by clients. Secondary data is usually relatively inexpensive and fairly easy to locate. Its availability is often examined prior to the collection of primary data; it can help define the parameters of the latter.

Marketing research falls into three research design categories. Exploratory research gives insight into the general nature of a problem, the possible decision alternatives and the relevant variables that should be considered. It is usually used when hypotheses are vague and tends to be flexible, unstructured and qualitative. Descriptive research is used when the problem is clear and specific. Descriptive design attempts to determine the frequency with which something occurs or the extent to which two variables are related. Thus this provides a more detailed representation of an aspect of the **market** environment. The third category, causal research, shows that one variable causes or determines the values of other variables.

Surveys are widely used in primary research. The two types of surveys are cross-sectional, which target many different respondents at a single point in time, and longitudinal, which track similar respondents at different points in time. These types of surveys are used to collect a wide variety of information. The types of information collected include respondent background characteristics, personality traits, attitude and lifestyle measures, and product-related variables (such as usage, determinants of brand choice, situational factors and intentions).

Research into consumer behaviour is particularly important in the tourism industry, and much progress in this respect has been made. This has assisted in product design, destination **development** and marketing, among other things. Consumer behaviour research draws from the disciplines of **psychology**, **sociology** and **economics**. Such research studies consider the effects of many variables such as **motivation**, personality, attitudes, culture and **lifestyle** upon consumers' buying behaviour.

Further reading

Churchhill, G. (1987) *Marketing Research: Methodological Foundations*, 4th edn, Hinsdale, IL: Dryden Press.

Kinnear, T. and Taylor, J. (1987) *Marketing Research: An Applied Approach*, 3rd edn, New York: McGraw Hill.

Kotler, P. (1994) *Marketing Management: Analysis, Planning and Control*, 10th edn, Englewood Cliffs, NJ: Prentice Hall.

ROGER CALANTONE, USA

mass tourism

Mass tourism refers to the steady stream of large numbers of **tourists** to holiday destinations. This movement began to develop in the 1960s, with growing affluence, longer holidays and cheaper **transportation** in and from industrialised countries. Initially, this tourist flow visited sun, sand, and sea destinations. Much of it was seasonal (for example, Mediterranean **resorts** in summer and the Canary Islands and the Caribbean in winter). Some, like Hawaii, remained popular throughout the year. Since the 1980s, more frequent holidays, inexpensive **package tours** and growing interest in **cultural tourism**, **nature-based tourism**, and **sex tourism** has influenced this pattern. There is now an increased tourist volume heading to Mediterranean, Southeast Asian and Pacific destinations.

Mass tourism generates considerable income and **employment**. At the same time, the sheer volume of tourists places the physical **environment** and culture of **destination** communities under great pressure. Generally the hosts cope remarkably well, but still mass tourism has some unpleasant repercussions. Some notorious examples are the rampant building and consequent environmental degradation along the Spanish coast, the debris littering the slopes of the Alps and Himalayas, the damage to chambers inside the Egyptian Pyramids, and the increase in child prostitution in Southeast Asia.

This form of tourism also commoditises culture and human relations (see **commercialisation**). There is little occasion and time for contact between **host and guest** populations. This dehumanises both. Traditional norms of **hospitality** cannot be maintained in the face of a constant flow of tourists, who become faceless customers and to whom **natives** are merely curious objects. Obviously, scale is an important factor. Discomfort caused by crowding is more keenly felt where the tourist volume is disproportionate to the **local** population, with popular small islands being particularly vulnerable. In certain destinations, locals now complain that tourists peer through their windows, enter their houses uninvited and look down on their customs. Measures to control tourist **impact** include stricter **planning** and building controls, limiting visitor flow, better guiding and leaflets on local culture/etiquette. Local residents protest threatened values by covert actions, hiding from tourists, communal celebrations, coupled occasionally even with overt aggression. Unavoidably, as the volume of mass tourism grows, so will its unpleasant consequences.

See also: ghetto

Further reading

Boissevain, J. (ed.) (1996) *Coping with Tourists: European Reaction to Mass Tourism*, Providence, RI and Oxford: Berghahn Books.

Smith, V. (ed.) (1989) *Hosts and Guests: The Anthropology of Tourism*, 2nd edn, Philadelphia: University of Pennsylvania Press.

JEREMY BOISSEVAIN, THE NETHERLANDS

mass tourism, organised

Organised **mass tourism** grew in the latter half of the century as a consequence of consumption shifts and changes in **marketing**, vertical and horizontal integration of the industry, **pricing** and **transportation**. It has been marked by tourist influxes, product **standardisation**, **market segmentation** (see also **segmentation, a priori**; **segmentation, a posteriori**), spatial polarisation, an oligopoly of producers and **middlemen**, dependency and external control, and an overexploitation of human and natural resources.

YORGHOS APOSTOLOPOULOS, USA

master plan

The master plan represents a national or subnational **planning** approach to tourism whereby the **supply** and **demand** elements are integrated into a single scheme. Such planning exercises are undertaken at infrequent intervals, due to the commitment of money and expertise involved. For the same reason, master plans are generally led and funded by the public sector or international agencies. They comprise a number of approaches and stages but commonly will include defining goals, aims and objectives, a research phase, strategy formulation and development, articulation of a physical plan and implementation.

Master plans emerged in the mid-to-late 1960s as planners began to appreciate the complex interrelationships involved in tourism **development**. In response, they attempted to design planning approaches which incorporated not only the external influences upon tourism development but also the regional context and the indirect and induced effects of this process. This 'integrated approach' was termed the master plan and is encapsulated in Lawson and Baud-Bovy (1977). A defining feature of this approach is the movement away from traditional technical planning approaches by the development of a feedback and monitoring procedure and the overt inclusion of both products and markets into the plan.

Opinion is divided as to the effectiveness of the master plan as a planning technique. Supporters point to the integration of all of the elements related to tourism development, the logical, sequential approach and the comprehensive consideration of its **impacts**. On the other hand, detractors state that the plans are rigid, often ambitious and fail to allow for updating and changing circumstances. In some respects the master plan is out of tune with new thinking in planning in the 1990s, as it has been overtaken by the need to involve communities and adopt a systematic, flexible, long-term and continuous planning process.

References

Lawson, F. and Baud-Bovy, M. (1977) *Tourism and Recreation Development*, London: Architectural Press. (A detailed elaboration of the PASOLP approach.)

Further reading

Gunn, C.A. (1994) *Tourism Planning: Basics, Concepts, Cases*, London: Taylor and Francis. (Explains the general procedures of tourism planning.)

Inskeep, E. (1991) *Tourism Planning: An Integrated and Sustainable Development Approach*, New York: Van Nostrand Reinhold. (A comprehensive review of tourism planning.)

—— (1994) *National and Regional Tourism Planning, Methodologies and Case Studies*, London: Routledge. (A useful set of case studies, many using master planning.)

CHRIS COOPER, AUSTRALIA

mature market *see* senior market

media

Tourism media comprise the channels through which messages are targeted at tourists by such sectors as **tour operators**, airlines, hotels, national tourism organisations and the like. These media can first be classified according to the stage of the trip. At the pre-trip stage, where the accent is on **escape** from the home environment, **sales**

promotion tends to be dominated by tour operators (destination outsiders). At the on-trip stage, where earlier communication is reinforced, the emphasis is on renewal of the **tourist**; here the message senders are usually destination insiders (such as national tourism organisations). Then there is the post-trip stage, that of reintegration into the society of origin, where returning tourists become promoters to potential tourists (often their friends and relatives). The process is hence cyclical, as indeed is tourism itself, following a ritualistic pattern of emancipation, **animation** and reincorporation.

The media of tourism may additionally be categorised by the nature of the sensory appeal (for example, the text and pictures of brochures relating to the sense of sight, a taped narrative to the sense of hearing). Some even evoke gustatory, olfactory and tactile responses, while others have combined effects (**television** and **videos** with auditory and visual communication). Within these categories, the media of tourism can be further broken down according to their principal sources. Thus print media, for instance, are largely literary (such as travellers' tales) or informational (newspaper accounts), whereas audio media are experiential (word-of-mouth), informational (**tour guides**) or electronic (**radio** call-in programmes).

The effectiveness of any given tourism medium is predicated mainly on the credibility of the sender. **Advertising**, for example, has less trustworthiness than an unbiased documentary. However, in maximising credibility, considerations of cost also have to be borne in mind. Undoubtedly the least expensive and most effective medium of tourism is word-of-mouth communication, which several **surveys** reveal to be the main persuasive source in tourists' choice of destinations. Yet surprisingly little research has been conducted in this area. There is also growing evidence to support the increasing importance of tourism communication over the **Internet** and the multimedia possibilities offered through it.

Further reading

Dann, G. (1996) *The Language of Tourism: A Sociolinguistic Perspective*, Wallingford: CAB International (See especially Chapter 6.)

GRAHAM M.S. DANN, UK

media planning

Media planning is the process of selecting those media which deliver the key messages of the **advertising** campaign to the target group. Three key media decisions are involved. The first decision to be made is selection of media categories (newspapers, **magazines**, **radio**, **television**, cinema, billboards and **Internet**) and the allocation of the advertising budget to these categories. Key determinants of this decision are costs of the categories (television commercials are expensive, radio costs less) and size of the budget (small budgets are inadequate for television), and advertising objectives (the acceleration of a car is best demonstrated on television or on a cinema screen).

The second is choice of appropriate newspapers, magazines, television networks or radio stations within the different media categories. The third decision is planning of media schedules. At this time, the media planner has to determine the size, colour, number and insertion times of ads within the chosen print media, and to determine the length, number and time schedule of radio, television and cinema commercials and banners on websites. The second and third decisions are normally based on information about readership and radio usage **surveys** in private households and television peoplemeters (all household members in the audience panel push buttons on a remote control when they start or stop viewing, programme selection and viewing time are recorded) and counting visits on websites.

The most important criteria to evaluate and select media are the reach in the target group (for example, 20 per cent of all Germans who plan summer holidays in Spain may at least once spot an ad in a given magazine); frequency (the number of times an average member of the target group may be exposed to an ad when it is inserted more than once in a particular media), often called 'opportunity to see/hear/read'; gross rating points (reach multiplied by frequency); and cost per thousand gross rating points within the target group. Media planning and analysis continue to be important in successful tourism **marketing** and promotion, especially since it is the image/perception of a place or **destination** which attracts attention and generates tourism in its direction.

Further reading

Sissons, J.Z. and Petray, E.R. (1981) *Advertising Media Planning*, 6th edn, Chicago: Crain Books. (Discusses the derivation of media objectives and plans from marketing and advertising objectives, as well the media planning process and problems involved.)

Sissors, J.S., Lehew, H.D. and Goodrich, W.B. (1981) *Media Planning Workbook – With Discussions and Problems*, 4th edn, Chicago: Crain Books. (Includes many practical examples of media schedules.)

HELMUT KURZ, AUSTRIA

media, recreational

Recreational **media** as a specialised outlet for communication has two connotations. The first covers all categories of media, mass and personal, print and electronic, that offer specialised coverage on recreational activities. These include travel guides and information brochures, media commentaries on recreational events (active and passive), publicity and promotional materials. The second includes categories of media that provide recreational **experience** for the users. **Leisure** activities involving reading, watching television or listening to music or sitcoms on the radio are some of the daily activities which are traditionally dependent on the recreational media. More recently, a number of computer-based leisure applications have encouraged further **demand** for recreational media products and services. Such new pursuits include random surfing of the **Internet**, e-mail, pay television services, friendship circles, psychic networks, talk-shows on radio and television, and video and computer games.

With the **trend** pointing towards shorter working hours and longer paid **vacations**, the recreational media industry is likely to continue to grow further as the demand for information on destinations and holiday activities increases. However, with problems of overcrowding becoming more acute, both en route to and at **vacation** sites, more families may choose to spend their leisure **time** at home; thus, the demand for domestic entertainment and recreational services will grow and this will lead to further expansion of recreational media as a source of entertainment for the community.

See also: information technology; sales promotion

Further reading

Gray, A. (1996) 'Behind closed doors: video recorders in the home', in P. Marris and S. Thornham (eds), *Media Studies: A Reader*, Edinburgh: Edinburgh University Press, 327–36. (Offers a description of how the video recorder affects contemporary family life.)

Sorlin, P. (1994) *Mass Media: Key Ideas*, London: Routledge. (Carries a useful section on entertainment discussing the manner in which recreational media influences public tastes and preferences in leisure activities.)

LATIFFAH PAWANTEH, MALAYSIA

meeting planner

A meeting planner is anyone who performs the services necessary for planning, running and controlling a meeting function. In the past, this position was filled primarily by secretarial and assistant staff members. Since the 1980s, however, meeting planners have become increasingly more professional. In many cases they hold executive ranks within their organisations.

Meeting planners can be segmented into two main types, with subcategories in each. The primary distinction is between internal and external planners. Internal planners work for the organisation that is sponsoring the meeting. For example, an association or corporation may well have executive level staff positions dedicated solely to meeting planning. External meeting planners come from a variety of sources. Some, known as independent meeting planners, perform the planning functions on a contractual basis with their clients. However, many other vendors now offer meeting planning services. For example, many **destination** management companies (DMC) and tour and travel organisers will actually plan meetings in addition to their other services. In fact, the distinction between DMCs and independent meeting planners is becoming increasingly blurred.

Both internal and external planners fall mainly into association planners and corporate. The difference between these two segments is obvious. A less understood subsegment of the corporate market is the **incentive** planner. Although not as widely known, this segment is very lucrative because the purpose of the meeting is to reward the participants. Therefore, per person budgets are usually higher than for other meeting types. Increasingly important is the planner who creates special events, which may include sports venues, celebrations and other functions. These planners must produce high impact functions that entertain as well as impress the participants.

No matter which segment or type, meeting planners must always determine the purpose for the meeting or event. In the case of traditional meetings, this is more than likely to educate the participants. The planner has the responsibility for creating an educational atmosphere that is conducive to the transfer of knowledge. Several professional organisations exist for the **education** of planners. In the United States, the primary ones are Meetings Professionals International and the Professional Convention Management Association. Both groups are dedicated to promoting **professionalism** through education and standards of performance.

See also: convention and visitor bureau; convention business

Further reading

Polivka, E. (ed.) (1996) *Professional Meeting Management*, Birmingham, AL: Professional Convention Management Association.

TED ABERNETHY, USA

meetings business

The sector which provides facilities and services oriented toward the **attraction**, hosting and management of meetings held by corporations, agencies and affinity groups is known as the meetings business or market. Major meeting types include business (such as conferences), education and training, trade and consumer shows. Convention centres, hotels and **resorts** cater to this sector, with professional **meeting planners** usually being involved.

DONALD GETZ, CANADA

mega-event *see* event; festival

merchandising

Merchandising is an ambiguous term and process used in the **marketing** literature in many ways. It relates to where, when and how to publicise or sell a product. This process refers to, among other things, promotional measures taken in packaging, distribution and point-of-sale activities or **pricing**.

See also: advertising; marketing; sales promotion

JOSEF A. MAZANEC, AUSTRIA

merger

The objectives of mergers in the **airline** industry are similar to those in other economic sectors. The primary motivation is to achieve synergism through the process of consolidation, where the combined firm's value exceeds that of the individual companies. Airline mergers seek to increase economic performance by providing increased **economies of scale**, greater financial strength, and increased market share.

Mergers may be identified as horizontal, vertical, congeneric or conglomerate, depending on the types of companies that combine. A horizontal merger occurs when two airlines (or other firms in the same sector, such as hotel companies) combine. Mergers are classified as vertical when an airline, for instance, acquires a **catering** company that supports its main air service activity. Congeneric mergers, on the other hand, take place when an airline acquires a business related to the tourism **industry**. Finally, there are the conglomerate mergers where airlines acquire completely unrelated businesses.

Before airline deregulation in the **United States**, mergers among airlines were limited to troubled carriers seeking solvent partners to avoid bankruptcy. Many governments, including that of the United States prior to deregulation, perceived mergers as the first step toward illegal monopolies and rarely approved them. After airline deregulation, merger strategies became more popular. Industry restructuring forced airlines to focus on the bottom line and seek synergism to survive in the free market environment. This merger and acquisition trend continues under the more politically acceptable designation of alliance formation, wherein ownership need not be transferred.

Given the current globalisation of **transportation**, airline alliance formations will likely continue, with a resultant creation of mega-carriers. Debates persist as to whether mergers strengthen or weaken a carrier. A review of the literature shows no correlation between mergers and profit realisation. Furthermore, the merger experience of US carriers has been rather problematic. Merging different corporate cultures often is more difficult than originally thought, and it appears that internal growth may be a much more effective way to expand an airline. No reason exists to suggest that mergers and acquisitions serve the interest of shareholders, employees or the public at large. There appears to be sufficient evidence that airline merger and acquisition strategies are not an effective way to achieve corporate greatness.

Further readings

Banfe, C. (1992) *Airline Management*, Englewood Cliffs, NJ: Prentice Hall. (An overview of practical aviation management, with pp. 72–8 and 167–9 focusing on merger and acquisition issues.)

Ott, J. and Neidl, R.E. (1995) *Airline Odyssey*, New York: McGraw-Hill. (Offers an analysis of the changing airline industry.)

DAVID B. VELLENGA, USA
WILLEM J. HOMAN, USA

metaphor

Metaphorical or metonymic **images** involve a word or phrase applied to or associated with an object or action as a substitute for it. Both are forms of linguistic/cognitive shorthand used to simplify and share understanding of complex or unfamiliar phenomena. 'Surfacing' these images helps to indicate underlying perceptions, thoughts and **attitudes** to the phenomena in question, but **gender** or cultural bias may influence images.

See also: iconography; marker; semiotics

ROBERT BROTHERTON, UK

methodology

An orderly arrangement of ideas is generally referred to as methodology. In the social sciences, methodology may be defined as a set of regulated procedures undertaken to test, validate and generate hypothesis and **theory**. Techniques such as observation and interviewing are various means by which research is conducted within a given methodological framework. The multidisciplinary character of tourism means that many different theories, methodologies and techniques are employed by researchers.

ROBERT B. POTTER, UK

Mexico

In 1996, Mexico was the seventh largest tourism **destination** worldwide, with over 21.7 million international **tourists** arriving and contributing $6.89 billion dollars to the economy. This figure made the country only the sixteenth earner in terms of **international tourism** receipts, mainly because of its relatively cheap prices as a result of monetary devaluation. When domestic tourism revenues are added, total tourism revenues for 1996 were expected at $18.1 billion dollars. The industry as a whole generates almost 12 per cent of Mexico's GDP, making an important contribution to the Mexican economy. It is estimated that one in every ten jobs in Mexico is related directly or indirectly to tourism, with 2.2 million jobs in 1996. In addition, approximately 15 per cent ($8.1 billion for 1996) of the total capital investment in **infrastructure**, installations and equipment is attributed to the industry, as well as 10 per cent ($50 billion for 1996) of the governments' total tax revenue.

The success of Mexico in the international tourism scene is justified. The country possesses a rich diversity of attractions in terms of natural **resources** going from Caribbean and Pacific beaches to mountains and snowy volcanoes, and from deserts to dense rain forests traversed by whitewater rivers and accommodating a large biodiversity. In addition, its cultural **heritage** is marked by ethnic diversity and the mix of cultures, ranging from important pre-Columbian civilisations like the Mayas, Aztecs and Olmecs that inhabited the territory to the Spanish conquerors who made it an important colonial centre. This mix is reflected in a large number of archaeological **sites**, many colonial buildings, an extensive variety of arts and handicrafts, a rich **cuisine** and colourful religious and folklore festivals. In addition, Mexico City, one of the most populous cities in the world, provides the cultural, artistic, and **shopping** conveniences of any large metropolis, as do many other large cities.

However, this vast supply of attractions has been largely underexploited because the government has focused its tourism **development** strategy to cater to the sun and beaches market, resulting in planned integrated beach **resorts** at selected destinations. The prime example of this type of development is Cancun (a hand-picked location built into a resort city from the ground up) that grew from 300 inhabitants in 1974 to more than 250,000 in 1995. Under the Mexican system, these resorts are known as 'megaprojects', and they are the responsibility of FONATUR (Fondo Nacional de Fomento al Turismo), a special government agency established in 1974 (although its predecessor had been created as early as 1956) by the special Ley de Fomento al Turismo (Law of Tourism Promotion), in order to foster the creation of this industry.

The country has a long history of government involvement in the promotion of tourism that goes back to the last century. In 1927, immediately after the Mexican Revolution, the road route Mexico City–Cuernavaca–Taxco–Acapulco was inaugurated, and with this one the most important tourism routes in Mexico to this date was created. Acapulco emerged as an important destination. The decade of the 1940s is known as the golden era of tourism in Mexico. At the end of the Second World War, with Europe and the Orient heavily damaged, Mexico emerged as an alternative destination, especially to its North American neighbours who constituted the bulk of tourism demand at the time. Furthermore, the president at the time, Miguel Aleman, considered a national tourism hero in Mexico, gave priority to the sector by building roads and airports, signing the first federal tourism law and creating the first tourism school; he also promoted the development of Acapulco as a beach resort popular with international tourists and jet-setters. In the 1960s, President Lopez Mateos advanced the First National Plan of Tourism Development and the National Tourism Council, headed by the now ex-president Aleman. At present, tourism in Mexico is the responsibility of the Secretariat of Tourism, created in 1974, the same year as FONATUR.

The main destinations of Mexico are Cancun, Acapulco, Ixtapa, Puerto Vallarta and Mazatlan for the sun and beaches segment. Monterrey and Mexico City constitute important destinations for the business segment, while Tijuana is famous for border-crossing tourism. The Baja California area is emerging as a destination for **adventure tourism**, in addition to beach resorts. The Yucatan area is famous for its Mayan archaeological sites; while Mexico City has some nearby Aztec archaeological sites. The Bajio area, which includes Guanajuato and San Miguel Allende, as well as Mexico City, Guadalajara, and Oaxaca, have many examples of colonial architecture.

In terms of **infrastructure**, the country is well endowed, especially compared to other developing countries. A recent build, operate and transfer legislative programme in infrastructure has significantly improved the road system, with modern highways connecting the main destinations. In addition, domestic and international airports at the main destinations are served by many national and foreign carriers. Many Mexican ports are routine stops for cruise lines. In terms of accommodation there were 342,000 hotel rooms in 1991, ranging from rustic cabins to luxury hotels. There are more than twenty international hotel corporations presently operating in Mexico. It should be noted that Mexico has no officially endorsed hotel **classification** system, although plans for implementing one are underway.

Some of the problems facing tourism development in Mexico are the same as those which face the country as a whole. Poverty is a big issue, with more than 50 per cent of the population living below the poverty line. The political system is plagued by corruption and scandals. Political uncertainty has been added by a process of democratisation undertaken by the government. This could change the political equation of the country, where government has been dominated during the past five decades by the Parido Revolutcionario Institucional. Adding to this is the emergence of guerrilla movements, mainly in the south and the impoverished indigenous populated sierra areas, like Chiapas, Oaxaca and Guerrero. Although these outbursts are small and presently do not represent a threat, they have been well publicised by the international press, scaring many international tourists away. Particularly within the industry, the main problem is that the model of sun and beaches and megaprojects has resulted in highly concentrated infrastructure building and the development of enclave destinations, with benefits that not always trickled down to the population but instead stayed with the developers and government officials. Moreover, the change in the worldwide tourism **demand**, with the focus shifting into **ecotourism** and adventure tourism, has given rise to stiff competition from other Latin American countries like **Ecuador**, **Peru**, **Costa Rica** and Belize. Recently, the government recognised this trend and is now promoting some small new projects, some of them located inland, away from the coastline.

ISABEL ZARAGOZA, MEXICO
MARIANO ROJAS, MEXICO

middleman

Middlemen are intermediaries between consumers and principals. The distribution channel in tourism consists of, for example, **travel agencies**, **tour operators** and handling companies. **Information technology** has begun to change the nature of the middleman. Threats of 'disintermediation' emerge as consumers are empowered to make arrangements directly with tourism suppliers. Opportunities arise as middlemen can improve their services and add more value.

See also: culture broker; entrepreneurship

DIMITRIOS BUHALIS, UK

migration

Migration is the permanent, or relatively permanent, relocation of an individual or group to a new, usually distant, place of residence and **employment**. **Tourists** are not usually regarded as migrants because they visit for less than a year. However, tourism promotes migration in at least two major ways. First, positive experiences gained in a place as a tourist coupled with the personal relationships which are established during the temporary stay may lead to a subsequent permanent move for social reasons, employment or on retirement. Second, the employment opportunities afforded by a growing tourism industry may attract migrants seeking jobs. Sometimes, the **seasonality** of this industry may result in short-time migrations of tourists, such as snowbirds, and of temporary workers, as in some ski **resorts**. The migratory habits of wildlife, such as whales, birds and butterflies, may constitute the basis of attractions when large numbers of individuals are concentrated in specific locations at particular times of the year.

GEOFFREY WALL, CANADA

minorities

A minority is a segment of a local population who, by some cultural, economic, political, or religious **marker**, are fewer in number relative to other groups of the population. They are differentiated and are formally or informally assigned a separate social position (see **marginality**; **ethnic group**). Ethnic minorities are often used in national tourism **advertising** because of their attractive traditions, like the Sami in Finland. They may indeed be the main reason for tourists visiting a destination, yet the minorities themselves may not be the major beneficiaries of the resulting business.

VALENE SMITH, USA
VERONICA LONG, USA

model

A model is an attempt to identify key variables in a situation and the relationship that exists among them. Model building in tourism seeks to understand a complex relationship and to aid the **management** of a place or process. For example, in econometric **forecasting** of tourism flows, the purpose is to help estimate future numbers of **tourists** so as to permit informed decisions.

CHRIS RYAN, NEW ZEALAND

modernity

Modernity is a temporal construct which periodicises a number of societal developments which evolved within industrial Europe and America after the First World War. Its defining features included technological mechanisation, mass communication, **internationalisation**, secularisation and avant-garde culture. Some writers see **tourism** as quintessentially a modernist phenomenon, others as postmodernist, while still others emphasise its roots in traditional travel forms (see also **postmodernism**; **prototypical tourism form**) .

A.V. SEATON, UK

modular programme

Modular programmes consist of learning experiences in which the totality of an experience is divided into independent components (termed modules). The design and content of each module is structured so that students may combine different modules according to their academic and/or professional needs, as well as the time available to them. Modular designs are commonly used to facilitate the delivery of an institution's programmes so as to maximise the cost-effectiveness of instructional programming.

See also: education; training

J.R. BRENT RITCHIE, CANADA

Morocco

Morocco, with an area of 710, 850 square km and 26 million population, features a diverse **climate**. This is due to its Atlantic and Mediterranean coastal fronts (over 3,500 km) and to the mountain chains which form its high boundaries. The geological structure of this North African country comes from the movement of tectonic plates: over two thousand million years ago, the Anti-Atlas and part of the Sahara were formed. The tertiary era saw the development of western and eastern Meseta and the High and Middle Atlas mountains, as well as the Rif mountains. Evidence of the ancient population can be found in the countless prehistoric **relics**. Since then, the Phoenicians, Carthaginians, Byzantines, Romans and Vandals succeeded one another, before the arrival of the Arabs in the eighth century. Today, Morocco is a constitutional monarchy.

The country is appealing to **tourists** because of its tradition and its quality of life. It is divided into four large **regions** : the coast dominated by fishing and manufacturing industries; the plains and plateaux of the northwest, the agricultural and commercial regions; the Rif and Atlas mountains where pasture and commercial forestry dominate; and the desert, where mining industries and the oasis culture co-exist. Morocco is located near main European tourism **markets**. It offers several touristic combinations, including the tour of fortified towns, crossing the Moroccan Atlas mountains, and the Big South. The tour of imperial towns comprises visits to Marrakech, Rabat, Meknes and Fez, as well as a short trip to Casablanca, the economic capital of the kingdom. Hiking, horse riding, mountain climbing, paragliding, hang-gliding and rafting are some of the activities sought by tourists. Tourism plays a fundamental part in Morocco's economic and social growth. It brings foreign currency, provides jobs and encourages regional development. In fact, tourism continues to be the second largest source of foreign currency, after money sent home by Moroccans living abroad.

SOUAD HASSOUN, MOROCCO
AHMED NAIM, MOROCCO

motel

The North American forerunner of the European **budget hotel**, offering inexpensive, 'no frills' **accommodation** in convenient roadside locations, motels are characterised by generous parking provision and informality. Originally 'tourist courts', groups of small individual cabins providing simple 'stopover' accommodation, their expansion was fuelled by the development of the interstate highways. Today they are predominantly franchised units targeting the business traveller.

KEITH JOHNSON, UK

motion sickness

Motion sickness is caused principally by conflicting sensory information. The motion usually has a rhythmic or erratic character. The inner ear monitors direction of motion. The eyes monitor where the body is in space (upside down, for example) and sense of balance or equilibrium. Skin pressure receptors, such as on the feet, tell the brain which part of the body is in touch with the ground. Muscle and joint sensory receptors tell which part of the body is moving. The central nervous system processes information from the four systems. Motion sickness is caused when the brain receives conflicting stimuli.

In mild cases, symptoms include headache and general discomfort. If motion continues, abdominal discomfort, nausea and dry mouth become acute leading to cold sweating and vomiting. It may be accompanied by depression and apathy. Feelings of lethargy may persist for hours after the motion has ceased. There are many types, including sea, air, car, train, swing, flight simulator and space sickness. It can occur during travel or on rides in an amusement/**theme park**. Sickness is most likely to occur due to turbulence, or travel in a small plane or boat. Several factors influence susceptibility. Children under two rarely experience motion sickness. Children three to twelve years old are the most susceptible. Females are more susceptible than males of the same age.

Tourists with a history of motion sickness can minimise symptoms with prophylactic use of anticholinergics or antihistamines. These are contraindicated in pregnancy, and may cause drowsiness, which will be exacerbated by **alcohol**. Anticholinergics suppress sweating and thus are suitable for use in hot climates. Skin patches worn behind the ear for up to three days are available. Symptoms may be alleviated by remaining in the most stable area of the vehicle, limiting **food** and alcohol intake, breathing plenty of fresh air, lying down or reclining and closing the eyes. Reading during travel accentuates the problem. If susceptible passengers sit where they can see the horizon, close eyes or travel at night, the problem can be lessened.

Motion sickness can add to a general feeling of malaise that may be associated with such common problems as jet lag, climate or altitude change, and the like. Motion sickness can ruin a tourist's enjoyment and may also impact on other passengers. Medical practitioners and biologists interested in neurophysiology, neurochemical and neuroanatomical mechanism research this condition. A number of sites on the **Internet** exchange information and treatments.

See also: health

ROBYN BUSHELL, AUSTRALIA

motivation

According to the sociologist Max Weber, motivation lies at the core of human behaviour. Consequently, the study of motivation is central to any social scientific undertaking since it provides understanding, explanation and prediction. It goes beyond the *how* questions of description to the *why* questions of interpretation and causality. As one domain of interpersonal activity, tourism is no exception to this general observation. Indeed, 'why do people travel?' is probably the most fundamental issue in tourism research today.

However, given that tourism is also a many-faceted phenomenon requiring multidisciplinary treatment, clearly the answer provided to the vast and complex question of international movement (motivation as movement being derived from the Latin *movere*), will be predicated on the perspective of the analyst. Economists, for example, may emphasise the dynamics of consumer choice over competing pursuits and their effects on scarcity, price and profit. Historians might wish to highlight the fact that patterns of travel go in periodic trends from the pleasure motives of Ancient Rome to the religious motives of the medieval pilgrimages and the educational motives of the **Grand Tour**. Geographers, environmentalists, political scientists and anthropologists will all have their own contributions to make to the mosaic of meaning underpinning reality. Each is a partial version of the truth. No single discipline can claim a comprehensive monopoly to the whole. Nevertheless, in tourism research, if there are two disciplines which perhaps stand out as having generated the greatest discussion on motivation, they are those of **psychology** and **sociology**.

At the risk of oversimplification, psychologists focus on the individual personality and intrinsic motivation, including cognitive and affective motives which are self-directing, autonomous and non-deterministic. Pearce's (1982) research is a good example of this position. Steering a middle course between psychoanalytic theories, which attempt to account for all spheres of human endeavour in terms of sexual and aggressive instincts, and those which exclude activities on the grounds that they are simply the result of habit and reflex, Pearce argues that this macro-perspective should be replaced with mini-theories which target specific groups and populations. Moreover, within this micro-framework, he maintains that a fruitful motivational **paradigm** for understanding tourism is one which combines **self-actualisation** with **achievement** and attribution theory. More specifically, Pearce contends that tourism research should address the future-oriented tasks of its subjects, be open to the possibility of multi-motives, examine the views of individual tourists rather than indulge in post hoc interpretative glosses of the researcher, and tackle the problems of empirical measurement. Drawing initially on Maslow's work, Pearce believes that the enhancement of self through tourism is the zenith of a hierarchy which requires progressive **satisfaction** from the basic physiological needs of hunger, shelter and so on, to those of safety, love, esteem and self-fulfilment. Later he develops these ideas in relation to a tourist career path which is based on stages of the life cycle.

By contrast, the sociological treatment of the tourist places the accent on society rather than on the individual, on extrinsic rather than on intrinsic motivation. Following the sociologist Emile Durkheim, Dann (1981) demonstrates that the prevailing conditions in the tourist's home environment are ultimately the dispositions for travel. Thus, the effects of an anomic existence (perceived normlessness and meaninglessness) (see **anomie**), may act as motivational push factors that persuade people to seek temporary respite in another society which is deemed to be less fraught with such problems (see also **push–pull factors**). Tourism provides the means for this **escape**. Similarly, persons who suffer from lack of belonging in their own society, chiefly through denial of status advancement, may journey to those parts of the world (often developing countries) where they are treated with greater respect. It is further possible to design a continuum (see **continuum model**) ranging from predominantly anomic to ego-enhancement motives, and to identify social profiles of toursits which, to a greater or lesser extent, measure up to these ideal typical constructs.

An extension of the sociological framework is provided by Sharpley (1994). He points out that there are several concrete social influences conditioning the decision to travel. He identifies the family, reference groups, social class, the surrounding culture and the workplace as the most important of these. The latter is particularly significant since it is conducive to compensatory and spillover effects in various types of travel. Thus, boring or monotonous jobs may lead to the quest for excitement, stress may result in a search for relaxation, and the regimentation of the office or factory may encourage the pursuit of **ludic** activities elsewhere (see **play**; **regression**). Alternatively, type of occupation, especially if self-rewarding, may stimulate similar interests on a **vacation** (such as the history teacher travelling to an archaeological **site**). As far as the influence of home culture is concerned, Parrinello (1993), in a postmodern context, has noted the tremendous **impact** of **consumerism** and product **advertising** on the decision to travel. For her, citizens of most developed countries live in a tourism-saturated environment and are daily bombarded with **images** of abroad.

Whereas at first glance psychological and sociological approaches may appear diametrically opposed, in many respects, they can be regarded as complementary. Whether one begins or ends with the individual or society, essentially both are intertwined. Indeed, if points of convergence rather than differences are stressed, several worthwhile advances in tourism motivational research can be identified. The first sign of mutual progress is a shared concern for theoretical awareness. Even though there are intra-paradigmatic differences attributable to schools or perspectives within each discipline, some sort of working consensus can be achieved with respect to terminology. In this regard, operational definitions can be agreed upon as to what exactly constitutes motivation, and how this varies from the allied concepts of aspiration, intention, reason, purpose, satisfaction, aim and goal. Thus, most psychologists and sociologists would concur that when tourists are asked to complete embarkation forms by stating the purpose of their trip in terms of business, pleasure or visiting friends and relatives, these categories, although possibly useful to the compilers of **statistics**, do not represent the full motivational meaning of a trip. Similarly, they would agree that many **surveys** which measure differences between expectation and reality are in fact dealing with satisfaction rather than motivation.

Another realm of theoretical convergence is the appreciation that motives, while reasonable to one individual, may be far from rational or logical to another. Hence, a person who, quite reasonably to himself or herself, seeks self-identity through the rigours of **wilderness** travel may be considered quite irrational by those who wish no more than relaxation by the hotel swimming pool. Furthermore, both may be regarded as distinctly odd by members of a destination **community**. What is meaningful and reasonable for one person may be motivationally senseless for another, especially when the other is a researcher. The important point is to realise how situations are defined, and then seek to monitor rather than to judge them.

A further area of theoretical consensus resides in the common belief that tourist motivation is future-oriented. From the psychologist's point of view, the **experience** should be seen in the context of long-term goals. From the sociologist's standpoint, Alfred Schutz's 'in-order-to' motivation is regarded as imaginatively *projected* action: as if it had already taken place as part of a meaningful *project*. That is why the verbal and pictorial discourse of tourism advertising is frequently couched in the future perfect tense, and why anticipation of a holiday is so important.

At the methodological level, both psychologists and sociologists emphasise the significance of an emic perspective. Pearce, for instance, has shown that personal accounts of negative and positive tourism experiences can reveal inferable motives, whereas the checklist approach of the structured survey only indicates superficial agreement with the a priori motivational categories that the investigator feels are worthy of inclusion. Many sociologists agree, to the extent that recent motivational research has employed projective tests whereby subjects linguistically frame their own responses to destination images.

Methodologically, too, psychologists and sociologists both appreciate that tourists may be unwilling, or even unable, to articulate their real motives for travel. Shared realisation of this difficulty has prompted such innovative research strategies as 'systematic lurking' and 'conversation sampling', which focus on the rarely investigated but nonetheless highly germane factor of word of mouth communication at all stages of a trip, including the ever more salient domain of pre-trip anticipation. Such techniques can be particularly useful in tapping the increasingly evident **nostalgia** motivation in contemporary tourism, one that is correspondingly difficult to unearth through the traditional methods of interviewing and questionnaire completion.

Finally, psychologists and sociologists agree that tourist motivation, like all motivation, is a highly subjective issue. The problem thus becomes how to study the realm of the subjective in an objective or dispassionate manner. For some (including constructivists and critical theorists), this is a non-issue, since value freedom in research is neither possible nor desirable. For others, subjectivity is explored not in its individuality but in its typicality. According to this view, one is not so much concerned about why John Doe goes on holiday as to why the type of person John Doe represents decides to take a vacation. Such a realisation has led to several typologies in tourism research, and these certainly constitute a useful starting point in understanding the tourist. However, typologies in themselves are heuristic, rather than explanatory, devices and more still remains to be done at both the theoretical and methodological level if the basic question of why tourists travel is to be adequately answered.

References

Dann, G. (1981) 'Tourist motivation: an appraisal', *Annals of Tourism Research* 8: 187–219.

Parrinello, G.L. (1993) 'Motivation and anticipation in post-industrial tourism', *Annals of Tourism Research* 20: 233–49.

Pearce, P. (1982) *The Social Psychology of Tourist Behaviour*, Oxford: Pergamon.

Sharpley, R. (1994) *Tourism, Tourists and Society*, Huntingdon: Elm Publications.

GRAHAM M.S. DANN, UK

motivation, intrinsic

Intrinsic **motivation** is an important concept in social psychological theory and research on **leisure** and free time **behaviour**, and has been used in explanations of children's **play**, **tourism** behaviour, **leisure satisfaction**, serious and forced leisure, and leisure and **health**. Along with perceived **freedom**, intrinsic motivation is one of the most important psychological characteristics of leisure. An engagement is more likely to be experienced as leisure when it is intrinsically motivated and the rewards for participation come from engaging in the **activity** itself. When an activity is engaged in primarily because of external rewards or obligations, it is extrinsically motivated and less likely to be experienced as leisure (Neulinger 1974).

According to self-determination theory (Deci and Ryan 1991), people are most likely to be intrinsically motivated when they are in free choice situations and behaviour is engaged in out of interest. When intrinsically motivated, people seek out optimal challenges and the internal psychological rewards are based on the innate needs of competence, self-determination and relatedness. Relatedness refers to the need to feel loved and connected to others. In addition to rewards, extrinsic motivation to participate in an activity can result from threats of punishment, evaluation, deadlines, obligations and imposed goals. Research suggests that if an originally intrinsically motivated activity is then extrinsically rewarded or regulated, the overjustification effect can occur. When people are rewarded or obligated to travel, play games or volunteer, their behaviour can become overjustified. In other words, they may begin to attribute their involvement to extrinsic motives and lose interest in the activity other than for the reward.

Extrinsic rewards do not always lead to an overjustification effect. If a person is not intrinsically motivated to start with, then a reward cannot hurt because the behaviour could not become overjustified. Further, if receiving a reward is not dependent on participating in an activity, then participation is likely to be experienced as self-determined and the reward perceived as a bonus rather than as a bribe. For example, people can **experience** their work as intrinsically motivating, and activities that involve rewards, obligations and commitments at times can be experienced as leisure.

References

Deci, E.L. and Ryan, R.M. (1991) 'A motivational approach to self: integration in personality', in R. Dienstbier (ed.), *Nebraska Symposium on Motivation, Vol. 38, Perspectives on Motivation*, Lincoln: University of Nebraska Press, 237–88.

Neulinger, J. (1974) *Psychology of Leisure*, Springfield, MA: Charles C. Thomas.

ROGER C. MANNELL, CANADA

motive manipulation

A process by which one's reasons for **behaviour** are changed constitutes motive manipulation. Tourism **motivation** is often associated with the need to **escape**, relax, learn and to be with and meet people. A number of different processes (including **marketing**) may be employed to encourage potential **tourists** to recognise the importance of a particular need and/or to emphasise the importance of one motive over another.

DANIEL R. FESENMAIER, USA
JIANN-MIN JENG, USA

motor coach tourism

Motor coach tours, as the most popular method of tour **transportation**, comprise an important segment of the tourism industry. This may be offered by itself or paired with scheduled **airline** flights to provide local **sightseeing** and transfers. Approximately 80 per cent of all **tours** in the United States are by motor coaches. This market size has been an encouraging factor in the planning and building of access areas at popular tour destinations such as **theme parks**, entertainment facilities and megamalls. Motor coach operators have a common goal of generating revenue from rental or **charter** of buses owned by their companies, combined with other tour elements by the operator.

Most vehicles used for these group tours accommodate from 45–8 **tourists** and are used primarily on trips of less than 800 miles. Modern comforts on board include large observation windows, video monitors, fully equipped restrooms and cocktail bars. Use of minicoaches is a relatively new practice in this business. These vehicles transport up to 25 tourists with amenities and comfort level comparable to the full-size coaches. They have the advantages of manoeuvrability, as well as lower amortised cost per passenger than the larger motor coach. A third type, with 76 seats, has an articulated midsection which gives it the same turning radius as a standard bus. Because of the Americans with Disabilities Act, most motor coaches in this country offer full **accessibility** for persons with disabilities (see **handicapped**). All motor coach tours are typically open to the public, but they may be organised for particular groups. **Tour wholesalers** form **partnerships** with motor coach companies, a symbiotic relationship which makes use of the company assets while providing the wholesaler an additional outlet for sales.

GINA K. McLELLAN, USA

mountaineering

The term mountaineering refers to a tourism activity taking place in a mountainous **environment**. Early developments of mountaineering for pleasure purposes started in Europe in the eighteenth and nineteenth centuries. One of the primary objectives at that time was to climb the highest mountains in Europe. Later, similar ventures were pursued in other parts of the world such as the Himalayas, Andes and Rocky Mountains. While originally very much an elite activity, numerous mountain clubs started to be established from the mid-nineteenth century, predominantly of German origin with the main area of activity in **Austria**. An ever-increasing network of mountain huts was built in the middle and higher elevations throughout the European Alps and to a lesser extent elsewhere in the world. These serviced huts served two main purposes. First, they served as a base for the mountain climbers who were primarily interested in scaling specific peaks or climbing difficult rock faces. Second, they were convenient stops for those who were hiking from hut to hut for several days or weeks. Another purpose of the huts was as rest stops for day hikers who ascended from the villages and resorts for a day trip to the huts and their immediate environments.

The rapid increase in mountaineering since the Second World War has resulted in great pressures on the environment as millions of **tourists** try to climb the European Alps or enjoy the scenery. Soil compaction and erosion, damage to vegetation and disturbance to wildlife are but a few physical effects. The increase in demand also led to an expansion of the **accommodation** facilities, accompanied by an improvement in the standards and available facilities such as running hot water, gondolas and luxury accommodations. This phenomenon of serviced huts is probably the primary distinguishing criterion to mountaineering in other parts of the world, where such facilities rarely exist and are even frowned upon. On a world scale, mountaineering is obviously less important than beach-oriented tourism. Nonetheless, for some **destination** countries such as Austria or Nepal, mountaineering provides one of the tourist's primary activities and attractions. This form of tourism also includes a competitive element, namely the competition of human against **nature** and/or oneself, quite frequently resulting in serious accidents.

Further reading

Unsworth, W. (1992) *The Encyclopedia of Mountaineering*, London: Hodder and Stoughton.

MARTIN OPPERMANN, AUSTRALIA

multidimensional scaling

Multidimensional scaling (MDS) is a class of techniques for discovering the latent structure underlying a set of observations and is used to derive or to confirm a scale in a spatial configuration. It provides a way to visualise the structure of observations (such as perceptual mapping). Multidimensional scaling has a very close relationship with unidimensional scaling techniques (like the Guttman scale). However, MDS is used to derive an R-dimensional spatial configuration rather than a one-dimensional (linear) configuration. Furthermore, each axis in a multidimensional space is considered a uni-dimensional scale.

Multidimensional scaling derives a spatial configuration of objects based on proximity data. These data can be measures of spatial distance (the actual distance between different cities) or judgements of similarity (or dissimilarity) among objects (the number of contacts among people). Metric MDS refers to proximity data with either an interval or ratio property, whereas non-metric MDS refers to analysis based on rank-ordered data. Recent developments in MDS such as INDSCAL and INCLUS have extended two-way estimations to three-way solutions which allow one to estimate individual differences or temporal variations.

In order to best interpret the MDS configuration, rotations are useful. Procedures such as linear regression can be used to regress the coordinate scores of each object (or independent variable) onto those original proximity data (dependent variables). A high coefficient regardless of sign suggests the interpretation of meaning for each co-ordinate. This procedure provides results similar to the structural coefficients provided in canonical correlation or the factor scores in factor analysis. Multidimensional scaling has many applications in tourism research. For example, it is used to describe consumer visitation patterns, to classify **leisure** activity types, as well as to explore the **market** position of alternative destinations.

See also: correspondence analysis; marketing research; perceptual mapping

Further reading

Gartner, W. (1989) 'Tourism image: attribute measurement of state tourism products using multidimensional scaling technique', *Journal of Travel Research* 28(2): 16–20. (Provides an example of using MDS to understand images of tourism products and the underlying attributes.)

Kruskal, J.B. and Wish, M. (1978) *Multidimensional Scaling*, Sage University Paper on Quantitative Applications in the Social Sciences, series no. 07–011, Beverly Hills, CA: Sage Publications. (Presents a detailed discussion on the multidimensional scaling technique and its statistical properties.)

Moscardo, G., and Pearce, P.L. (1986) 'Visitor centres and environmental interpretations: an exploration of the relationships among visitor enjoyment, understanding and mindfulness', *Journal of Environmental Psychology* 6: 89–108. (Discusses an example of the use of MDS in clustering visitor centres.)

JIANN-MIN JENG, USA
DANIEL FESENMAIER, USA

multidisciplinary education research *see* education, multidisciplinary

multinational firm

A multinational firm, also known as a multinational corporation (MNC), can be defined as a corporation which has expanded its business(es) internationally. Such firms generally have subsidiaries strategically located around the world in areas where their businesses are concentrated. In the past decades, **hospitality** firms enjoyed tremendous growth opportunities and profitability. However, as the external environment moved from relatively stable to one that is increasingly changing at a faster pace, the industry operators realised that they no longer can manage their businesses the way they used to. By the late 1980s the domestic hospitality or tourism **market** showed signs of saturation. The demand curve for its products/services began to level off.

The hospitality industry in general is in the maturity stage of its **life cycle** and is experiencing limited growth domestically. Competition has become fierce, suppliers and customers have been gaining power, and the supply of labour has been low. Thus, many firms are seeking viable ways to survive the economic hardship that they are facing. As a result, the industry has turned to global expansion and thus the emergence of multinational firms as a means of continual growth and survival. Just as there are opportunities that arise from the global community, there are also new sets of challenges facing tourism executives. It is a considerable task for any given organisation to function effectively in the domestic market given today's dynamic environment. It will be more of a challenge for a MNC to compete successfully in the international markets as it faces a complex set of environmental conditions. Thus, it is imperative that multinational firms weigh the costs and benefits of expanding their businesses to certain countries due to the high level of risks the firm could face. Some of the benefits that tourism multinational firms enjoy are potential market share, continual growth in **profit** and return on **investment**. Some of the costs that multinational firms face are training costs of unskilled labour, global customer acceptance of the **product** and/or **service**, increased risks due to political unrest, availability of qualified **management**, and a host country that has the **infrastructure** for tourism development.

See also: economies of scale; joint venture; risk

Further reading

Jones, P. and Pizam, A. (1993) *The International Hospitality Industry*, New York: Wiley. (Discusses prevalent issues occurring at the international level of the hospitality industry and gives examples of multinational firms and their operating environments.)

ELIZA CHING-YICK TSE, CHINA

multiplier effect

Multiplier effects refer to an economic concept that was conceived in the nineteenth century and developed throughout the early period of the twentieth century, but not formalised until the work of John Maynard Keynes in the 1930s. The concept is now universally accepted amongst economists and applies to changes in exogenous **demand** for any industry's output, and is thus not solely related to tourism activity. Within the context of tourism multiplier effects are those economic **impacts** brought about by a change in the level or pattern of tourism expenditure. The term 'multiplier' is derived from the fact that the value of expenditure is multiplied by some estimated factor in order to determine the total economic impact. The multiplier effect can be estimated by using ratios that reflect either the *direct plus indirect* effects or the *direct plus indirect plus induced* economic effects of tourism spending.

The direct economic effects are those that occur at front-line tourism-related establishments. Therefore, when **tourists** spend their money in hotels, **restaurants**, **transportation** and communication services and retail outlets, for example, this will create direct income, output, government revenue and **employment** effects, as well as requiring some direct imports of goods and services. The direct effects are generally less than the volume of tourism receipts because of the expenditures that immediately leak out of the economy under study (see **leakage**; **economics**).

The indirect economic effects are those subsequent effects as a result of the direct economic effects. For instance, when the tourist spends money in a restaurant, the restaurant will spend some of the money it receives on food and beverage supplies, some of it on transport, heating and lighting, accountancy and other business services, and so on. All of these subsequent activities are classified as indirect effects, as are those economic effects created as the suppliers to these other industries find the demands for their services increasing. It is often the case that these subsequent demands for goods and services within the local economy result in further demands upon the front-line tourism-related establishments, such as when an intermediate supplier increases its demands for hotel and food and beverage services. In this case there will be further subsequent rounds of spending and this will continue, with the amount of money circulating getting smaller at each successive round of activity as money leaks out of the economy in the form of savings and imports, until the amount of money circulating in the economy as a result of the initial tourism spending becomes negligible.

The induced effects occur because at the direct and indirect levels of economic impact, income will accrue to residents of the local economy. Some of this money will be saved and leak out of the system, but some of it will be spent on goods and services within the local economy and this will generate further rounds of economic activity. This additional activity and its subsequent effects reflect the induced effects of the initial change in tourist spending.

The multiplier effect refers to the sum of these different levels of impacts. They can be positive, such as when the level of tourism spending increases, or they can be negative, such as when the level declines. However, it is important to note that the volume of tourism spending can remain unchanged but its distribution amongst the different economic sectors may change, and this can bring about a change in the economic impacts of tourism spending.

A distinction may be drawn between partial and complete multiplier effects. The former refers to the multiplier effect associated with a single productive sector (industry) of the economy, so that if the demand for that specific industry's output changes then the total impact, taking account of the secondary effects generated by changes in its output needs, can be estimated. However, the latter (complete) multipliers refer to the economic impacts of changes in the level and distribution of tourism spending across all sectors of the local economy. They are, in effect, a weighted average of the partial multipliers where the weights are determined by the distribution of spending.

There are a range of methods that can be used to calculate multiplier effects, including export base theory **models**, Keynesian multiplier models, ad hoc multipliers and **input–output analysis**. The export base theory and Keynesian multiplier models are no longer utilised because the high degree of aggregation makes the models redundant for **policy** purposes. The method used to estimate the multiplier effects will, to some extent, determine the values of the multipliers. For instance, the ad hoc multiplier models generally estimate multiplier ratio values that are approximately thirty per cent lower than input–output multiplier coefficients for the same economy. This is because the latter includes all of the sectors of the economy rather than just those that are clearly linked to tourism activity.

Regardless of the methodology used, there are a number of different multiplier effects that can be attributed to changes in tourist spending. The first, transactions/sales multipliers, refers to the change in the volume of transactions as a result of one unit change in tourism expenditure. The second, output multipliers, relates to the effect of one unit of expenditure upon the level of output within the economy. The third, income multipliers, refers to the shift in local income created as a result of a change of one unit in tourism spending. The income encompasses all its forms, including wages, salaries, profits, rent and interest, but does not include those which are repatriated to persons or businesses outside the economy in question. The fourth, employment multipliers, refers to the effect upon the number of jobs available that are supported by a given amount of spending. These multipliers are generally expressed with respect to changes in spending greater than unity because of the magnitude of the multiplier values. Fianlly, government revenue multipliers, refer to the changes in government revenue from all sources created by a unit change in tourism spending.

It should also be noted that changes in tourism spending also create predictable changes in the volume of imported goods and services. These can be estimated in the same way as any of the other economic indicators noted earlier. The multiplier concept is an invaluable tool for use by those involved in the policy formulation and planning of tourism **development**. Multiplier values will provide information relating to human resource requirements, government revenue, imports and income level changes that are essential if tourism is going to be developed and maintained in an optimal fashion.

Further reading

Fletcher, J.E. (1993) 'Input–output analysis', in S. Witt and L. Moutinho (eds), *The Tourism Marketing and Management Handbook*, 2nd edn, Hemel Hempstead: Prentice-Hall, 480–5.

Fletcher, J.E. and Archer, B.H. (1991) 'The development and application of multiplier analysis', in J. Fletcher (ed.), *Progress in Tourism and Recreational Research*, London: Francis Pinter, 28–48.

JOHN FLETCHER, UK

museum

Museums are institutions for the collection, preservation, exhibition and explanation of cultural and natural phenomena. Typically they focus on culturally defined branches of knowledge such as **art**, **history**, **religion**, **geography** and natural history. These institutions overlap with natural, cultural and ethnic interpretative centres and ecomuseums; with places of entertainment such as wax museums, Disneyland or Universal Studios; with zoological and horticultural gardens; and with preserved **landscapes** such as archaeological sites, architectural monuments and natural parks and reserves.

Museums and tourism have much in common both intellectually and historically. In their Western forms (modern Asian tourism stems more directly from pilgrimages), both tourism and museums started as privileges of the nobility and upper classes as post-Renaissance phenomena for 'knowing the world', in the forms of the **Grand Tour** for northern Europeans, and cabinets of curiosity where the European elite accumulated, classified and displayed natural and artificial wonders gathered in their expansionist world.

The mercantile system increased wealth so that both tourism and museums broadened their clientele. More rich people travelled and museums became the property of the rich as well as the nobility and, starting with the Ashmolean Museum at Oxford in 1683, of educational and scientific bodies. In the nineteenth century, the industrial revolution saw the decline of aristocracy and the growing wealth of entrepreneurs. Tourism became more routinised and, with the advent of steam-powered **transportation** and the commercial successes of Thomas Cook, it expanded its geographical scope and was opened up to the majority of the middle classes. In the meantime, museums became national, regional and municipal institutions, open to all the public and increasingly allied to the democratisation of education. The twentieth century expanded on these trends. Universal education was extended to most of the world. Since the Second World War, greater affluence and better transportation have enabled most middle-class people to travel internationally. Museums have multiplied tenfold to one hundred-fold worldwide, and they have become essential ingredients of much middle-class tourism.

Common values of the middle **class** are expressed in both museums and tourism in general save the most hedonistic: education, to increase understanding of the cultural, ethnic, historical, natural and more diversities of the world; conservation of the past for its own sake and to maintain continuities between the past and the present; aesthetic appreciation and sense of the wonderful and the special; status enhancement, for the tourist *vis-à-vis* their home reference group and for museums as markets of place, status, taste and enlightenment in their local community and the larger world; and entertainment and the provision of a relaxed social atmosphere, in contrast to the workday world.

Museums fall into two types with respect to their functions for tourism. The first may be called 'world museums'. These are deemed to be important ingredients of the **heritage** of the middle classes of the world. In addition, they are **markers** which stand for and serve as essential components of the touristic appeals of their home cities. Such museums are primarily **art** museums, like the Louvre in Paris, the Metropolitan and MOMA in New York or the Ueno Park in Tokyo. Such museums are great assets as attractions, for they are thought to contain some of the best of mankind's cultural productions and are a 'must see' for cultural tourists (see **cultural tourism**). They are exemplars of their fields, and they provide a set of values or criteria against which other collections may be judged.

The competition for **tourists** is often played out in efforts to raise regionally famous museums to such world visibility, such as the Gulbenkian in Lisbon or the new Museum of Modern Art in San Francisco. The same could be said for natural history or other comparable museums, whereby the world class institutions such as the South Kensington museums in London or the American Museum of Natural History in New York have a global status that newer entries such as the Ring of Fire Museum in Osaka or the Exploratorium in San Francisco may envy. Only rarely have regional institutions been able to leap into the world class category, as did the Canadian Museum of Civilisation of Ottawa or the Museo Nacional de Anthropologia in Mexico City.

The second type, the local or regional museum, serve as a touristic guide to the historical, geographical, natural historical or ethnic characteristics of a region. These museums help define the identity of a region, or, in the case of many **Third World** nations, the nationality. They frequently serve as interpretative centres for tourists who want to quickly grasp the key characteristics of a locality. Though nominally educational, such museums are under pressure to be informally didactic, or even 'info-tainment,' to present information in a light-hearted way that does not appear to the consumer to be 'learning.'

Recent trends in world tourism show a sharp increase in 'cultural tourism' as opposed to more hedonistic or recreational kinds. Museums, along with art galleries, theatres, film festivals and other cultural assets, play a large part in attracting tourists. Economic calculations show that each 'cultural' **attraction** has a large monetary value, by bringing in visitors who spend money on these as well as on beach, natural feature, sports event and recreational attractions.

Further reading

Ames, M.M. (1992) *Cannibal Tours and Glass Boxes: The Anthropology of Museums*, Vancouver: University of British Columbia Press. (A critique of contemporary, especially anthropology, museums as possible sites for hegemonic class-based world views.)

Graburn, N. (1977) 'The museum and the visitor experience', in L. Drager (ed.), *The Visitor and the Museum*, Berkeley, CA: Programs Planning Committee, Museum Educators of the American Association of Museums and the Lowie (Hearst) Museum, pp. 5–26. (Discusses multiple roles of modern museums, as awe-inspiring houses of the sacred and inspirational, as institutions of informal learning about the social and natural world, and as gathering places for locals or visitors.)

Kaplan, F. (ed.) (1994) *Museums and the Making of 'Ourselves': The Role of Objects in National Identity*, London and New York: Leicester University Press. (Concerns the identity-creating and maintenance roles of regional and national museums in the First, Third and Fourth Worlds.)

Richards, G. (ed.) (1996) *Cultural Tourism in Europe*, Wallingford: CAB International. (Demonstrates the rise of cultural as against other forms of tourism in contemporary Europe, and the crucial role of museums as icons and attractions in tourism circuits.)

NELSON H.H. GRABURN, USA

music

Music has an intrinsic relationship to tourism as an **attraction**, as a marker of status and **lifestyle**, and as the bearer of messages and **motivation** to travel. At the simplest level, musical events and performances are almost ubiquitous attractions for **cultural tourism** of all kinds, from the famous summer music festivals of Europe, to the many jazz events of North America, or the stereotypic steel band, rumba and carnival music of the Caribbean. With the spread of post-colonial diaspora, musical events are no longer confined to their places of origin, but may partake in the mélange of metropolitan attractions, such as African music in Europe and Japan, or Indonesian *gamelan* bands in North America and elsewhere.

Furthermore, music is internationally recognised as an essential part of cultural and lifestyle expressions. Even with special events, tourists stereotypically expect certain music to accompany their experiences of cultural places, such as waltzes in Vienna, jazz in New Orleans or rumba in Cuba (Daniel 1995). The tourism industry uses these associations in its **advertising** and in providing authenticating **markers** for **resorts** and target areas. These place associations are also part of a large system of **class** and status markers (see **prestige**), with musical tastes being ranked according to systems of connoisseurship indicating the class, status and lifestyles of associated locales. Powell's (1988) pioneering research has shown the way to analyse a more subtle power of music in providing, through lyric and **image**, messages for tourism. She shows how pop music of the 1960s, 1970 and 1980s provided models for **youth tourism**, specific to age and **gender** groups, giving 'permission' to visit new destinations in a cultural framework of **alternative tourism**.

References

Daniel, Y. (1995) *Rumba: Dance and Social Change in Contemporary Cubs*, Bloomington, IN: University of Indiana Press. (Examines how a lower class dance has been celebrated, professionalised and commoditised for foreign tourists.)

Powell, A. (1988) 'Like a Rolling Stone: notions of youth travel and tourism in pop music of the sixties, seventies and eighties', in N. Graburn (ed.), *Kroeber Anthropological Society Papers* (67–8): 28–34. (Shows how pop music provides models for youth tourism.)

Further reading

Lie, J. and Abelmann, N. (1992) book review of *Kanko to Ongaku* (*Tourism and Music*), *Annals of Tourism Research* 19: 609–12. (Covers many aspects of the interactions between tourism and ethnic and popular music.)

NELSON H.H. GRABURN, USA

myth

Myths are stories in which the sacred and the paradigmatic are intertwined. They speak of desires, beliefs and the ambivalence and contradictions of life in the world. Dufour's (1978) scholarly study of the French weekend examines some of the ancient mythological connotations associated with **weekend** tourism. Dufour argues that, through the use of such fundamental symbols as water, sun, earth, fire and sky, weekenders are in fact engaged in quasi-divine contemplation of themselves, their natures and their place in the world (see also **symbolism**). His most fundamental hypothesis is that the tourist is 'l'homme du jeu', or 'l'homo ludens'.

Barthes (1983) argues that the experiences of the Eiffel Tower by Parisian citizens and tourists are essentially totemic. Not only does the tower denote Paris and its inhabitants, but it also connotes multiple mythological associations between the prehistorical (the river from which early man emerged), the medieval and the modern worlds. A collection of essays on tourism mythology (Selwyn 1996) reveals a wide range of themes underlying destinations. Israeli walking tours mythologically reconstruct the origins of the state, Japanese coastal tourism thrives off myths of sexy, pearl-diving women, while tours in the English West Country feed off nostalgic longings for the life of rural communities long gone (see **nostalgia**).

Romantic Western tourists like to believe that Nepal is a 'Shangri-La', while in fact Nepal is one of the world's poorest countries with high levels of infant mortality and poverty. To borrow Dufour's words, myths are at once 'liberating' in the sense that they are vehicles for imagining the possible, and 'desirable' in that they lead perception away from the pragmatic realities of history and political economy. In these senses, tourism myths reveal one sort of truth in the process of concealing another. Myths lie at the very heart of **tourism motivation**, fuelling dreams and fantasies. Despite some seminal work, the area remains relatively understudied and presents future researchers with a rich seam of research possibilities, in which tourism sheds light on mainstream sociological, religious, economic and political questions. Tourism myths suggest not so much how the world really is but how one would really like it to be.

References

Barthes, R. (1983) *The Eiffel Tower*, New York: Hill and Wang. (Analyses the Eiffel Tower as a totemic object for both Parisians and tourists.)

Dufour, R. (1978) 'Des mythes du loisir/tourisme de weekend: alienation ou liberation', *Les Cahiers du Tourisme*, Serie C, No.47, Aix-en-Provence: Centre des Hautes Etudes Touristiques. (Examines links between weekend tourism in France and ancient mythological themes.)

Selwyn, T. (1996) *The Tourist Image*, London: Wiley. (Offers a collection of largely anthropological essays on the subject of tourist mythology.)

TOM SELWYN, UK

national character

While complex and under threat from **globalisation** and **nationalism**, national character can be a valuable and stable **marketing** tool for tourism. It is the stereotypical personality that most **host and guest** populations hold in their mind when they think of a particular country or community of people. It is directly related to the culture of a country and is reflected in how its citizens view themselves, as well as in how they are viewed by the rest of the world. In reality, most countries are very complex, with national characters that have formed over a long period of settlement and nation building.

The overall national character of a country is most easily seen in its international relations and the personalities of its institutions. Thus, some countries are aggressive in international diplomacy while others work quietly behind the scenes. Some are characterised as logical and business oriented, while others are romantic and artistic. These **stereotypes** are more fundamental than the daily changes in a country's political or economic circumstances. While tourism **images** of a country as a **destination** may change rapidly as a result of sudden misfortunes, the underlying character of the nation changes very slowly. This make national character an important part of a country's **marketing plan**.

The process of accurately generalising the complexities of a country's character is difficult, and can be offensive to **minorities** who may see it as ethnic **nationalism**. However, positive images that reflect the national character of a country play a major role in their promotion as destinations (see also **promotion, place**). It is efficient to be able to portray the character of a country in a single image or icon, such as the maple leaf for Canada or the Eiffel Tower for France (see **ethnography**; **marker**). Tourism marketers hope that the more stable national character is what will attract tourists through changing political and economic fortunes.

There is some concern that distinct national characteristics are being acculturated (see **acculturation**) and displaced by **globalisation**. Tourism is accused in this process because of its tendency to modernise traditional societies and Anglicise everyone else. The governments of some countries (such as France and Indonesia) are responding to this by requiring that elements of the national culture, especially language, not be displaced by foreign influences on signs and in the **media**. At the same time, knowledge of a nation's character is important for increased **international understanding** and **cross-cultural management** in tourism.

See also: myth; semiotics; sociology

Further reading

Crompton, J.L. (1979) 'An assessment of the image of Mexico as a vacation destination and the influence of geographical location upon that image', *Journal of Travel Research* 17(4): 18–23. (Uses semantic differentials to characterise different areas in Mexico to derive an overall image of the country.)

Lew, A.A. (1992) 'Perceptions of tourists and tour

guides in Singapore', *Journal of Cultural Geography* 12(2): 45–52. (Compares the attraction and overall country image of tourists and residents.)
Pizam, A. (1995) 'Does nationality affect tourist behaviour?', *Annals of Tourism Research* 22(4): 901–17.
World Tourism Organization (1978) *Tourist Images*, Madrid: WTO.

ALAN A. LEW, USA

national park

National parks are officially designated areas of a substantial size, established by a central government in which natural **resources** and processes are protected by legislation. Such parks are usually afforded **management** structures and processes to ensure the continued protection of the resource base. **Recreation** and tourism are often activities that are licensed or permitted subject to the requirements for nature **conservation**.

The first national park bearing this distinct title, the Yellowstone National Park in the United States, was designated in 1872. This park, designated at a time of rapid land **colonisation** of the North America Continent, was established to fulfil three objectives: to prevent exploitation of wildlife, areas of natural beauty and the **environment**; to enable visitors to derive enjoyment from their contact with the protected area; and to promote scientific study of the natural systems and ecosystems within the national park. These three themes have, by and large, defined the general concept of national parks up to the present, including parks that are designated as biosphere reserves and world **heritage** sites.

Other nations of the English-speaking world were relatively quick to adopt similar measures. **Australia**, **Canada**, **New Zealand** and **South Africa** had established national parks by the end of the nineteenth century, and **India** followed in the first decade of the twentieth century. In Europe, the governments of **Germany**, **Russia**, **Sweden** and **Switzerland** had all established some sort of national parks prior to the First World War. In the United Kingdom and the Netherlands, national parks had been established by private rather than governmental initiative (Allin 1992). The concept was also spreading to colonised nations. The Dutch government followed the American example by establishing the Udjung Kulon Reserve in Java early in the twentieth century. American influence has also been cited by those responsible for national park development in England, **Japan**, Sweden and Switzerland. By Yellowstone's centenary there were 1,000 national parks throughout the world that meet the International Union for the Conservation of Nature's (IUCN) criteria. In 1969, a formal IUCN resolution defines national park as:

> A relatively large area where one or several ecosystems are not materially altered by human exploitation and occupation, where plant and animal species, geomorphologic sites and habitats are of special scientific, educative and recreative interest or which contains a natural landscape of great beauty; and the highest competent authority of the country has taken steps to prevent or eliminate as soon as possible exploitation or occupation in the whole area or to enforce effectively the respect of ecological, geomorphologic or aesthetic features which have led to the establishment; and visitors are allowed to enter, under special conditions, for inspirational, educative, cultural and recreative purposes.

This definition also included several recommendations for certain areas that should not be designated as national parks. Included here are nature (or scientific) reserves where access is limited to those with special permission; reserves operated by lower order governmental authorities (like state, provincial or local government) and private reserves; inhabited and exploited areas that may have protected status but in which **recreation** and tourism **development** take precedence over the conservation of ecosystems (IUCN 1975).

The requirement for designation at the national level in the establishment of national parks can be problematic, especially in federations and such situations where land ownership is vested at the provincial or state levels. In these situations, **public participation** at local levels and the demonstration of local level economic **benefits** is now inevitably required to support a proposition for new park designations. In other societies, such as the Polynesian Pacific, where 'native communal land' rights exist and where land ownership is inalienable, the concept of a 'national park' is similarly meeting new challenges.

The original 'Yellowstone' model of protecting wild lands (see **wilderness**) also has only limited applicability in heavily populated areas, or where **landscapes**, such as in Europe and the United Kingdom, present evidence of human modification. Some of these landscapes, while not meeting the strict criteria of national parks *per se*, still hold considerable merit when the broader objectives of **sustainable development** and the retention of biodiversity are considered. Situations where **indigenous** groups may at times harvest traditional resources on a sustainable basis also present a challenge to a strict definition. In such situations, national parks can be seen as reserves to protect cultural as well as natural processes.

While the national park concept has served **nature** and the world well, its application is inevitably limited. By definition it cannot apply to those landscapes (or seascapes) of special quality where resident populations and their resource use patterns are integral but have materially altered their naturalness. Thus the strict definitions of national parks, born in largely frontier societies at the end of the nineteenth century, were found lacking in their application in places with higher population densities and longer histories of human habitation. The recent trend is to see national parks as one part of a spectrum of 'protected landscapes' (Lucas 1992) and part of wider action on sustainability (see **Agenda 21**). When the concept of a national park was first introduced, it represented a seminal step in defining a system of protected areas. Today, national parks are recognised as the second category in the IUCN's wider **classification** of protected areas, which ranges from nature/scientific reserves, to protected land (and seascapes), to multiple use management areas. Within this classification, human habitation and **recreation** and **leisure** activities assume a greater significance *vis-à-vis* the goals of strict nature conservation.

As unimpeded **sites** become rarer and the progress of national park establishment slows, the search for alternative mechanisms to achieve the goals of nature conservation has broadened. Thus the IUCN has recognised a separate category for nature conservation, that of protected landscapes/seascapes (IUCN Category V) where people are a permanent part of the landscape and day-to-day activities affecting it in a harmonious and sustainable relationship. Such places importantly offer opportunities for public enjoyment through recreation and tourism within the normal **lifestyle** and economic activity of these areas (IUCN 1985). The concepts of protected natural landscape and national parks are closely related. Protected landscape has been a key **management** tool for conservation in Europe for forty years, as announced in the Lake District Declaration of 1987. In some countries, such as Japan and the United Kingdom, they are still officially called 'national parks' in spite of a call by the IUCN in 1969 requesting governments to retain the title 'protected landscape' for areas which have national park management objectives.

National parks are, where appropriate, sometimes cross-listed in other international classifications and systems of protection. Important among these are biosphere reserves and world heritage sites. Biosphere reserves are established under the Man And the Biosphere programme serviced by UNESCO in Paris. The principle aim of such areas is a conservation role reinforcing the conservation of genetic resources and ecosystems and the maintenance of biological diversity. Similarly, the World Heritage Convention (1978) – the convention concerning the Protection of the World Cultural and Natural Heritage – aims to identify, list and protect natural and cultural properties of outstanding universal (that is, world as opposed to 'national') value for the benefit of all people. Traditionally, natural and cultural sites have been considered separately by IUCN and ICOMOS (International Council on Monuments and Sites), respectively, and the important interplay between nature and culture has tended to be neglected. Notwithstanding, many national parks meet the requirements for world heritage site status (such as the Galapagos Islands), and together with many cultural sites and monuments (including designated temples of the Kathmandu Valley and the Great Wall of China) form a specific form of **attraction**. World Heritage Sites are administered under a separate UNESCO jurisdiction, the World Heritage secretariat.

National parks and other protected areas have special significance for tourism. They stand as national icons or symbols of a **destination**. They are attractions in their own right, and are important sites not only for nature conservation but for a range of recreational activities. In more recent times they have become closely associated with new forms of tourism, **nature tourism** and **ecotourism** in particular. Because of their pre-eminence in tourism, there is almost inevitable debate about the amount and form that tourism may appropriately take in such situations. The literature reports concerns of overuse, trampling and other environmental **impacts**, and considerable discussion exists about recreational **carrying capacity** and other management requirements.

In response, management practices have variously attempted user education, nature interpretation, the establishment of use fees, the granting of permits and concessions, and other forms of rationing. Currently, management practices focus on environmental monitoring, impact assessment (see **impact assessment, environmental**) and defining **limits of acceptable change**. As tourism has grown in size and scope, the management of some national parks has recently required that facilities that support (and encourage) high levels of use be removed to outside park boundaries.

References

Allin, C.W. (1992) *International Handbook of National Parks and Nature Reserves*, New York: Greenwood Press.

International Union for the Conservation of Nature (1975) *United Nations List of National Parks and Equivalent Reserves*, Morges: ICUN.

—— (1978) *Categories, Objectives and Criteria for Protected Areas*, Morges: IUCN.

DAVID G. SIMMONS, NEW ZEALAND

National Recreation and Park Association

The National Recreation and Park Association (NRPA) goes back to 1898 when some leading superintendents of parks from Massachusetts met in Boston and formed the New England Association of Park Superintendents. Between 1898 and 1965 several organisations were formed, including the American Recreation Society which was founded in 1946. With the emerging emphasis on outdoor recreation during the late 1950s and early 1960s, new opportunities came for blending **recreation** and **park** philosophies. Coupled with increasing competition among recreation and park related organisations, this process in 1965 created incentives for the National Recreation Association to merge with American Institute of Park Executives, American Recreation Society, and the National Conference on State Parks to form the National Recreation and Park Association.

NRPA promotes the interests of the park and recreation movement through public information, political advocacy, research and professional development. Through its divisions and programmes, the association strives, among other things, to build public awareness of the role of physical fitness in health, encourages recreation among the elderly, and promotes standards for recreation services for the **handicapped**. Its National Therapeutic Recreation Society is working to improve professional qualifications and standards. Its publications include *Dateline: NRPA* (bimonthly), ***Journal of Leisure Research*** (a quarterly), *Recreation and Parks Law Reporter* (quarterly) and *Legal Issues in Recreation Administration* (quarterly). The headquarters of NRPA is located in Alexandria, Virginia, United States.

TURGUT VAR, USA

National Restaurant Association

Since its founding in 1919, the mission of the National Restaurant Association (NRA) has been to protect, educate and promote the **food** service sector and its 9.4 million employees. Membership in this association provides a linkage to more than 30,000 members worldwide, representing 175,000 food service outlets including **restaurant** and **fast food** outlets. Its membership includes a variety of businesses, professionals and the academic community associated with the food service sector. Members of the association receive, among other things, the monthly magazine *Restaurant USA* and the *Washington Weekly Newsletters*. The NRA is known for it annual 'Restaurant Show' held is Chicago. Its first convention in Kansas City, when the association was formed, brought about 68 attendees from 17 states. Today the convention is one of the largest in the world, attracting more than 100,000 participants from many countries, with more than 19,000 exhibitors displaying their products and services. The show, designed for all levels of **hospitality** management, also features a wide range of educational programmes, **management** clinics and presentations. The Education Foundation of the NRA is a non-profit organisation whose mission is to enhance the **professionalism** of the food service sector through **education** and **training**. It organises a wide array of training seminars designed to keep foodservice operators educated and current. It administers the largest career assistance programmes of its kind in this field. Scholarships include undergraduate, professional management development programme, teacher work–study grants, graduate fellowships and industry assistance grants. The headquarters of the association is located in Chicago, Illinois.

MAHMOOD KHAN, USA

National Tour Foundation

Chartered in 1982, the National Tour Foundation (NTF) was created to promote **tour operator** travel services in North America. Its mission is to enhance quality of **education** programmes and processes in providing travel services, to improve the **quality**, scope and use of research and information for **planning**, developing and **marketing** tourism **products**, and to create awareness of tourism's value. Its headquarters is located in Lexington, Kentucky, USA.

MARIA FUENMAYOR, USA

national tourism administration

The core function of an NTA is to provide central government with poicy advice relating to tourism. Alternatively it may exist as a small policy unit within a larger ministry or department. Elsewhere, policy-making may be undertaken by a larger multi-functional national **tourism organisation** (NTO). In such a case, the terms NTA and NTO are often used synonymously.

Further reading

Elliot, J. (1997) *Tourism: Politics and Public Sector Management*, Routledge, London.

Pearce, D.G. (1992) *Tourism Organisations*, Longman, Harlow.

DOUGLAS G. PEARCE, NEW ZEALAND

nationalism

The term 'nation' is normally used to refer to a given set of people thought (by themselves and/or others) to be bound together by the sharing of such common and distinctive characteristics as territory, language, **religion**, cultural traditions, history and somatic features. Nationalism refers to the sense of belonging to the same nation.

Unlike their nineteenth-century forebears, who regarded nations as part of the natural order of things, more recent scholars have stressed the fact that nations are made rather than born. For Gellner (1983), both nations and nationalism arose as a direct consequence of the conjuncture of capitalism, printing and the need in the modern world for a coherent unit of communication and administration. In similar vein, both Hobsbawm and Ranger (1983) and Anderson (1983) have regarded the nation as an imagined or invented community, and nationalism as the vehicle (linguistic, cultural, literary, mythological) to express that imagination; both nations and nationalisms emerging from specific political and economic circumstances.

It is, precisely, by way of the **images** and imaginings spawned by the tourist industry that nations and nationalism enter tourism studies, specifically through their association with the idea of '**heritage**'. This is a term which has been adopted, if not appropriated, by tourism planners and managers to describe some tourist sites as embodiments of supposedly 'essential' or 'basic' national characteristics. Thus, for example, parts of the English tourist circuit are made up of such monuments to nostalgic British nationalism as Buckingham Palace, the Tower of London and Oxbridge colleges. Other parts of the UK tourist itinerary are attended by complimentary sorts of imagery such as souvenir shop postcards consisting of photographs of bull dogs, men with bowler hats waving umbrellas, of London 'bobbies', and of people (who often seem to be placed on bridges over the river Thames) with either breasts or beer bellies adorned with Union Flags. All of this paraphernalia makes the tourist appear as a sort of voyeur specialising in what are imagined to be nationalist totems. Indeed, there are schools of thought amongst tourism advertisers and compilers of brochures that using the 'nationalist' card can yield dividends, especially in the case of a destination such as the UK, whose weather patterns disqualify it from playing the 'sun, sea and sand' card too boldly.

To tinge a site with nationalist connotations may, as in the British case for example, amount to little more than to add a touch of frothy glamour to tourist products. In contexts of real nationalist rivalries in shared or neighbouring territories, on the other hand, some 'heritage' sites (national memorials, religious buildings, relics of past conflicts, museums, for example), used by the tourist industry as tourist attractions can and do take part in altogether more serious nationalist struggles for hearts and minds. In this sense the economics and politics of contemporary tourism are, in various parts of the world, giving nationalist sentiments a new sort of fillip.

References

Anderson, B. (1983) *Imagined Communities: Reflections on the Origin and Spread of Nationalism*, London: Verso.

Gellner, E. (1983) *Nations and Nationalism*, Oxford: Blackwell.

Hobsbawm, E. and Ranger, T. (1983) *The Invention of Tradition*, Cambridge: Cambridge University Press.

TOM SELWYN, UK

natives

In the past, people who were objects of the tourist **gaze** were considered as natives. The term now has derogatory connotations, implying that these **indigenous** people have a lower culture, while the **tourists** from the West have political **power** and superior knowledge. But, today, in a transnational global era, the distinction is less clear since former natives now travel as tourists in turn to the centres of Western power.

See also: ethnic group; minorities; professional native

EDWARD M. BRUNER, USA

nature

Nature is undeveloped resources including water, vegetation soil and wildlife that support and attract tourism activities. These **resources** in nature influence tourism activities as **attraction** features, settings or pristine areas. As attraction features such as waterfalls, the resources themselves bring in tourism because people enjoy experiencing natural wonders. In settings such as the rain forests, the organisation and arrangement of resources brings in tourists because it is ideal for certain activities like **hiking** or bird watching. The undisturbed condition of areas such as uninhabited tropical islands, jungles and inland waters, has a special appeal to tourists, hence has the potential to generate income from them, increasing the incentive for the host **community** to protect its resources.

The value of nature as it relates to tourism includes aesthetic, ecological and ethical components. Aesthetically, the value of nature is a collection of resources that creates visual, auditory and other sensory effects which can be experienced by tourists. They visit areas to witness these effects first hand in an authentic (see **authenticity**) experience that permits them to explore the mystery and unknown elements of nature. Ecologically, the value of nature is for its own sake, where it is seen as more than a collection of resources and involves an interrelated, interconnected set of functions and processes composing a greater ecosystem. Ethically, the value of nature is in protecting the natural resources and processes, and in preventing **impacts** that tourist activities may cause. Protecting nature and avoiding negative **impacts** is a common element of many **ecotourism** and **nature tourism** definitions. The difference between these definitions is in part due to how this environment is interpreted. This in nature-based tourism represents an aesthetic value of using the resources of the non-manmade **environment** to support tourism activities whereas the nature in ecotourism represents ecological values of ecosystem protection, and ethical values of enhancing or maintaining the balance between tourism use and natural system integrity.

Further reading

Armstrong, S.J. and Botzier, R.G. (1993) *Environmental Ethics: Divergence and Convergence*, New York: McGraw-Hill. (Chapter 3 compares the aesthetic and ecological values of nature.)

Nelson, G. (1994) 'The spread of ecotourism: some planning implications', *Environmental Conservation* 21(3): 248–55. (Pages 48–250 describe the ethical components and main principles of ecotourism and nature-based tourism.)

TRACY FARRELL, USA

nature tourism

The goal of experiencing flora and fauna in natural settings is regarded as nature-based tourism, often used synonymously with **ecotourism**. Increased awareness of the **environment** and greater **accessibility** to remote regions have made this a fast growing form of tourism. Whale-watching, visits to rain forests and to habitats of large mammals compliment the more traditional activities of bird-watching and observing scenery in **national parks**.

CHARLES S. JOHNSTON, NEW ZEALAND

nature trail

A path through a natural area featuring indigenous biological and/or geologic features is described as a natural trail. Interpretative signs may be present, or self-guiding brochures provided to enhance the educational value of the experience. Generally, nature trails are short loops requiring at most a few hours to complete, and are popular in public parks and wildlife preserves.

TOM BROXON, USA

nearest neighbour analysis

Nearest neighbour analysis produces a quantitative measure of whether a point pattern (where the points represent discrete entities such as hotels or other tourism businesses) is clustered, uniform or random. The technique is used to formulate and test hypotheses regarding forces affecting the **location** of tourism facilities.

STEPHEN SMITH, CANADA

need, recreational

A common **classification** of human needs is a threefold typology identifying primary, derived and integrative needs. Primary human needs are the most basic and relate to such necessities as food and defence. Derived needs refer to such things as education and language, while **recreation** can be classified as an integrative need. In affluent Western societies, primary needs are generally well satisfied so that many citizens are in the fortunate position of being able to give some of their **time** to the **satisfaction** of less essential, integrative needs. Many governments around the world, for example, have legislated for regular holidays and **weekend** breaks from work, a **trend** that recognises the values and benefits to be derived from non-work time. The academic literature, however, is deeply divided over the question as to whether people can be said to have a genuine 'need' for recreation. Some argue that periodic **escape** from a stressful urban setting is essential for psychological well-being. Others note that there is no clear evidence that a lack of recreational opportunities or the presence of an overstimulating environment has ever contributed to physical or mental stress, and conclude that it is impossible to put precise figures or predictions of such claims.

In the twentieth century, recreational planning (see **planning, recreation**) in affluent, industrialised cities has often involved the identification and use of normative criteria for judging whether an area is disadvantaged in terms of available recreational opportunities, or whether people are having sufficient non-work time. Standards relating to hectares of open space per thousand population in metropolitan areas are of this kind, and in the literature this has come to be known as a measure of 'normative need'. However, **critics** have pointed out that in most instances such 'standards' have no basis in sound research, are simply used for no other reason than that they have been used before elsewhere, and should not be taken as a measure of any kind of 'need'. What can be said, however, is that differences between areas in terms of open space provision (or some other parameter) can indeed be quantified, and this can be used as a measure of 'comparative need'. Finally, while 'felt needs' (sometimes known as latent **demand**) are not easily measurable, 'expressed needs' (in the form of actual recreational **behaviour** patterns) certainly are. What is always called for is careful identification and definition of the kind of 'need' being discussed.

See also: behaviour, recreation; demand, recreational; participation, recreation

Further reading

Bradshaw, J. (1972) 'The concept of social need', *New Society*, 30 March: 640–3.

Mercer, D.C. (1973) 'The concept of recreational need', *Journal of Leisure Research* 5: 37–50.

DAVID MERCER, AUSTRALIA

neo-colonialism

Many international analysts believe that tourism readily recreates the dependencies of colonialism (see **colonisation**) and can become a form of leisure **imperialism** or hedonistic neo-colonialism. Such ideas stand as powerful **rhetoric** to illustrate the ties between core and periphery areas, and to describe the potential loss of control which a removed/isolated host region/state may face *vis-à-vis* controlling foreign or elite interests.

KEITH HOLLINSHEAD, UK

Nepal

Nepal, a mountain kingdom of 22 million people, remained isolated until the Chinese invasion of Tibet forced Nepal to seal its border with the latter and encourage stronger ties with India. The result was the country's first motor road from the Indian border to the capital, Kathmandu in 1950. The Chinese invasion also created a massive inflow of Tibetan refugees who would support themselves by creating handicrafts for the fledgling tourism industry.

Initially tourists consisted of mountaineers attracted to the Himalayan peaks, the highest in the world. Today, **trekking** is a major tourism form. The government struggles with how to balance the fragile mountain **environment** and the need for the revenue trekking permits and the resulting **employment** generate. Litter, deforestation and social **impacts** are getting increased attention. Budget tourists often attracted by the drugs available have been largely replaced by luxury tourism built around trekking, the wildlife **parks** in southern Nepal and the rich cultural artefacts of this predominantly Buddhist society. Temples and antiquities have inspired massive international interest in **heritage** protection. Many art treasures were smuggled out of Nepal during its early years of tourism. These losses encouraged the **Pacific Asia Travel Association** to make heritage preservation awards. More recently, the United Nations Development Programme has assisted in Nepal's tourism reorganisation and **planning**.

Tourism became the leading **industry** in Nepal in 1983, but it has often suffered from its dependence on its politically unsettled neighbours for **airline** flights and overland links. India has on occasion stopped flights to Nepal. Nevertheless, by 1996 Nepal was attracting 404,000 **tourists** and generating $130 million in tourism receipts. Recently, the Ministry of Tourism and Aviation was strengthened by assuming more policy planning and regulatory roles. A Nepal Tourism Board, consisting of representatives of various sectors of the industry and the major government agencies with which they interact, was created to further coordination and cooperation between government and the private sector. 'Visit Nepal Year – 1998' has been promoted around the need for **ecologically sustainable tourism**. Its website furthers this goal by specific suggestions for tourists concerning the ecology and culture of the country.

See also: jungle tourism; smuggling

Further reading

Richter, L. K. (1989) *The Politics of Tourism in Asia*, Honolulu, HA: University of Hawaii. (Chapter 8 deals with Nepalese tourism development.)

LINDA K. RICHTER, USA

Netherlands, the

Tourism has been an important business for the Netherlands since the 1960s. The coast, the flowers, the cities and the cultural appeals make the country an attractive **destination** all year around. One problem is the congestion in summer: April and May are important months for the flower tourists, and July and August are busy because of the short period of time in which the Dutch have their school holidays.

The Netherlands, with a population of 16 million, has good **infrastructure** with a well-equipped one-terminal concept airport at Amsterdam Schiphol. Due to its limited permit to grow to 44 million passenger movements within the next few years there is a discussion about growth and transfer passenger acceptance. The environmental and commercial issues were on the political agenda in 1997 and 1998. Roads are well maintained and plentiful, though there is congestion during rush hour. Railway infrastructure is to be improved to allow high-speed traffic on mid-haul distances (including Paris, Berlin, Frankfurt and Munich). The seaports of Rotterdam and Amsterdam are well equipped for cruise ships, with the latter city scheduled to open new facilities soon.

The 2,400 hotels in the Netherlands and countless numbers of bungalow **parks** with indoor subtropical swimming facilities can cater for 6.6 million foreign **tourists**. Some 60 per cent of the Dutch population take one or more holidays per year, including 12 million holidays abroad and 15.8 million domestic holidays. Important **attractions** are the **museums** in Amsterdam, the miniature town of Madurodam in the Hague, the Efteling **theme park** in Kaatsheuvel, the canal cruises in Amsterdam and the woods in central Holland (Veluwe). In the north and southwest there are important sea and lake areas for boating. The Germans, who are appreciated guests in the Netherlands, frequent these areas in particular.

The tourism administration is organised in several bodies. The government tourism office, with a budget of $55 million, provides promotion for **inbound** tourism. A separate organisation for domestic tourism promotion promotes the Netherlands to the Dutch people. The regional promotional offices develop new **products** in their region and promote these via the **industry**. The government only wishes to co-fund ventures if new **markets** are developed, and provides in principle a lump sum of about 50 per cent to the two agencies. The local tourism promotion offices are funded by local hotels and local government. The policy of the government is to leave the industry to the market mechanism. **Tour operators** provide about 30 per cent of the business; the remainder arrive independently. Over 70 per cent of foreign tourists to the Netherlands are from the **United Kingdom**, Belgium and **Germany**.

Further reading

Peerenboom, J.P. (1997) *Travelmarketing*, 's Gravenhage: Hotelschool The Hague T &C.

van Dok van Weele, A. (1996) *Werken aan concurrentiekracht het tocristisch bedleid tot 200*, 's Gravenhage: SDU.

INEKE WITZEL, THE NETHERLANDS

neurocomputing

Neural networks (NNW) are approximate computer representations of (very) elementary brain functions. They originate from artificial intelligence and neurophysiological research. Neurocomputing is inspired by what is known about the neuro-physiological structures in humans and animals. In imitating nature, NNWs share the advantages of parallel data processing such as fault tolerance (the ability to learn from incomplete or noisy information), making inferences and generalising on novel data. Neurocomputing **models** have to be trained; they learn by being exposed to examples. Once the network has been trained it is ready for use in fields such as **forecasting**, **classification** and optimisation. NNWs today predict financial time series, rate the creditworthiness of customers, classify consumers, screen satellite data, recognise speech and **images**, help accelerating the transmission of bulky data, and extract rules from observing sample cases for an expert system.

NNWs consist of data processing units (neurones), arranged in layers and largely interconnected. Each unit performs very simple and fundamental computational functions. However, the connectivity and the parallelism make the network a powerful classifier and pattern recogniser. In conventional computing systems the information is stored in data files which are separated from the programs retrieving old data and generating new data. NNWs have a distributed memory that is implicitly contained in the network's weight structure.

Many neural network operations may also be implemented with well-established mathematical and statistical methods. NNWs, however, are still operative where conventional **methodology** fails, because a problem becomes analytically intractable or runs out of computer resources. NNWs handle nonlinearities and interactions which often plague the classical methods. Neurocomputing helps analysts to keep up with the explosion of data. One may think of scanner data, the information generated by automated checkin/checkout systems, or by computerised reservation and travel counselling systems (see **computer reservation systems**; **decision support system**; **reservation**). The model offers promising applications, including classification of **tourists** into market segments.

Further reading

Andersen, J.A. and Rosenfeld, E. (eds) (1988) *Neurocomputing: Foundations of Research*, Cambridge, MA: MIT Press. (A collection of the seminal articles from a variety of disciplines.)

Caudill, M. (1993) *Neural Networks Primer*, 3rd edn, San Francisco: AI Expert. (An introduction for readers lacking a background in mathematics.)

Haykin, S. (1994) *Neural Networks. A Comprehensive Foundation*, New York: Macmillan. (An in-depth introduction for readers seeking a rigorous treatment.)

Mazanec, J.A. (1992) 'Classifying tourists into market segments: a neural network approach', *Journal of Travel & Tourism Marketing* 1: 39–59. (An example of an NNW application in tourism marketing research.)

JOSEF A. MAZANEC, AUSTRIA

new product development

New product development involves the formation and/or modification of tourism **products**. It is a research-driven, sequential and disciplined process which recognises that tourism **markets** and destinations constantly evolve (see **life cycle**). It can either be achieved by acquisition of products or by **planning** new ones, but both mechanisms ensure the constant adjustment of tourism products to meet changes in market **demand**.

CHRIS COOPER, AUSTRALIA

New Zealand

New Zealand lies in the South Pacific Ocean, 1,600 km east of **Australia**. It comprises a North and South Island and numerous smaller ones, with a total land area of 270,500 square km. Although highly urbanised, with a resident population of 3.6 million, the population density is generally low. Its touristic potential has been recognised since the first European arrivals (early to mid-1850s), and this has led to the development of the first government-sponsored national tourism office in 1901. From its early beginnings with explorers and mountaineers, and some **package tours**, New Zealand tourism entered the modern era with the arrival of jet engined aircraft in the late 1950s and the subsequent breaking of distance barriers to major **markets**. Growth in international tourist **demand** has averaged 9 per cent over much of the past two decades, ahead of both Pacific regional and world tourism **trends**. Australians, Japanese, North Americans and United Kingdom residents comprise about 60 per cent of all arrivals; one-half of these originate from Australia. There is strong contemporary growth from emerging Asian economies.

Major tourism products include **nature**, **adventure tourism** and cultural experiences. In the more temperate North island, volcanic activity, **indigenous** Maori culture and water-based activities are major attractions. The South Island, dominated by the Southern Alps and associated glaciers and fjords, offers considerable scope for walking, tramping and **skiing** and a range of adventure (bungee jumping, rafting, 'flightseeing' and so on) and **nature tourism**. Tourism is a significant sector of the New Zealand economy. The 1.5 million international arrivals contribute 5.5 per cent of gross domestic product. Government plans have recently shifted away from a long-term target of 3 million tourists per annum, to a figure of $4.5 billion in **foreign exchange** (10 per cent of current GDP). New Zealanders themselves are prolific tourists with up to one million departures per year.

A government-funded New Zealand Tourism Board was established in 1991 to develop, implement and promote strategies for tourism, while an Office of Tourism and Sport provides advice to a Minister of Tourism. The Department of Conservation (which includes administration of thirteen **national parks** and maritime parks) manages the 30 per cent of land vested in the Crown, and is a significant agent in tourism **development** and **planning**. Industry concerns are advanced by the New Zealand Tourism Industry Association. At the site level, Territorial Local Authorities administer specific developments, and they are influenced by the Resource Management Act of 1991 which charges them with facilitating the **sustainable development** of resources.

DAVID G. SIMMONS, NEW ZEALAND

non-discretionary income

Income can be defined in a variety of ways including nominal and real, disposable, discretionary and non-discretionary. The first one is the income recorded in national accounts and is generally expressed in terms of current prices. This covers such sources as wages, salaries, profits (dividends) rent and interest. It is generally the definition of income used in **impact** studies and **multiplier effects** because it corresponds to national **accounting** practices. On the other hand, real income (after the effects of inflation have been taken into account) is a term generally used to describe the purchasing power of an individual or the residents of an economy.

Disposable income refers to that income available to an individual after mandatory deductions for such things as taxes and social insurance payments. However, it should also take into account any transfer payments such as income support or unemployment benefits. Therefore, it is national income less 'net' **tax** payments. It should not be confused with discretionary income, for it takes no account of voluntary commitments that may have been entered into by an individual.

Discretionary income refers to that over which an individual has some choice in how it is to be spent. Many of the personal services such as tourism are associated with relatively high income elasticities of **demand**. That is, as income levels go up, the increase in demand for tourism-related goods and services grows more than proportionally. This is in part explained by the fact that the amount of discretionary income increases as income levels increase.

Non-discretionary income refers to that disposable income over which the recipient has no choice in terms of expenditure. Examples of this category include all enforced and long-term spending such as mortgage payments, pensions, health schemes and so on. They also include part of the expenditures that are made on essential purchases such as that amount of spending on heat, light and food necessary for survival. Thus, income required to meet prior commitments may be considered to be non-discretionary for the individual has no discretion as to how that income may be spent. The higher the proportion of non-discretionary income out of total income, the less scope there is for additional spending upon tourism, unless income levels increase overall.

JOHN FLETCHER, UK

non-governmental organisation

A non-governmental organisation (often referred to as an NGO) is defined by the United Nations as any formal association that is neither a **government** nor hopes to replace a government or its officials, is funded from voluntary contributions and is not involved in for-**profit** activity, and does not engage in or advocate violence. The NGOs must support the goals of the United Nations or other governmental agency that recognises them. A number of such organisations, due to their scope, have direct involvement in tourism.

ALAN A. LEW, USA

non-profit organisation

Non-profit organisations are those with a dominant service or welfare, as opposed to **profit**, orientation. In tourism, these may include organisations providing a service, including national, regional or local tourism offices, tourism **marketing** bureaus, **conservation** bodies or trusts owning or managing **attractions** (like national park authorities), and welfare organisations such as **social tourism** operators and other charitable bodies.

RICHARD SHARPLEY, UK

Norway

The long alpine coastline and fjords surrounded by steep mountains with waterfalls are chief **attractions** of Norway's natural scenery. The country's tourism **industry** started in the romantic period of the nineteenth century. Today, Germans and Scandinavian neighbours dominate inbound tourism, which is concentrated to the summer season when the midnight sun is visible north of the Arctic Circle.

JAN VIDAR HAUKELAND, NORWAY
JENS KRISTIAN STEEN JACOBSEN, NORWAY

nostalgia

Nostalgia is a feeling of loss or anxiety about the passage of time, accompanied by a desire to **experience** again some aspects of the past. Although nostalgia started out as the morbid disease of **homesickness**, a spatial displacement cured by going home, it has come to mean a sentimental awareness of temporal disorientation to be countered only by symbolic **time** travel. During the past hundred years, nostalgia has emerged as a major **motivation** for tourism (Dann 1996). **Modernity**, the belief in progress and rational solutions to problems, automatically highlights both present imperfections and lost of the past (Lowenthal 1985). Postmodernity, the dissolution of boundaries, of high–low distinctions and of colonially organised societies, creates a pastiche in which often incongruous fragments may be incorporated for present interests (Graburn 1995) (see also **postmodernism**).

Attempts to compensate for nostalgic loss result in different types of tourism. Personal nostalgia, the awareness of ageing or of becoming marginal in a changing society, may spur a return to earlier stages of life, as simple as visiting the places of one's youth or childhood or enjoying a reunion with long familiar friends and relatives, or perhaps trying to behave or be treated as in past times by engaging in sports, having a 'second honeymoon' or being pampered as in childhood. Even 'fat farms' and **health** clinics may be nostalgic efforts to return to former physical states.

Equally important are social and environmental nostalgia, born of the fear that society is changing too fast and for the worse, or that present human civilisation is too artificial and is damaging **nature**. Historical nostalgia is an enjoyment of the past, along with pride in the achievements of one's forbears. It is reflected in the ubiquity of **heritage** tourism and the growing phenomenon of historical re-enactments. Stronger forms of social nostalgia lead one to seek simpler and presumably more authentic lives of rural or exotic peoples, to represent a more humane past. Nostalgia, precipitated by the crisis of confidence in man's ability to manage the world, is manifested in the search for and immersion in nature, devoid of other people, representative of how the earth was before humans despoiled it.

See also: regression

References

Dann, G. (1996) *The Language of Tourism: A Sociolinguistic Perspective*, Wallingford: CAB International. (See 'Ol' Talk: the Register of Nostalgia Tourism', pp. 218–28.)

Graburn, N. (1995) 'Tourism, modernity and nostalgia', in A. Ahmed and C. Shore (eds), *The Future of Anthropology: Its Relevance to the Contemporary World*, London: Athlone Press, 158–77.

Lowenthal, D. (1985) *The Past is a Foreign Country*, Cambridge: Cambridge University Press.

NELSON H.H. GRABURN, USA

novelty

The desire for change or something new, as an alternative to the daily living environment, has often been highlighted as a major ingredient of tourist **motivation**. The **quest** for difference forms part of such curiosity. The pursuit of **strangeness**, as a counterpoint to the familiarity of home, is also linked to this motivational need.

GRAHAM M.S. DANN, UK

occupancy rate

A tourism business occupancy rate refers to the number of **airline** seats or the units of **hotel** room space sold. This **demand** is usually measured as a percentage of available seats or space occupied for a given period of time. It is calculated by dividing the number of occupied rooms/seats by the total number available for sale during the same period.

DAVID G.T. SHORT, NEW ZEALAND

occupational safety

Occupational safety is one of the most controversial pieces of federal **legislation**. The purpose of this act has been to centralise regulation of labour and workplace for employees in the United States. The Occupational Safety and Health Act of 1970 created three new government agencies, the Occupational Safety and Health Administration, the Occupational Safety and Health Review Commission and the National Institute of Occupational Safety and Health. The first one has the responsibility for formulating and enforcing regulations for on-the-job **safety**. Almost all US tourism businesses and their operations are influenced or even regulated by various occupational safety standards framed by this and other institutions.

ROBERT H. WOODS, USA

off-road vehicle

Off-road vehicle is a term used for both automobile-like vehicles, such as jeeps, and for recreational vehicles such as mountain bikes, dune buggies and snowmobiles. The latter group are also known as all-terrain vehicles. They rose in popularity during the 1980s, but unsafe design led to a decline in **demand** for three-wheelers. Off-road vehicles increase **accessibility** to various **attractions** and **landscapes**, but their usage has sometimes had adverse environmental **impacts** often resulting in conflict over resource use.

CHARLES S. JOHNSTON, NEW ZEALAND

operating cost

The costs relating to production can be broken down in a variety of ways. For fixed tourism costs, for example, factors of production are the same whether one or more units of the **product** (such as a **hotel** room) are sold. Variable costs have a tendency to decrease initially, as economies of large-scale production are enjoyed and then, at some level of output, increase as diseconomies of large-scale production are encountered. The notion of operating costs relates to those costs that need to be covered in order to break even.

See also: accounting

JOHN FLETCHER, UK

opportunity cost

The opportunity cost of a resource needs to be considered in the context of a proposed use of it. As such, the opportunity cost is the **profit** that would be foregone by not using the **resource**, such as a tourism **attraction**, in the most profitable alternative use that is feasible. Hence, if the profit from the proposed use is not at least equal to the opportunity cost, the proposed use is not economically justified.

SIMON ARCHER, UK

opportunity set

Opportunity set refers to a combination of destinations/places which a **tourist** actually considers for a particular trip. The size (number) and elements (places, destinations) of the opportunity set may vary substantially among individuals and trips. One's knowledge of destinations, past tourism **experience** and personality, the nature of the trip (length, membership, costs and so on) and related constraints influence the components of the opportunity set.

JIANN-MIN JENG, USA
DANIEL R. FESENMAIER, USA

optimal arousal

Psychological (see **psychology**) arousal below or above a person's optimal level is experienced as unpleasant, while increases or decreases toward the optimal level are experienced as pleasant. Optimal arousal is an important **motivation** for **leisure** and tourism. Vacationers with a high need for arousal are more likely to seek out **novelty**, whereas those with low arousal needs choose more predictable experiences.

ROGER C. MANNELL, CANADA

organisation culture *see* culture, organisation

organisation design

Organisation design can be defined as decisions about such configurations as formal and informal structures, systems and processes. Structures are the patterns of relationships among groups and individuals. Systems are comprised of elements, including social and physical elements, in structured and continually interacting relationships. Processes are specifically designed sequences of steps, activities or operational methods. An organisation's strategic intent must be embedded in its design. Its ultimate goal is to complement and support strategic objectives through aligning all the components. Organisational designers consider the symmetry between its two aspects: the effectiveness of the design in matching its strategic intent, and the design's impact on individuals, group relationships and the political dynamics of the organisation.

In general, there are five primary dimensions to consider in the design or redesign of an organisation. The first four dimensions include centralisation (locus of control), complexity (division of labour), formalisation (rules, procedures, and instructions) and configuration (shape of role structure). The last organisational dimension, flexibility, is a reflection of the firm's ability to alter its organisational design in the context of a changing environment, and is dependent upon the overall level of integrity of these four dimensions.

Four key steps are involved in developing a full sequence of organisational design activities. The first, initial evaluation, includes procedure for a thorough assessment of the organisation. The primary goal is to identify whether or not the organisation's intent matches its effectiveness. The second is strategic design. If the organisational assessment has reached the conclusion that reshaping is necessary, then the reconfiguration of an organisation's strategic design is in order. This reconfiguration involves an extensive analysis of the necessary information, and the integration of appropriate patterns for connecting strategy and objectives, which includes the earlier five organisational dimensions. The third step is operational design. Once the strategic design is approved, specific considerations are focused upon. These include business processes, work flows, information flows, resources allocation and communications.

The proper fit of these elements is essential in order to shape an active, well-functioning organisation. The degree of transformation involved in a new operational design is determined by the degree of alteration in strategic design. The final step is implementation, which involves the actual execution of new practices inserted into existing organisational practices. For any organisational design to be effective, appropriate monitoring should be coupled with consistent **management**. The tourism research community has had progress in relation to this theme and its application in various sectors of the industry.

Further reading

Nadler, D. A. and Tushman, M.L. (1997) *Competing by Design: The Power of Organisational Architecture*, New York and Oxford: Oxford University Press.

Pugh, D.S., Hickson, D.J., Hinings, C.R. and Turner, C. (1969) 'The context of organisation structures', *Administrative Science Quarterly* 14: 91–114.

KYUNG-HWAN KIM, KOREA

Organisation for Economic Cooperation and Development

The Organisation for Economic Cooperation and Development (OECD), was transformed from an earlier status in 1961 to achieve a sustainable economic **growth** and **employment** and a high standard of living in member countries; to contribute to sound economic expansion in member as well as non-member countries; and to contribute to the expansion of world trade on a multilateral, non-discriminatory basis in accordance with international obligations. This agency has a long history of being involved in the area of tourism. Its headquarters are in Paris, France.

BONG-KOO LEE, USA

Organisation of American States

The Organisation of American States (OAS) is the world's oldest regional organisation, dating back to the First International Conference of American States, held in Washington, DC from October 1889 to April 1890. This meeting approved the establishment of the International Union of American Republics. The Charter of the OAS was signed in Bogotá in 1948, and entered into force in December 1951. It has been amended several times, the latest amendment being in 1996. The OAS currently has thirty-five member states. In addition, it has granted permanent observer status to thirty-seven states, as well as the **European Union**.

The basic purposes of the OAS are to strengthen the peace and security of the continent; to promote and consolidate representative democracy, with due respect for the principle of non-intervention; to prevent possible causes of difficulties and to ensure the settlement of disputes that may arise among the member states; to provide for common action on the part of those States in the event of aggression; to seek the solution of political, juridical and economic problems that may arise among them; to promote, by cooperative action, their economic, social and cultural development; and to achieve an effective limitation of conventional weapons that will make it possible to devote the largest amount of resources to the economic and social development of its members.

In June 1996, in recognition of the importance of tourism in the hemisphere, in order to strengthen the tourism group of the organisation, and their activities in this industry, the Secretary General of the OAS created the Inter-Sectoral Unit for Tourism, responsible for matters directly related to tourism and its development in the Western hemisphere. The Inter-American Travel Congress was established earlier, in 1939, to develop tourism in the Americas, conduct technical studies and maintain contact among government organisations and the private sector. The headquarters of the OAS is located in Washington, DC, USA.

TURGUT VAR, USA

orientalism

Orientalism is a potent **discourse** which systematically defines India/the Middle East/the Far East as **other** than, and generally inferior to, the West.

The Orient – after Edward Said – is a socio-political, military, scientific and imaginative imperial production. It was largely framed as a disciplined practice of cultural hegemony by British/French/American trade and political interests in the nineteenth century. Since, arguably, the world's tourism industry has historically been considerably platformed upon ethnocentrism generated within Western/colonialist practices, it tends to accentuate established images of the East as Other: an exotic, enticing, but peripheral (and frequently misconstrued) place.

KEITH HOLLINSHEAD, UK

orienteering

A competitive sport of Scandinavian origin, orienteering involves cross-country running by **map** and compass. Orienteering attracts male and female participants of all ages, and has a strong family component. Orienteers travel widely within their own countries and throughout the world in following their sport, sometimes as organised orienteering **tours**. The world's largest orienteering event, the Swedish Five-Days, attracts over 15,000 participants.

See also: sport, recreational; sports tourism

DAVID HOGG, AUSTRALIA

other

Othering is the imaginary construction of different/alien people by external individuals who remain marginal (yet powerful) in that **encounter** with their exotic 'others'. The othering of foreigners tends to deny these others of genuine identifications as they are conspiratorially (but often unconsciously) appropriated. The **management/development** practices, and the narratives of tourism, regularly further the capture/destruction of others.

See also: anthropology

KEITH HOLLINSHEAD, UK

outbound

Outbound (or outgoing) tourism is defined as tourism involving residents of a country travelling to another country. Outbound **tour operators** offer **package tours** abroad. They either operate the **tours** themselves, or they commission the services of an **inbound** operator to handle **local** arrangements at the **destination**. The country from which the **tourists** originate is known as the generating **market** or country.

LIONEL BÉCHEREL, UK

outdoor recreation

Outdoor recreation covers any diversionary activity taking place outside of enclosed structures, in a variety of settings ranging from an urban **park** to a **wilderness** area, and it may include organised games, non-competitive group activities and individual pursuits. In recent years, planners and developers associated with governmental agencies have defined outdoor recreation as being 'resource-based' or dependent on some nature-based characteristic that is **compatible** with specific activities. In turn, the latter have been generally categorised as either land-based or water-based. The **facilitation** of these **activities** may require some modification of the **environment** such as the construction of sports playing fields, the installation of lifts and aerial trams to transport skiers to the tops of slopes, the designation and marking of trails for wilderness hikers, or the **zoning** of water bodies to segregate incompatible activities as swimming and water skiing.

A key ingredient in most outdoor recreation has long been the pursuit of aesthetic values and therapeutic benefits perceived to be derived from existing, though temporarily, in a natural setting. Although it is difficult to evaluate these benefits in terms of psychological and social importance due to cultural difference, a common characteristic is the apparent need for individuals to **escape** on occasion from the demands and stresses of modern everyday life. Participants in outdoor recreation activities are often motivated to become active in the promotion of the **conservation** of natural resources because the viability of their favourite pursuits depend on the continued existence of natural or near-natural environments. Popular activities include walking for pleasure, **hiking**, **camping**, fishing, hunting, motor and sail boating, wind and wave surfing, swimming, picnicking, sunbathing, mountain and rock climbing, **skiing**, snow-boarding, skating, nature **photography**, bird watching, golfing, tennis and the various team sports that require out-of-doors playing fields (see also **sport, recreational**; **sports tourism**).

As populations have become more affluent and more removed from the basic necessity of struggling for survival, the **demand** for appropriate spaces in which to engage in outdoor recreation has increased. This has presented challenges for resource managers who often are charged with the allocation of scarce resources among competing uses. They must resolve conflicts between resource extractive industries, conservationists, preservationists and, increasingly, developers of commercial facilities seeking to exploit recreation resources. In addition, new activities arrive on the scene as technological innovators create and **market** new devices ranging from mountain bikes to jet skis. These in turn lead to **development** of specialised places for their use, which are often in **conflict** with existing activities. In order to accomplish their multifaceted tasks, modern outdoor recreation resource managers must draw on the professional skills of several disciplines in the natural and social sciences and be versed in their practical application.

MARY LEE NOLAN, USA
TOM BROXON, USA

outsourcing

Outsourcing, according to Barret (1995), is a process or operational alternative which gives an organisation the option of employing an external supplier(s) to perform a function normally undertaken in-house (such as **catering** which is provided externally to many tourism businesses). This action may involve transfer of people and machinery, as well as **management** responsibility to the supplier(s). Reasons most frequently given in favour of outsourcing are **economies of scale** and specialist skills, and hence the resulting **efficiency** and savings.

References

Barret, P. (ed.) (1995) *Facilities Management: Towards Best Practice*, Oxford: Blackwell Science.

FRANCISCO SASTRE, SPAIN
IMMACULADA BENITO, SPAIN

overseas office

Overseas tourism offices are branches of national tourism administrations maintained in foreign countries that are primarily involved in **marketing** their destinations. The specific types of marketing efforts conducted by them vary by country and are greatly influenced by the total budget available for operations. Developed countries such as Australia, Japan and the United Kingdom have extensive networks of overseas offices, while developing countries tend to use their consular offices or outsource the tourism marketing function to sales representative firms (see **outsourcing**).

The typical strategy for establishing overseas tourism offices is to locate them in the country's major tourist-generating markets. For example, the Australian Tourist Commission maintains offices with full-time staff in the country's seven major markets: Japan, United States, New Zealand, United Kingdom, Germany, Hong Kong and Singapore. The purpose of such offices is to create awareness of and stimulate greater interest in the country as a tourism destination. This is accomplished primarily through **sales promotion**. The total funds available for the office's marketing in a specific country must be apportioned between consumer (individuals and groups) and travel trade marketing (retail travel agents, **tour operators**, incentive planners, convention/**meeting planners** and corporate travel managers). The promotional activities undertaken include **advertising** to build relationships through cooperative marketing programmes, sales calls, travel trade **sales**

promotions, **education** and **training** and thus to penetrate various markets and constituencies.

There is an assortment of such promotional activities which stimulate greater interest and the desire to learn more about countries through 'collateral' materials (such as brochures, visitor guides and **maps**), audio-visual materials (including films, **videos**, CD-ROMs and slide presentations), and information through personal computers on **World Wide Web** sites. Fulfilling requests for information has been the traditional role of overseas tourism offices. They are expensive marketing organisations to maintain, and national tourism administrations frequently close some offices or combine them to represent several countries. Some, such as those in Scandinavia, have joined forces and operate multi-country rather than single-country offices. There is considerable discussion today in tourism about the future role of these offices in the era of global information dissemination through the **Internet**. As potential tourists increasingly use the World Wide Web to gather information, the traditional role of overseas tourism offices may be redefined or diminished.

Further reading

Morrison, A.M., Braunlich, C.G., Kamaruddin, N. and Cai, L.A. (1995) 'National tourist offices in North America: an analysis', *Tourism Management* 16(8): 605–17. (Discusses operation of overseas tourism offices in North America.)

ALASTAIR M. MORRISON, USA

P

Pacific Asia Travel Association

The Pacific Asia Travel Association was founded as a not-for-profit agency in Hawaii in 1951. PATA was established to promote the Pacific Asia region's destinations and to provide information, research, **education** and **training** programmes for its members. Organisationally, it is divided into four divisions, Pacific, Asia, Europe and Americas, with chapters located in many countries. Its membership consists of more than 2,100 worldwide tourism establishments, including over 100 government, state and city tourism bodies, and such tourism businesses as hotels, airlines and cruise lines. Additionally, its chapters have some17,000 tourism representatives as their members. Major activities of the association and its chapters include regional and international conferences, educational and promotional workshops, its Travel Mart which brings buyers and sellers together, and publication of documents supporting its goals. PATA's headquarters was recently moved from San Francisco to Bangkok, Thailand.

CHOONG KI LEE, KOREA

Pacific Rim

This is a collective term for nations in the Pacific region which foster economic cooperation and reflect a common outlook in international relationships. Initiated in **Japan** in the 1960s as a response to world trade focus, including tourism, shifting from Europe to rapidly developing economies of East Asia, the term incorporates the **United States**, **Canada**, **Australia**, **New Zealand**, Japan, **Indonesia**, **Malaysia**, **Singapore**, South **Korea**, **China**, **Hong Kong**, **Taiwan**, Brunei, the **Philippines** and **Thailand**.

NGAIRE DOUGLAS, AUSTRALIA

Pacific Tourism Review

Pacific Tourism Review deals with changes in **outbound** and **inbound** tourism patterns occurring in the wider Pacific area and their associated effects on the region's economies and environments. A refereed interdisciplinary journal, it aspires to advance excellence in tourism research, promote high-level **education** and nourish cultural awareness in various tourism sectors by integrating industry and academic perspectives. First appearing in 1997, it is published quarterly by Cognizant Communication Corporation (ISSN 1088–4157).

RENE BARETJE, FRANCE

package tour

A package tour is an **inclusive** form of travel organised by intermediaries or **middlemen**. It represents a bundle of tourism goods and services, marketed as one particular **product** or brand and sold at an inclusive price. It is often defined as a prearranged combination of not fewer than two items, such as **transportation** and **accommo-**

dation, offered for sale for a period of more than twenty-four hours (Middleton 1994).

Package tours were originated in the 1850s by Thomas Cook. The idea evolved rapidly after the 1950s when air transport made travel quicker and more affordable. Package tours can be independent for individuals, but frequently are organised for groups. They are categorised by the main transportation used, such as coach, train, boat or airplane (both scheduled and charter). **Tour operators** or wholesalers sell their package tours directly or through a retailer. **Travel agencies** retail tour operators' packages and create independent package tours for individuals. In Europe, package tours refer mainly to overseas destinations, travelling on charter flights for a fixed period, typically one to two weeks.

Package tours offer several benefits to both consumers and the **industry**. For the former, tours maximise **security** and offer guidance, while facilitating the information seeking and **reservation** processes. The optimisation of operations reduces costs, often at the expense of **quality**, making travel affordable. For the principals, package tours enhance their **distribution** and promotion mix, and reduce operational **risk** and **seasonality** problems. However, they present several disadvantages, mainly because they encourage **mass tourism**. They stimulate large-scale **development** and jeopardise the sustainability of **resources** of destinations. Intermediaries also tend to lead consumers to products and destinations where they can maximise their **profit margins**. Tourists' contact with the destination and hosts becomes artificial or staged. Intermediaries concentrate bargaining **power** which they utilise to reduce principals' prices, resulting in low profitability and economic **benefits** at the destination (Buhalis 1994).

Future **trends** in this area include the improvement of **quality** and the **growth** of thematic tours. However there is a trend towards individualisation, and towards greater **satisfaction** of consumers' needs. This is facilitated by **information technology**, which enables consumers to identify and package suitable products themselves and thus threatens a certain degree of disintermediation in the **distribution channel**.

References

Buhalis, D. (1994) 'Information and telecommunications technologies as a strategic tool for small and medium tourism enterprises in the contemporary business environment', in A. Seaton *et al.* (eds), *Tourism: The State of the Art*, Chichester: Wiley, 254–75.

Middleton, V. (1994) *Marketing in Travel and Tourism*, 2nd edn, London: Butterworth-Heinemann.

Further reading

Cooper, C., Fletcher, J., Gilbert, D., Shepherd, R. and Wanhill, S. (eds) (1998) *Tourism: Principles and Practice*, 2nd edn, London: Addison-Wesley.

Laws, E. (1997) *Managing Packaged Tourism*, London: International Thomson Publishing.

Sheldon, P. (1994) 'Tour operators', in S. Witt and L. Moutinho (eds), *Tourism Marketing and Management Handbook*, 2nd edn, London: Prentice Hall, 399–403.

DIMITRIOS BUHALIS, UK

paid vacation

The growth, during the twentieth century, of the practice of employers providing their employees with **vacation** with pay has been a key component in the expansion of tourism in Western societies. Paid vacations have fuelled the growth in volume of tourism as well as its **democratisation**. Until the early twentieth century, paid vacations of one or two weeks per year were generally limited to the professional middle classes. For manual workers, the practice was limited to a few individual employers or industrial sectors.

The critical period for the acceptance of the principle of paid vacations came between the two world wars. In the United Kingdom, two main phases can be identified as the years immediately after 1917, and immediately before and during the Second World War. These phases coincided generally with trade cycle booms, but also required constant pressure from workers' unions. Whereas in the early 1920s one million manual workers had paid holiday provision, by 1938 this had risen to four million and by 1945 ten million workers had two weeks of paid holiday. The 1938 Holiday With Pay Act effectively recognised this process.

Elsewhere in Europe, paid vacation entitlement was increasingly enforced by legislation during the interwar period, spreading throughout central and eastern Europe. In the midst of a national strike, France introduced a compulsory two-week paid holiday, the *congé payée*, in 1936. This movement was less pronounced in the United States, where pressure for legal entitlement to paid vacations did not develop on the same scale as it had in Europe. Instead, length of paid vacation was, and still often remains, a part of individual or collective contracts. This reflects, to some extent, the resistance in the United States for the state to intervene in workplace relationships. In general, the process of increasing growth in paid variations is not inevitable. While for many people in the world this provision remains unattainable, changing work practices and **employment** rights in the developed world are slowing down expansion and, in some cases, reversing the trend.

See also: social tourism

Further reading

Cameron, G.C. (1965) 'The growth of holidays with pay in Britain', in G.L. Reed and D.J. Robertson (eds), *Fringe Benefits, Labour Costs and Social Security*, London: Allen and Union. (A standard, detailed source on the subject.)

Cross, G. (1990) *A Social History of Leisure Since 1600*, State College, PA: Venture. (Discusses paid variations in the United Kingdom, Europe and North America.)

JOHN TOWNER, UK

paradigm

Paradigms are fundamentally the worldviews (value windows) through which things are seen, and known. Since 1970, some huge shifts have occurred in the basic beliefs and assumptions that are held about **nature**, reality and humanity, particularly regarding the conduct of social science. The conventional paradigms of science have been challenged ontologically (the perceived nature of reality), epistemologically (the relationship between the knower and the known) and methodologically (approved ways to carry out investigations). The term 'paradigm' itself is commonly associated with the pioneering work of Thomas Kuhn (1970), despite the fact that he supposedly used it in a score of different ways himself. Indeed, the paradigm concept can mean many things to many people.

In 1989 an important Alternative Paradigms Conference was held in San Francisco to clarify the parameters of what some believe are the three leading paradigms that have succeeded positivism in social science: postpositivism (objectivist, but critico-realist, lenses on the world), critical theory (dialogic and transformative outlooks that seek to eliminate false consciousness of some particular kind or kinds) and constructivism (relativist views that identify multiple but contextual truths in and of the world). The debates about the ethics, the goodness criteria and the values involved as knowledge accumulates under each of these paradigms have been comprehensively reproduced in Guba (1994). Some social scientists consider that the focus of the 1989 gathering was reductionist in celebrating just three so-called 'master' paradigms, and claim genres like **structuralism** and neo-Marxism are also worthy of the term 'paradigm', while still others would apply the concept even more flaccidly for an infinity of interpretative positions variously grounded in **race**, **class**, **gender** and the like.

In 1996, the **International Sociological Association**'s Research Committee on Sociology of Tourism held an important paradigmatologie in Finland to explore the manner in which competing worldviews empower and/or delimit understanding in tourism. The ISA text that emanated from that symposium (Hollinshead and Graburn, forthcoming) should prove to be a highly valuable examination of both the manner in which different paradigms exercise hegemony over the **industry** and the fashions in which different paradigms are respectively influential in interpreting heritage, culture and nature, as they interface with the tourism phenomenon.

References

Guba, E. (1990) *The Paradigm Dialog*, Newbury Park, CA: Sage.

Hollinshead, K. and Graburn, N. (eds) (forthcoming) *Shifting Sands: Established, Emergent, and Emerging Worldviews in the Social Sciences*, New York: Cognizant Communication Corporation.

Kuhn, T.S. (1970) *The Structure of Scientific Revolutions*, 2nd edn, Chicago: University of Chicago Press.

KEITH HOLLINSHEAD, UK

paradise

Paradise is regarded as a place of exceptional happiness and delight, represented in the Western imagery as an ideal or enchanted garden such as the Garden of Eden. The continued **quest** for the earthly paradise has a long history (Manuel and Manuel 1973). The image of paradise suggests pristinity, simplicity and primitiveness, on the one hand, and unlimited wish fulfilment, enjoyment and **fun** on the other. As such, it is one of the most powerful metaphors for the promotion of destinations. Initially used primarily in the promotion of 'pristine' **nature** sites, such as beaches and 'enchanting' small islands, the **metaphor** has been applied to a widening range of diverse destinations with an ever more tenuous relationship to nature: there are **skiing** and golfing paradises, sex paradises and even **shopping** paradises, in the metaphorical sense of places offering complete customer **satisfaction**.

However, even as commercialised attractions become paradises, nature paradises undergo **commoditisation**. Remote, initially 'pristine' and 'untouched' beaches and islands become vacation **resorts**, providing tourists with an increasingly more complete, sophisticated (and expensive) touristic **infrastructure** of hotels, **restaurants**, shopping areas, **bars** and other facilities. Such paradises, integrating natural amenities with modern, often luxurious facilities, become principal attractions of **mass tourism**. **Commercialisation** may be accompanied by a studied, playful primitivism, as for example in the Club Mediterranée resorts. However, seekers of more 'pristine' paradisiac sites and simpler vacationing **lifestyles** move further afield onto ever more remote beaches and smaller islands, thus frequently serving as spearheads of touristic penetration.

With the **growth** of mass tourism, such paradises become not only commercialised but also commoditised: a package-like stereotype of a nature paradise is constructed and advertised as any other commodity with uniform, standardised ingredients. The image mostly consists of waving palm trees, calm beaches, quiet lagoons, hills clothed in rich vegetation and surf pounding on coral reefs, beyond which lies a deep blue sea. Voluptuous native women frequently complete the image (see also **natives**).

However, with the **development** of tourism, a marked discrepancy tends to emerge between the 'paradisiac' destinations so advertised and their actual reality. As natural sites become despoiled in highly developed tourism areas, entrepreneurs start to establish contrived paradises, isolated enclaves in which the stereotyped image of paradise is artificially recreated in order to correspond to the expectations of **tourists**, who have been beguiled by the promotional literature.

References

Manuel, F.E. and Manuel, F.P. (1972) 'Sketch for a natural history of paradise', *Daedelus* 101: 83–128.

Further reading

Ballerino Cohen, C. (1995) 'Marketing paradise, making nation', *Annals of Tourism Research* 22(2): 404–21.

Cohen, E. (1982) 'The Pacific islands from utopian myth to consumer product: the disenchantment of paradise', *Les Cahiers du Tourism* Série B, 27.

Lewis, L.S., and Brissett, D. (1981) 'Paradise on demand', *Society* 18(5): 85–90.

ERIK COHEN, ISRAEL

paratransit

Paratransit refers to a class of public passenger **transportation** that is more flexible and personalised than conventional forms such as **rail** or bus. It includes services of taxicabs, jitneys, buses and vans which may operate on irregular routes and/or schedules. Paratransit also includes prearranged ride-sharing services of van pools and car pools. These and other transportation modes are put into creative uses in many tourism destinations.

JOHN OZMENT, USA

park

A park is a land or water area that has been reserved for purposes which include **recreation** and tourism. Amusement parks, **theme parks**, state parks and **national parks** share this characteristic. However, as **protected areas**, state and national parks also seek the **conservation** and even the **preservation** of natural and cultural **heritage**, orientations that distinguish them from their park kin. These seek to achieve conflicting public goals: managing recreation and tourism opportunities for tourist **satisfaction** and managing the natural **environment** to maintain or to restore ecological integrity.

Tourism had been a factor in establishing the earliest national parks. Canada's Banff National Park, for example, was established in 1885 by the federal **government** as a vacation **destination** for tourists on the transcontinental railway. Yosemite National Park, however, was established in 1872 to protect the scenic beauty of the Yosemite Valley from tourism and other **economic development**. Tourism in parks was recognised early as both boon and blight. Parks continue to be destinations, as sources of **nature** and its **interpretation**, as settings for **adventure tourism** and **escape**, and as opportunities for business and **profit**. Tourism to parks expanded through the post-Second World War period, fuelled by **growth** in **camping** at first and later, by **differentiation** in camping itself and other **activities** such as bicycling.

Park **management** has become complex. Parks have assumed an educational function to interpret natural and cultural **heritage** to tourists. The presentation of natural and cultural themes helps them to understand particular features in parks and to appreciate their **roles** in protecting species, habitats and diversity. Where parks have been established in areas claimed as home territory by aboriginal people, interpretation of cultural themes has helped **tourists** to understand how they lived prior to contact with European cultures.

Protection of **biological diversity** and ecological integrity becomes an urgent goal as agriculture, forestry, mining and tourism invade other areas. Parks and protected areas such as the Serengeti National Park and the adjacent Maasai Mara Game Reserve afford a refuge for many migratory African wildlife species. **Australia**'s Great Barrier Reef Marine Park protects the world's largest reef complex. Managers recognise, however, that both are significant tourism attractions. This complexity demands a new generation of **planning** and management frameworks. While all stem from **carrying capacity**, some such as the **recreation opportunity spectrum** integrate use and protection. Others such as the **Limits of Acceptable Change** build consensus among stakeholders in park decision making.

Further reading

Banff-Bow Valley Study (1996), *Banff-Bow Valley: At The Crossroads*, summary report of the Banff-Bow Valley Task Force (Robert Page, Suzanne Bayley, J. Douglas Cook, Jeffrey E. Green and J. R. Brent Ritchie), prepared for the Honourable Sheila Copps, Minister of Canadian Heritage, Ottawa, Ontario, 76 pages. (Illustrates the conflicts between protection and tourism in Canada's oldest National Park.)

Keiter, R.B. and Boyce, M.S. (1991) 'Greater Yellowstone's future: ecosystem management in a wilderness environment', in R.B. Keiter and M.S. Boyce (eds), *The Greater Yellowstone Ecosystem: Redefining America's Wilderness Heritage*, New Haven, CT: Yale University Press, 379–415. (Discusses the potential of ecosystem management in managing people's use of the natural environment in Yellowstone National Park.)

Killan, G. (1993) 'Creating the "gospel according to parks": 1967–1978', *Protected Places: A History of Ontario's Provincial Parks System*, Toronto: Dundurn Press, 239–87. (A historical examination of a policy-based approach to managing protection and use in the parks system of the Canadian province of Ontario.)

ROBERT J. PAYNE, CANADA

participation *see* public participation

participation, recreation

Detailed investigations of recreational participation have now been undertaken at the local, regional and national levels in numerous countries around the world. Sometimes at considerable **cost**, the national studies have surveyed thousands of people (usually adults) in depth about their recreational preferences and participation rates, either in the form of specialist surveys or as part of wider 'quality of life', **health** or **time** budget studies. The rationale for such expensive social science exercises is that they provide essential data for **infrastructure** and **service** provision, and that the promotion of active recreational participation in particular has positive benefits for **community** health.

Examples of these ambitious, large-scale studies are national surveys carried out in New Zealand, the *New Zealand Recreation Survey*, published in 1984, and the 1991 Life in New Zealand Commission Report. The first of these used trained interviewers who questioned 4,011 New Zealanders over the age of ten about their involvement in 220 activities over the previous twelve months. The second utilised a similar-sized sample, but used a mail-out questionnaire targeting only people over the age of 15 years and restricted itself to involvement in fifty-four home and away-from-home activities over the preceding four weeks. Because such surveys typically find small overall participation rates in most activities, it is common to group them into 'clusters'. Thus in the case of the first study, the five clusters identified were cultural pursuits, sporting activities, home-based **leisure**, interest groups and 'other recreational activities'. It is difficult to make meaningful comparisons among countries (or even within the same country between different dates) on the basis of such surveys because different methodologies and definitions have been used and the times of recall vary considerably, thus casting doubt on the reliability of findings.

In addition to the enormity of **statistics** provided by such surveys (most of which highlight the overwhelming importance of home-based **recreation**) for most countries, now have excellent data are available on participation in individual sports, both in schools and the wider community. Such data are relatively easy to collect. Sports participation statistics in particular highlight the strong influence of hallmark events like the Olympic Games and the global **media** in promoting certain sports rather than others. However, the publicity given to sport tends to mask the fact that in any society most people, and not just the young and the elderly, are not active participants in sporting activity.

See also: behaviour, recreation; demand, recreational; need, recreational; recreation

Further reading

Department of Sport, Tourism and Recreation (1985/6) *Recreation Participation Survey*, Canberra: DSTR.

Perkins, H.C. and Cushman, G. (eds) (1993) *Leisure, Recreation and Tourism*, Auckland: Longman Paul.

DAVID MERCER, AUSTRALIA

partnership

Partnership is defined as arrangements devoted to some common end among otherwise independent organisations. In tourism, this relationship refers to all cooperative activities sustained between the private and public sectors, or even strategic alliances practised within the **industry** itself. This concept is not new in tourism **development**. Traditionally, governments provide physical, regulatory, fiscal and social frameworks, as well as basic touristic **infrastructure** such as roads, airports and communications. In turn, the private sector has been responsible for developing the tourism industry, covering the whole range of such goods and services as **tour operators**, **travel agencies**, **transportation** and communications, **hospitality**, **restaurants**, **recreation** and more.

Local tourism authorities, **boards** and other organisations (for example, the public–private association Maison de France) are responsible for coordination and unification of both municipal and individual initiatives within a given area or country.

Nevertheless, in recent years governments from the economically developed countries have tended to opt for disengagement from tourism in favour of both local authorities and the private sector. This **trend** will probably lead to new forms of public–private sector partnership in tourism. In emerging countries or those in an economic transition stage (Central and Eastern Europe, for instance) tourism activity is undertaken entirely by the public sector (WTO 1996). Most commonly, public–private sector partnerships are found in the areas of tourism promotion and **marketing** functions. However, **training** provision for operators, **quality** improvement, **accessibility**, **environment**, attractions and infrastructure are some of the other partnership programmes undertaken in local tourism development.

In general, partnerships in tourism share certain defining characteristics (Long 1994). They seek a better collaboration and cooperation between public and private organisations located in a certain tourism area; they are action-oriented, focused on implementing initiatives rather than research and **strategy formulation**; and they normally include development, marketing, information and environmental advisories. Although the rationale for governmental involvement in tourism is changing – due to constraints on public sector budgets and a changing political and socioeconomic climate – the importance of partnership approaches in tourism development is still crucial (WTO 1996).

References

Long, P. (1994) 'Perspectives on partnership organisations as an approach to local tourism development', in *Tourism: The State of the Art*, Chichester: Wiley, 442–52.

WTO (1996) 'Towards new forms of public–private sector partnership: the changing role structure and activities of National Tourism Administrations', Madrid: World Tourism Organization.

Further reading

Bonham, C. and Mak, J. (1996) 'Private versus public financing of state destination promotion', *Journal of Travel Research* 35(2): 3–10.

Selin, S. and Chávez, D. (1995) 'Developing and evolutionary tourism partnership model', *Annals of Tourism Research* 22(4): 814–56.

MARTINA GONZALES GALLARZA, SPAIN

pastoral care

The pastoral care in tourism arose as the response of the church and an attempt to salvage as much as possible from the wreckage. **Tourists** are castigated for blatant disregard of Christian morality, but measures are being introduced to help believers to behave in accordance with religious teaching. Since the pastoral ministry cannot be separated from the life of humanity and the problem of time, it is quite logical that pastoral care should deal with the tourist as a part of a completely new area of the pastoral ministry in the world. The basis of the idea of the pastorisation of tourism is found in the organisation of religious ceremonies for believers spending their holidays in certain tourism destinations.

See also: religion; religious attraction; religious festival

BORIS VUKONIĆ, CROATIA

patrimony

Originally an individual inheritance from a father or other ancestor, patrimony now has the extended meaning of any aspect of the present deriving from the past. The term is frequently used to describe collective and often national patrimony as much as individual patrimony, and by extension, is considered as an endowment made by the present to future generations. In tourism, selected aspects of the past are commodified into **heritage** products.

G.J. ASHWORTH, THE NETHERLANDS

payroll cost analysis

Payroll cost analysis is based on a ratio, which is in turn a percentage derived by dividing total payroll and related expenses by total revenue for operating departments. For **service** departments, it is analysed on a comparative basis. The largest expense for most tourism operations is labour cost. Payroll is wages and salaries paid to employees. Related expenses are payroll taxes and employee benefits.

See also: accounting

STEPHEN M. LEBRUTO, USA

peace

As tourism increases, the importance of making it an instrument for peace becomes critical. Following the Second World War, the European Economic Community was established to reconcile the former enemies. Its premise was that as people came to know each other there would be less likelihood of war. A cornerstone of the European Economic Community's **policy** is **freedom** of **travel** and minimising frontier controls. Similarly, as the 'cold war' thawed, USSR General Secretary Gorbachev and US President Reagan declared in their joint statement following the 1986 Geneva Summit that: 'There should be greater understanding among our people and to this end we will encourage greater travel'. Recently, the **benefits** of tourism in reconciling former enemies have been acknowledged in the Peace Accord between Jordan and Israel and the Palestinians and Israel.

The **growth** in student, cultural and professional exchanges and international sporting events give people an appreciation of differences and similarities among nations, including shared aspirations. Annually, over ten thousand international conferences facilitate the exchange of concerns, ideas and opportunities to work together. Rapidly emerging forms of alternative visits including **ecotourism**, and these forms can foster a global ethic which reveres the dignity and interconnectedness of all people. In the process of forging common bonds, **international tourism** also increases respect for **nature** and helps individuals to discover themselves as well.

By nurturing this discovery, tourism has the potential to be a global peace industry. To facilitate the process by which **tourists** can become ambassadors of peace, the **Ecumenical Coalition on Third World Tourism** and many other tourism groups distribute a **code of ethics** for tourists. The International Institute for Peace Through Tourism distributes this credo of the Peaceful Traveller:

> a journey with an open mind and gentle heart to accept with grace and gratitude the diversity I encounter; revere and protect the natural environment which sustains all life; appreciate all cultures I discover; respect and thank my hosts for their welcome; offer my hand in friendship to everyone I meet and act upon them by my spirit, words and actions; and encourage others to travel the world in peace.

Further reading

D'Amore, L.J. and Jafari, J. (eds) (1988) *Tourism: A Vital Force for Peace*, First Global Conference, Vancouver.

LOUIS J. D'AMORE, CANADA

People on the Move

People on the Move publishes studies on issues concerning the various groups of people who are in one way of another involved in the phenomenon of human mobility, including refugees, migrants, nomads, circus and fair people, seafarers, civil aviation workers and passengers, foreign students, **tourists** and **pilgrims**. It produces documents concerning laws or other questions of particular significance and reviews publications related to the theme of human mobility. First appearing in 1971, it is published three times yearly by the Pontifical Council for the Pastoral.

RENE BARETJE, FRANCE

perception *see* advertising; anticipation; freedom, perceived; image; perceptual mapping; marketing; risk, perceived; satisfaction

perception, environmental

Environmental perception refers to the process by which humans organise and interpret elements of their **environment** into a meaningful picture of their world or 'life-space'. An individual's perceived environment consists of **images** derived from interaction between what is selectively scanned from that environment and the individual's scheme of **values**, past experiences, expectations, **motivations** and needs. It is the perceived or subjective environment, rather than objective 'reality', which is of greatest significance in explaining human environmental **behaviour**. It is from within this perceived environment that alternative courses of action are selected; alternatives which are seen as optimal from a perceived and limited range of options and the perception of the outcomes of the choices made.

Environmental perception is basic to an understanding of tourism **decision making** and why **tourists** choose particular settings and activities. Tourism behaviour is discretionary and tourists are free to choose experiences according to how they perceive opportunities, filter environmental stimuli, interpret information and establish preferences. A predisposition towards tourism is translated into participation in it through a choice mechanism heavily dependent upon perceptions (or personal mental constructs) of opportunities and experiences on offer.

Perception operates over several dimensions and various scales in tourism decision making and initial mental constructs may be confirmed or revised as a result of further spatial search and learning. Information levels, as well as the ability to use that information (including personality characteristics, aversion to **risk** and so on) also help structure evaluative beliefs and mental images concerning the nature and **quality** of anticipated experiences. Predictions regarding tourism behaviour would have greater validity if more attention was given to environmental perception and to attitudes and motivations affecting decision making. This would help explain why certain destinations and **activities** are favoured, and how and why **alternative tourism** opportunities are ranked.

Information sources and the credibility of the information itself are key issues in the choice of tourism settings. Information helps structure the images of the environment to which tourists respond. This is part of the rationale for tourism promotion and, indeed, for all forms of persuasive advertising. Information flowing from an (objective) environment or potential destination is filtered through the perceiver's set of preferences, values and cultural interpretations of place meaning. These in turn are open to manipulation by external influences. Inadequate information or misinformation impinge upon environmental perception and constrain the process of discriminating between alternative tourism settings and experiences.

See also: environmental aesthetics

Further reading

Garling, T. and Golledge, R. (eds) (1993) *Behavior and Environment*, Amsterdam: North Holland.

JOHN J. PIGRAM, AUSTRALIA

perceptual mapping

Perceptual maps are multidimensional instruments in which **products** (goods, **services**, destinations, offers and ideas) are represented as points in the map. Perceptual mapping is a useful tool because of its pivotal role in designing **marketing** strategy. Tourists' purchases are primarily driven by their perceived value of products, which is a subjective reality, as seen by the target consumers. A tourism business can proactively differentiate its products from the competition and create a higher perceived value relative to its competitors. This may be accomplished by communicating its offer, including benefits, price and other aspects of the offer, to its target consumers. Such maps are also found to be useful in identifying gaps in the marketplace and designing and positioning products to fill those gaps, **market segmentation** and competitive analysis.

There are basically two types of analytical procedures available for producing perceptual maps: factor analytic techniques and **multidimensional scaling** of proximity (similarity or dissimilarity) data. Both procedures require collection of easily obtainable primary data from target consumers via **survey** research methodology. The data requirements for **factor analysis** tend to be more intensive than for multidimensional scaling. Such procedures are useful if some of the **attributes** of a **product** are difficult to scale, or are hidden and difficult for the tourists to articulate.

Further reading

DeSarbo, W.S. and Manrai, A.K. (1992) 'A new multidimensional scaling methodology for the analysis of two-way proximity data in marketing research', *Marketing Science* 11(1): 1–20.

DeSarbo, W.S., Manrai, A.K. and Manrai, L.A. (1994) 'Latent class multidimensional scaling approaches: a review of the recent developments in the marketing and psychometric literature', in Bagozzi (ed.), *Advanced Methods of Marketing Research*, Cambridge, MA: Blackwell Publishers, 190–222.

Manrai, L.A. and Manrai, A.K. (1993) 'Positioning European countries as brands in a perceptual map: an empirical study of determinants of consumer perceptions and preferences', *Journal of Euromarketing* 2(3): 101–29.

AJAY K. MANRAI, USA
LALITA A. MANRAI, USA

performance appraisal

Evaluating the performance of an employee or manager is always a difficult process, especially in service industries such as tourism. No matter what is said about one's performance, the employee receives the appraisal as nothing more than ambiguity. Indeed, performance appraisals are subject to human emotions, judgements and errors, and hence are unlikely to be completely objective. Thus it is not surprising to learn that in a recent **survey**, 70 per cent of employees indicated that their performance appraisals failed to provide them with a clear picture of what was expected of them.

Performance appraisals can fill many different needs in organisations. Most of these needs fall into one of two categories, improving work performance or aiding in making work-related decisions. For instance, appraisals can be used to provide feedback to employees. When used in this manner, the primary purpose is probably to either reinforce or encourage performance or to help employees develop their careers. Appraisals can be used to support personnel decisions. As such, performance appraisals help in separating poor performers from good ones. Therefore, this process can be used in promotion, discipline, **training** or merit decisions, as well as in establishing goals or objectives for training programmes, for validating selection (and other processes) or diagnosing organisational or departmental problems.

When developing a performance appraisal system, managers must first recognise what kind of performance is to be appraised. The three principal types of ratings used are trait-based, behaviour-based and results-based. The first type is used primarily for the purpose of assessing the personal characteristics of employees. As a result, information on subjects as company **loyalty**, communication ability, **attitude** toward superiors, ability to work as a team and decision-making ability are most relevant. The second is used for evaluating employee behaviours, including their friendliness toward customers, their helpfulness, how often they thank customers for their patronage and the like, all important traits for front-line tourism employees. In this **industry**, the behaviours displayed by employees towards **tourists** and other employees is often just as important as their actual ability to perform specific tasks. The last type helps to measure output or production.

ROBERT H. WOODS, USA

performance indicators

Performance indicators are comparative measures which provide a systematic means of assessing the progress of an enterprise, organisation, project or programme, towards meeting stated goals and objectives and relative to established and accepted levels of **achievement**. They are commonly used to monitor and evaluate, in quantitative terms, the success or otherwise of an undertaking.

Forming judgements about performance involves weighing multiple, competing and sometimes contradictory objectives and measures, and subjective, value-laden components that change over time. Moreover, the process of performance monitoring is multifaceted and should go beyond the mere achievement of intended results. Attention to unintended effects, to flexibility and responsiveness to **change** internally and externally, to appropriateness and relevance to the social context and operational environment, and to the pursuit of sustainability are all important aspects of a project's performance. These take on added significance for tourism developments where the potential for **impacts** on the **environment**, both negative and positive, is a recurring reflection on performance.

Conventionally, measurement of performance can be grouped under three basic elements: economy, **efficiency** and effectiveness. The first category is related to the achievement of standards and the resources needed and the measures required to reach these levels. However, the relationship is not straightforward. Choices and trade-offs among socioeconomic goals, environmental sustainability, community interests and public policies and institutional structures can blur the picture and make the selection of performance indicators complex. Again, this is often typical of tourism undertakings where commercial realities do not always mesh with **community** expectations and **compliance** requirements. Efficiency indicators are also concerned, in part, with economic performance, but also with technological and allocative efficiency, and accountability. Sustainability, or the persistence of benefits over time, is obviously important for tourism projects. Effectiveness as an element of performance measurement is about outcomes, not necessarily outputs. This distinction is of particular relevance to tourism where precise definition of the desire product may not be possible. Rather, effort in improving performance should be directed towards upgrading the processes in place. This is far more difficult to measure on an individual project basis, but may be of much more widespread and lasting benefit to the tourism industry as a whole.

Further reading

Cook, T., Van Sant, J., Stewart, L. and Jamie, A. (1995) 'Performance measurement: lessons learned for development management', *World Development* 23(8): 1303–15.

Mosse, R. and Southeimer, L. (1996) *Performance Monitoring Indicators Handbook*, Washington: World Bank.

JOHN J. PIGRAM, AUSTRALIA

performance standard

Organisations must establish consistent performance criteria and reward systems. Without these, the opportunity for considerable conflict exists within the system. In the absence of common performance standards employees in different parts of an organisation may attempt to attain disparate goals. For instance, without consistent performance standard sales and marketing personnel may place all of their emphasis on attracting business. While front desk managers or attendants in tourism, for example, may concentrate on serving a specific type of guest, these two goals are obviously in conflict. Reward systems are useful in establishing performance standards that are consistent. In the final analysis, such systems must motivate individual employees to work more productively; but, unfortunately, not all of them have the same wants and needs. As a result, not all compensation programmes work for all employees.

Motivation theories fall into two principle types: content and process theories. The former, more important to understanding performance standards, relate to how organisations can encourage employees to work. The latter theories approach employee behaviours from a different perspective. Those that are widely acknowledged are the expectancy, equity, and reinforcement theories. The former assumes that people are always motivated. The goal of managers is to get workers to focus their motivation on organisational goals.

According to this theory, workers are most productive when they believe that there is a high probability that their efforts will result in desired rewards. An understanding of the term valence completes the expectancy theory, which essentially is the strength of an individual's personal preference for a particular outcome. This theory is based on the assumption that employees all ask two questions of their work: 'what am I giving' and 'what do I receive in return for what I give (compared to others who give the same)?'

Employees create mental ratios about their work situations to determine whether these conditions are met. Equity can occur only when employees believes that outputs given to them are equal to their inputs in relation to what they see others doing and receiving. The reinforcement theory, which is intuitively logical, is useful in understanding how people respond to performance standards. It assumes that people are conditioned to behave in certain manners because of past responses to similar situations. Employees respond according to how they are treated. If the response of a manger to an employee's behaviour is positive, then the employee will perform the same way again. In contrast, if the manager's response is negative, the employee will likely either behave differently the next time or avoid the actions that caused the previously negative response. In general, the theory and practice of performance standard has received much attention in the tourism management field, including exploring the work and performance of employees who have direct and daily contacts with tourists as well as those who are performing their tasks in the 'back of the house'. Because of the nature of tourism, this subject also connects to intercultural challenges and opportunities, relating to both employees and tourists coming from all over the world.

See also: performance appraisal

ROBERT H. WOODS, USA

personal selling

This is one of the promotional tools available to the **marketing** function of an organisation or company. Personal selling is a marketing technique in the promotional mix along with **advertising**, **public relations**, **direct marketing** and **sales promotions**. Personal selling involves direct interaction between the customer and company with the intent of selling a product or service.

In tourism, personal selling is a common marketing strategy for booking business meetings, conventions, trade shows and **incentive** travel, as well as for working with **travel agencies**. Personal selling is a less commonly used technique for selling **vacations** and tourism **products** to families or independent tourists. In a service-based organisation or company where the nature of the product is a human interaction, almost every employee is a salesperson. In a product-based business, human interaction may only occur between a professionally trained salesperson and the customer. Thus while service organisations have more opportunities to sell, there is also more need for sales **training** among all employees.

Personal selling involves identifying potential customers and building business relationships with them. A starting point to selling is identifying potential customers by reviewing past customer lists, directories and telephone listings. Next, a salesperson begins to conduct research by understanding past purchase patterns, future product needs and financial conditions. Then, a salesperson will contact this potential customer by phone or in person to arrange for a meeting. This first personal contact provides the opportunity for mutual understanding. Products which satisfy needs begin to be identified, such as facilities and **hospitality** for an end-of-the-year party, convenience **food** service for an **airline**, or **advertising** placement for a travel agency. A next step for the salesperson is to define the potential of the account and then to strike a sale. After a sale is made, follow-through is needed which insures that the customer is fully satisfied with their purchase (see **cognitive dissonance**). The sale does not end here; instead it starts all over again through the maintaining of relationships (see **loyalty**) and the development of new sales opportunities. Salespersons are compensated with base and commission pay structures. Rewards are offered to individuals in personal selling by reaching or exceeding sales quotas and might include cash, time off, vacations, electronics or gift certificates.

Further reading

Assael, H. (1993) *Marketing Principles and Strategy*, 2nd edn, Forth Worth, TX: The Dryden Press.
Belch, G.E. and Belch, M.A. (1995) *Introduction to Advertising and Promotion: An Integrated Marketing Communications Perspective*, 3rd edn, Chicago: Irwin.
Morrison, A.M. (1989) *Hospitality and Travel Marketing*, Albany, NY: Delmar.

CHRISTINE VOGT, USA

Peru

The largest South American country on the Pacific Ocean, Peru was home to many ancient civilisations which were later conquered by the Inca to form the basis of their empire. The Spanish arrived in the sixteenth century. In 1821 Peru became an independent nation. Physically, it is divided into three distinct areas: the coast, one of the most arid deserts in the world; the Andes mountain ranges, with many peaks over 6,000 m; and the lowlands jungle, with thinly populated dense tropical forests. Peru has famous archaeological sites of international interest (Nazca Lines, Cuzco and Machu Picchu, Lord of Sipan, Chan Chan citadel). Its coasts, washed by the Humboldt ocean current, have attracted sport fishermen for decades. Music, both Spanish romantic and Andean, has promoted Peru all over the world.

Hotels are to be found in all cities and close to attractions. Many need to be renovated to be able to compete with the new developments, and require more modern facilities and design. Enturperu, the official chain of hotels, has been privatised. An official star **classification** system has been set up by the Dirección Nacional de Turismo to encourage loyal competition and clarity in information. In air **transportation**, Peru officially supports an open skies **policy**. Insufficient traffic volume and distance to the originating markets keep tariffs comparatively high, hence limiting development. Aeroperu, the national **airline**, has been privatised but its practical monopoly on most international connections limits competition and discourages improvements in **service**. The road network is extensive but lacks proper maintenance. Recently, a loan from the International Development Bank helped put back in proper service the stretch of the Pan-American Highway that runs along Peru's coast.

The newly acquired **political stability** and climate of opportunities through privatisation is allowing markets to recover. An independent government office, Promperu, is dedicated to promoting internationally the **image** of Peru, its assets and opportunities. International agencies are supporting the firm, and in some cases swift decisions by **government** are expected to have very positive **impacts** and consequences for the **future** of the economy.

Arrivals grew at a steady 10 per cent per annum until 1988. After this, tourism suffered from the much publicised cholera outbreak and persistent problems with guerrilla violence, reducing arrivals to a low of 216,000 in 1992. Only after new government policies and privatisations had been effected did tourism recover to 600,000 arrivals in 1996, still well below its evident capacity of attraction.

LLUIS MESALLES, SPAIN

phenomenology

Fundamentally, phenomenology amounts to the study of how the lived world is conceived and appreciated by individuals. Phenomenologists maintain that all people are architects of their identities and activities, becoming the creators of the social realities they live out. Such explanations of the lived world thus tend to privilege theories of mental determinism over those of material determinism (Hirschman and Holbrook 1992: 38). Phenomenology emerged in the early 1900s alongside the other leading subjectivist philosophy, existentialism. Later, Husserl's work pivoted on the power of conscious thought to perceive the external world and produce intentional objects as phenomena of consciousness.

Following Husserl, Schultz platformed his investigations not on isolated subjects but on the meeting point between subjects, such as the cultural realm of intersubjectivity. Partly inspired by him, Berger and Luckman (1973) looked for the legitimations by which individuals (as members of social groups) justify the world. In Berger and Luckman's terms, it is important to examine the processes by which individuals (in these preferred social groups) engage in universe maintenance to construct preferred lifeworlds around themselves.

Phenomenological approaches to understanding are steadily becoming important in studies of tourism. While strong empirical logico-positivist/Cartesian approaches to understanding tend to be mechanistic and dualistic (focusing on componential and reductionist predictions for third-person theoretical accounts about things), phenomenological inquiry tends to be contextual, in-the-world inquiry (focusing on holistic and thickly described accounts of first-person experiences) (Thompson *et al.* 1989: 137). In this regard, the constructivist-*cum*-interpretivist accounts of Goodman's irrealism, Gergen's social constructionism and Guba and Lincoln's constructivist paradigmatic approach (Schwandt 1994: 125–30) may thus loosely be labelled phenomenological styles, since each seeks to analyse the intricacies of lived **experience** from the viewpoint of those living it, each grasping for the actor's definition of things (Verstehen) within the context of that particular lived reality amongst the possibility of multiple realities. Clearly, the prospect of using such heavily emic approaches across the different populations and affinities of the developed/developing/undeveloped world of the tourism encounter is immense.

See also: anthropology; psychology; sociology

References

Berger, P. and Luckman, T.L. (1973) *The Social Construction of Reality*, London: Penguin.

Hirschman, E.C. and Holbrook, M.C. (1992) *Postmodern Consumer Research: The Study of Consumption as Text*, Newbury Park, CA: Sage.

Thompson, C.J., Locander, W.B. and Pollio, H.R. (1989) 'Putting consumer experience back into consumer research: the philosophy and method of existential phenomenology', *Journal of Consumer Research* 16: 133–46.

KEITH HOLLINSHEAD, UK

Philippines, the

Tourism's potential and pitfalls are evident in its evolution in the Philippines. Despite over 7,000 islands, beaches, mountains and a culturally diverse citizenry, arrivals in 1995 were still under 1.5 million. Erratic **development**, political unrest and corruption have wasted opportunities to build on the many tourism assets. Lacklustre development following the devastation of the Second World War and frenzied **hotel** building during the martial law era of President Marcos (1972–86) left the nation with an underused tourism **infrastructure**, huge debts and a variety of environmental and social problems. The failure by President Marcos to follow a phased plan of tourism development cost the Philippines dearly.

The logistics of an archipelago, the absence of a tradition of **domestic tourism** and **pilgrimage** and severe poverty have made **international tourism** always the focus of **planning** and promotion. Insufficient **demand** and capital shortages resulted in fewer than 145,000 **tourists** in 1972. Within months of declaring martial law, Marcos created a cabinet-level Department of Tourism and launched a building spree that added twelve five-star hotels to Manila in eighteen months. The tourism infrastructure was built with guaranteed loans from the **government**'s pension system. So desperate was the government to fill the new hotels that mega-conventions were courted, veterans and Filipinos living abroad were given subsidised travel to the Philippines and **sex tourism** and paedophilia tours proliferated. By 1980, tourism's linkages to the discredited regime had created a backlash against the **industry**. Dissident groups bombed conventions and set fire to luxury hotels (see **terrorism**). Tourism plummeted and many of the hotels went broke, leaving the government their reluctant owner.

Under President Corazon Aquino (1986–92), the sites associated with the 'People Power' revolution that ousted Marcos became attractions. Suggestive **marketing** was halted and efforts made to halt **sex tourism** and sell government-owned hotels. Few faulted Aquino's priorities, but political unrest continued to deter tourism **growth**. **Political stability** under President Fidel Ramos allowed tourism to flourish as part of an integrated national development effort. Curbing child **prostitution** and environmental issues such as beach and reef protection are just a few of the concerns on the Philippine tourism agenda.

See also: crime; impact, environmental; risk analysis

Further reading

Richter, L. (1989) *The Politics of Tourism in Asia*, Honolulu, HA: University of Hawaii Press. (See Chapter 3 for tourism policies during the Marcos and Aquino administrations.)

LINDA K. RICHTER, USA

photography

Photography is ubiquitous in tourism. Wherever tourists are, either as individuals or on a group tour, there are still or **video** cameras. Universal advice given by most tour agencies to their clients is to take more film than they think will be needed, and to test cameras before leaving home. Do not be disappointed photographically on your trip of a lifetime, they counsel.

The most frequently cited reason for the prevalence of photography in **tourism** is that the tourists want a memento and record of their trip, so they collect photographs as they collect **souvenirs**. The photograph shows that they were there, at the **site**, which accounts for the plethora of unimaginative photos of sometimes awkward-looking persons standing in front of a famous monument. Often, the photographs are arranged at home in a temporal sequence and projected, if they are slides, or assembled in an album as prints. With the exception of a few professional photographers and lecturers, the audience for the photographic record of the tour is usually quite limited, consisting of the tourists themselves and their immediate family and friends.

There is evidence suggesting that each wave of tourists take pictures of the same famous images that are found in guidebooks, **postcards**, travel **brochures** and such well-known periodicals as *National Geographic*. Rather than seeking the original and the unique, most tourists (with the exception of some serious photographers) take pictures of images that are already familiar and have become common icons in popular cultures on the visited site. Indeed, many tour **itineraries** have marked photo points that are notably scenic, and most tourists take pictures from the same locations.

Photographic visualisation is a key sensory mode for the apprehension of the object of the tourist **gaze**. In a sense, the camera lens stands between the tourist and the object, creating a space or distance that protects him or her from the unfamiliar and the strange, especially in **Third World** tourism. Photography mediates the **experience**. In African, Southeast Asian and other Third World tourism, tourists may be said to control the exotic through photography as a mediating **technology** of seeing. Roland Barthes and Susan Sontag have even suggested that photography may be viewed as an aggressive act (Chalfen 1987). On a **safari**, the camera is a substitute for a gun, as the tourist shoots a picture or captures an image. As the colonialists of old, the modern tourist controls and domesticates the wild through photography.

References

Chalfen, R. (1987) *Snapshot: Versions of Life*, Bowling Green, KY: Bowling Green State University Popular Press.

EDWARD M. BRUNER, USA

piety

Involving shrines and celebrations, travel for piety typically entails people visiting one or several sites, usually located within a universal religious tradition (but note also the significance of other belief systems such as animist and quasi-religious/secular). Devout acts may be performed both during the journey as well as the sojourn at the **destination**(s).

See also: pilgrim; pilgrimage; religion; site, sacred

JOHN EADE, UK

pilgrim

A pilgrim is one who journeys to a sacred place, such as a **shrine** or a centre of his or her **religion**. While it is often assumed that the pilgrim's **motivation** is primarily devotional, many travel for practical reasons, such as in quest of a cure for illness or the fulfilment of a wish, or because of a religious obligation (such as in the Moslem **hajj**). Pilgrims mostly travel in groups, which encourages a state of *communitas*, devoid of normal social divisions and restraints, and leads to much closeness and camaraderie (Turner 1973). This claim, however, has been contested by some researchers (Sallnow 1981).

Most pilgrims in the contemporary world come from the more traditional and conservative social strata, especially the peasantry. Religious members of the middle classes tend to practice religious tourism, a looser and less confining mode of travel than the **pilgrimage**. Modern tourists are sometimes perceived as secular pilgrims in search of authentic experiences, a secular surrogate of the sacred, which they hope to **encounter** in the course of **sightseeing** trips. However, while the 'motivation of tourists may resemble that of pilgrims, the direction of their travel differs: rather than journeying to their own religious centres, they often seek **authenticity** in other, remote, non-modern places and cultures, or in the remnants of other ages. Tourists, like pilgrims, often mix a serious quest with **recreation**, **play** and **fun**; but insofar as they tend to focus exclusively on such activities, the resemblance between tourists and pilgrims passes away.

References

Sallnow, M.J. (1981) 'Communitas reconsidered: the sociology of Andean pilgrimage', *Man* 16: 163–82.

Turner, V. (1973) 'The center out there: pilgrims goad', *History of Religions* 12(3): 191–230.

Further reading

Eade J. and Sallnow, M.J. (eds) (1991) *Contesting the Sacred*, London: Routledge.

Morinis, A. (ed.) (1992) *Sacred Journeys: The Anthropology of Pilgrimage*, Westport, CT: Greenwood Press.

Smith, V.L. (ed.) (1992) 'Pilgrimage and tourism: the quest in guest', *Annals of Tourism Research* 19(1).

Wagner, U. (1977) 'Out of time and place – mass tourism and charter trips', *Ethnos* 42(1/2): 38–52.

ERIK COHEN, ISRAEL

pilgrimage

Pilgrimage is travel inspired by religious reasons towards holy places (elements of the geographic **environment**, holy mountains, **sites** of revelations or the activities of the religious founders, **shrines** containing **relics** of saints or worshipped likenesses, and so on). Such a trip can last several years. Special forms of this journey include the pilgrimages of the sick and dying.

See also: piety; pilgrim; shrine

ANTONI JACKOWSKI, POLAND

pilgrimage route

The **pilgrimage** route is the road leading the **pilgrim** towards the holy place, to the meeting with the sacrum. As a rule the routes are fixed; some are almost 20,000 kilometres long. In the Middle Ages, an entire system of paths was formed starting in Europe and leading to Santiago de Compostella (*magnum iter sancti Jacobi*), Rome, the **Holy Land** and other sacred places. Sea routes were also used for pilgrimage, as well as river routes such as the Nile and Ganges. There was also a system of caravan trails leading to Mecca.

ANTONI JACKOWSKI, POLAND

pilgrimage site

The **pilgrimage** site can be defined as the **site** of holy objects, a place that serves as the goal of religiously motivated journeys. Some places are credited with miraculous healing powers. The usual structure of pilgrimage sites of religious significance contains historic and/or artistic importance. There is a relatively large number of local and regional sites that do not attract the attention of many secular **tourists**; this is usually the case when there are no buildings of architectural or artistic value present. The largest and best known of such shrines in Europe include San Sebastian de Garabandal in northern **Spain** and San Damiano in the Po valley in **Italy**.

BORIS VUKONIĆ, CROATIA

placelessness

Placelessness refers to the absence of distinguishing characteristics of places rendering them non-specific, and thus non-memorable, which may lead to a disorientation of those experiencing them. It exists particularly in locations designed to handle efficiently large numbers of people in a standardised manner, such as **airport** terminals and large hotels, and hence is especially prevalent in **mass tourism**.

G.J. ASHWORTH, THE NETHERLANDS

planning

Planning is a process of determining appropriate future action through a sequence of choices or organising the future to achieve certain objectives. Planning occurs at a wide variety of scales from individuals making plans for their vacations, to **destination** areas plotting future strategies to achieve **community** goals, to states charting futures for the tourism **industry**, and to international organisations preparing their own future activities and assisting countries and others to look ahead and prepare the way for desirable **change**. Mostly it is used in the context of forms of urban and regional planning from local to national levels.

Planning can have many different focuses. It can have an economic, social or more comprehensive orientation, it can be primarily concerned with land uses or **infrastructure** such as **transportation** facilities, electricity, water supply and waste disposal, it may focus upon **parks** and **protected areas** or the manpower required by an economic sector, it can be directed specifically at tourism or it can view it as part of a broader set of activities. In addition to the planning of tourism as a whole, there is a substantial literature which is directed at its particular aspects such as transportation, parks and protected areas, conventions and festivals.

Planning has come to mean more than the preparation of static planning documents with increasing attention being paid to the processes by which decisions concerning possible desirable futures are made. Thus, there has been growing concern to complement the **roles** of experts with local inputs gained through **public participation** and recognition that, although theoretically desirable, it is not possible to be truly comprehensive and that flexible, incremental approaches provide more scope for adjusting to changing circumstances and taking advantage of opportunities.

Tourism planning, like planning itself, has evolved over the years. Getz (1991) has identified five tourism planning traditions, each with its own associated concepts, methods and biases. The first, boosterism, is little more than the promotion of **development** and thus is not really planning at all. The second tradition views tourism as an **industry**, analogous to other ones, and has a predominantly economic focus with an emphasis on development and **marketing**. The third stresses the spatial aspects of tourism and physical resource planning based upon careful resource analysis and notions of **accessibility**. **Community planning** requires that such places should take control of the planning process, set their own goals and plan accordingly, using concepts like **community planning** and social **carrying capacity**. The fifth tradition, an integrated and systematic approach, suggests that goals, policies and strategies should be based upon a fuller understanding of how the tourism system works.

In developing countries, although there has been an increase in the number of jurisdictions incorporating a tourism component into more general plans, this has often been based primarily on control of development through building codes, **zoning** systems, environmental impact assessment and the like which are not specific to tourism. Further, there has been a recent trend to turn more authority and responsibility over to the private sector. In contrast, many less-developed countries have prepared sophisticated tourism plans, often with the aid of outside experts and finance, as a means to display a sense of vision and thereby attract **investment** in what has often been viewed as a growth industry.

See also: product planning

References

Getz, D. (1991) *Festivals, Special Events, and Tourism*, New York: Van Nostrand Reinhold. (Discusses planning and assessment of special events.)

Further reading

Gunn, C.A. (1979) *Tourism Planning*, New York: Crane, Russack and Co. (A classic which discusses attractions, services, facilities, transportation, information as well as planning processes.)

Heath, E. and Wall, G. (1992) *Marketing Tourism Destinations: A Strategic Planning Approach*, New York: Wiley. (Applies marketing concepts to the planning of destination areas.)

Inskeep, E. (1991) *Tourism Planning: An Integrated and Sustainable Development Approach*, New York: Van Nostrand Reinhold.

Murphy, P. (1985) *Tourism: A Community Approach*, New York and London: Methuen. (A comprehensive discussion of tourism planning from a community perspective.)

GEOFFREY WALL, CANADA

planning, environmental

Environmental **planning** is a generic term covering a wide range of related activities, from site planning to national **policy** making, from the activities of individuals through non-governmental organisations to those of United Nations Development Organisations and the World Bank. Most successful destinations, for example, depend upon clean physical surroundings and protected environments, as well as on their particular social, economic or cultural attributes. Destinations that do not protect these are suffering a decline in **quality** and tourist use. However, because the word 'planning' has so many popular meanings, it is necessary to consider in each case what is meant when environmental planning is discussed.

It is possible to distinguish the planning endeavour on the basis of environmental policy development and implementation, differing levels of specificity related to area-wide planning, particular types of techniques adopted to control land use, and/or other community issues with the environmental one. Each of these has its own concerns and has spawned differing techniques and requirements for action since the development of modern town planning approaches in the wake of the eighteenth-century European experience with industrialisation and urbanisation. The recent but increasing concern with the wider biophysical **impacts** of human settlement and development has led to the explicit use of the word 'environmental' in association with the term planning, but the pedigree of planning is far older.

Environmental planning provides the basis for achieving integrated, controlled and sustainable human activity within the earth's biosphere. Planning is carried out in accordance with a systematic process of setting objectives, environmental **survey** and analysis (audit), formulation of a plan with recommendations for the control of **development**, and implementation, followed by continuous **management**. Planning takes place at a variety of levels, ranging from the macro national and regional to the micro destination and site levels. At the **local** level, determination and adoption of facility development and design standards are essential to ensure that tourism development is, for example, appropriately sited and designed with respect to local environmental conditions and desired character. However, even though local plans may be prepared independently, it is essential that they fit into the wider context of national and regional environmental policy and plans. The macro level of environmental planning provides the framework for developing activities at the community level.

Tourism development policy and physical structure planning, which indicates major attractions and activities, sites to be developed or designated **protected areas**, major tourism market segments, gateways, **regions** or zones, and **trans-**

portation and support networks, is part of and constrained by the requirements of national and regional environmental plans. At the community, **resort** and development area levels, environmental plans integrate tourism into overall desired development patterns (this level should ideally also include consideration of community wishes and **community** participation in **decision making**). At the site planning level, the emphasis is less on environmental planning policy and more on how the location and layout of buildings and structures, engineering design, parking, landscaping, recreational facilities and related uses conforms to sound environmental planning.

Therefore, environmental planning can fulfil economic, social and cultural needs while maintaining the ecological integrity of a given area. It can assist in providing for today's tourism while protecting and enhancing the same opportunities for the future. However, it also involves making hard political choices based on complex social, economic and environmental trade-offs. It requires vision that encompasses more than just local regulation of polluting activities and/or **environmental rehabilitation** after the event, and must be based upon the following to provide an effective tool. All of this must be part of an overall approach to **sustainable development** strategies for a region or a nation. It should be cross-sectoral and integrated, involving **government**, **industry**, citizens' groups and individuals for the widest possible benefits.

Good information, research and communication on the nature of environments and potential developments is essential, especially for local people, so that all can participate in and influence the direction of development and its effects as much as possible in the individual and collective interest. Integrated environmental, social and economic planning analyses should be undertaken prior to the commencement of major project development, with careful consideration being given to alternatives and the ways in which they might link with existing uses, ways of life and biophysical considerations. Throughout all stages of planning and development, a careful assessment, monitoring and environmental mediation programme should be conducted in order to allow local people and others to take advantage of emerging opportunities or respond to environmental change.

The deterioration of environmental conditions worldwide is of major significance. Certainly, the planning of all tourism development must include consideration of its potential **impact** on the social, economic and environmental qualities of host areas. Recent **growth** in the techniques and concerns of environmental planning ensure that economic development is not now an exclusive goal of developers and host communities, and a proactive environmental protection stance is fast emerging on all sides. Environmental planning can ensure that development objectives, visitor desires and environmental protection are compatible goals for any community, at any level.

See also: codes of ethics, environmental; conservation; ecologically sustainable tourism; environmental engineering; environmental management systems; environmental management, best practice; legislation, environmental; planning, environmental.

Further reading

Gunn, C.A. (1994) *Tourism Planning*, 3rd edn, Washington: Taylor & Francis.

World Tourism Organization (1993) *Sustainable Tourism Development: Guide for Local Planners*, Madrid: WTO.

MALCOLM COOPER, AUSTRALIA

planning, recreation

Recreation **planning** aims at shaping a desirable recreational **environment** based on the society's preferences. In **outdoor recreation** planning, the process begins with an **inventory** of **resources** which are evaluated for their potential uses. Based on this assessment of **site** capabilities, a physical plan is decided upon, involving the allocation of site facilities and the construction of access routes to the sites.

There are three popular planning approaches. First, single-site planning is characterised by a supply orientation and incrementalism which, in the long run, usually leads to overcrowding and

displacement of original users. Second, **carrying capacity** planning is based on a permit system which imposes limits on the use of facilities to avoid overcrowding, which may lead to environmental damages and erosion of the quality of recreational **experience**. Third, **recreation opportunity spectrum** planning pays attention to the changing structure and composition of the user population which may result in corresponding changes in use patterns. This approach incorporates the concept of social succession, which suggests that recreational preferences will change following the displacement of original residents by recent migrants. Before such succession occurs, planners may introduce a range of new recreational activities which would allow original users to mitigate negative impacts of **change** while continuing to enjoy preferred activities. The third approach allows for a comprehensive area planning while offering a greater variety of recreational opportunities.

As a society undergoes changes, both in demographic structure and social values, the patterns of recreation behaviour also change. One of the goals of recreational planning is to create desirable futures based on an anticipation of these changes. This calls for **forecasting** of recreational demand. There are three notions of forecasting: positive forecasting which extends the past into the future by extrapolation, normative forecasting which takes into account society's future values, and prescriptive forecasting which begins with the establishment of values based on experts' opinions. One such approach is the **Delphi technique**.

Since the early 1990s, following a widespread awareness of the shortcomings in the traditional advocacy approach in recreational planning, there has been a noticeable shift in approach in favour of participatory planning. The new orientation calls for more public debates and citizen involvement in the planning process. This stage marks a shift from supply to **market** orientation in recreational planning. With expansion in the tourism industry, recreational sites, originally designed for local **community** use, have begun to cater to touristic needs and consequently have become part of the tourism attractions.

Further reading

Lieber, S.R. and Fesenmaier, D.R. (eds) (1983) *Recreational Planning and Management*, State College, PA: Venture Publications.

Van Lier, H.N. and Taylor, P.D. (eds) (1993) *New Challenges in Recreation and Tourism Planning*, Amsterdam: Elsevier.

SULONG MOHAMAD, MALAYSIA

play

Play is a behavioural disposition, characterised by pleasure, enjoyment, freedom, and spontaneity, which elicits engagement by participants, and which is manifest in a variety of different forms. Since play is a behaviour so ubiquitous in humans, mammals, and birds, one can assume it serves some evolutionary function. Assuming that play does have survival value, the challenge is to delineate what its properties, functions, benefits, and consequences are.

A feature that is almost unanimously acknowledged to be the hallmark of play is that it is intrinsically motivated behaviour. The designation intrinsic or extrinsic motivation is closely connected with the question of the function, purpose, and the goal of certain patterns of subjective experience. Intrinsically motivated behaviour is centred within the individual and occurs in the absence of any external force or event instigating it. In contrast, extrinsically motivated behaviour is centred outside of the individual, leads to aims outside itself, and is more a means to an end. Play is regarded as such because it is activity performed for its own sake and in which pleasure is inherent in the activity itself.

Play is characterised by attention to means rather than ends. Attention to means allows for the creation of new combinations of behaviours, which themselves may lack the efficiency of acts that are employed to achieve a goal. This may be characterised by a lack of economy of movement. Without having to worry about efficiently achieving some completed product, the individual is able to dismantle established instrumental behavioural sequences and reassemble them in new ways. This characteristic helps distinguish play from other intrinsically motivated behaviours directed toward the attainment of specific goals (enjoyable work) as well as from those that seem aimless or unfocused.

Play is regarded as organism-dominated, rather than stimulus-dominated. In other words, it is governed by the individual rather than the environment. Play serves to introduce stimulation and maintain an optimal state of arousal when little challenge or information flow can be extracted from the environment. In this reasoning, play is contrasted with exploratory behaviour in which the question for the player is 'what can I do with this object?' while for the explorer it is 'what is this object and what can it do?' The latter is about external events, about learning something; the former is about personal desires, about pretending to be in charge of the something that has already been learned. The attribution by players that internal wishes determine play, rather than physical parameters of the external environment, is an important outcome and is one that holds throughout the life span.

This organism-dominated character often yields its nonliteral description. Play behaviours are not serious renditions of the activities they resemble. For example, it is easy to detect the difference between fighting and play fighting in both humans and animals. Similarly, the usual functional meanings of objects can be dispensed with within play and the individual explores new potential meaning by treating the object as if it were something else. Thus, play has been regarded as possessing an 'as if' representational set, focusing on its imaginative and pretence qualities. This becomes important in the acquisition of the ability to engage in subjective thought processes, and in the development of a sense of autonomy, individuality, and competence or power.

There are two general perspectives about the value of play to the individual and society. The first argues that play serves to prepare the individual for the future. In this sense, it is predominantly what children do. These preparatory theories argue that play prepares the child for adult life, that the young are born with immature or imperfectly honed mechanisms and skills that need to be developed and finely tuned. Selection has favoured pressure-free periods of time in childhood during which the subroutines of adult skills can be acquired through observational learning and imitation. Play also serves a critical social function: to reduce social distance between individuals and to develop intimate social bonds that may approach what some have termed 'love' in peer relationships and between parents and offspring. Evolutionists, biologists, developmental psychologists, and feminists all subscribe to this 'preparatory' approach to play.

The second rhetoric about the value of play is a more pessimistic one. It posits that individuals play to adjust to their present life circumstances and to the stressful situations in which they find themselves. The psychoanalytic explanation for play and the phenomenological nature of the **flow** experience are prime examples of this approach, in which the impetus for play is to master anxiety and conflict: what has been suffered in ordinary experience must be purged. This way, play serves to restore the individual to a more pleasant state of mental health. Thus, in contrast to the first theme, individuals do not grow as a result of their play, instead they strive for contentment. It is in this sense that tourism has been regarded as play or a form of it, with tourists playfully throwing themselves into the **liminal** and **ludic** moods. While some have attempted to show and develop the relationship between tourism and play, the theme has remained mainly underdeveloped. The behaviour and activities of tourists have often been the subject of analysis, but without explicit reference to the concept of play. Some researchers believe that it is precisely the study of this relationship which can make the tourist culture meaningful. Understanding the degree and level of interaction of tourists (or the absence or artificiality of it) with the host population is another topic which would benefit from this study, with applications both in the realm of concept and practice.

Play is widespread, encompassing, and often indistinctive from other forms of behaviour. However, it is often argued that play is a separable category of behaviour by having distinguishable play places (like playgrounds, playrooms, sports arenas, **gambling** boats, **parks**, etc.) and play times (like recess, game nights, vacations, guided tours). To play seems to be the easiest thing to do. However, to define play is quite the opposite. Its rich kaleidoscope of many facets and forms renders it one of the most remarkable and empirically challenging of the vast realm of behaviours and experience both in everyday life and in touristic playgrounds now spread and popular worldwide.

Further reading

Cohen, E. (1985) 'Tourism as play', *Religion* 15: 291–304.

Huizinga, J. (1955) *Homo Ludens*, Boston: Beacon Press. (The seminal text on man as playful being.)

Pellegrini, A.D. (ed.) (1995) *The Future of Play Theory*, Albany, NY: State University of New York Press. (Discusses theory and research about play, its appearance, forms, functions, definitions, and benefits.)

LYNN A. BARNETT, USA

pleasure periphery

The term 'pleasure periphery' has been borrowed from international development studies, where notions of 'core' and 'periphery' relate to the fundamental inequalities inherent in North–South relations. In contemporary tourism studies there is increasing interest in the relationship between the **pleasure tourist**-generating countries of the metropolitan 'core' and the tourist-receiving countries of the pleasure periphery.

TOM SELWYN, UK

pleasure tourist

Even though people travel to visit friends and relatives more frequently than for any other reason, the study of tourism has focused mainly on those who travel in pursuit of pleasure. Ironically, although the pleasure tourist has been constituted as its principal subject, the study of tourism has paid little attention to the nature of pleasure itself.

TOM SELWYN, UK

Poland

A country in Middle Europe with 39 million inhabitants and a land area of 312,000 square km, Poland has borders with Germany, Czech Republic, Slovakia, Ukraine, Byelorussia, Lithuania and Russia. It has the Baltic sea on its north, with 524 km of coast, and Carpathian Mountains in the south with Tatra and its highest peak, Rysy, on the Slovakian border. The capital, Warsaw, has 1.7 million inhabitants. Other major cities include Cracow (the former capital), Lodz, Wroclaw, Poznan and Gdansk. Prince Mieszko I unified several Slav territories and accepted Christianity, and his son Boleslaw Chrobry became the first king of Poland. In the fifteenth century, Poland (unified with Lithuania) was one of the largest powers in Europe. By the end of the eighteenth century the country had been divided among Russia, Prussia and Austria and disappeared from the map of Europe. Independent from 1918 to 1939, after 1945 Poland became a socialist country and part of the Soviet bloc. In 1989, as a result of the 'Solidarity' movement, Poland became once more independent and stood as an example to the other socialist countries.

Tourism in Poland had begun to develop by the end of the nineteenth century, first with mountain tourism in the Tatra. By 1918 the country was gaining in popularity for its pleasure and **spa** destinations. In 1920 the Orbis Travel Agency and in 1929 the Polish Airlines LOT were created. In 1990 the number of incoming **tourists** reached 18 million and in 1994 reached 75 million, accounting for 18.5 million overnight trips. In 1994 the national tourism **development** plan was accepted by the cabinet. Tourism is the subject of research, **education** and **training** in Poland. The first annual meeting of the **International Academy for the Study of Tourism** took place in 1989 in Zakopane. The main tourism **attractions** of Poland include cultural objects in Cracow, Warsaw and Gdansk, and the well-known pilgrimage centre of Czestochowa and the old salt mines in Wieliczka. The appeal of the Tatra mountains, the Masuria lakes, the National Park of Bialowieza and the Bieszczady mountains continue to attract still a larger volume of **international tourism** to this country.

KRZYSZTOF PRZECLAWSKI, POLAND

polar

Polar regions lie poleward of the tree line (roughly 60° latitude) where the average temperature of the warmest month remains below 50° Fahrenheit. In the north, stunted slow-growing tundra vegetation (mosses, sedges and lichens) provides sparse cover over heavily glaciated bedrock and permafrost (permanently frozen ground and ice). Long sunny summer days, the midnight sun, a carpet of wildflowers and **outdoor recreation** opportunities attract **adventure tourism** including fishing, **hiking** and canoeing/rafting. There is limited road access in both Europe and North America. Winter tourism is usually limited to festival events including watching the *aurora borealis* and migratory polar bears. The rich marine life in the Arctic Ocean supports unique **indigenous** cultures such as the Alaskan–Canadian–Greenland Inuit and the Scandinavian Saami, whose ancestral **lifestyles** are now a focus for **ethnic tourism**. In the Siberian Arctic, summer flooding on the deltas of several northward flowing rivers creates vast marshes and shore access is difficult. Here, inland tribes such as the Tungus, Samoyed and Chulcchi domesticated reindeer centuries ago and used these animals for sledging. The polar region is a fragile **environment**, and especial care must be given to spatial **development** and land use.

The Antarctic continent, fifth largest in size (5,000,000 square miles) lies entirely within the polar region, and is the coldest, driest and windiest area on earth. More remote, it is accessible only during the Austral summer, by ice-strengthened cruise ships and by limited air services. It is particularly noted for sea mammals and penguins, as well as dramatic glacial scenery. Antarctica is geographically unique because it is a mirror **image** of the Arctic (a land mass surrounded by seas) and no single nation holds sovereignty over any land in this vast terrain. Following two centuries of progressive exploration and consequent land claims, the International Geophysical Year (1957–58) established the need for an international accord and the Antarctic Treaty was signed on 1 December 1959. Austral summer cruise tourism began in 1958 and increased slowly but steadily; in the mid-1990s, some 6,000 passengers cruised its coast. A land-based touring operation, Adventure Network International provides DHC-6 Twin Otter flights to the South Pole and aerial support for explorers and adventure tourism skiers.

In the absence of sovereignty and thus a total lack of **policy** or **planning**, in 1989 three Antarctic **cruise line** operators plus Adventure Network International founded the International Association of Antarctic Tour Operators and established two sets of guidelines, one for tour operators and another for Antarctic visitors. Comparable standardised regulations were then signed in Madrid at the 1991 Antarctic Treaty Consultative Meeting, and are known as the 'Protocols on Environmental Protection to the Antarctic Treaty'. Subsequently, the US National Science Foundation has sponsored on-board monitors for each sailing to ensure the protocols are observed.

See also: Antarctic tourism; Arctic tourism

Further reading

Smith, V. and Splettstoesser, J. (eds) (1994) 'Antarctic tourism', special issue of *Annals of Tourism Research* 21(2).

VALENE L. SMITH, USA

policy

Policy is a key concept in **government**, political and business studies. Business policy refers to the guidelines that managers use in making decisions. In the study of tourism, however, the majority of researchers use the term with respect to government, state or public policy. There is no universally accepted definition of public policy. Its analysis may be generally described as whatever governments choose to do or not to do. This description covers government action, inaction, decisions and non-decisions as they imply a deliberate choice among alternatives. For a policy to be regarded as public policy, at the very least it must have been processed, even if only authorised or ratified, by public agencies. Thus tourism public policy is whatever governments choose to do or not to do with respect to tourism.

Although closely related to the fields of **political science**, **political economy** and public administration, public policy is regarded as a separate field. It is an important area of scholarship that generates much debate, literature and research, because of the highly applied and political nature of its conclusions. Public policy is not independent of the political processes which, including its processes, outputs, outcomes and analysis, cannot be value free. Interest in public policy research has grown rapidly since the 1960s. This began in the United States and the United Kingdom as social scientists were attracted to the applied, socially relevant, multidisciplinary, integrative and problem-directed nature of policy analysis, particularly in areas such as health, welfare, housing, **crime**, **transportation**, **economic development**, **energy** and **environment**. However, tourism has only recently emerged as an obvious commitment and important consideration in the public sector.

As in the case of political science, tourism policy analysis has historically not been a substantive subject of scholarly inquiry. Nevertheless, the importance and relative influence of tourism, and particularly its economic impacts, are reflected in its dramatic growth in global political prominence since the 1970s. As interest in the **development**, promotion and **impacts** of tourism becomes integrated into the machinery of all levels of government in both developed and developing countries, increased attention is being given to the formulation, outputs and outcomes of tourism policy. There is little agreement about how tourism public policies, as a relatively new area of study, should be analysed and the reasons underpinning such research. As a result, analysis of tourism policies is often constrained by the lack of consensus concerning definition of such fundamental concepts as tourism, **tourist**, and the **industry**; the lack of recognition given to tourism policy-making processes and the consequent lack of comparative studies; the lack of well-defined analytical and theoretical frameworks; the limited amount of quantitative and qualitative data; and concerns by policy makers that studies may draw negative conclusions on government activity.

Tourism public policy studies are usually undertaken in order to understand the causes, consequences and appropriateness of policies, decisions and actions. Therefore, its study offers the opportunity to examine many topics which are relevant to the industry, government agencies, interest groups, **destination** communities and researchers working on the boundaries of related disciplines such as **economics**, **geography**, **history**, political economy, political science and **sociology**. These topics include the political nature of the tourism policy-making process; **public participation** in the tourism **planning** and policy process; the sources of **power** in policy making; the exercise of choice by bureaucrats in complex policy environments; the institutional arrangements surrounding tourism, particularly with respect to the activities of national and regional organisations; attempts by interest groups, including business and environmental groups, to influence the process and outcomes of policy; the evaluation of the impacts of policies and the **efficiency** and effectiveness of the policy-making process; and perceptions as to the effectiveness of policies.

Tourism policy analysis is an activity for which there can be no fixed programme. Public policy **theory** serves as the basis for explaining decision-making and policy-making processes, and for identifying the causal links among events. However, the importance, use and relevance of particular public policy theories often rest on the research philosophy and worldviews of the analyst or those who designed the study. Different theoretical perspectives such as pluralist, elitist, Marxist, corporatist and public choice, while not mutually exclusive, conceptualise the policy process in distinct ways. Theories can also be distinguished from one another by their level of analysis, and by the methods they employ in studying policy. Each perspective thus differs in its assumptions about political conflict, the appropriate level of analysis and research methods. This does not mean that policy studies is an anarchic field. Rather, policy analysis is akin to an 'art' or 'craft' in that the overall structure of the argument of a policy study will determine its acceptability and attractiveness to its intended audiences. Within the conceptualisation of policy as an art, debate is perceived as a highly positive contribution to the formulation of appropriate policies as it provides a basis for selecting from alternative policy directions. Within such a conceptualisation, policy studies explicitly acknowledges the role of competing values, interests and assumptions in the analysis and formulation of policy and the production of policy knowledge.

Despite the lack of a dominant approach to policy studies, policy research can be broadly described as being based on either prescriptive or descriptive frameworks of analysis. Prescriptive or normative **models** seek to demonstrate how policy making should occur relative to pre-established standards and offer a guide to future policies. Descriptive models document the way in which the policy process actually does occur. The majority of tourism policy literature is highly prescriptive. This may be because of ignorance of public policy studies and the politics of tourism or it might reflect a desire not to be perceived at odds with government and industry sponsors of tourism **education** and research. For example, the policy dimensions of **community planning** and **development**, **growth** management, **destination** management and sustainable tourism have generally been based on prescriptive models of tourism.

A descriptive approach is preferred when exploring a new subject in a particular area of tourism policy. Descriptive (positive) models give rise to explanations about what happened during the decision-making and policy-making processes. They help analysts understand the effects that choice, power, interests, institutional arrangements, perception and **values** have on the nature of the policy-making process. Descriptive studies indicate that tourism policies are formulated and implemented in dynamic environments where there is a complex pattern of decisions, actions, interaction, reaction and feedback. The complexity of the policy process emphasises the importance of analysing different stages of the policy process and different levels of analysis. Three levels of analysis are usually identified: the micro-level of decision-making within organisations, which sees organisations as highly dynamic political entities; the middle range analysis of policy formulation and implementation (the meso-level), although it is noted that policy formulation and implementation are difficult to separate on a consistent basis because policy is often formulated as it is implemented and vice versa; and macroanalysis of political systems including examination of the role of the state in tourism. It is the interaction among levels which is particularly significant and problematic. The relative lack of descriptive policy studies of tourism at national, regional and local levels has been a major impediment in developing effective models of planning and development strategies which are accepted by the various stakeholders. In other words, although prescriptive models are deductive, one cannot deduce in the absence of prior knowledge.

Several sub-fields of tourism policy studies are also gradually developing, including the study of interest (lobby or pressure) groups, policy making in federal systems, tourism as a government response to economic restructuring in rural and urban areas, regional tourism development policies, urban re-imaging strategies, public–private partnerships, tourism and cultural policies including the **commoditisation** of **heritage** for tourist consumption, the hosting of hallmark events, the values and **ideology** of participants in policy making, tourism and environmental policies including **national parks**, and policy monitoring and evaluation. Although policy studies of tourism have for long received only limited attention by government, practitioners and academics, the practical significance of understanding why policies and plans fail is leading to greater recognition of the need to understand the policy-making process.

Further reading

Edgell, D. (1990) *International Tourism Policy*, New York: Van Nostrand Reinhold. (A prescriptive approach towards tourism policy.)

Hall, C.M. (1994) *Tourism and Politics: Policy, Power and Place*, Chichester: Wiley. (Places international policy within the wider political context of tourism.)

Hall, C.M. and Jenkins, J. (1995) *Tourism and Public*

Policy, London: Routledge. (The most comprehensive text on tourism and public policy.)
Hall, C.M., Jenkins, J. and Kearsley, J. (eds) (1997) *Tourism Planning and Policy in Australia and New Zealand: Cases, Issues and Practice*, Sydney: Irwin. (Integrates tourism planning and policy issues, with overview chapters on tourism policy in international, nature area, rural area and urban area context.)
Richter, L.K. (1989) *The Politics of Tourism in Asia*, Honolulu, HA: University of Hawaii Press. (An analysis of tourism policy and public administration in ten Asian countries.)

C. MICHAEL HALL, NEW ZEALAND

political development

The linkage between tourism and political development has become an important subject of social science research in recent decades. As a subfield of **political science**, the study of political development became prominent in the period following the Second World War when numerous territories were struggling for political independence and **economic development**. This period has also been characterised by substantial **growth** in **mass tourism**, with much of it directed toward those same developing countries.

Political development can be defined generally as the refinements that a political system makes in its political institutions and processes to improve its capacity to encourage and accommodate new demands and functions. The objective of political development is to keep the political system responsive and effective. Tourism is a cluster of human activities associated with the desire and ability of people to **travel** outside their home environment. Such travel places demands upon political institutions in a given polity for regulation of both **inbound** and **outbound** tourism and for adjustments in various aspects of the society.

It is difficult to study political development without also examining concurrent development in social and economic processes and institutions. Changes can be directly related to tourism, but in the larger context they may be prompted by other events or pressures. Separating tourism's impact from the general forces of modernisation is usually difficult. Political development's minimal task is to maintain order in society or channel **conflict** in manageable ways. As the complexity of a society increases, the goals of political development become more complicated. Political development turns into the process whereby society refines its political institutions to handle more diverse and challenging tasks. Tourism can create new demands for more effective **government**. Unfortunately, some policies may contribute to political decay rather than development if the system is unequal to the pace or type of tourism or if the government is corrupt.

Tourism has its **critics**, who argue that the **industry** is the source of much political and cultural friction, and that it has a negative effect on **indigenous** values and moral character. At the same time, tourism is seen by some as a catalyst for positive political and cultural **change**, often prompting political institutions to embrace new ideas, to be more open to diverse **values** and to accommodate new technologies. Both views offer elements of truth. The influence on political culture may depend on the system and the values of the researcher. In closed political systems, the desire for economic benefits from tourism may be stronger than the willingness to protect the *status quo* in terms of political culture. In such a case, persons who believe in more open political systems would likely say that tourism has a positive effect on political development, but this would be a view driven by predisposed political values.

There are a number of issues of political development that are greatly affected by tourism. **Domestic tourism**, for example, is a critical input to national integration. It helps overcome tense regionalism based either on geographic or cultural identities. In a similar way, **international tourism** encourages political development of nation-states in terms of their conduct of foreign affairs, and promotes an international political climate conducive to travel and cultural exchange. International tourism helps identify shared values and common political needs among otherwise diverse nations. The development of contemporary international **law** and its corollary political institutions has been a major factor in the growth and sophistication of international tourism in the modern world.

Tourism is also linked to political needs of regimes and the jurisdictions they represent. The **demonstration effect** may encourage populations to embrace new ideas, new technologies or new political values. Because tourism is a very important means of earning **foreign exchange**, governments have used the industry as a way of opening the door to even greater foreign **investments** in their country. If the industry flourishes, this demonstrates to the outside world a sense of legitimacy and stability that encourages more travel and **investment**.

Serious and enduring criticisms of the impact of tourism upon political development and political culture have been particularly harsh from developing regions of the world. Many social scientists see a negative demonstration effect which highlights the contrast among the wealth, **leisure** and appearance of **tourists** and the poverty of the indigenous population. As international tourism grew in the post-Second World War period, the populations which could afford travel were concentrated in Western Europe and North America. Later, as the Japanese and other Asians began to engage in outbound travel, they also sought vacations in the warmer climates of the **Third World**, particularly the South Pacific region. In the Caribbean, for example, the contrast between the white tourists and the non-white population made tourism a highly visible activity. Moreover, the tourists were affluent, they came mainly from Europe and North America, and the institutions that made their trips possible were mostly foreign-owned. Caribbean economists cited this situation as evidence that tourism, along with other types of foreign-owned industry, was a form of neo-colonialism and neo-**imperialism** since it returned its profits to the metropolitan country and since those who benefited from the industry were mainly nationals of the generating country. This critique of tourism in developing countries gradually eroded as local investors, politicians and workers began to benefit substantially from it. The critique also encouraged political institutions, and the political culture in general came to integrate the demands of tourism with the needs of the local population. The extent to which the industry has been a catalyst for such political development has varied among the different political systems.

See also: globalisation; ideology; political socialisation

Further reading

Brohman, J. (1996) 'New directions in tourism for Third World development', *Annals of Tourism Research* 23(1): 48–70. (Updates on contemporary issues of tourism development in developing regions.)

Jackman, R.W. (1996) *Power without Force*, Ann Arbor, MI: University of Michigan Press. (Basic text on political development as a field.)

Palmer, C.A. (1994) 'Tourism and colonialism: the experience of the Bahamas', *Annals of Tourism Research* 21(4): 792–811. (Examines how tourism is often seen as inherited from colonialism.)

HARRY G. MATTHEWS, USA

political economy

Political economy describes the examination of the relationship among political and economic policies, institutions and structures and their influence on **development**, societies and individuals. Often described by conservatives as being 'radical', political economy distinguishes itself from **economics** by its explicit recognition of the political and ideological dimensions of analysis, **policy** and **theory**.

Political economy examines the formal relations within capitalism that express the real **power** relations among individuals, groups, institutions (including the state) and culture. Political economists focus on the organisation of capitalism as an economic and cultural phenomenon and the manner in which it has undergone successive transformations which include the penetration of peripheral areas by global capital, such as **Third World** peripheral regions in which tourism development is controlled by First World capital, and the ever-increasing **commoditisation** and incorporation of culture and places within the capitalist system through **globalisation**. Analysis may also occur in the context of a specific place in which political and economic relations are identified in terms of **class** relations.

The idea of class is integral to Marxist or neo-Marxist approaches to political economy. Marxist analysis has often been misunderstood or misrepresented in mainstream economics which promotes itself as value-free science. The Marxian notion of the dialectic or interconnectedness between culture and economy is essential to the critical study of **heritage**, **image**, place and power in tourism, although this does not occur within mainstream studies in this field.

Issues of **identity**, **representation** and exclusion are significant contemporary foci for tourism-related political economy, recognising that it is the ruling class which controls the form and content of historical recreations and tourism **landscapes**, legitimising itself by projecting its own contemporary sociocultural values upon the past. A political economy approach is also applicable to the production of tourism knowledge. For example, this encyclopedia can be interpreted as an exercise in academic power and legitimisation which excludes or ignores certain ideas, values and interests from its text, thereby selectively representing the phenomenon of tourism.

See also: centre–periphery; political science

Further reading

Britton, S.G. (1982) 'The political economy of tourism in the Third World', *Annals of Tourism Research* 9: 331–58. (Studies political economy in developing countries.)

—— (1991) 'Tourism, capital and place: towards a critical geography of tourism', *Environment and Planning D: Society and Space*, 9: 451–78. (Places tourism analysis in the context of contemporary capitalism.)

C. MICHAEL HALL, NEW ZEALAND

political science

Scholars and political scientists have been slow to recognise that the enduring questions of politics are central to tourism issues. Political science is the study of politics in all its myriad forms. Many issues can be highlighted by first, looking closely at a well-known definition of politics, and second, looking at the way political science as a discipline is organised to teach and research these topics.

There are many definitions of politics, but one that is widely used and adapted forms the title of Harold Lasswell's *Politics: Who Gets What When and How* (1936). This definition pithily focuses attention on who has the **power** to control the distribution of **resources** and under what conditions or in what pattern it is dispersed. It concentrates attention on such issues as *who* those most affected positively or negatively by tourism **policy** are; whether elites, interest groups and the general public are involved in making decisions; and whether there are monitoring mechanisms to discern the **impact** of tourism on individuals and groups. For example, special events like the Olympics and World Fairs may have some obvious **benefits** for those who participate and attend and for the businesses that they patronise, and in terms of **prestige** for the hosting nation and city. But there are also those who bear the costs of tourism. There are evictions, increased **taxes**, **security** measures, traffic, crowding and **crime**. All create winners and losers. Within the same political unit, there may also be a great variation in who benefits along lines of **class**, **race** and **gender**.

What is to be distributed is often the subject of political controversy. It may be new attractions, beach **development**, a hotel **training** institute, or a noise abatement ordinance regulating airline takeoffs over urban areas. It can even mean **vacation** policies or a nationally declared holiday. *When* and *how* ask the political questions about the circumstances, pace, financing and scope of tourism decisions. Governments, investors and other groups recognise that the timing, sequencing and care with which decisions are reached have much to do with the level of support they enjoy. Who participates in the making of key decisions about tourism policy: the **World Tourism Organization** or some other international body, the nation, the state or province, or local authorities? Or are key decisions primarily the result of individual, corporate or even transnational private interests?

What procedures and processes exist for legitimating these decisions? For example, is there a referendum on a hotel **tax**, a **zoning** meeting on a proposed **resort**, an environmental **impact** statement on planned development, or opportunities for national tourism organisations to coordinate policy with **energy**, immigration, labour and commercial policies? Opening up a tunnel for tourists near a Muslim holy **site** in Jerusalem in 1996 led to deadly riots because the action was taken without consultation of those in the area. Those who feel negatively affected may not be able to exert persuasive power at the bargaining table, but the resulting political instability can threaten the **peace** and cripple tourism.

Lasswell's definition was later amended to also ask, 'and who already has what?' That addition asks political questions central to tourism **planning** and **development**. It does not take for granted that power is evenly distributed, but looks at who is advantaged or disadvantaged by the *status quo*. Political scientists explore who thinks the most in terms of social class and gender, or gains from the existing or proposed structuring of tourism. Policies that consciously put attractions in areas of high unemployment or see that new roads also serve the **transportation** needs of non-tourists are considering how development can serve to enhance economically a depressed area. Often, **enclave tourism** is adopted with just the opposite perspective: to keep tourists away from the poor or the traditional who might force policy makers to assume the difficult task of integrating tourism into a plan of overall development.

This writer has argued that the definition of 'Who Gets What When and How' should not only include 'and who already has what', but also 'and who cares?' Political distributions of power may be of academic interest to political scientists, but it is the struggles around the *status quo* that affect the policy-making process and shape the evolution of tourism in a particular setting. For example, the sovereignty movement in Hawaii, aboriginal claims in Australia and fundamentalist Muslim dictates toward women in Egypt, the Maldives and Afghanistan are all instances in which the *intensity* of group beliefs is impacting tourism. **Risk analysis** not only is used in planning for site preparation and design but also for assessing the prospects for **political stability**, labour peace and cultural conflicts.

Political science as a discipline is organised into a number of subfields which could, but rarely do, focus on tourism. Major subfields include comparative government, international relations, public administration and political thought. Comparative politics usually examines nations along several dimensions including political structures, culture, leadership and public policy. A few political scientists have compared national tourism policies in terms of planning, taxation, import politics, **political development**, labour relations and leisure time policies, and the role and status of national tourism organisations. The role of **government** in the promotion, development and ownership of tourism components (such as airlines, hotels and **national parks**) has also been studied. More recently, political studies of the problems associated with tourism, including **prostitution**, environmental damage, loss of local control and crime, have concentrated attention on the many different ways governments have responded to the potential and pitfalls associated with tourism development.

Comparative studies have also called attention to the fact that developed countries usually have a major base of **domestic tourism**. Citizens are both hosts to guests and tourists themselves at home and abroad. Less-developed countries have a typically different experience with tourism. Citizens, except for a narrow elite, tend to serve as labourers in the tourism industry, as hosts but not as guests. **Dependency theory** as well as other **political economy** models, Marxist analysis and market theories have been used to prove the political relationships associated with the asymmetry between developed and developing countries-with respect to tourism. Comparative political studies of corruption and crime have also touched on tourism-related issues of prostitution and **smuggling**.

International relations is a subfield in which tourism **roles** can be salient. Tourism is often a barometer of relations between countries and a harbinger of increased aid from developed to developing nations. When relations improve, tourist exchanges are one of the earliest indications. Such tourism can be a 'confidence-building', low-risk step to other relationships. Similarly, one of the first casualties of worsening relations is a decline of tourism. Countries may reach a point where they forbid citizens to visit certain other countries. Others may refuse them entry. Bilateral and multilateral treaties and the burgeoning field of tourism **law** are growing around such issues as **terrorism**, **airline** rights, smuggling, **investment** and **health** issues (see also **treaty**). Multinational organisations like the United Nations, **UNESCO**, the **World Bank**, the International Monetary Fund and the **World Tourism Organization**, as well as nongovernmental organisations like the Economic Coalition on Third World Tourism, transnational corporations and even religious organisations are also involved in international labour relations, historic site **preservation** and environmental lobbying.

Public administration is centred around the **budgeting**, organising, personnel and overall implementation of policies which increasingly has required an attention to tourism. Growing numbers of governments at all levels have sought **economic development** from an expanded tourism base. Agencies are much more involved in promotion, **planning** and in some cases actual ownership of tourism facilities, as well as the collection and **auditing** of taxes collected from the private sector. Some governments have also used their public sector to burnish the images of controversial regimes through government-instigated tourism. Public administration as a subfield has paid little attention to the growing **employment** and proliferation of departments in public sector tourism, though there have been a number of case studies that have examined industry's role in a particular jurisdiction.

Political **theory** is another subfield that could raise issues germane to tourism but rarely does. As St Augustine suggests, 'The world is a book. He who stays at home reads only one page.' Scholars note the various intellectual influences on philosophers, but often equally relevant are their travels and experiences in other societies. Alexis de Tocqueville's classic *Democracy in America* was a product of his nineteenth-century journey through the new nation. Thomas Jefferson, Mahatma Gandhi and Karl Marx were but a few thinkers profoundly affected by their travels. Some have argued that explorations and subsequent travel accounts premised on the 'backwardness of the **natives**' helped to legitimise imperial adventures and justify the **colonisation** of millions throughout the world.

Further reading

Hall, C.M. (1994) *Tourism and Politics*, London: Wiley. (Explores the politics of tourism with special reference to the Pacific.)

Matthews, H. and Richter, L.K. (1991) 'Political science and tourism: the state of the art,' *Annals of Tourism Research* 18(1). (Reviews and organises political studies of tourism and identifies research gaps.)

LINDA K. RICHTER, USA

political socialisation

Political **socialisation** is the process by which one acquires familiarity with political institutions and a sense of one's **role**, place and **values** with respect to politics. The family, **education**, work, the media, social **class**, age and **religion** are among the elements that shape such socialisation. Tourism also affects the acquisition of political values. Few have actually considered how tourism has shaped political awareness, although the school fieldtrip, the religious **pilgrimage** and the diffusion of ideas through **outbound** travels are commonly acknowledged to be influential. Even the family **vacation** can convey political information, though such outings are seldom mentioned as an agent of political socialisation.

Though government may influence many of the institutions affecting socialisation, the degree of control varies among societies. Moreover, groups and individuals may have very different socialisation experiences within the same society. White South Africans, for example, have had quite a different orientation to political **power** than have blacks. The symbols of sovereignty, the political memories and milestones will differ. Political and cultural conditions will affect if, how and where tourists travel. Women, for example, may not be allowed in some nations to travel alone or without a veil, or drive a car. Minority ethnic and racial groups may be unwelcome in some destinations,

accommodation may be unavailable, or **security** may be problematic. Others may have trips aborted or ruined by political **terrorism**. Political **attitudes** toward other cultures or even other regions may affect both tourists and those who serve them.

Key events like **war** and conquest spawn political memorials like the Hiroshima Memorial and the Alamo. Political infrastructure like the Washington Monument, the Kremlin and the British Parliament are also tourism attractions. Tourism can become an antidote to **ethnocentrism** or be used by groups and **government** as a way to reinforce cultural **identity**, instil pride, preserve political memories and interpret them to others. Enormous debate, public expenditures and effort are directed at controlling what is remembered, what is covered up and how sites should be interpreted to tourists.

See also: political science

Further reading

Horne, D. (1984) *The Great Museum*, Melbourne: Pluto Press. (A comparison of European Museums and what and who they choose to immortalise.)

Matthews, H.G. and Richter, L.K. (1991) 'Political science and tourism,' *Annals of Tourism Research* 18: 120–35. (Details the political socialisation impact of tourism.)

LINDA K. RICHTER, USA

political stability

Political stability has a peculiar relationship to **international tourism**, which is often seen by local officials as a means to develop political stability. They argue that international tourists will infuse the local economy with both money and new jobs. Both will provide governments at all levels with increased revenues. Such funds will allow more public needs to be met, reducing popular pressures, thereby stabilising the political system. Others disagree. They note that the presence of leisured, affluent tourists with expensive tastes may create a **demonstration effect** that can itself be destabilising, demoralising workers and creating demands for luxuries that the **government** cannot provide. In some countries, the presence of these outsiders may also offer an opportunity to dissident groups to get their agendas before the world press by kidnapping tourists, rioting or other acts (see **terrorism**) that cannot be ignored because foreigners are involved.

Because tourists are courted by many competing destinations, the **industry** in a particular nation is critically dependent on the perception of political stability. Tourism may often contribute to political stability, but this is also a precondition for tourism's **sustainable development**. Because of this, some public officials are tempted to suppress a wide range of dissent, fearful that it might be seen by international tourists as signs of 'unrest'. Such crackdowns can contribute to the very political opposition they were designed to thwart.

The concept of political stability is itself open to a variety of definitions using a multitude of criteria. One observer's assessment of a large rally might be that it signalled 'unrest'; another might see it as the sign of a healthy democracy. Who has the **power** to define what is or is not political stability can greatly affect the industry. Many countries like Fiji, **Kenya**, **Thailand** and the **Czech Republic** that are dependent on overseas tourism dollars are precisely those which lack the **resources** to shape the ways in which their own events are interpreted abroad. They do not control CNN, Asahi Shimbun, the US State Department or the **American Society of Travel Agents**, all of which are continually assessing the political stability of countries.

Further reading

Enloe, C. (1990) *Bananas, Beaches and Bases*, Berkeley, CA: University of California Press. (Chapter 2 discusses factors contributing to problems associated with tourism.)

Richter, L. (1992) 'Political instability and tourism in the Third World', in D. Harrison (ed.), *Tourism and the Less Developed Countries*, London: Belhaven, 35–48. (Examines specific types of instability and their impact on tourism.)

CYNTHIA ENLOE, USA

pollution management

Pollution can be defined as deterioration of part of the **environment** due to the occurrence of substances or processes of such types and in such quantities that the environment cannot assimilate them before they cause damage. Some would assert that this definition does not go far enough, and that any discharge of effluents or emissions pollutes the receiving environment in that it changes the state and perhaps the **quality** of that environment. Realistically, all that can be done is to reduce pollution to the minimum possible, in socioeconomic terms, control the types and levels of pollutants acceptable, and determine selectively the location where certain pollutants are to be released. Pollution management is expensive, both in terms of technology and procedures called for, and tradeoffs may be required in production and capacity levels.

Although the scale of most tourism operations does not always lead to an identifiable pollution problem, they can be the source of liquid or gaseous substances which are potentially a hazard to **health** and the environment. Among such sources are the discharge of sewage into bodies of water and the ocean; emissions from heating and refrigeration units; discharge of hazardous substances through the sewerage or drainage system; vehicle emissions; odours and spills; and noise and light pollution. In many situations, practices which are potentially polluting are controlled by regulation. Even where this does not apply, good customer and **community** relations call for a mode of operation in a tourism establishment which will minimise the release of harmful or undesirable substances into the environment.

Effluents and emissions can be reduced to a practical minimum by phasing out the use of hazardous substances, such as chlorine bleaches in pools, leaded petrol, toxic detergents and so on, substituting cleaner technologies, installing treatment and filtration facilities, and adopting acceptable procedures for storage, use and disposal of hazardous substances. Relatively simple amendments to operating practices should reduce or eliminate nuisance to neighbouring environments.

It is important that measures to manage pollution in a tourism establishment be subject to an **environmental auditing** system. Monitoring and follow-up are essential to ensure that the measures are effective and self-regulation is as important as mandatory inspections by a regulatory agency (see also **regulation, self**). Pollution management is an integral part of best practice environmental management in the **industry**. It need not be expensive, and is cost-effective in terms of long-term operational savings and guest relations.

See also: codes of ethics, environmental; education, environmental; legislation, environmental

Further reading

International Hotel Association (1995) *Environmental Action Pack for Hotels*, Paris: UNEP.

JOHN J. PIGRAM, AUSTRALIA

pornography

Writings, pictures and films designed to stimulate sexual excitement are termed pornography. **Fantasy**, including sexual, is a major determinant of tourism, and **destination** differences in the **accessibility**, legal and social acceptability of pornography can lead to 'porno-tourism', motivated by the purchase or experience of pornography. The distinction is increasingly being made between acceptable sexual **fantasy** and consumption, and unacceptable sexual exploitation.

G.J. ASHWORTH, THE NETHERLANDS

portfolio model *see* marketing

Portugal

Located on the Western side of the Iberian peninsula, Portugal has a territory of 35,574 square miles, including the mainland and the Azores and Madeira islands. With a population in 1997 of 9,943,000, Portugal is one of the oldest independent countries in Europe, harking back to the thirteenth century, and was well known many centuries earlier. From the fifteenth to the nineteenth centuries, Portugal was one of the main imperial **powers** of Western Europe and it kept some colonies (Angola, Mozambique, Macau) until after the 1974 revolution that brought democracy back to the country. The Portuguese economy has been on a quick pace to **development** since the country joined the European Community (now the **European Union**) on 1 January 1986, with an average rate of growth of 4.5–5 per cent over the last few years. In 1995, it reached a per capita product of $9,740, just on the international divide between developed and developing countries.

International tourism is one of the major exports of the country, with international arrivals contributing about 4–5 per cent in 1996. There are no accurate estimates for the strength of Portuguese internal tourism, though it can be guessed that it adds an extra 1–1.5 per cent. According to the **World Tourism Organization**, in 1994 there were over 9.1 million international arrivals (2.8 per cent of the European market), generating $3.8 billion (2.1 per cent of the European market); this volume in 1997 reached over 10.1 million (with European market share remaining unchanged) generating over $4.3 billion (2.0 per cent of the European market). Thus, although there has been a slight upward trend in international arrivals, Portugal's market share in Europe has remained stable of late and there are no prospects for significant changes in the near future. In this sense, Portugal is a mature **destination**. In 1996 it ranked twenty-third in the world for international arrivals and twenty-ninth for international receipts.

Even though the country borders the Atlantic, the evolution of Portuguese tourism has followed what might be called a Mediterranean pattern. **Resort** vacations (sun, sea and sand) became the main touristic staple after the inception of **mass tourism** in the 1970s. Old fishing communities developed nearby beaches, high rise condos and hotels crowded the coastline, and European **tour operators** provided a constant **supply** of touristic stays. Even though Portugal learned some lessons about the need for environmental protection and **planning** from neighbouring **Spain**, it was unable to avoid the pitfalls of overcrowded spaces and strained facilities, above all in the popular Algarve region in the south of the country.

The **infrastructure** for mass tourism has increased significantly over the last twenty years. There has been an impressive growth of 52.5 per cent in **hotel** rooms and of 59 per cent in bed places in the country between 1985 and 1996. The corresponding total numbers for the latter year are 91,094 and 208,205. Heavy investments in the road network as well as improved **airport** facilities also had a share in making Portugal a more accessible destination. This in turn explains the success of Portuguese tourism in enriching its touristic **product** mix. There has been a clear trend to develop affluent sports and golf resorts close to the coastal areas. The country today has over fifty golf courses, with a significant concentration in the Algarve region (see also **sport, recreational**; **sport tourism**).

Portugal's history shows in its cultural attractions and destinations. Some eight old cities and areas of the country have been classified by **UNESCO** as world-class cultural preserves. Many visitors visit Lisbon for **weekend** minitrips in order to enjoy the quaint charm of the city (the Alfama, the St George Castle, the Carmo and the Alta areas in town, so celebrated by, among others, the great Portuguese writer Eça de Queiroz), her architectural monuments (above all the Manuelino Torre de Belem and Mosteiro dos Jeronimos) and some excellent restaurants. Porto, Portugal's second city, and Coimbra have also become more popular. Expo 98 confirmed Portugal's ability to host world class mega events. The Fatima **shrine** has become a magnet for Catholic **pilgrims** the world over. Even the old rural region of Alemtejo has started to attract itinerant tourists, mainly from neighbouring Spain. **Ecotourism**, **agrotourism** and farm stays complete the increasing offer of Portugal.

Portugal's main **markets** are found in Europe. Spain, the **United Kingdom** and **Germany**, in that order, account for over half of touristic arrivals while the same countries in different order (United Kingdom, Germany and Spain) provide for most hotel nights. Overseas visitors are another significant, though still small, part of foreign flows (in 1997 there were around 235,000 arrivals from the **United States**, 107,000 from **Brazil** and 43,900 from **Japan**). This market concentration is one of the problems for Portugal's touristic **development** and makes it too dependent on the demand from a handful of foreign countries. Another problem arises from the fact that **inbound** tourists have a low level of expenditure in the country as compared with other European destinations. In 1996 the average stay of foreign visitors was seven days, but the receipts per arrival stabilised around $438 (as compared with $1,900 in **Denmark**, $1,537 in **Sweden**, $1,155 in **Germany** and $683 in **Spain**).

Portuguese vacationers account for a significant, though yet unknown, amount of tourism's contribution to the GNP. Their proportion is still small relatively to other European countries, as only 27 per cent of Portuguese citizens take a vacation away from home. Their average per capita expenditure on **vacation** is largely smaller than that of foreigners, around $112 for a seven-day vacation. This segment, however, is expected to grow in numbers and in expenditure within the next years as the Portuguese become more affluent.

Further reading

DG Turismo (1997) 'Análise de Conjuntura', Boletim n. 24, Lisbon.

World Tourism Organization (1998) 'Europe', Tourism Market Trends Series, Madrid: WTO.

JULIO R. ARAMBERRI, USA

postcard

Postcards are cards for mailed messages, often printed by governmental postal services. They were first introduced in Austria in 1869. Picture postcards, covered on one side with a photo or drawing and printed privately, were introduced in 1894. The text on the latter was initially limited to 'Greetings from...' but later a limited space for correspondence was provided on the obverse. In the early years of the twentieth century a postcard craze spread in Europe and America, and postcards became big business, as well as collector's items.

Early picture postcards depicted primarily local views and events; however, following the growing demarcation of **tourist** spaces and the staging of attractions, postcards began increasingly to depict touristic **sites** and sights, such as vacationing **resorts**, beaches, costumed natives and a variety of appealing facilities rather than the ordinary flow of local life (Albers and James 1986). As such, postcards, like posters, advertisements, guidebooks and travel **videos**, help to form an (often stereotyped) **image** of destinations, and to motivate prospective tourists to visit them. While they engender expectations prior to the trip, they also serve as standards with which actual experiences are compared, and as **souvenirs** of the trip or proofs of 'having been there'.

Postcards are a useful but minor source of historic documentation. They are probably more important as a ubiquitous form of imagery, lending themselves well to **content analysis** and **semiotic** research. For such purposes, four principal components can be distinguished in most postcards. The first is a pictorial **representation**: a photograph, painting, drawing, or caricature covering one side of the postcard. The topics, actual contents and composition vary widely. Historically, however, a general trend can be observed, from contextualised, metonymic representations of ordinary life, to decontextualised, conventionalised (though often mystified) metaphoric representations of destinations, attractions and facilities (Albers and James 1983). Contemporary postcards in particular tend to be flashy, spectacularised and highly embellished, especially if they present popular destinations; their very artificiality possibly impairs their credibility as realistic representations for modern, sophisticated tourists.

Second, the pictorial representations are normally accompanied by a caption, which names, localises, describes or classifies, and sometimes interprets or extols, the pictorial representation. Captions initially tended to be brief, and became gradually more elaborate. They are sometimes written in the language of the tourists' countries of origin rather than the national language. Third, if the postcard has been mailed, it contains on the obverse a message, which may or may not relate to the pictorial representations; if it does, it may supply significant information on the attitude of the tourist to the depicted topic (Baldwin 1988). Fourth, if mailed, the postcard also includes an address to which it has been sent, and which could serve as a clue of the flow of communication about people and destinations from tourists to their relatives, friends, neighbours or colleagues.

Studies of postcards have focused primarily on the analysis of representations in a body of postcards with a common, mostly ethnic topic. However, there are no studies of the institutional context within which postcards are produced and distributed, and especially, purchased. Regarding their purchase, some important questions would be who buys which postcards and for what purpose – for mailing, as a **souvenir** or as a collector's item – and if mailed, which kinds of postcards and accompanying messages are sent to different sorts of addressees.

References

Albers P. and James, W. (1983) 'Tourism and the changing image of the Great Lake Indians', *Annals of Tourism Research* 10(1): 128–48.

—— (1986) 'Travel photography: a methodological approach', *Annals of Tourism Research* 15: 134–58.

Baldwin, B. (1988) 'On the verso: postcard messages as a key to popular prejudices', *Journal of Popular Culture* 22(3): 15–18.

Further reading

Cohen, E. (1995) 'The representation of Arabs and Jews on postcards in Israel', *History of Photography* 19(3): 210–20.

ERIK COHEN, ISRAEL

post-industrial

The origins of **mass tourism** are found in the 1920s and 1930s with collective outings to the seaside by factory workers. However, the radical expansion of tourism into the world's pleasure peripheries (see **pleasure periphery**) is normally associated with a post-industrial age defined by the rise in importance of the service sector and the shift in emphasis from production to consumption.

TOM SELWYN, UK

postmodernism

The term 'postmodern' was first used in the 1970s to describe a new kind of architecture which recycled styles from the past and combined them in sometimes incongruous ways as surface decoration. The term was quickly assimilated into a variety of fields, including tourism studies, as a name for current cultural and political phenomena. The characteristics of postmodernity are **nostalgia**, a 'lack of depth' in its **art**, architecture and **social relations**, a valorisation of surface appearances, a failure to distinguish between originals and fakes, and an assertion that there is no difference between truth and non-truth (that in postmodern aesthetic, domestic and civic life, there are only 'truth effects'). Proponents of postmodern cultural theory have argued that its politics, art and commodity production have become fully integrated into a single system. This integration effectively blocks social change or reorganisation that might be based on a critique of the *status quo*. Thus postmodernity marks the death of the critical or free human subject and the end of history.

The global growth of tourism is strongly associated with the spread of postmodern culture. Tourists take paths that have been marked out for them in advance. They are characterised by the superficiality of their understanding of the peoples and places they visit. Those who play host to tourists are more interested in making money than in preserving the specificity and integrity of **local** cultures. The aggregate of attractions on a regional or global base resembles a postmodern pastiche of incongruously connected objects and events. The most recently constructed large-scale commercial attractions in Las Vegas and the proposed new addition to Disneyland are copies of other destinations. 'The Paris Experience' and 'New York, New York' **hotel** casinos in Las Vegas promise tourists a simulation of the kind of **experience** they might have if they visited Paris or New York. 'The Luxor', built in the shape of a pyramid, promises a simulated Egyptian experience, complete with a boat ride down the 'Nile' in its lobby. 'Redwood Forest' and 'Wild River' rides are planned for the 'California Fantasy' coming to Disneyland. The popularity of these attractions may confirm the postmodern thesis that there is no difference between originals and copies, at least when viewed from touristic and commercial standpoints.

See also: commercialisation; hyperreality; staged authenticity

DEAN MacCANNELL, USA

power

Power is a central concept within social science which has had only limited application in studies of **policy** and **decision making** in tourism. Power is all forms of successful control by subject A over subject B, whereby the former secures the latter's compliance. Power cannot be said to have been exercised unless compliance has occurred. The examination of the application of power in tourism assists in identifying who **benefits** from tourism and how.

The highly political nature of studying this concept means that the use of power is tied to a given set of value assumptions which predetermine the range of its application. Three different approaches or dimensions may be identified in its analysis, each focusing on different aspects of a decision-making process: a one-dimensional view emphasising observable, overt **behaviour**, **conflict** and decision making; a two-dimensional view which recognises decisions and non-decisions, observable (overt or covert) conflict, and which represents a qualified critique of the behavioural stance of the one-dimensional view; and a three-dimensional view which focuses on decision making and control over the political agenda (not necessarily through decisions), and which recognises observable (overt or covert) and latent conflict.

Although not stated explicitly, different concepts of power have underlain different approaches to and formulation of problems in tourism **development**, **planning**, **policy** and **political science**. **Models** of tourism planning have usually assumed a one-dimensional, pluralist conception of power in which all members of a **community** are regarded as having equal access to the decision-making process. Two-dimensional perspectives of power have tended to be realised in studies of tourism planning which have identified the importance of possible alternative decisions or paths of tourism development being deliberately ignored or excluded from policy agendas. Three-dimensional views of power in tourism are utilised in studies of the **political economy** of tourism which recognise that institutions, culture and capital have a political and ideological dimension and highlight the **social relations** within the consumption of tourism services and production of knowledge.

Further reading

Hall, C.M. (1994) *Tourism and Politics: Policy, Power and Place*, Chichester: Wiley. (Identifies the exercise of power with respect to the state, development processes, policy and international relations.)

Norkunas, M.K. (1993) *The Politics of Memory: Tourism, History, and Ethnicity in Monterey, California*, Albany, NY: State University of New York Press. (Examines the relationships among power, class and tourism representation.)

C. MICHAEL HALL, NEW ZEALAND

precautionary principle

The precautionary principle is essentially about how to act responsibly in the face of uncertainty and lack of full scientific knowledge of the likely environmental **impact** of a given **development** decision, or that anticipatory action should be taken to prevent or abate a future environmental harm or **risk**. The precautionary principle does not mean that all developments with uncertain environmental impacts should not go ahead, as that would be to forgo benefits for current and future generations without justification. However, all options need to be identified and explored when considering significant developments with unpredictable future consequences. Tourism developers and their host communities can incorporate the precautionary principle into their **planning** by adopting **management** approaches. One approach is strategic, in that an attempt is made to anticipate problems. Another is adaptive, whereby the extent of uncertainty in many areas is recognised; account is taken of economic, social and environmental information at the beginning of the planning process, and development occurs in an exploratory fashion. Still another approach is cautious in implementation, so that when outcomes are uncertain policies and programmes can be modified as new information becomes available.

Practical **policy** instruments that can be used to achieve implementation of this principle involve communities in the development of tourism **growth** strategies that set targets for such objectives as the maintenance of **biological diversity**, and impose constraints on the forms of development that are permitted. Safe minimum standards, shifting the onus of responsibility and burden of proof with respect to environmental impact towards those who wish to initiate developments that may adversely impact on the environment, improving resource **pricing** mechanisms, introducing policies that make users pay for the full costs of providing access to resources, and polluters paying in full for the costs they impose on others, are all examples of such instruments. The precautionary principle concept is relatively new, and several issues remain unresolved for tourism developers. These include incorporation of the principle in extending **benefit–cost analyses** of its developments; the refinement of methods of resource pricing to take adequate account of the principle; and importantly, simply agreeing to divorce economic growth from environmental degradation by not allowing developments to go ahead without precautions where adverse impacts may occur.

See also: ecologically sustainable tourism; planning, environmental

Further reading

Boer, B., Fowler, R. and Gunningham, N. (1994) *Environmental Outlook: Law and Policy*, Sydney: Federation Press.

MALCOLM COOPER, AUSTRALIA

preservation

The non-use of **resources**, when applied in natural areas, refers to limited **development** for the purpose of saving species and **wilderness** for the **future**. Sometimes used interchangeably with **conservation**, particularly in North America, preservation is used in a more restricted sense where management of natural areas only extends as far as preventing unnatural interference to the natural resources, which are central tourism attractions worldwide.

ROSS K. DOWLING, AUSTRALIA

prestige

Like its counterpart term, 'status', prestige is linked to both the history and contemporary practice of tourism. Throughout the centuries some **tourists**, as reflected in their writing and in broad sociological accounts, have emphasised the social rewards of travelling to fashionable locations. The equation underlying this prestige relationship is relatively simple. Since travel in earlier centuries was expensive and time-consuming, only those wealthy and affluent members of society could engage in such **conspicuous consumption**. Additionally, select locations were identified as particularly prestigious because, in addition to the expense of reaching these settings, an appropriate introduction to an inner circle of respectable and knowledgeable personnel was also required. In contemporary society, expensive and fashionable places still confer prestige on the **tourist**, partly through the **pricing** of special **resorts** and partly through the knowledge of what is fashionable.

There are a number of signs, however, that the neat relationship between tourism sites and prestige is changing. Now that its products are becoming increasingly differentiated at the same **location**, it is no longer particularly impressive to report on tourism to any named **site**. Instead, how one travels, how one stays and what experiences are achieved are gaining in value. Postmodern theorists such as Urry (1990) argue that for many tourists these prestige location relationships have largely disappeared as new consumers learn to try on tourism identities and vary their experiences.

The enduring importance of the prestige notion in tourism is supported by three research directions. The prestige of and the fashionability of locations are central to Butler's tourism **product** life cycle model, as little prestige is attached to those locations which are in decline. Furthermore, different **market** segments will tire of the same location at different points in its product **life cycle**. Recent research in the language of tourism reinforces the importance of prestige-related images and text in tourism marketing. Finally, tourists who have been questioned on the tales and stories they tell other tourists have been shown to discuss their visits quite selectively. Indeed, tourists show a particular preference for managing their **identity** so that their prestige and personal status is highlighted in their holiday stories.

References

Urry, J. (1990) *The Tourist Gaze: Leisure and Travel in Contemporary Societies*, London: Sage.

Further reading

Pearce, P.L. (1991) 'Travel stories: an analysis of self-disclosure in terms of story structure, valence and audience characteristics', *Australian Psychologist* 26(3): 172–5.

PHILIP L. PEARCE, AUSTRALIA

pricing

Pricing represents the published or negotiated terms of the exchange transaction of goods or services between producers and consumers. It is an economic concept in which each party (consumers and producers) in the transaction aims to maximise their **satisfaction**. For their part, consumers attempt to buy at a price which maximises their perceptions of **benefits** and **value** for money as they choose from competing products on offer. Producers or service providers aim to reach or exceed targeted objectives of sales volume or revenue and to maximise profits and/or **return on investment**. In the long term, prices should be set at a level which reflects an organisation's costs, anticipated profits and the degree of **market** competition.

Successful tourism providers understand that pricing is an essential activity and, in practice, a complex one for several reasons. First, segmented selling of tourism products results in differential pricing which can vary from the published price, especially if tourism products are selected as a 'package' of priced options. More importantly, pricing forms the main tactic in managing trade fluctuations, because the **product** is not able to be stored for the future. Further, it is difficult to expand the capacity of a **hotel** or an **airline** in the short term. **Yield management** is the process by which tourism providers try to maximise revenue by manipulating the demand/price relationship. At times of high demand, published prices are quoted while at slacker times, extra trade can be encouraged by attractive pricing. In general, there is a high price elasticity for discretionary segments such as the **leisure**, **recreation** and holiday markets. This means that customers for these tourism products are sensitive to the prices charged and producers need to operate pricing strategy with care.

Further reading

Fay, C.T., Jr, Rhoads, R.C. and Rosenblatt, R.L. (1971) *Managerial Accounting for the Hospitality Service Industry*, 2nd edn, Dubuque, IA: Wm. C. Brown, 363–81. (Describes and classifies pricing decisions.)

Harris, P. (1992) *Profit Planning*, Oxford: Butterworth-Heinemann, 88–105. (Discusses methods of pricing products and services in hospitality and tourism.)

Middleton, V. (1994) *Understanding the Marketing Mix in Travel and Tourism*, Oxford: Butterworth-Heinemann, 95–104. (Explains the issues in pricing for tourism products.)

Muller, T.E. (1991) 'Defining segments in an international tourism market', *International Marketing Review* 8(1): 57. (Demonstrates, through empirical research, how the identification of different segments will allow the application of different marketing and pricing strategies.)

RICHARD TEARE, UK
HADYN INGRAM, UK
GAVIN ECCLES, UK

principal components analysis

Principal component analysis is a multivariate statistical technique designed to order and simplify the relationships between a large set of variables. Therefore, it is considered to be a data reduction technique. It is most often described as a type of **factor analysis**, but sometimes is presented as an alternative to it. This technique takes the correlations between a large number of variables and uses these to derive a smaller set of factors. Factors represent hypothesised underlying variables which explain the responses given to a set of the variables measured. It is a technique particularly suited to the investigation of underlying patterns in data such as large numbers of **attitude** statements or rating scales.

All factor analyses share the same basic steps. First, a correlation matrix is produced for all the variables under investigation. These correlations are then examined in order to extract factors which explain the variance in one or more of the variables being analysed. These resulting factors are then usually rotated to produce the clearest possible interpretation of the factors.

Principal components analysis is a particular method for extracting factors, a common technique used in tourism to study, for example, vacation **motivation**, **destination** choice or host attitude. Principal components are always orthogonal, that is, independent of each other, and the first factor always explains the largest amount of the variance; the second factor explains the next largest amount of the variance, and so on. Principal components are usually contrasted with common factors, which are the other main extraction method used in factor analysis. Principal components are derived from the data, whereas common factors reflect causal relationships which the researcher has hypothesised to exist between the variables. This distinction means that this analysis technique is strictly a data reduction and exploration technique.

Further reading

Lui, J.C., Sheldon, P.J. and Var, T. (1987) 'Resident perception of the environmental impacts of tourism' *Annals of Tourism Research* 14(1): 17–37. (Offers an example of the use of principal components factor analysis in an investigation of resident attitudes towards tourism.)

Moscardo, G., Morrison, A.M., Pearce, P.L., Lang, C. and O'Leary, J.T. (1996) 'Understanding vacation destination choice through travel motivation and activities', *Journal of Vacation Marketing* 2(2): 109–22. (Provides an example of the use of principal components factor analysis in an investigation of tourism motivation.)

Stevens, J. (1986) *Applied Multivariate Statistics for the Social Sciences*, Hillsdale, NJ: Lawrence Erlbaum and Associates. (The chapter on principal components which discusses this technique is relatively free of mathematical equations.)

GIANNA MOSCARDO, AUSTRALIA

product

The concept of a tourism product is, at first glance, deceptively simple. A tourism product presumably is whatever one buys while away from home. Indeed, from a **marketing** perspective, one can define any product as 'anything that can be offered to a **market** for attention, acquisition, use or consumption that might satisfy a need or want. It includes physical objects, services, persons, places, organisations, and ideas' (Kotler 1984: 463). In the context of tourism, the most common types of products arguably are services and places, although tourists will also use physical objects, persons, organisations and ideas.

Marketers, however, tend to view tourism products – and services in general – as being fairly complex phenomena. Some authors suggest tourism products consist of a number of components or layers. One **model** suggests that tourism products consist of two parts: a 'tangible' good or **service**, and its symbolic value. Another model suggests there is a core component (such as an **airline** flight) and peripheral components (such as **reservations** and baggage handling). A slightly more complex model posits three levels: a 'formal' commodity the customer is seeking (usually conceived as specific benefit the visitor is seeking, such as a relaxing weekend), a 'core' commodity actually being sold (such as a room at a **resort** and access to a golf course), and an enhanced commodity that consists of the core commodity plus value-added features (such as ambience or free drinks). An even more complex model suggests tourism products consist of attractions, facilities, access, **images** and price. Numerous other models have been proposed, and are summarised in Smith (1994).

A different perspective of tourism products forms part of the basis of **satellite accounts**. In this context, tourism products are those commodities for which a substantial part of **demand** comes from visitors. The range of such commodities is wide, but they can be grouped into five sectors: **transportation**, **accommodation**, **food** services, **recreation** and entertainment, and other tourism services such as **travel agencies**. A complication in the empirical measurement of the demand for tourism products is that they are also consumed by non-tourists. **Local** residents eat in restaurants, non-tourism travellers (based on the **World Tourism Organization** definition of tourism) travel by air or **rail**, temporary workers live in hotels, and so on. Further, tourists also consume non-tourism products in the course of an **experience**. For example, they often purchase food from grocery stores or clothing from retail stores. Neither groceries nor clothing are tourism products because the portion of demand for these products from tourists is minuscule.

Partly because of this, tourism is not considered to be an **industry** in the conventional sense: there is no homogeneous product nor a common production process However, drawing from the work of numerous authors who have explored the fundamental nature of tourism products, Smith (1994) argues that in fact, one can hypothesise the existence of a single generic tourism product: the experience of tourism. He describes a five-layer model of the ideal generic tourism product. This ideal product consists of a 'physical plant', which might be a **site**, a natural **resource**, a built facility or equipment (such as a cruise ship), service, **hospitality** (the style or **attitude** with which service is delivered), the **quality** of perceived freedom (the sense that the visitor has some degree of choice in the consumption of the tourism product) and the quality of involvement (the ability to relax, not worry, to focus on the experience without being concerned about other issues).

Smith also proposes a generic production process that gives rise to the tourism product. The process begins with primary inputs such as land, water, agricultural produce and building materials. Through additional inputs of labour and capital, these are transformed into intermediate inputs. They are facilities such as attractions, restaurants, **resorts** and airports. The 'physical plant' level of the generic product is developed at these two stages. Facilities, with additional inputs of labour – especially service and hospitality – and capital, are processed into intermediate outputs or services. Services include all the commodities normally associated with tourism: performances, conventions, meals, overnight accommodation, festivals and events, and so on. This level – services – is the one which normally represents the tourism product. However, Smith suggests that there is an additional step in the production process: the consumption of the service by a tourist who combines individual services into an overall experience. The combination of the experience of the individual services into a holistic tourism or trip experience completes the production of the generic tourism product. This model makes the consumer an integral part of the production process. In the fullest sense, tourism products do not exist until a trip to the point of production and 'assembles' the service components into the final product. Ultimately, tourism product and production are inseparable.

See also: management; product life cycle; product planning

References

Kotler, P. (1984) *Marketing Management: Analysis, Planning, and Control*, 5th edn, New York: Prentice Hall.

Smith, S.L.J. (1994) 'The tourism product', *Annals of Tourism Research* 21: 582–95.

Further reading

Medlik, S. and Middleton, V.T.C. (1973) 'Product formulation in tourism', in S.F. Witt and L. Moutinho (eds), *Tourism Marketing*, vol. 13, Berne: AIEST. (Conceptualises tourism products as an overall experience consisting of five components.)

Middleton, V.T.C. (1989) 'Tourist product', in S.F. Witt and L. Moutinho (eds), *Tourism Marketing and Management Handbook*, Hempel Hempstead: Prentice Hall, 573–6.

STEPHEN L.J. SMITH, CANADA

product life cycle

The concept of a **life cycle** has been transferred from biological sciences to **economics** and adapted to uses in business management. It describes the evolution of a **product** in time, passing through different stages which require different **marketing** strategies. The application to tourism is not without criticism, because its **product** is in the first place the sum of experiences (from the tourist point of view) and the range of facilities and attractions in a **destination** area (from the suppliers side). The latter application has inspired much empirical research in tourism. Product life cycle offers a generally accepted framework to study and to manage the **development** of a destination over time, the physical, social and economic changes which are induced by a growing tourist activity.

Further reading

Cooper, C. (1995) 'Product lifecycle', in S.F. Witt and L. Moutinho (eds), *Tourism Marketing and Management Handbook*, 2nd edn, Hemel Hempstead: Prentice Hall, 342–50.

MYRIAM JANSEN-VERBEKE, BELGIUM

product planning

Product planning is the process that tourism organisations and destinations use to identify, evaluate and select alternative approaches to developing or modifying their physical plant or facilities, festivals and events, equipment, **services**, **infrastructure** and **package tours**. Product **planning** and development is considered to be one of the four Ps of marketing or the **marketing** mix along with promotion, place and price.

To understand tourism product planning, one must first study the **product**. Tourism scholars have yet to agree on its definition. Most experts concur that the product includes the physical plant, natural and cultural resources and sites, built facilities (hotels, restaurants, built attractions), equipment and **infrastructure**. There is also agreement that **service** and **hospitality**, although not physical items, are part of the tourism product. Disagreement exists on its more intangible aspects such as **destination** image, **product positioning** and **branding**. Many practitioners, especially those involved in destination marketing, would include planning that leads to the development of new package tours. In essence, the product being planned varies according to the type of tourism organisation involved.

The process of product planning should be based on research and a thorough **SWOT analysis**. This research and analysis identifies product gaps in the competitive marketplace and the need to modernise or otherwise improve the current product. Customer **surveys** and focus groups, competitive analyses, and the use of expert opinions may be employed to assist in identifying potential product changes. Specific techniques such as feasibility studies may be used to determine the potential viability of new projects such as hotels, **resorts** or attractions. Outside consultants may be contracted to conduct these studies. Another dimension of this process in tourism is the extent of involvement of those outside of the organisation responsible for the planning. At the destination level, there should be broad **community** and **industry** participation and input into the planning process. Individual corporations tend to maintain their competitive advantages.

Further reading

Morrison, A.M. (1996) *Hospitality and Travel Marketing*, 2nd edn, Albany, NY: Delmar Publishers. (Provides a comprehensive description of product development in Chapter 10.)

Smith, S.L.J. (1994) 'The tourism product', *Annals*

of Tourism Research 21(3): 582–95. (Explains the elements of the tourism product.)

ALASTAIR M. MORRISON, USA

product positioning

Product positioning and **market segmentation** are strategic issues of the **marketing** planning process. Deciding on the target position of a product brand, the marketing manager determines the key attributes of the product brand or service and the main contents of **advertising** messages. Positioning always refers to one or more particular groups of customers (market segments), thus making the two domains of marketing strategy dependent on each other. Targeting an offer to meet the customers' needs requires prior information on how the consumers react to the attributes of the brands in a product class. In tourism and **hospitality**, the concept of positioning has been adopted for all major application areas including destinations and individual businesses like hotels, **travel agencies**, **tour operators** or **theme parks**. In addition to the typical **perceptual mapping** exercises, the tourism sectors have acquired experience in analysing consumer preferences with conjoint measurement techniques.

Positioning analysis centres on the consumers' evaluative processes involving perceptions and preferences. Consumer **behaviour** theory provides the hypotheses linking the perceptual and preferential position of a product brand (or a business as a whole) to the expected brand choice reactions. Various techniques of multivariate analysis assist in locating these positions by processing empirical data. Perceptual mapping techniques are common in the diagnostic phase of a positioning study. They are particularly useful in deriving a spatial representation of the brand evaluations and in communicating these visualised results to managers. Positioning models may go one step further in adopting an additional normative component which aims at optimising the mix of product features. As positioning analysis operates with the perceptions and preference judgements of consumers, the perceptual findings must be translated into the language of product and service engineers to prompt innovations and improvements in product/service design.

See also: master plan

Further reading

Carmichael, B. (1992) 'Using conjoint modelling to measure tourist image and analyse ski resort choice', in P. Johnson and B. Thomas (eds), *Choice and Demand in Tourism*, London: Mansell, 93–106.

Fenton, M. and Pearce, P. (1988) 'Multidimensional scaling and tourism research', *Annals of Tourism Research* 15: 236–54.

Fodness, D.D., and Milner, L.M. (1992) 'A perceptual mapping approach to theme park visitor segmentation', *Tourism Management* 13: 95–101.

Urban, G.L. and Hauser, J.R. (1993) *Design and Marketing of New Products*, 2nd edn, Englewood Cliffs, NJ: Prentice-Hall.

JOSEF A. MAZANEC, AUSTRIA

productivity

In its simplest form, productivity is an economic concept, defined as the ratio of output to input, shown by the mathematical formula:

$$\text{Productivity of an operation} = \frac{\text{Value of what is produced}}{\text{Value of production cost}}$$

Overall productivity can be regarded as the sum of partial productivities. In other words, both the upper term (numerator) and lower term (denominator) in the above relationship can be broken down as follows.

$$\text{Productivity of an operation} = \frac{\text{Value of components} \times \text{A+ B+ C etc}}{\text{Value of costs (labour, materials, energy, etc)}}$$

This relationship can be further reduced by considering only one part of the numerator (to give the productivity in respect of one component) or only one part of the denominator (to give the productivity in respect of one input). The latter is most commonly done, usually by considering labour productivity at the expense of other inputs. This is also a justifiable approach for industries such as tourism and **hospitality**, where material and other costs are arguably small compared with the labour input.

Productivity is fairly easy to conceptualise if it is concerned with manufacturing industry, or even with the national product of a whole **region** or country. However, it is not clear whether it is an appropriate concept to apply to services such as hospitality and tourism. One problem is that services do not produce a product, and the value of their contribution to wealth is unknown. Jones and Hall (1995) suggest that productivity implies a manufacturing **paradigm**, which is not applicable to **service** industries. However, Heap (1995) proposes that productivity is appropriate for tourism and hospitality so long as it reflects customer needs.

Two parallel concepts, **efficiency** and effectiveness, help to explain the role of productivity in the provision of services. Efficiency refers simply to the most cost-effective use of resources, so that high efficiency would seem to bring high productivity. However, the use of resources is not necessarily productive, if it does not also satisfy the customer or the **tourist**. Effectiveness refers to the way in which a service operation meets its objectives (of satisfying customers and achieving repeat business). Thus, in order to be productive, a service industry needs to be both efficient (in order to achieve short-term financial/cost-effectiveness goals) and effective (in order to achieve short-term financial/cost-effectiveness goals) and effective (in order to achieve longer term goals of customer **satisfaction** and competitiveness).

References

Heap, J. (1996) 'Top-line productivity: a model for the hospitality and tourism industry', in N. Johns (ed.), *Productivity Management in Hospitality and Tourism*, London: Cassell, 2–18.

Jones, P. and Hall, M. (1996) 'Productivity and the new service paradigm, or service and the neo-service paradigm?', in N. Johns (ed.), *Productivity Management in Hospitality and Tourism*, London: Cassell, 227–40.

NICK JOHNS, UK

professional native

'Professional native' is a somewhat critical term for a local person who assumes the role of **culture broker** by interpreting the **indigenous** culture for foreign tourists, one who usually seeks payment for services rendered. Professional natives need an understanding of the tourists' language and culture and offer themselves as experts on the traditional culture, but in their own society they may occupy marginal positions, lacking **prestige**, **power** or expertise.

See also: entrepreneurship; natives

EDWARD M. BRUNER, USA

professionalism

The concept of professionalism cannot be easily separated from the study of ethics. The term 'ethics', which means character, is derived from the Greek word 'ethos'. In its basic dimension, the study of professionalism is concerned with 'taking the correct action' in a situation. This stance has roots in Immanuel Kant's categorical imperative, which implies that an individual's actions should be based, not on rightness or wrongness, but on 'good will'. At an optimum outcome, behaving in a professional manner involves acting in a normative manner that results in a positive **service** occurrence.

Professionalism in the tourism **industry** is highly sought after by service providers in lodging **accommodation**, **travel agencies**, tour companies, airlines, **food** service, **transportation**, entertainment and attractions, as well as in education and tourism-related associations. The very nature of the industry requires that an individual, immediate work group and organisation act in a socially responsible manner. Given this view, the desire to act in a professional manner has become a primary mandate for tourism service providers. However, professionalism is not easily defined. Clearly, the degree of professionalism within an occupation is rooted in the **values** and norms that individuals bring into an occupation, are exerted by individuals within the context of the immediate workplace, and are developed from societal values, proscriptions and prescriptions.

Researchers have attempted to define professionalism by determining the factors that when taken collectively are seen as indicators of accepted **behaviour**. These factors encompass a **code of ethics**, presence of formal and informal workplace sanctions, reaction to **community** sanctions, presence of a reward system that reinforces ethical behaviour, an educational degree or specialised **training** within an appropriate discipline, dedication to the highest level of customer service possible, behaving in a socially responsible manner within a public service setting, a high degree of **prestige** and a high degree of competence within the field. In essence, these factors have been shown by research conducted from the 1970s to the 1990s to advance an individual in an occupation to a professional standing. Overall, the cumulative combination of these factors has advanced the **image** of the individual working in the tourism industry. The net result is that the individual and the profession as a whole are viewed as credible by the consuming public.

Further reading

D'Amore, L. (1993) 'A code of ethics and guidelines for socially and environmentally responsible tourism', *Journal of Travel Research* 31(3): 64–6.

Hall, S. (1993) *Ethics in Hospitality Management*, East Lansing, MI: Educational Institute of the American Hotel and Motel Association.

Kwansa, F. (1992). 'A conceptual framework for developing a hospitality educators' code of ethics', *Hospitality Research Journal* 15(3): 27–39.

Sheldon, P. (1989) 'Professionalism in tourism and hospitality', *Annals of Tourism Research* 16(1): 492–503.

RANDALL S. UPCHURCH, USA

profit

The residual of revenues minus costs of an enterprise is known as profit. This may be expressed on a per unit basis, such as profit per cover (**restaurants**), segment or total enterprise (departmental, all **hotel**), unit of time (annual, monthly), or level (gross, net). Key profit determinants are unit price level, volume of units sold or guests served, and level of fixed and variable costs.

LEE M. KRUEL, USA

profit centre

A profit centre is an operating department, in tourism businesses and elsewhere, which generates revenues and has related expenses. The activities of a profit centre are detailed on a departmental schedule which provides information on revenues, payroll and related expenses, **cost** of goods sold, other expenses and departmental income. The charges against revenues include only a limited amount of expenses that are traceable to the department.

STEPHEN M. LEBRUTO, USA

profit margin

Profit margin refers to the percentage ratio of net **profit** to total revenue generated in tourism or other production units. It measures the overall relative operating **efficiency** of the business (sales generation and profit control) from one operating period to another. It is computed after all costs are deducted from sales. Key determinants are costs relative to selling prices. High volume operations tend to have low profit margins, while low volumes have high margins.

LEE M. KRUEL, USA

profit sensitivity analysis

Profit sensitivity analysis measures the affect of changes to dependent variables when changes occur in the related independent variables. For example, if a hotel's rooms department experienced an increase in its fixed costs, an independent variable, the additional rooms required to be sold to absorb this increase, a dependent variable, could be computed by dividing the fixed cost increase by the contribution margin of the rooms department. Contribution margin is the selling price of a room minus the variable costs of selling it. This assumes that these two other independent variables, the selling price of the room and its variable cost, remain constant

Profit sensitivity analysis is used in cost volume profit analysis and breakeven point analysis. It involves identifying all costs as fixed or variable. Fixed costs are those that remain constant over a relevant range of activity. Costs such as property **tax**, interest expense, **hotel** management salaries and depreciation are common examples of fixed costs. Variable costs are those that remain constant on per unit basis but change in total relative to sales. Examples of variable costs are those of **food** or beverages sold, and perhaps management fees and other similar expenses that could be determined as a percentage of sales. The remaining costs are mixed costs. The majority of costs in a **hospitality** operation are mixed, having both a fixed and variable component. Examples are repair, maintenance and utilities. In order to perform breakeven point analysis and cost–volume–profit analysis, mixed expenses must be broken down into their fixed and variable components. Once the fixed and variable components are identified of all costs, sensitivity analysis provides the opportunity to measure the affect of the change on dependent variables as a result of changes in independent variables.

There are at least three ways that mixed costs can be segregated into their fixed and variable components: the high-low two-point method, a scatter diagram and regression analysis. Both the high-low and scatter diagram methods are rough approximations. The preferred method of segregating a mixed cost into its fixed and variable components is through the use of regression analysis, which computes the best straight line through an array of data.

Further reading

Schmidgall, R. (1997) *Hospitality Industry Managerial Accounting*, 4th edn, East Lansing, MI: Educational Institute of the American Hotel and Motel Association.

STEPHEN M. LEBRUTO, USA

profit variance analysis

Profit variance analysis is a **management** analysis tool providing departmental managers with information relative to revenues and expenses directly under their control. The three types of variances are revenue, expense and variable labour. Variances occur when the actual amount is greater than or less than a predetermined budgeted amount. A variance is either favourable or unfavourable. Revenue variances are favourable when the revenue exceeded the standard to which it was being compared. Expense or variable labour variances are favourable when the expense or variable labour were less than the standard to which they were being compared.

Most of the variance in revenue occurs because of differences in price and volume. A **restaurant** may budget revenue of $100,000 based on a forecast of 10,000 covers (volume) times a forecast of check average of $10 (price). The actual revenue may be $88,000. The restaurant served 11,000 customers at an average check of $8. The revenue variance would be $12,000 unfavourable. Price variance is computed by taking the budgeted volume (10,000) and multiplying it by the difference between the actual price and the budgeted price (–$2). The price variance was $20,000 unfavourable. Volume variance is the budgeted price ($10) times the difference between the actual volume (11,000) and the budgeted volume (10,000). The volume variance was $10,000 favourable.

Most of the variance in expenses occurs because of differences in costs and volume. A **hotel** may budget guest room supply costs of $20,000 based on a forecast of 10,000 rooms sold times a forecast cost of $2 per room. The actual expense may be $18,000. The hotel sold 8,000 rooms at a cost of $2.25 per room. The expense variance would be

$2,000 favourable. Cost variance is computed by taking the budgeted volume (10,000) and multiplying it by the difference between the budgeted cost ($2) and the actual cost ($2.25). The cost variance was $2,500 unfavourable. Volume variance is the budgeted cost ($2) times the difference between the budgeted volume (10,000) and the actual volume (8,000). The volume variance was $4,000 favourable.

Most of the variance in variable labour expenses occurs because of differences in volume, rate and **efficiency**. A hotel may budget room attendant wages of $35,000 based on a forecast of 10,000 rooms sold with average hourly wage of $7 and an allowed time of thirty minutes to clean a room. The actual variable labour expense may be $36,000. The hotel sold 8,000 rooms with an average hourly wage of $7.50 and averaged thirty-six minutes to clean a room. The variable labour expense was $1,000 unfavourable. Volume variance is the budgeted rate ($7) times the difference between the budgeted time (5,000) and the allowable time for actual output (4,000). The volume variance was $7,000 favourable. Rate variance is computed by taking the budgeted time (5,000) and multiplying it by the difference between the budgeted rate ($7) and the actual rate ($7.50). The rate variance was $2,500 unfavourable. Efficiency variance is computed by taking the budgeted rate ($7) and multiplying it by the difference between the allowable time for actual output (4,000) and the actual time (5,000). The efficiency variance was $7,000 unfavourable.

Further reading

Schmidgall, R. (1997) *Hospitality Industry Managerial Accounting*, 4th edn, East Lansing, MI: Educational Institute of the American Hotel and Motel Association.

STEPHEN M. LEBRUTO, USA

promotion *see* sales promotions

promotion mix *see* marketing; marketing mix; sales promotions

promotion, place

Communities and regions can be promoted as destinations and as places in which to live, meet, invest, relocate or make purchases. Many regions actively market themselves through development agencies, while **community** services play an important supporting role. Overall environmental quality and **safety**, as well as attractive **lifestyles**, feature prominently.

Place promotion has at least three dimensions: **image** formation and communication, public **policy** and **marketing** activities (Gold and Ward 1994). Marketers attempt to create and disseminate positive messages which will be received as valid, believable, distinctive and appealing. Re-imaging is often needed to deal with bad publicity or to reposition an area (see **positioning**). Incentives are often given to attract investors, **media** coverage and special events. New attractions are justified, and **infrastructure** must be continuously improved. Civic amenities and design enhancements are important. Whole districts are created with an orientation toward **tourists**, and commercial and industrial zones are planned and serviced for investors. Community **recreation** services can be programmed to attract visitors.

Place promotion is typically the responsibility of trade and **economic development** offices, convention and visitor bureaus or tourist offices, and in some cities, offices of sports, culture and special events (see also **sports tourism**). Non-traditional tools are used, such as facilitating **film** or **television** productions; mega-events and high-profile 'hallmark' events are sought and developed because of their publicity value. Critics charge that place promotion commoditises communities, especially in the absence of **public participation**. Problems can also occur when mixed messages are sent owing to a multiplicity of actors. If promotion occurs without **planning**, there is a risk of exceeding local **carrying capacity**, and if promotion is indiscriminate, incompatible or harmful segments might be attracted. Those engaged in place promotion might be unable to influence actual development of **infrastructure** and **supply**, resulting in a **quality** –expectations gap. With these factors in mind, Ashworth and Voogd (1994) believed that place **management** is a more appropriate strategy.

References

Ashworth, G. and Voogd, H. (1994) 'Marketing of tourism places: what are we doing?', in M. Uysal (ed.), *Global Tourist Behaviour*, New York: Haworth, 5–19.

Gold, J. and Ward, S. (eds) (1994) *Place Promotion: The Use of Publicity and Marketing to Sell Towns and Regions*, Chichester: Wiley.

Further reading

Ashworth, G. and Goodall, B. (eds) (1990) *Marketing Tourism Places*, London: Routledge.

Ashworth, G. and Voogd, H. (1996) *Selling the City*, New York: Belhaven/Columbia.

Kotler, P., Haider, D. and Rein, I. (1993) *Marketing Places: Attracting Investment Industry and Tourism to Cities, States and Nations*, New York: The Free Press.

DONALD GETZ, CANADA

promotion, puns in

The play on words associated with some forms of tourism **advertising**, typically those promoting destinations (such as 'Bermuda shorts'), **accommodation** ('suite inspiration') and transport ('Europe with a trained eye') are examples of promotional puns. This technique is also used as an attention grabbing caption device in travelogues (such 'Thailand Fling') and in slogans ('Cotton on to Burnley'). Commentators are divided as to the effectiveness of such **humour**.

GRAHAM M.S. DANN, UK

property management system

Property management systems are a part of **hospitality** information systems. They were the first evidence of the integration of mechanical systems into the **hotel** sector. Developers and providers of this **technology** are constantly expending efforts to make systems more user-friendly, cheaper and faster. Hardware and software refinements have facilitated this development. Most property management systems are now run on a microcomputer. They first promoted guest interaction by providing **video** checkout, which has improved this **service**. The continuous improvements in this general area will lead to self-service check-in. Newer systems have the availability of sorting more data and providing built-in interfaces, such as credit card processing.

There are many key elements that almost all property management systems possess. Hotels do not need to be of any particular size to take advantage of the benefits of a system, although some are clearly better suited for different types and sizes of properties. Their operation provides the front office **accounting** function and keeps records from the guest's point of view. Along with this, most systems are capable of maintaining a detailed guest history for a significant period of time. Reports are some of the most valuable processed information that is generated from property management systems, which also provide a housekeeping status report. This report details the rooms that are occupied, vacant, clean, dirty and out of order. Property management systems are also collectors of revenue data from other operating departments, organised and reported to management on a report detailing the enterprise's revenue. Most property management systems have mechanical interfaces to the other sources of revenue in the hotel, such as **food** and beverage, telecommunications and other such items.

All these systems are under the control of **management**, and as such, rates and availability of rooms can be changed at management's discretion. This is important as management attempts to maximise revenues for a particular period of time. Most property systems have a reservation feature which provides a constant update of the status of the property. The most important information to hotel managers is **occupancy rate** and average rate on a daily and cumulative basis. This information system provides these data virtually instantaneously.

STEPHEN M. LEBRUTO, USA

prostitution

The purchase of commercial sexual services is regarded as prostitution. Tourists, freed from their normal social controls, seek such services, which may also not be permitted in their countries of origin. Indirectly, prostitution and associated **pornography** outlets contribute an important element of excitement and **fantasy** within 'red-light' nightlife districts that are significant tourism **attractions** in many cities (such as Amsterdam's 'Wallen', Hamburg's 'Reeperbahn' and the Rue St Denis in Paris).

See also: eroticism

G.J. ASHWORTH, THE NETHERLANDS

protected area

Protected areas preserve or conserve geological and physical features, ecosystems and flora and fauna habitats, including tropical forests, deserts, wetlands and lake and ocean systems that are unique or representative examples of the diversity of species and **landscapes**. Protected areas are critical elements in the tourism system, with the **experience** of natural and cultural environments an important tourism **motivation**. In many countries, tourism has been developed and promoted with much reliance on protected areas, including **wilderness** areas or **national parks**, some of which are unable to withstand even small numbers of tourists.

Various **management** approaches have been developed for use in protected and other areas. These approaches include the analysis of biological and social **carrying capacity**, **limits of acceptable change**, **recreation opportunity spectrum**, visitor impact management and the visitor activity management programme. Despite their widespread critical evaluation and use in many countries, each approach has inherent limitations, such as balancing the diverse values and interests of individuals and agencies within **resource** constraints, which make their implementation by resource managers difficult.

Protection of the environment should be an essential prerequisite of tourism **development**. However, as the **growth** of commercial tourism, including **ecotourism**, continues to create tensions between this industry and protected area managers and interests, it is unfortunate that many countries often lack the resources to undertake appropriate management strategies. In addition, problems of value and interest conflicts are exacerbated by the lack of research into the relationship between tourism and protected areas, which then limits the ability of managers to adopt proactive **policy** and **planning** approaches and thus follow **precautionary principles**.

Further reading

Graham, R. and Lawrence, R. (1990) *Towards Serving Visitors and Managing Our Resources*, Proceedings of a North American Workshop on Visitor Management in Parks and Protected Areas, February 14–17, Waterloo: University of Waterloo. (Provides both a comprehensive overview of tourism in selected protected areas and discussions of a number of related tourism management approaches.)

Leslie, D. (1986) 'Tourism and conservation in national parks', *Tourism Management* 7(1): 52–6.

Pigram, J.J. and Jenkins, J.M. (1989) *Outdoor Recreation Management*, London: Routledge. (Gives particular attention to the adequacy of the resource base to provide a quality environment for sustained recreational use.)

Shackley, M. (1995) 'The future of gorilla tourism in Rwanda', *Journal of Sustainable Tourism* 3(2): 61–72.

JOHN M. JENKINS, AUSTRALIA

prototypical tourism form

The term prototypical tourism form refers to early forms of **tourism** that took place before the industrial revolution. Only when **technology** and urbanisation allowed the masses the time and resources to travel did modern tourism emerge. Such prototypical forms were dominated by the leisured classes (**spa** towns, **Grand Tour**, the Olympic games of ancient Greece). All travelled mainly for non-leisure purposes, often religious (see **pilgrimage**).

DUNCAN TYLER, UK

pseudo-event

Pseudo-events are inauthentic, manufactured **sites** or 'happenings', predominantly focused on mass tourists in guided groups that thrive on the contrived, often being inattentive to the 'real' world around them. Usually pseudo-events are enjoyed by tourists staying in the environmental bubble of placeless-style hotels, which insulate them from meaningful contact with the **local** population. In recent years, pseudo-events have become ever more fanciful, grandiose and staged.

See also: hyperreality; staged authenticity

KEITH HOLLINSHEAD, UK

psychographics

The term 'psychographics' is used in tourism market research as a summary description for identifying **tourists** sharing attitudes, **values**, motives and activity preferences. It is a contrasting term to demographics, which profiles them according to such variables as age, **gender**, nationality and income. Psychographics are being used increasingly in the **market segmentation** field usually as a supplement or in combination with demographics analyses.

PHILIP L. PEARCE, AUSTRALIA

psychology

Psychology concentrates on the **behaviour** and **experience** of individuals. As the scientific study of human behaviour, psychology can claim to be over 120 years old, tracing its epistemological foundation to the laboratory research of Wundt in 1879. The study of human behaviour in a non-systematic fashion has a much longer history and several strands of contemporary theorising draw on earlier philosophical and intellectual traditions. Social psychology, one of its several distinct branches, addresses the behaviour of individuals as influenced by the groups to which they belong. Environmental psychology, an even more recent branch, considers the influences of the physical setting on human behaviour. The principle contributions of psychology to tourism come from social and environmental psychology although the broad topic of **cognition** which refers to human thinking and information processing has provided some important conceptual tools for tourism research and analysis.

Some of the prominent topic areas in psychology which have been adapted and developed for tourism study are reviewed here, but psychology research has perhaps played its most important role in a general or superordinate way. The methodological rigour and the data collecting and appraising style which defines the scientific approach to human behaviour is central to the work of many tourism researchers. Much work is conducted in the area of analysing tourism markets, tourist **satisfaction** and needs. At times this work has been conceptually limited as the work has had an immediate and highly applied focus. Nevertheless the design of **survey** questions, the use of **attitude** scales, the nature of sampling and the statistical treatment of the data are all contributions from the research world of psychology in general as well as from the boundary areas of social psychology and **sociology**. In the 1930s, 1940s and 1950s substantial research and analysis of survey techniques, fieldwork methodologies and statistical treatment of data were developed in the United States and Britain in such content areas as surveys of political attitudes, television watching, the war effort, education and racial segregation. Contemporary tourism study benefits in a major way from the methodological outcomes and knowledge acquired during these years of **innovation** and evaluation in survey and social research.

For some researchers, the starting point of study is with the psychology of the tourist and in particular with the question of what motivates people to **travel**. The study of tourism motivation has evolved as a significant interest with theoretical developments, including Plog's (1988) allocentric/psychocentric approach, optimal arousal theory and Pearce's (1988) travel career ladder. All such motivation theories have to deal with certain critical issues including the ease of communicating the theory to other users, measurement and assessment issues, the dynamic nature of motivation, the operation of multiple motives in tourist behaviour and the role of intrinsic versus extrinsic motivation. The study of tourism motivation is a basis for enhancing **market segmentation** work which is widely conducted by consultants for major marketing organisations. Researchers with a psychology background have also played a strong role in identifying and understanding emerging market segments in tourism **demand**. Some specific groups who are distinguished by motivational profiles include ecotourists, cultural tourists and young budget travellers or backpackers. Another prominent topic which is linked to psychology research is tourism **decision making**. Typically researchers have adopted a set theory approach to this theme, but increasingly the value of heuristic or rules of thumb approaches to tourists' decisions are being considered by researchers. This line of work has direct antecedents in mainstream psychology research on cognition and decision making.

Tourism as an experience can influence tourists' attitudes and change their behaviour. Psychology research on persuasive communication and attitude change has provided an important resource for researchers studying the creation of **destination images** and evaluating tourism communication activities. One concept in particular, that of mindfulness has been usefully applied to studies of tourist reactions to interpretative settings and communications. Langer (1989) and colleagues first identified the mindfulness–mindlessness distinction in a series of laboratory and field experimental studies with the definition that mindfulness is 'active information processing in which the individual is fully engaged in creating categories and drawing distinctions' (Langer 1989). This active information processing is stimulated by novel environments, well-organised displays and themes, personal relevance and multisensory activities and is the basis for changing tourists' attitudes to the places and people they visit.

Social psychology has a special focus on the interaction of individuals within groups and the application of ideas such as group dynamics and social situations analysis to tourism have also been prominent. There are **social situation** analyses of **farm tourism**, of guided tours, of **youth tourism** and of cross-cultural encounters. An emerging area of particular interest is cross-cultural interaction with its attendant stimulus for the tourist as well as its difficulties. Concepts such as **culture shock** and cross-cultural **training** have been borrowed from the psychology literature to enhance the understanding of these tourism situations. In particular, the development of training approaches such as the cross-cultural assimilator extend earlier psychology research on intercultural communication education.

Environmental psychology has relevance to tourism studies both at a conceptual level and in the direct application of existing research findings. Concepts used to define such terms as place and territory are important in understanding attractions and dealing with such issues as crowding and privacy in settings. Psychological perspectives on crowding have been valuable in repeatedly emphasising that the tourism motivation affects the perception of crowding. This set of findings is important for the **management** of natural **resources** and **leisure** settings because it highlights the need to consider the tourist type and the nature of activities in the setting rather than simply dealing with crowding through the imposition of numerical counts, ceilings or thresholds. Additionally, environmental psychology takes a strong perspective on designing this natural setting with people in mind. Such approaches have been useful in looking at such issues as seat design in aeroplanes, resort layout and the use of equipment in leisure activities.

At the broadest level, social and **community** research in psychology has developed concepts such as social representations which can be defined as the shared network of attitudes held by a group about a significant topic. This broad appraisal of how individuals and larger units view the world has been of initial use in understanding tourism community relationships and offers further promise for investigating links between social **identity** issues and the appraisal of **future** tourism developments in emerging destinations. In some respects, psychologists have not been as prominent in tourism study as geographers, anthropologists and economists. The strong hypothetico-deductive systems of empirically based psychology research has many virtues including rigour in designing studies and considerable analytical acumen. Regrettably, the demands of the inherently social and context-dependent phenomenon of tourism have often been seen as unmanageable for researchers with a predisposition to conduct laboratory work. Nevertheless, the study of tourist behaviour is vital to a sound understanding of tourism and the continued development of psychologically-based studies for understanding the entire phenomenon of tourism is vital for customer satisfaction and management success.

References

Langer, E.J. (1989) *Mindfulness*, Reading, MA: Addison-Wesley.

Pearce, P.L. (1988) *The Ulysses Factor: Evaluating Visitors in Tourist Settings*, New York: Springer Verlag.

Plog, S. (1974) 'Why destination areas rise and fall in popularity', *Cornell Hotel Restaurant and Administration Quarterly*: 55–8.

Further reading

Ross, G.F. (1994) *The Psychology of Tourism*, Melbourne: Melbourne Hospitality Press.

PHILIP L. PEARCE, AUSTRALIA

public goods

Products or services provided and indiscriminately distributed by the state to all members of the society are collectively known as public goods. This is in contrast to private goods, which are dispensed by **markets**. Public goods such as highway signs serve tourists. Some public goods are differently distributed, like police forces focus attention on restless areas. Impure public goods, like parks and recreations areas, lie at particular locations.

MARJA PAAJANEN, FINLAND

public health

The new public health integrates human well-being into the ecosystem. Public health has its origins in mid-nineteenth century Europe, introduced to deal with urban problems of air and water quality, food-borne infection (see **food-borne illness**) and diseases arising out of poor hygiene and poverty. The most important solutions to these problems came from changes to social conditions rather than advances in medical technology. Environmental health, preventative medicine and social reform were key components of this notion of public health.

In more modern times, despite changes in the pattern of illness from infectious **disease** to **lifestyle**-related illnesses, this conception of public health began to fade. By the 1970s, the emphasis shifted from 'public' to 'individual' **risk** factors. The link between social change and public health weakened. Less emphasis was placed on the interdependence of health and the social and physical environments. Individuals were seen as responsible for their lifestyles and health risk factors.

In the 1980s, the World Health Organisation and other authorities recognised alarming trends related to a reduction in health equity. The Ottawa Charter of 1986 marked another reorientation of public health away from the dominant notion of the individual, simplistic 'cause–effect interventions' and surveillance approaches, towards a more complex environmental and social **model**. This ecological approach served to emphasise the interconnections among humans, their physical and social environments and their health.

Tourism is a modern phenomenon linked to a number of positive causal factors, including a rise in the standard of living, desires to improve quality of life, increases in **recreation** and **leisure** pursuits, and the desire to reduce the stress of modern living. As part of that improvement, it is logical that successful tourism should be closely linked with health promotion. The World Commission on Environment and Development report entitled *Our Common Future* (also known as the Brundtland Report) (1987) stressed the need to reconcile **economic development** with the resource endowment of the natural world and equity in human well-being. The challenge posed by the report was to strive for **sustainable development**, though this admirable objective has been clouded somewhat by the ambiguity of the term 'sustainable'. Health is clearly part of the rhetoric of ecologically sustainable tourism, but is yet to inform **policy** and **planning** adequately.

For tourism **development** to be environmentally responsible, and if the use of the term 'health' in tourism research is to embrace a more ecological and broader approach, it is necessary to consider biohistory, or the study of human situations in the history of life on earth. This begins with understanding biological processes, economical principles and the sensitivities of biological systems, of which humans are a part. Many argue that the future well-being of humanity depends on satisfying both the health and well-being needs of both the biosphere's ecosystem and human beings. People in developed countries (high-energy societies) are generally well off in terms of life expectancy, though their quality of life could arguably improve; whereas those in the developing world are far worse off. The biosphere is necessarily of global concern, due to the universal ecological **impact** of humans, the combined effects of an extraordinarily high material standard of living and a massive increase in world population. A key question is whether it is possible for human beings to lead healthy and enjoyable lives in a society characterised by a steady rate of **resource** and **energy** use. Tourism, unfortunately, usually encourages high energy usage in developing as well as developed countries in a manner that is not sustainable.

Therefore, the conceptual framework for consideration of health and tourism interactions should include wellness and illness, promotion and prevention. It should not be confined only to reporting on diseases, accidents and health risks related to tourism. Wellness promotion and illness prevention considers the broad interactions of physical, mental, social, economic and environmental factors, with the policy and planning frameworks which impact on stresses related to quality of life. The Brundtland report, in its broadest senses, is the strategy for sustainable development which aims to promote harmony among human beings and between humanity and **nature**.

General concepts relevant to health and tourism include issues of equity and access, in particular relating to the treatment of illness and injury in developing countries or remote locations, to anthropocentric versus biosphere attitudes and values which underpin approaches to human health, to individual well-being versus **community** health, to specialist versus holistic integrated human health with environmental/ecosystem health, and to sociocultural impacts. Tourism specific concepts include guest versus host well-being, and links between wellness and quality of life of host community with tourist **satisfaction**, as well as issues of **safety** and risks.

Further reading

Brown, V.A. (ed.) (1989) *A Sustainable Healthy Future: Towards an Ecology of Health*, Australia: La Trobe University and Commission for the Future. (Proceedings of a World Health Organisation workshop aiming to set a new agenda for public health policy.)

Clift, S. and Page, S.J. (eds) (1996) *Health and the International Tourist*, London: Routledge. (Deals with health problems and preventative measures from the perspective of tourists.)

World Commission on Environment and Development (1987) *Our Common Future*, New York: Oxford University Press. (Discusses global issues of sustainable development.)

World Health Organisation (updated annually) *International Travel and Health: Vaccination requirements and health advice*, Geneva: WHO. (Prescriptive information on immunisation requirements, prophylactic measures and other health advice for tourists abroad.)

ROBYN BUSHELL, AUSTRALIA

public participation

In democratic societies, it has become increasingly recognised that individuals should be able to have input into the decisions that are impingeing upon them. Public participation is the act of providing input into such decisions and it has become part of the conventional wisdom in **planning** as a whole, as well as in much tourism **development**, at least in the Western world, where **local** involvement is widely advocated and may even be mandated. It is argued that public input can help to ensure that the adverse consequences of developments for residents are minimised, that support for initiatives is present, and that it can smooth the managerial path. However, there is little agreement on how best to proceed. Public hearings, open houses, questionnaire surveys and court actions all constitute means of ascertaining public opinions with a potential to influence decisions. The appropriateness of these and other methods of acquiring public input will vary with the time and money available, the culture and the degree to which there is a willingness to share **power**. Arnstein (1969) developed a **classification** of types of involvement which she called a ladder of citizen participation in decision making, extending from manipulation through therapy, informing, consultation, placation, **partnership** and delegated power to citizen control. She argued that the first three were not true participation, that the middle two were forms of tokenism, and that only the latter three were true participation in that they involved some delegation of power to the public. The importance and process of involving **community** members and stakeholders in tourism **planning**, **development** and operation have gained popularity in recent tourism studies.

See also: community planning

References

Arnstein, S. (1969) 'A ladder of citizen participation', *Journal of the American Institute of Planners* 35: 216–24.

Further reading

Elder, P.S. (ed.) (1975) *Environmental Management and Public Participation*, Toronto: Canadian Environmental Law Research Foundation.

Simmons, D.G. (1994) 'Community participation in tourism planning', *Tourism Management* 15(2): 98–108.

GEOFFREY WALL, CANADA

public recreation *see* park; recreation

public relations

Public relations is the **management** of communications and relationships between an organisation and its public to establish goodwill and mutual understanding. Some experts simply call it 'reputation management'. The target groups of public relations activities are several. First, there are employees, who should receive information (by newsletters, magazines, the home page on the worldwide web or at annual meetings) about the company's objectives, strategies and basic ethic guidelines so that they can understand them and identify with them. **Service** personnel in hotels and restaurants, for instance, should know about the importance of tourism for the company or the whole country.

Second, stockholders, banks and investment fund managers should find their investment decisions confirmed by regular reports, newsletters or at annual stockholders meetings (investor relations). Third, suppliers, distributors and customers should be regularly informed by direct mail, press releases, websites and **television** infomercials in order to keep them loyal. Fourth, local residents, pressure groups (like consumers) and the inhabitants of the countries in which the company markets its products should receive honest information, especially when something goes wrong (environmental pollution, mass layoff of personnel and so on). In this **industry**, the providers of services of the national tourism **board** should promote understanding of its importance to the country's welfare, especially when **mass tourism** occurs. Fifth, politicians, **media** managers and journalists, who select and sometimes even distort and distribute information to the public should be targets for public relations. In this case, public relations turns into 'lobbying' for the objectives of a company or a whole industry.

The effectiveness of public relations campaigns is normally measured in two ways: counting the length and number of press articles and television reports ('clippings'), and measuring the organisation's **image** before and after a special public relations campaign by conducting surveys in the relevant target groups. The major problem of this method is that the image of a company is also influenced by other communication instruments (including **advertising**, **sales promotions** and sponsoring).

See also: communication mix; marketing

Further reading

Center, A.H. and Jackson, P. *Public Relations Practices – Managerial Case Studies & Problems*, 5th edn, Englewood Cliffs, NJ: Prentice Hall. (Includes several public relations case studies.)

Gregory, A. (1996) *Planning & Managing a Public Relations Campaign: A Step by Step Guide*, London: The Institute of Public Relations.

Newsom, D., Turk, J.V. and Kruckeberg, D. (1996) *This is PR – The Realities of Public Relations*, 6th edn, Belmont, CA: Wadsworth Publishing Company. (Gives an overview of public relations, objectives and activities and its ethical and legal foundation.)

HELMUT KURZ, AUSTRIA,

publicity *see* public relations

purchasing

Purchasing is one of the stages in the **service delivery system**, in which raw materials are purchased, converted to saleable goods and sold, generating income for further purchases. The purchasing operation involves identifying the best sources of **supply**, making suitable arrangements with suppliers, drawing up purchasing specifications, placing orders and ensuring payment.

According to Odgers (1985), setting up a purchasing operation in a tourism or **hospitality** business consists of four phases: identifying who is responsible for purchasing; establishing standards for the materials and equipment which must be purchased calculating or estimating the quantities required; and evaluating different prices and products in order to find the most suitable for needs. The individual responsible for purchasing materials or small equipment is usually an operational manager, for example, a head chef in a **hotel**. In the case of specialised or capital equipment for specific projects, engineers or architects are often involved.

The specific/evaluation cycle for purchasing materials or equipment is similar to that involved in other evaluation processes, such as choosing a contractor or selecting a job candidate from a short list. The intended use of the material or equipment is identified and a purchase specification is drawn up which contains two lists. One of these lists the essential characteristics which the intended purchase must possess, while the other shows desirable features which the purchase ideally should possess.

The next step of the evaluation process is to obtain details of available products which might fulfil the purchase criteria from suppliers. It is important to guide suppliers towards providing information which relates to the lists of essential and desirable criteria on the purchase specification. Their lists can then be compared with the specification and the best match determined in terms of value for money. The purchase specification may only be concerned with the tangible **quality** of the purchase. However, it is frequently also dealing with factors related to the volume required and availability. Frequently it is necessary to trade such features off against one another. For example, reliability of supply may have to be bought at the **cost** of a small loss of product **quality**.

Further reading

Johns, N. (1995) *Managing Food Hygiene*, London: Macmillan.

Odgers, P. (1985) *Purchasing, Costing and Control*, Cheltenham: Stanley Thornes.

NICK JOHNS, UK

purpose

Purpose of tourism normally refers to different basic categories in the context of statistical surveys, such as holiday, **shopping**, **gambling** and business. It is also equated by some with **motivation**, including physical (**health**, sport), cultural (**art**, **religion**), interpersonal (**visiting friends and relatives**) or status/**prestige** (ego enhancement) reasons.

See also: motivation

RICHARD SHARPLEY, UK

push–pull factors

An early **paradigm** for understanding tourist **motivation** is the push–pull model. This is based on the distinction between factors which encourage individuals to move away from their home setting through tourism (push factors) and those attributes of a different place which attract or 'pull' them towards it.

Push factors are evident at the individual or social level, or as a combination of both. **Anomie**, for instance, relates to a breakdown in society itself and may be manifested in escalating **crime**, increased drug addiction, decline in law and order or, in general, high rates of social disintegration. Those who perceive their own environment in such negative terms become disposed to seeking out alternative places where defects of this nature are considered to be minimal or non-existent. Poor **health**, on the other hand, is usually an individual push factor which influences the decision to relocate temporarily to more curative and benign milieux, while feelings of **nostalgia** may be experienced personally and collectively, inclining people to look for alternatives to the unbearable present and dreaded future in locations reminiscent of the selective good times of the past. On the other hand, pull factors refer to the qualities of the destination area which are either natural (such as the climate or topography), derived (the warmth of the people) or contrived (**theme parks** or hotels).

For analytical purposes, push factors precede pull factors both logically and temporally, since the decision whether or not to travel is prior to a specific choice of destination. In practice, however, such **decision making** may be virtually simultaneous. It follows from the above that the most effective forms of tourism promotion are those which attempt to match the pull factors of the destination with the push factors in the client (matching **supply** and **demand**, including **target marketing**). Thus the urge to satisfy curiosity in the potential **tourist**, for instance, can be matched with novel and exotic experiences in faraway places.

See also: promotion, place

Further reading

Dann, G. (1977) 'Anomie, ego-enhancement and tourism', *Annals of Tourism Research* 4: 184–94.

—— (1981) 'Tourist motivation: an appraisal', *Annals of Tourism Research* 8: 187–219.

GRAHAM M.S. DANN, UK

Q

qualitative research

Qualitative method draws a distinction between subjective or humanistic research techniques and 'formal' or mathematical models, known as **quantitative methods** or research. While partisans often assert the inherent superiority of one research style over the other, each has a legitimate role to play and neither should be considered inherently subservient or weaker to the other.

Within business research, quantitative methods were long considered to be superior and the true means of gathering information and testing hypotheses. In this intellectual environment, qualitative research was often used in order to make preliminary assessments or to forge a testable hypothesis for future quantitative research. Eventually, business researchers discovered situations where qualitative research is more effective than its counterpart; when decisions must be made immediately, for example, qualitative research can be readily conducted while quantitative research can typically be more time-consuming.

More significantly, business researchers have learned that certain qualitative research tools (such as the focus group) are able to provide findings that more quantitative techniques cannot deliver. In a focus group, subjects are brought together to discuss an issue. Supervised by a skilled facilitator, the group brainstorms the topic and often comes up with findings that the researchers could not have discovered using other more formal methods. The focus group experience has convinced many researchers that qualitative methods are superior to quantitative methods in a variety of circumstances.

In recent years, business scholars (such as those in **marketing** and consumer research) have developed a variety of qualitative techniques that generally borrow research methods from the humanistic disciplines and the social sciences. In marketing, for example, ethnographic methods, inspired by anthropological fieldwork, have come into vogue (see **ethnography**). In addition, techniques closely related to literary criticism and cultural history have been embraced. Much of this work stems from the subjective orientation of deconstructionism and poststructuralism. This work has proved to be especially useful when investigating unique aspects of diverse groups of people.

By its nature, much tourism research is especially amenable to qualitative techniques. Researchers, for example, often seek to evaluate how a tourism opportunity will impact the host community of a region. Quantitative techniques may not take the unique feelings of specific people into account, a fact that can result in future problems being unrecognised. Humanistic-oriented qualitative research, however, may be more effective in such circumstances. Tourism research has a long tradition of embracing various qualitative techniques (such as those that derive from sociocultural **anthropology**) and the results of this qualitative investigation have been impressive and appropriate.

Today, tourism professionals are increasingly interacting within the framework of business. As a result, their research needs to be considered as legitimate by the standards and frameworks of business scholars and practitioners. On the one hand, this situation demands that tourism researchers master and become familiar with quantitative methods (even if only to argue against using these methods in particular circumstances). Tourism specialists should be aware that some of their business colleagues might discount the value of qualitative methods even though these tools have long been a staple in tourism research.

Tourism, as a field of investigation and as a profession, needs also to use the strengths and legitimacy of qualitative and humanistic research traditions. Today, due to the embrace of qualitative research by business scholars, it is no longer necessary to defend this choice. Both quantitative and qualitative research methods are significant and legitimate. All researchers must be able to defend study techniques they embrace, including their decision to use (social) specific research strategies in a particular circumstance. In the final analysis, on many occasions qualitative research techniques are indeed the method of choice.

See also: Delphi technique

ALF H. WALLE, USA

quality

Quality, like beauty, lies in the eye of the beholder. Definitions of quality have gradually evolved from a focus on the product-based attributes of the manufacturing sector to the intangible characteristics of quality found in tourism sectors. However, quality in the manufacturing sector differs significantly from quality in tourism. In manufacturing businesses, a tangible **product** is produced where replication is often possible. Customers of these products experience less **risk** as they can always return items if quality is lacking. The goods are produced in an environment separate from the point of sale and unsold items can usually be stored.

In tourism, the product is often intangible and quality is not apparent until after it is consumed. The consumer has no opportunity to return the product if quality is inferior. Its perishable nature also makes it impossible to store it, and production and consumption often occur simultaneously with the consumer being present during the process. The high level of human input into the product makes replication more difficult, particularly when greater amounts of customisation are possible in the product.

Quality guru W. Edwards Deming set out fourteen points for the **management** of quality in the manufacturing sector (Dobyns and Crawford-Mason 1994). The main issue apparent in these points is that producing quality is a continuous process or series of processes, with a wide range of factors influencing both the processes and the final outcome. Understanding the processes and adding, changing or removing any steps that hinder the process are key requirements for achieving the goal. For a quality end product to be obtained, it is essential to recognise what quality means to the consumer in industry sectors, individual businesses and specific transactions. In an AT&T study (Sanes 1996), 70 per cent of customer **satisfaction** was found to be dependent on customer **service** rather than the tangible product.

Quality has been defined as zero defects or defections (Reichheld and Sasser 1992), but still may be defined by the customer. Quality is conformance to standards, with its composite as a function of its component. The International Organisation for Standardisation defines quality as 'the totality of features and characteristics of a product or service that bear on its ability to satisfy stated or implied needs'. Others believe that quality is a combination of outcome and processes, including internal and external conditions, and it is obtained when the expectations and needs of customers are met.

Many tools exist to assist organisations along the road to superior quality. Those firmly committed to this goal may wish to look at quality best practices and benchmarks for their own organisations. **Benchmarking** is a technique where standards are set in key metrics defined by the organisation. Examples of quality best practices can come from the same industry or from other industries. Companies in the retail sector may apply and adapt best practices from the **hotel** or other tourism sectors, and vice versa. There are many areas where quality can be benchmarked. In human resource management, best practices for attracting, selecting, inducting and **training** and developing employees can be used. In the service area, quality practices such as unconditional guarantees, measurement of customer satisfaction, employee satisfaction and mystery **shopping** evaluations, as well as specific standards for complaint resolution, can be applied.

Quality is important in relation to its impact on the bottom line and an organisation's profitability. Reichheld and Sasser (1992) studied customer profitability over time. With higher levels of customer retention as a result of superior quality, an organisation benefits from profits due to: increased purchases and higher balances, reduced operating costs, referrals and a price premium. British Airways found that customers were willing to pay 5 per cent more for superior quality (Weiser 1995). A study by Reichheld and Sasser of different companies found that a 5 per cent reduction in defections increased profits between 25 per cent and 85 per cent, depending on the type of company and industry sector.

By eliminating inferior quality, organisations avoid the costs associated with poor production. These costs may be internal in relation to the cost of having to redo something, facility downtime and loss of morale, which in turn results in high employee **turnover** and lower levels of employee **marketing**. External costs relate to that of compensating customers for poor quality and the resulting loss of their **loyalty**. The customer experiences monetary losses related to the poor quality and non-monetary costs which can be more damaging than the former situations. Non-monetary costs relate to inconvenience, a customer's time and possible emotional responses.

To obtain information on poor quality, organisations require listening posts to capture complaints in order to avoid defections. Up to 95 per cent of unsatisfied customers can be won back if the complaint is resolved quickly and to their satisfaction (Cottle 1990). If quality is determined by the sum of numerous experiences as was found in tourism, any dissatisfaction with a single attribute of the product or services may have a negative impact on the quality of the overall experience (Pizam *et al.* 1978). Thus, attention to the individual components of a tourism experience is critical in producing quality outcomes for satisfied tourists.

References

Cottle, D.W. (1990) *Client-Centered Service: How to Keep Them Coming Back for More*, New York: John Wiley & Sons.

Dobyns, L. and Crawford-Mason, C. (1994) *Thinking about Quality*, New York: Times Books.

Pizam, A., Neumann, Y. and Reichel, A. (1978) 'Dimensions of tourist satisfaction with a destination area', *Annals of Tourism Research* 5: 314–22.

Reichheld, F.F. and Sasser, Jr, W.E. (1992) 'Zero defections: quality comes to services', in C.H. Lovelock (ed.), *Managing Services*, Englewood Cliffs, NJ: Prentice-Hall, 250–8.

Sanes, C. (1996) 'Employee impact on service delivery', *Management Development Review* 9(2): 15–20.

Weiser, C.R. (1995) 'Championing the customer', *Harvard Business Review* November–December: 113–16.

MARGARET ERSTAD, UK
EDUARDO FAYOS-SOLÀ, SPAIN

quality, environmental

Environmental quality is a multifaceted concept which relates to the condition of the natural or biophysical resource base as much as to **environmental aesthetic** aspects of structures, traffic congestion, residential neighbourhoods, depreciative human behaviour and human welfare. The various dimensions of environmental **quality** make agreement on a definition elusive. Apart from the highly subjective appreciation of what constitutes human values, the quality of an **environment**, which in most aspects may be satisfactory or better, could be made unliveable by deficiencies in one key area (such as air quality). There also appears to be clear evidence that perception of environmental quality is sharpened as standards of living rise. Similarly, awareness of environmental deterioration is heightened with affluence, as are efforts to attain and maintain desirable levels of environmental excellence.

Moreover, there appears to have been a shift from needs to wants as essential components of environmental quality. Simply, subsistence-oriented cultures, food and shelter were fundamental to the quality of the environment. In the more aware, sophisticated and technologically-based societies of the industrialised world, environmental quality is equated with quality of life. This in turn assumes access to a range of creature comforts and amenity resources which are wanted to enrich human welfare well beyond basic needs of existence.

In tourism, environmental quality can act as a major **attraction** or an impediment. The environment of a **destination** which is seen to be impaired by deficiencies in physical attributes (such as water quality), or shortcomings in the social sphere (like **crime**), will have a negative effect on **tourist** numbers and **satisfaction**. Conversely, a tourism facility or destination which strives to build and maintain an attractive, functional and secure environment of high quality can anticipate increased patronage. The challenge then is to ensure that greater numbers do not lead to a consequent deterioration in environmental quality.

Evaluation of the quality of environments for tourism is perhaps even more susceptible to subjectivity. Whereas most environments probably have some potential for tourism, this is not easy to establish with any precision because of the personalised manner in which settings are selected and experienced. Generalisation and interpersonal comparisons are of doubtful validity and the multiple characteristics of a tourism environment make disaggregation and evaluation of environmental quality a risky undertaking.

Despite the essentially subjective nature of the variables involved, much attention is being devoted to developing useful measures of environmental quality. However, once again, achieving agreement on what constitutes superior quality and the many aspects which call for assessment make the task problematic. The best that can be achieved is the application of a specific set of environmental indicators to a discrete attribute of the tourism environment (such as noise levels and traffic congestion). Even then, care needs to be taken that a simple additive process is not applied to generate a composite representation of environmental quality, especially when intangible factors such as aesthetics can blur the overall setting.

State of the environment reports are now being produced by many industrialised countries and such reporting is an obligation for member nations of the **Organisation for Economic Cooperation and Development**. This body uses a pressure-state-response model for its reporting system. The **model** is based on the concept of causality. Human activities, such as tourism, exert pressures on the environment and change its state or condition. The changed state of the environment leads to responses by way of **policy** initiatives, legislative reforms and changes in public behaviour. These responses complete the **cycle** by influencing tourism activities which exert pressure on the environment.

Clearly, monitoring the quality of the environment and the way it responds to pressures is fundamental to determining whether current patterns of tourism are sustainable. A common approach, noted above, is the development of a specific set of indicators for each matter of environmental concern. Such indicators need to be scientifically credible, easily understood, cost-effective, and serve as a robust, sensitive indicator of environmental change and potential problems. In terms of the OECD model, indicators can be developed to measure the pressures on the environment arising from tourism; the condition or *state* of the environment (that is, the environmental quality as a consequence of tourism); and the extent and effectiveness of responses to environmental concerns.

In the context of a tourism facility, an indicator of increased environmental pressure might be the introduction of motorised pleasure craft on to a confined waterbody. Indicators of the consequent state of the environment might include the presence of air and water pollutants, noise nuisance and erosion of foreshores. Indicators of response could in this case encompass a range of measures including speed restrictions, licensing and **zoning**. The effectiveness of societal responses to pressure and changes to environmental quality should in turn be subject to appropriate **performance indicators** and with reference to established base-line conditions. By reporting regularly and systematically on the state of the environment and responding appropriately, the tourism industry can help ensure its long-term sustainability.

See also: codes of ethics, environmental; ecologically sustainable tourism; environmental aesthetics; environmental auditing; environmental compatibility; impact assessment, environmental; perception, environmental

Further reading

State of the Environment Advisory Council (1996) *Australia, State of the Environment, 1996*, Collingwood: CSIRO Publishing.

JOHN J. PIGRAM, AUSTRALIA

quality, transportation

Providing **transportation** service can be an extremely challenging endeavour. First, the production of any **service** can be more difficult to manage than the production of goods, since services are largely intangible. In the production of a pure service, such as in transportation, there is no physical item to examine prior to purchase or use, so **service quality** is not as readily observable as in the case of goods. Customer perceptions are extremely important, making it necessary for the provider to perform the service to the expectations of the customer.

Second, transportation service must be performed on a continuous basis. Unlike the production of goods, there can be no inventory of services to draw from to meet the needs of customers when **demand** increases. If transportation capacity is not available to meet the needs of customers at the time and place required, the opportunity to fill that need is lost forever. Additionally, production of transportation services necessarily involves **management** of people and equipment from a distance. Managers typically are not in the same geographic **location** as the transportation equipment or the employees operating it. This makes it very difficult for them to know how well the service is being provided and to make the necessary adjustments to ensure that the service conforms to the needs of the customer.

As the transportation sector continues to mature in a deregulated environment, there has been an increasing emphasis on service quality competition. Today, managers are recognising that improved service quality can be used as an important tool to differentiate their service offering from that of other providers. This **differentiation** leads to enhanced customer **satisfaction** and improvements in market share and profitability and is true for both passenger and freight service. There are many dimensions to these services, but here the various dimensions of transportation service quality can be grouped into four general categories: speed, availability, dependability and communication.

Tourists are concerned with the speed of transportation service. Not only are long trips typically very tiring, but in today's fast-paced economy, the value of the traveller's time plays an important role in the choice of transportation mode. Naturally, faster service is considered to be of higher quality and usually commands premium fares. For example, taking a long distance trip by bus or private automobile may be less expensive than flying, but airlines are generally much faster. Tourists will be more refreshed after a few hours of flying than after days of driving or riding, and more time is available to enjoy **leisure** visits or conduct business associated with the trip. In freight service, carriers that provide fast service typically charge a premium price for their services, such as for overnight parcel delivery services.

Availability is also an important dimension of the service quality. Tourists prefer carriers which offer frequent and more convenient schedules to/from the desired origin and **destination**. Thus, carriers with more frequent service offerings to/from preferred locations on a given day have a distinct advantage over those with schedules of lower quality service. If tourists have to drive great distances to connect with an airline, bus or train, the carrier may be at a significant disadvantage relative to carriers that provide service to more communities.

A carrier that has appropriate transportation equipment at the right locations and at the right time is another dimension to availability. Those that always have needed equipment available for pickup and delivery of passengers and/or freight shipments permit more flexibility in customer transportation decision making, thus providing an important competitive advantage. Dependability is still another important dimension of service quality. Business travellers must depend on the transportation provider to arrive on time so that they can schedule business meetings without wasting valuable time. Service failures, such as late arrivals or cancellations, can result in significant costs, or worse, passengers may miss important meetings altogether.

Although **leisure** tourists may be less time sensitive than their business counterparts, service failures also lead to a lower service quality since holiday trips may be shortened or impossible to complete. Thus, transportation providers who offer high quality, dependable service can expect to enjoy a competitive advantage over lower quality service providers. Shippers of freight frequently want their shipments delivered to a particular place at a specific time, such as for a **restaurant** that uses 'just-in-time' **purchasing**.

Communication is yet another significant dimension of transportation service quality. A carrier's ability and willingness to communicate with tourists or shippers may be extremely important in terms of the customer's perception of service quality. Even though a service failure may occur, if the customer is made aware of the problem in advance, alternate arrangements can be made. When no advance notice of problems is given, they or freight may be stranded for long periods of time. This may cause serious problems leading to substantial costs beyond the transportation expenses, including the possible loss of **future** business from the customer.

The carrier's ability and willingness to communicate with its customers may mean that a tourist or shipper can avoid surprises, reducing the potential for having to deal with unexpected problems. Generally, the more information that the carrier can communicate to the customer, the more effective and satisfactory the service becomes. Thus, transportation service is a multifaceted concept, and in order to ensure that the service quality meets the needs of customers, managers must carefully measure and control all dimensions.

See also: transportation pricing

Further reading

Parasuraman, A., Zeithaml, V. and Berry, L.L. (1985) 'A conceptual model of service quality and its implications for future research', *Journal of Marketing* 49(Fall): 41–50. (Suggests that service quality is best defined in terms of gaps that exist between customer expectations of service and their perceptions of the actual service provided.)

Gourdin, K.N. and Kloppenborg, T.J. (1991) 'Identifying service gaps in commercial air travel: the first step toward quality improvement', *Transportation Journal* 31(1): 22–30. (Discusses commercial airline service and identifies opportunities for managers to improve service quality.)

JOHN OZMENT, USA

quantitative method

Historically, the sciences and the business fields have tended to favour quantitative methods. In statistical analysis, the researcher gathers data and evaluates it in order to locate **trends** or correlations. The researcher begins with a 'null hypothesis' which presupposes no patterns exist; thus, a flipped coin is assumed to have an equal chance of turning up 'heads' or tails'. By flipping the coin a number of times the researcher gets a sample of observed phenomena which is used to determine the odds that the observed **behaviour** actually resulted from chances. The researcher then applies a 'decision rule' to the observed data and uses it to either accept or reject the null hypothesis; the decision rule takes the form of stating the odds that the observed phenomena could be the result of chance. Although somewhat dictated by research traditions in specific disciplines, the researcher is free to choose a specific decision rule.

A variety of statistical tools have been developed for a wide number of research situations which are supported by computer programs. In a simplistic example, the researcher may want to know if having live entertainment in a **hotel** or beach **bar** leads to higher **profit**. The researcher could compare the profits from a random sample of nights when there was entertainment and when no entertainment was available. All variables except entertainment should be held constant. Using statistical analysis, the researcher could point to a possible relationship between entertainment and profits. Still, the researcher would have to demonstrate that the two samples were truly representative. Thus, if the weather was very bad on every night when live entertainment was presented, the public may have stayed away because of the weather and thus the results of the statistical analysis would not reflect reality.

Statistical analysis must of course be used with care. For example, the question may be as simple as whether a hundred-year-old man has a greater chance of living one more year than a five-year-old boy. The former has a century-long track record of living one more year, so statistically, the odds of him surviving are very high. The five-year-old, on the other hand, has a shorter track record, so his survival rate will not be predicted to be as high. But in fact, the boy has a better chance of survival. This can be determined by comparing samples of five-year-olds and hundred-year-olds and noting their survival rates. Therefore, although the proper method of analysis is self-evident in this heuristic example, researchers must take care to insure that similar but less obvious flaws do not creep into their work.

Researchers (or those commissioning research) are urged to spend adequate time and effort deciding exactly what they want to know and why they need it. Although much quantitative research can be routinised and performed at a reasonable price, developing questionnaires and determining what sample populations are to be tested is crucial. Researchers and clients who have not adequately considered these issues will spend more money on research, and typically they will receive less useful results.

Quantitative methods are a staple of business investigation, including tourism. But as the latter becomes more intertwined with modern business and its methods and practice, tourism researchers will need to master quantitative concepts and skills. While quantitative research provides powerful tools, investigators should keep in mind that it represents but one option of analysis and that none of is techniques are foolproof.

See also: qualitative research

ALF H. WALLE, USA

quest

A quest is the act of seeking or searching for something often associated with a sense of **adventure** and the unknown. The great explorers of the past set out to identify and map new trade routes, travel to sacred religious sites for spiritual enlightenment, unearth ancient relics and mysteries, acquire new colonies and collect information. By analogy, modern-day **tourists** are searching for something which they cannot find in their home environment. Identifying the tourist as a modern-day **pilgrim** on a quest, MacCannell (1989) suggests that the **motivation** for tourism is the search for **authenticity** in other times and other places which are significantly different from everyday life. While the religious **pilgrim** travels to pay homage at a particular sacred site (see **site, sacred**), the modern-day tourist visits a large number of **sights** as specific attractions.

At the heart of the quest lie the **souvenirs** and symbols collected to verify the **experience** and, in some cases, prove the quest was a success to others. The more exotic the **location** or uncommon the memento, the greater the **prestige** which is awarded the trip. Beyond the collection of physical artefacts, a quest can be a voyage of **exploration** and **self-discovery**. Cohen (1979) stresses the need to understand the 'spiritual centre' of different tourist types (see **typology, tourist**). Some identify more strongly with their home culture. However, others feel alienated and have the desire to seek out and understand alternative societies in order to locate and submerge their elective spiritual centre within a new cultural environment.

In another context, Almagor (1985) examines the concept of 'vision quest' with respect to the touristic **experience** of **nature**. Here, 'quest' relates to the practice of setting off into the **wilderness** in search of visions through fasting and solitude in order to communicate with 'the beyond'. Animals seen in the visions become the dreamer's guardian spirits. Those who take part in wilderness outings and encounter wildlife are equated with the **indigenous** populations of North America who engage in the 'vision quest'. The touristic adoption of an animal encountered in the wilderness to later symbolise or represent their existential quest with nature may possibly be equated with the 'vision quest' itself (see **symbolism**).

References

Almagor, U. (1985) 'A tourist's "vision quest" in an African game reserve', *Annals of Tourism Research* 12(1): 31–47.

Cohen, E. (1979) 'A phenomenology of tourist experiences', *Sociology* 13: 179–201.

MacCannell, D. (1989) *The Tourist: A New Theory of the Leisure Class*, 2nd edn, New York: Schocken Books.

Further reading

Smith, V. (1992) 'Introduction: the quest in guest', *Annals of Tourism Research* 19(1): 1-17.

DAVID J. TELFER, CANADA

quick service *see* fast food

R

race

Racial **identity**, including appearance, language, **religion** or other ascribed characteristics, affects the lives of millions of people who are forced to define and even defend themselves in relation to it. Identity and the privileged meanings of it form the social core of all human groups. Humans exert control over others by utilising socially constructed images and terms in language and **behaviour** to create hierarchy and legitimise actions. Thus, race is a cultural notion applied and used in specific ways to structure, legitimise or reproduce political and economic conditions and relationships.

Tourism **development** has had far reaching effects on the construction of race and **experience** of racism. Racial constructs are often used to legitimise efforts to gain control of land, labour and resources of peripheral peoples. **Resources** formerly used to support traditional activities and sustain the household and community are redefined as commodities supporting **growth** in the national and global economy. Traditional beliefs, actions and endeavours are termed 'primitive' or 'backwards', and imposed tourism projects and activities represent 'progress'. Resident peoples are described in racial and subservient terms, legitimising their low status and place in the socio-economic hierarchy of an **international tourism** economy.

Tourism based on the **exploration** of exotic differences can stimulate cultural revitalisation and enhance the relative position and **power** of **minorities**. Racism can be a tourism commodity, with tourists exploring, for example, the historical experience and struggles against slavery and segregation. More typically, the sharp contrasts in the tourism experience – between residents and visitors, those who serve and are served, and those with money and those who work for wages – underscores the various divisions in society, emphasising and at times exacerbating racism and other forms of inequality.

See also: commoditisation; ethnic group; ethnic tourism; exoticism

Further reading

Bolles, A.L. (1994) 'Sand, sea and the forbidden', *Transforming Anthropology* 3(1): 30–4. (Analyses race, class and gender dimensions of Jamaican tourism, challenging the notion of the 'oppressed' as a passive victim.)

Harrison, F.V. (1995) 'The persistent power of "race" in the cultural and political economy of racism', *Annual Review of Anthropology* 24: 47–74. (A critical review of anthropological discourse on race.)

Olwig, K.F. (1985) *Cultural Adaptation and Resistance on St. John: Three Centuries of Afro-Caribbean Life*, Gainesville, FL: University of Florida Press. (A history of St John, including the impacts of National Park and related tourism development on demography, identity and race relations.)

BARBARA ROSE JOHNSTON, USA

radio

Radio is an underrated medium of tourism information and promotion (see **promotion, place**). As an **advertising** medium, its main virtues are economic reach of niche audiences, high repetition and mobility (including being able to reach audiences in the car, in different rooms of the house or outdoors). Creatively, its appeal to the imagination can compensate for its inability to deliver a visual image.

A.V. SEATON, UK

rail

Rail **transportation** of passengers, including **tourists**, dates from nearly two centuries ago. Today, a number of rail systems, including some devoted purely to providing a nostalgic **experience** for tourists, are found worldwide. High-speed rail travel was developed in the 1960s, providing timely long-distance travel. The first Shinkansen line in **Japan** opened in 1964, linking Tokyo and Osaka with trains travelling at speeds of up to 210 kph (130 mph). Since then, additional Shinkansen 'bullet trains' have linked Tokyo and other Japanese cities. This country is now investing over $3 billion in the development of 'maglev' (magnetic levitation) systems, which will permit speeds of more than 300 mph.

The French National Railroads began its TGV (train à grande vitesse – 'train of great speed') service between Paris and Lyon in 1981 at speeds of about 320 kph (200 mph). TGV service expanded to Marseille in 1983 and, at a cost of several billion dollars, was extended to several of **France**'s European neighbours including the **United Kingdom**, **Italy**, **Spain** and **Germany**. The latter is developing a high-speed rail service similar to the TGV, except that freight will be carried as well as passengers. Germany plans to invest more than $60 billion in rail **infrastructure** by the year 2000, including investment in maglev projects. The government is partnering with private enterprise to link Hamburg and Berlin with a 400 kph (250 mph) maglev system.

To permit high-speed service such as that of the Shinkansen or TGV, trains require new or upgraded tracks, tunnels, bridges, signalling systems and special power units. In addition to Japan, France and Germany, countries such as the United Kingdom, Italy, Spain, **Sweden** and **Korea** all have committed billions of dollars to the development of high-speed rail systems. For high-speed rail service in the **United States**, Amtrak was authorised to upgrade the Northeast Corridor between Washington DC and Boston at a cost of $900 million. Completion was scheduled for 1999, permitting train speeds of up to 240 kph (150 mph). The government's National Maglev Initiative identified twenty-six potential high-speed corridors. Several US groups are interested in developing maglev systems, but Congress is reluctant to fund development projects.

Further reading

Hollings, D. (1997) 'Europe's railways in the 21st century', *Travel and Tourism Intelligence* 4: 4–22. (Discusses current and future rail systems in terms of expansion, technology, and ownership.)

JOHN OZMENT, USA

ratio analysis

Ratio analysis is the comparison of related numbers resulting in a single figure. The purpose is to increase the understanding of information reported in financial statements. For example, the comparison of the net income of a company to its owner's equity results in the return on this equity, which is often more significant than simply the amount of net income or the amount of stockholders' equity by themselves.

Ratios express financial information in different ways including percentages, per unit basis, turnover and coverage. For example, the comparison of the **cost** of food sold to the sales is expressed as a cost percentage. The comparison of total food sales to the number of customers served is the average service check, a per unit basis expression. The comparison of total sales to average accounts receivable results in the accounts receivable turnover. Further, the comparison of current assets to liability results in the current ratio, a coverage expression.

Managers, creditors and investors use ratio analysis. Ratios help managers to monitor the operating performance of their operations and evaluate their success in meeting a variety of goals. By tracking a limited number of ratios, managers are able to maintain a fairly accurate perception of the effectiveness and **efficiency** of their operation. Creditors use ratio analysis to evaluate the solvency of **hospitality** and **tourism** operations and to assess the riskiness of future loans (lenders) and the extension of future credit (suppliers). Creditors sometimes use ratios to express requirements for operations as part of the conditions set forth for certain financial arrangements. Investors use ratios to evaluate the financial performance of a company. They are keenly interested in financial returns as they relate to their investments.

Ratios are classified by the type of information they provide. Five common **classifications** are liquidity, solvency, activity, profitability and operating ratios. Different user groups tend to focus on different classes of ratios. Investors and creditors focus primarily on solvency and profitability ratios. **Management** focuses primarily on operating and activity ratios and to a lesser extent the other categories. The overall result of using ratio analysis is to increase the understanding of financial information. Each ratio is most meaningful when it is compared to a standard and viewed over a period of time. The standard for management is often the budget. Ratios computed for the current **accounting** period should be compared to those computed for the past periods to detect trends. Ratios used by investors and creditors generally are compared to industry averages and the enterprises' historical ratios, since detailed budgeted financial information is generally not available to these user groups.

In short, ratio analysis permits users to receive more valuable information from financial statements than they could receive from reviewing the absolute numbers reported in the documents. Vital relationships can be monitored to determine solvency and risk, liquidity, operating performance and profitability. A combination of ratios can be used to efficiently and effectively communicate more information than that provided by financial statements from which they are calculated.

RAYMOND S. SCHMIDGALL, USA

rebirth

A return to the womb is a symbolic theme usually employed by the language of tourism promotion in reference to **nature**, and the maternal presence evoked by valleys, caves, lakes and so on (see also **promotion, place**). In tourism publicity, rebirth also connotes the emergence of a new person and the opportunity for **self-discovery**.

GRAHAM M.S. DANN, UK

reception

Reception refers to both the act of greeting the **guest** and the area of the hotel, or similar tourism settings, where guests are welcomed by front-line staff. Friendly, courteous, sincere and prompt reception gives the guest a good first impression. Reception areas are where guests are registered or are provided with orientation.

DEBORAH BREITER, USA

recreation

Recreation is considered as a pleasurable, socially sanctioned **activity** that restores the individual, concomitant with the **experience** of **leisure**. Considerable debate exists about the nature and meaning of recreation and its relationship to tourism (*Annals of Tourism Research* 1987). While this debate is ongoing, there appear to be continued calls for merging of the fields of enquiry, from their originally different ideological beginnings: leisure and recreation as welfare; tourism as business (Moore *et al.* 1995). Stephen Smith (1992) traces the origins of the term to the Latin *recreare*, to renew or to be re-created. The concept of restoring the individual (often, historically, 'to return to work') still pervades most contemporary definitions. While these terms emphasise a human and at times ideological or spiritual perspective, a resultant tension has evolved among definitions emphasising one's experience and one's engagement in activity. Other important definitional criteria centre on whether or not recreation and/or the experience of leisure can occur during time spent on obligatory or extrinsically motivated activity, such as within **employment**.

Despite such debates, there is general agreement on some of the basic elements of recreation. Neumeyer and Neumeyer (1958) summarise these as following: that recreation is an individual or collective activity that can occur during leisure, that it must have some element of intrinsic value (although they argue that it may also have extrinsic **value**, a view not necessarily shared by others), that the primary motive is the **satisfaction** arising from participation in an activity, and that social stimulation and cultural influences shape specific forms of recreation.

As well as definitional debates, the recreation literature reports various attempts at **classification** of recreational activities, which they have been subject to review. In his comprehensive review Shivers (1981) noted the restrictive limits most authors impose on their classifications and definitions. He summarised these as addressing five key concepts: when recreation occurs (during leisure time); why (intrinsically satisfying); how (freely chosen); what (physical activity); and its social acceptability. Shivers then set out to challenge or question each of these five defining concepts. He argues, for example, that for the first concept, recreation can occur through work or obligatory time spent with one's family, and for the second that some forms of recreation, such as sports or some hobbies, can be extrinsically rewarding (see also **sport, recreational**; **sports tourism**). **Freedom** of choice may similarly be absent in some recreational contexts, such as in activities prescribed in a therapeutic context. For the latter two concepts, Shivers also notes that some recreations, such as watching **television** or reading, might be termed sedentary or passive, and that 'social acceptance' might also be an unnecessarily restrictive criterion and present paradoxes: sexual activity might be recreation if undertaken with one's spouse, but is not recreation if it is with a prostitute.

Given the definitional problems associated with recreation, more recent literature has tended to focus on the **benefits** derived from recreation and to the various managerial contexts in which recreation occurs: commercial, **community**, physical, outdoor and therapeutic. What these debates do indicate is that there is active and ongoing debate about the nature of and conditions for recreation, a debate that is similar in many ways to the definitional debates surrounding tourism. Of interest in these debates is the relationship between recreation and tourism.

Tourism has a special relationship with recreation and leisure. At the simplest level, tourism is most often a freely chosen activity, that occurs in discretionary time and that involves discretionary expenditure. This approach has led some researchers to view tourism as a separate subcategory of leisure, distinguished primarily by its differing spatial (travel) and temporal (involving at least one night's stay) arrangements. These additional elements result in a clearer focus on tourism's industrial elements (**transportation** and **accommodation** especially), although the distinctions are not so clear for the public or commercial recreation sectors. Conversely, business tourism does not present the same levels of intrinsic worth or of free choice implicit in the above definitions of recreation, and seeks to confound its relationship with tourism.

When tourist **behaviour** is studied, much of the findings are about tourists' activities and satisfaction. In terms of satisfaction with their **experience**, it is often a **destination**'s 'foundation resources' (public and common property goods) and recreational attributes that are to the fore. Similarly, when reported, tourists' activities comprise an extensive list of recreations: **shopping**, entertainment, passive (**sightseeing**) and active forms of **outdoor recreation** (canoeing, **camping**, rafting, **trekking**, **skiing**, sports and so on). Opportunities for recreation and associated benefits are also major elements in **marketing** and **image**-making of destination areas.

Crompton (1979), in his analysis of the 'motivations for pleasure vacations' makes the important observation that seven of the nine tourism motivations arising from his study are 'sociopsychological' in origin and closely match those defined elsewhere for recreationists. Indeed, the general categories of recreation motives enacted through tourism may explain much of the push factors in tourist motivation (**escape**, relaxation, **prestige** and social interactions), while 'cultural' motives (**novelty** and **education**) may help explain pull factors shaping specific **destination choice** (see **push–pull factors**). While **motivation** is multifaceted and may change over the course of a single trip, Crompton cautions that much tourism may thus simply be diversionary and destination-specific. Several contributions on 'recreational travel' represent a deliberate integration, and again reinforce the 'push' (in terms of 'seeking' personal and interpersonal rewards) and 'pull' ('escaping' personal and interpersonal environments) motives for travel. The common goal is the re-creation or restoration of the individual.

Leiper's (1995) work follows a similar vein. He advocates that tourism is 'a special form of leisure', albeit with some dimensions that raise it above daily recreation. In his analysis, the physical nature of withdrawal and return and the strong temporal and physical demarkers of touristic movements (departure, travel to and travel from destinations, and return) promote this heightened distinction. Moore *et al.* (1995) further advance the argument for a behavioural approach to the study of recreational travel (that is, tourism) as a contemporary **career** among one of everyday life's central 'planes'. The other central planes of contemporary life (at least in the developed world) are family and work, each with its own career structure. Others have argued that both satisfactory recreation and tourism experiences require a match between existential (subjective reality: personal evaluations) and structural (environmental or structural reality: industry programming) elements, and this remains an important caveat for tourism developers and providers of services to tourists.

While the debate about recreation and tourism will continue, there can be little doubt that recreational activity is a central construct in tourist choice and **decision making**. Thus the interrelationships between leisure, recreation and tourism seem set to continue to occupy a central place in tourism (and recreation) research for some time, whatever the disciplinary (**geography**, **sociology**, **psychology**, **management**) or interdisciplinary focus.

References

Annals of Tourism Research (1987) 'Interrelationships of leisure, recreation and tourism', special issue, 14(3).

Crompton J.L. (1979) 'Motivations for pleasure vacation' *Annals of Tourism Research*, 6(2): 408 – 24.

Leiper N. (1995) *Tourism Management*, Collingwood, Victoria: TAFE publications.

Moore, K., Cushman, J.G. and Simmons, D.G. (1995) 'Behavioural conceptualisation of tourism and leisure', *Annals of Tourism Research* 22(1): 67 – 85.

Neumeyer, M.H., and Neumeyer, E.S. (1958) *Leisure and Recreation*, 3rd edn, New York: The Ronald Press.

Shivers J.S. (1981) *Recreation and Leisure Concepts*, Boston: Allyn and Bacon.

Smith S.L.J. (1992) 'Recreation', in *Dictionary of Concepts in Recreation and Leisure Studies*, New York: Greenwood, 253–7.

DAVID G. SIMMONS, NEW ZEALAND

recreation behaviour *see* behaviour, recreation

recreation business district

The **recreation** business district denotes an area of a city that is largely influenced by tourism activities such as hotel **accommodation**, attractions, **shopping** opportunities and tourism retailers (see **urban tourism**). This notion often forms a sub-area of the central business district and is synonymous with it in tourism-dominated **resorts**.

See also: recreation

MARTIN OPPERMANN, AUSTRALIA

recreation centre

Within a **resort**, a **recreation** centre is a complex of sport, **health**, entertainment and social activity spaces for guests (see also **sport, recreational**; **sports tourism**). Common facilities are swimming pools, spas, exercise equipment and multi-purpose rooms. Many communities have recreation centres that are used by both residents and tourists, and in these cases issues of differential user fees, preferential bookings, conflicting activities and overcrowding can arise.

See also: community recreation; resort

DONALD GETZ, CANADA

recreation education *see education, recreation*

recreation experience

A **recreation** experience is a type of human **experience** which is based on the realisation of intrinsically motivating, voluntary engagements during non-obligated times. This is more than just participating in recreational activities such as **camping**, since the importance of the term comes from reward or benefits of participation in the activity.

The intrinsic benefits result from specific outcomes as a result of the experience, such as an enhanced understanding of **nature**, a greater appreciation for the job of the **park** ranger and a greater **development** of outdoor skills. While an engagement in a recreation activity is important to establishing the conditions for the recreation experience to happen, it does not guarantee it, because intrinsic benefit must also be gained.

Classifying experiences that offer recreational meaning and **value** is somewhat difficult, but one meaningful method includes sports and games, hobbies, **music**, **outdoor recreation**, mental and literary recreation, social recreation, arts and crafts, dance and drama. Identifying and measuring recreation experiences focus on the individual and their perception of the **satisfaction** which the experience has produced. There are two major ways of identifying response to specific phenomenon. One is to observe the person's **behaviour**, and the other is to get the person to describe verbally what they did and how they felt about it.

Which strategy is best depends on the nature of the activity, but because evaluation of the experience depends on the participant being able to describe the experience and the feelings produced by it, some method of asking for a description verbally (that is, through an **interview**) usually yields the best results. Participant observation is less likely to be successful because much of what is said about the activity would have to be inferred by the observer from facial expressions, mannerisms and so on, and these are not nearly as reliable. Once the different recreation experiences have been evaluated, then it is possible to measure the importance of these experiences in terms of monetary worth, their psychological value or their intellectual worth.

See also: expectation

Further reading

Brown, P.J. (1983) 'Defining the recreation experience', R.D. Rowe and L.G. Chestnut (eds), in *Managing Air Quality and Scenic Resources at National Parks and Wilderness Areas*, Boulder, CO: Westview Press.

Driver, B.L. and Tocher, S.R. (1970) 'Toward a behavioural interpretation of recreational engagements, with implications for planning', in B.L. Driver (ed.), *Elements of Outdoor Recreation and Planning*, Ann Arbor, MI: University Microfilms.

CAROLYN M. DAUGHERTY, USA

recreation manager

The recreation manager is the individual responsible for the **management** of an organisation or section for the provision of recreation **experience**. Roles at executive level primarily include **decision making** and **leadership** through application of diverse administrative and management concepts (**corporate strategies**, **management accounting**, **marketing**, performance indicators, strategic **planning**, total quality management and so on) to effectively utilise the organisation's human and material resources (see **asset management**).

CHRISTINE L.H. LEE, AUSTRALIA

recreation need *see* need, recreational

recreation opportunity spectrum

The **recreation** opportunity spectrum, developed by the US Forest Service as a tool for managing **outdoor recreation** and tourism on US National Forest lands, integrates these with other land uses in forest management plans. Since it constitutes a practical response to the ecological and social dimensions of **carrying capacity**, it appeals to managers who must balance outdoor recreation and tourism with **biological diversity** protection.

The spectrum framework centres on the idea of opportunity, a combination of three components of people's touristic involvement: activities, settings and experiences. An opportunity is the **product** of an agency's **resources** and **management** practices. The recreation opportunity spectrum's capability to provide an **inventory** of the **supply** of opportunities is its most significant feature. It segments and **maps** land areas into six opportunity classes or zones: **wilderness**, semi-primitive non-motorised, semi-primitive motorised, roaded natural, rural and urban. These opportunity areas, with particular physical (such as large size), social (few encounters with other people) and managerial (legally designated wilderness area) characteristics, can support some kinds of activities and experiences, but not others.

In analysing the **impacts** of other activities on opportunities, this research resolves conflicts among recreational activities and between opportunities and non-recreational land uses, thereby surmounting **displacement** effects. **Management** decisions affect the existing supply of opportunities. The spectrum framework can simulate the shift in opportunities likely to occur, for example, by constructing a road into a roadless area before that road is actually constructed. Managers can then examine **development** scenarios before committing themselves to practices with unforeseen effects on opportunities.

User **demand** for opportunities can be joined with **supply** in the recreation opportunity spectrum framework. Just as supply is expressed in activity, setting and experience terms, so too is demand. Researchers have demonstrated that people have preferences both for activities and for settings and experiences. In establishing standards for outdoor recreation and tourism settings, this research tool combines two types of diversity: that which exists in people's interests, preferences, activities and expectations, and that found in **nature**. Both diversities are meaningful in **planning** and managing opportunities for outdoor recreation and tourism. In planning for outdoor recreation and tourism in **parks** and other **protected areas**, the tool enables managers to work towards achieving two conflicting goals: sustainability (see **sustainable development**) and user **satisfaction**.

Further reading

Clark, R.N. and Stankey, G.H. (1979) *The Recreation Opportunity Spectrum: A Framework for Planning, Management and Research*, Gen. Tech. Rep. PNW-98, 32 pages, Seattle: USDA Forest Service, Pacific Northwest Forest and Range Experiment Station.

Driver, B.L., Brown, P.J., Stankey, G.H. and Gregoire, T.G. (1987) 'The ROS planning system: evolution, basic concepts and research needs', *Leisure Sciences* 9: 201–12.

ROBERT J. PAYNE, CANADA

recreation participation *see* participation, recreation

recreation planning *see* planning, recreation

recreational carrying capacity *see* carrying capacity, recreational

recreational demand *see* demand, recreational

recreational geography *see* geography, recreational

recreational media *see* media, recreational

recreational sport *see* sport, recreational

recreational tourist *see* tourist, recreational

recruitment

Recruitment can be defined as the process of identifying qualified employees and encouraging them to apply for positions in an organisation. The process begins by reconciling the **demand** and **supply** for personnel. This personnel reconciliation process determines how many employees are needed and what qualifications they should have.

Much of the work in recruiting is done prior to actually placing any ads in newspapers or posting notices in employee lounges. This 'pre-recruiting' stage of the recruitment process begins by reviewing the information contained in job analyses, job descriptions and job specifications. Activities undertaken at this point include reviewing job analysis, job description and job specification information for currency and applicability; identifying applicable laws and regulations, determining the message intended to be conveyed to applicants, deciding what other questions to be asked during recruiting, determining whether to recruit internally, externally or both, selecting recruiters, implementing recruiting methods and strategies, establishing criteria for evaluating the pool of applicants, and establishing criteria for evaluating the recruiting method. Costs per hire, number of contacts made, acceptance–offer ratios and salary requested, rates all vary depending on the type of method used. Prior to beginning a recruiting programme, tourism companies need to establish acceptable rates for each of these and other evaluation criteria. The process continues until desirable candidates are identified and encouraged to apply. Preference is increasingly given to those who have completed various phases of **hospitality** and tourism **training** and **education**, now popularly offered worldwide.

ROBERT H. WOODS, USA

recycling

Recycling is the process of recovering materials from waste and reprocessing them into useful **products**. Recycling is now increasingly practised at the domestic household level, in the wider community and as a component of commercial **waste management** and minimisation. Tourism establishments are well suited to recycling programmes, which are often perceived by guests as more important than waste reduction.

Ample opportunities for recycling are presented by tourism operations. Onsite recycling can also be practised through the composting of organic wastes (food/vegetation), composting toilets for human wastes and reuse of grey water on gardens and in the toilet system. Successful and efficient recycling calls for separation of wastes and adequate recycling **infrastructure**. Collecting widely dispersed waste materials for recycling and getting them back into the production system can be a difficult task. Lack of an efficient system for collection, transportation and storage of recyclable materials can be a significant barrier to recycling, no mater how strong and enthusiastic the commitment. This becomes even more of an impediment at isolated tourism establishments, such as remote **nature tourism** or **ecotourism resorts** and island **resorts**, where longer term storage and periodic despatch to centralised treatment facilities is necessary.

A basic deterrent to recycling is **cost**. Larger tourism undertakings can operate at a sufficient scale to install appropriate infrastructure for recycling. Smaller independent concerns may not generate sufficient wastes of sufficient diversity to justify the costs of separation, storage and transport. Moreover, the returns from disposal of recyclable materials may not offset the cost of recycling. Again, older tourism establishments may not have the space to install recycling infrastructure.

Commercial business realities dictate that the pursuit of recycling policies for their own sake, even as a gesture to guest relations, is difficult to justify. Recyclables are seen as components of the waste stream which have a value. In the absence of an adequate and accessible **market** for reprocessing recycled materials and satisfactory returns, support for recycling will always be qualified. **Legislation** which mandates this action should also require longer term performance-based contracts for recycling operators and stable prices. As the **technology** improves and the **quality** of recycled material is upgraded, a stronger market should develop for the use of recyclables in the tourism industry. This **trend** should be strengthened by the emergence of a more environmentally aware and demanding guest clientele seeking a greener tourism product (see **environment**; **green tourism**).

JOHN J. PIGRAM, AUSTRALIA

region

The concept of the region is fundamental in **geography** and is of particular value in gaining an understanding of the special nature of different places and areas, especially to help with **decision making** in **planning**. It is a classificatory device used to differentiate one area from another according to the presence within the bounded area of selected criteria and characteristics. The principal features of a region are its defining boundary, its internal common characteristics, and the degree of difference between those characteristics inside and beyond the boundaries. The purpose of defining the region is to create a differentiated area which is amenable to **planning** and **development**, to organisation and **government**, and to study and analysis. The complexity of the region will be determined by the nature and scope of the characteristics used to secure the desired level of homogeneity.

There are two basic types of region. A formal region is defined by the presence of characteristics such as natural phenomena, particular flora or fauna, weather or climatic conditions. A functional region is defined for organisational purposes such as governmental jurisdiction, professional or trading, or for the convenience of recognising **spatial interaction** or linkages in economic activity and trade. The criteria or characteristics used to define regions may be physical/natural (such as mountainous areas, river catchments), cultural/social (such as territory occupied by particular tribal groups, collections of people with consistent socioeconomic characteristics or **ethnicity**), economic (such as shared trading relationships, or a dominant form of economic activity such as mining), political (such as a nation/country, an electoral division, or an association of countries), any combination of such characteristics, or the qualities of **image**, **identity** or attractiveness.

In the case of tourism, regions may be defined for convenience of **marketing**, organising and **planning**, or to give a spatial reference to tourism **infrastructure** development. For economic feasibility studies, a region may be defined as the hinterland or catchment area for potential users of a **resort**, recreational **shopping** centre or sports venue (see also **sport, recreational**; **sports tourism**); and a region may be used as the spatial frame of reference for the consideration of the geographical extent of environmental impacts of particular tourism developments.

See also: destination; regional organisation

Further reading

Gunn, C.A. (1994) *Tourism Planning*, Washington, DC: Taylor and Francis. (Chapter 5 examines some of the spatial patterns and criteria for defining tourism regions.)

Smith, S.L.J. (1989) *Tourism Analysis*, Harlow: Longman Scientific and Technical. (Chapter 7 examines some principles of regionalisation and describes a number of regional analytical techniques.)

MICHAEL FAGENCE, AUSTRALIA

regional organisation

A regional tourism organisation is a governmental or non-governmental entity responsible for tourism and which represents a specific geographic area within a state, province or territory. This unit usually encompasses several communities and local **government** districts. It is common practice for such offices to be established as non-profit organisations that receive their funding partly through grants from the state, provincial, or territorial governments and partly through industry memberships.

Several countries, such as Australia, Canada, New Zealand, the Netherlands and the United Kingdom, have established regional offices as a third tier of **destination** marketing organisations after the national tourism administration and state, provincial or territorial tourism organisations. These offices are more likely to be found where there exists a structure of regional government agencies or councils. The United States is one major country without a comprehensive network of such offices in most of its states. Instead, responsibility falls to visitor and convention bureaus and other **local organisations**. Their network establishment often follows the completion of regional tourism strategies or plans within states, provinces or territories.

The typical roles of regional offices include **marketing**, **planning**, research and tourist servicing, with the former tending to be the primary activity. Their marketing focus tends to be domestic, as they target their own and other state markets. It is unusual for them to have their own exclusive international marketing programmes. Rather, they tend to work within larger state or national organisations in cooperative international marketing programmes. Regional offices are often involved in tourist servicing through the operation of travel **information centres** within their regions. They may also play a role in the **development** of regional tourism strategies or plans. The main distinction between these and state offices tends to be in the former lack of **policy** making and regulation roles, and in their operation of membership programmes. Despite these distinctions, there is some confusion as to what exactly constitutes a region. Some would argue that this represents several states, provinces or territories within one country. Others see a region as being composed of several countries in a part of the world such as the Asia Pacific region and might characterise the **Pacific Asia Travel Association** as a regional tourism organisation. There is a definite need to clarify this confusion.

Further reading

Pearce, D.G. (1992) *Tourist Organisations*, Harlow: Longman. (A study of tourism organisations in six countries.)

ALASTAIR M. MORRISON, USA

register

A register is a way of speech which varies according to the status of the speaker, medium of communication and topic. In the language of tourism promotion, it is topic which principally distinguishes its varieties (see also **promotion, place**). There are thus different registers for tourism which focus on **nostalgia**, **gastronomy**, **health** and the **environment**, and these in turn display differences of style and imagery.

GRAHAM M.S. DANN, UK

regression

A regression is a return to maternal dependency often evoked by tourism promotion (see **promotion, place**). The symbolic regression to childhood can be called forth via discursive returns to **nature**, rurality, roots, innocence and the like. The nostalgic **rhetoric** emphasises the safety and protection of travelling to destinations which are reminiscent of the past. Coterminously, it is grounded in such **myths** as Heliopolis and the Fountain of Youth.

GRAHAM M.S. DANN, UK

regulation, self

The issue of regulation is about how standards for controlling environmentally damaging activities are achieved through **legislation**. Since environmental issues are complex, specialised, changing and requiring quick responses, environmental legislation – the primary Acts of Parliament – tend to create the general framework and strategy for dealing with issues, allowing for the development of further, detailed **rules** at a later stage by delegated authorities. Traditionally this delegation has been given by statute to the **government** and the detailed rules thus created within guidelines set out in the statute are known as regulations. However, while it is the government which formally promulgates regulations, it is the minister responsible for the subject concerned who has the carriage of them. In turn, it is the public servants under the minister who consult experts in the field and draft the regulations. One advantage of this is that theoretically the detailed standards can be altered more quickly as need dictates without being delayed in the formal procedures of parliament.

Other statutory procedures exist which have developed in recent times into more subtle and accommodating processes for allowing a degree of negotiation, user input and even self-regulation. The starting point is the **licensing** procedures under various resource legislation. For example, environment protection authorities grant authorisations to undertake environmentally sensitive, polluting activities on the basis of flexible responses to the need to achieve a certain limit on the flow of pollutants into the environment. Higher production of pollutants may be allowed, for instance, over the first six months on the basis of an undertaking of safe storage in return for the clean-up of past usage areas to new, stricter standards and **development** of procedures to deal with stored waste and maintain output thereafter at or below the usual standard. Environment protection legislation explicitly recognises this flexibility by encouraging firms to undertake voluntary environmental audits and for entering into environment performance agreements and environment improvement programmes of a contractual nature between the company and the authority. The original **model** for these consensual agreements can be seen in **heritage** agreements which allowed property owners to become willing protectors of heritage **sites** in return for some government assistance. Such agreements are now used even for vegetation protection, and are the forerunners of agreements and management plans which encourage rural landowners to become involved in and initiate the protection of their properties against degradation in cooperation with government agencies. During recent years, tourism has more clearly understood its relationship in respect to this theme and the resulting **benefits** when policing its own plans and operations.

See also: conservation; precautionary principle; preservation

DUNCAN HARTSHORNE, AUSTRALIA

regulatory agency

All countries have **laws** that influence the tourism **industry**. Many of these laws are administered by regulatory agencies. For instance, in the United States a number of agencies have responsibility for the **quality** of **food**. The United States Department of Agriculture regulates the grading and inspection of foodstuffs; the Food and Drug Administration regulates food labelling and packaging, adulteration and misbranding; and the United States Department of Commerce operates a voluntary **grading system** and inspection process for fish and fish products.

See also: Federal Aviation Administration

PETER JONES, UK

relationship marketing

Relationship **marketing** is the business process of establishing, developing and maintaining mutually beneficial relationships with customers – and also with suppliers and other stakeholders – on a long-term basis. The objective is to learn about customers and to develop and maintain a continuos relationship which does not expire after, for example, the **hotel** guest checks out. This is done through the use of database marketing and provision of customer-oriented services.

MARCO ANTONIO ROBLEDO, SPAIN

relics

Relics are objects venerated because of their association with a saint, martyr or even a holy place. These objects are often used in religious **rituals** (prayer books, breviaries, rosaries, crosses and the like), and religious tourists (see **pilgrims**) may purchase them or replicas as **souvenirs**. Relics often attract pilgrims or become attractions (such as pieces of the cross, Muhammad's footprint and the skull of St. John the Baptist). In the Middle Ages, the competitions between churches in Europe was so fierce that relics were often manufactured or even stolen from one church and taken to another.

BORIS VUKONIĆ, CROATIA

religion

Numerous religions in the world cannot be described in a simple definition. Most generally, religion is an organised system of beliefs, ceremonies, practices and worship that centre on one supreme god or deity, or a number of gods or deities. Religion is primarily an **attitude** towards the world, and everything is seen in this respect. Faith is not reason, and thus God cannot be created by reason, nor can faith be explained by it. A believer of one faith may have the same or similar **experience** as a believer of another, yet followers of each express themselves in different ways on the rational, emotional, moral and every other plane. Almost all people who follow some form of religion believe that a divine power created the world and influences their lives.

A general intensification of religious belief seems apparent in many parts of the world, accompanied by a weakening of belief in the established church. Apart from the eight major religions (Judaism, Christianity, Buddhism, Islam, Hinduism, Shinto, Daoism and Confucianism) there is an enormous number of beliefs, cults, myths and sects in the world. To be religious no longer means to blindly follow all that the church says and does. Today, religion can be a personal belief in an entire system, or in the meaning of this or that **ritual**. A certain degree of **alienation**, which is so characteristic of the modern societies of the developed world, is reflected in a similar way; by the springing up of new meets and mythologies, assuming specific contents and often adopting dubious values. Theologians have put forward the thesis that it was in religion that free **time**, rest and travelling were discovered, so it is logical that these should become topical and necessary themes of religious teaching and even pedagogy.

Religion has found the starting point for its perspective on tourism as a form of free time in the need to explain theologically its meaning in human life and to provide the ethical principles on which tourism should rest, just as it previously expounded the meaning of work for human life as a whole and defined the ethical principles of work. In ideological views, the attempt is made to rely on original texts in the Bible, the Qur'an and other sacred texts. These views are founded on the claim that the role of tourism is to provide people with a chance to become familiar with the natural world, with animate and inanimate nature as God's creation. They are thus able to use their free time for their own spiritual enrichment, even their moral renewal, by exploring the ultimate cause and meaning of their existence. Moreover, the 'myth of the weekend', at least in Christianity, enters into the concept of the 'seventh day' because this is the day of rest in the Biblical **image** of the creation of the world.

In the Christian **interpretation**, it is impossible to suppose that people will find their realisation in **leisure**, since that would mean a split personality. According to this view, free time and leisure are a unique and unified time given to people by God, which should thus be used to serve God. Leisure time, a part of free time in which people will express their most intimate inclinations and devote themselves only to that which satisfied them completely, is the ideal time for people to find the peace they need to give themselves to God and receive Him.

The main connection between the religions and tourism can be seen in **pilgrimages**, the religiously motivated journeys which have been an important part of most religions. Almost every major religion requires its followers to go to holy places. Depending on the degree of their religious belief, people are prepared to undertake journeys covering shorter or longer distances, and sometimes very long ones to follow their religious need or perform an act designated by their religion. This religious nucleus is persistent enough on a global scale to overcome **class**, national, ideological, age, professional or any other affiliation.

Theories of tourism consider this movement as one whose participants are motivated either in part or exclusively for religious reasons. Religious tourism most often appears in three forms: as a **pilgrimage**, a continuous group and individual visit to religious shrines; as large-scale gatherings on the occasion of significant religious dates and anniversaries; and as tours of and visits to important religious places and buildings within the framework of a tourist **itinerary**, regardless of the time of the tour. The most popular pilgrimage destinations in the world are Rome, Lourdes, Compostela, Loretto, Fatima, Einsiedeln, Medjugorje, Czestochowa, Guadeloupe and others for Christians; Mecca and Medina for Muslims; Varanasi (Benares), Allahabad, Lumbina, Leshan and Mandalay for Buddhists and Hindus; Lhasa for believers in Tibetan Buddhism; and Jerusalem for Christians, Jews and Muslims (see **Holy Land**).

Religious ceremonies and commemoration days, the climatic location of the pilgrimage **sites** and the work calendar of the population are the main reasons that religious tourism is bound to a certain **seasonality**. Religious tourism is closely connected with **cultural tourism**. For pilgrims, a free day is often planned in the programme so that they can also make trips into the surrounding tourism area. Religious tourism has also political aspects. Numerous religious places are also national sites. Some authors have tried to determine how and to what extent the phenomenon of pilgrimage and tourism differs. They have even attempted to establish certain similarities between these phenomena and to find arguments to support the thesis that tourism as a kind of pilgrimage of modern civilisation. The more serious forms of tourism, where the motives of the journey are more substantial than pure **recreation** and entertainment, are analogous to the ecstatic forms of pilgrimage in their spiritual meaning for the **tourist**, but the symbolic language in which tourists are obliged to express their pilgrimage is different. On their journeys, they always move towards destinations which are a kind of symbol of their wishes and needs, just as **pilgrims** do when they head towards the **shrine** to which the pilgrimage is being made.

Religious tourism is certainly a clearer concept in its secular **interpretation**. From the perspective of tourism, the religious motive is only one among many which impels tourism movements. Consequences for all categories involved in the process are important: the travellers-believers, the providers of services and the space (**region**) in which such movement takes place or toward which they are directed. The **theory** of tourism considers that a part of religious tourist's **behaviour** is simply activities which serve to fulfil basic religious needs.

Theological explanations of the concept of religious tourism have a similar point of departure, but there are differences here between religions. There are those who deny totally such intermingling of the religious and the profane. They attribute tourism as applied to the concept of religion as not even mentioned anywhere in Buddhism, for instance, and Islam avoids this concept or tries to distance itself from it, while Roman Catholicism, although it does not accept explicitly the possibility that a religiously motivated journey may bear the attribute of the tourist, does not totally deny it. However, theologians are reluctant to talk about religious trips as a specific form of tourism. On the contrary, those religious teachings in which the stance toward tourism has been expressed in a relatively strong and well-defined manner, as is the case with the Catholic church and Islam, advocate quite clearly the standpoint that adopting the concept of religious tourism would mean accepting the idea that religion can have another meaning and goal apart from that of faith. In other words, theologians deny that religious motives can be called touristic. By their opinion, the fact that believers on religious journeys have to satisfy their biological needs is not a sufficient argument for the whole phenomenon to acquire a touristic (i.e., profane) rather than a religious character. And on the other hand, the fact that they may have religious needs does not mean that tourists should be seen primarily as believers.

The Catholic church advocates 'the religious and moral dangers of tourism' and the 'heavy responsibility' of the participants, which they undertake in contact with other individuals and environments, as well as the conflicts that this can lead to. But at the same time, the Catholic Church recognises the fact that tourism helps to lessen various prejudices among people and leads to mutual respect among nations, as well as creating the objective conditions for the spiritual elevation of individuals. The idea of pastorisation of tourism was born in Christianity and contained in the organisation of religious ceremonies for believers spending their holidays in a certain destination. This idea in its highest form leads to the spreading of religion and the promotion of its teaching. In this sense, tourism is fertile ground for such work because in their free time people are relaxed, they accept debate and they are prepared to meditate and to gain new insights.

Islamic theologians find it difficult to reconcile Islam with the 'Western contents' of tourism, stressing rather its spiritual and social dimensions. They apostrophise especially the negative impacts of the **development** of modern tourism that are reflected in a marked sexual permissiveness, **pornography**, voyeurism, nudism and so on, as well as in certain forms of **supply** that are opposed to Islamic tradition and religious teaching (as regards to **food**, drink, **gambling** and the like). But on the other hand, in Islam, **hospitality** is a major theme in all holy books, writings and teachings. The attitude of Islam towards hospitality was to be expected, considering the fact that the **hajj** is one of the fundamental commandments for an Islamic believer. Buddhism is much more tolerant in this respect. This is reflected especially in the different ways in which foreign tourism is treated in some Buddhist countries.

The standpoints of the tourism theory are somewhat different. In religious tourism, the dominant sacred content of the journey is important, but it is also important that other, so-called touristic, contents should be present at such a destination. Therefore the very fact that the tourist is a believer is not sufficient for such a person to be called a religious tourist. Tourists who are religious are simply manifesting their personal conviction. Such tourists do not join touristic movements impelled by religious motives; they use their religious needs and rituals in the same way usually done in their permanent place of residence. They also demand that certain religious contents be included in the obligatory range of touristic supply amenities, but these contents or buildings are not crucial to their decision to travel to a certain destination, although they may affect their final decision.

The most visible connection between tourism and religion is the thousands of sacred buildings that are frequented by tourists. The reason for their interest is increasingly to be found in the cultural content or historical value of the sacred building, rather than its original religious purpose. These contents are determined by their function in religion. What attract (religious) tourists are pilgrimage shrines, defined as places that serve as the goals of religiously motivated journeys from beyond the immediate locality; religious attractions, in the form of structures or sites of religious significance with historic and/or artistic importance; and festivals with religious associations.

There are enormous numbers of objects which have a religious meaning, and are thus used in religious rituals (prayer books, breviaries, rosaries, crosses and more), which tourists keep as religious souvenirs. In modern tourism one is witnessing a large-scale vulgar commercialisation of religious motives and their use on the most varied objects, which thus become symbols of certain religious sites or content. Today there are probably no orthodox theologians or other theorists who would deny the economic impacts of religious tourism. Believers had to be accommodated and catered for, and they bought various objects as souvenirs of their stay in the place of pilgrimage as well as other kinds of goods and food. This represented a constant source of income for the local population. Rome was probably the first world shrine which not only felt the economic benefits of pilgrimage but also undertook certain activities to increase the impact. It is thought that there were over two million pilgrim and religious tourists every year.

It is hard to escape the impression today that in most of the places of pilgrimage in the world, the profane impact is more and more on a par with the religious impact. In the religions that are more 'hardline' or conservative in their requirements for the strict observance of all the religious duties of their adherents, such benefits are no longer denied.

There is no reason to believe that the religious motive for travelling will weaken. Man has been given reason and free will, so acceptance of God's law is a question of individual conscience. This view of religious freedom presupposes freedom from any religious pressure. It argues in favour of the view that religious tourism will become increasingly individualised, and also that the visits to the religious places will develop more or less with the same intensity as in the past.

Further reading

Eade, J. and Sallnow, M.J. (eds) (1991) *Contesting the Sacred*, London: Routledge.

Smith, V.L. (1992) 'Pilgrimage and tourism: the quest in guest', *Annals of Tourism Research* special issue 19(1).

Vukonić, B. (1997) *Tourism and Religion*, London: Routledge.

BORIS VUKONIĆ, CROATIA

religious attraction *see* attraction, religious

religious centre

Religious centres are primarily places where a certain **religion** 'started' or where prophets were mostly located when teaching the religious topics. The most popular religious centre in Christianity is Rome and the most important in the Islamic world is Mecca. In some religions there are many religious centres of different significance and importance.

Religious centres developed in two basic directions: as closed centres, accessible only to priests and other religious teachers or as places of **pilgrimage** for large numbers of people. Centres can be classified according to their importance into religious centres of a global or international significance and those of regional or **local** significance, of which there may be dozens for any one particular religion.

There are also specific places and smaller or greater areas where believers come because of their religious content or character. These locations, often called pilgrimage centres or holy places, cannot be directly included among the sacral structures or buildings visited by both religious and pleasure tourists, but should be considered the spatial framework in which such contents are located.

Further reading

Bhardwaj, S.M., Rinschede, G. and Sievers, A. (eds) (1994) 'Pilgrimage in the Old and New World', *Geographia Religionum*, vol. 8, Berlin: Reimer Verlag.

Rinschede, G. (1992) 'Forms of religious tourism', *Annals of Tourism Research* 19: 51–67.

—— (1999) *Religionsgeographie*, Braunschweig: Westermann.

Rinschede, G. and Bhardwaj, S.M. (eds) (1989) 'Pilgrimage in the United States', *Geographia Religionum*, vol. 5, Berlin: Reimer Verlag.

Stoddard, R.H. and Morinis, A. (eds) (1997) *Sacred Places, Sacred Spaces: The Geography of Pilgrimages*, Baton Rouge, LA: Geoscience Publishing.

GILBERT RINSCHEDE, GERMANY

religious festival *see* festival, religious

religious souvenir *see* souvenir, religious

remote sensing

Remote sensing is the gathering of information about the earth's surface from a point above the surface. This may be done by aerial photography from an airplane, or by digital image scanners in low and high-altitude satellites. Remote sensing data is used in geographic information systems for **management** of natural resources and land use **planning**, all used to guide tourism **development** and operation.

ALAN A. LEW, USA

rent

The practice of rental or hire allows the use of spaces or durable objects, facilities and services for a limited period of time. Rented facilities are usually advertised with published fees. Payment for using some facilities, including **hotel** rooms and condominiums, is often included in the price of the **package tours**. In many **resorts**, cars, boats, bicycles or even ski boots are available for rent.

TANJA MIHALIČ, SLOVENIA

repeat tourist

Tourists who return to a **destination** or purchase products and services from the same purveyor on a regular basis are known as repeat customers. Marketers refer to them as being part of the active market segment; that is, they have purchased the product in the recent past and intend to do so again in the near future. Repeat tourists may be drawn to an area because of cultural appeals that exist there. They are responsible for perpetuating the culture, as they initially learn about it on their first trip and reinforce it on subsequent visits. When and if the culture changes, **displacement** of the repeat tourists will take place as they search for alternative destinations with which they are comfortable.

There is no simple answer to the question of what makes someone a repeat tourist. Seasonal (second) homeowners are some of the most visible and frequent repeat tourists to an area (see **second home**; **summer cottage**). Owners of recreational vehicles or motor homes may exhibit characteristics of seasonal homeowners, but, due to their increased mobility, they may choose to relocate if cultural, political or environmental conditions **change**. Small towns or villages located in proximity to larger urban areas have great potential to cater to repeat tourists by offering an alternative touristic **experience** to that found in the population centre. Regardless of the conditions leading to persons becoming repeat tourists, it is important to understand the characteristics, including behavioural patterns, of this group as they are generally the foundation for the rest of an area's tourism. Word-of-mouth **advertising** – or organic image formation – has long been recognised as the most effective means of image development. Therefore, long-term development of touristic products, including **destination** development, requires careful attention be paid to the characteristics, needs and expectations of repeat tourists.

Repeat tourists may also be repeat customers for a business. All tourism-dependent businesses rely to an extent on repeat business, but it is more critical for some than for others. Smaller businesses generally must have higher **profit margins** to offset the lack of sales volume in order to maintain profitability. To cultivate repeat business, **quality** customer **service** becomes a critical part of **value added** to the product. Depending on the type of business, name recognition, preparation of favourite **food** and **reservation** of favourite rooms are some of the characteristics of a quality service programme developed especially for repeat customers. Large businesses also recognise the value of repeat consumers. **Airline** frequent flyer programmes, with various recognition and benefit levels, are commonplace in the industry. Similarly, many multinational **hotel** chains offer frequent stay programmes. Frequent buyer programmes are increasingly being adopted by many different types and sizes of tourism-dependent businesses, as recognition of the value of repeat tourists/customers is acknowledged.

WILLIAM C. GARTNER, USA

representation

Representation is the order of appearance of a thing or event, according to conventions. Representations range from the authentic/real to the inauthentic/unreal. Under postmodernity, representations and signs are increasingly consumed instead of actual entities. The **technology** of simulation is increasingly used to frame touristic **pseudo-events**, and mechanical/electronic/kinematic means are deployed to construct new sign values.

See also: authenticity; staged authenticity

KEITH HOLLINSHEAD, UK

research and development

Research and development (R&D) refers to a continuous process of examining consumer needs and wants, evaluating experiences, and improving targeted **products** and **services** through application of research findings. As such, it is an inherent and essential part of **marketing**. At the **destination** level, there is often a gap between promotion and research agencies and the tourism **industry** which is expected to develop products.

DONALD GETZ, CANADA

reservation

A reservation request is an advanced communication to the tourism or **hotel** operator from the customer (or through a third party like a **travel agency**) specifying their requirements. Reservations and the techniques for processing them vary from one tourism business to another. Some companies have developed sophisticated computerised reservation systems. Alternatively, independent **restaurants** and small operations might use only a traditional reservation book.

DAVID G.T. SHORT, NEW ZEALAND

residential recreation

Residential **recreation** can be defined as publicly-financed or sponsored facilities and programmes that enrich the life of the total **community** by providing opportunities for the use of **leisure** time through a wide variety of year-round activities for all groups, regardless of age, **race**, sex, economic or social status. Most communities include the provision for recreational opportunities of some type in the approval process for new residential neighbourhood developments. Opportunities to enjoy both planned recreation programmes as well as impromptu recreation opportunities are the goals of many residentially-based recreation programmes.

The US National Recreation Association has classified three categories of recreation space found in a typical residential setting. These are the playlot, the neighbourhood playground and the playfield. Each type fulfils a special function in the design of neighbourhoods and groups of neighbourhoods. The playlot is for children of pre-school age. It is often operated in an undeveloped piece of property adjacent to neighbouring single family homes. It serves approximately 30–60 families, and ranges in size from 1,500 to 2,500 square feet. All facilities and equipment are designed and arranged for small children between six and fourteen years of age. Most children in this age group prefer to play in areas three to four blocks from home. The preferred **location** for a playground is adjacent to a **community recreation** centre or an elementary school where supervised recreation is possible. The facilities should contain equipment suitable for informal by-play as well as courts for various organised games.

The playfield may serve four or five residential neighbourhoods. Minimum size is about fifteen acres, with one-half acre per 1,000 population. The facilities include those in the neighbourhood playground but have additional space for such sports as football, baseball, a swimming pool, an outdoor theatre and a recreation building. Night lighting is provided. Some residential neighbourhoods, by virtue of their sites and specialised markets (such as retirees), can be oriented around such amenities as golf courses, marinas, ski areas and equestrian facilities. All these resources of various **development** types assume special importance in tourism promotion and operation, especially in respect to the family **market** with younger children.

See also: community attitude; community development

Further reading

Eisner, S. and Gallion, A. (1993) *The Urban Pattern*, New York: Van Nostrand Reinhold.

Russell, R.V. (1982) *Planning Programs in Recreation*, St Louis, MO: The C.V. Mosby Company.

Urban Land Institute (1990) *Residential Development Handbook*, Washington, DC: The Urban Land Institute.

CAROLYN M. DAUGHERTY, USA

resort

A definition of resorts should include at least three features: they are small geographic units or areas that offer an array of touristic **attractions** and **services**; their population, at least during the tourism season, is mostly made up of transients or **visitors**; and their economy consists in a high percentage of transactions where the **tourist** is one of the parties. Most touristic exchanges are conducted in resorts. Often resorts develop in clusters and form touristic zones usually branded and marketed together (for example, Cannes, Nice and St Tropez are part of the Côte d'Azur).

The concept of resorts has evolved over time and their typology is virtually endless. Old resorts usually were outgrowths of pre-existing communities that adapted to the demands of modern tourism, such as fishing ports that spawned new urban areas with high-rise apartment units around a beach, or mountain villages now surrounded by ski lodges and other related facilities. These resorts tend to have a distinct morphology where the **leisure** areas feed on business and market districts located in the old towns.

More recently, resorts have been developed in a planned way to become self contained touristic units. Some of them were carefully blueprinted from the beginning by private companies, including Disneyworld in Florida, Hilton Head in South Carolina and many others. On the other hand, Cancún in Mexico started in the 1970s as part of a government plan to provide a better economic future to the depressed Yucatan peninsula. Once the infrastructural facilities were put in place, the Mexican government entrusted the **development** of touristic facilities to private firms. The Cancún model has been followed in other **regions** of the world.

Some of the new resorts, such as the Club Med villages and its many imitators, often follow an enclave model. Not only are they smaller and more specialised than the older resorts, but they also tend to keep their attractions exclusively reserved to their clients and to be off limits for most of the local population. Physical boundaries, all-inclusive meals and entertainment plans to keep guests inside and payments made in beads or resort money are all traits that tend to cut the tourists off from the surrounding social environment. In this way, exchanges with the host communities are reduced to the bare minimum; usually, meaningful social interactions take place only among the **tourists** themselves, and the carefree **lifestyle** that their promotional literature depicts becomes their central **attraction**.

See also: ecoresort; resort club; resort development; resort hotels; resort morphology

JULIO ARAMBERRI, SPAIN

resort club

A resort **club** is a brand of all-inclusive holiday, featuring exclusive 'club' resort destinations which are sometimes grouped together under a common theme. Club members may participate freely in all resort activities, including sports and entertainment (see also **sport, recreational**). The term is also applied to designated areas within particular **resort** settings, where access to special facilities is restricted to valued clients or 'club' members.

BRIAN KING, AUSTRALIA

resort development, integrated

Integrated **resort development** involves the simultaneous creation of recreational facilities, hotel **accommodation** and real estate. The project is conceived on an extremely large scale, requiring considerable financial and technical resources often out of reach of a single individual, or even the local **community**. Hence the developer may be a large metropolitan-based company (such a transnational hotel chain) or a consortium formed especially for the purpose of creating the resort. Whether a company or an individual, a single developer is generally responsible for the project. Due to such external ownership, major financial **benefits** leak out of the area. Nevertheless, communities may find remaining benefits, occurring through sale of land and an increase of employment, to be attractive. The latter comes in two phases, first through the needs for construction jobs and later through the need for staffing.

Communities are sometimes able to bargain for a percentage of the revenues obtained from use of the recreational facilities.

Following from the nature of ownership, a unity of purpose in resort **planning** and construction enables integration to occur. The **development** can be designed so as to maximise harmony between the natural setting and the human facilities. The proper ratio of facilities to accommodation can be specified (see **facilities management**). The balanced development resulting from effective overall planning may enhance financial success. Recreational features that are unprofitable in the short term can continue to operate because rapid return from other areas brings in sufficient capital. The large scale of **investment** typically mandates that facilities be of first-class quality. Costs of development are thus offset by higher fees to customers.

In Europe, integrated ski **resorts** and marinas are common, while elsewhere coastal resorts focused on water and golf are often the most typical form (see **golf tourism**). These are sometimes referred to as 'mega-resorts'. A luxury hotel and golf course are built, initially, to attract potential real estate customers. The clientele is relatively wealthy, hence the greatest return accrues through the sale of up-scale condominiums and other types of housing.

Integrated resorts enjoyed a development boom in the West during the early and the middle of 1980s. This was later extended to the Pacific region through the injection of capital from Japanese and **Pacific Rim** development concerns. Changes in **tax** laws in the United States, increasing complexity of environmental legislation and the recession in Japanese real estate market curtailed the proliferation of integrated resorts (see **legislation, environmental**). More recently, changes in **vacation** patterns in the **United States** have led developers to downsize integrated resorts and build them closer to their urban-based clientele. Caribbean destinations such as Puerto Rico have thus been chosen by developers.

Further reading

Pearce, D. (1989) *Tourist Development*, 2nd edn, Harlow: Longman, and New York: Wiley.

CHARLES S. JOHNSTON, NEW ZEALAND

resort enclave

Self-contained tourism complexes are also known as resort enclaves (see **resort hotel**). Usually associated with **sun, sand, sea and sex** in **Third World** settings, resort enclaves typically offer amenities such as tennis, golf, scuba diving, and horseback riding, and are popular with affluent Western mass **tourists**. Because of foreign ownership and low levels of interaction between enclave tourists and the **indigenous** population, there are many controversies over economic benefits as well as social impacts.

See also: resort development, integrated

KLAUS J. MEYER-ARENDT, USA

resort hotel

Resort properties which provide **tourists** with accommodations are categorically known as resort hotels (see **resort enclave**). Recreational facilities may include swimming pools, tennis courts, golf courses and gymnasiums. Typically there is access to other **attractions** such as beaches, lakes, ski slopes, shops, nightclubs or natural areas that provide enjoyment for guests. Facility **quality** is generally designated by number of 'stars'.

ONG LEI TIN, SINGAPORE

resort morphology

Resort morphology, or form, entails the delineation of a resort area into its functional land use components for better understanding of the spatial patterns, processes and **impacts** of **recreation** and **tourism**. The uniqueness of urban morphologic patterns in resort areas was recognised by tourism geographers and others as early as the

1930s, but not until the 1970s did detailed urban and economic analysis generate a considerable body of academic literature. Today, the dissecting of the component parts of a resort (or of any urban area that displays resort functions) to better understand their linkages and evolution has become a small but important research thread among planners and tourism geographers.

The key component of a resort's morphology is what has become known as the 'recreational business district'. This zone of concentrated tourism activity has long been recognised, particularly in highland and seaside resorts. In the 1930s, it was noted that English seaside **resorts** took on a lineally elongated urban form because commercial tourism activity developed along the shorefront, the initial touristic **resource** and **attraction**. Later studies of British and North American shorefront resorts refined identification of the land use components of a resort's morphology by describing zones of 'frontal amenities' or 'honky-tonk' districts as well as discrete non-touristic commercial zones such as 'downtown' and zones of more modest accommodations. In studies of New Jersey resorts, the term 'recreational business district' came into use to distinguish the locus of a town's basic economic activity (recreation and tourism) from the non-basic economic activity characteristic of the central business district. Several morphologic 'models' of European and American seaside resorts were developed in the 1970s, and both similarities (fishing piers, amusement piers) and differences (promenades versus boardwalks) were identified.

Subsequent case studies of resorts around the world have been made to verify conformity to – or deviation from – the early 'models' of resort morphology, especially in coastal resorts. Although many of the case studies were in English-speaking countries, preliminary evidence indicated a high degree of morphologic similarity. This suggests diffusion of resorts from an Anglo-American hearth. By examining resort morphology for various 'time slices', the historical element has also been incorporated into evolutionary models of resort morphology developed for Gulf of Mexico and Southeast Asian resorts. Morphologic analyses of non-coastal resorts have been relatively few, in spite of the fact that mountain resorts in particular would lend themselves well to this type of comparative study.

Further reading

Jeans, D.N. (1990) 'Beach resort morphology in England and Australia: a review and extension', in P. Fabbri (ed.), *Recreational Uses of Coastal Areas*, Dordrecht: Kluwer Academic Publishers, 133–48.

Meyer-Arendt, K.J. (1993) 'Morphologic patterns of resort evolution along the Gulf of Mexico', in K. Mathewson (ed.), *Culture, Place and Form: Essays in Cultural and Historical Geography*, Geoscience and Man, vol. 32, Department of Geography and Anthropology, Baton Rouge, LA: Louisiana State University, 311–23.

Smith, R.A. (1992) 'Beach resort evolution: implications for planning', *Annals of Tourism Research* 19(2): 304–22.

KLAUS J. MEYER-ARENDT, USA

resource-based amenity

Sand, sea, surf and sun are resource-based amenities. **Tourists**, probably instinctively, find natural **resources** like oceans and seas, spectacular landforms, reliably warm and dry or cold and snowy climates, lush vegetation and abundant wildlife appealing. Many destinations are oriented toward resources such as the strandline between land and water, tropical rainforests, deserts, savannahs, mountains, **wilderness** and animal preserves.

LISLE S. MITCHELL, USA

resource evaluation

Resource evaluation refers to the process of determining the suitability of various **resources** for use in tourism, and is an important component of its **planning** and **development**. This covers the development of an **inventory** of the resources available for tourism (see **resource inventory**). Such resources include attractions, sites of cultural, historical and/or natural interest, and facilities available for **transportation**, **accommodation** and activities. Once such an inventory has been compiled, the next stage in the planning process is to evaluate these resources in terms of their suitability for tourism use (see **sustainable development**; **ecologically sustainable tourism**).

A number of criteria have been established for evaluating resources for tourism. The physical and social **carrying capacity** of a resource must be determined. The aim of such evaluations is to assess the levels and types of touristic use which can be sustained by the resource. This usually involves an assessment of the cultural, historical and natural significance of a **site** or area. There are two aspects that must be considered: first, the importance to local, regional or national **environment** or culture, and second, the appeal or attractiveness for tourists. Evaluations of significance and **carrying capacity** should include assessments of resident or local **community** use of a resource and their perceptions of appropriate tourism use. Planners also consider the needs and preferences of tourists. The experiences associated with a potential tourism resource can be assessed in terms of how well they match the needs of existing or desired markets. Another factor which must be considered in resource evaluation is the ease of access to a resource and the **infrastructure** and **investment** required to make tourism use viable.

See also: ethnography; geography

Further reading

Dowling, R. (1993) 'An environmentally-based planning model for regional tourism development', *Journal of Sustainable Tourism* 1(1): 17–37.

Gunn, C.A. (1994) *Tourism Planning*, 3rd edn, Washington, DC: Taylor & Francis.

GIANNA MOSCARDO, AUSTRALIA

resources

'Resources are not, they become', asserts a maxim from economic **geography**. This **commoditisation** substantiates **economics** as the primary arbiter of **value** in Western society. Therefore, it is reasonable to speak of 'human resources', 'cultural resources' and 'natural resources', meaning, in each case, something which has instrumental and, possibly, monetary value in **economic development**. A resource is valuable only because it produces wealth.

Natural resources figure prominently in tourism as well as in other economic activities. Resource inventories and **resource evaluations** determine the significance of potentials and hazards for various forms of tourism development. Elements of **nature** such as beaches or moose are highly valued because they attract tourists; others such as wolves and wetlands, lacking any perceived economic value, are ignored or eliminated. This dominant 'resourcist' **paradigm** has been attacked by critics who argue that instrumental value is not the only significant value which people attach to nature.

A growing multifaceted valuing of nature in the past forty years in Western society has weakened the resourcist position. It is now commonplace for tourists to visit sites of natural **heritage** such as **national parks** which promote both instrumental and heritage values. **Interpretation** programmes help tourists understand and appreciate why such sites are significant in non-monetary ways. Where tourism is an important industry in fragile environments such as **Australia**'s Great Barrier Reef, the cloud forests of Costa Rica or Canada's Arctic, environmental codes of ethics and environmental audits have been adopted to illustrate that natural heritage values also matter to the tourism **industry** and to governments.

The instrumental valuing of nature has been further challenged formally by **Agenda 21**, the international Convention on Biological Diversity. Signatory nations to the convention endorse the protection and maintenance of biological diversity within their territories. Agenda 21 attempts to force a shift away from natural resource management toward ecosystem **planning** and management, a new paradigm in which ecological values are as important as instrumental. **Green tourism** or more commonly **ecotourism**, depending on definition and practice, seems to recognise this change in paradigm, incorporating an array of values in which ecological value occupies a high position. Government natural resource tourism agencies have been slow to adopt the new paradigm. It is testimony, some commentators argue, to the postmodern character of Western society (see **postmodernism**) that the resourcist paradigm continues to guide **decision making** even as the new ecological one grows in influence.

Further reading

Gauthier, D.A. (1993), 'Sustainable development, tourism and wildlife', in J.G. Nelson, R. Butler and G. Wall (eds), *Tourism and Sustainable Development: Monitoring, Planning, Managing*, Department of Geography Publication Series Number 37, Waterloo, Ont.: University of Waterloo, 97–109.

Koppes, C.R. (1988) 'Efficiency, equity and esthetics: shifting themes in American conservation', in D. Worster (ed.), *The Ends of the Earth: Perspectives on Modern Environmental History*, Cambridge: Cambridge University Press, 230–51.

Rolston, H. (1986), 'Values gone wild', *Philosophy Gone Wild: Essays in Environmental Ethics*, Buffalo, NY: Prometheus Books, 118–42.

ROBERT J. PAYNE, CANADA

responsible tourism

The **demand** for responsible tourism results from more than twenty years of experience which has demonstrated that **mass tourism** has many negative effects on host communities. This is true wherever it is located, but the impacts have been particularly harsh in those places where there is a great distortion in the distribution of **resources** and **power**. In such communities, local people do not share the huge profits made from tourism, and are normally worse off economically because of the inflationary pressures of imported goods and services as well as the consumerist mentality which accompanies tourism. They are also disadvantaged by the capital costs borne by their governments for **infrastructure** and promotion at the expense of both essential services and the **alienation** of land and water from public use to touristic needs.

Cultural and religious practices are often cheapened into attractions. **Prostitution** has frequently developed, further eroding the spiritual and moral basis of community life. Fragile coastal areas, atolls and islands have been turned into **resort** areas at great environmental **cost**. Resorts and golf courses make heavy demands on scarce water resources, requiring polluting pesticides. **Ecotourism**, which has been hailed as a step toward **ecologically sustainable tourism**, has in many places degraded the very environment it is promoted as preserving.

People privileged enough to travel, especially to countries in the **Third World**, have a responsibility to those who cannot go away yet must bear the costs of tourism. Responsibility begins when people make a choice to travel or not, and continues when they decide how and where to go. Responsibility entails minimising the negative impacts by choosing to use local facilities, learning appropriate cultural **behaviour** and being environmentally sensitive. Tourists can seek positive relationships with their hosts.

It should not be assumed that **alternative tourism** forms are necessarily responsible. The backpacking and adventure tourist often paves the way for opening up new areas to mass markets, leaves little economic return to the hosts, and is insensitive to the culture. Responsible tourism developers are those who, before they make their investment, ensure that they are considerate of the aspirations of host people and work in **partnership** with them. Similarly, travel agents should ensure that the products they sell do not undermine the dignity and rights of local people. Travel boycott by some tour developers of the repressive government of Burma and the denunciation by the **World Tourism Organization** of child prostitution illustrate consideration of the powerless. The **Ecumenical Coalition on Third World Tourism** provides a **code of ethics** to guide sensitive tourists.

PETER HOLDEN, THAILAND

restaurant

A restaurant is typically defined as an establishment or property where refreshments or meals are served. There are many different types of restaurants depending on the **market** served (mass, family, up-scale and so on), concept or theme (such as ethnic or dinner house) **product** range (type of menu and number of items), service style (quick service/**fast food** or full service) and price.

There are various ways in which restaurants may be categorised. The American National Restaurant Association has developed, in conjunction with other organisations, the idea of 'concept groups'. The quick service segment is subdivided mainly on the basis of core product, into 'concept groups' of chicken, doughnut, Mexican, pizza and so on. Mid-scale restaurants are made up of seven concept groups: cafeteria, casual dining, family-style, hotel, steakhouse, speciality (seafood, ethnic) and varied menu. The up-scale segment has concepts described as casual dining, high-check (high-priced), moderate check (moderately priced) hotel, speciality and varied menu.

An alternative **classification** was also developed in the United States in 1992. The restaurant typology explicitly recognises the **growth** and significance of multi-unit restaurant chains. It proposes five types of restaurant, each of which has distinctive menu characteristics and operational features. These include quick service outlets, offering low price, speed and consistency; mid-scale restaurants, based around menu choice, value and comfort; moderate up-scale, with delivery ambience, flexibility and a 'fashion statement'; up-scale outlets which deliver an experience, style and ambiance; and business dining, based on location, price and value.

The restaurant sector is a significant part of the tourism **industry**. The average American or Frenchman eats out about four times a week. In other countries such as the UK and Australia this figure is lower, but the trend is towards a significant increase in dining away from home. In 1995, the US restaurant industry had sales of over $180 billion. Of great importance is the growth of restaurant chains in all segments of the market. Chain restaurants such as McDonalds, TGI Fridays and Pizza Hut are becoming global brands, with thousands of outlets and sales in excess of $1 billion. Despite this there continues to be a large number of independently owned and operated units serving local markets.

Further reading

Jones, P. (1996) 'Restaurants', in P. Jones (ed.), *Introduction to Hospitality Operations*, London: Cassell, 122–37. (Defines restaurants and outlines the sector in the United Kingdom.)

Walker, J.R. (1996) 'The restaurant business' in J.R. Walker (ed.), *Introduction to Hospitality*, Engelwood Cliffs, NJ: Prentice Hall, 154–79. (Discusses the development and classification of US restaurants.)

PETER JONES, UK

return on investment

Return on **investment** is a ratio that measures the net **profit** or net cash flow generated by an investment in tourism or other businesses divided by the average investment. It is expressed in percentage terms. The average investment is computed by summing the amount of the investment at the beginning of the period measured and the ending amount of the investment of the period measured, and dividing it by two.

STEPHEN M. LEBRUTO, USA

revenue forecasting

Revenue **forecasting** is the estimating of revenues for a future period. Qualitative **forecasting** methods include **market** research, juries of executive opinion, sales force estimates and the **Delphi technique**. **Quantitative methods** of revenue forecasting are both time series and causal. Time series approaches are naive methods, the application of a smoothing constant, and a decomposition time series. Causal methods include regression analysis and econometrics. These and other research methods are used extensively in recent tourism studies.

STEPHEN M. LEBRUTO, USA

rhetoric

In the late twentieth century, rhetoric is being reappreciated as the inventive political craft of discrimination and polemic, not merely as the art of composition. In tourism as elsewhere, rhetoric is increasingly found to be that powerful and pragmatic inducement by which rhetors – be they individuals, organisations, professions or movements – harness the paralinguistic ambiguities of language to persuade particular audiences.

KEITH HOLLINSHEAD, UK

risk

As tourism is all about fulfilling people's travel desire, both the **industry** and tourists try to minimise the risk involved in visiting a given **destination**. In order to do so, providers of destination facilities must ensure the **safety** and **security** of their guests.

Tourists' invulnerability can be assured if the entire **infrastructure** and supporting **services** at a destination adhere to high standards of safety. These include regulation of **food** quality to prevent any kind of food poisoning (see **food-borne illness**); strict **sanitation** regulations to prevent the communication and spread of diseases; controlling and monitoring the use of hazardous building materials; the application of explicit restrictions on potentially dangerous tourist activities; the introduction of dedicated and carefully designed facilities for the disabled; the introduction of safety regulations to **transportation** facilities; and the provision of adequate and highly competent **health** services. As yet, the level of introduction and enforcement of such standards varies from one destination to another. This means that some destinations are still far from offering tourists a risk-free environment as far as safety is concerned. Consequently, tourists still have to seek information regarding the level of safety of a given destination. They also need to take various precautions to avoid potential safety hazards once on-site.

Controlling the security situation at a destination is also highly important in order to minimise the perceived and/or the actual risk involved in tourism. Security situations such as occasional **terrorism** aimed directly or indirectly at tourists, warfare, social unrest and political instability might all be perceived as high risk as they could endanger tourists' lives. In such circumstances, they feel reluctant to visit affected destinations.

The level of perceived risk is related to the extent of those violent activities; the frequency of violent activities; the geographical proximity of such events to destinations; and the duration of each security event, including the way the **media** coverage of such events is featured. Many destinations that had suffered from safety and/or security situations have already incorporated new policies, regulations and enforcement measures to reverse the emerging risky **image**. However, it takes careful **planning** and aggressive **marketing** efforts to convince tourists that the actual risk has been minimised to a reasonable level.

YOEL MANSFELD, ISRAEL

risk analysis

The tourism industry is under increasing pressure to identify potential threats to life and property in and around tourism facilities and, through risk analysis, to determine what can be done to eliminate or reduce the **risk** to acceptable levels. **Resorts**, for example, may face sociopolitical, economic and/or physical hazards. Economic factors, such as shortages of **food**, **transportation**, housing and skilled labour, can have a devastating effect on tourism operations. Political factors, such as unstable governments, changes in government **leadership** and violence targeted against hotels or **tourists**, can pose serious threats. Employees may be killed or injured, facilities may be damaged and customers may be frightened away.

Risk analysis is frequently divided into risk assessment and risk management processes. The former involves defining the nature of the risk, including its probability of occurrence and likely intensity, and measuring its potential impact on customers or employees or others. The risk is usually expressed as a probability of fatality or property loss. Risk management involves developing a plan to reduce the level of political, economic and/or physical risk. Assessments identify the *de minimis risk*, meaning those risks that are so negligible that nothing need be done to control them, so that resources can be focused on the major hazards.

Tourism facilities are often vulnerable to environmental hazards. Locations along coastlines, in the mountains, along rivers and lakes, in wooded areas and so on may increase the danger from hurricanes, earthquakes, floods, fire and other disasters. Assessing the risk requires examination of 100-year and 500-year flood levels, hurricane cycles, seismic activity estimates and other predictions of the occurrence and intensity of disasters. Political risk is assessed in terms of the history of political instability, judgements concerning the level of government support for the industry, and other factors that may change over time as the political situation in the nation changes. Of particular importance is the capacity of tourism businesses and governments to protect tourists and residents from natural disasters and political threats.

See also: emergency management

Further reading

Drabek, T. (1994) 'Risk perceptions of tourist business managers,' *The Environmental Professional* 16: 327–41. (Examines how tourism executives interpret risk and act to reduce threats to their facilities and customers.)

Kunreuther, H., and Slovic, P. (1996) *Challenges of Risk Assessment and Management*, Thousand Oaks, CA, Sage Publications. (Provides a comprehensive guide to risk analysis and suggests ways for firms to reduce risk.)

WILLIAM L. WAUGH, USA

risk, perceived

Perceived **risk** includes the anticipated hazard, natural or social occasions, which is always greater for risk-averse than for risk-taking **tourists**. On the **supply** side, it refers to currency uncertainty perceived by international operators, political and economic instability for foreign investors, default risk for creditors and stock return volatility for shareholders, which is composed of market-related systematic risks and firm-specific unsystematic risks.

ZHENG GU, USA

rite of passage

Rites of passage or transition rites are anthropological terms for rites where a person is transferred from one status into another, like in initiation rites and puberty rites, or for rites that are enacted when crossing the boundaries between the profane and the sacred or the natural and the supernatural, as in secularisation, rites of desacralisation and rites of purification. In transition rites, the **ritual** subjects go through phases that are called preliminal, liminal and postliminal (see **liminality**). The first is the normal profane state of being, the second is sacred, anomalous, abnormal and dangerous, and the third is the normal state of things to which the ritual subject re-enters after the transition. The liminal is a state and a process in the transition phase during which the ritual subjects pass a cultural area or zone that has minimal attributes of the states preceding or following the liminal, and where the norms and sanctions of the society do not necessarily apply. Graburn (1989) has shown how tourism can be understood as a journey to the sacred in an analogy with transition rites.

The stage in tourism that resembles the liminal in rites of passage could be called the liminoid or quasiliminal. This is produced and consumed by individuals while the liminal is believed by the members of society to be of divine origin and is to its nature anonymous. The liminoid is also fragmentary compared to the liminal. Often, elements of the liminal have been separated from the whole to act individually in specialised fields like art. In **art**, popular culture, entertainment and tourism products that promise to remove the consumer from the everyday experience are made for consumption by individuals and groups. They promise a transition into a stage that resembles the liminal for a limited time span. This liminal stage of the transition rite, the stage where social structure partly loses its significance, has inspired many anthropologists studying tourism.

See also: ludic; play; sacred journey

References

Graburn, N.H.H. (1989) 'Tourism: the sacred journey', in V. Smith (ed.), *Hosts and Guests: The Anthropology of Tourism*, 2nd edn, Philadelphia: University of Pennsylvania Press, 21–36.

Further reading

Gennep, A. van (1960) *The Rites of Passage*, Chicago: The University of Chicago Press.

Turner, V. and Turner, E. (1978) *Image and Pilgrimage in Christian Culture*, New York: Columbia University Press.

TOM SELANNIEMI, FINLAND

ritual

Consideration of the relationships between tourism and ritual has been as central to the **anthropology** of tourism as the study of ritual has been to anthropology itself. In sociocultural anthropology, rituals are normally understood as stylised performances with a communally understood symbolic architecture. Such performances may, as **rites of passage**, mark life crises such as birth, marriage and death, signalling the transition of persons as they move from one social status to another, and may involve families or collections of families. Alternatively, rituals may involve larger social collectivities: whole villages in the case of feasts commemorating patron saints, representatives of entire religious communities in the case of **pilgrimages**, nations and international diasporas in the case of such festivals as Passover or Christmas. Although, by prescribing particular patterns and regimes of fasting, feasting, movement, decoration and so on, most rituals make use of the human body, their purpose is precisely to enable participants to transcend the physical world of individual bodies and to shift their focus to a level of engagement and identification with wider social, spiritual and moral structures.

In studies of tourism, these ideas have been used in at least the following four ways. First, the origins of tourism have been traced to traditional ritual occasions, including totemic events (see **myth**) and **pilgrimage**. Second, attention has been drawn to the structural similarities of ritual and tourism as a whole. Third, the ritualistic aspects of particular types of tourist **behaviour** have been identified. Fourth, and on a slightly different level, the place of local ritual events as factors in attracting tourists to particular destinations has been examined.

Several writers have drawn attention to the similarities between the structure and function of traditional totemic ritual and those of tourism. McCannell (1976), for example, has argued that tourism **sites** are subject to the same kind of veneration and 'sacralisation' as traditional totemic sites. Following this view, the similarities between contemporary tourism and traditional pilgrimage have been emphasised. Both involve the making of journeys to sacred or quasi-sacred places. Furthermore, both tourists and pilgrims temporarily place themselves on the margins of everyday society, adopt particular codes of behaviour, and (as is known of pilgrims from *The Canterbury Tales*) use the journey to meet old friends, make new ones, tell stories and reflect on wider social issues and cultural values.

Following Van Gennep's (1960) classic text, the idea of tourism itself as a rite of passage has been explored by several writers in tourism studies. A tourist passes through rites of separation, which may involve crossing a frontier, transition, staying for a period away from home in unfamiliar surroundings, and of (re)incorporation, involving crossing back into and picking up the threads of life at home but in a changed state. Awareness of such structural formations directs attention to another characteristic feature of tourism, elegantly symbolised by Lucy's Florentine adventures in E.M. Forster's *A Room with a View*, namely the saliency of transgression, or the temporary crossing by individual travellers of the boundaries and prohibitions of the everyday. This in turn has pointed to intriguing semiological lines of enquiry which have set out from the central symbolic role in tourism of coasts, mountain passes, rift valleys, rivers and other such territorial 'edges', all of which ambivalently inhibit and invite 'crossings'.

Several specific features of the tourist behaviour and **experience** are clearly ritualistic in character. The buying and, more importantly, the later display of **souvenirs** is one example. Souvenirs are objects which stir memory and longing for the places, times, people and relationships they represent. Their symbolic power derives from the associations they call up and is at once totemic and fetishistic. Then there is the repetitive nature of much of tourism. This is not confined simply to the repeating of visits to much loved places, or the calendrical repetitions flowing from the arrangement of holidays at the same time each year, but also involves the repetition while on holiday of familiar routines. There are other examples too, ranging from the ritualised consumption of food and drink to the behavioural rituals of beach and disco.

Further, ritual events, together with the buildings and other artefacts which mark them, have always constituted central features of **attraction** to travellers and tourists. Thus, for example, the front covers of recent editions of two well-known guidebooks featured churches. Lonely Planet's *Guide to Mediterranean Europe* displayed a photograph of the dome of a church in the Greek island of Santorini, while the Rough Guide's *Poland* opened with an illustration of one of the country's sumptuous baroque cathedrals. Indeed, from the latter book's text, the tourist learns that churches are at the top of the list of tourism **sites** in Poland. Yet it is not only religious buildings which are presented as attractions. Aware of the touristic importance of 'culture' more generally, the Polish authorities are presently encouraging cultural festivals on the model of traditional Catholic feast days. Moreover, Boissevain's (1992, 1997) seminal work on Maltese festivals in particular and revitalised European rituals in general provides ample ethnographic evidence that, far from declining, the attraction of such ritual events to contemporary tourism is increasing.

Although, in the various ways described, tourism is embedded in ritual, it does not follow that it necessarily gives rise to senses of collective harmony and accord. Tourists in cities such as Jerusalem thrive of the ritual and mythic qualities (see **myth**) to be found there. But, as Crick (1996) points out in the case of the Sri Lankan town of Kandy, the relationships and processes engendered by tourism in such 'resplendent sites' are seldom anything but discordant.

References

Boissevain, J. (ed.) (1992) *Revitalising European Rituals*, London: Routledge. (Collection of essays on the subject of the resurgence of ritual events in Europe, many associated with tourism.)

—— (1997) 'Ritual, tourism and cultural commoditization in Malta: culture by the pound?', in T. Selwyn (ed.), *The Tourist Image*, London: Wiley. (An examination of the touristic significance and local management of patronal festivals in Malta.)

Crick, M. (1996) *Resplendent Sites, Discordant Voices*, Switzerland: Harwood Academic Publishers. (Ethnography of the social, economic and political processes of tourism in Kandy, Sri Lanka.)

McCannell, D. (1976) *The Tourist: A New Theory of the Leisure Class*, New York: Schocken. (Classic text in the sociology of tourism which examines the ritual structure of modern tourism.)

Gennep, A. van (1960[1908]) *The Rites of Passage*, Chicago: University of Chicago Press.

TOM SELWYN, UK

roles

A role can be defined as a set of behavioural expectations associated with a particular position (such as host, guest, father and wife). Every situation has a number of specified roles which provide the individual with a fairly clear **model** for interaction. Roles involve a great number of expectations about the actions, beliefs, feelings, attitudes and **values** of the person holding that role. These roles can be seen as encompassing the duties or obligations or rights of the social position. Roles become institutionalised when certain values, expectations and behavioural patterns are exhibited by the **community** at large. People generally endeavour to harmonise with the expected and shared norms or values of their societies.

A role can be defined by the **rules** that apply to an occupant of a position. Cultural norms and rules affect role development processes. These roles in themselves are culture-bound, and often are determined by the community or the culture with which they surround themselves. People tend to make judgements about appropriate and inappropriate **behaviour** by expecting certain ones in certain roles. Those that diverge from what we expect can be quite unsettling.

In cross-cultural situations, such as tourists proceeding to an unfamiliar culture, cross-cultural encounters with members of the host community are often a major source of stress due to the person not knowing the rules and roles that apply to these episodes in the receiving culture. Familiar behaviours in the ways of their own society may even be seen as ambiguous and inexplicable in their new surroundings. In other words, the social role can be influenced by different rules in different environments. **Social situation** analysis is an example of social skills **theory** that can be applied in investigating of cross-cultural interaction. Social situation analysis can be used in such cross-cultural tourism contexts to identify systematically components which influence the outcome of the encounters or particular social situation.

See also: cross-cultural studies; cultural tourism; culture shock

Further readings

Argyle, M., Furnham, A. and Graham, J.A. (1981) *Social Situations*, Cambridge: Cambridge University Press.

Bochner, S. (ed.) (1982) *Cultures in Contact: Studies in Cross-Cultural Interaction*, International Series in Experimental Social Psychology, vol. 1, Oxford: Pergamon Press.

Cushner, K.C. and Brislin, R.W. (1996). *Intercultural Interactions: A Practical Guide*, Cross-Cultural Research & Methodology, vol. 9, London: Sage Publications.

Y.J. EDWARD KIM, AUSTRALIA

romanticism

Contemporary romantic **attitudes** towards the people and places on the tourism map have their origins in the aristocratic travel of the nineteenth century, a period shaped by both **imperialism** and **orientalism**, when the romantic movement in literature, **music** and the **arts** was also at its height. On an (almost) different level, many types of holiday are quasi-synonymous with romance.

TOM SELWYN, UK

room night

This term refers to the stay of one **hotel** guest for one night. It can also be used in reference to group booking. For example, a group of pleasure or business tourists may book 150 rooms for each of four consecutive nights, thereby using 600 room nights. Other terms that are used interchangeably are bed night, guest night and guest day.

DEBORAH BREITER, USA

route system

The routes that carriers offer are an important component of the **transportation** product. The ability of such a firm to offer **service** on routes between desired origins and destinations is part of a carrier's transportation service quality. Availability of this service includes the points served and the frequency on specific routes. This is true for long-distance travel and also for carrier service within **resort** or urban areas. The characteristics and issues surrounding route systems are arguably most prominent for airlines, especially after deregulation.

Carriers may offer a number of different types of routes, sometimes in combination. The simplest route network is a line system, either linear or grid. In the former network a series of two or more points served are essentially connected in a straight line. The latter system resembles an intertwining network of individual routes in a number of different directions with multiple points served.

For some grids, individual route segments cross over one another without using the same airports or terminals.

Hub-and-spoke routing systems are those that have one or more main airports with a number of smaller ones connected to the hub via 'spokes' like in a wheel. Service on the 'spokes' is often provided by smaller aircraft, either jet or turbo-prop. Often, a regional airline offers service on the 'spokes' and has a **partnership** or **code sharing** agreement with a major airline. Where a route system has more than one hub, service between them typically consists of frequent, high-capacity jet service.

The main reason why a carrier might use some form of linear system is the simplicity of the route network, but often with increased expenses and limitations on revenue due to the limited network. With a hub-and-spoke system, a carrier gains **efficiency** through centralisation of many functions. Perhaps the biggest gain, however, is the potential increase in market power at the hub airport. As open skies (or equivalent) and deregulation spread, transportation managers will find routing decisions even more critical to carrier success.

Further reading

Bania, N., Bauer, P.W. and Zlatoper, T.J. (1998) 'U.S. air passenger service: a taxonomy of route networks, hub locations, and competition', *Logistics and Transportation Review* 34(1): 53–74. (Examines the evolution of airline route networks after deregulation in 1978.)

Hanlon, P. (1996) *Global Airlines: Competition in a Transnational Industry*, Oxford: Butterworth-Heinemann. (Chapter 4 provides an in-depth discussion of airline route networks and the next chapter focuses on hub-and-spoke operations.)

FREDRICK M. COLLISON, USA

rules

Rules are shared beliefs which dictate which **behaviour** is permitted, not permitted or may be required in some situation. Rules are generated in **social situations** in order to coordinate the behaviour of interactors so that goals can be attained and needs satisfied. All this relates to various sociocultural studies in tourism, especially **host and guest** relationships.

Y.J. EDWARD KIM, AUSTRALIA

rural recreation

The term rural recreation refers to recreation that takes place in a rural environment rather than an urban setting (see **urban recreation**; **urban tourism**). Participants are either rural residents themselves, day visitors from urban areas, or tourists staying in rural areas (see **rural tourism**). The major activities are any form of **outdoor recreation**, such as **hiking** and biking, **sightseeing**, water-based activities, **nature** appreciation and visiting **heritage** sites.

MARTIN OPPERMANN, AUSTRALIA

rural tourism

Rural tourism uses the **countryside** as a resource. It is associated with the search by urban dwellers for tranquillity and space for **outdoor recreation** rather than being specifically linked to **nature**. Rural tourism includes visits to national and state **parks**, **heritage** tourism in rural areas, **scenic drives** and enjoyment of the rural **landscape**, and **farm tourism**.

In general, the most attractive rural areas for tourists are those which are marginal for agriculture, often located in thinly populated, isolated and less-favoured upland regions. It has long been popular in northern Europe, and is now spreading to Eastern and Central Europe. It is often found in mountainous areas such as the Pyrennees or the Rockies, where it was originally associated with winter sports but now is a year-round activity (see also **sports tourism**). Tourism offers an additional source of income, especially for women, and is important in reducing the rate of rural depopulation. Tourism **investment** may preserve historic buildings, and traditional activities such as village festivals may be brought back to life by tourist interest.

Abandoned buildings in declining villages or on farms may become **second homes** for urban dwellers. Farmers may modernise their unused buildings for letting to tourists, as in the French *gite* system, or develop part of their land for caravan sites or recreational activities. Such developments bring renewed prosperity to poor rural areas, but can also destroy the very attributes of the landscape which first attracted tourists. The growing presence of urban people changes the social character of villages. Peak inflows of cars and caravans cause traffic jams on narrow country roads and hinder movement of animals. Traffic pollution, uncontrolled pets and walkers who leave gates open can hurt livestock and standing crops. Congruence of seasonality for agriculture and tourism also causes friction because of competition for workers. Thus the balance of the costs and **benefits** of rural tourism is not always positive, but in many areas tourism is seen as inevitable.

Further reading

Bouquet, M. and Winter, M. (eds) (1987) *Who from their Labours Rest? Conflict and Practice in Rural Tourism*, Aldershot: Avebury.

Coppock, J.T. and Duffield, B.S. (1975) *Recreation in the Countryside: A Spatial Analysis*, London and Basingstoke: Macmillan.

Canoves, G. and Garcia-Ramon, M.D. (1995) 'Mujeres y turismo rural en Cataluna y Galicia: ?La nueva panacea de la agricultura?' *El Campo* 133: 221–39.

Unwin, T. (1996) 'Tourist development in Estonia: images, sustainability and integrated rural development', *Tourism Management* 17(4): 265–76.

JANET HENSHALL MOMSEN, USA

Russia

The emergence of new active markets in Central and Eastern Europe has been one of the most remarkable recent developments in **international tourism**. Russia, the most important and dynamic tourism **market** in this part of the world, quite deservedly has been at the centre of attention of the international tourism industry. The country's transformation into a market economy has given an impetus to accelerated **development** of the tourism market. Fewer and easier frontier and customs formalities and doing away with limits previously imposed on the amount of foreign currency allowed to be taken out of the country created favourable conditions for satisfying huge deferred demand for travel abroad. If in the early 1980s around 2 million foreign trips by Russians were registered annually, in 1996 this figure reached more than 21 million. But it should be noted that the number of foreign destinations for the Russian citizens has somewhat increased, and now includes the former Soviet countries and Baltic States. The fact that these countries maintain close trading, industrial and cultural links with Russia accounts for the bulk of outbound tourist traffic from the country.

In 1996, the Russians made 13 million trips to the former Soviet countries with 90 per cent of these journeys made for the **purpose** of **visiting friends and relatives** and for other private purposes. At the same time, the progress of economic reforms, expansion of international contacts and growth of economic activity of regions have lead to a tremendous increase in travel of foreign and Russian businessmen across the country. The unique natural and climatic **resources** of Russia and its great cultural and historic **heritage** open up infinite possibilities for large-scale development of tourism and its transformation into a booming **industry**. In the past two years, overall foreign arrivals in Russia increased threefold. Some 13 million foreigners now visit Russia annually, while the former Soviet Union received no more than 8 million at best.

Domestic tourism has been staging a comeback (in the late 1980s the overall number of domestic trips reached 60 million per annum). With the Russian **outbound** tourism boom subsiding, internal demand has been switching to national destinations and **resorts** which have become more competitive in the last few years. The most famous and demanded regions from the point of view of development of tourism are Moscow and St Petersburg. They are the traditional tourist and business centres which attract up to 80 per cent of tourists, including Suzdal, Vladimir, Kostroma and Yaroslavl. These ancient Russian cities which form the 'Golden Ring' are much in demand among Russian and foreign tourists. The main advantages of this region are the closeness to Moscow and St Petersburg, well-developed transport communication with other cities of the central region, and unique architectural monuments which have survived in remarkable condition.

Owing to its unique natural resources, the Siberia and Far East region became quite famous thanks to active promotion in the 1980s. The region possesses excellent tourism potential for **ecotourism** and **adventure tourism**, hunting and fishing tours and **safaris**, as well as the Trans-Siberian railway between Europe and Asia. The Northern Caucasus and the Black Sea region has a big tourism potential thanks to its famous balneological resorts, national **parks**, Alpinism centres and mountain **skiing**. For example, Sochi, one of the world-famous sea **resorts**, is a traditional place where people from Russia, the former Soviet countries and the Baltic States spend their vacations.

Federal and local authorities provide all-round support to the revival of domestic tourism former volumes. This positive trend is reflected in the 1995 approval by the Russian government of the Federal Programme of Tourism Development in Russia and of similar local programmes enacted in a number of Russian **regions**. This programme envisages large-scale renovation of existing **accommodation** properties and construction of new hotels. **Investments** required to carry out the projects included in the programme are estimated approximately at one billion dollars (in 1995 prices). The programme provides for different types of government **incentives** to attract private investment. Special credits, **tax** rebates and subsidies are expected to effectively stimulate potential investors.

It should be noted that the tourism market in Russia has in a short period of time overcome many difficulties and adopted a more universal form of development. Of major importance was the formation of the legislative basis for the development of national tourism industry. Rapidly growing demand for foreign tourism has resulted in setting up a huge number of companies. It is estimated that now there are around ten thousand **tour operators** and **travel agencies**, with nearly seven thousand of them fully licensed. It is known that the tourism trade is a magnet that draws many incompetent and sometimes unscrupulous people, and this is true not only of Russian but of foreign tourism companies as well. Russia is literally flooded with different tourism **products**, many of which are far from being good **quality** as regards programmes of stay and standards of service offered. Some newcomers to Russian tourism are obviously not interested in the overall strategic planning of their activities within the framework of the national industry.

The Russian government has been taking steps to put the national tourism industry in order and make this a respectable business. Stricter controls have been introduced to make tourism companies meet their obligations to clients and partners, including withdrawal of licence as a last resort. A new federal law, 'On the Basic Principles of the Tourism Activities', is expected to be passed shortly by the Russian Parliament.

Directly interested in steady development of the tourism trade, tour operators and travel agents have been pooling their efforts together within the framework of different professionals. Membership of the National Tourism Association includes the whole spectrum of the national tourism industry, including tour operators, travel agents, air carriers, hotels, insurance companies, museums and other tourism enterprises. This national association has been actively involved in drafting legislation and other standard-setting documents as well as development programmes. It lobbies tourism development projects and promotes **education** and professional **training** in this field.

NATELA SHENGELA, RUSSIA

S

sacred journey

In common usage 'sacred journey' could be considered as another term for a **pilgrimage**, but a strict theological explanation demonstrates that there is a difference between them. Pilgrimage in the modern world is usually partly a religious journey and partly touristic. The **sacred journey** has no secular or touristic characteristics; it is devoted to religious reasons for travelling or, simply, it has only a religious component. In the **anthropology** of tourism, the modern practice of tourism has sometimes been described as a sacred journey because certain characteristics of its journeys are analogous with the stage of **rites of passage**, especially the alternation of profane and sacred time and the liminal stage in the **ritual** process.

BORIS VUKONIĆ, CROATIA

safari

A safari is a fully equipped, guided and catered journey or expedition for hunting, **exploration** or scientific investigation, originally associated with East Africa. Increasingly, packaged tours to eastern and southern Africa as well as areas of natural resource elsewhere are given this name. The primary purpose of the modern safari is seldom hunting, but rather the viewing and photographing of wildlife.

TOM BROXON, USA

safety

Tourists might be exposed to various safety hazards if the **infrastructure** and supporting services in a **destination** are not up to acceptable standards. High safety levels are achieved if the quality of **food**, **sanitation**, building materials, tourist activities, **transportation** facilities and **health** services are all strictly incorporated into the tourism industry.

See also: risk; security

YOEL MANSFELD, ISRAEL

sales force management

The effort of recruiting, **training**, motivating, tracking, and evaluating individuals who are selling or representing a **product** or **service** for an organisation or company constitutes the parameters of sales force management. It can be managed by product line, geographic territory or customer. Management efforts focus on maximising revenues, minimising expenses related to selling, and building customer and employee loyalty.

CHRISTINE VOGT, USA

sales promotions

Sales promotion is one of the promotional tools available to marketers of an organisation or company. It is a **marketing** technique in the promotional mix along with **advertising**, **public relations**, **direct marketing** and **personal selling**. These are marketing activities which provide an extra value or incentive to either the consumer or the trade intermediaries. Their cost is often paid by the selling organisation or business. However, sometimes for more **product** or **service** consumers are willing to pay slightly more for the incremental value.

In the tourism industry, sales promotion is a common marketing strategy for both consumers and trade. Sales promotions are offered to both in advertisements, press releases, **direct marketing** and sales offers, and are aimed at boosting sales at a typically slow revenue period or in the shoulder seasons. The focus on the consumer includes samples of products, coupons, premiums, refunds, rebates, contests and bonus packs. Free airline tickets for cruise reservations, two-for-one attraction admission or buy-one-ticket-get-one-free, or a special Sunday brunch for repeat resort customers are among typical examples of efforts to attract additional customers. Trade sales promotions include trade allowances, point-of-purchase displays, contests and dealer incentives, **training** programmes, **trade show** representation and co-operative **advertising**. Examples of these include a cruise company paying travel expenses for representatives of a **destination** marketing group to attend a trade show in an effort to sell both cruise experiences and port-of-call excursions, a hotel company offering a reduced room rate to an airline that is packaging air–land itineraries so that the airline has a larger **profit margin**, and a car rental company and an **attraction** entering into a cooperative advertising programme.

Sales promotions are typically offered for times when business volume is lower than peak months. Often they are called 'blitzes' because of the immediate response that is desired. Other occasions for sales promotions include the launch of new products (see **product planning**) by offering samples and coupons to create awareness for it, attending a trade show and offering a special-priced **vacation** package plus a drawing for a free vacation, and sponsoring a major event and creating a commemorative product.

Further reading

Assael, H. (1993) *Marketing Principles and Strategy*, 2nd edn, Forth Worth, TX: The Dryden Press.

Belch, G.E. and Belch, M.A. (1995) *Introduction to Advertising and Promotion: An Integrated Marketing Communications Perspective*, 3rd edn, Chicago, IL: Irwin.

Morrison, A.M. (1989) *Hospitality and Travel Marketing*, Albany, NY: Delmar.

Rothschild, M.L. (1987) *Marketing Communications*, Lexington, MA: D.C. Heath and Company.

CHRISTINE VOGT, USA

sanitation

Sanitation involves 'the creation and maintenance of hygienic and healthful conditions' (Marriott 1994). Within **food** service and similar establishments in business, the major concern is with the development of systematic procedures for the protection of human **health**. Hazards to health may include pathogenic micro-organisms together with chemical and physical contamination (see **food-borne illness**). The term sanitation covers all stages in the food service system which ensure this food safety. These stages include purchasing, delivery, storage, preparation, cooking, holding, service, cleaning and waste disposal.

In order to control the growth of food pathogens, it is necessary to identify how they get into the food chain and how a hazard can be eliminated through the destruction of the organism or the control of its rate of growth. Food pathogens may enter this chain as a contaminant. In addition to the initial contamination of raw materials, this may result from contact with equipment, food preparation surfaces, employees, pests or cross-contamination from other foods.

Many companies now use a systematic procedure for identifying and managing risk, known as hazard analysis and critical control points. This procedure is divided into a number of logical steps: (1) risk assessment (review all processes in order to identify possible hazards); (2) determine critical control points (identify stages in the process where hazards can be controlled); (3) establish target levels and tolerances in order to control the risk at each point; (4) establish procedures for the monitoring and controlling of all critical control points; (5) identify corrective action for situation outside the defined tolerances; (6) establish a formal system of record keeping of all monitoring stages; and (7) establish conformance testing and audit procedures to validate the system.

References

Marriott, N.D. (1994) *Principles of Food Sanitation*, 3rd edn, New York: Chapman and Hall.

Further reading

Cliver, D.O. (1990) *Foodborne Diseases*, San Diego: Academic Press.

HCIMA (1995) *Hazard Analysis and Critical Control Points (HACCP)*, Technical Brief No. 5, London: Hotel & Catering International Management Association.

DAVID KIRK, UK

satellite account

A tourism satellite account describes the structure of the industry in a nation, measures its economic size, and serves as an information system to collect and interrelate statistics describing potentially all quantifiable data related to it. The term 'satellite' refers to the fact that the account is a subset of a nation's input–output accounts, which detail the values of each commodity produced and consumed by each separate industry.

A complication is that tourism commodities, such as **restaurant** meals, are consumed by both tourists and residents; the former also consume non-tourism commodities such as clothing. One cannot simply identify any set of industries and aggregate their statistics to describe a nation's tourism activity. Using data derived from consumer and business surveys, a satellite account identifies the portion of outputs from both tourism and non-tourism businesses consumed by tourists.

When fully developed, this satellite account consists of four levels: financial data related to the **supply** and demand of commodities; activities supported by these financial flows, such as tourist numbers; characterisation data of those activities such as tourist demographics; and **planning** and **policy**-related data such as the rates of tourism business failures. A significant feature of the account is that it is governed by the rigorous rules of national accounting systems. This ensures that its statistics are as credible, consistent and balanced as data for any other industry.

The concept of a tourism satellite account first appeared in the mid-1980s. Macro-economists in Canada, France, Spain and Sweden developed their own interpretations of the concept. The so-called 'Canadian model' is the most inclusive interpretation in that it incorporates all four data levels described here. Satellite accounts are the 'cutting edge' of tools to measure the structure and size of tourism in a country and add credibility to the field.

See also: balance of payment; economics; industry

Further reading

Lapierre, J. and Hayes, D. (1994) 'The tourism satellite account', *National Income and Expenditure Accounts*, Second Quarter 13–001: xxxiii–xiviii. (Explains the principles and conventions used in developing a satellite account and presents core data.)

National Task Force on Tourism Data (1987) *Tourism Satellite Account: Working Paper No.3*, Ottawa: Statistics Canada.

Smith, S.L.J. (1995) 'The tourism satellite account: perspectives of Canadian tourism associations and organisations', *Tourism Economics* 1: 225–44. (Describes the relevance and limitation presented by the account in the context of individual organisations.)

STEPHEN SMITH, CANADA

satisfaction

Motivation, personality, **attitude**, expectations, perceptions and stereotyping are significant variables in setting goals, influencing choices and decisions about destinations, **accommodation**, activities and other elements important in planning a holiday. These factors also influence tourists' **behaviour** and are determinants of satisfaction. This can be investigated under the rubric of several disciplines, including concepts and theories of **economics**, **psychology**, **sociology**, **anthropology** and ethnographic domains of enquiry. All contribute to understanding the individual, group, crowd, **community**, society and place. In turn, the complexity of interactions that determine levels of satisfaction include the nature of involvement, the role of risk, stress, frustration, fulfilment and **anticipation**.

The psychology of tourism, including attitudes of tourists and residents, and consumer behaviour and values, are among important themes contributing to an understanding of this subject. Concepts used to explain tourist behaviour linked to satisfaction include Maslow's theory of human needs, which is one of the best known theories of motivation. Personality constructs of tourist behaviour such as allocentric/psychocentric model (see **allocentric**) relate satisfaction to marketing purposes, by segmenting markets according to personality profiles. Allocentric tourists prefer exotic destinations and unstructured holidays, and psychocentrics prefer familiar destinations and packaged tours. Stress models such as the irritation index seek to explain the cumulative effects of tourism development over time. As a community is stressed by tourism, poor host–guest relationships can be expected. Destination life cycle theory further explains impacts on these relationships and on tourist satisfaction. Expectancy theory which relates to motivation in the work environment has relevance to customer satisfaction. If tourism does not provide adequate rewards and **motivation** to staff, this will translate to a mismatch with customer expectations of an industry in which the notion of **service quality** is paramount.

Anthropological theories of **acculturation** and others involving tourist–host transactions and the development of service economies explain adaptations and sociocultural changes which accompany the evolution of tourism systems in response to tourists' needs. The service economy is focused on providing for the needs of transient, leisured strangers. Satisfaction varies considerably on whether the tourist is seeking **authenticity** or staged experiences. **Commoditisation** of host–guest relationships is central to the loss of authenticity. Many techniques are employed to research satisfaction, either as **marketing research** to enhance economic viability, or from more academic perspectives of understanding the host–guest relationship and sociocultural impacts.

Further reading

Pearce, P.L. (1982) *The Social Psychology of Tourist Behaviour*, Oxford: Pergamon.

Ross, G.F. (1994) *The Psychology of Tourism*, Elsterwick: Hospitality Press.

Ryan, C. (1995) *Researching Tourist Satisfaction*, London: Routledge.

ROBYN BUSHELL, AUSTRALIA

satisfaction, customer

In social psychology and **marketing**, customer **satisfaction** represents the positive result of the consumption of goods and services. Customer satisfaction results when the tourist's expectations are met. It is linked to other factors of behaviour such as motivations, attitudes and the service encounter. While some of these depend on the performance of the industry, many (such as weather) are outside its realm of operation.

MARTINA GONZALES GALLARZA, SPAIN

scale of development

Scale reflects both the size and rate of tourism **growth** and **development**. This can be measured by infrastructural development, capital investment, tourist arrivals or **employment** generated. It is usually argued that the social, cultural and environmental **impacts** or costs increase with scale and that local control over development declines, although this last feature depends on the initial ownership of resources and on start-up costs.

DAVID WILSON, UK

scenic drive

A scenic drive is a route designated for recreational travel by car because of the attractive vistas which it affords. As a verb, it refers to the taking of a journey by car primarily to view the scenery en route. Such routes are commonly signposted or marked on a **map** and thus identified as attractions in themselves as well as being links to a number of points of interest. In some cases, as in the case of the Blue Ridge Parkway, a 469-mile scenic road between Shenandoah National Park in Virginia and Great Smoky Mountains National Park in North Carolina and Tennessee, USA, the road may have been purpose-built for recreational travel with winding routes, pull-offs, landscaping and restricted speed limits.

GEOFFREY WALL, CANADA

seamless service

On multisector air journeys involving transfer connections at intermediate points, airlines may participate in **code sharing**, **franchising** and block bookings to make interline connections appear as if they were online connections. This cooperation among airlines to provide an apparently single connected service is referred to as 'seamless'. Airlines will often advertise their alliances and combine their livery, painting their aircraft in the colours of the participating companies.

TREVOR H.B. SOFIELD, AUSTRALIA

seasonality

Seasonality refers to temporal fluctuations in the volume of tourism. Normally recurring and often regular, they involve tourist numbers and phenomena related to such fluctuations in receipts, visitation numbers, **occupancy rates** and bed nights. Seasonality is one of the most distinctive features of tourism in many parts of the world, and is generally viewed as a major problem facing the industry. In many areas, considerable efforts have been made to reduce the seasonal changes in tourism, which have been blamed for low returns on **investment**, problems in retaining staff, problems in accessing capital and problems in overuse and underuse of the capacity of physical plant.

Seasonality can be categorised into two primary types: natural and institutional. The first is caused by temporal changes on an annual basis in the natural world, such as the four seasons, and corresponding changes in temperature, precipitation, sunshine and hours of daylight. These natural changes, which are primarily climatic in nature, increase in severity as one moves further from the equator. Much of the traditional temporal patterns of tourism reflect seasons in the Northern Hemisphere, because most of the world's tourism originates in the developed countries and are located there.

The second form of seasonality in tourism, institutional, is caused by human decisions and relates to what were often traditional temporal variations in the patterns of human activity and inactivity. This form varies much more widely across the world than the natural form, reflecting cultural diversity and beliefs. It is the accumulated result of religious, ethnic, cultural and social factors, and varies from patterns reflecting natural phenomena to historical inclinations. The most common form of institutionalised seasonality is the formal holiday, derived in most cases from holy days of rest, common in form if not in date to most of the world's religions. Such breaks from work are normally of short duration and occur at similar but not exact times each year. Of more significance for tourism are the long holidays which reflect more recent institutional decisions and subsequent legislation. The two major ones are school holidays, now enshrined in the concept of holidays with pay in most industrialised countries. Such holidays are normally in the summer season but increasingly are available at unspecified times during the year, thus reducing one element of seasonality which was extremely strong at the time of the introduction of industrial holidays in the late nineteenth century.

There are other less significant forms of seasonality in tourism. One is seasonality of a social nature whereby specific tourist activity is dominated by social factors and constraints such as fashion. In earlier years, specific and sharply defined seasons existed for participating in certain activities and for visiting facilities, such as taking the waters at **spas**, or hunting. Such seasons normally involved small numbers of elite tourists and are of less significance in contemporary tourism. Seasons relating to sporting activities are now more common and reflect to a degree climatic and related conditions which may be a requirement of the specific activity, such as **skiing**. A great deal of the seasonal pattern of tourism can be explained by inertia and tradition. People take holidays at specific times because that is when vacations have been taken historically.

Seasonality is viewed as a concern in many areas and in the industry because it results in uneven loading on facilities. Most elements of the tourism infrastructure have to be large enough to accommodate peal numbers and are therefore unused and unproductive for large periods of the year. While it may be that in a few areas it is possible to use labour for tourism which is employed in other activities in the non-tourist season, in most cases this is not feasible as the season often corresponds to times of peak demand for labour for agriculture and other resource-related activities.

In two respects, seasonality can be viewed as a beneficial feature. In the case of the environment, the non-tourist season allows for vegetation and wildlife to recover from the demands of tourism use. In the case of residents of destinations, the periods without tourists allow them a 'normal' life for part of the year. In such situations, some services may be discontinued or reduced because of lack of demand due to reduction in tourist numbers.

Efforts to reduce seasonality have been introduced in many areas, and include lengthening the main season of visitation, establishing alternative seasons based on other attractions, diversifying and broadening **markets**, staggering holidays to spread domestic tourism over a longer period, creating off-season attractions such as festivals and special events, and economic incentives such as differential taxation and pricing. Almost all of the steps taken have been related to destination areas, and few initiatives have been taken in origin regions. Perhaps for this reason, most attempts have not had lasting success. In many cases, while tourist arrivals have increased, numbers have also increased in the peak season, suggesting that seasonality is a complicated and deep-seated characteristic. In areas where there has been a change in seasonal patterns, at least parts of the change appears to have come about through changes in the areas of origin of tourists, such as additional holidays and changes in tastes.

Seasonality also has spatial components, and is more accentuated in rural and remote regions and is less problematical in urban centres. Such destinations have more non-seasonal attractions and more business **travel** and, in many cases, are less vulnerable to climatic changes. As well, many of their infrastructural features operate year-round, particularly those relating to accessibility. The reasons for current patterns of seasonality have been little explored and may well relate more to the motivations and behavioural attributes of tourists than to innate climate or historical characteristics in destination areas.

See also: climate; destination; marketing; vacation

Further reading

Bar On, R.V. (1975) *Seasonality in Tourism*, London: Economist Intelligence Unit. (Reviews seasonality in a number of countries.)

Butler, R.W. (1994) 'Seasonality in tourism: issues and problems', in A.V. Seaton (ed.), *Tourism State of the Art*, Chichester: Wiley, 332–9. (Discusses origins and causes of seasonality in tourism and their policy implications.)

RICHARD BUTLER, UK

second home

Second homes are properties maintained for **leisure** purposes which are not the principal residences of their owners. They date back to classical times when the social elite retreated for pleasure and privacy to other residences, often to escape unpleasant summer temperatures in their normal city home. Owning a town house as a principal residence and a rural estate for leisure use, often related to sporting activities, became a feature of the elite of many countries. Examples include the royal summer residence of Balmoral Castle in Scotland and Vanderbilt's 'Breakers' in New England. Numbers of second homes increased rapidly in the twentieth century, especially in North America where rural scenic land was cheap and available. Second homes now range from palaces to shacks, used from a few days each year to semi-permanent retirement dwellings. As numbers of second homes increased drastically with the democratisation of leisure and increased affluence, the scale of most new second homes shrank, and in North America they were appropriately called **summer cottages**. However, they still remained for the most part 'purpose-built for an inessential purpose'.

Many properties formerly used only in the summer season have been converted for leisure use on a year-round basis, reflecting the rise in popularity of winter sports. Traditionally located in relatively remote areas and often in a waterfront setting, second homes are now found in a great variety of environments, corresponding to changing leisure tastes and preferences. They have become more widespread and are now found integrated into golf course developments, conventional resorts, winter sports areas and even urban centres. In the last two decades, variations on the traditional second home have included boats, hobby farms, canal barges, apartments and condominiums, some in the form of **timeshare** units. Single-purpose second home communities have been established, often in conjunction with other leisure facilities such as golf and water sports developments, especially in North America and the Mediterranean. Air travel and improved accessibility generally have allowed second homes to be located a long distance from the owner's permanent residence.

Acquisition of conventional residences for use as second homes has caused negative reactions in many rural areas, particularly in Western Europe where there is competition for housing in high amenity areas. In some jurisdictions, legislation has been passed to prevent non-residents acquiring property for use as a second home, and limits have been placed on the amount of use of second homes to prevent them being used on a permanent basis, often with political implications and reactions.

RICHARD BUTLER, UK

security

Security refers to a perceived and/or actual invulnerability of tourists considering visiting a given **destination**. It deals with manmade potential or actual activities that **risk** the lives of tourists and/or their possessions. Activities that might jeopardise tourists may include **crime**, social or political unrest, terrorist activity and/or warfare. Security in tourism has been regarded as one of the major concerns of both tourists and industry. Examination of the relationship between this and security encompasses four basic aspects: crime committed against tourists and **safety** measures to prevent or minimise it; evaluation of the risk involved in visiting destinations affected by security problems, namely **terrorism** and **war**; monitoring the impact security situations could have on global, national and regional patterns of tourist flows; and defining market strategies to deal with the short-term and long-term consequences that emerge in the wake of a given security situation.

Crime and terrorism aimed against tourists are causing the most evident concern to all concerned parties. Terrorist activities are aimed at attracting the public's attention to certain political and/or ideological interests. In pursuit of such goals, terrorists attack targets that either symbolise opposing interests or can dramatically damage the basic interests of a given regime. By attacking tourists, terrorist groups achieve immediate, dramatic and wide-range media coverage while disrupting an important economic resource. Sporadic and infrequent attacks normally cause a minimal damage to the industry, as tourists' memory tends to fade after a while. An instance of such terrorist activities against tourists and tourism installations is the infrequent Irish Republican Army bombings in the West End of London. On the other hand, frequent waves of terror create a long-lasting image that causes many tourists to avoid such troubled areas altogether.

Crimes against tourists are primarily a result of an existing or perceived socioeconomic gap between hosts and guests. Thus, destinations have become an 'arena' where lower-class hosts are demonstrating their anger and jealousy (through crime) against rich guests, who are using their wealth to pursue their eternal search for enjoyment. Law enforcement agencies, together with owners and operators of tourism and hospitality facilities, are in part responsible for lack of prevention of crime against tourists. If tourism is all about merging dreams with reality, the industry faces a major dilemma. On the one hand, it is expected to supply the tourist with a crime-free environment. This means high costs, a high level of cooperation and organisational measures, which are not always within reach. On the other hand, too many evident crime-preventive measures could be regarded as admitting that a crime problem does exist in a given **destination**. Thus, such measures might become counterproductive, deterring tourists as a high-risk message is conveyed.

Tourists, state tourism agencies and providers of products all share a common interest in reducing both the actual and perceived risks involved in visiting troubled areas (see **risks, perceived**). One of the most important facilitators of minimised risk is available information. Up-to-date, comprehensive and accurate information about the level of risk and security at destinations allows tourists to make a better choice of destination. However, quite often they are not exposed to such relevant information by a reluctant industry that wants to create a positive image of its products. The printed and electronic media, on the other hand, tend to overplay the risk involved in visiting destinations affected by present or past security problems. Thus, risk evaluations made by tourists cannot be accurate as they rely on either biased information produced by the industry or on the media. Only in a few countries, such as the United States, do governments provide their citizens with updated risk evaluation.

Security situations negatively affect the propensity of potential tourists to visit and stay in an affected destination. As a result, hotel occupancy rates drop sharply, small businesses supporting the local industry collapse, many workers lose their jobs and locals lose their faith in tourism as a viable industry, and corporations divert their investment capital to safer places. In such circumstances, both the public and private sectors lose expected income and profits. A limited number of studies have attempted to evaluate the impact of security situations on the tourism performance of affected destinations. Findings show that the outcomes most detrimental to the affected industry occur in the short run. Those destinations which adopt proper crisis management measures manage to regain **inbound** tourist flows. The pace and intensity of the recovery process is directly correlated with the severity level of security events, their frequency, the extent of media coverage, the geographical proximity of destinations to troubled areas, the geographical distance between the generating markets and the insecure destinations, and the level of attractivity of the affected destinations (Mansfeld 1996: 266).

There is a need to provide the industry with means to regenerate tourist flows, and thus reduce the damage caused by security situations. Local and national governments have to take a leading role in the reactivation of inbound tourist flows. In countries and regions with a long history of security problems, proactive contingency plans are executed. Once the situation calms down, public tourism organisations and the private sector join forces to change the media's attitude towards the **conflict** and its implications. Already substantial funds are invested in various promotional activities among the generating markets in order to regain tourist flows. In countries with high attraction potential, such intensive promotional campaigns succeed, helping the recovery of affected destinations.

References

Mansfeld, Y. (1996) 'Wars, tourism and the 'Middle East' factor', in A. Pizam and Y. Mansfeld (eds), *Tourism, Crime and International Security Issues*, Chichester: Wiley, 265–78.

Further reading

Tarlow, P. and Muehsam, M. (1996) 'Theoretical aspects of crime as they impact the tourism industry', in A. Pizam and Y. Mansfeld (eds),

Tourism, Crime and International Security Issues, Chichester: Wiley, 12–22.

YOEL MANSFELD, ISRAEL

segmentation *see* market segmentation

segmentation, a posteriori

A variant of **market segmentation** used in tourism and other businesses, where the number, size and structure of sub-markets is initially unknown. Instead, it is hypothesised that a **market** contains sub-groups of individuals who are homogeneous in terms of activities, motives, attitudes or other **psychographics** and behavioural characteristics (see **behaviour**). Subsequent analysis (see **cluster analysis**; **discriminant analysis**; **neurocomputing**) assists in checking this assumption and examining its consequences for purchasing behaviour.

JOSEF A. MAZANEC, AUSTRIA

segmentation, a priori

A variant of **market segmentation** where a dominant characteristic of buyer subgroups is fixed in advance. It is predetermined by the **marketing objective**, which defines a priority of the target segments to reach via selective market operation (such as first-time visitors or travellers on a short trip). Subsequent analysis (see **automatic interaction detection**; **discriminant analysis**) assists in finding correlates (such as demographic, socioeconomic and other **behaviour** criteria) of the main classifier for better reachability through mass media and distribution channels.

JOSEF A. MAZANEC, AUSTRIA

self-actualisation

In the tourism **motivation** literature built on Maslow's ideas, the complex motivation of self-actualisation is identified. This term refers to tourists who experience peace, inner harmony and profound **satisfaction** with themselves and the places they visit. It is a sophisticated mental state reflecting personal fulfilment, and only a limited number of tourists strive for or achieve such feelings.

PHILIP L. PEARCE, AUSTRALIA

self-discovery

One outcome of vacationing may be self-discovery, that is the generation of new insights about the holiday taker's own personality and **behaviour**. Self-discovery, like **self-actualisation**, is best described as an incidental outcome of tourism rather than a motive which actively drives behaviour. Occasionally, tourism operations focus on self-discovery tours aimed at those seeking life-changing or personally insightful experiences.

PHILIP L. PEARCE, AUSTRALIA

semiotics

Semiotics incorporates all forms and systems of communication as its domain. The central idea in semiotics is a conception of the sign which is defined as a bond between a signifier and a signified. Examples are many: they include the bond that exists between a sound (signifier) and its meaning (signified) for words spoken in a given language, or between marks on the page and meaning in the case of written language, or the conventional agreement that the colour red (signifier) stands for danger (signified) in some communities. Semiotic research involves the study of conventions, codes, syntactical and semantic elements, and logic; in short, all the mechanisms which serve to produce and obscure meanings and to change meanings in sign systems. Semiotics is the only method of inquiry into the nature of communication which does not necessarily presuppose intersubjective agreement as a condition for 'meaningful' communication. Communication occurs across cultural and language boundaries, between humans and animals or humans and machines and so on. Semioticians assume that signs can be exchanged among multiple subjectivities.

The complexity of communications involving more than one kind of mind is the domain of contemporary semiotic study. Thus, semiotics is well adapted for the study of communication and behaviour in tourism settings, where assumptions of shared values, beliefs and language among hosts and guests may be untenable.

Contributions to tourism research which have made intensive use of semiotic methods include Dean MacCannell's chapter on 'The Semiotics of Attraction' (1989), Bennetta Jules-Rosette's book (1984), and several of the papers collected in the special issue of the *Annals of Tourism Research* on 'The semiotics of tourism' (1989). Several of the papers in the special issue of the *Annals* describe interactions between Third World hosts and tourist guests from Europe and North America. When these tourist–host interactions are subject to close semiotic description, it is evident that the tourists and their hosts do not meet as representatives of 'modernity' and 'tradition', respectively. Semiotic analysis of the interactions reveals that the interactions between Third World tourists and their hosts are framed 'as if' the tourists are modernised and the hosts are traditional or even 'primitive,' but this is mainly a superficial agreement or 'definition of the situation'. Closer examination dispels all stereotypes. In several studies, 'primitive' and 'peasant' people are revealed to be motivated mainly by economic rationality, while the tourist participants in the same exchanges appear to inhabit a world shaped mainly by myth and superstition. Similarly, Jules-Rosette discovered Africans who carve 'traditional figures' for the tourist trade spend their spare time carving replicas of such things as telephones and Western style men's suits for their own amusement. MacCannell argued that the global system of attractions, the Taj Mahal, Mount Kilimanjaro, the Egyptian Sphinx, the Eiffel Tower, the Pyramid of the Sun and so on, constitutes a special class of signs intended to communicate across every cultural and historic boundary.

The American pragmatist philosopher, Charles Sanders Peirce, determined there are three classes of signs. Each class is based on differences in the relationship between the signifier and the signified, and each is relevant to the study of tourism attractions and communication involving tourists. First, iconic signs depend on a bond of resemblance between the signifier and the signified. For example, a photograph of a famous sight is iconic to the extent that the sight can be recognised in the photo. Much tourism imagery found in brochures and guidebooks is of the iconic type. The saturation of the media with iconic images is necessary to the **motivation** to travel to specific destinations. Tourists exchange iconic signs for a view of the 'real thing'. Second, indexical signs or indices are produced by the direct action of that which they represent. For examples, Friday's footprint in the sand was Robinson's first sign that he shared the island with another human being. The presence of crowds of tourists is indexical of the popularity of a given **attraction** or **destination**. Third, symbolic signs or symbols are arbitrary and conventional, and they require community consensus on proper meanings. The large statue of a woman standing in New York harbour is, by convention, a symbol of Liberty. It is an example of a symbol which has attained cross-cultural significance.

Any object or idea can be represented by each of the three types of sign. For example, 'liberty' can be signified indexically by the broken bars of a cage, or iconically by someone dancing naked on a tropical beach. A problem with using semiotic analysis for the study of tourism is that it tends to produce ontological insecurity. Within semiotics, everything is either a signifier or a signified. There is no realm of 'reality' which is separate from and merely modelled by signs. Everything is signs or parts of signs and everything stands for something else. Nothing can simply stand for itself, not even so-called 'natural' things. Flowers signify passion, rocks durability, and so on. Thus, living in a world of signs imposes upon the human subject a condition of being forever in flux, involved in exchanges and transformations.

The restless, incessant searching of tourists worldwide can be read as a symptomatic exposure of the semiotic basis of human existence after the erosion of older fictional grounding of the human subject (gender, ethnicity, geographical roots, work or profession and so on). Further, there is very good reason to believe that once 'the tourist' becomes a kind of universal identity, it will not be possible to return to older more grounded, specific versions of the human subject. The global system of attractions arises out of the naming of things. But no matter how much the tourist wants to identify with the attraction, by having a photograph taken beside it or by telling the story of the visit to it, he or she is always blocked by the protective railing, the velvet cord, the 'official' viewpoint. Once an object, place or event is named as 'an attraction', it can never be experienced for what it is. The Statue of Liberty is not liberty. Visiting the Statue of Liberty is not to be free. The system of attractions blocks identification with the very values the attractions are supposed to embody. The tourist can approach, but never fully enjoy, never become 'one with' the values encoded in the attraction.

References

Jules-Rosette, B. (1984) *The Messages of Tourist Art: An African Semiotic System in Comparative Perspective*, New York: Plenum Press.

MacCannell, D. (1989) *The Tourist: A New Theory of the Leisure Class*, New York: Schocken Books.

—— (ed.) (1989) special issue on 'The semiotics of tourism', *Annals of Tourism Research* 16(1).

DEAN MacCANNELL, USA

senior tourism

Defined variously by age, individuals as young as 50 may be included in the senior market, and ages 55, 62 and 65 also serve as benchmarks. Tourism **behaviour** of the US mature market has been found to be similar to the rest of the public, with somewhat lower rates of business and outdoor recreation and higher rates of entertainment-related travel.

MICHAEL A. BLAZEY, USA

service

Service can be seen simply as those points of interaction between service providers, normally the employees of an operation, and their customers. A broader description would include all the elements that go to make up a complete service package or **experience**, which might include, as in a hotel, a complex mixture of products and services. The nature of the service act means that operations have a number of particular characteristics that affect their management. It is possible to apply a range of different perspectives to service management focussing on the **encounter** itself, the management of the operation, marketing and the role of services in an economy.

Services are often described in contrast to goods. While the latter are tangible items that can be created and then sold or used later, a service is intangible and perishable. It is usually created and consumed simultaneously. Services can be grouped into five areas: business services such as consulting, finance and banking; trade services such as retailing maintenance and repair; infrastructure services such as communications and transport; social/personal services such as restaurants, hotels and health care; and public services such as education and government. It is possible to distinguish between goods and services by identifying the percentage of the price that represents the cost of physical goods in the purchase. Operations will then range on a continuum from high goods content purchases such as self-service groceries to high service content purchases such as hotels or consultancy. Placing businesses along this continuum soon reveals that there is not a clearcut distinction and that the majority of purchases actually represent a package of goods and services.

A service package consists of four interrelated elements that must provide a consistent **image** to the customer. First, facilitating goods are the materials purchased or consumed by the buyer or the physical items that an operation uses during service delivery. For a **restaurant**, this would include the food and drink consumed by the customer as well as the cutlery, crockery, china and linen that it uses during service. Second, explicit services are the readily observable or sensual benefits that the operation delivers. For a restaurant, these could include not only the presentation, aroma and taste of the food but also the speed of delivery, the accuracy of the order, the menu range and so on. Third, implicit services are psychological benefits that derive to the customer from using a particular service. For a restaurant, this would include the aesthetic appeal of the restaurant decor, the feeling of care provided by the treatment from the service staff, the status derived from visiting a prestigious operation and the like. Fourth, supporting facility is the physical **environment** that must be in place before a service can be offered. For a restaurant, this would include the physical structure of the building, the internal decor including all the furnishings and fittings both front and back of house as well as the car park and external landscaping.

The special characteristics of service operations mean that their management should be treated differently from that of other production systems. These characteristics have been described by many authors under slightly different terminology, but four are particularly common. The first characteristic is intangibility. Services are often described as being intangible, or at least a lot less tangible than goods. While the purchaser of a product has the opportunity to see, touch, hear, smell or taste it, this is not the case for pure services, which may be better described as ideas or concepts rather than things. The tourism sector, however, does not consist solely of pure services but of hybrids that combine **product** and **service** features. **Hospitality** operations, for example, do not just consist of the service performance and the intangible factors that affect this interaction. A large part of hospitality consists of tangible product elements. On the product side, there are the tangible elements of the **food**, drink or **accommodation** itself, but there are also the intangible elements of the built environment. On the service side, there are the intangible elements of the friendliness or care offered by the hospitality provider, but at the same time, it is possible to identify tangible service elements such as the time taken to deliver the service or the effectiveness of the service performed.

The second characteristic is heterogeneity. While manufacturing operations may be able to guarantee the consistency of the output from the production process, there is a great deal of variability in the output from a single service operation or indeed from a single service employee, largely brought about by the combination of the intangible nature of the service and the presence of the customer at the point of production. It is difficult to ensure consistent quality from the same employee from day to day, and harder still to get compatibility between employees; yet this will crucially affect the nature of the service that the customer receives. Although a customer may expect some variability in the service offered, the range of tolerance on the product side seems to be much lower.

The third characteristic is simultaneity. Services cannot be moved through distribution channels. Customers usually must come to the service facility or the provider must be brought to the customer, as in the case of home delivery pizza. A service business can only operate a limited geographical area, and while a customer may travel many thousands of miles to a particular tourism destination, the location of a hotel within a resort may be crucial. For the same reason, services cannot be counted, measured, inspected, tested or verified before sale for subsequent delivery to the customer, and this places a premium on the **quality** assurance function of the operation.

The fourth characteristic is perishability. Services cannot be stored, so removing the buffer of an inventory can be used to cope with fluctuations in customer demand. An airplane seat is a very perishable product. Empty places cannot be stockpiled for a busy day sometime in the future. Once a seat has been left empty, the potential revenue from the occupation of that space is lost. If the demand for tourism services did not fluctuate, then it would be possible to set the capacity of the service at the level of demand, and few scheduling or out-of-stock situations would occur. Unfortunately, tourism and hospitality services face dramatic and volatile changes in demand on an annual, daily and even hourly basis, and managing capacity becomes one of the major preoccupations of management.

Services can be considered from a variety of angles, each of which has its own distinct focus, with its own literature and research base. First, the service economy theme focuses on the role of service industries within the economy. Key concerns are definitions of service industries and service occupations, the changing role and importance of services within developing and developed economies and the impact of service productivity on the economy as a whole. As tourism represents over 10 per cent of the world's GDP, it has a major part to play in the global economy. Second, the service encounter theme focuses on the interaction between the customer and the service provider and the ways in which this can best be managed. Service encounters play a significant part in the customer's overall perception of the quality of the tourism experience, and yet are probably the most difficult and variable elements of the package the customer receives. Topics here would include considering the service encounter from a sociopsychological perspective, the role of boundary-spanning employees, queuing, communication and control, and the detail of individual verbal and non-verbal exchanges. Third, the service operations theme focuses on explaining the ways in which managing a service operation is different from managing a manufacturing operation and, over the past twenty-five years, has become a major branch of operations management. Fourth, in a similar way to the operations perspective above, the service marketing theme focuses on development from the mainstream of marketing; service marketing is now considered to be a significant area of study in its own right. Finally, the service management theme focuses on providing an integrating framework that can provide a link between all disciplines as they are applied in the service context.

Tourism is a complex amalgam of a range of different service operations that link together to provide the complete experience. Understanding the ways in which service is delivered and the ways in which service operations are managed are of fundamental importance to ensuring the continuing success of the tourism industry.

See also: service quality; service delivery system

Further reading

Fitzsimmons, J.A., and Fitzsimmons, M.J. (1994) *Service Management for Competitive Advantage*, New York: McGraw-Hill. (One of a growing number of texts on service operations that is easily accessible but well grounded in theory.)

Glynn, W.J. and Barnes, J.G. (eds) (1985) *Understanding Services Management*, Chichester: Wiley. (A collection of chapters from eminent writers in the field of service management.)

Johnston, R. (1994) 'Operations from factory to service management' *International Journal of Service Industry Management*, 5(1): 49–63. (Traces the development of operations management and the emergence of interest in service operations management.)

Silvestro, R., Fitzgerald, L., Johnston, R. and Voss, C. (1992) 'Towards a classification of service processes' *International Journal of Service Industry Management* 3(3): 62–75. (Examines the manufacturing process typology and, following a critique of existing service process types, proposes three service archetypes.)

ANDREW LOCKWOOD, UK

service delivery system

A unique aspect of any service industry, to which tourism belongs, is the service delivery system. This term covers those aspects of the operation which take place at the time that a customer or client comes to the service point and requests a specific **service**. The delivery system is designed to provide the customer with the required variety of services and with the minimum delay. A common feature of delivery systems is the customer line or queue. A key aspect of any system is to control the length of queues.

Hospitality and **tourism** services, among others, can be differentiated on the basis of variety offered and the volume of customers processed. This can vary from a high-volume operation offering a very restricted range of products and services to a lower volume operation where every customer receives a unique or highly customised service. Another key variable in any service operation is the nature of the service contact. High contact has an extra level of social interaction associated with staff, while a low-contact service requires the staff to perform largely technical skills. Each requires a different approach, with a high level of contact requiring sophisticated social skills training, together with a good level of **product** knowledge, whereas a low level of contact requires simpler customer care and technical training. A low-contact service may also be suitable for substitution by information technology or self-service.

There are a number of different ways of organising service delivery systems, depending on the mix of products and services on offer. In a pure service, the process can only begin when the customer places an order. Where there is a mix of products and services, the customer order may initiate production or stocks of finished materials can be held awaiting the placing of an order.

See also: commissary system; fast food

Further reading

Levitt, T. (1976) 'The industrialisation of service', *Harvard Business Review* 54: 63–74.

Sasser, W.E. and Fulmer, W.E. (1990) 'Creating personalised service delivery systems', in D.E. Bowen, R.B. Chase and T.G. Cummings (eds), *Service Management Effectiveness*, San Francisco: Jossey Bass.

DAVID KIRK, UK

service quality

Service quality is a measure of how well the **service** delivered meets customer expectations, resulting from comparing these with the actual performance on both the outcome and the process dimensions of the service. From the provider's perspective, delivering service **quality** means conforming to or exceeding these expectations consistently.

See also: management; service delivery system

ANDREW LOCKWOOD, UK

sex tourism

Sex and tourism are linked in several ways, and sex tourism is one variant within a range of possible relationships. It can be defined through a study of **motivation** or by focusing on the organisations involved. When sex tourism is defined as holiday making where the primary motive is to experience relatively short-term sexual encounters, it is linked to **prostitution**, here considered as a commercial and short-term transaction involving the explicit provision of sexual services in return for payment, in cash or kind. However, much sexual interaction occurs among tourists themselves, particularly the young and 'single'. Holiday makers also meet locals from the 'host' society, especially men, at entertainment centres, and temporary liaisons are common. These need not reflect equality, for young men in some societies may regard conquests over tourists as personal (but temporary) recompense for real or imagined political oppression. The prognosis for such relationships is usually poor.

When affluent tourists visit developing countries, interaction between them and their hosts is rarely on equal terms. From the earliest days of **colonisation**, stereotypes of colonised peoples – and their sexual proclivities – were widespread in the West and are still expressed in tourism promotional literature. Colonialism (and/or the presence of armed forces from overseas, in wartime or during periods of 'rest and recreation') also led many colonised people to believe all white visitors were wealthy, a view reinforced by the growth of international tourism. Such stereotypes contributed to the **commoditisation** of many aspects of human relations, which was already occurring on an unprecedented scale.

Just as 'true love' has its material ramifications, the commoditisation of sexual relations is rarely total. Many men and women in developing countries may quite literally 'wait' for a spouse from overseas to provide them with a passport to a more developed society, but such relationships are not necessarily totally instrumental. Partners need not be the same age, and studies from the Gambia and the Caribbean indicate a deliberate search by middle-aged white women for younger black men, to be either temporary holiday lovers or permanent partners in their home country. Female prostitution in Southeast Asia has its equivalents elsewhere. In such circumstances, the distinction between the exploiter and exploited is unclear, and we must differentiate between individual and institutionalised **power** relationships.

Although sex workers in cities of the developed world also target tourists, in circumstances relatively little publicised in the literature, the links between tourism and prostitution are most notorious in Southeast Asia. Even here, though, other factors (including the Vietnam War, a perceived need for the tourism dollar and culturally specific views of the subservience of women) must be considered. Such caveats should discourage simplistic assertions that 'tourism causes prostitution' or women 'are forced by poverty' to sell their bodies. Even in Thailand, where 'sex tourism' is highly commoditised, studies demonstrate considerable room for ambiguity in the relationships of female prostitutes with their *farang* clients. But prostitution is undoubtedly extensive. In Thailand in 1988 there were up to one million working adult female prostitutes, mostly catering for Thai clients, and more with previous involvement in the industry. The sex industry is also growing in South Korea and the Philippines. This is a matter of both moral debate and practical concern, given the spread of AIDS. Although tourism does not cause prostitution, sex tourism undoubtedly exacerbates the situation.

According to this perspective, this is not a form of 'special interest' tourism but an initially instrumental relationship of **host and guest**, changeable over time and place and often occurring within the same organisational form and **infrastructure** as other types of tourism. However, if one brackets the problematic issue of tourists' motives, the nature of prostitution in such places as Thailand might be considered an extreme version of 'sex tourism' for two reasons: first, it is often quite specifically organised, across national boundaries, to bring in self-selected parties of (in this case) male clients for the primary purpose of meeting male or female prostitutes, and second, there is an increasingly organised trade in child prostitution (allegedly to reduce the risk of contracting AIDS), which has prompted universal condemnation. By shifting attention from the morality of adult prostitution to the institutional arrangements through which prostitution is perpetuated, the possibilities of control become clearer. Indeed, this has been the approach of several prominent organisations opposed to child prostitution, including the **Ecumenical Coalition on Third World Tourism** and the Campaign to End Child Prostitution in the Third World, with support from such organisations as Tourism Concern in the United Kingdom and elsewhere. By the end of 1996, more than a dozen countries had passed legislation allowing for the prosecution of nationals for offences committed against children while overseas. The details vary from country to country and, because of the practicalities, prosecutions are likely to be few. Nevertheless, such legislation may reinforce public opinion and prompt stern and consistent action against offenders in countries where offences occur, which is ultimately where the remedies lie. It is in the long-term interests of the tourism industry that such attempts should succeed.

Further reading

Black, M. (1995) *In the Twilight Zone*, Geneva: International Labour Office. (A study of child workers in the hotel, tourism and catering industry.)

Cohen, E. (1993) 'Open-ended prostitution as a skilful game of luck: opportunity, risk and security among tourist-oriented prostitutes in a Bangkok soi', in M. Hitchcock, V.T. King and M. Parnwell (eds) *Tourism in South-East Asia*, London: Routledge, 155–78 (An account of a group of prostitutes working from bars and coffee shops in Bangkok.)

Hall, C.M. (1992) 'Sex tourism in South-East Asia,' in D. Harrison (ed.) *Tourism and the Less Developed Countries*, Chichester: Wiley, 64–74. (A review of the evolution of tourism-oriented prostitution in the region.)

Meyer, W. (1988) *Beyond the Mask*, Saarbrücken: Verlag Breitenbach. (A detailed account of prostitution in Thailand.)

Truong, Thanh-dam (1990) *Sex, Money and Morality*, London: Zed Books. (A study of prostitution and tourism in Southeast Asia.)

DAVID HARRISON, UK

Seychelles

Tourism in the Indian Ocean microstate of the Seychelles is a tightly controlled, internationally focused, up-market industry with policies to protect extensive marine parks and nature reserves (40 per cent of the total land area), distribute tourists throughout the main islands, and prevent over-development (there is a ceiling of 4,500 beds). Numbers have oscillated around 100,000 annually since 1990. The industry contributes about 50 per cent of GDP. Cruise tourism was a new growth area in the mid-1990s.

DAVID WILSON, UK

shopping

The growing importance of retailing in **destination** areas and resorts is a reflection of the role of shopping in the tourist **experience**. In terms of **time** budgets and expenditures, shopping can now be considered as an important tourist activity. The actual economic benefits have long been underestimated. Although shopping is not commonly mentioned as a prime motive or a key factor in the destination choice, recent surveys of tourist **behaviour** patterns indicate the actual time budgets spent on shopping. The rapid expansion of the city trip market and **short breaks** strongly supports the tourismification of the retail sector.

The importance and success of catering to tourists' shopping proclivities is manifest in tourism shopping villages, urban shopping areas in destinations and resorts, museum shops and, not least, in retail trade development in amusement parks and other attractions. Potential tourists are now being offered arrangements for shopping trips in which the shopping opportunities are promoted as the core product. Examples are Christmas shopping trips to London or New York and tours by coach or train to visit the German Christmas markets.

The attraction of tax-free shopping in airports (such as London, Amsterdam, Frankfurt, Dubai, Hongkong or Bangkok) and on international ferryboats persists, despite the decreasing competitive advantages. Cross-border shopping is appealing to many, and generates important flows of consumers. Mega-shopping malls such as West Edmonton Mall (Canada), the Mall of America (USA) and Centro Oberhausen (Germany) are becoming important destinations even beyond the domestic tourism market.

Tourists' propensity to shop varies according to their cultural background and the range and nature of shopping opportunities in the destination area. Japanese tourists are frequently catered for, as they have a reputation of being big spenders. This aspect of behaviour strongly supports the development of the souvenir industry. The production and marketing of craft **souvenirs** is a global response to this tourism market, and raises questions about **authenticity** and cultural convergence.

Tourism retailing offers interesting prospects for many locations, in particular the 'exotic' places and for small business enterprises, but also brings a serious risk of **standardisation**. The theming of festival markets, the upgrading of street markets, the redevelopment plans for **heritage** sites and urban waterfront areas to become the carriers of **urban tourism** all are based on very similar concepts of the **product** mix. In order to survive the trend towards convergence, there is a need to safeguard the competitive advantages of uniqueness by emphasising cultural links with the host community and its traditions. This applies to the product mix as well as to the characteristics of the shopping environment and its functional integration into the local tourism system.

Further reading

Anderson, L.F. and Littrell, M.A. (1995) 'Souvenir purchase behaviour of women tourists', *Annals of Tourism Research* 22(2): 328–48.

Butler, R.W. (1991) 'West Edmonton Mall as a tourist attraction', *The Canadian Geographer* 35(3): 287–95.

Getz, D. (1993) 'Tourist shopping villages: development and planning strategies', *Tourism Management*: 15–26.

Kent, W.E., Shock, P.J. and Snow, R.E. (1983) 'Shopping: tourism's unsung hero(in)', *Journal of Travel Research* 21(4): 2–4.

Jansen-Verbeke, M. (1994) 'The synergy between shopping and tourism: the Japanese experience', in W. Theobald (ed.), *Global Tourism: The Next Decade*, Oxford: Butterworth Heinemann, 347–62.

MYRIAM JANSEN-VERBEKE, BELGIUM

shrine

A shrine is an object or place sacred to a **religion**. The term may also refer to a place of national or patriotic importance. A shrine originally meant a box or chest that contained holy objects. It later came to mean the place where such a container is kept. It is also any structure built on a place considered holy because some significant religious event happened there. In addition, shrines may be built to honour a saint or a virtue. In many countries in the world, there are national shrines built to honour the memory of national heroes or battles for freedom. The best known monument of this type is the Tomb of the Unknown Soldier, created after the First World War by the governments of the Allied countries to honour the memory of many unidentified soldiers killed in battles.

Shrines are organised in a similar way in most places. Apart of the main sacred place within the shrine, there is very often a bustling town, a more or less highly developed commercial centre where visitors reside, relax and shop. The shops sell everything that a **tourist** or **pilgrim** would need, as well as all kinds of religious souvenirs and other goods. The shrines become attractions and thus the reason for large numbers of tourists, believers and non-believers alike, sacred or secular, who are drawn to them.

For religious persons, shrines are places where many miracles are said to have been performed. Hundreds of thousands of ill and disabled men and women have made pilgrimages to such shrines; many have left their crutches in the church as tokens of their healing. These visitors to shrines usually distinguish themselves from tourists, who are seen as not deeply interested in the religious meanings and rituals taking place within the shrine. But at night, for example, little difference can be observed between pilgrims and tourists in the way they relax. Moreover, in many of the popular shrines tourists are presented to the essential meaning of the shrine.

BORIS VUKONIĆ, CROATIA

sight

Places and structures which are considered worth seeing are known as sights. Natural sights such as beaches, lakes, forests, rivers, waterfalls and mountains are visited for their beauty. Cultural/historic sights and **shrines** are mostly buildings and monuments, the results of human culture and history at work, and are appreciated as symbols of mankind's greatness. Built sights such as **theme parks** are mostly created for entertainment and commercial value.

HUBERT B. VAN HOOF, USA

sightseeing

The term sightseeing refers to the act of visiting a location or locations during the course of a journey or as the point of the journey to look at objects. These objects will range in type from works of art and buildings, including national monuments as well as those of architectural interest, to people involved in some form of performance or display about their culture, present or historical, or in the conduct of their daily lives. In this latter respect, which can also apply to the static matters, the objects involved in sightseeing can be both part of the official tourist itinerary and even outside of it. Regardless of the situation, what ties them all together is that they are seen as 'representations' of the destinations, the peoples of the destinations, and thus of the touristic experience itself. The majority of the objects that are displayed or become the focus of tourist attention do so because they are listed in **guidebooks**, **itineraries**, or mediated to the tourist through the likes of tour operator representatives.

'To do a bit of sightseeing' is an integral part of a holiday for the mass tourist populations in the **sun, sand, sea and sex** category. However, there has not always been an association between travel/tourism and sightseeing. As Adler (1989) points out the sight has been conceived in different ways during the development of tourism, and the importance of visualisation is also linked to changes in philosophical and cultural conditions in Europe. In the very early days of travel more emphasis was given to engaging in discourse by word than the need to look at objects.

In more recent times, the emphasis on sightseeing can be connected with the idea of legitimacy of the experience especially in the idea of going to see for oneself. Then to capture that on film (static or mobile) enables the experience to be revisited again. If individuals or groups are also incorporated they too become part of the spectacle. Seeing for oneself and becoming part of the spectacle serves to legitimise the experience. The importance of sightseeing or 'gazing' (see **gaze**) on spectacles, performances or objects has according to John Urry (1990, 1995) become a central, if not the central aspect of the act of tourism itself. Engaging in the act of the gaze effectively positions the tourist as consumer of the things they gaze upon.

References

Adler, J. (1989) 'The origins of sightseeing', *Annals of Tourism Research* 16: 7–29.

Urry, J. (1990) *The Tourist Gaze: Leisure and Travel in Contemporary Societies*, London: Sage.

—— (1995) *Consuming Places*, London: Routledge.

HAZEL ANDREWS, UK

sign

Semiotics, the science of the life of signs in society, defines a sign as the product of a symbol (or signifier) – anything from a word, sound or physical gesture to a physical object – and the associative meaning it carries within a shared cultural universe (its 'signified'). For a **tourist**, for example, a beach might denote (indicate, imply, mean) **freedom** and pleasure, while for a former soldier active in Normandy in 1944, a beach might appear as a symbol denoting war, death and victory. Either way, when it is linked to a meaning, a beach is a sign. Language as a whole is constructed from such denotative relations between words, things and significations.

Moreover, at a higher level, there is a second order of meanings into which signs are routinely woven into the more complex textures of everyday life. This is the level which anthropologists and semiologists refer to as the mythological and/or **ritual**. Here, signs connote more subtle, covert and powerful ideas and messages. The advertising industry depends upon such connotative processes. Thus, for example, designer luggage or sports equipment are likely to be understood as signs connoting the lifestyles of the rich and famous who have the world at their fingertips. As such, they would be expected to exert powerful influences on readers and viewers exposed to them in magazines and other visual **media**.

Tourism has its own life of signs, the study of which is gathering momentum. It is a field which repays study, partly because the signs which form its subject matter inhabit overlapping worlds. Thus, for example, the **interpretation** by tourists of signs for **paradise** – tropical islands, verdant landscapes, fertile gardens and so on – are likely to be shaped to some extent by references to religious ideas and values, while signs of **freedom** – images, for example, of quantities of nubile bodies and/or mountains of food (both connoting unfettered choice, a fashionable contemporary version of the notion of freedom) – clearly have roots and references which are at once political and economic as well as commercial. Recent studies of the **heritage** industry have concentrated on the ways in which the **interpretation** of historical signs is shaped and moulded not only by the political and cultural fashions of the time, but by the social positions, dispositions and intentions of the persons or groups doing the interpreting. Such studies have important implications for the understanding of such phenomena as nationalism, racism and identity formation. More generally, studies of tourism and its signs contribute uniquely valuable insights into the nature of contemporary culture.

Further reading

Crick, M. (1989) 'Representations of sun, sex, sights, savings, and servility: international tourism in the social sciences', *Annual Review of Anthropology* 18: 307–44.

Dann, G. (1996) *The Language of Tourism*, Oxford: CAB International.

Hawkes, T. (1977) *Structuralism and Semiotics*, London: Methuen.
Leach, E. (1976) *Culture and Communication*, Cambridge: Cambridge University Press.
Lowenthal, D. (1998) *The Heritage Crusade and the Spoils of History*, Cambridge: Cambridge University Press.
Selwyn, T. (1996) *The Tourist Image: Myths and Myth Making in Tourism*, Chichester: Wiley.

TOM SELWYN, UK

significant omission

Just as the pictorial and verbal content of advertisements leads to expected **behaviour**, so too can important items which are missing. Thus photographs of **resorts** bereft of children send a controlling message that, for optimal pleasure, tourists should leave their offspring at home. Non-target groups may also be omitted (such as fat people or minorities), along with the undesirable aspects of a **destination** or **heritage** site.

GRAHAM M.S. DANN, UK

Singapore

With a population of about 3.1 million, **tourist** arrivals in Singapore reached 7.3 million in 1996. In the same year, exports of travel services reached $11.1 billion, or about 5 per cent of total exports of goods and services. The maturity of Singapore as a **destination**, the strong Singapore dollar and rising competition from neighbouring countries imply very stiff **market** conditions for the city-state. Thus, it has repositioned itself to be a tourism capital so as to tap into its pivotal role in the **region** and capitalise on its international business hub network. To become such a capital, the Singapore Tourism Promotion Board has defined a 'Tourism Unlimited' strategy, which reflects the traditional role as well as new concepts of a tourism business centre and hub under the 'New Asia–Singapore' label. It is repackaging traditional attractions to cater to more sophisticated and demanding tourists under a number of themes. These include 'Ethnic Singapore', 'Mall of Singapore', 'The Night Zone', 'Nature Trail', 'Singapore Heartland', 'Theatre Walk' and 'Rustic Charm'.

An all-round strategy includes building a tourism industry cluster, configuring new tourism space in the region, twinning and packaging complementary projects with partners for success, and more. Besides marshalling a multi-agency approach involving the public and private sectors within Singapore, international companies of world-class standing are encouraged to set up their overseas headquarters in Singapore. The government is investing in superior **infrastructure** to encourage activities related to meetings, incentive travel, conventions and exhibitions. However, obvious constraints to 'Tourism Unlimited' lie in limitations in physical size and labour. Information technology has been used to improve productivity and enhance a global information network. A pool of competent managers and tourism manpower has to be nurtured to inculcate professionalism and sustainability in the industry. Niche areas include catering to older tourists or single Japanese females.

At the same time, any negative sociopolitical or environmental impacts need to be minimised, especially when Singaporeans in their land-scarce country share many facilities and recreational amenities with tourists. Other countries in the region are also concerned with negative and sensitive cultural and environmental aspects of the industry, and Singapore works together with their efforts. Overall, the prospects for tourism in this country are good as Asians become more affluent and eager to travel. As a key sector, tourism is also actively promoted by the government and ties in well with overall development goals and strategies. There is a lot of scope and potential for economic cooperation in Asia Pacific in tourism.

LINDA LOW, SINGAPORE

site

A site is literally any location in space. Sites in tourism are usually individual attractions, whether indoor or outdoor, assembled to comprise a trip. Such site 'construction' may be performed by tourism intermediaries or, more often, by tourists themselves. Sites are frequently also **sights**, combined in the tourism activity of **sightseeing**.

G.J. ASHWORTH, THE NETHERLANDS

site analysis

A **site** is a specific real estate holding usually identified by an address. After a firm has chosen a **location** or trading area in which it wishes to locate its business, it must select a site from one or more alternatives available. This is the evaluation of one or more sites for the purpose of conducting a profitable business. Since changing sites is costly, this selection is an important medium to long term decision for most companies.

Site analysis is especially important in the case of **service** industries like tourism, because the customer must be present when and where the service is rendered. From the perspective of the consumer, good sites must be easy to find and access. Those which are located in well-known business areas or streets are more easily found than those off the main routes, but they must also be attractive and safe from **crime**. Good sites are highly visible (for example, street corners), or can be made so with signage. If customers arrive on foot, sites near subway or bus stops are preferable. Those arriving by private car require nearby parking with easy access and egress. They will often shop for several products or services in one trip to save time and money, or enhance their enjoyment. Consequently, sites located in areas with many merchants are often preferable to those which stand alone. Impulse sales are a direct benefit of locating in sites which maximise both visibility and traffic.

In addition to factors which influence consumer preference, sellers are concerned with operating, competitive and regulatory factors. A potential site must be large enough to accommodate longer term business goals, or allow for future expansion into adjacent space. Operating costs are much higher for sites that are not connected to public utilities, sewer and telephone communication lines. By-laws may restrict the use of a site, exterior or interior renovations and the size or type of signage permitted. Fire, health and handicap access regulations may vary from one site to another and contribute to renovation costs. Some firms locate near their competitors (for example, hotels) while others prefer to keep their distance. In general, the best sites command high rents or property values.

Site analysis can be done qualitatively through discussions and consensus. **Quantitative methods** are also available to assist in making optimal decisions, such as multiple linear regression approach. The Huff model uses the proximity of consumers to competing sites to select the best location. A relatively simple breakeven analysis can also be used to select the best site based on fixed and variable cost estimates, expected sales volume and profit targets.

Further reading

Huff, D.L. (1996) 'A programmed solution for approximating an optimal retail location', *Land Economics*, August: 293–303.

Kimes, S.E. and Fitzsimmons, J.A. (1990) 'Selecting profitable hotel sites at La Quinta Motor Inns', *Interfaces* 20(2): 12–20.

JOACHIM BARTH, CANADA

site, sacred

Almost all **religions** define certain places to be holy, and travel to them is often recommended or even obligatory. A holy place is a sacred building (church, mosque or temple) located at a certain **site**, or sometimes a religious statue, tomb or painting or some other religious object (such as the Kaaba in Islam). The sacred site can be defined as a religious nucleus in a greater holy area of religious significance with historic and/or artistic importance. Very often religious ceremonies or festivals are celebrated at these sites at a specific time or on specific occasions, and at those times the presence of **tourists** may be limited or banned. A distinction is made between shrines, holy places and sacred sites.

BORIS VUKONIĆ, CROATIA

sites, biblical

Biblical **sites** are those holy places named in the bible which became the sacred places for believers of two religions: Judaism and Christianity. Mostly located in Judea and Galilee, they have become the most popular **pilgrimage** destinations.

The bible, also known as the Holy Scripture, is the most sacred book of the Jewish and Christian religions. It is also the most widely read and influential book in history. Both religions consider it to be the Word of God, and they base their most important beliefs, ceremonies and holidays on it. The bible consists of the Old and New Testaments (Jews accept only the Old Testament, so it is known as the Hebrew Bible). The first generation of Christians preserved memories of Jesus's teachings, deeds, and Crucifixion largely by word of mouth. The New Testament was written to testify new views of Christian faith born during the second century AD. The church also wanted to preserve the authentic story of Jesus's life and death in writing for future generations of Christians.

The bible views God as the chief character of the events it describes, founding evidence of God's presence in the history of old Israel, the life of Jesus Christ, and the development of the early church. The bible refers to many events in the history of ancient Israel and the early church. Such events include the deeds of kings and the journeys of apostles. There are many historical events covered in the bible that represent the only or best information about a particular period. Therefore, many of the places mentioned in the Old Testament are also archaeological sites in the Middle East. These sites are often not directly connected to the Jewish or Christian religions, but bear evidence of, for instance, the Assyrian or Sumerian cultures.

From the oldest times, pilgrimages – that is, visiting the sacred places – became a vital part of the religious life. Places of pilgrimage are sites of apparitions, healings, miracles or omens. These were ordinary places in the **landscape**. Later, people added to these spots symbolic objects representing the divinity, altars for sacrifices and offerings, and then buildings, 'houses of gods' and shrines protecting the altars and statues. Pilgrimages to such holy places are simply a way for a believer to learn about the earthly life of Jesus and about the lives of the saints, with the aim of strengthening faith. Catholics, especially in the Middle Ages but also during many centuries until today, made pilgrimages in solemn procession to the Church of St Peter in Rome and St James in Santiago de Compostela in Spain.

The most popular biblical place visited by millions of people every year is Jerusalem, with the holiest sites of Jews, the Western Wall (also known as Wailing Wall), which was part of the holy Temple during biblical times; and of Christians, the Church of the Holy Sepulchre, which is believed to stand on the hill of Calvary (or Golgotha), where Jesus was crucified and buried. Jerusalem is also a holy place to Islam, and the city's holiest Muslim **shrine** is the Dome of the Rock. There are many other sites in Jerusalem and in other cities of old Palestine, which are the sites connected with the bible. One of them is the Rock. Jews believe that on this Rock (the same from which Muslims believe Muhammad rose to heaven with the angel Gabriel and spoke with God) Abraham, the leader of the ancient Hebrews, prepared to sacrifice his son, Isaac, at God's command. Another one is the Via Dolorosa (Way of Sorrows), which is believed to be the route over which Jesus carried His cross to Calvary. On Good Friday, a huge crowd of pilgrims retraces the route to the Church of the Holy Sepulchre which rises at the end of Via Dolorosa.

Other places connected with Jesus's birth and life, as named in the bible, are also among the most popular sites for believers to visit and gather. They include Mount Sinai (where Moses declared the Ten Commandments), Bethlehem (where Jesus was born), Nazareth (where Joseph, Mary's husband was raised and where they settled with Jesus after Herod died), Cana (where Jesus's first miracle took place), the Mount of Olives with the Gethsemane garden (where Jesus was arrested), the River Jordan (with the water from the river where Jesus was baptised), and many more.

BORIS VUKONIĆ, CROATIA

skiing

Skiing may well be more than 5,000 years old. The word 'ski' itself is from a Norwegian word referring to a snowshoe (*skilober*) that was used by the natives of Northern Europe. Skis dating back to 2500 BC have been found in the Altai Mountains of Siberia, and they bear startling resemblance to those in use today. One of the first written accounts of skiing appears in the sagas, the classic literature of the Viking period. Skiing was so much a part of Viking life that a god and goddess of skiing, Uller and Skada, were objects of worship. Skiing was used as a mode of travel and as a means to prove men's physical prowess, but it was not used in warfare until 1721, when a ski company was organised in the Norwegian army.

It was in the late 1830s that skiing as a sport first entered the picture, again in **Norway**. Speed skiing and ski jumping soon became popular. The spread of Nordic skiing to the rest of the world can be attributed to two major causes: the emigration of Norwegians and the publication of Fridtjof Nansen's book describing his 1888 expedition on skis across the southern tip of Greenland. Nansen's book popularised the idea of skiing; the Norwegian emigrants contributed their knowledge of the equipment and techniques. They first travelled to Germany in 1853, to Australia in 1855, to North America in 1856, to New Zealand in 1857 and to Switzerland in 1868. Most were in search of a better life through farming, forestry or mining, but they brought with them a recreational sport which has transformed the winter season throughout the world (see also **sport, recreational**).

Winter sports are one of the fastest growing tourism sectors in the world, and their growth during the last hundred years has been remarkable. (Skiing originated as Nordic (cross-country) skiing. Then, following its introduction into central Europe, it became an almost exclusively alpine (downhill) sport. Today, skiing like many other forms of **recreation** is feeling the influence of new **technology**. Snowboarding (which uses one larger more elliptical board which holds both feet) has become the most recent fad of the young and agile. A new invention called the 'ski key' lets skiers keep track of their vertical miles skied in order to win prizes, and even alerts them to phone messages from their office.

Further reading

Gamma, K. (1983) *The Handbook of Skiing*, New York: Alfred Knopf.

Rosenberg, D. (1995) 'High tech skiing' *Newsweek* 125(3).

Scharff, R. (1978) *Encyclopedia of Skiing*, New York: Harper and Row.

CAROLYN M. DAUGHERTY, USA

Slovenia

Slovenia, a central European country on the Adriatic Sea, has a population of 2 million and an area of 20,253 square km. Its proclamation of independence in 1991 caused a brief war against the Yugoslav Army. Slovenian inbound tourism was ruined, more by the negative image of war than by real war horrors. Slovenian spa, mountain and sea destinations and cities are now promoted as the Green Peace of Europe.

TANJA MIHALIČ, SLOVENIA

small business management

The main preconditions for small business **management** result from the dependency on a specific micro-social and micro-economic environment, the dominant role of an entrepreneur, the influence of the **life cycle** phase into which the firm has developed, and the management methods which are being used. These four interdependent variables form a configuration which characterises the concrete strategic position of an individual firm. Small business management is designed to analyse the concrete configuration of variables, to elaborate goals for further development and to adopt measures to move the firm towards a new configuration.

The micro-social environment of small businesses – which is almost universally typical to most tourism operations, including hotels, **restaurants** and **travel agencies** – is characterised by close relationships to family members (in the case of a family business), friends and stakeholders. Entrepreneurs and representatives of their environment (business partners as well as local authorities) often form loosely coupled systems which can at least partly compensate the asymmetry of information, **power** and **risk** for small firms on their markets. Small business owners are not necessarily entrepreneurial. However, **entrepreneurship** is assumed to form a decisive factor of success for small firms, characterised by traits such as the need for achievement, internal locus of control, risk-taking propensity and the like.

Most firms are founded as small firms, and develop later through several phases of a life cycle such as growth, decline, in-depth **innovation** of **technology**, products, markets and succession, or close down. Strategic small business management has to identify the specific characteristics of each stage of development and the optional paths for further development. General management methods require specific adaptations for small firms. Management styles and tools need to fit the individual configuration which is made up by the environment, the personality and knowledge base of the entrepreneur, the stage of development of the organisation (taking into account limited **resources**) and the level of management techniques which has already been achieved. Strategic goals in small firms tend to emerge according to the characteristics of a particular configuration rather than being designed following the results of a **SWOT analysis**. Operative management in small firms is based on more direct communication and less bureaucracy.

Further reading

Landstroem, H., Frank, H. Veciana, J. (ed.) (1997) *Entrepreneurship and Small Business Research in Europe: An ECSB Survey*, Aldershot: Avebury.

Storey, D.J. (1994) *Understanding the Small Business Sector*, London: Routledge.

JOSEPH MUGLER, AUSTRIA

smuggling

Smuggling is the illegal cross-border trade that occurs through official and unregulated locations. Tourism's role in smuggling is rarely studied despite numerous reports of this activity by tourists, who sometimes falsify the nature, value and use of what is being traded across official entry or exit points. For example, European rules permitting the free movement of goods across frontiers have enabled tourists to import legally large amounts of **alcohol**, ostensibly for personal consumption but actually to resell in their home countries. Regardless of its nature and place, this action by tourists occurs as the result of state regulations that create an excess of supply or demand in the **market**. Therefore, smuggling is as diverse as economic life and not only includes drugs, endangered species, prostitutes, babies, precious metals and antiquities, but also extends to other valuable goods.

Tourism is often associated with the **commoditisation** of the distant past. Antiquities are souvenirs that embody the **experience** of that past. Collecting them is now a booming business that has grown in response to the escalating value of priceless artefacts. Poorly guarded museums in less-developed countries have become easy targets for those seeking to sell **relics** to collectors posing as tourists. International investors touring these countries have developed sophisticated trade networks by financing national looters of treasured artefacts and smuggling them to either Europe or North America.

Exchange policies in many developing countries also favour smuggling activities related to the tourism industry. Increased foreign exchange receipts from tourism put upward pressures on the exchange rate. If the national currency is overvalued, a tourist can gain a margin of profit by exchanging hard currencies in the parallel **markets** dominated by smugglers and speculators. Under flexible exchange rates, upward pressures from tourism may translate into increased imports and reduced legal exports from non-tourist sectors of the economy. Smuggling exports make up for the decline in official exports. Holding foreign currency provides freedom of trade to those unwilling to be dependent on the vagaries of most national currencies in developing countries.

Further reading

Bhagwati, J. (1981) 'Alternative theories of illegal trade: economic consequences and statistical detection', *Weltwirtschaftliches Archiv* 117: 409–26. (Classic discussion of different types of smuggling and their links with social welfare.)

KISANGANI N.F. EMIZET, USA

social control

Since tourism is such a massive and rapidly expanding phenomenon, it is necessary for the industry, **destination** authorities and host populations to exert control over its negative sociocultural and environmental **impacts**. This constraint is effected by placing limits on touristic behaviour through the language of brochures, travelogues, notices, posters, **guidebooks** and so on, while at the same time trying to preserve individual **freedom**.

GRAHAM M.S. DANN, UK

social interaction

Social interaction is commonly identified as a tourism **motivation**. Early research on the latter incorporated the underlying hypothesis that an individual has a need for love and affection and the desire to communicate with others. Further explorations revealed enhancement of kinship and friendship relationships and the facilitation of varied and increased social interaction as motives which directed vacation choice and **behaviour**.

It is important to recognise the different levels of the need for social interaction while away from home. At one level is the more intimate need for interaction with family, perhaps in an attempt to develop togetherness; at another level is general socialisation with others for reasons such as the development of new friendships or relationships with the opposite sex; and at still another is the fulfilment of more educational goals and needs expressed through the desire to interact with the host **destination**. Explorations of the **novelty** dimension in motivation research emphasise that some tourists (such as drifters and more recently young budget groups) prefer to seek excitement and novelty by engaging in direct contact with a wide variety of new and different people, while others just prefer to observe the lives of **local** people.

Another important consideration is that social interaction occurs within a cultural setting and culture – as a reflection of a whole way of life – affects social behaviours directly and indirectly. People often have difficulties when they move across cultures, and the growth in tourism means that increasing numbers of them will have extensive social interaction in cultures other than their own, pointing to an increased need for cross-cultural studies. The analysis and management of tourists' social experiences is important, since they are often discussed when tourists meet each other or return home; negative experiences are often retold, thereby reinforcing negative images of the host country. While much research in the field of social **psychology** has concentrated on understanding the processes and structural characteristics of social interaction, tourism research has been primarily limited to its identification as a tourism motive.

Further reading

Argyle, M., Furnham, A. and Graham, J.A. (1981) *Social Situations*, Cambridge: Press Syndicate, University of Cambridge. (Discusses the analysis of social situations.)

Dann, G. (1977) 'Anomie, ego enhancement and tourism', *Annals of Tourism Research* 4(4): 185–194. (Explores tourism motivation.)

Pearce, P.L. (1990) 'Farm tourism in New Zealand: a social situation analysis', *Annals of Tourism Research* 17(3): 337–52. (Applies social situation analysis to a tourism setting.)

LAURIE MURPHY, AUSTRALIA

social recreation

Leisure activities of short duration designed to promote **socialisation** in a casual, non-competitive environment are regarded as social **recreation**. Four commonly identified types are parties, clubs, eating events (picnics, banquets and barbecues) and visiting. Where activities have other purposes, such as competition in children's party games, the activity would be considered social recreation if the socialisation factor is considered to be the most important.

MICHAEL A. BLAZEY, USA

social relations

Social relations between **tourist** and host, tourist and tourist, and tourist and representatives of the industry have been found to be one of the more important determinants for **satisfaction** with holidays. Reasons for social relations include the desire to meet new people and discover new things about their way of life, the need for status and ego enhancement, and a desire to relax with 'interesting' people. However, resultant satisfaction from such meetings will depend upon the tourist types, the need for interaction with others, the context where the social contact occurs, and the nature of the social interaction that happens. Additionally, the presence of others for this relationship is not in itself a determinant of satisfaction; those relations have to be enjoyable. Likewise, the very existence of others with whom it is possible to interact might be a source of dissatisfaction for the tourist who does not want company. Some may wish for isolation in natural places. On the other hand, given that any tourist maintains social skills, it is possible that even in **wilderness** settings a chance meeting with a like-minded person might create a positive **experience** from the social interaction. The relationship between social relationships and positive holiday experience can be described many ways. For example, if the tourist desires social contact, and it is present and found to be rewarding, then the tourist will be satisfied. If others are not present, the tourist will not be satisfied. There is also a position where a social relationship is sought and is present but is found unrewarding, thereby creating dissatisfaction.

In host–tourist relationships, the determinants of social relations within situations can be identified as the numbers of tourists, the structure within which meetings take place (scripted or unscripted occasions), host–guest ratios, the **motivations** of both hosts and guests, culture gaps, nature of **location** (urban or rural) and the destination's life-stage. One theory, that of the Irritation Index, argues that initially tourists are welcomed, but as they grow in numbers and the destination grows in size, host communities may at best only tolerate tourists. At the end, it can be noted that social relations may be positive or negative, may be fragmented in the sense that not all members of the host community may be involved in host–tourist encounters, and may be of social significance in playing a role in creating demonstration and **acculturation** effects.

See also: demonstration effect; social interaction; social situation

Further reading

Doxey, G.V. (1975) *A Causation Theory of Visitor–Resident Irritants: Methodology and Research Inference*, paper given at San Diego, California, The Travel Research Association Conference no 6, TTRA, 195–8.

Ryan, C. (1991) *Recreational Tourism: A Social Science Perspective*, London: Routledge.

CHRIS RYAN, NEW ZEALAND

social situation

A generic term, social situations describes fully the interaction of people in routine ways in specific settings. They possess nine features: goals, **rules**, **roles**, repertoire of elements, sequences of **behaviour**, concepts, environmental settings, language and speech, and skills and difficulties. Applications in tourism include host–guest, tourist–guide and backpackers' interactions.

LAURIE MURPHY, AUSTRALIA

social tourism

Social tourism is an extremely diverse and complex phenomenon, and its meanings vary depending on the time periods and countries under discussion. A partial list of terms which may be synonymous with social tourism gives a good idea of the breadth and complexity of social tourism: it may include tourism for workers, for families, associations (often employees of one organisation), for personal development, associated with a vocation, to promote social cohesion, on a non-profit basis, or in the public interest. Social tourism is usually defined in terms of the objectives pursued, the methods employed for achieving them and the outcomes of participation.

The following criteria are identified as being among the most fundamental underpinnings of this concept and practice. First, social tourism recognises the basic right of all, irrespective of their social, financial or geographical situation, to have leave from work and to have vacations. It reflects concerns for the availability and accessibility of vacations at reasonable, affordable prices. Second, it acknowledges the importance of leisure and holidays as exceptional occasions in the physical and cultural development of individuals, promoting their **socialisation** and their integration into their **community** of workers as well as the broader society. Third, it reflects the concern of its proponents that social tourism should be an instrument of economic development, a means of managing and enhancing the national territory and, at the same time, of preserving the natural and human environments of **destination** regions. Today, these objectives are often called lasting or sustainable development.

In countries with a high standard of living, the history of social tourism can be divided into three periods: a period of emergence after the Second World War, the two decades of very strong growth in the 1960s and 1970s, and a period of contraction of opportunities and participation, and of questioning, re-evaluation and redefinition of objectives and basic concepts. This led to the 1996 'Declaration of Montreal' at the Congress of the Bureau International du Tourisme Social, which constituted a new 'charter' for social tourism. It validated the globalising of social tourism and the spread of the movement from the European countries (France, Belgium, Italy, Portugal and Switzerland in particular) in which it had originally taken root, to the Americas, particularly the industrial countries of Canada and the United States and the developing country of Mexico. It recognised the great diversity of organisations and regulations for social tourism, as well as different philosophies and facilities which had arisen in very different geographical and political contexts associated with **internationalisation** of the phenomenon.

Social tourism policies are equally heterogeneous with respect to the involvement of the state, local community groups and **employment** associations. They concern the priority given to the various ways of assisting people to take holidays (the financial support given to people for holidays and, for example, support for wholesome vacations or **vacation** vouchers as in Switzerland and France). Further, they are also concerned with the provision of basic **infrastructure**, especially that of specialised lodgings, **accommodation** for families and associations, holiday villages, campgrounds, youth hostels and other types. In countries most committed to this policy, this last type of action has resulted in the establishment of important lodging concentrations in destinations managed by true 'employment associations'. Thus in France, which is the uncontested leader in the field of social tourism, associations control more than one-fifth of the accommodation capacity which is not in private ownership, for a total of about 900,000 beds (of which one-third are for camping and caravans, 260,000 are beds in family lodging and 240,000 are beds in youth hostels).

GEORGES CAZES, FRANCE

socialisation

Socialisation is the process of interaction through which people learn social norms and rules of **behaviour** in a given social group, complementing enculturation, **acculturation** or transmission of a society's cultural ideology and identities. Socialisation is a primary factor shaping touristic interactions of **leisure**, consumption and producer/

provider behaviours that establish and cross-cut ethnic, **gender**, **class** and racial boundaries.

See also: political socialisation

MARGARET B. SWAIN, USA

Society and Leisure/Loisir et Société

Society and Leisure/Loisir et Société is a multidisciplinary refereed journal dedicated to the study of **leisure** and **recreation**. It publishes articles based on either fundamental or methodological research, aiming at both disseminating the results of empirical research and initiating critical debates about the relationships between free time and social change. Its issues are thematic and usually are divided into three main sections: articles dealing with the theme of the issue and forming its most substantial part, non-thematic articles addressing current scientific issues, and book reviews. First appearing in 1978, it is published twice yearly by Les Presses de l'Université du Québec (ISSN 0705–3436).

RENE BARETJE, FRANCE

sociocultural change *see* change, sociocultural

sociolinguistics

Sociolinguistics is a relatively new sub-discipline within linguistics which examines the interrelationships between language and society. Since 'language is not a single code used in the same manner by all people in all situations' (Trudgill 1983: 32), sociolinguistics studies the varieties of language which are attributable to the different sociocultural characteristics and contexts of speakers and listeners.

Applied to tourism, sociolinguistics constitutes a **paradigm** or multi-theoretical perspective based on the premise that tourism communication is a form of language. Many contributors towards an understanding of tourism in **sociology** and **anthropology** either tacitly or openly acknowledge such a linguistic component. Some works, especially those dealing with the **semiotics** of attractions, are grounded in **markers** or representations of a **sight**. Whether these signs of reality are off-sight (such as travel books) or on-sight (like notices), they nevertheless speak in terms of the 'must sees' of sightseeing. Tourism thus develops a moral **rhetoric** of **authenticity** complete with its own glossary.

Those scholars who conceive of tourism as a sacred journey are very much aware of the icon-filled messages of **maps** and other forms of publicity which connote the ritualistic obligations of travel. Others who take a different approach, founded on the distinction between familiarity and strangerhood, nevertheless have a fascination with the **discourse** of tour advertisements and how particular words are used to frame the experience of novelty as an alternative to **mass tourism**. Dann (1996) has extended this treatment to travelogues and the verbal techniques employed by **travel writers** to reduce the unfamiliarity of foreign destinations for their readers. Those who adopt more of a postmodern ludic perspective for an appreciation of contemporary tourism still see the need to explore the discourse of the **media** through which themes are created for cultural consumption.

However, those who assume constructivist positions in tourism research are perhaps the closest of all to the sociolinguistic paradigm. Writers such as Hollinshead (1993), for instance, regularly fill their analyses with terms such as narrative, discourse, rhetoric, idiom, speech, talk and storylines, as different players in various touristic scenarios vie for their version of the truth and the power to articulate it. This 'invention of culture' position is necessarily processual and conflictual, since it assumes that definitions of situations are competitive in nature, particularly those which relate to history and heritage.

Finally, there are several recurring tourism communication encounters which can benefit from a sociolinguistic treatment for a fuller understanding of the asymmetry underlying verbal exchanges which take place. For example, according to Cohen and Cooper (1986), when tourists and **destination** people interact on a temporary basis, there is considerable linguistic accommodation that ensues on account of perceived status differences and the resultant need to respectively talk down or up to each other. As other researchers have shown, tour guides, also represent another instance of sociolinguistic patterns of superordination and subordination when, from a position of privileged knowledge, they attempt to 'secularise' a sight for visitors. Indeed, so ubiquitous are linguistic encounters in tourism that they may be considered as extensive as the phenomenon of tourism itself. Hence the need for sociolinguistic analyses.

References

Cohen, E. and Cooper, R. (1986) 'Language and tourism', *Annals of Tourism Research* 13: 533–63.

Dann, G. (1996) *The Language of Tourism: A Sociolinguistics Perspective*, Wallingford: CAB International.

Hollinshead, K. (1993) 'The truth about Texas', unpublished Ph.D. dissertation, Texas A & M University.

Trudgill, P. (1983) *Sociolinguistics: An Introduction to Language and Society*, London: Penguin Books.

GRAHAM M.S. DANN, UK

sociology

The sociology of tourism is a target field or specialty of sociology. It applies approaches, theories and **methodologies** to the study of touristic phenomena, and of their wider social significance and consequences. The sociology of tourism concerns itself specifically with the **motivations**, roles and social relations of tourists, the structure and dynamics of the tourist system and of touristic institutions, the nature of attractions and their representations, and the **impact** of tourism on destinations and host societies. Some researchers (MacCannell 1976; Urry 1990) also see in tourism an avenue through which the wider problems of modern (and postmodern) society could be examined. The field has a relatively brief history; although some German sociologists analysed tourism in the 1930s, the first publications in English appeared only in the 1960s. Nevertheless, the intellectual perspective and theoretical approach to tourism underwent significant transformations over this short period. The early work in the field has been reviewed in Cohen (1984); the main theoretical approaches have been examined in Dann and Cohen (1991).

The principal intellectual issue in the field is the relationship between Western tourism and **modernity** (or postmodernity). In comparison, the transformational impact of tourism on host settings – though of considerable practical importance – has rarely become a focus of much intellectual concern (Bruner 1991). The controversy regarding the nature of modern tourism was initiated by a biting critique of tourism by an American social historian, Boorstin (1964). Within a broader critique of contemporary America, he contrasted the traveller of the past with the modern tourist, claiming that the former submitted to the tribulations of travel to achieve authentic experiences, while the latter thrives on easily attainable, inauthentic 'pseudo-events', contrived by locals for touristic consumption. Boorstin thus sees in tourism just another domain in which the superficiality and inauthenticity of modern life is manifested.

The first sociological paradigm for the study of tourism, which informed the **discourse** of the topic throughout the 1970s and 1980s, was proposed by MacCannell (1973, 1976). In studied opposition to Boorstin's conception of the tourist, he sought to formulate a distinctly sociological approach to tourism, disengaged from and contrasting with its common sense image. MacCannell conceived of the tourist as an alienated modern, who – in contrast to Boorstin's view – engages in a serious quest for authentic experiences, in remote, non-modern places and pre-modern times. Rather than a superficial consumer of pseudo-events, the tourist is construed as a secular **pilgrim**, whose quest for **authenticity** is analogous to the quest for the sacred in simpler, pre-modern societies. MacCannell, however, stresses that in contrast to the latter, the 'sacred' of modernity (in the Durkheimian sense of symbols of 'society') is split up, manifesting itself in a multiplicity of attractions. The secular tourist's visit to the attractions is analogous to the religious pilgrim's homage to **sacred sites**. However, as tourists proliferate at a

destination, their quest is frustrated through the emergence of a '**tourist space** ' within which the locals 'stage authenticity' (see **staged authenticity**), in that they construe spurious attractions which are then represented or promoted as 'real'. The tourists' quest is thus frustrated, since they are unable to penetrate the 'fronts' and 'false backs' with which they are confronted and reach into the back, the 'real' life at the destination.

MacCannell's **paradigm** engendered considerable controversy and criticism as well as empirical research. Central to it was his concept of 'authenticity'. The discussion focused on three principal topics. The first is the empirical question of the extent to which tourists are in fact motivated by a quest for authenticity; this appears not to be generally the case, even though such a quest may be a central cultural theme of modernity (Trilling 1972; Cohen 1988). The second is a critical examination of the concept of authenticity; it was claimed that the concept is socially constructed, and hence may have a different connotation for various types of tourists (Cohen 1988). Rather than to posit the concept as given, some researchers thus turned to the empirical examination of the tourists' own perceptions of authenticity. Such an operationalisation of the concept of authenticity weakens the analytical power of MacCannell's paradigm. The third is a shift away from authenticity as the focal topic in the sociological discourse of tourism, which occurred under the joint impact of gradual changes in the nature of tourism and the employment of new, especially postmodern theoretical approaches to its analysis. This is the trend which characterised the 1990s.

Under emergent 'postmodern' conditions, the notions of 'beginnings' or 'origins' are being denied their aura of sacrality. In turn, the centrality of authenticity as a cultural theme has been weakened. Consequently, 'post-tourists' (Urry 1990: 93ff) tend to engage more readily in the playful enjoyment of explicitly contrived attractions rather than in a serious **quest** for authenticity in some unmarked back regions of their destination. This cultural shift has led to the growing popularity and proliferation of new kinds of attractions, most of which are sanitised simulations of reality (Jules-Rosette 1994), such as large-scale amusement parks, emblematically represented by the Disneylands, **theme parks**, living museums, reconstructed prehistoric environments and renovated **heritage** sites. To many of these, the concept of 'authenticity' can no longer be unequivocally applied. Indeed, some were treated as 'authentic copies' (Eco 1990: 20) or 'authentic reproductions' (Bruner 1994). While facilitating an effortless and playful postmodern enjoyment of surfaces, such simulated sites in fact lead to a progressive segregation of tourism from the daily reality of destinations and to the erosion of the boundary between tourism and ordinary **leisure** (Cohen 1995; Moore *et al.* 1995). This process culminates in the application of the technology of '**virtual reality**' to the simulation of **travel**, which the 'virtual tourist' can enjoy in his or her home environment.

These trends have been accompanied by a shift in the theoretical focus of the sociological discourse of tourism to another of the leads in MacCannell's (1976, 1989) early work: **semiotics**. Ironically, however, the adoption by some students of tourism of the more radical views of 'postmodern' semioticians such as Baudrillard (1981b) and Eco (1990) led to the ultimate subversion of the basis of MacCannell's paradigm. The contemporary tourist travels in 'hyperreality'. This is a world of simulacra, of fakes which appear more real than the original, or of endlessly reproduced copies, without origins. Representation is all, reality is problematised. Tourists engage in the interpretation and consumption of signs of this problematised reality, with them being perceived as 'consumers of signs' (Urry 1990: 83 ff). Commodified tourist objects and activities are said to possess 'sign value' (Watson and Kopachevsky 1994: 645–7), rather than use value or exchange value. A new research programme, if not a paradigm, for the study of tourism in terms of a 'political economy of [touristic] signs' (Baudrillard 1981a, 143–163; Watson and Kopachesvsky 1994) thus emerges.

The semiotic turn in the sociology of tourism engendered a growing concern with the construction of touristic representations, **images** and performances (Dann 1966; Selwyn 1996). Significantly, however, such constructions are frequently contested, by bearers of contrasting cultural and political perspectives and interests (Bruner 1996). Dovetailing with, and partly drawing upon semiotics, is Urry's (1990) Foucaultian study of the tourist '**gaze**'. Stressing the primacy of the visual in tourism, he analyses the social organisation and authorisation of the tourist's 'gaze' and of its consequences for the constitution of its object. Though his empirical material is drawn primarily from British resorts, it appears that his argument applies primarily to **sightseeing** rather than vacationing tourism, where a total sensual emersion rather than a dominant visual **experience** appears to be the rule.

The prevailing contemporary theoretical concern with touristic images and representations contrasts sharply with the more practical concerns of sociologists with the ecological, social and cultural **impacts** of tourism upon that very reality which postmodernists tend to 'problematise'. The rapid growth of tourism and its progressive penetration into more remote and less developed regions of the world is perceived as a serious threat to the survival of those natural environments and ethnic cultures and **lifestyles** which became popular destinations of **mass tourism**. The concern with this threat engendered a variety of ideas and proposals to soften the impacts of tourism, under such headings as **alternative tourism**, sustainable tourism (see **sustainable development**), green tourism and **ecotourism**. Sociologists engage in both the promotion of such concepts as well as in their critical examination.

The sociology of tourism thus presently moves in contrasting, but in some sense complementary, directions. On the one hand, it engages in a theoretical analysis of tourism in terms of signs and representations, accompanied by a marked lack of concern with its concrete consequences. On the other hand, it manifests a growing practical concern with the impact of tourism on destinations 'marked' by those very signs and representations. The growing theoretical sophistication of the field, even though involving a 'critique' of tourism on an abstract level, however, appears to contribute little to the resolution of the practical problems engendered by the accelerated growth and expansion of mass tourism.

References

Baudrillard, J. (1981a) *For a Critique of the Political Economy of the Sign*, St. Louis: Telos Press.

—— (1981b) *Simulacres et Simulation*, Paris: Ed. Galilee.

Boorstin, D.J. (1964) *The Image: A Guide to Pseudo-events in American Society*, New York: Harper & Row.

Bruner, E.M. (1991) 'Transformation of self in tourism', *Annals of Tourism Research* 18(2): 238–50.

—— (1994) 'Abraham Lincoln as authentic reproduction: critique of postmodernism', *American Anthropologist* 96(2): 397–415.

—— (1996) 'Tourism in Ghana', *American Anthropologist* 98(2): 290–304.

Cohen, E. (1984) 'The sociology of tourism: approaches, problems and findings', *Annual Review of Sociology* 10: 373–92.

—— (1988) 'Authenticity and commoditization in tourism', *Annals of Tourism Research* 15(3): 371–86.

—— (1995) 'Contemporary tourism: trends and challenges', in R. Butler and D. Pearce (eds), *Change in Tourism*, London: Routledge, 12–29.

Dann, G. (1996) *The Language of Tourism: A Sociological Perspective*, Oxford: CAB International.

Dann, G. and Cohen, E. (1991) 'Sociology and tourism', *Annals of Tourism Research* 18(1): 155–69.

Eco, U. (1990) *Travels in Hyperreality*, San Diego: Harcourt Brace Jovanovich.

Jules-Rosette, B. (1994) 'Black Paris: touristic simulation', *Annals of Tourism Research* 21(4): 679–700.

MacCannell, D. (1973) 'Staged authenticity: arrangements of social space in tourist settings', *American Journal of Sociology* 79(3): 589–603.

—— (1976) *The Tourist: A New Theory of the Leisure Class*, New York: Schocken Books.

—— (ed.) (1989) 'Semiotics of tourism', *Annals of Tourism Research* 16(1): 1–18.

Moore, K., Cushman, G. and Simmonds, D. (1995) 'Behavioural conceptualisation of tourism and leisure', *Annals of Tourism Research* 22(1): 67–85.

Selwyn, T. (ed.) (1996) *The Tourist Image*, New York: Wiley.

Trilling, L. (1972) *Sincerity and Authenticity*, London: Oxford University Press.

Urry, J. (1990) *The Tourist Gaze: Leisure and Travel in Contemporary Societies*, London: Sage.

Watson, G.L. and Kopachevsky, J.P. (1994) 'Interpretations of tourism as commodity', *Annals of Tourism Research* 21(3): 643–60.

ERIK COHEN, ISRAEL

soft tourism

Originating in Alpine Europe, 'sanfer tourismus' is a term used to describe forms of tourism that have low or minimal negative impacts on the physical and sociocultural **environment** in **destination** areas. Generally considered the antithesis to **mass tourism**, it is associated with small-scale tourism development that optimises the benefits to local communities.

See also: alternative tourism; responsible tourism; sustainable development

RICHARD SHARPLEY, UK

South Africa

After many years of underperformance due to political difficulties in the country, a new era has dawned for South African tourism, particularly after the first all-inclusive elections in 1994. The publicity surrounding the peaceful transition to democracy, the stature of former President Nelson Mandela, the hosting of major events such as the Rugby World Cup in 1995 and the implementation of 'Explore South Africa', the destination marketing campaigns based on the country's tourism strengths such as the environment, wildlife and cultural diversity, ensured a growth rate of 8.7 per cent in 1997 and 15.2 per cent in 1998. This is well above the world average growth of 2.4 per cent for 1998, as reported by the World Tourism Organization.

During 1998, South Africa hosted more than 5.6 million tourists, of whom 4.2 million were from Africa. Europe constitutes the largest part of the overseas market (some 66 per cent), followed by North America (about 14 per cent) and Asia (almost 10 per cent). Market surveys indicate that the main reasons for visiting South Africa centre on scenic beauty, the opportunity to experience political change, abundant wildlife choices and the rich heritage of cultural diversity. Although still performing well below potential, the impact of tourism on the economy of South Africa is already significant. During 1998 tourism contributed an estimated 8.2 per cent to the economy, was the third largest earner of foreign exchange and employed more than 730,000 people.

South Africa is richly endowed with a wide diversity of natural attractions, ranging from the largely unspoilt coastline of 3,000 kilometres to an abundance and variety of animals and plants. This reinforces its image of being host to 'the greatest wildlife show on earth'. The 'big five', namely lion, elephant, rhino, leopard and buffalo, and whales are abundant in South Africa. There are also vast conservation areas such as those controlled by South African National Parks, a leading conservation body. Particularly since 1994, the rich cultural diversity of South Africa has been further developed and promoted as a key tourism attraction. South Africa's depth and diversity of attractions are underpinned by excellent tourism facilities and infrastructure. Its road networks, transport infrastructure and health services are in line with the First World standards. An indicator in this regard is that during 1998, sixty-three airlines had scheduled flights to the country.

ERNIE HEATH, SOUTH AFRICA

souvenir

Souvenirs are material objects which serve as reminders of people, places, events or experiences of significance in a person's biography. Some authors distinguish souvenirs from mementoes. The term souvenir is used for commercial artefacts, particularly those acquired in the course of a journey, as reminders of places visited or of particular travel experiences. Similarly, mementoes are acquired non-commercial objects, such as stones or dry flowers. Souvenirs have a personal value for the individual which is typically much higher than their often negligible market value. The circumstances of their acquisition endow souvenirs with a metonymic or metaphoric significance: bottled water from the Jordan river is a metonym signifying the Holy Land for Christians, while a miniature representative of the Church of the Holy Sepulchre is a metaphoric symbol for it. Souvenirs are seen as one expression of the tourist's **quest** for the 'authentic' (see **authenticity**). If so, such a quest is more immediately satisfied by metonymic souvenirs (the Jordan water is 'real' water from the river) than by metaphoric ones, though the latter may be symbolically significant, even if they are not 'authentic'.

The meaning of souvenirs may be wholly individual, or shared by a collectivity. Mementoes are normally only individual, but so too are some commercial souvenirs: a standard object, such as a sun hat, may become a souvenir of a personal **experience**, for example as a reminder of a particularly hot trek in the tropics. A cross acquired on a **pilgrimage** to Jerusalem has an additional shared collective meaning, as a symbol of the pilgrim's Christian **identity**.

There are some important differences between the kinds of souvenirs acquired by different types of tourists. Authenticity-seeking tourists will tend to acquire 'real', often exotic, artefacts as reminders of their encounter with the 'other' or with the 'extraordinary'. Less authenticity-prone tourists will acquire artefacts which are less strange, so they can be displayed or used in everyday life. For example, instead of a used native costume, tourists may purchase a pillow cover with 'typical' native designs, but in colours fitting their living room. Tourists enjoying the playful reversal of everyday life on their trip may acquire funny, childish or 'tacky' artefacts, such as loud T-shirts or risqué postcards, expressing the un-serious, ludic nature of their **vacation** (Gordon 1986).

While souvenirs have a personal significance for individuals, the emergence of **mass tourism** encouraged the mass production of souvenirs. Three modes of such production can be distinguished: industrial (such as T-shirts, key rings, bags, postcards, miniaturised attractions and similar inexpensive, machine-made products), crafts (moderately priced pottery, embroideries and carvings produced in forms, designs, decorations or colours 'typical' of local styles, or depicting typical local scenes, customs, occupations or people), and arts (relatively expensive, unique works by reputable local painters or sculptors, representing local motifs).

It should be stressed that none of these modes of production serves exclusively as souvenirs. The industrially produced objects may be acquired for the tourist's own use or for presents; crafts may be bought purely for their attractiveness or as collectors' items, particularly when they are exported and reach international markets, where their purchase is unrelated to any personal experiences of the customers. While works of art may principally be acquired for their aesthetic value and rarely serve as 'mere' souvenirs, in many destinations, some artists are explicitly oriented to the souvenir market, producing large numbers of works in a similar style on which the same scenes are frequently repeated. In such instances, the borderline between 'tourist crafts' and 'art' is practically obliterated. Research on souvenirs has focused mainly on their origins and development in relationship to the **local** culture. Some studies exist which inquire into their mode of production and **marketing**. However, the study of their significance for the ultimate consumers is still at an early stage.

References

Gordon, B. (1986) 'The souvenir: messenger of the extraordinary', *Journal of Popular Culture* 20(3): 135–146.

Further reading

Cohen, E. (ed.) 'Tourist Arts', special issue of *Annals of Tourism Research* 20(1). (Of particular importance are two articles: S. Littrell, 'What makes a craft souvenir authentic?', and Shenhav-Keller, 'The Israeli souvenir'.)

ERIK COHEN, ISRAEL

souvenir, religious

The term religious **souvenirs** refers to a group of objects which are, by their content or shape, reminiscent of religious teaching, figures or **sites**. There is an enormous number of objects that have a religious meaning, and are thus used in religious **rituals** (including prayer books, breviaries, rosaries, crosses and so on), which religious tourists keep as touristic souvenirs; on the return from the journey, these souvenirs are then used regularly when performing religious rituals. There are also a large number of objects with religious characteristics (pictures of saints, of the crucifixion, of the Virgin, or various motifs from Biblical and other religious stories, women's scarves with the same pictures, religious motifs on garments or articles of everyday use, and so on). In modern tourism, the **commercialisation** of religious motifs and their use on the most varied objects means that they become, for the **pilgrim** as well as for the regular **tourist**, symbols of a certain sacred site or content (see **site, sacred**). In other words, they become religious souvenirs. The commercial mass production of these objects is stimulated further by the large numbers of pilgrims gathering in religious centres worldwide and representing an enormous and willing mass of purchasers of souvenirs of this kind.

BORIS VUKONIC, CROATIA

spa

A spa is a **resort** which provides mineral or thermal water for drinking or bathing. Although the phenomenon dates back at least to Greek and Roman times, the word is taken from the name of a town in Belgium which provided medical cures in congenial surroundings. Originally having a mix of health and pleasure functions, spas reached their zenith in Europe from the end of the sixteenth century to the beginning of the nineteenth century when places such as Bath and Tunbridge Wells in the United Kingdom became fashionable pleasure resorts. Although tastes have changed, spas still function in many parts of the world (such as Hungary, where there is an extensive network of spas) and **health** resorts have seen a recent resurgence in popularity. The transformation of spas into pleasure resorts, together with the **Grand Tour**, and the emergence and growth of coastal resorts constitute the beginnings of tourism as one knows it today.

Further reading

Lawrence, H.W. (1983) 'Southern spas: source of the American resort tradition', *Landscape* 27(2): 1–12.

Patmore, J.A. (1968) 'The spa towns of Britain', in R.P. Beckinsale and J.M. Houston (eds), *Urbanisation and its Problems*, Oxford: Blackwell, 47–57.

Towner, J. (1996) *An Historical Geography of Recreation and Tourism in the Western World 1540–1940*, Chichester: Wiley, Chapter 4.

GEOFFREY WALL, CANADA

space allocation

Space allocation refers to the space required for a total facility such as a **restaurant**, or the space within it for different allocations such as dining room or production kitchen. In this case, this is usually expressed as space per seat. For instance, the total space needed for a table service restaurant is 2.5 m^2 per restaurant seat, with 2 m^2 for the dining area, 0.9 m^2 for the kitchen, 0.3 m^2 for the cloakrooms and 1.2 m^2 for the bar area.

PETER JONES, UK

Spain

According to the **World Tourism Organization**, Spain is the number two **destination** in the world, just after the United States, occasionally slipping to third position in close competition with France. The country began its tourism development as a mass destination in the 1950s thanks to a well-endowed resource base in natural and cultural amenities, very favourable relative prices and its proximity to the most important European markets of origin.

In that first stage of the Fordian Age of Tourism, when standardised mass-produced tourism services were offered to non-discriminating consumers, Spain had a limited quantity of facilities and those that did exist were wanting in conveniences and sophistication, even by the standards of the time; tourism **infrastructure** and public services were also lacking. However, as early as 1956 the number of international arrivals to Spain had reached 2.7 million, and it was clear that tourism demand was not hindered by those shortcomings.

Because of close cooperation between the national industry and international tour operators, investment in tourism has been enormous since then, resulting in practically continuous growth in tourist numbers and revenue. The number of international arrivals to Spain was 62 million in 1996, of which an estimated 42 million were tourists. **Domestic tourism** has also grown in parallel, with a demand for both international holidays (some 4 million in 1996, with Spain occupying the position 10 in the ranking of tourism origin countries) and national tourism services. At present, it is estimated that 45 per cent of the tourism demand in Spain is international, while the remaining 55 per cent is from holidaying Spaniards themselves. The Spanish stock of tourism capital is estimated to be worth over $300 billion and includes over 1.3 million **hotel** beds, another 5–6 million non-hotel tourist beds, over 58,000 restaurants and tens of thousands of other tourism entertainment installations. The main markets for Spanish tourism are in Europe, with tourists coming mainly from **France**, **Germany**, the **United Kingdom**, **Portugal**, **Italy** and Belgium. Long-haul arrivals from the **United States**, **Japan**, Latin America and other origins are on the rise, but still represent a minor segment of demand.

Spain is endowed with rich and diversified **resources**, although tourism has concentrated largely on the peninsular coastline and its two archipelagos (Balearic and Canary Islands). Therefore, the potential resource base for tourism is still considerable, with the interior of the country largely undiscovered by **international tourism** and providing for authentic cultural experiences and controlled visits to many natural reserves.

The **impact** of tourism on the Spanish economy is enormous, accounting for an estimated 9–10 per cent of the GDP and some 11–12 per cent of total **employment**, and being one of the fastest growing activities. Of course, these figures are not evenly distributed across the whole country but vary considerably among the regions. It has been said that 'Spain is not a tourism power... some of its regions are'; Catalunya, Andalucía and Valencia, all coastal regions, together with the Canary islands and the Balearic archipelago, account for over 65 per cent of total demand. Other coastal regions in the Mediterranean (Murcia) or the Atlantic (Galicia, Asturias, Cantabria and the Basque Country) are experiencing a strong increase in tourism demand and responding with new tourist facilities and modern infrastructures, while adopting high standards of environmental quality.

The role of government in tourism **development** has been active since the 1950s and even before, but has changed much in scope. Because of the importance of tourism in Spain's balance of payments as a foreign exchange generator, central government assumed early the lead of promotional activities in foreign markets, with varying degrees of coordination with private initiative. Foreign **tour operators** did most of the actual selling abroad, with the Spanish industry concentrating on the product side within the country and adding little in the way of international distribution. Government intervention in tourism production was limited to legislation concerning the quality of services (consumer information and protection) and isolated investments in high-category lodging, adapting national **heritage** buildings (castles, monasteries and so on) to hotel use and managing the resulting product, the *Paradores de Turismo*.

In the beginning of the 1980s, and because of the political devolution process, the policy powers in tourism were handed to the regions, the so-called *Comunidades Autónomas* (Autonomous Communities). These have almost full powers to control tourism in their respective areas and to promote, even abroad. For several years, between 1982–3 and 1990, many chose to exert that capacity with little coordination with the central government or other *Comunidades Autónomas*. Later, at the beginning of the 1990s, the importance of policy coordination in achieving **efficiency** was recognised.

The tourism policy of the *Comunidades Autónomas* in the 1980s concentrated on the promotional side, while regional governments were consolidating and assuming their tourism capabilities. Product policies and quality issues received less attention at a time of high inflation, with the result of a loss in competitiveness in tourism markets. By the end of the decade, Spain as a whole was rapidly losing market share and facing severe problems of **image** and performance of its tourism industry.

In 1991, after the preparation of a White Paper on Spanish tourism and serious concern in the private and public sectors about the future of the industry, a new government decided on the preparation of a coordinated business strategy with the *Comunidades Autónomas.* After months of conceptual work and continuous discussion with the tourism business leaders, the trade unions and the governments of the autonomous communities, a tourism policy plan, called *Plan Futures,* was agreed by all and signed by the central government and the regions in June 1992. Its first period of application was 1992–5 and its second stage rang from 1996–9.

The *Plan Futures* sets a framework for the product, communication, distribution and environmental tourism policies of all decision makers in Spanish tourism, at central, regional and local level, in the public as well as in the private sector. This plan is meant to be a *horizontal* instrument of economic policy, not intending to force action at any concrete location or in any specific activity, but creating a range of economic, financial and legal instruments for use by the tourism industry. Public budgets at all levels of government are supposed to back the implementation of the plan, and use of European funds is also possible within its frame; partnership action with the private sector is central to its philosophy.

The *Plan Futures* has sought to improve the competitiveness of the Spanish tourism industry by acting in three main areas: coordination (national and international, vertical and horizontal), modernisation (**technology**, human resources, process re-engineering, and the like), and product quality and diversification, promotional and environmental quality (nature and built). Thousands of applications for partial funding of projects within these areas are processed each year by the central and regional governments. In the first stage of *Futures,* all final decisions on funding were taken at central level, after screening by the *Comunidades Autónomas.* Today, the funds are distributed among them and the applications are considered by the corresponding regional government. Only projects of a multiregional or national scope are evaluated at central level.

Worthy of specific mention within *Futures* is the action taken on *Planes de Excelencia* (environmental quality). In the six years of operation of *Futures,* some twenty tourism destinations have been chosen for integral quality action. The Spanish Department of Tourism, together with the regional and local government and tourism business associations, took the initial lead in defining needs and adopting priority actions. Other government departments and individual businesses later joined in.

This policy, and a favourable situation in Mediterranean and global markets in the last years, has resulted in high and stable rates of growth for Spanish tourism (There was a 12 per cent increase in foreign receipts in 1997). Several mechanisms for public–private partnership have been created to facilitate cooperation between them at all levels. 1997 saw accelerated growth of the tourism private sector in the country, the adoption of a more detailed *Plan de Estrategias en Materia Turística* (strategic plan in tourism) and the celebration of a National Conference in Tourism, bringing together some 2,000 decision makers. Although the problems of the late 1980s have not been entirely solved, the Spanish tourism industry seems to be more ready for the challenges and opportunities of the coming years.

Further reading

Congreso Nacional de Turismo. Conclusiones y Medidas Adoptadas (1997) Madrid: Ministerio de Economía y Hacienda. Secretaría de Estado de Comercio, Turismo y Pequeña y Mediana Empresa.

Futures: Plan Marco de Competitividad del Turismo Español 1992–95 (1992) Madrid: Ministerio de Industria, Comercio y Turismo. Secretaría General de Turismo.

Futures: Memoria de Ejecución del Plan Marco de Competitividad del Turismo Español 1992–95 (1992) Madrid: Turespaña, Instituto de Turismo de España.

Futures: Plan Marco de Competitividad del Turismo Español 1996–99 (1996) Madrid: Turespaña, Secretaría General de Turismo.

Informe Anual del Movimiento Turístico en Fronteras (Frontur) 1997 (1998) Madrid: Ministerio de

Economía y Hacienda. Secretaría de Estado de Comercio, Turismo y Pequeña y Mediana Empresa, Instituto de Estudios Turísticos

Plan de Estrategias y Actuaciones de la Administración General del Estado en Materia Turística (1997) Madrid: Ministerio de Economía y Hacienda. Secretaría de Estado de Comercio, Turismo y Pequeña y Mediana Empresa.

EDUARDO FAYOS-SOLÀ, SPAIN

spatial interaction

In tourism, spaces travelled and occupied by tourists are often clearly marked, at specific hotels, **sites**, monuments, roads and advertised attractions. Indeed, **guidebooks**, in highlighting these spaces, advise tourists where to go, and what to see and experience. However, most of the spaces occupied by **indigenous** peoples are neither marked nor visited by tourists. Interactions between tourists and locals occur mostly in patterned ways in pre-selected spatial settings.

EDWARD M. BRUNER, USA

special interest group

Special interest tourism is a controversial label which is often used mistakenly to describe a form which is in some way preferable to **mass tourism**. However, it is not synonymous with terms such as **alternative tourism**, ethical travel, **appropriate tourism** or **ecotourism**, since the special interest type is not necessarily either responsible or sustainable (see **responsible tourism**; **ecologically sustainable tourism**). Special interest tourists are motivated not so much by desire to be responsible tourists as by the desire to pursue a particular interest, hobby or activity, be it a sport such as scuba diving or golf, an outdoor activity such as mountain climbing or bird-watching, a cultural or **heritage** interest such as folk music or period architecture, or an educational pursuit such as European history or archaeology. They choose a **destination** that somehow offers a better or unique way of enjoying that special interest, and/or they choose a tourism experience that facilitates the enjoyment of something they are passionate about.

While independent travellers can be special interest tourists, organised special interest groups are both more visible and more relevant to the industry. Indeed, there are travel agents and **tour operators** who cater exclusively for special interest groups, often designing and marketing their products in ways that appeal to a narrow segment or segments of the travelling public. Others respond to requests from organised groups on a one-off or recurring basis.

The characteristics of special interest tour groups and their members have been casually observed but seldom investigated in any systematic way. Issues relating to the provision and delivery of organised study tours are addressed regularly in international conferences known as The Global Classroom, a biennial assembly of non-profit and commercial educational travel providers. Despite the lack of research, the growth of special interest tourism is presumed to be more rapid than other forms, attributed mainly to the increased sophistication of the public and the resulting demand for more focused and high-quality tourism experiences.

See also: adventure tourism; cultural tourism; ethnic tourism; historical tourism; nature tourism; pilgrimage

Further reading

Weiler, B. and Hall, C.M. (1992) *Special Interest Tourism*, London: Belhaven. (Includes review chapters on tourism interest groups, along with case studies of each segment.)

BETTY WEILER, AUSTRALIA

sponsored event

Sponsored events are those which receive monetary payment or other support in return for specified benefits. Sponsors seek publicity, sales, relationships with target segments and on-site **hospitality**. Events gain additional resources, expertise and augmented marketing. Sponsorship has accelerated growth and variety in the events sector, including creation of media events aimed at broadcast audiences (see **event marketing**).

DONALD GETZ, CANADA

sport, recreational

The concept of recreation sport includes a field of physical **activities** organised and planned so as to give a **tourist** or individuals within a group a sense of **satisfaction** that is refreshing and enlivening. Out of this participatory **experience**, people find renewed strength to face daily life. Unlike career sports, recreational sports do not normally involve occupational obligation and pecuniary rewards.

JOSEPH KURTZMAN, CANADA
JOHN ZAUHAR, CANADA

sports tourism

Baron Pierre de Coubertin, the father of the modern Olympic Games, gave inspiration to the development of sports tourism in the belief that it would bring people together, thus contributing to a better understanding among peoples and nations. Coubertin's concept gave sports tourism its vitality and impetus for pursuit of business **entrepreneurship**, and financial profitability.

Sports tourism has been defined as 'the use of sports as a vehicle for tourism endeavours', in which activities encompass the physical context of running, jumping, walking, racing, throwing, shooting, hitting and the like. The primary qualifier for the concept lies in tourist participation or attendance at a predetermined sports activity. Sports tourism has also been delineated along the lines of activity categories having direct relationships to tourism (sports events, attractions, resorts, tours, cruises) and are influenced by inherent elements such as history, **destination**, **policy**, sociocultural characteristics, **marketing** and economic impact.

Many unique sports activities are associated with sports tourism; for example, if a particular camel race attracts tourists, then the race would be considered a sports tourism activity. Both the participants and spectators might well be sports tourists. This example, viewed as such, could be associated with motor sports racing, horse racing and more.

Certain organisational factors are characteristic and prevalent within sports tourism. They include resources utilised to stage an activity within a sports environment, and certain dependency on tourists by sports organisers.

Sports tourism activities take place within a recreational and/or competitive environment. The nature of the physical activity *per se* implies competition between individuals and groups at varying levels of skill and ability. For the competitor, this process includes self-challenge through self-imposition. A self-challenge would be exemplified in trekking in Nepal, a location where would-be participants come from outside the region to perform their professional best in the attempt to reach the top. For the spectator, the appreciation of skill and ability mastered by the participant brings about satisfaction, enjoyment and applause. This is well demonstrated in the Olympic Games. For tennis resorts, a person could be a spectator or avid participant. Further, in the sports halls of fame or museums, the touristic role is confined to that of onlookers.

Sports tourism has shown to provide significant economic impact, not only from the hosting of major sports events but also from the development of sports resorts and sports attractions. Most recently, increased marketing of sports tourism by countries, regions and communities suggests potential for improved tourism receipts. Sparked by the popularity of the cruise industry, the sports celebrity cruise has become an important element of sports tourism. Each of these advancements contributes to the impact of sports tourism. The importance of the sports tourism field is now well recognised and has given rise to a professional association, the Sports Tourism International Council.

JOSEPH KURTZMAN, CANADA
JOHN ZAUHAR, CANADA

staged authenticity

Postmodern tourists revel in constructed experiences from computer assisted virtual travel to entire cities made to resemble theme parks, such as Las Vegas. Classically, tourists were concerned about the truth of their experiences: this is the exact place where the Magna Carta was signed, this is the true Crown of Thorns, this is unspoiled nature, and so on. The issue of **authenticity** inhabits the tourist **experience**. One finds melancholic or joyful denials of the possibility of authentic experience on the part of postmodernists, or a continued striving for the authentic on the part of ecotourists and others who refuses to give up the **quest** for the real, the true, the authentic.

Researchers abandon all efforts to provide expert philosophical or other judgement based on external criteria regarding what is or is not authentic. In the place of absolute and external criteria of authenticity, a social constructivist approach is proposed. Since almost every place that tourists visit provides a local answer to the question of authenticity, an **anthropology** of actual practices of staging authenticity can be substituted for absolutist determination of what is authentic and what is not.

Even those attractions that might qualify as authentic according to an external criterion (like Leonardo's Last Supper or the Grand Canyon) depend upon stagecraft for their presentation: they are marked off from their surroundings, protectively separated from their viewers, framed, provided with special lighting and landscaping, and explained by guides. The presentation apparatus of tourism always intervenes between the tourist and the attraction. The apparent authenticity of any attraction is socially constructed, whether or not it is real. Even those tourists who seek authentic or real experiences mainly witness the staging apparatus. The industry that grew around the desire for direct experience of real other peoples, places and things blocks the possibility of an authentic experience in its drive to supply such experiences: in the place of an authentic native ritual is found a performance of an 'authentic native ritual' for tourists. What the tourists see, even under conditions where the greatest care has been taken to preserve the authenticity of an attraction, is still the touristic version of the natural, historical or ethnological original.

Thus artificially constructed postmodern tourism environments (such as the New York, New York and Paris Experience hotels in Las Vegas) are a logical outgrowth of the original tourist desire for authenticity and the staging of authentic experiences that occurred in the classical phase of the development of global tourism until about 1975. Most of the original attractions are now entirely superseded by their staging apparatus and/or their artificial reconstructions and simulations. Actually, there is a progression of increasing apparent authenticity in the presentation, or staging of attractions. The first stage is simply setting up a distinction between ordinary everyday occurrences and something that is worthy of a touristic visit, the marking of an **attraction** that separates it from its context. Already, the experience is framed as much by the institutions of tourism as by the desire of the tourist. On site, the tourist will find aesthetic cues suggesting what kind of experience one is supposed to be having: solemn and uplifting, or fun, or edifying. In its most developed forms, attractions are designed to totalise the tourists' experience, such as to immerse them in the everyday life of a reconstructed sixteenth-century village where they can engage in dialogue with living enactors. Alternatively, attractions may provide the tourists with privileged but highly controlled access 'behind the scenes', to the rooms where actual executions took place, or a famous person was born.

A prevalent form of staged authenticity is the 'work display', which includes factory tours, historical enactments of the everyday work activities of **pilgrims** or **museum** reconstructions of outmoded manufacturing techniques, to name only a few. There is a common trajectory to the **development** of former sites of work into destinations. First, a place of work becomes a display for tourists, like a beach where small fishing boats were hauled out, nets repaired, today's successes and failures discussed and tomorrow's activities planned. The entire scene is susceptible to becoming an object of touristic consumption, to be taken in as an example of the 'picturesque'. As soon as development for tourism begins, it is no longer necessary for fishing or related activities to continue, so long as some of the boats, nets and fishermen remain photogenically arrayed as a reminder of their former purpose. Eventually, the picturesque elements are selectively integrated into the decor of beach bars and discos which retain a 'traditional fishing village' theme. The staging of authentic tradition appeals to tourists who feel a need to be reminded of the existence of simpler forms of work while they are on **vacation**, or they may need to have hard work made to appear pretty and leisurely to assuage their guilt. Thus, a place of work is made into a place of **leisure** for tourists. The nature of the actual work that occurs in the village has changed. The fishermen, or their children, are now integrated into the global economy as service workers for tourists. The actual work that is taking place is masked by the work display and the staging of authentic **traditions** for touristic consumption. The repetition of this scenario places staged authenticity on the cutting edge of the global transformation of traditional economies for tourism.

The distinction between a pseudo-attraction and a real one is unsustainable. So-called real attractions are subject to the same staging requirements as pseudo-attractions, and some places, such as Las Vegas, come to be so much associated with tacky glitter for tourists that fakery becomes 'the real' of the place. European tourists to the United States expect to see impersonal cityscapes, barren superhighways, litter, street crime, fast food chains, **theme parks**, trailer parks and Las Vegas-style show business glitz. They would be disappointed if they did not see such things, because for Europeans these 'tacky' and 'pseudo' attractions are what is most authentically American.

Since 1980, tourism researchers have mainly turned away from questions of presentation and staging of attractions. Instead, they have focused on psychological questions of whether tourists are motivated by a desire for authentic experience, or are just as happy with touristic shows which they know are staged for their entertainment. Theoretical developments in other fields, especially the emergence of postmodern theory and its concern for the simulacrum and hyperreality, have absorbed debates about real versus fake that started in tourism studies.

DEAN MacCANNELL, USA

standardisation

Standards are the norms with which **products** or **services** must comply. Specified in documents which determine how work must be carried out, they can be defined as the objectives which must be reached to satisfy clients' needs, desires and expectations. The fulfilment of these objectives, provided that they are articulated as requirements, lead the company to deliver **quality**.

Standards must be relevant and simple; that is to say, they must describe the phenomenon as fully as possible by being detailed, precise, specific and easily understandable. For a well-established definition, it is necessary to consider: the final result expected by the client, client satisfaction and the need to be fulfilled or the client's initial expectations; the characteristics and components; the work process and the responsibilities of the organisation's members; the global objective of the organisation and thus its chosen strategy; and the means by which these standards can be achieved (human means, methods and procedures through which a degree of excellence can be achieved, materials and equipment to help provide the service and raw materials) (Horovitz 1986: 54–6).

A product can be defined as a physical good or service. It often consists of both elements, thus identifying a physical good–pure service continuum which clearly affects the perception of quality. This dualism is present in the tourism sector through the unique characteristics by which services are identified, such as heterogeneity, intangibility, simultaneous production and consumption, and perishability. Heterogeneity is linked to the ability to vary the result of a service, and the very nature of a tourism service can vary according to the producer, the client or the moment when the service is required. It is fundamentally this characteristic which hinders standardisation. However, a definition of standards and the means which allow standards to be achieved is essential in the **quest** for quality. In fact, both the absence of standards for the execution of work and errors in its delimitation are considered key causal factors of the so-called *gap 2* in the theoretical and methodological development of model of service quality proposed by Zeithaml *et al.* (1990).

At least three distinct components of the total quality concept can be recognised in all tourism sector activities: a technical or physical quality (what is delivered), a functional or interactive quality (how it is delivered), and a corporate quality or **image**. To use as an example the services carried out in an **hotel** (Saleh and Ryan 1992), the technical characteristics would include the tangible elements of the hotel's reception area (check-in desk, seats, brochures), rooms (beds, television, toilets), restaurants (food, drink, table-ware), and the like. The functional dimension would relate to courtesy, speed, empathy and so on, and would in essence result from the interaction between the clients and the tangible elements, the clients and the service providers, or the clients and other clients. Furthermore, the third component would be the **image** that the client has of this type of hotel establishment, which is fundamental given the intangible nature of the service. To guarantee high levels of quality in this service, the clients' expectations of the previously mentioned elements must be converted into standards; that is, converted into specific norms, and that will depend on the extent to which the executable work or tasks can be standardised and converted into routine activities (Zeithaml *et al.* 1990).

Standardisation is a basic mainstay in the theory of daily work improvement, and forms the central axis of the shift towards improvement. Levitt (1976: 66–8) indicates three possibilities for standardisation or industrialisation of service which are applicable to the tourism sector. One, with hard **technology**, substituting those systems with more personal contact and human effort for machines and equipment, allows the effort to be concentrated on those tasks which require a higher level of personal service (principally those tasks related to 'how it is delivered' or functional quality). The tourism industry has a large number of activities that can be standardised in this way, thus the standards are effectively established and fulfilled, and the service is improved. Second, with soft technology, work procedures are modified and perfected. Although this often implies some sort of modification to tools, the essential characteristic is within the system itself, where routines are specifically designed to produce the desired effect. In tourism, it is possible to standardise some aspects of the service, for example, the package deals offered by tour operators. Third, there are hybrid methods which allow the simultaneous use of both technologies.

Both the automation required to maintain a high level of consistency in the tourism industry and the development of programmes that allow improvement to operational processes constitute the basis for structuring important elements in the service provision. Despite the inherent difficulties in the industry, standardisation is essential given that only in this way will it be possible to establish objectives that must meet or surpass such standards with the aim of achieving client satisfaction.

References

Horovitz, J. (1986) 'La non-qualité tue', *Harvard-L'Expansion* 53–61.

Levitt, T. (1976) 'The industrialisation of service', *Harvard Business Review* September–October: 63–74.

Saleh, F. and Ryan, C. (1992) 'Conviviality – a source of satisfaction for hotel guests? An application of the servqual model', in P. Johnson and B. Thomas (eds), *Choice and Demand in Tourism*, New York: Mansell, 107–22.

Zeithaml, V.A., Parasuraman, A. and Berry, L.L. (1990) *Delivering Quality Service*, New York: The Free Press.

IRENE GIL SAURA, SPAIN

state park *see* park

statistics

Formally, 'statistics' describes both a field of study in mathematics, and certain quantitative characteristics of a sample. The former sense includes descriptive statistics, the organisation and summary of quantitative information from a collection of actual observations of a phenomenon, such as tourism demand, and inferential statistics, the generalising from a sample of actual observations to a larger population. Statistics in the latter sense is used to describe a sample of observations quantitatively. The mean, median and mode are used to identify the centre or middle set of values observed. Range, standard deviation and variance indicate the spread of the observations.

Statistics are gathered on the **behaviour** of tourists and other travellers (in relation to tourism demand, for example) for **market segmentation**, selecting target markets, monitoring tourist trends, determining the effectiveness of marketing programmes, **forecasting demand** and estimating the economic impact of tourism. Tourism demand data of most value include tourist characteristics (such as place of residence, **gender**, age, level of education, occupation, household income) and trip characteristics (including origin, **destination**, duration, purpose, transport, accommodations, related expenditures).

Tourism supply statistics are usually collected from establishments, the single physical locations where productive activity takes place. These include places of **accommodation**, **transportation** terminals, eating and drinking places, sites of recreational, cultural and sports activities, and offices of other services such as **travel agencies** and **tour operators**. Statistics are gathered on location, capacity, quality rating (stars, crowns or other rating terms), amenities, prices, business receipts, employees, wages and salaries paid, taxes paid and net income.

To be of greatest use, tourism statistics should be comparable across space and time. Unfortunately, this has not been the case. Even national tourism administrations do not agree on how to define basic terms that ensure comparability among countries' data on international tourism behaviour. However, the **World Tourism Organization**, an affiliate of the United Nations, is working to standardise tourism terminology and classifications throughout the world, through training manuals and technical seminars.

See also: quantitative methods; travel survey

Further reading

Cunningham, S. (1991) *Data Analysis in Hotel & Catering Management*, Oxford: Butterworth-Heinemann. (Application of statistical principles to data available to hospitality managers for solving operational problems.)

Goeldner, C.R. (1994) 'Travel and tourism information sources', in J.R.B. Ritchie and C.R. Goeldner (eds), *Travel, Tourism and Hospitality Research: A Handbook for Managers and Researchers*, 2nd edn, New York: John Wiley & Sons, 81–90. (A guide to English-language sources of current data on travel, tourism and hospitality.)

WTO (1995) *Collection and Compilation of Tourism Statistics*, Madrid: World Tourism Organization. (Discusses objectives and principles of conducting national tourism surveys of inbound or outbound international arrivals.)

DOUGLAS C. FRECHTLING, USA

stereotype

Borrowing a term from printing **technology**, social scientists refer to simple, inaccurate and change-resistant ideas about the characteristics of other groups of people as stereotypes. Rather than being formed with individual experience, stereotypes about different groups tend to be learned second-hand. Because they create expectations for encounters between different kinds of people, stereotypes are a natural subject of interest to scholars of tourism.

Brewer's (1984) research on interactions between **local** shopkeepers and Mexican and non-Mexican tourists in Baja, a Californian resort town illuminates how the shopkeepers' stereotypes serve several functions. First, the stereotypes they hold help the locals to make sense of their tourists' **behaviour**. Second, business transactions are conducted with the expectation that the parties will behave in predictable ways. The merchants' stereotypes of non-Mexicans and Mexicans are relative rather than fixed, contrasting with one another.

One of the clearest examples of stereotypes shaping tourism is found in the American Southwest, where large numbers of European-Americans visit tribal lands. There, non-native understanding of American Indians has been organised into stereotypes by media, especially Hollywood movies that had no particular knack for accuracy or diversity in their portrayals of Native Americans (O'Connor 1980). Two reciprocally titled articles, 'How "they" see "us"' (Evans-Pritchard 1989) and 'How "we" see "them"' (Laxson 1991), emphasise the negative and oppressive stereotypes of American Indians held by tourists, but also the empowering character of Indians' stereotypes of obnoxious tourists in preserving local privacy even as they leave, thinking that they really know the **indigenous** people.

Stereotypes tend to be negative attributions, but they may also be positive. The transient character of tourism dictates that long-term, in-depth relationship-building between tourists and hosts is rare. The research mentioned above suggests that in the relative absence of such relationships, stereotypes help to guide fleeting encounters by injecting some predictability into interactions, with each attempting to meet the expected needs of the other. To a certain extent, hosts and guests alter their behaviour to fit the stereotypes assigned to them.

References

Brewer, J.D. (1984) 'Tourism and ethnic stereotypes: variations in a Mexican town', *Annals of Tourism Research* 11(3): 487–501.

Evans-Pritchard, D. (1989) 'How "they" see "us": Native American images of tourists', *Annals of Tourism Research* 16(1): 89–105.

Laxson, J.D. (1991) 'How "we" see "them": tourism and Native Americans', *Annals of Tourism Research* 18(3): 365–91.

O'Connor, J.E. (1980) *The Hollywood Indian: Stereotypes of Native Americans in Films*, Trenton, NJ: New Jersey State Museum.

ROLAND S. MOORE, USA

strangeness

Strangeness is the opposite of familiarity. It signifies that which is unknown, or even incomprehensible, in terms of the accustomed categories of one's 'thinking as usual'. Moderate strangeness in another society can be overcome since a newcomer can be socialised into it. Radical type, on the other hand, is beyond human intelligibility, and hence cannot be overcome. Such strangeness is experienced as profoundly ambivalent; whereas it is uncanny, threatening and dangerous, it is also – in contrast to the familiar – exciting, alluring and fascinating. This ambivalence is expressed in the contrasting symbols through which radical strangeness is frequently represented (the demonic and the divine). In the cosmological imagination of many, particularly premodern peoples, the familiar, ordered 'cosmos' is surrounded by primal 'chaos', a strange, uncreated and hence incomprehensible expanse, inspiring fear. It is populated by threatening, uncategorizable beings, such as demons and monsters. However, within or beyond these chaotic surroundings is hidden the numerous, divine Other, the source of all creation. **Wilderness**, such as primeval forests and deserts, these concrete symbols of the chaotic surroundings of the ordered, inhabited world, therefore attract mystics and hermits, who seek in them an encounter with the divine.

Exposure to moderate strangeness may also provoke some ambivalence. However, the extent to which it attracts or repels individuals depends on the intensity of their exposure to it, and the degree of their prior preparation and experience with similar encounters. For an unprepared, inexperienced person, exposure to even a low degree of strangeness may be threatening: for example, travelling alone in a strange modern city without the knowledge of the local language. By contrast, a well-prepared, experienced individual may find a solo unassisted crossing of Antarctica an attractive challenge.

Unmitigated exposure to strangeness can cause a disorienting and incapacitating **culture shock**. The force of this shock may be lessened by the intervention of such factors as adherence to a group, professional assistance, of the shelter of an 'environmental bubble' in which individuals are provided with some familiar aspects of their home environment. This bubble offers a level of protection and **security** which enables them to function in strange surroundings.

Tourists generally desire to experience some degree of **novelty** and strangeness on their trips: to see people and **landscapes** differing from those at home, to smell and taste other **cuisine** and enjoy another **climate**. Nevertheless, they are frequently scared of exposure to the more extreme manifestations of strangeness, such as, for example, volcanic moonscapes or savage **natives**.

Such strangeness, to remain enjoyable, has to be experienced from the **security** of some familiar shelter, ameliorated by a touch of familiarity, or even demarcated as 'unreal', as a show which is not part of reality. Thus, tourists can observe the Arctic wilderness from the comfort of low-flying planes or cruiseliners, without being exposed to the severity of the elements. They can savour local cuisine in tourist **restaurants**, where the taste has been adapted to suit western palates; and they can playfully enjoy cannibalistic rites, aware that no real killing of human beings is involved.

Whereas some tourists are readily prepared to expose themselves to the strangeness of the host environment – especially if they avidly seek 'authentic' experiences – others tend to confine themselves to the protective familiarity of **resort** hotels, **restaurants** and guided tours, ever afraid of exposure to their unfamiliar surroundings. Indeed, the preferred relationship between familiarity and strangeness in tourists' travelling styles has been employed as the basis for a typology of tourists, ranging from 'organized mass tourists' to 'drifters' (Cohen 1972).

References

Cohen, E. (1972) 'Towards a sociology of international tourism', *Social Research* 39(1): 164–82.

Further reading

Furnham, A. (1984) 'Tourism and culture shock', *Annals of Tourism Research* 11: 41–57.

ERIK COHEN, ISRAEL

stranger

Strangers are people who are spatially close to one, but are socially or culturally remote. The classical approaches to 'the stranger' as a social type deal with individuals who either sojourn permanently in the host setting (Simmel 1950; Siu 1952) or who are newcomers to it, and intend to settle in it (Schuetz 1940). Even those sociologists who deal explicitly with temporary strangers refer primarily to expatriates, students or foreign workers, whose stay in the host setting is relatively extended. Travellers, who in Simmel's words, 'came today and leave tomorrow', have mostly been excluded from the **discourse**. Stranger theory has up to now contributed little to the study of tourism, and it is thus necessary to build a conceptual bridge between the sociology of the stranger and of tourism.

The **sociology** of the stranger dealt primarily with two major issues: the relations of strangers with the host society or their position in it, and the process of their transition from strangeness to familiarity within the host society. The former relates to the social aspect of the strangers' strangeness, the problem of their '**acculturation**'. Within the latter aspect, two further dimensions can be distinguished: cognitive (strangers' comprehension of the cultural patterns, the values, rules and customs of the hosts), and normative (their identification with these values, rules and customs). It is important to note that the three dimensions of the strangeness, the social, cognitive and normative, do not always overlap; there may be a disjunction between the social and the two cultural dimensions. Strangers may acculturate, but, owing to the nature of the host society's rules of membership, still not be accepted. In this case they remain as 'acculturated outsiders'. There may also be a disjunction between the two cultural dimensions: strangers may comprehend the hosts' cultural pattern, but not identify with it, a situation typical of some intercultural roles, like the implanted spy, the anthropologist and the missionary.

Tourists do not seek acceptance in the host society, except as 'guests', and neither do they seek to acculturate, even though they may show interest in the host culture. In these respects, they in some degree resemble the other disjunctive intercultural roles. Tourists frequently seek an explanation of the host **behaviour** or customs, in order to understand them, however superficially, but they only rarely seek to identify with the host culture, except playfully. It should be noted, however, that an opposite trend can be found among some counter-cultural tourists: they are often 'infatuated' by the hosts' culture, which they seek to 'experience' but do little to understand.

Newcomers, and strangers sojourning in a host setting, typically experience some degree of **culture shock** in their encounter with it, and feel uncomfortable or unable to function normally. Therefore, strangers of a common background tend to create their own 'environmental bubble' within this setting, establishing their own associations, schools, services, places of worship and neighbourhoods which resemble the surroundings with which they are familiar. Travellers, who may suffer even more from direct exposure to strangeness than more permanent strangers, do not have the opportunity and means to establish an 'environmental bubble' by themselves. Personal guests are provided the equivalent of such a bubble by the home of their hosts. Impersonal guests, such as tourists, make use of a commercialised bubble consisting of hotels, restaurants and a variety of travel and personal services provided by the local tourism industry. Various tourist types can be distinguished by the extent to which they confine themselves to the familiarity of that bubble, or tend to expose themselves to the strangeness of the host surroundings.

References

Schuetz, A. (1944) 'The stranger: an essay in social psychology', *American Journal of Sociology* 49(6): 499–507.

Simmel, G. (1950 [1907]) 'The stranger', in H. Wolff (ed.), *The Sociology of George Simmel*, London: Free Press, 402–8.

Siu, P.C.P. (1952) 'The sojourner', *American Journal of Sociology* 58: 34–44.

Further reading

Harman, L.D. (1988) *The Modern Stranger: On Language and Membership*, Berlin: Mouton de Gruyter.

Yoshida, T. (1981) 'The stranger as God: the place of the outsider in Japanese fold religion', *Ethnology* 20(2): 87–99.

ERIK COHEN, ISRAEL

strategic business unit

A strategic business unit, in tourism as well as other establishments, is the primary generator of cash flow for that firm. It is where resources are concentrated in order to produce revenues and incur costs associated with the production of goods and services that make up the firms core competitive methods.

Further reading

Pearce, J.A. and Robinson, R.B. (1994) *Strategic Management*, Boston: Irwin, 344–6.

MICHAEL D. OLSEN, USA

strategic marketing

Strategic marketing involves the analysis, **planning**, implementation and control of an organisation's efforts over the longer term to satisfy customer needs and wants in the context of a competitive microenvironment and a set of macroenvironmental conditions and trends. Strategic marketing is a central sub-element of an organisation's or corporation's search for direction. It is undertaken as part of the organisation's overall planning process. Together with other functional plans concerning finance, production/operations and human resource management, strategic marketing plans identify how a corporation and its various strategic business units are to achieve their long-term vision and objectives

Whereas **marketing management** deals with an organisation's activities to develop, implement and direct marketing programmes designed to achieve designated business goals, strategic marketing focuses on the formulation of those goals, and the means and timing of realising their achievement. The interplay of three forces, known as the strategic 3 Cs – the customer, the competition and the corporation – is central to strategic marketing. Effective plans must deliver better value to customers (or tourists), while capitalising on corporate strengths and addressing weaknesses, and differentiating the organisation's product effectively from its competitors. Trends in the broader macroenvironment (demographic, economic, sociocultural, technological, ecological and political/legal) influence the application of these principles at any point in time.

Strategic marketing plans connect the organisation's present competitive position to its desired future state in a deliberate, planned execution of a coordinated set of specific steps. It recognises the organisation's unique competitive advantages and **resources**, and deploys them in a way which ensures that it can sustain these competitive advantages over the long term. In short, it deals with where, how and when the business is to compete. Strategic marketing has become vital in today's aggressive, rapidly changing business environment, such as in tourism, where consumer tastes are dynamic, competitors are quick to respond to market opportunities and the array of different needs evident in a more heterogeneous market, and environmental changes seem to be occurring more frequently.

See also: corporate strategy

Further reading

Cravens, D.W. (1987) *Strategic Marketing*, Chicago: Irwin.

Day, G.S. (1984) *Strategic Market Planning: The Pursuit of Competitive Advantage*, St Paul, MN: West Publishing Company.

Luck, D.J., Ferrell, O.C. and Lucas, G.H., Jr (1989) *Marketing Strategy and Plans*, Engelwood Cliffs, NJ: Prentice-Hall.

GEOFFREY I. CROUCH, AUSTRALIA

strategic planning

As the tourism market becomes saturated and firms attempt to adapt and succeed in an increasingly dynamic and turbulent competitive environment, there is an increased emphasis among managers on the use of strategic planning and decision making. To some scholars and practitioners, the term strategic planning connotes a process that represents part of strategic management. The latter refers to the broad overall process, while strategic planning involves the formulation phase of total management activities.

Strategic planning becomes one of top management's major tools for coping with changes in the **environment**, to gain a **competitive advantage** and to achieve long-term survival and success. Through this process, an organisation determines its objectives, generates and evaluates alternative strategies, develops and maintains an optimal 'fit' or 'co-alignment' between the deployment of an organisation's resources and structure and the opportunities in its changing environment, and monitors the results of the implemented plan. The nature of strategic planning is for the long term, usually three to five years. It is not a static process; it evolves as the size and structure of the company changes. Thus, to achieve the ultimate goal of a firm, the devised strategic plan should be continuously evaluated and revised as it is carried out.

The concept of strategy can be traced back to the Old Testament and the *Art of War* by the Chinese strategist Sun Tzu, which is still widely read, especially by military personnel. The adoption of the concept by the business community came after the Second World War. The 'rise' of strategy can be traced to the 1960s, when scholars at the Harvard Business School articulated the concept of strategy as a tool to link together the functions of a business and to assess a company's strengths and weaknesses against competitors. Strategic planning came to prominence and became a feature not only of the corporate organisation but also of the business school curriculum.

However, in the early 1980s, facing global competition, companies turned away from strategic planning and began to focus on operational improvement and total quality management. By the late 1980s, corporate America began massive downsizing and re-engineering of operations to increase efficiency and productivity. In the 1990s, after years of downsizing, companies began to focus on how to grow. This time strategic planning came back, but with a difference. It is no longer perceived to be an abstract, top-down approach with corporate-level top management initiating the strategy formulation process by calling upon divisions and functional units to formulate their own strategies as ways of implementing corporate-level strategies. Now, companies are urged to democratise the process by delegating strategic planning activity to teams of line and staff managers from different disciplines (Byrne 1996). Strategic planning and total quality management have received increasing attention in recent tourism studies.

See also: strategic formulation; strategy marketing

References

Byrne, J.A. (1996) 'Strategic planning: it's back', *Business Week*, 26 August, 46–51.

ELIZA CHING-YICK TSE, CHINA

strategy formulation

Strategy formulation is the process followed by tourism and other organisations to develop their strategic plan. It should be viewed as an iterative rather than linear process. It begins with the scanning of the business environment in order to identify threats and opportunities. This effort is designed to detect both long-term and short-term trends affecting the business. This step in the process is then followed by the creation/evaluation of the mission statement which defines what business the firm is, or plans to be, and in what environmental domain it will compete.

Following the mission statement, the firm decides on the competitive methods it will chose in order to take advantage of the opportunities in its domain environment. The competitive methods are viewed as the primary value-producing activities of the business. A close match must be achieved between the opportunities in the environment and the competitive methods chosen. Once completed, the firm will then assess its strengths and weaknesses to determine if it has the **resources** and capabilities to properly implement and execute the chosen methods. This assessment will determine whether the firm will be able to realise the overall strategy it has chosen through its selection of competitive methods.

Once the firm has identified its strengths and weaknesses, it is then ready to set long-term and short-term objectives. Long-term objectives are considered to be one year or more in the future. The objectives should be designed so as to enable the firm to overcome its weaknesses and take full advantage of its strengths. Each objective should be tied directly to a particular competitive method or set of methods. The objective must identify the physical, financial and human resources necessary to implement and execute the chosen strategy.

Once the objectives are finalised and approved, and resources allocated, implementation begins. This process involves the actual utilisation of resources for the successful execution of processes and activities associated with each competitive method. While implementation is technically not a direct part of the formulation process, it is the link to strategy evaluation. Evaluation involves the assessment of the success of each competitive method in adding its targeted value to the firm. Results of this evaluation are used to continually cycle through the formulation process.

See also: corporate strategy

Further reading

Hofer, C.W. and Schendel, D. (1978) *Strategy Formulation: Analytical Concepts*, St Paul, MN: West Publishing.

Lombardi, D. (1994) 'Chain restaurant strategic planning', *Cornell Hotel and Restaurant Administration Quarterly* 35(3): 38–40.

Reid, R. and Olsen, M.D. (1981) 'A strategic planning model for independent food service operators', *Journal of Hospitality Education* 6(1): 11–24.

MICHAEL D. OLSEN, USA

structuralism

Borrowed from linguistics, anthropological structuralism is a mode of analysis focusing on the 'grammar' of social relations and ideological formations. Steeped in this tradition, and inspired by the studies of myth and totemism by Levi-Strauss and others, the structuralist tradition in tourism studies owes most to the writings on contemporary mythologies by Roland Barthes (see **connotation**).

TOM SELWYN, UK

substitution

Substitution typically implies that for a pair of 'things' (such as x and y) that replace each other, there exists a measure specifying an individual's willingness to *interchange* x with y, or a measure giving the degree to which an aggregate (like population) is 'willing' to interchange x with y. For instance, x and y may be different trips, activities, **accommodation** types and/or quality; the choice may involve choosing a weekend trip or concert; a cruise or tour, and a 5-star or 4-star **hotel**.

Substitution has many different meanings. For individuals, various correlation measures used confuse substitution with complementarity (doing x to some degree results in y) or preferences (doing x and y meet different needs). Much of what is written is not quantitative. 'Perfect' substitution or interchangeability, whether referring to (x,y) or (y,x) does not matter, is implicit in much of what is written (Wyman 1982). When substitution is used in relation to **regression** models, other than individual conjoint models, for example, substitution is an aggregate concept. Properties of the aggregate are often, without justification, ascribed to each individual. Aggregate substitution recognises such trade-offs as price or location, and considers consequences of forced substitution. Trade-offs identified need not imply individuals' willingness to substitute.

Aggregate substitution measures may reflect **displacement** from activities, facilities/services or **adaptation** to change. User **conflict** or resource conflict can relate to change processes involving substitution, but more often **displacement** should be considered. Substitution at the segment or individual level need not show up in factors, regression coefficients or other population level measures. Aggregation masking substitution is expected with segments with conflicting preferences. It is recognised that substitution is important in forecasting change or assessing the consequences of change. However, studying **supply** change, changing **spatial interaction** of populations, changing individual preferences and the like can yield valid information about apparent substitution. The existence of constraints and non-homogeneous supply can also lead to an incorrect impression about substitution occurring. Non-substitution factors must be controlled and factored out for results to be valid.

References

Wyman, M. (1982) 'Substitutability of recreation experience', *Leisure Studies* 1(3): 277–93.

Further reading

Anderson, D. and Brown, P. (1984) 'The displacement process in recreation', *Journal of Leisure Research* 16(1): 61–73.

Beaman, J. (1975) 'Comments on the paper "The substitutability concept: implications for recreation research and management", By Hendee and Burdge', *Journal of Leisure Research* 7(2): 146–52.

Brunson, M. and Shelby, B. (1993) 'Recreation substitutability: a research agenda', *Leisure Sciences* 15(1): 67–74.

JAY BEAMAN, CANADA

summer cottage

A summer cottage is a **second home** used on a seasonal basis usually during the summer, particularly common in North America and Scandinavia. These properties may be user-owned or rented on a short-term basis. Many summer cottages have been converted to year-round **leisure** use or permanent residences, frequently as retirement homes. Generally found in high amenity settings, often on waterfront locations, they range in scale from shacks to mansions.

See also: seasonality

RICHARD BUTLER, UK

sun, sand, sea and sex

The expression 'sun, sand, sea and sex' is normally used as a mnemonic to denote **mass tourism**. It is associated (rightly) with the environmental threats of ill-planned free market mass tourist **development**. The 'unholy quartet' has also been used by some commentators, nostalgic for a time when travel was the exclusive domain of the upper and middle classes, to cast pejorative aspersions on mass tourism itself.

TOM SELWYN, UK

sunlust

The term 'sunlust' coined in the early 1970s to describe tourism that is resort-based and motivated by the desire for rest, relaxation and '**sun, sand, sea and sex**'. Important features of sunlust tourism are **climate**, comfort and familiarity of **accommodation** and cuisine. It is normally associated with mass, package tourism as opposed to independent, explorer-type travel.

See also: ludic; liminality; play

RICHARD SHARPLEY, UK

supply

Supply, in economic terms, is a schedule showing the amount of a **product** that will be made available for purchase at various price levels. Its opposite, **demand**, is a schedule showing the amount of a product purchased at various price levels. The intersection of the two is an equilibrium point where the amount produced and purchased at a certain price level equal each other. This microeconomic view is vitally important to planning at the individual firm level, including tourism dependent businesses. Because the experiential tourism product is, in effect, produced and consumed at the same time, it is highly perishable. Oversupply of its marketable physical components (such as **hotel** rooms) is a costly situation. On the other hand, too few of the physical components lead to lost sales that cannot be made up at some time in the future.

Most economists would argue that supply of tourism products is dependent on demand. That is not entirely the case in other industries. A farmer can use high-yield farming practices to produce additional corn that is then sold at a **market**-determined price. Even if the majority of farmers adopted high-yield practices and demand remained constant, an individual farmer's revenue could increase if the cost of new farming practices were less than the increased revenue from the additional product produced. Since tourism is a **service**-oriented business, it is very difficult to employ new 'high yield practices'. Demand for tourism products is to a large extent dependent on factors outside of the individual manager's control. For example, the **image** an area projects may have more to do with demand than the products of any one business within the **destination** area. However, there are some things businesses and the communities in which they are located can do to increase demand for the product and, consequently, supply.

Clustering similar type businesses, a process known as organic bunching, can lead to a synergistic effect where total sales would exceed the sum of what could be obtained by a number of individual businesses working independently. Developing a theme adopted by each business and represented in their building design has also been shown to increase demand. In the tourism season, lengthening strategies could also be employed. Developing enough of the physical components of the tourism product to serve a high-use season can be a risky strategy especially when those components are expensive to construct and maintain (like accommodations). **Destination** areas with historically high and low-use periods (including alpine **resorts**) are continually searching for touristic offerings, such as off-peak festivals, which allow for season lengthening. When such strategies are employed, supply shifts from a microeconomic perspective to that of a macroeconomic one. The conglomeration of businesses within a region or destination area forms the physical components of its tourism supply base.

Tourism supply is a function of an area's natural and socioeconomic characteristics as well. In fact, many tourism businesses may be better classified as secondary attractions or supporting businesses. Often, it is an area's culture and/or natural resource that form the true attraction supply component of tourism. In some cases, it can even be the ephemeral quality of faith that supplies the attraction in places such as Medugorje, Croatia or Lourdes, France.

Similar to business clustering described above, attraction clustering might serve to draw tourists thereby creating demand. Las Vegas (USA) is an example of this type of phenomenon. In this case, supply of the physical components merges to create demand for a product that is part **gambling** and part **fantasy**. This phenomenon, where supply creates demand, is known as Say's Law.

Businesses that offer services including non-profits (including tourism associations) and educational institutions are also part of the supply picture. However, classifying a business as part of the tourism industry or some other sector is problematic. Tourism-dependent businesses often serve residents, and vice versa. There is no standard industrial classification code for tourism as there is for almost all other business pursuits. That is one reason that when tourism supply is discussed, it is its component parts rather than the supply of a particular product which are examined. However, attempts to account for the value of tourism to an area, related to the purchase of the component parts, is an active area of study.

Different **accounting** systems using existing frameworks have shown promise not only for identifying tourism-dependent businesses but for determining their combined economic impact. These **satellite accounts**, as they have become known, are at the forefront of research in this area. Other areas of research pursuit dealing with the supply of tourism product components include locational analysis for new businesses. This is again a microeconomic perspective, and although numerous **models** exist for retail firm location, very few of these models are applicable for tourism-dependent firms where the market base may be a wide geographic area, even the world. However, certain methods have been used to determine the relative supply of certain tourism resources within a sub-unit of a much larger geographical area. This **supply** side analysis (Smith 1987) is useful for formulating **development policy** with respect to the type of tourism most likely to succeed within a defined development zone.

Defining supply as it relates to tourism is not an easy task. What analysts are comfortable with is describing the supply of the physical components of the product provided by businesses and the attractions which draw people to the area. The combination of these form a tourism product which is individually distinct.

References

Smith, S. (1987) 'Regional analysis of tourism resources', *Annals of Tourism Research* 14(2): 254–73. (Describes a procedure for defining tourism regions on the basis of resource patterns.)

Further reading

Gartner, W.C. (1996) *Tourism Development: Principles, Processes, Policies*, New York: Van Nostrand Reinhold. (Part IV of this book deals with tourism demand/supply.)

Smith, S. (1995) *Tourism Analysis*, 2nd edn, Essex: Longman. (Examines the more important quantitative methods utilised by planners, researchers and consultants.)

WILLIAM C. GARTNER, USA

survey

Surveying is a method of gathering information directly from a sample of residents or tourists, with the objective of inferring certain trip and personal information about a larger host or guest population. It is an extremely cost-effective way of determining the characteristics and behaviour of a larger population, such as the residents of a nation or their visitors. It is widely used to gather **statistics** for **marketing research**, **planning**, programme evaluation and **impact** assessment in tourism.

The conduct of a valid survey is a complex task, and certain elements are key to its success. First, the target population, the larger group the survey should represent, must be carefully defined. This is most often the residents of a country or tourists visiting an **attraction** or **destination**, but may include all passengers on an airline or all guests at a **hotel**. Next, a sampling frame – that is, a list of all units in the population, such as all households in a country – or rules for identifying all such units, such as a computer programme that randomly generates household telephone numbers, must be specified.

The survey planner must determine how many respondents to interview (the sample size) and where to interview (the venue). For statistical reasons, larger samples produce more reliable results than smaller ones. However, a sample size of 1,500 to 2,000 is adequate for many tourism research purposes. Popular venues for surveys include national frontiers, **transportation** modes, places of **accommodation** and households.

The rules for selecting the respondents to be included in the sample must be specified. These rules must allow every one in the target population to have a known, non-zero chance of being selected for the sample. If either the interviewer determines who is interviewed, or respondents choose whether to be included or not, the validity of the survey is compromised. The questionnaire or interview form must be carefully designed to avoid ambiguity and administered to the sample by trained interviewers. Better surveys pretest the sample selection procedures, the interview process and the questionnaires in advance to remove obstacles to the smooth conduct of a valid final survey.

Further reading

WTO (1994) *Conducting and Processing a Visitor Survey, Instructional Materials*, Madrid: World Tourism Organization. (Outlines procedures for planning, organising, conducting, processing and analysing a survey of visitors to a destination country or other area.)

—— (1995) *Collection and Compilation of Tourism Statistics*, Madrid: World Tourism Organization. (Discusses objectives and principles of conducting national tourism surveys of inbound or outbound tourists and tourism.)

DOUGLAS C. FRECHTLING, USA

survey, guest

One method of collecting data on tourist characteristics is represented by a guest **survey**. This kind of **marketing research** addresses the **tourist** at the **destination**. Therefore, it emphasises other aspects of **behaviour** than a **travel survey** based on household samples.

Guest surveys are administered to gather in-depth knowledge about tourists' needs, spatial mobility, consumption of natural, transport, and cultural facilities as well as their **satisfaction** levels. Scanning the tourists with regard to their demographic (see **demography**), socioeconomic and **psychographic** characteristics provides information for tourism **marketing** and effective **advertising**. Guest surveys are conducted in different ways. Front-end surveys are used for collecting a limited set of trip characteristics that are not subject to change by the trip itself. Possible variables are **length of stay**, type and means of **accommodation**, principal destination, travel party, geographic origin, travel preparation and organisation.

Surveys during the stay focus on variables such as tourist activities, interests, attitudes, benefits sought and **satisfaction**, as well as spending patterns. During-visit surveys may be administered unsupervised (with paper and pencil or interactively by computer interface) or supervised by personal interviewers (see **interview**). Whereas the unattended version leads to low field costs in combination with low response rates and a biased representativeness, personal interviews have substantially higher costs but offer the advantage of receiving more reliable answers and shortening the field process. For particular purposes such as diagnosing spending patterns, **time budget** allocation or **leisure** participation, it is appropriate to apply diary surveys that keep records of the tourist's daily activities. Before leaving the country, tourists may be invited to deliver different information about their trip. This type of research is called exit surveys, and is preferably conducted at departure lounges of **airports** or harbours.

See also: causal model; expectation; guests; image; impacts; marketing research; quality

Further reading

Chadwick, R.A. (1994) 'Concepts, definitions and measures used in travel and tourism research', in R.B. Ritchie and C.R. Goeldner (eds), *Travel, Tourism and Hospitality Research*, New York: Wiley, 65–80. (Discusses measurement concepts and different survey approaches in tourism research.)

Frechtling, D. (1994) 'Assessing the impacts of travel and tourism: measuring economic benefits', in R.B. Ritchie and C.R. Goeldner (eds), *Travel, Tourism and Hospitality Research*, New York: Wiley, 367–91. (Examines different approaches to estimating travel expenditures.)

Veal, J.J. (1992) *Research Methods for Leisure and Tourism: A Practical Guide*, Harlow: Longman.

ANDREAS ZINS, AUSTRIA

sustainable development

In 1989, 'Our Common Future', the report of the United Nations World Commission on Environment and Development (commonly called the Brundtland Commission after the Norwegian chair) was published. Although the concept has other precursors, this report brought sustainable **development** to prominence and defined it as development that meets the needs of the present without compromising the ability of future generations to meet their own needs. As a minimum, it was suggested that sustainable development would require the maintenance of ecological integrity and diversity, the meeting of basic human needs, keeping options open for future generations, reducing injustice and increasing self-determination. Thus, it encompasses a long-term perspective and notions of equity among individuals, between present and future, and between humans and other organisms.

Apparently addressing the tension between economic development and environmental preservation, and perhaps partially because of its imprecision which permits a multitude of interpretations, sustainable development has received widespread acceptance by governments and representatives of most economic sectors. The tourism industry has endorsed the concept, but often from a narrow perspective; sustainable tourism is thought of as tourism in a form which can maintain its viability in an area for an indefinite period of time. Unfortunately, issues of sustainability often lie in the competition between economic sectors for the scarce **resources** of land, water, energy and waste assimilation capacity. It would thus be wrong to assume that the perpetuation of tourism, of necessity, leads to sustainable development more broadly conceived. It follows that a comprehensive perspective is required so that sustainable development in the context of tourism might be defined as that which is developed and maintained in a form such that it is viable over an indefinite period, and does not degrade or alter the physical and human **environment** to an extent that it prohibits the successful development and well-being of other activities and processes.

Given the imprecision of the concept, which can be used to describe a philosophy, a process, a **product** or a plan, it should not be surprising that it has proven difficult to identify means of implementation or to assess the sustainability of particular tourism initiatives. Much effort is being directed, with mixed success, to the identification of sustainable development indicators. The **World Tourism Organization** has suggested a number of national indicators which span economic, environmental and social domains but also recognises that the manifestations of tourism are so varied and occur in such diverse environments that site-specific indicators will need to be developed and adopted.

See also: ecologically sustainable tourism; environmental valuation

Further reading

Hunter, C. (1997) 'Sustainable development as an adaptive paradigm', *Annals of Tourism Research* 24(4): 850–67.

Nelson, J.G., Butler, R. and Wall, G. (eds) (1993) *Tourism and Sustainable Development: Monitoring, Planning, Managing*, Waterloo: Department of Geography Publication Series No. 37, University of Waterloo.

Wahab, S. and Pigram, J.J. (eds) (1997) *Tourism Development and Growth: The Challenge of Sustainability*, London: Routledge.

Wall, G. (1997) 'Is ecotourism sustainable?', *Environmental Management* 21(4): 483–91.
World Tourism Organization (1993) *Indicators for the Sustainable Management of Tourism*, Madrid: International Working group of the Environment Committee of the World Tourism Organization.

GEOFFREY WALL, CANADA

Sweden

Tourism in Sweden is 80 per cent domestic. In terms of **inbound** tourism, Germany and Norway are the main markets. The 8.9 million Swedes have a high tourism propensity, with imports twice the size of exports. Main **outbound** destinations are the other Nordic countries, Spain and Greece. The national tourism board is a company jointly owned by the state and the tourism industry.

LARS NYBERG, SWEDEN

Switzerland

Switzerland is a traditional tourism country, which in the early nineteenth century earned a reputation as 'the playground of Europe'. The alpine charms of the world's oldest living democracy, in a country which welcomed foreigners, proved above all irresistible to the English. The fact that the Swiss at that time also happened to be innovative pioneers in such key areas as the development of mountain railways and grand hotels also helped. In 1844 Karl Baedeker, creator of the celebrated travel guides, noted that: 'Switzerland unquestionably has the best hotels in the world.'

The Swiss maintained a near-monopoly of international tourism based on alpine hotels right up to the outbreak of the First World War. The number of **hotel** overnight stays in 1914 was greater than it is today. In the period between the wars, Swiss hotels lost considerable market share. It was not until the 1960s that Swiss tourism enjoyed a renaissance. With major improvements in the living standards of the Swiss themselves, many of whom had moved from the countryside to the towns and cities, a new domestic market began to grow up to rival the tradition of international tourism with its luxury hotels. Less service-orientated and more price-conscious, this new type of tourism required an entirely different kind of **accommodation**. The end of the Second World War again brought new opportunities with the democratisation of **skiing**, which allowed Switzerland to position itself as a winter sports **destination**. Since then, Swiss tourism has tended to cater for all four seasons.

This restructuring, both exogenous and endogenous, coincided with an unexpected boom period in tourism. In most cases, the new **demand** concentrated on the 'flagship' **resorts** of the earliest day of Swiss tourism, St Moritz, Interlaken, Locarno-Ascona, Montreux and so on, all of which can be easily reached by rail. The next round of democratisation, courtesy of the automobile, brought an expansion of tourism in terms of the total surface area covered. However, the new tourism resorts that sprang up after the Second World War continued to revolve around winter sports. The need to come to terms with the external effects of a continuous tourism growth which lasted right up to the end of the 1970s presented the authorities responsible for tourism and area **planning** with one of their greatest challenges. It was at this time that a new comprehensive tourism policy began to evolve, designed to encourage qualitative as opposed to quantitative growth. This policy today forms the basis of the Swiss concept of tourism.

The arrival on the world scene of new destinations, and an ever-greater competition in terms of price and **quality** in the international **market**, resulted in a period of stagnation for Swiss tourism, beginning in the early 1980s. The efforts of Swiss promoters were hampered by the high exchange rate of the Swiss franc. Moreover, the mainly small and medium-sized enterprises which are the backbone of its industry had difficulty adapting to the increasingly industrial scale of **international tourism** structures. The prolonged recession throughout Europe produced a real slump in the industry at the beginning of 1990s. Since that time, however, this crisis seems to have been resolved. New government measures designed to encourage a more innovative and united approach to the international market, on a quasi-industrial scale, have succeeded in improving the international competitiveness of Swiss tourism.

If revenues from cross-border tourism are used as the yardstick, then Switzerland remains one of the top ten tourism countries in the world, and has been so for quite some time. This country also holds the record for the highest **value added** per employee. With approximately 75 million overnight stays, 2.5 million beds in hotels or other forms of accommodation including second homes, and a great variety of experiences on offer in some unique destinations, Switzerland will remain one of the world's most important tourism countries in the foreseeable future.

PETER KELLER, SWITZERLAND

SWOT analysis

Strengths, weaknesses, opportunities and threats (SWOT) analysis is a synthesis technique used to identify potential management opportunities for a business, property or organisation as part of **strategic planning**, **management** or **marketing**. The technique is applied systematically to summarise the major SWOT that have been examined from other analyses of the business, such as **market**, site and competition analyses. A matrix is developed that allows for cross-referencing information from various analyses to determine common themes. The major and recurring themes build to become the SWOT of the business.

To illustrate the nature of the analysis, examples of this analysis are presented here in the context of a **community** better positioning itself for tourism. *Strengths* are characteristics that give the community competitive advantages such as a unique recognisable historical district or close proximity to natural and historical attractions. *Weaknesses* are aspects of the business that have performed poorly or not to expectations. For a community, this may be a poorly organised **hospitality** sector, or little available **accommodation**. *Opportunities* are usually the result of a combination of strengths and weaknesses. For example, with a limited accommodation capacity and a unique historical district there may be opportunities to develop charming **bed and breakfast** inns in the **community**. *Threats* are more difficult to identify since many threats are usually external factors (economic, environmental, and political) that cannot be controlled by the community, but must be considered in the analysis process. In the case of the community, they may include a weak regional economy or restrictive historical legislation.

The adaptability of the SWOT analysis makes it a valuable strategic tool for all types of tourism businesses and organisations from a resort to a national tourism organisation. In addition, it can be applied at a micro level to one business function like marketing, as well as at a macro level to address the strategic positioning of an organisation.

See also: strategic planning; strategy formulation

Further reading

Bryson, J. (1990) *Strategic Planning for Public and Non-Profit Organisations: A Guide to Strengthening and Sustaining Organisational Achievement*, San Francisco: Jossey-Bass Publishers. (Describes the SWOT process in more detail in the context of organisational restructuring.)

ROBERT A.G. WONG, CANADA

symbolism

A symbol is something considered as representing, typifying or recalling something else by association in fact or thought or because it has comparable qualities. The same symbol may represent something concrete and something abstract. Thus the Great Wall of China is an icon of and for China. The first word signifies the greatness of a culture with an unbroken **heritage** stretching back five thousand years. The second word signifies the strength and endurance of that culture. Combined, the three words encapsulate all that is Chinese, moving the symbol from the concrete (a wall built for military purposes) to the ideas, philosophies and histories of that nation and its people. Globally, the Wall, as one of the wonders of the world, symbolises all humankind's heritage. Touristically, it is a compelling symbol for visitation both by Chinese (akin to a **pilgrimage**) and by international tourists. Symbolically, it has elements of both the sacred and profane.

Symbols take many forms – colours, sounds, **signs**, shapes, actions, linguistics – but the meanings of these forms may vary across cultures. Thus red is commonly associated with danger, but to Chinese it also represents happiness and good luck. In some Arabic societies a burp demonstrates appreciation of a fine meal; in Western societies it is a symbol of poor manners. The varied interpretation of the same symbol is contextual, and symbols are an inherent part of the **socialisation** process of a culture.

Tourism utilises symbolism in many different ways, functionally signifying other realities such as signage (words and pictures) to indicate facilities available at a **destination** or **attractions** in a **theme park**. Referentially tourism employs the specialised symbol 'language' of maps and charts, which are themselves symbolic representations of cities or countries. Promotionally it uses **iconography** (famous sites, buildings, people or places) to represent an entire country. In terms of imaging, the sun is perhaps the most widely applied symbol in tourism. The **sign** (a circle with radiating lines) represents the object (the sun) but is interpreted as an abstract feeling (happiness), and also as every destination imaginable, from the tropical islands of the Caribbean to the ski slopes of **Switzerland**. Tourism thus moves symbolism beyond the duality of signification and representation to a triadic relationship because of the intervention of a third factor, the interpretant presentation of sign and object. **Images** are put in place of reality, as presentations of reality. The tourism symbol becomes **hyperreality**.

See also: semiotics

Further reading

Dann, G.M.S. (1996) *The Language of Tourism: A Sociological Perspective*, Wallingford: CAB International. (Discusses the use of symbolism, especially through linguistics, in tourism.)

Norbert, E. (1991) *The Symbol Theory*, London: Sage Publications. (Essential to an understanding to the theory of symbols.)

TREVOR SOFIELD, AUSTRALIA
SARAH LI, AUSTRALIA

systems theory

Systems theory is useful in many types of research and in many fields of study, for it provides distinctive ways of understanding (Emery 1981). While its best-known application has been in the development of computer technologies, it has profoundly shaped thinking and practice in **economics**, **psychology**, **management**, ecology and many other areas. Systems theory is especially valuable in multidisciplinary education on applied fields. For all these reasons, its relevance to tourism studies and tourism industries is pervasive and diverse.

Systems *thinking* can be distinguished from systems *theory*. The former is a way of looking at things (concrete systems) or ideas (abstract systems) which attempts to be holistic and cohesive. The search for holistic vision means attempting to view a whole entity rather than focusing on some parts, ignoring others. The search for cohesion means attempting to understand how the parts are arranged and how they function in combination, as a system. Systems *thinking* has been evident for two thousand years in the work of isolated scholars including Aristotle, Ptolemy, Copernicus, Galileo, Ibn Khaldun and Vico.

Systems *theory* is newer. It was created in the 1930s and continues to develop. It formalises systems thinking, as a quasi-discipline, an organised body of knowledge, for the study of systems. Several applied sciences have grown from it, including cybernetics, information theory, **game theory**, decision theory, topology, **factor analysis**, systems engineering and operations research. Central to all this is general systems theory, devised by Bertalanffy (1972) and others.

The distinctive purpose of general systems theory is to deal with any thing or any idea that seems complex. The aim is to reduce the complexity. The value is that after complexity has been reduced, things and ideas become easier to understand, analyse and manage. The strategy for achieving this is, broadly, breaking down a whole (thing or idea) into its elements, and then identifying the crucial attributes of elements to see how they are connected. Element, attribute and systems hierarchy are basic concepts in systems theory, where these terms have particular meanings. An element is a component which must exist for the system in question to exist. For example, useful models of whole tourism systems have tourist(s) as an element, since without tourists there can be no tourism. They also have **destination region** and origin (or generating) region as two more elements, by the same logic.

Attributes are features or characteristics of an element that makes it part of a system, linking it to the other elements. For example, urban congestion and pollution are attributes of many of the world's traveller-generating regions; those features influence many residents to seek temporary **escape** via tourism, a behavioural process which involves those residents becoming tourists, visiting places beyond their home city, places which are destinations.

Elements forming a system do not do this because of their immanent qualities but by their arrangement in that system, where they are not significantly connected except with reference to the whole. That axiom indicates a distinction between relationship thinkers and system thinkers. If the former sees beautiful scenery, they imagine that the region is (or should be) a destination because of its beauty, the two facts seem related. Systems thinkers understand that beautiful scenery is an immanent quality which, on its own, does not and cannot make a region a destination, nor can advertising or business investment (Leiper 1995: 30–1). For a place to become a successful destination, a number of connected elements are necessary.

Certain systems can be seen in hierarchies. In principle, every system has superior and subsystems. The latter are derived by analysing elements and attributes in their fundamental components. 'Whole system' is a heading for the superior system in a field. Common concepts of systems theory include whole systems and subsystems, framework and clockwork systems, inputs and outputs, environments, feedback, entropy, negative entropy, proliferating variety, dynamic systems and homeostatic systems, equifinality, interdisciplinarity and multidisciplinarity. While all those concepts are mentioned in the tourism literature, to date few have been explored in depth.

Tourism generally involves notably open systems, so environmental interactions are a major interest. In some places, environmental impacts are reduced, or altered, by closing off parts of the system, reducing its open qualities. One strategy for this is integrated resorts. A number of researchers have devised simple framework and clockwork models of systems. Theoretical research on feedback and proliferating variety in relation to tourism has been conducted by Leiper (1995: 307–19); this has helped analysis of part-industrialisation, which has implications for all elements of tourism systems and their environments.

Systems thinking's pervasion of traditional disciplines has been a counter balance to the main approach in modern academia. That approach has allowed a rapid expansion of knowledge in specific physical and social sciences. But it involves concentration on analytical fact-finding and narrow theory building. Countering this fragmentation, systems theories are serving as a unifying bridge, a means of co-ordinating multidisciplinarity education and research.

See also: system, tourism

References

Bertalanffy, L. von (1972) 'General systems theory: a critical review', in J. Beilshon and G. Peters (eds), *Systems Behaviour*, London: Open University Press, 29–49.

Emery, F. (ed.) (1981) *Systems Thinking*, Harmondsworth: Penguin.

Leiper, N. (1995) *Tourism Management*, Melbourne: RMIT Press.

Further reading

Churchman, C.W. (1979) *The Systems Approach*, New York: Laurel Press.

Espejo, R. and Harnden, R. (eds) (1989) *The Viable Systems Model: Interpretations and Applications of Stafford Beer's VSM*, Chichester: John Wiley.

Fox, K. and Miles, D.G. (eds) (1987) *Systems Economics: Concepts, Models and Multidisciplinary Perspectives*, Iowa State University Press.

NEIL LEIPER, AUSTRALIA

T

Taiwan

Taiwan is a country with diverse cultural and natural **resources**. Most **inbound** tourism is generated from its neighbouring countries and North America. In 1999 over 2.4 million international **tourists** visited Taiwan, with business trips and participation in conferences having a giant share. Elimination of entry visa for many nationalities in 1994 has contributed to the growth of **international tourism** in this country. Because of its economic strength, Taiwan's outbound tourism is also sizeable, with **Hong Kong**, mainland **China**, **Japan** and **Singapore** favoured for pleasure and business purposes. **Domestic tourism** is also well developed in Taiwan, with weekend holidays and festivals adding much to its popularity.

YEONG-SHYANG CHEN, TAIWAN

target marketing

An enterprise or organisation focusing its activities to one or several **market** segments is exercising **target marketing**. The first step of this strategic process is **market segmentation**. Consumers or tourists are subdivided into meaningful groups with respect to criteria such as number of trips, **destination** preferences, intentions, choice of **accommodation** or **tour operator**. The criteria for determining market segments depends on the type of business (such as hotels, **restaurants**, airlines, tour operators and car rentals). However, it can be organised in two different ways: classifying tourists by one key variable (like age, purpose, amount of spending, **special interest group**), or arranging homogenous groups of tourists by typical profiles of **psychographic** or behavioural characteristics (including activities, benefits sought, **motivation**, **lifestyles** or **vacation** styles).

During the second step, each segment has to be evaluated according to normative criteria derived from organisation goals, consumer needs, environmental constraints and competitive forces. The outcome of this stage can be a selection of one or more segments to be served. Products and services are commonly adjusted (see **product positioning**) to fit the targeted segments. The last step of this process refers to the **adaptation** and coordination of the whole **marketing mix**. Communication, distribution and pricing instruments have to be integrated into the **product** decisions and additional business activities.

See also: attitude; event marketing; perception; strategic marketing; typology, tourist; values

ANDREAS ZINS, AUSTRIA

tax

Taxation is a system of raising money for public expenditure, and taxes are the amounts paid by individuals or organisations to local, regional and national government agencies in accordance with such a system. There is a variety of forms of taxes, and distinctions are made between direct and indirect types, as well as between personal and corporate. Direct taxes are levied on income or property, while indirect taxes are those levied on expenditure (in the form of sales tax or value added tax which are collected by the seller but ultimately borne by the purchaser).

Historically, taxes have been a major source of **conflict** between governments and the governed. Such conflicts led, for example, to the English Civil War in the seventeenth century which resulted in the dethronement and execution of King Charles I, and to the American War of Independence in the late eighteenth century. One of the key principles of democracy is 'no taxation without representation', that is, taxes must be enacted into **law** by a democratically elected legislative body.

The most common direct personal taxes are the personal income, property and capital transfer taxes. Income tax may be payable to municipal, **local** and national governments, and in federal countries to the foregoing as well as to state, cantonal or provincial government. Individuals owning unincorporated businesses (as sole traders or in **partnership**) are liable for personal tax on their business **profit** or share of it. Property tax on real estate is usually payable to local or municipal government. Capital transfer taxes typically take the form of inheritance tax on the estate of a deceased person, being paid to the national government. Individuals also suffer indirect taxes as purchasers of goods or services which are subject to sales or **value added** tax. They may also be liable to pay other indirect taxes such as tax on transfers of real estate.

The most common direct corporate taxes are those on business profits and on property (real estate). Indirect corporate taxes include those on share issues and on transfers of real estate. Businesses do not normally suffer sales taxes, since these are passed onto the ultimate individual customer. In the case of value added tax (VAT), normally a business has to pay the government value added tax on its sales (output VAT) less value added tax on its purchases (input VAT). Whether in economic terms the burden of such taxes falls onto the business or on the consumer depends on the state of competition in the **market** and the price elasticity of **demand** for the good or **service** in question. These determine the extent to which the business in its **pricing** may pass the burden of the tax onto the consumer or have to absorb the tax by reducing its **profit margin**. In many countries, specific indirect taxes are levied on tourist expenditure (such as hotel bills). This may be done in order to help pay for **infrastructure** costs incurred by government in connection with tourism **development**. One of main arguments made in favour of tourism development is its ability to generate large tax earnings for governments, which in some countries amounts to billions of dollars.

SIMON ARCHER, UK

taxi

A taxi is a fast, flexible but relatively expensive mode of **transportation** frequently used to transfer **tourists** from airports or railways stations to their **accommodation**. They are particularly useful for travelling around unfamiliar city streets or for transporting luggage. Taxis are very popular with business tourists (see **business travel**). To avoid financial and other malpractices, taxicabs are often licensed by the **government**.

ANNE GRAHAM, UK

teacher education

Both professional and academic experts in tourism have identified the quality of educators and the **quality** of the educational material they use as being among the most important determinants in the success of an education system. They have further identified a number of issues affecting the success of efforts to educate tourism educators. Structural issues involve a range of contextual factors which create somewhat unique problems for those responsible for tourism **education policy**. These include the late arrival of tourism as a field of education and **training**, a lack of industry consensus on the need for education, the diverse nature of tourism/hospitality education and training, the multiple educational demands of a rapidly growing industry, the lack of institutional structures to support tourism education, a shortage of positions for educators, and finally, the lack of advanced level programmes to properly train tourism educators.

In addition, there are a range of professional issues that must be taken into account by individuals interested in pursuing a career in tourism education.

The most significant of these are the lack of clear **career** path for tourism educators, and the strong **conflict** between the demand for a strong academic training as well as practical experience. When combined, they create powerful forces that impact heavily upon both current and future educators. In addition, the need to develop a specialised disciplinary expertise, while achieving a broad interdisciplinary understanding of tourism, must be addressed by the committed educator. These, along with pressure to gain international experience while at the same time demonstrating a strong **local** commitment, creates strong pressures on the instructor/scholar. Such pressures, added to a lack of well-developed supporting teaching materials, results in a strong challenge for administrators who seek to support the educational goals and efforts of future faculty members.

Despite the complexity of the problem, a number of **policy** guidelines for tourism education administrators have been identified. These include the need to provide well-defined **leadership** and organisational responsibilities for teacher/educator training programmes. Such leadership is essential to ensure that other components of the system realise the importance and seriousness of tourism education. Another critical principle underlying efforts to prepare future educators is the need to ensure that their training is sensitive to the needs of the industry. At the same time, it is essential to stress that academic programming in tourism should not be totally driven by industry. In addition to their pragmatic responsibilities, the educators of tomorrow must be prepared to provide intellectual and conceptual leadership to a rapidly evolving industry.

J.R. BRENT RITCHIE, CANADA

technology

Technology, despite its fundamental and widespread influence on individuals and societies, is difficult to define precisely. A simple but popular view of technology revolves around its physical manifestations, machines and equipment, and it is not difficult to appreciate why this is so when one considers the myriad of impressive examples of that manifestation (space shuttles, jet aircraft and satellite communications systems, to name but a few). Even mundane examples, such as lightbulbs and telephones, have such a dominant influence on the way we live that they readily support this popular view.

However, technology can be defined as the systematic knowledge and action applicable to any recurrent activity, which introduces an awareness of the human element in the design, choice and application of those physical artefacts and in the creation and direction of the organisations in which they are used. Thus management systems are also forms of technology. An all-encompassing definition which introduces an element of changing conditions is more useful, and therefore the following is recommended: technology is a flexible repertoire of skills, knowledge and methods for attaining desired results and avoiding failures under varying circumstances. Reference to such a definition does not automatically result in the utilisation of machines or equipment, but indicates the need to consider relevant objectives and the various ways those objectives might be achieved, which may or may not involve the most modern and impressive equipment on display at the trade fair or advertised in the media.

Technology, particularly the machines and equipment variety, has undoubtedly played a large part in establishing the general conditions in society leading to the creation of **demand** for tourism. For example, consider the creation of wealth through the application of technology to manufacturing and the creation of more **leisure** time, as working life becomes more efficient and thus less time needs to be spent in the workplace. Technology also provides the means to satisfy that demand. For example, the provision of jet aircraft can quickly, safely, comfortably and relatively cheaply carry large volumes of people to previously inaccessible places, where they can stay in hotels built to the highest standards of comfort and **efficiency** and equipped with every modern amenity, including communication installations which allow visual contact with business colleagues thousands of miles away; plus restaurants offering foods which are never out of season, cooked on low energy stoves, with all processes and payments being monitored, controlled and paid for through sophisticated computer-based systems.

While it provides the means to create, stimulate and then satisfy tourism demand, it is necessary to be aware of possible infatuation with physical technology. There is the danger that because it is available, we feel it must be used. Does a fifty-room, **budget hotel** need the same front office sophistication as a five-hundred room, five-star **hotel** ? Does a remote scenic beauty spot need an access highway and hotel and resort developments? Technology is a powerful tool, but its use should always be thoroughly considered from all perspectives.

Further reading

Pine, R. (1987) *Management of Technological Change in the Catering Industry*, Avebury.

RAY PINE, CHINA

telemarketing

Telemarketing includes a variety of direct actions which, in an integrated and systematic manner, utilise information and telecommunication technologies together with management systems in order to optimise the promotional mix used by a company to reach its customers. Its main application for a tourism company is for reservations, customer care and complaint and account management.

MARCO ANTONIO ROBLEDO, SPAIN

television

Television is internationally the major mass medium, reaching up to 96 per cent of some populations every week. Its combination of sound, movement, and colour gives it great impact. Since the 1980s, there has been an astonishing growth of specialist television travel programmes in the developed world. It is increasingly used as an **advertising** medium by large tourism organisations, though it is unaffordable for smaller ones.

A.V. SEATON, UK

Teoros

Teoros: Revue de Recherche en Tourisme acts as a medium to promote transfer of knowledge from the academic and research community to the francophone practitioners in the field. It is mainly aimed at agents responsible for the tourism **development** in Québec, Canada, and to tourism instructors and students. With its issues being mostly thematic, *Teoros* favours contributions from tourism students, but contributions from all other sources are sought. First published in 1982, it appears three times per year (ISSN 0712–8657).

RENE BARETJE, FRANCE

terrorism

Terrorism may be defined as the systematic use of terror as a means of coercion (Wall 1996). It is a systematic and persistent strategy practised by a state or political group against another state, political or social group through a campaign of acts of violence, such as assassination, hijacking, the use of explosives, sabotage and murder, with the intention of creating a state of terror and public intimidation to achieve political, social, or religious objectives (Ezzedin 1987; Wahab 1996). The extent to which terrorism may be distinguished from **crime** and political violence is debatable. Crime tends to be perpetrated by individuals or groups for their own, selfish motives, whereas terrorists tend to act on behalf of others as well as themselves. Further, the legitimacy of terrorist acts tends to be related to the political perspective of the commentator. That is, one person's freedom fighter is another person's terrorist.

Buckley and Klemm (1993: 193) have suggested that:

> The problem with any kind of civil unrest is that unfavourable images are beamed across the world so that even those who are not afraid of terrorism will be discouraged from taking a holiday there. It is not so much that the area is dangerous; more it does not look attractive.

Terrorist incidents receiving media coverage did have an almost immediate effect on British people's willingness to holiday in Northern Ireland, but, conversely, if the level of incidents declined, potential tourist attitudes improved.

Even though most of the **supply** components of the tourism **product** are to a large extent within the control of the destination, if there is any concern over the **security** situation, **demand** for that **destination** is at **risk**. As the Economist Intelligence Unit (1994: 70) asserted: 'International tourism has shown itself to be susceptible to concerns over political instability and risks to personal safety. At its most extreme, outbreaks of military conflict are able to destroy established tourism sectors in very short order'. In less extreme circumstances, adverse publicity or short term political disruption can wipe out several years of **growth** and **development** in tourism and cause a destination to lose several more years of trend growth until either the cause of the disruption is dealt with or tourists' memories fail them.

Studies of the economic impact of terrorism on tourism are few, but the estimated effects can be very substantial (Gamage *et al.* 1997). Three categories of the causes of disruption to international tourism can be identified (Economist Intelligence Unit 1994): fundamental, long-term disruption such as has occurred in Lebanon, Northern Ireland, Sri Lanka, Uganda and the former Yugoslavia; continuing volatility/uncertainty in tourism destinations such as in Egypt, India, Israel, Jamaica, Kenya, Peru, the Philippines and Turkey; and short-term, single event disruption such as in China, Fiji, Florida and, due to occasional terrorist attacks, the UK.

See also: war

References

Buckley, P.J., and Klemm, M. (1993). 'The decline of tourism in Northern Ireland: the causes', *Tourism Management*, June: 184–94.

Economist Intelligence Unit (1994) 'Occasional studies: the impact of political unrest and security concerns on international tourism', *EIU: Travel and Tourism Analyst* 2: 69–82.

Gamage, A., Shaw, R.N. and Ihalanayake, R. (1997) 'The cost of political upheaval to international tourism to Sri Lanka', *Asia Pacific Journal of Travel Research* 2(1): 75–87.

Wahab, S. (1996) 'Tourism and terrorism: synthesis of the problem with emphasis on Egypt', in A. Pizam and Y. Mansfeld (eds), *Tourism, Crime and International Security Issues*, Chichester: John Wiley & Sons, 175–86.

Wall, G. (1996) 'Terrorism and tourism: an overview and an Irish example', in A. Pizam and Y. Mansfeld (eds), *Tourism, Crime and International Security Issues*, Chichester: John Wiley & Sons, 143–58.

Further reading

Ryan, C. (1993) 'Crime, violence, terrorism, and tourism – an accidental or intrinsic relationship?' *Tourism Management*, June: 173–83.

ROBIN SHAW, AUSTARLIA

testimony

Testimony is touristic promotion of a **destination** or **hotel** by recognisable spokespersons (like Australia and Crocodile Dundee), association with the rich and (in)famous, and the verbatim accounts of satisfied customers. Testimonial **rhetoric** as an authenticating device can also be found in the narratives of **tour guides**. As a type of word of mouth communication, testimony rates high in terms of credibility.

GRAHAM M.S. DANN, UK

Thailand

Thailand has a distinctive place in the political history of Southeast Asia in general and in terms of the **evolution** of tourism within the nation. Thailand was never colonised as its neighbouring countries were, and thus Thai culture developed without the discontinuities associated with **imperialism** or later independence struggles. However, Thailand was not immune from the desperate poverty, urban decay or threats to the **environment** that have affected its neighbours. This country of 61 million is comparable in land area to France, but its economic condition is that of a **Third World** country.

Thailand was an absolute monarchy until 1932. Ironically, it was the efforts to democratise that led to the persistent involvement of the military in Thai politics. Coups against civilian governments have been common for over sixty years, and military regimes have been frequent. For the most part, these transfers of powers have not been bloody. Thus, despite the general truism that political instability depresses tourism, this industry has flourished alongside the political upheaval and can even be said to have developed as a consequence of regional instability.

The Tourism Organisation of Thailand began in 1960 as a minor tourism **marketing** arm of the **government**. The country was at that time receiving only 100,000 tourists a year. That changed as the Vietnam War escalated and nearby Thailand became a rest and **recreation** base for American troops and allies. Bars, nightclubs and **prostitution** (officially illegal) skyrocketed. The Thai tourism base moved from its largely cultural orientation to a more tawdry but lucrative orientation on sex and recreation.

By the mid-1970s, when the war ceased, Thai tourism had more than one million arrivals annually and was no longer dependent on the soldiers. However, it retained an emphasis on **sex tourism** in the major cities catering to business tourists and unattached males. During most of the time since the war, over 75 per cent of the tourists to Thailand have been males. This dependence on male tourists has created a dilemma for the Thai government, which is trying to find a way to diversify its tourism **market** without jeopardising its receipts from this industry.

The country's abundant attractions have become well known. Culturally oriented tourists can visit some of the world's most attractive Buddhist temples. The hill tribes of northern Thailand have become both a **destination** and a production centre for quality handicrafts. Thai silks, gems and other specialities have a ready market in the growing tourism trade. In the last fifteen years, however, the major tourism **growth** has come in the southern beach communities. As **infrastructure** has improved, Pattaya and Phuket are but two of the cities to experience tremendous growth.

Tourism has become increasingly important to the economy, surpassing rice in 1982 as the leading industry. Growth has been quite impressive. In 1996, Thailand welcomed over seven million visitors, a 3.6 per cent increase over 1995. Tourism receipts were up nearly 132 per cent to $8.6 billion over the year before. Thailand ranks thirteenth among the world's nations in the amount earned annually from tourism. These figures do not include the growing receipts from **domestic tourism**. Often, developing nations are assumed to have little or no domestic tourism. That is not the case in Thailand. By 1987, there were more than 8.7 million domestic tourists in the country. Their contribution to the economy has not been precisely measured. Domestic tourism does not bring in **foreign exchange**, but by 1983 the government listed it as one of its major **planning** goals.

Since the Third National Social Development Plan (1972–6), tourism **development** has been a central part of economic planning. The Tourism Authority of Thailand, established by Parliament in 1979, has overall responsibility for this industry, but the private sector successfully resisted efforts of this agency to significantly regulate the industry. In recent years, the Thai government has privatised its national airline, Thai International, and has allowed the private sector both incentives and considerable freedom in tourism development. That is in keeping with the post-Cold War ideology of deregulation, decentralisation and privatisation, but it has been a mixed blessing for orderly tourism development.

Tourism has major problems; the same deregulation the industry fought against has resulted in ruinous hotel-building sprees. Crippling traffic congestion and air pollution in Bangkok threatens future tourism development, particularly large-scale **convention business**. Opposition to government controls has made once beautiful seaside cities into planners' nightmares. Environmental degradation is coupled with enormous social problems associated with tourism.

The incidence of **AIDS** and other sexually transmitted diseases has soared. The government was reluctant to acknowledge the spread of these until recently, fearing it would discourage tourism. Increasingly, it has launched education campaigns about the **disease**, but sex tourism has sought to deal with the challenge by bringing poor women from the hills, refugees from Burma and younger girls into the industry. Beyond its AIDS education campaign, the government has taken several steps to enhance **safety** and a quality tourism experience for tourists. The country is making a concerted marketing effort to attract more families and female tourists, so that Thailand's **image** as a sex destination is diluted by images of the nation's culture and scenic attractions.

Moreover, since **crime** and corruption are also associated with tourism, Thailand has established a separate tourism police to control problems threatening its most important industry. Finally, the government passed in 1992, and amended in 1993, a bill which enforces tough standards on tourism companies and their guides. It requires annual permits and imposes substantial fines for non-compliance. This bill brings Thai tourism into line with the types of consumer protection commonplace in most of the developed world.

Further reading

Cohen, E. (1983) 'Hill tribe tourism,' in J. McKinnon and W. Bhruksasri (eds), *Highlanders of Thailand*, Kuala Lumpur: Oxford University Press, 307–25. (This and some other articles by Cohen offer an anthropological perspective on Thai tourism.)

Elliott, J. (1983), 'Politics, power and tourism in Thailand,' *Annals of Tourism Research* 10: 377–93. (Provides an introduction to the political environment of Thai tourism.)

Richter, L.K. (1989) *The Politics of Tourism in Asia*, Honolulu, HA: University of Hawaii. (Chapter 4 discusses the evolution of Thai tourism policy.)

LINDA K. RICHTER, USA

thanatourism

Thanatourism or 'dark tourism' is travel to locations associated with death and disaster. It includes visits to battlefields, murder and atrocity sites, places where the famous died, graveyards and **war** memorials, **relics** and reconstructions of death. Thanatourism is an old, widespread and often controversial form of tourism which was, in the winter of 1996, the subject of a special issue of the *International Journal of Heritage.*

A.V. SEATON, UK

theme park

Theme parks are capital intensive, highly developed, self-contained recreational spaces which invariably charge admission. The entertainment, rides, speciality foods and park buildings are usually organised around themes or unifying ideas such as a specific period in **history** or a particular geographic **region**. These themes are crucial to the operation of the parks as they create a feeling of involvement in a setting which is in stark contrast to daily life. A distinction can be drawn between commercial theme parks, which are well described by the theming and entertainment elements mentioned, and outdoor museums or historic theme parks, which may be less commercial in emphasis and have goals in **heritage preservation** and public education. Zoological gardens, while sharing some of the characteristics of theme parks, usually have important conservation and species breeding goals and should be studied and understood separately from theme parks. The scale of the theme park phenomenon in global tourism is impressive. Most North American cities of over 1 million population have a theme park, and many have two. Several new theme parks are being built in Asia, and while some are closely modelled on North American ventures, a number of distinctive Asian influences and styles in architecture, **food** and style of entertainment are emerging.

The best-known and most successful theme parks are those operated by the Disney corporation in Orlando, Los Angeles and France. There are significant chains of theme parks such as the Six Flags parks as well as variations of the Sea World concept in several states of the United States and Australia. Theme parks dependent on or related to movie themes are widespread as are those built on produce and food themes. Theme parks evolved from the earlier widespread amusement parks, essentially collections of rides associated with **fairs** and carnivals. The contribution of the Disney organisation to the tourism industry has been described as sanitising the amusement park for the middle classes with scrupulous attention to cleanliness, visitor comfort and quality **services**. For the open air museums and historic theme parks, imaginative interpretive programmes have reduced the boring **museum** image of such attractions, and the commercial viability of parks of this style has improved.

Much of the research done on theme parks has concentrated on visitor **satisfaction** with facilities as well as understanding the market segments and drawing power of individual sites. The construction and subsequent difficulties of the Paris Disneyland park has received considerable research attention. The need to adapt the American theme park concept to **local** cultural styles, particularly in areas such as eating behaviours and perceived attractiveness of themes, has been noted in the research literature. With large-scale investments by brewing companies, publishing houses, **film** companies and **hotel** groups, the future of theme parks in the **landscape** of tourism seems assured.

PHILIP L. PEARCE, AUSTRALIA

theory

A theory is a set of connected statements used in the process of explanation. The nature and status of theories in tourism will vary from one philosophy of social science to another. Every theory and philosophy of the social sciences presupposes an ontology which is the set of things to which a theory ascribes existence. Therefore, ontology is described as a 'meta-theory' which seeks to answer the question of what must the world be like for knowledge to be possible.

Three broad ontological positions can be distinguished within the philosophy of the social sciences: classical empiricism/positivism, transcendental idealism and transcendental realism. Under positivism, a theory comprises a set of hypotheses and constraining conditions which, if validated empirically, assume the status of universal laws. These coherent linked statements provide the basis for further research from the known (theory and law) to the unknown (hypotheses). In the philosophy of idealism, there are no universal theories; each individual has sets of individual theories which serve as the basis for human action. In the philosophy of realism, a theory is a means of conceptualising a framework within which reality is apprehended. The test of a theory to someone using it is thus its coherency and adequacy, rather than its empirical adequacy under positivism.

The construction of tourism research philosophies and theories has not been extensively studied. The majority of research and journals in tourism implicitly adopt an empiricist/positivistic philosophy, particularly in **economics**, **management**, **marketing** and **psychology**, although theory construction is poorly formulated. The philosophy of idealism has been expressed most strongly in historical analyses, and has also been influential in cultural studies which attempt to place tourism within a postmodern perspective. The philosophy of realism underlies much theory generation in geographical, **planning** and **policy** studies, and has been extremely influential in recent accounts of tourism public policy and planning being founded on craft notions of the adequacy of planning and policy arguments. Such examinations of the production of knowledge have stressed that the ideas, theories and structure of the study of tourism have developed in response to complex social, economic, ideological, political and intellectual stimuli. Therefore, the case for understanding the changing nature of tourism studies contextually parallels the case made by realists for appreciating all human activity; the operation of human agency must be analysed within the constraining and enabling conditions provided by its **environment**.

Further reading

Ellis, J. (1989) *Against Deconstruction*, Princeton, NJ: Princeton University Press.

C. MICHAEL HALL, NEW ZEALAND

Third World

Third World is a loosely defined pejorative term used to denote developing countries, or those lacking sufficient human and natural **resources** in order to provide an adequate standard of living. Most of the Third World countries are in Africa, Asia and Latin America. Their economies, including revenues derived from tourism, are locked in to denote relationships with the so-called developed First World. All seek foreign exchange through this growing industry.

VALENE L. SMITH, USA
VERONICA LONG, USA

time

A **change** in the perception and organisation of time is a principal characteristic of the transition from ordinary life to tourism. The latter can be seen as a **leisure** activity or an activity in time free from obligations. Indeed, one of the major varieties of tourism is called '**vacation** ', a period of vacant or free time. However, tourist time differs from ordinary leisure, in that it is a special, extraordinary time, anticipated and cherished as a major break of everyday routine, and often remembered as a highlight of one's biography.

The break in time between routine and tourism has been conceptualised in two contrasting but related pairs of terms. Graburn (1977) conceives of touristic time as akin to the sacred time in premodern societies, and juxtaposes it to profane time, thus emphasising the reverberation of religious or festive motifs in the touristic experience. Others, following Victor Turner, conceive touristic time as essentially a limited reversal of the ordered, organised nature of time in everyday life (Wagner 1977) This liminal or liminoid time (see **liminality**; **rites of passage**) is akin to timelessness: it is free from the chronological divisions and of the associated 'timetables' of daily routine. This conception comes close to that of 'flow', the loss of a sense of time experienced by those deeply engrossed in an activity.

Ironically, the high value attached to the often limited time people dispose of for tourism, may in practice countervail its liminal character. Tourists frequently seek to accomplish as much as possible on their trip; though at leisure, they are in a hurry and pressed for time. Routine, everyday modes of temporal organisation thus penetrate tourism, especially in its more organised institutionalised forms such as group **tours**. In contrast, some alternative forms of tourism, such as drifting, purposely avoid fixed travel timetables and **itineraries**.

Tourism and time are also related in another respect, namely, the temporal nature of tourist attractions. A pervasive motif of modern tourism is the **quest** for the authentic, often conceived in temporal terms as the original, pristine, primitive or antique. In that respect, travel to remote attractions is not only a trip in space but also back in time, as, for example, a visit to a 'primeval forest' or a 'stone age' tribe. Indeed, the value of sights visited, artefacts seen and **souvenirs** purchased is often appreciated in terms of their antiquity.

In contrast, postmodern tourism is often oriented in the opposite temporal dimension; the **future**, vicariously experienced in **theme parks** or in fantastic voyages in **virtual reality**. Both the past and the future orientations of tourism are appreciated owing to their contrast to the present, or as a mode of escape from it. However, the ultimate **escape**, time travel, is not yet available, although according to some it could become a realistic possibility sometime in the third millennium.

References

Graburn, N.H.H. (1977) *The Sacred Journey*, in V.C. Smith (ed.), *Hosts and Guests*, Philadelphia: University of Pennsylvania Press, 17–31.

Wagner, U. (1977) 'Out of time and place: mass tourism and charter trips', *Ethos* 42(1/2): 38–52.

Further reading

Cohen, E. (1986) 'Tourism and time', *World Leisure and Recreation* 28(6): 13–16.

ERIK COHEN, ISRAEL

time budget

Time budgets are an established social scientific research technique which provides for a systematic record of a person's use of **time**. They describe the duration, sequence and timing of a person's **activities** for a given period, usually of between a day and a week. When combined with the recording of the **location** at which activities occur, then the record is referred to as a space–time budget.

Time budget studies provide for the understanding of spatial and temporal **behaviour** patterns which may not be directly observable by other research techniques, either because of their practicality or their intrusion into individual privacy. Such studies are usually undertaken through the use of detailed diaries which are filled in by participants. The research method was initially used in the urban and transport **planning** field to assist in the understanding of urban travel behaviour, and in audience research by **advertising** and **media** organisations. In a number of countries, ratings for **television** and **radio** programmes utilise time budget studies to evaluate the number of people watching or listening to individual stations or programmes. Topics investigated by tourism researchers using time budgets include spatial search behaviour, activity patterns, attitudinal information, **change** in tourist's mood and activities, relative mobility among different users and the relative time given to different activities.

The use of time budgets generates a number of methodological issues: what is to be recorded; how is it to be obtained, particularly with respect to the selection of respondents; how long a time period should be used in assessing behaviour and activities; and how should the record be analysed and presented. Time budgets have the potential to assist in the **differentiation** between actual and perceived tourist activity, as well as observing behaviour that would otherwise be unavailable to the researcher. As well as contributing to tourism **marketing**, such information may also lead to improved management strategies.

Further reading

Anderson, J. (1971) 'Space–time budgets and activity studies in urban geography and planning', *Environment and Planning* 3(4): 353–68. (Provides an early review of the subject.)

Debbage, K.G. (1991) 'Spatial behaviour in a Bahamian resort', *Annals of Tourism Research* 18: 251–68. (Discusses the influence of length of stay on the overall mobility patterns of tourists.)

Pearce, D.G. (1988) 'Tourist time-budgets', *Annals of Tourism Research* 15: 106–21. (Reviews methodological issues involved in undertaking tourist time-budget research.)

C. MICHAEL HALL, NEW ZEALAND

timeshare

The timeshare concept entitles a purchaser to occupy a unit in an urban or resort project for a specified period of time each year over a number of years. This practice began in Europe in the 1960s, and gradually developed mostly among vacationers, but problems soon arose which contributed to its slowing down. Timesharing moved to the United States, and two timeshare exchange companies were created in 1974–5 (Resort Condominiums International and Interval International). It swiftly became a worldwide practice. The concept is based on adding the dimension of time to the dimension of space so that the accommodation unit is divided into further units of time.

Timeshare enables the owner or member to purchase a guaranteed accommodation (for a certain period, on a week unit basis) at a resort every year for a specified number of years, or in perpetuity. It also gives owners the chance to exchange their periods for accommodation owned by others at different resorts in the same country, or in another country (in case the **resort** is affiliated with a worldwide timeshare exchange system). This enables an individual to hedge against the inflation to **vacation** costs by contracting for the purchase, which usually takes the form of either an ownership (fee simple) interest in the resort property or a prepaid lease or license (right to use) for the accommodation at a resort for a specified number of years.

The cost of a timeshare unit depends upon several factors, including its **location**, the country, **accessibility**, degree of luxury, size of **accommodation**, season required, available recreational and other facilities. Timeshare acquires its importance from several advantages and criteria, including high utilisation for the tourist unit all over the year, increasing the number of tourist arrivals and nights through the exchange programme, and nourishing the individual tourism movement. As for the timesharers, buyers pay for the time they use; they pay a one-time reasonable fee for the unit plus a yearly maintenance fee for the unit.

The exchange programme is now cited as a primary **motivation** for purchase. Not all exchange systems are alike. They have different philosophies and operate under different performance and affiliation standards. Resort Condominiums International is still the leading exchange company worldwide. The other famous international companies are Interval International, Whippy and Club Mediterranean. There are leading associations aiming to embrace the timeshare industry worldwide, such as the American Resort and Residential Development Association in the United States and Timeshare Council in UK. Recognising the **growth**, **impact** and problems that might ensue from the application of timesharing, many countries now have statutes or regulations to control this growing industry.

WEASEL ABU-ALAM, EGYPT

tipping

Tips, or gratuities, are a traditional element of pay for many employees within the **hospitality** and tourism industries. They are discretionary payments made by the customer directly to the employee in recognition of the **quality** of the **service** provided. There are, however, some cultural/national variations in tipping customs; tipping is expected in some cultures and not in others. Tipping must not be confused with service charge, which is imposed by the employer as a means of increasing revenue, and may or may not be redistributed to all employees.

Tipping is most prominent where there is a direct and personal service to the customer, and so its significance to the employee may vary from one sector of the industry to another and from one job to another. Hotels and **restaurants** are the most likely businesses where tipping may form a substantial proportion of take-home pay, whereas bars, pubs, clubs and contract **catering** outlets may attract few gratuities. The job categories where tipping may be at a high or varying level are in **food** and beverage service, luggage porters and **concierge**, whereas back-of-house positions and departments such as housekeeping normally attract few tips.

Although in many cases the individual employees keep their own tips (and handle their own income tax declaration), there are also systems of pooling the amounts, sometimes known as the tronc, to be divided out at regular intervals on some pre-agreed basis. A points system often prevails where senior and long-service staff may receive a larger share of the pool. Such a system may also include the non-service workers, in recognition of their contribution. If such a pooling arrangement exists, it is necessary to appoint a senior member of staff to be the official supervisor of the pool, to ensure propriety and to deal with taxation issues. In addition to cash being given by customers as a tip, it is increasingly the case that an amount may be added to the bill by customers and thus paid as total settlement of the bill via cheque or credit card. Here, the onus rests with the employer to redistribute these non-cash tips as additional pay through the normal wages system.

Further reading

Boella, M., Calabrese, M., Goodwin, C. and Goss-Turner, S. (1996) *Catering Questions & Answers: Employment Law*, Kingston-upon-Thames: Croner Publications Ltd.

STEVEN GOSS-TURNER, UK

tour

Once defined as any journey from one place to another, a tour is now commonly used in two distinct senses; to describe either a day trip or **excursion**, or any touristic journey involving a period of **travel** and overnight stay. The tourism industry uses **package tour** or inclusive tour to describe an **itinerary** put together by a **tour**

operator or other supplier, usually incorporating at least three elements, **transportation**, **accommodation** and transfers, although some packages may include additional services such as excursions. Historically, organised tours can be traced to at least 1500 BC, and were common at the time of the Greek and Roman Empires, especially to Egypt. In the seventeenth and eighteenth centuries, the **Grand Tour** was an essential element in the education of the aristocracy and later the rich merchant classes in the United Kingdom and Germany, entailing long cultural tours to France and Italy, sometimes for as long as three years.

Thomas Cook is usually credited with devising the first modern **package tour** taking a group by charter train between Leicester and Loughborough in 1841. Today's **mass tourism** market developed in Britain shortly after the end of the Second World War, based on the use of charter aircraft. By buying facilities (aircraft seats and **hotel** rooms) in bulk, tour operators gained from **economies of scale** and were able to offer low prices, the convenience of all elements booked simultaneously under one roof, and a guarantee of consistent **quality**. After inclusive tour prices were legally permitted to undercut scheduled air fares at the end of the 1950s, the Northern European market for sun, sea and sand package tours to the Mediterranean countries expanded rapidly; the flow of British tourists was soon joined by the Germans, Dutch, Scandinavians and others seeking guaranteed sunshine at low prices. In North America, the market similarly grew for package tours to Florida and Caribbean destinations (mainly for winter sun), and to Europe for cultural tours.

In the 1990s, the **demand** for **package tours** widened to incorporate activity and cultural programmes. Mass market tours are now to global destinations, with Europeans increasingly visiting North America, the Far East and Australia. Tour organisers offer packages for independent or group travel, using either charter or scheduled airlines, as well as coach tours, **rail** tours and self-drive programmes. A feature since the mid-1990s has been the **growth** of the all-inclusive package, providing on-site entertainment and unlimited **food** and drink at the **resort** complex.

Further reading

Holloway, J.C. (1998) *The Business of Tourism*, London: Pitman. (Describes the organisation and institutionalisation of tourism, including the modern package tour.)

Laws, E. (1997) *Managing Packaged Tourism: Relationships, Responsibilities and Service Quality in the Inclusive Holiday Industry*, London: International Thomson Business Press. (Explains the business relationships involved in the organisation of tours to ensure customer satisfaction.)

Yale, P. (1995) *The Business of Tour Operators*, Harlow: Longman. (Offers a detailed explanation of the procedures involved in setting up and running tours.)

J. CHRISTOPHER HOLLOWAY, UK

tour guide

A number of terms are in use to describe those whose responsibility it is to shepherd and inform groups of tourists. Courier is most commonly applied to describe the role, although other terms used include tour leader, tour captain, tour escort, tour manager and tour guide. The latter term is more correctly used to describe one whose principal task is seen as imparting information. Consequently, tour guides are more likely to be highly educated and formally trained, although few countries actually require them to possess a licence to practice. In the United Kingdom, Registered or Blue Badge guides are those who have passed examinations organised under the aegis of the official tourism boards. Most are members of the Guild of Guide Lecturers, whose role is to promote the professional status of the guide and to safeguard their interests in the tourism industry. Their services are usually offered on a freelance basis, and tour organisers are under no legal obligation to employ only licensed guides. However, at many key **heritage** sites such as noted cathedrals, it is the practice of institutions to restrict guiding to those employed specifically for this role, partly to ensure accurate and satisfactory information is professionally imparted, and partly as an aid in restricting the number of tourists admitted to very popular and frequently congested sites.

The origins of the role can be traced at least as far back as ancient Greece. At that time, guides fell into two categories, the Periegetai, or 'leaders around' and the Exegetai, or 'explainers'. Herodotus, writing around 490 BC, noted the gullibility of travellers and their exploitation by many clearly less than professional guides. Later, the guide fulfilled the role of tutor to those on the **Grand Tour** in the seventeenth and eighteenth centuries, conducting their charges while pointing out objects of interest.

The role of a guide suffers from being largely a seasonal occupation (apart from those key year-round tourism centres such as London), and from offering little **career** progression. Thus, it is often seen as a temporary or part-time job, attracting teachers, actors and others with a good knowledge of foreign languages, often a key criterion for **employment**. Increasingly, the role is linked to that of entertainer or 'animateur' at heritage sites, with guides offering historical **interpretation** of the **site** while acting out **roles** in appropriate period costume. The role attracts those with both acting skills and local knowledge.

Further reading

Cohen, E. (1985) 'The tourist guide: the origins, structure and dynamics of a role', *Annals of Tourism Research* 12: 5–29. (Examines the guide's role from a sociological perspective, based on research among Thai guides.)

Holloway, J.C. (1981) 'The guided tour: a sociological approach', *Annals of Tourism Research* 8(3): 377–402. (Describes and analyses the way in which guides interpret their roles in coach tour settings.)

Pearce, P.L. (1984) 'Tourist–guide interaction', *Annals of Tourism Research* 11: 129–46. (Analyses the interaction between tourists and guides.)

Schmidt, C.J. (1979) 'The guided tour: insulated adventure', *Urban Life* 7(4): 441–67. (Examines the function of the guided tour and the circumstances under which guides enhance this function.)

J. CHRISTOPHER HOLLOWAY

tour leader *see* guided tour; tour guide

tour operator

Tour operators are business organisations that combine **transportation**, **accommodation** and other **service** suppliers in **package tours** that are then sold through a **distribution** channel to the public. The terms tour operator and **tour wholesaler** are often used interchangeably, and are treated here as having identical meanings.

The origins of the tour operator can be traced back to 1841, when Thomas Cook chartered a train and organised an **excursion** to attend a temperance meeting in the United Kingdom. However, it is in the period from the early 1950s that tour operators have developed to become a significant part of the travel industry. Advances in air travel have enabled the transportation of large numbers of people quickly over large distances at relatively moderate prices, and this factor, together with increases in disposable income and leisure time, has resulted in the sustained **growth** of the sector.

The tour operator is involved in the **planning**, contracting, coordinating, reserving and the actual operating of tours. Tours are publicised through the use of brochures and sold to the public, either directly or through an intermediary such as a retail **travel agency**. The **package tour** (sometimes termed the inclusive tour or air inclusive tour) may include such items as airline flights to and from a **destination**, transfers between the airport and a **resort hotel**, accommodation and **catering** arrangements at a resort, the services of a company representative to deal with problems and inquires and optionally the provision of insurance cover and excursions whilst at a resort.

Tour operators offer benefits to both customers and suppliers of tourism services. To customers, they are able to offer packages at prices below those that could be arranged by the individual, because they are able to purchase services in bulk quantities at discounted prices which they are then able to pass on. Access to tourism information is also made easier through the tour operator's brochure. To the suppliers of services such as hotels and airlines, tour operators are able to assist in marketing products and to guarantee viable volumes of custom.

NIGEL G. EVANS, UK

tour wholesaler

The terms **tour operator** and tour wholesaler are often used interchangeably, and some writers such as McIntosh *et al.* (1995) do not distinguish between the two terms. Mill and Morrison (1985), however, regard tour wholesalers as being distinct from tour operators in two respects. First, the former are involved with the planning, preparing, **marketing**, reserving and possibly also the operation of package tours, whereas strictly speaking, the latter are involved only in the actual operation of the package itself. Second, the wholesaler does not sell directly to the public but sells through retail intermediaries such as **travel agencies** or outlets owned and operated by the tour wholesaler, whereas the tour operator sells either through an intermediary or directly to the public.

In common usage, the distinction between the two terms has ceased to exist and companies frequently vary their distribution channels, sometimes choosing to use intermediaries and at other times dealing directly with the public according to **market** conditions. Consequently, it is neither common practice nor particularly useful to distinguish between them. Tour wholesalers are, therefore, business organisations that combine **transportation**, **accommodation** and other service suppliers into **package tours** that are then sold through a distribution channel to the public.

See also: tour operator

References

McIntosh, W., Goeldner, C.R. and Ritchie, J.R.B. (1995) *Tourism Principles, Practices and Philosophies*, 7th edn, New York: Wiley.

Mill, R.C. and Morrison, A.M. (1985) *The Tourism System, An Introductory Text*, Englewood Cliffs, NJ: Prentice-Hall.

NIGEL G. EVANS, UK

tourism

As the scope and range of topics covered in this encyclopedia reveal, tourism is indeed a challenging multisectoral industry and a truly multidisciplinary field of study. To reveal and understand both its manifest and hidden dimensions, much has been written on this subject. While earlier studies through the 1960s mainly focused on its economic contributions, present efforts define and treat it as a whole, whether as an industry, a phenomenon, or both. To frame this comprehensive focus, during recent years holistic treatments and definitions have gained popularity. For example, tourism is defined as the study of man (the tourist) away from his usual habitat, of the touristic apparatus and networks responding to his various needs, and of the ordinary (where the tourist is coming from) and nonordinary (where the tourist goes to) worlds and their dialectic relationships. Such conceptualisations extend the frame beyond the earlier trade-oriented notions or definitions mostly devised to collect data and calculate tourist arrivals, departures, or expenditures. Significantly, it is this holistic view which accommodates a systemic study of tourism: all its parts, its interconnected structures and functions, as well as ways it is influenced by and is influencing other forms and forces relating to it. The purposefulness of the emerging landscape of knowledge, as demonstrated in this volume's coverage and through new studies regularly published in now over forty research journals in this field, as well as other media, all point to the more scholarly horizons ahead, but without failing to recognise that it is tourism as an industry – with its perceived and documented socioeconomic costs and benefits – which has brought all this worldwide academic attention and popularity to the forefront.

See also: definition; industry

Further reading

Jafari, J. (1989) 'Structure of tourism', in S. Witt and L. Moutinho (eds) *Tourism Marketing and Management Handbook*, New York: Prentice Hall, 437–442.

—— (2000) 'The scientification of tourism', in V.

Smith and M. Brent (eds), *Hosts and Guests Revisted: Tourism Issues in the 21st Century*, New York: Cognizant Communication.
Jafari, J. and Pizam, A. (1996) 'Tourism management', in M. Warner (ed.), *International Encyclopedia of Business and Management*, London: Routledge, 4903–12.

JAFAR JAFARI, USA

Tourism Analysis

The aim of *Tourism Analysis: An Interdisciplinary Journal* is to promote a forum for practitioners and academicians in the fields of **leisure**, **recreation**, tourism and hospitality. The journal encourages researchers to examine tourism issues from a functioning systems approach. Its scope includes behavioural models (quantitative–qualitative), decision-making techniques and procedures, estimation models, demand–supply analysis, monitoring systems, assessment of **site** and **destination** attractiveness, new analytical tools and research methods related to these and other themes. As an interdisciplinary refereed (double blind) journal, it aims to appeal to scholars, professionals and students in this field. First appearing in 1996, it is published quarterly by Cognizant Communication Corporation (ISSN 1083–5423).

RENE BARETJE, FRANCE

Tourism Economics

Tourism Economics covers the business aspects of tourism in the wider context. It takes account of constraints on **development**, such as social and **community** interests and the sustainable use of tourism and **recreation resources**, and inputs into the production process. Articles deal with hotel and restaurants businesses, **merchandising**, attractions, transport, entertainment and tourist activities, as well as economic organisation of tourism at micro and macro levels. In addition to full-length articles, the journal accommodates shorter contributions such as research notes. Article submissions are subject to a double blind review. There will be a minimum of two referees for each paper. First published in 1995, it is published quarterly by In Print Publishing (ISSN 1334–8166).

RENE BARETJE, FRANCE

Tourism Management

Tourism Management: Research, Policies, Practice publishes original research, analysis of current trends and information on the **planning** and **management** of all aspects of tourism. This refereed journal (double blind) favours a multisectoral approach and multidisciplinary treatments, including **geography**, **economics**, **transportation**, hotels and **restaurant** management, **sociology**, **marketing** and **development** studies. *Tourism Management* aims to publish articles of interest to members of both the tourism industry and the academic community. Manuscripts submitted to this journal cannot be simultaneously submitted or published elsewhere, but translated studies, which have not been published in English, are considered. First appearing in 1980, it is published six times per year by Elsevier (ISSN 0261–5177).

RENE BARETJE, FRANCE

tourism organisations

Tourism organisations are established to foster and manage the **growth** and **development** of tourism through the pursuit of common goals by joint action. The term is generally used to refer to destination-based organisations consisting of the official administrative bodies responsible for tourism, those which bring together two or more members with interest in tourism or some public–private sector **partnership**. These organisations may have different goals, functions and structures and operate at a range of scales, from the local and regional through to the national (known as **national tourism administration** or NTA) and international (such as **Pacific Asia Travel Association** and **World Tourism Organization**).

The diverse and interdependent nature of tourism, the small scale of many operators, market fragmentation and the spatial separation of origins and destinations encourage joint activity. Interdependence leads to a need to coordinate the activities of the different sectors such as **accommodation**, **transportation** and attractions, while smallness and distance from fragmented **markets** may give rise to united action to achieve **economies of scale** and accomplish goals beyond the reach of individual operators. Diversity and interdependence also generate issues of **public goods** where by the benefits of activities such as destination promotion are enjoyed by a wide range of individuals and enterprises and thus, it is argued, are best undertaken jointly or by the public sector. Similar arguments result from the negative **impacts** which tourism may generate. However, organisations generally have little direct control over the type and **quality** of the **products** and **services** being offered at the destination and to be effective depend heavily on their ability to co-ordinate and take a **leadership** role.

Functions undertaken by organisations include **marketing**, visitor servicing, development, **planning**, research, policy making, regulation, **human resource development** and lobbying. Marketing, which is the most common among them, is seen to be one of the most direct means of fostering growth, a function where the benefits of joint action are widely recognised. Some organisations, through limited resources or a desire to specialise, concentrate on a single function. Others exercise a range of functions due to their greater resources, the diverse interests of their members or recognition of the interdependence of activities such as marketing and development, research and planning. Differences may also occur from one scale to another. Tourist servicing is most commonly undertaken by **local organisations** through the operation of visitor information centres.

Marketing, planning and research is frequently the domain of national and **regional organisations** due, for example, to their greater ability to develop and create stronger **images** in distant markets or undertake such activities more effectively at these scales. The number and type of functions undertaken may also reflect broader governmental **policies** of state intervention with differences occurring between developed and developing countries or between market-led and centrally planned economies. Likewise, the functions exercised may be influenced by the level of growth; development may be a more common **role** in emerging destinations, marketing emphasised when **demand** declines and planning when problems of saturation arise.

Organisations may also differ in the way in which they are structured and the extent to which they depend on public and private sector funding. Traditionally, the public sector has been the dominant and often sole member and organisations have taken the form of a government ministry or department run along bureaucratic lines, especially at the national and regional levels. In the late 1980s and 1990s, a trend developed towards greater public–private sector partnerships through the creation of more flexible structures such as foundations, **boards** and limited liability companies. Often there has been a separation of the more bureaucratic functions such as planning, policy making and regulation from more commercially oriented ones such as marketing. The latter requires greater freedom to operate in a competitive market place and which benefit from mechanisms, which allow more private sector input into **decision making** and facilitate cofinancing of the organisations' activities. Local bodies, which are closer to the day-to-day activities of businesses and in more direct contact with **tourists**, often have a much greater private sector membership but generally still depend on local authority grants and assistance.

The development of organisations at various levels with different goals, functions and structures, but also with a large degree of mutual interdependence, means that their collective network exists in most countries. Concerted promotional campaigns in the markets, for example, need to be complemented by the delivery of tourism information services and the development of appropriate products at the destinations and vice versa. The characteristics of the network, the number, nature and type of organisations that comprise it, and the relationships between these bodies may be mandated by statute, established on a voluntary, co-operative basis, or reflect the outcome of competitive forces and resource availability. Given the large degree of public funding which has supported many organisations, the network is often structured hierarchically and reflects the prevailing administrative units and political framework. While organisations based on functional tourism regions may exist, most follow the administrative boundaries of municipalities, counties, states or provinces. National bodies usually play a more significant role in centralised nations while regional tourism organisations may be stronger in federal systems. Strong formal and functional linkages may exist among them by way of marketing agreements and the operation of integrated reservations systems and databases, or alternatively, the network may consist of a loose **association** of disparate organisations.

There is no single best type of tourism organisation nor inter-organisational network. Each country must evolve a system which best reflects local, regional and national conditions. As these change, some evolution in their structure and function may also be expected. A successful tourism organisation might be considered to be one that is soundly established, well resourced, highly regarded and that is meeting its goals.

Further reading

Chow, D. (1993) 'Alternative roles of national tourism organisations', *Tourism Management* 14(5): 357–65. (A comparative analysis of the government tourism organisations for the top five Asia-Pacific destinations.)

Pearce, D.G. (1992) *Tourist Organisations*, Harlow: Longman. (A systematic and comprehensive cross-national study of tourist organisations in six countries offering a conceptual framework and analysing common patterns and processes.)

WTO (1994) *Budgets of National Tourism Administrations*, Madrid: World Tourism Organization. (A world-wide survey of national tourism organisations with an emphasis on promotions.)

DOUGLAS G. PEARCE, NEW ZEALAND

Tourism Recreation Research

Tourism Recreation Research, a multidisciplinary journal, publishes research studies on various recreational environments and attempts to propose appropriate strategies for sound **growth** and **development** of tourism in both developed and developing countries. This refereed journal regularly publishes special issues focusing upon popular and emerging themes. First appearing in 1976, it is published twice yearly by the Centre for Tourism Research and Development, India (ISSN 0250–8281).

RENE BARETJE, FRANCE

tourism, secular

'Secular' means not sacred or ecclesiastical, and secularism means the indifference to or exclusion of **religion**. The term 'secular tourists' thus represents those who seek only to satisfy their curiosity about a holy place, or perhaps about the **pilgrims** as well. Regardless of their motivations, all tourists visiting such attractions require some level of services. The larger, better known **pilgrimage** shrines of Europe draw varying numbers of secularly oriented tourists along with religious pilgrims.

BORIS VUKONIĆ, CROATIA

Tourism Society

The Tourism Society was established in 1977 in the United Kingdom. Its activities focus on the communication of information and knowledge, primarily through a combination of meetings and its quarterly journal, *Tourism*. It acts as an advocate for the industry by addressing current issues and expressing views on them to government, the public and private sectors. Further, it promotes contact amongst colleagues throughout the UK tourism industry and with professionals in this field internationally. Membership is around 1,300 and ranges from students, academics and consultants through lower and middle management levels in the private and public sectors to industry leaders. The Society also incorporates the Association of Tourism Teachers and Trainers as well as the Tourism Consultants Group.

ADRIAN CLARK, UK

tourism system

The concept of a tourism system was developed by applying core ideas from **systems theory**. Two levels of systems can be identified in these applications. First, there are models of whole tourism systems. These are an arrangement of all the elements deemed necessary for tourism to occur. Second, there are models for subthemes, perceived in systemic terms.

Gunn's (1988) model of whole tourism systems specifies five elements: information and direction, **tourists**, **transport**, attractions, **services** and facilities. In a diagram, this model has the first four elements in a circle, with the attractions element shown above the fifth element, the visual superiority signifying relative importance. Jafari's (1989) tourist model sets out six phases that occur in all normal trips and as such are elementary: they are corporation, expatriation, animation, repatriation, incorporation and omission. The model has this process in contexts of the ordinary world (routine) and the non-ordinary world (during trips). Leiper's (1995) model has a human element (tourists); three geographical elements representing roles that places have in all tourists' itineraries (generating **region**, **transit** route and **destination** region); and an organisational supportive element (tourism industries). Various kinds of environments, such as physical, social and economic, are shown around the system. One diagram for this model is ideographic, depicting itineraries; a second is abstract, depicting heuristic perspectives.

See also: education; industry

References

Gunn, C. (1988) *Vacationscape: Designing Tourist Regions*, 2nd edn, New York: Van Nostrand Reinhold.

Jafari, J. (1989) 'Structure of tourism', in S.F. Wit and L. Moutinho (eds), *Tourism Marketing and Management Handbook*, Englewood Cliffs, NJ: Prentice-Hall, 437–42.

Leiper, N. (1995) *Tourism Management*, Melbourne: RMIT Press.

Further reading

Leiper, N. (1990) 'Tourist attraction systems', *Annals of Tourism Research* 17: 367–84. (Shows how attractions can be studied systematically, as subsystems in all whole tourism systems.)

Romsa, G.H. and Blenman, E.H.M. (1987) 'The Prime Minister's dilemma', *Annals of Tourism Research* 14: 240–253. (Tourism systems dynamics help explain complex interrelationships affecting political decisions.)

Sessa, A. (1988) 'The science of systems for tourism development', *Annals of Tourism Research* 15: 219–35. (A systemic view of regional tourism.)

NEIL LEIPER, AUSTRALIA

Tourisme

Tourisme: Systèmes-Environnements-Cultures publishes in French original research, analysis of current trends and information on the **planning** and management of various aspects of tourism. The journal aims to emphasise the broadness and interrelatedness of the industry, of interest to both academics and practitioners. Articles may cover **geography**, **economics**, **law**, **transportation**, hotels and **catering**, **sociology**, **anthropology**, cultures, **marketing** and **development** studies. Translated studies not published in French before are considered for publication. First appearing in 1993, it is published annually by the Université de Toulouse le Mirail (ISSN 1250–5773).

RENE BARETJE, FRANCE

tourist

The term 'tourist' was invented as an extension of '**tour** ', which earlier had evolved to its modern sense of a trip for pleasure. While 'tour' has long meant a circular trip, its modern sense – a pleasurable trip – evolved only 270 years ago. Daniel Defoe's book, *A Tour Through The Whole Island of Great Britain* (1726), is an early **sign** of this modern meaning, and the book, an instant best-seller, helped promote it. Previously, **motivations** of travellers involved sombre attitudes to scholarship, exploration, politics, commerce and **religion**. As a pioneer of pleasure **travel**, Defoe retained interests in those subjects but imbued them with pleasure, which became an extra **motivation**. Because of this new **attitude** he was regarded as slightly eccentric, certainly unusual. In Japan, a near contemporary of Defoe, Basho, was another pioneer of pleasure travel; his book *Narrow Road to the Deep North* appeared in 1692. A generation after Defoe, the eccentricity became fashionable in Europe, for by 1745 the expression **Grand Tour** was widely used, referring to leisurely circuits of the Continent for high cultural purposes.

A generation later, in the 1770s, Adam Smith formed an opinion that the cultural ideals had been sufficiently eroded by pleasurable priorities to justify a new word, 'tourist'. Smith's neologistic device, 'tour-ist', symbolised persons making a **ritual** of quick visits to cultural sites but spending most of their time seeking pleasure. Thus in its beginning, 'tourist' was pejorative. Today, while it still carries various senses of inferiority (in form, style, **class**, standard, price, dependency and more) among many who use or hear it, those connotations are not universal.

Currently, three sets of meanings of tourist can be identified, each serving a particular context. The three are popular ideas, technical definitions and heuristic meanings. Failure to discriminate clearly among the sets can lead to confusion. Dictionaries report common popular meanings, but cannot report all in this set. Different persons use varied concepts and perspectives for distinguishing tourists, so that what constitutes a boundary between tourists and non-tourists cannot be specified in a precise manner that suits everybody.

Technical definitions are used for statistical data. An unambiguous statement allows everybody responsible for collecting, processing and using data to know what is included. Because popular ideas of tourists are diverse and subjective, official **statistics** cannot leave the demarcation to opinion. Widely followed technical definitions for 'international tourist' have changed over the years, but the following is a current version:

> For statistical purposes the term 'international visitor' describes any person visiting a country other than that in which they have their usual residence, but outside their usual environment, for a period not exceeding 12 months and whose main purpose of visit is other than the exercise of activity remunerated from within the country visited. International visitors include 'tourists' (overnight visitors) who stay at least one night in a collective or private accommodation in the country visited and 'same-day visitors'.
>
> (World Tourism Organization 1997: 3)

This **definition** is much broader than many popular ideas, for it counts as tourists those making trips for many purposes, including **vacation**, holiday, business, **pilgrimage**, conference, visiting relatives, study and so on. This scope should be heeded when interpreting statistical data. However, technical definitions are not authoritative prescriptions which must be adopted for the next, third context.

Heuristic meanings help learning. Formal studies or research on tourists' behaviour, and formal discussions or lectures, can be helped by a crafted description and ultimately a precise definition as to what is meant by 'tourist'. Such statements help focus thinking and remove ambiguities. Since technical definitions usually embrace many trip purposes, they are inappropriate for detailed discussions of tourist behaviour. Consequently, individual researchers should devise their own heuristic concepts, refined to definitions if precision is needed, to suit each project.

In that process, four criteria can be considered. First there is **itinerary**: domestic or international, or both. Second, minimum and maximum duration of trips can be indicated. Normally tourists are distinguished from day-trippers, for if the latter were regarded as tourists, studies of tourism would logically have to be biased towards the special nature of day-tripping, which is a far larger phenomenon. A third criterion (not essential) is minimum distance travelled. A fourth is distinctive **behaviour**, which can be indicated by saying that tourism revolves around **leisure**. Tourists' leisure involves recreational and/or creative experiences from features or characteristics of places visited. These are commonly called attractions. Their central elements are sights, sites, objects, events and other phenomena.

Once, many publications referred to 'the tourist' in ways implying that all tourists behave similarly ('Xanadu has much to offer the tourist'). This stereotype has been progressively abandoned. Diversity is now reflected in typologies, in acknowledgement of various **purposes** of trips, and in touristic categories such as adventure, cultural, pleasure, business, domestic and so on. Further diversity occurs in the degree to which different tourists depend on services and goods supplied by tourism industries. Highly dependent tourists are consumers relevant for business **marketing**, while independent or self-sufficient tourism occurs beyond or at the fringes of markets served by those industries. Meanwhile the complex and heterogeneous nature of tourists' behaviour can be explored using many social science and business disciplines or concepts, such as **anthropology**, **behaviour**, **anomie**, **attitude**, **escape**, **experience**, **fantasy**, **management**, **marketing**, **psychographics**, **psychology**, **regression**, **self-actualisation** and **sociology**.

References

World Tourism Organization (1997) *Recommendations on Tourism Statistics*, Madrid: WTO. (Sets out current UN-endorsed technical definitions relating to tourism, as a guide to statistics.)

Further reading

Leiper, N. (1995) *Tourism Management*, Melbourne: RMIT Press. (Chapters 1–3 discuss meanings of tourist, sociological and psychological insights to roles, images and behaviour.)

NEIL LEIPER, AUSTRALIA

tourist as child

The term 'tourist as child' refers to the personality state of illusory **freedom** and happiness, into which tourists are cast by the industry in order to reduce the otherwise harsh effects of the need to control them. Appeals are thus made to the ego's desire for unlimited pleasure through messages which take tourists back to the sun and **fun** of childhood.

See also: liminality; ludic; rites of passage

GRAHAM M.S. DANN, UK

tourist culture *see* culture, tourism

Tourist Review, The

Revue de Tourisme – The Tourist Review – Zeitschrift für Fremdenverkehr is the official journal of the Association Internationale d'Experts Scientifiques du Tourisme, with contributions published in English, French and German. It covers contemporary theoretical and practical issues on tourism and related fields, including academics, practising managers, consultants, politicians and students. It features full-length articles, reports, reviews, and news about the Association. First appearing in 1946, it is published quarterly by the Association Internationale d'Experts Scientifiques du Tourisme (ISSN 0251–3102).

RENE BARETJE, FRANCE

tourist, recreational

Recreational **tourists** are those who visit a **destination** primarily for purposes of **recreation**. This distinguishes them from those who travel for purposes of business, attending conferences, sporting events, or **visiting friends and relatives**. The recreational purpose may be 'playful' or serious, because the word **recreation** can be used in the sense of relaxing amusement or to restore or re-create, a notion which comes very close to that of **pleasure tourist**.

CHRIS RYAN, NEW ZEALAND

tourist space

A physically or socially demarcated area arranged for touristic visits is known as tourist space. The literature emphasises the covertly staged nature of such space as a manifestation of **staged authenticity**. A covert tourist space is made to appear as part of the normal lives of **destination** people, although it is in fact fabricated for tourists, such as the performance of a 'real' tribal ritual. An overt tourist space is explicitly marked off from the host's everyday reality, as for example in an environmental **museum**.

ERIK COHEN, ISRAEL

tourist trap

Tourist traps conjure up the negative **stereotypes** associated with tourism. They infer neon signs, cheap **souvenirs**, crowds, traffic and lots of advertising. These locations or destinations have a bright and shiny surface but little substance and a less beneficial price–value relationship for **tourists**. Usually the term implies deceit, cheat, **crime** or theft. Tourist traps often result from little or no **community planning** and/or standards.

ROBERT M. O'HALLORAN, USA

trade show

An exposition is a temporary marketplace intended to promote or sell products directly to the buyer. A trade show is an exposition open to attendees from a particular trade or profession. Consumer shows are expositions open to the public. Trade shows and expositions are used extensively for tourism **marketing**. They also generate money into local tourism firms and economies.

PATTI J. SHOCK, USA

trading area

The trading area is a geographically defined **region** from which a **service** firm draws its customers. Trading areas are often subdivided into primary and secondary zones and may contain **demand** generators which draw people into the areas from further afield. Such areas vary substantially in size for different kinds of firms. A highly reputed fine dining **restaurant** may draw its customers from a radius of fifty miles or more, while quick service restaurants often have trading areas less then one mile in diameter.

JOACHIM BARTH, CANADA

tradition

The variety of related meanings associated with the word tradition is the cluster around notions of delivering instructions and handing down ideas and sayings. Religious and/or cultural traditions are normally thought to be those customs and practices which are not only long-established but also in some way definitive of this or that religion and/or culture. In more general contemporary usage tradition carries with it connotations of authority and the legitimacy of that which is passed down from one generation to another. It is in its associations with authority and the authoritative that tradition enters the tourism field – for carrying this aura the idea of tradition is a powerful mobiliser. Tour operators, for example, will readily maintain that to 'see a bit of tradition' (a bull fight in Spain, a Life Guard in England) is to catch a glimpse of an essential part of Spanish or British culture. This is the point at which the term gives rise to a cornucopia of questionable assumptions about the nature of human groups and collectivities. Thus, it is seldom the case that a given people (or nation as in the example above) have a single set of traditions. In the case of contemporary Spain or Britain, there are clearly many different traditions at work, few of which are fixed, many of which contradict one another, and many of which are in the process of being disputed, renegotiated or invented (Hobsbawm and Ranger 1983). Yet how many picture postcards there are which confidently announce to the tourist that this group of old women weaving on the balcony and that old fisherman mending his nets are at the heart of the traditional culture of the destination? In the end, therefore, like many other terms found in the tourist lexicon the term tradition draws its potency to mobilise and excite precisely from the fact that it is poised between the supposedly scientific and the unashamedly romantic and mythic.

References

Hobsbawm, E. and Ranger, T. (eds) (1983) *The Invention of Tradition*, Cambridge: Cambridge University Press.

TOM SELWYN, UK

train *see* rail

training

Training and **education** have long been discussed as two distinct concepts and activities. Training can be described as the process of preparing or being prepared for a job, and is usually deemed to be vocationally specific. Dewey described education as a 'reconstruction or reorganisation of experience which adds to the meaning of experience, and which increases ability to direct the course of subsequent experience' (1916: 89–90). Thus, training is viewed as a means to teach a skill, whereas education fosters the development of the whole person without regard to practical application. In the case of tourism **human resource development**, distinguishing the two concepts is illogical and unproductive; education and training are intimately interwoven.

In tourism, the social and **market** context in which people work changes over time. Standards **change**, **technology** and economic climate influence **change** in the workplace; even the perceptions of client **service** expectations change. These changes relate to emerging skills and skill levels that impact the training needs of the workforce. Tourism professionals continually develop their knowledge base and skills in order to meet the demands of a dynamic and evolving work climate. This may be why the tourism training orientation in, for example, Canada is largely based on humanistic and technological ideologies.

A humanistic orientation promotes training as a process of personal development and **self-actualisation**. Humanists argue that training must be relevant to the learner, and exploratory with a focus on problem-solving skills, **innovation** and creativity. Emphasis is on process rather than product. Relative growth of the learner is more important than set goals or criteria to be obtained. The technological perspective is founded on **efficiency** and accountability on achieving results. Training is competency-driven; the emphasis is on what learners must be able to perform in the workplace to be considered competent. Learners are expected to master skills through a planned and contrived sequence of instruction. All learning objectives are based on criteria for performance and evaluation. The primary concern is to determine what is not yet learned, and to continue to focus on meeting these objectives. Often, technologists arrange training resources in a modularised format.

Traditional approaches to on-the-job training leaned towards the technological perspective, but training is becoming increasingly more focused on humanistic propositions. Principles from either or both pedagogic theories are prevalent in tourism training because of their practical and relevant applications to the workplace. The task of providing effective training poses a dilemma for many employers. Their primary concern is with employee competency; however, they may not be good trainers. Regardless of how well they know the job, bestowing the skills on a trainee is a very different task. By accepting and following the principles put forward by both humanists and technologists, employers enable the learner to engage in meaningful, relevant training tasks and thus the latter is made accountable for their own professional development. The problem with this accepted approach is the difficulty in succeeding to define essential prerequisites and learning hierarchies. Both require an analysis or understanding of the cognitive functions, which is beyond the scope of most employers.

Tourism education and training is not a constant. In order for employees to succeed in an ever-changing work environment, they must acquire transferable skills that provide opportunity and job mobility. Industry professionals must be taught the fundamental skills necessary for working in various occupations, including cross-occupational skills and essential skills. The need to retrain (upgrade skills) is equally as important as the life cycle of jobs and of specific skills associated with particular job change. Therefore, the currency of skills learned is a significant commodity.

To ensure that a tourism workforce remains competitive in the global marketplace, educators and employers must work together to meet the training needs of the industry. Specific technical skills/competencies can be readily acquired in the workplace; for example, an employee can learn how to serve a table, process a **reservation**, guide a **tour** or even prepare a lunch. Essential skills (those most in demand and core to any job) such as problem-solving skills, effective communication skills or customer relations skills are probably best taught in a formal environment before applying them to the workplace.

As one example, the Canadian tourism industry has since the mid-1980s been working to develop various instruments for employers and educators to support **human resource development**. Occupational standards, competency-based curricula (for the workplace and/or for the classroom), 'prior learning' assessment tools and essential skills profiles are examples of the tools and the types of **research and development** work that continues to evolve. Occupational standards are the core documents which influence effective human resources practices.

Such standards are statements that outline the knowledge, performance and attitudes required of an individual to be considered competent in an occupation. Standards provide the basis for curriculum and programme development, assist employers with defining training needs and staff development, help identify **career** paths and provide a basis for certification, as well as many other benefits that lead to improved service levels and the development of professionals in the industry.

Training in tourism will continue to evolve as the work climate changes. To meet the demands of a global **market** and ever-increasing service culture, tourism professionals will have to continue to seek meaningful ways to increase their repertoire of knowledge and skills. Formal education alone will not adequately meet the training needs of industry professionals. Employers will also have to develop their training skills in order to meet the demands of a dynamic and evolving work climate.

See also: curriculum design

References

Dewey, J. (1916) *Democracy and Education*, New York: Macmillan.

Further reading

McNeil, D.J. (1990) *Curriculum – A Comprehensive Introduction*, New York: HarperCollins.

Mondor, P. (1995) 'Articulating curriculum intentions for the Canadian tourism industry', Canadian Tourism Human Resource Council.

PHILIP E. MONDOR, CANADA

transactional analysis

Transactional analysis is a **theory** of **psychology** and interpersonal dynamics developed by Eric Berne in the 1960s, popularised with the publication of his and other books, and extensively used to train employees of tourism organisations in the 1970s and 1980s. Berne devised a simple terminology to replace psychology's technical jargon and new concepts which could be easily understood by lay persons. Tourism companies determined that transactional analysis was particularly useful for improving a frontline employee or manager's ability to handle difficult interpersonal situations with customers or employees. Transactional analysis is now included in many human relations and organisation **development** textbooks, and thus may be seen as a beneficial area of study for tourism professionals. For the purpose of tourism, the theory may be separated into four areas of study.

The first area is 'transaction'. A transaction is the basic unit of study and social intercourse which occurs when one person encounters another and says or does something to acknowledge the other person. The sender gives a transactional stimulus for which the receiver gives transactional response. A conversation between a tourism employee or manager and a customer or another employee involves a series of transactions. Most are either complementary or crossed. The latter causes trouble because an expected response is not received, thereby stopping good communication and setting off a number of ineffective transactions. Some crossed transactions are so common that they have been termed games and given distinct names such as 'uproar' or 'ain't it awful'.

The second area is 'ego states', or structural analysis of the personality. An ego state is a distinct mood with a consistent pattern of feeling and behaviour which resides in the brain like a tape in a video recorder. Each person has a similar parent, adult and child ego state which contain the 'taped messages' of a lifetime and differ based on each person's unique life experiences. Tourism professionals and customers may transact from only one of these ego states at a time, and this may be analysed and identified based on ego state-specific **behaviour** (such as gestures, body postures, facial expressions and words). For example, the adult ego state is the thinking, objective and rational part of the personality which has attentive posture and facial expression with eye-to-eye contact, relaxed gestures, an unemotional tone of voice which uses question words such as what, how, where, or factual words such as 'the flight takes off at 2:00 pm'.

The third area is 'improving crossed transactions'. Transactional analysis suggests that the tourism professional should remain in their adult ego state so that they are not 'hooked' into the parent or child by the person crossing the transaction. Adult questions may then be used to make the other person think. This usually results in their return to the adult. Other methods include agreeing in some way with what the person is saying and active listening.

The final area is 'additional concepts'. Other transactional analysis concepts such as strokes, time structuring, trading stamps and life positions can further the tourism professionals' understanding of themselves and others. For example, a life position describes a person's basic feelings about themselves and the healthy 'I'm OK, You're OK' position, the one that a successful tourism professional should rationally choose. People are born into this position, and it is reinforced by many 'OK' experiences with others. Although tourism professionals may have some 'Not OK' feelings, according to transactional analysis theory, these may be reasoned away by a person's adult ego state.

Further reading

'Using T.A. to keep things OK', (1978) *Hotel and Motel Management Journal*, April.

Wachtel, J.M. (1980) 'Transactional analysis training for the travel industry', *Annals of Tourism Research* 7(3): 455–71.

JEFFREY M. WACHTEL, USA

transit

Transit, also called mass transit and/or urban passenger **transportation**, refers to the movement of passengers within a city or metropolitan area. Transit systems are also important to rural areas, but light density traffic patterns make them inefficient and have limited their development. Transit systems may carry a mix of **local** residents for commuting and personal travel, as well as **tourists**. Transit consists primarily of buses, subways, and light **rail** systems.

Bus systems use motor-powered vehicles operating on fixed schedules along fixed routes, but not confined to fixed guideways. This provides flexibility to alter routes and service times to accommodate changes in demand and traffic patterns. Subways refer to high speed rail systems operating in tunnels or on elevated structures, separated from highways to avoid interference with traffic. They may use high-performance trains reaching speeds of up to 130 kilometres (80 miles) per hour, carrying up to 40,000 passengers per hour. Light rail systems, also called streetcars or trams, are electric railway systems that use single rail cars or short trains driven by overhead electric power lines. These supply an attractive option to cities that need more capacity than bus systems provide, but not enough to justify investing in expensive subway systems.

By the 1860s, many cities in the United States and elsewhere had horse-drawn streetcar systems. Electrification led to rapid expansion of streetcar systems during the 1890s, and the first underground system (subway) began operations in New York City during 1904. Automobile development in the early twentieth century offered individuals more flexibility and convenience, creating major problems for public transit systems. By the 1960s, the US government **investment** in streets and highways, accompanied by dramatic increases in automobile ownership and urban sprawl, led to a significant decline in ridership and the demise of virtually all privately-owned systems. To preserve public transit services, local governments took over operations of transit systems, with most in the United States heavily subsidised. It is believed that returning transit systems to the private sector would permit competition to generate efficiencies and reduce congestion, although not all would agree.

See also: transportation

Further reading

Cervero, R. (1992) 'Futuristic transit and futuristic cities', *Transportation Quarterly* 46(2): 193–204. (Provides an overview of potential transit developments in the twnty-first century.)

Taafe, E.J., Gauthier, H.L. and O'Kelley, M.E. (1996) *Geography of Transportation*, 2nd edn, Upper Saddle River, NJ: Prentice-Hall. (Chapter 6 provides an overview of urban transportation.)

JOHN OZMENT, USA

transportation

Transportation, in its simplest form, is the movement from one place to another of either people or goods (tangible products). Of course this **service** involves much more, including the amenities provided for passengers travelling between two or more points such as comfort, food and beverage service. Passengers may also judge a carrier on such transportation service **quality** elements as speed, availability (frequency of service and points served), dependability and **safety**.

For many destinations, transportation plays a vital role in the development of a viable tourism industry, both in the transportation of tourists to, from and within them, and in the transportation of goods (cargo) such as food and supplies to support tourism operations. Without ample and convenient access between tourists' origins and intended destinations, development of those locations would be nearly impossible.

There are five modes of transportation, based on the physical characteristics of the service offered and the right of way over which a transportation carrier operates. These are air, highway, **rail**, water and pipeline, arranged approximately in descending order of speed. With the exception of pipeline, each of the modes provides passenger transportation in some form. Some may argue that space transportation may constitute a sixth mode, at least in the **future**, although not all agree.

The air mode (see **airline**) consists of vehicles (aircraft) flying often at considerable altitudes, above ten kilometres (six miles). These aircraft may be 'heavier than air' (fixed-wing aircraft such as jets and non-fixed-wing aircraft such as helicopters, for example) or 'lighter than air' (such as blimps, balloons or dirigibles). Typical speeds for air carrier vehicles are from about 960 kilometres (600 miles) per hour for most jet aircraft to 2,170 kph (1,350 mph) for supersonic aircraft such as the Concorde (see **aircraft, supersonic**). Aircraft such as helicopters may provide passenger transportation over short distances of only a few kilometres or miles. The main **impact** of air transportation, however, is over longer distances, up to over 12,900 km (8,000 miles) for non-stop jet service. For the air mode, the right of way is generally not geographically constrained, at least as far as in-air routes go. Air traffic is often constrained, however, by the location of terminals (see **airports**) and the demands of air traffic control to fly in fixed lanes.

The highway mode consists of a number of types of vehicles including bus, taxicabs, car hire and private automobiles. Bus services can be found within urban areas and connecting origins and destinations of varying distances, while taxicabs are generally used only for trips of short duration. Automobiles, whether private or rented, can travel anywhere from a few blocks to thousands of kilometres or miles in the process of transporting a traveller. Some forms of highway transportation are constrained geographically, although the automobile can generally reach anywhere there is some type of road. With off-road vehicles, few locations are unreachable other than mountainous or very steep terrain. Even with buses, little in the way of terminal development is required when compared to air, rail or water transportation. In fact, with bus service it is possible to pick up or drop off passengers at nearly any point along the route served.

Rail transportation is confined to a limited right of way, since the carrier must travel where there is a set of railroad tracks. Often, railroad tracks are only found between and within urbanised areas where sufficient volumes of passengers can be found (see **transit**). Additionally, the number of terminals can be quite limited and far apart, especially in non-urban or rural areas. The speed of rail service can vary widely from over 300 kph (200 mph) for high-speed rail services to average speeds lower than 80 kph (50 mph).

The water mode is typically characterised by vehicles that move on the surface of the water, such as passenger ships, ferries (see **ferry**), or hovercraft. Ferries are generally used on shorter routes, including within urban areas (Seattle, for example) or in archipelagos (the Philippines or Indonesia, for instance). In some cases the mode operates below the surface, such as with submarines. These vessels are used primarily for viewing marine life and underwater scenes, rather than for transportation between an origin and a destination.

In recent decades, the most prominent sector of the passenger ship market was and continues to be the cruise lines, which offer far more than just transportation between two points. They may provide lavish entertainment, fine dining, **gambling** and other amenities to their guests. The right of way of the water mode may be constrained to navigable waterways, especially for inland routes, but all forms of this mode are constrained by the location of terminals (seaports or inland ports). The water mode does not provide very high speeds compared to air, rail or highway, with a typical cruise vessel operating at about 40 kph (25 mph).

All modes of transportation are regulated in some manner, be it economic or non-economic. Economic regulation focuses on the routes served, frequency and capacity of the service provided, and the fares or rates charged by carriers. Non-economic regulation, on the other hand, focuses on aspects such as safety, certification of carrier personnel and traffic control. Nearly all nations have some form of domestic transportation regulation, with international transportation regulation often being more complex.

Internationally, the degree of economic regulation varies by mode of transportation. International **airlines** have traditionally been heavily regulated, for both economic and non-economic reasons. Cruise lines, however, have seen little in the way of economic regulation, with a varying level of safety regulation depending upon where the ships are registered. The rail and highway modes have also been regulated internationally when vehicles pass between nations.

Economic regulation has been reduced in a number of transportation markets in recent years, beginning with domestic **airline deregulation** in the United States in the late 1970s. Since then, many other countries have deregulated their national transportation systems to some degree, including air, highway and rail. Internationally, economic regulation has been reduced in some markets, with the **European Union** being a prime example of including not only air but other modes as well.

Passenger transportation plays a critical role in determining the success or failure of nearly every segment of tourism. Without a reliable and economic form of passenger transportation to, from and within a destination, enticing tourists to visit that destination may be very difficult. In some cases, transportation becomes an identifiable part of the tourism **product**, such as the cable cars in San Francisco, double-decker buses in London, and gondolas in Venice.

The transportation carrier may also be a 'destination' in its own right, as has occurred with some cruise lines in areas such as the Caribbean, Mediterranean, and Mexican Riviera. Other forms of transportation that serve as 'destinations' include tourism railroads (typically steam-powered), luxury trains such as the 'Orient Express', and 'flights to nowhere' on the Concorde. In other cases, a transportation carrier may provide the only viable way of viewing a destination, such as the cruise lines serving the southeast coast of Alaska.

Technological advances in transportation over

the years have made a tremendous impact on tourism volumes. In the nineteenth century and before, means of transportation were limited, slow and often expensive. As a result, most destinations attracted only a fraction of the potential tourists that could be drawn to that location. The development of rail and highway transportation in the late nineteenth and twentieth centuries allowed for a much larger volume of travel.

The development of the air mode, along with the highway mode, has arguably had the greatest impact on travel, particularly for longer distances. The advent of the 'jet age' in air transportation ushered in a period of tremendous growth in long-distance travel, especially across the Atlantic and throughout the Pacific Basin, where travel distances tend to be quite long and often over water. Destinations such as Hawaii, Tahiti, Australia and New Zealand would not have experienced tourism growth to any substantial degree without the speedy and economical transportation afforded by jet aircraft. As the demand for tourism grows in the future, transportation managers and policy makers will face growing challenges. Without such efficient and effective systems, the expansion of tourism will be hindered.

See also: international aviation liberalisation; motor coach tours; route system

Further reading

Heraty, M.J. (1989) 'Tourism transport – implications for developing countries', *Tourism Management* 10(4): 282–92. (Discusses the role and importance of ground transport in developing nations' tourism industries.)

Hobson, P.S. and Uysal, M. (1993) 'Infrastructure: the silent crisis facing the future of tourism', *Hospitality Research Journal* 17(1): 209–15. (Examines future transportation scenarios as they affect the tourism industry.)

Page, S. (1994) *Transport for Tourism*, London: Routledge. (Evaluates the relationship between tourism travel and the transport industry and impacts on transport providers, managers and policy makers.)

World Tourism Organization (1994) *Aviation and Tourism Policies: Balancing the Benefits*, London: Routledge. (Discusses recent developments in aviation and tourism policies.)

FREDRICK M. COLLISON, USA

transportation, globalisation of

The emergence of the global economy is perhaps the most significant **trend** of the twentieth century. The interconnectedness and interdependencies of world trade and travel will preclude individual markets from remaining isolated from the rest of the world. As the globe continues to shrink, rapid communications and **transportation** linkages will become even more essential. A major contributor to this **growth** is the international aviation industry, which has benefited from the deregulation and liberalisation policies of the last decade.

Since the 1980s, the world has witnessed very rapid increases in international commerce. These gains are a result of many factors, including **technology** advances and transfers, **productivity** increases, freedom for former socialist states and the resultant market expansions, the establishment of free trade areas and economic unions, and more affluent consumers. As a result, producers and consumers need more frequent interaction and communication in order to provide responsive services. Often, cooperative agreements or alliances are made with international partners in order to facilitate smooth exchanges of goods and services.

International airlines have been prime beneficiaries of this phenomenal growth in world trade. These airlines have experienced tremendous increases in the volume of passengers and air cargo carried in recent years. At times, however, these airlines have been frustrated in expanding their global networks in a timely and efficient manner. Among the difficulties to be overcome are archaic governmental regulations and traditional bilateral approaches to international air services (see **international aviation bilateral**).

The major purposes of these economic regulations and bilateral agreements were (and are) to protect the national flag carriers and limit the competition faced from foreign flag airlines. Restrictions were placed on fares, routes and carriers in particular markets. Most nations also had strict prohibitions on mergers between home nation and foreign airlines. Additionally, the extent of foreign ownership allowed for a home nation airline has been limited, usually in the 25–49 per cent range.

As pressures for the globalisation of air services increased, many countries have relaxed their economic regulations and allowed more competition. Approaches to these changes have varied from total, immediate airline deregulation, as in the United States and Canada, to liberalisation in which the transition is made in a slower, gradual process, as in the **European Union**. Open skies agreements, which promote a free market approach to international air services agreements, have become more widespread throughout the world. These changes have allowed international airlines to enter into strategic alliances with each other. Such cooperative agreements are legal and often do not violate existing anti-merger and acquisition laws of the nations involved.

One intended outcome of these alliances is to allow **seamless service** in which passengers can buy a single ticket, check their baggage once and fly across the world via several airlines. The member carriers coordinate their schedules in order to minimise passenger waiting times and to increase load factors and economic viability of the service. Participating airlines also integrate frequent flyer programmes and airport lounge privileges, and may also combine elements such as aircraft maintenance, **purchasing**, **reservation** and **catering** functions. Economic rationales for these alliances include **economies of scale**, density and scope; declining unit costs; and offering a more competitive service in the marketplace.

An example of the largest global alliance to date (1998) is the STAR alliance, in which Air Canada, Lufthansa, Scandinavian Airlines System, Thai International, United Airlines and Varig coordinate their services across a wide range of customer service departments. This vast global network allows passengers and air cargo to reach almost anyplace in the world with the assurance of consistently high standards of service and safety.

The formation of such mega global alliances are not without their **critics**. Some observers feel that it still remains to be seen if such agreements will be profitable in the long term. They state that current alliances focus primarily on high levels of customer service and potential revenue increases. Significant cost reductions due to factors such as decreases in workforce levels, combined purchasing power, use of common and integrated computer reservation systems, and schedule coordination have not yet materialised in many cases.

In addition, corporate culture and personality clashes have occurred in some alliances. **Infrastructure** problems exist at some key hub airports, and difficulties thus arise in allocating gates and take-off and landing slots to new and existing airlines. Such problems, along with aviation policy differences between the United States and the United Kingdom, have stalled the proposed British Airways and American Airlines alliance. Finally, some observers fear that competition will decrease and fares will increase as a result of these alliances. Nevertheless, airlines throughout the world are attempting to position themselves in the most advantageous alliances. Global networks are changing rapidly. We can be confident that these trends will continue for some time.

Further reading

Gialloreto, L. (1988) *Strategic Airline Management: The Global War Begins*, London: Pitman. (Describes future global airline networks and the necessary changes to achieve them.)

Leonhardt, D. and Echikson, W. (1998) 'Taking a whack at airline alliances', *Business Week*, March 16: 106. (Points out the potential problems in airline alliances.)

Oum, T.H. and Taylor, A.J. (1995) 'Emerging patterns in intercontinental air linkages and implications for international route allocation policy', *Transportation Journal* 34(7): 5–27. (Provides an analysis of changes taking place in international air services.)

'S T A R alliance unlikely to impact on airline cost cutting' (1997) *Airfinance Journal*, June: 18. (Describes the world's largest global alliance to that date.)

DAVID B. VELLENGA, USA

WILLEM J. HOMAN, USA

transportation pricing

Many transport carriers rely on 'cost plus' methods in which a carrier finds the unit expense for a particular journey and adds a mark-up to provide for a profit. Alternatively, many carriers use a marginal costing approach which is concerned with the contribution (selling price minus marginal or variable cost) earned from each unit of sale. In both cases, the **pricing** approach is tempered by **market** conditions.

Fixed costs or overheads are wholly inescapable in the short-run, such as administration, **marketing** and reservations, leases on buildings and cost of utilities. These are paid even when the carrier has empty seats. The level of fixed costs may also be route-related as with crew salaries and expenses, fuel, maintenance, payments to regulatory bodies (air traffic control, customs and excise) and terminal operators who provide terminal facilities. Variable costs are normally linked to the number of passengers carried, such as with meals, baggage handling and ticketing.

However, pricing is not as simple as that since not all services will be full, and generally carriers who offer scheduled (published) services operate them no matter how many passengers wish to undertake the journey. Carriers need to determine how many seats on a service they could expect to sell in order to reach a breakeven point. In order to do this an airline, for example, will examine historic booking profiles to determine expected load factors.

Passenger load factors can be maximised by strategically releasing seats available at promotional fares into the **reservation** system, especially for **leisure** tourists. Normally business people require flexibility in their itineraries, but such flexibility comes with a price attached. The lower the price of a ticket, the less flexibility a passenger has in relation to refunds, changes and duration of stay.

On high-density routes competition will be greater, forcing down prices and in effect forcing up service levels. Carriers need to remain competitive by matching the prices of their rivals, offering special deals, **loyalty** schemes (see **frequent flyer programmes**), or adding value to their services to attract customers. Predatory pricing, or the undercutting of competitors to an unsustainable level, has been used previously to drive rivals out of the marketplace. High usage of equipment will reduce unit expenses and many carriers will try to maximise such use. However, this can mean that in the event of equipment breakdown a substitute vehicle may not always be available.

THRINÉ HELY, UK

transportation service quality *see* quality; transportation service

travel

Travel involves movement from place to place. This is a fundamental aspect of tourism, and in its absence there would be no tourism. Improvements in the ease of travel have greatly increased the magnitude of tourism and have influenced the forms which it takes. In fact, many forms of travel, such as walking, canoeing, rafting, horse-riding, **skiing**, driving for pleasure and snowmobiling, are tourism **activities** in their own right.

Although sometimes used synonymously with tourism, travel is a broader concept. There is a diversity of types of travellers, such as migrants or exchange students, not all of whom may be tourists. Both travel and tourism involve the movement of people between origins and destinations along connecting routes. However, depending upon the distance travelled, whether **borders** are crossed, **length of stay** at the destination and **motivations**, travellers may or may not be considered to be tourists. Furthermore, the application of such criteria may vary from situation to situation, making the distinction between travellers and tourists imprecise.

There may also be temporal and qualitative differences in the meanings ascribed to travel and tourism. Thus, there is a tendency for some to refer to people moving individually and in small groups, and in periods prior to **mass tourism**, as travellers in contrast to the more ubiquitous and larger scale of modern tourism. At the extreme, tourism is used as a disparaging term ascribed to others by those who claim superior motives and tastes who regard themselves as travellers.

GEOFFREY WALL, CANADA

travel advisory

Countries often inform their citizens about unstable political or social conditions existing in destinations around the world as a means of providing **safety** warnings. These advisories are not meant to deter **travel**, although they may have that effect, but rather to prepare potential tourists for situations they may not be familiar with and allow them to take appropriate caution.

WILLIAM C. GARTNER, USA

travel agency

Travel agencies sell inclusive tours, holiday, **transportation** tickets and other related products such as **accommodation**, car rentals, **attraction** tickets and insurance to the public. The suppliers (or principals as they are often known) of the products sold may include **tour wholesalers** or **tour operators**, and transportation, **car rental** and **hotel** companies.

A travel agency is an indirect form of **distribution channel** in that the agent acts as an intermediary (or **middleman**) between the potential tourists and the suppliers. Such a channel can be contrasted with a direct distribution channel, whereby the supplier communicates directly with the customer. These agents as retailers act on behalf of their customers in making arrangements of their choosing with the suppliers. Legally, however, they operate as an agent of suppliers, and are paid a commission by them for sales made. To tourists, the value of using this **facilitation** is that agents can offer advice and recommend the best **product** to meet their requirements. For suppliers, they represent a cost-effective distribution channel for their products.

Thomas Cook is credited with being the first travel agent when, in 1841, having chartered a train in the United Kingdom to attend a temperance meeting, he sold the **excursion** to the public. Before the Second World War, most agents predominantly sold ship and **rail** travel. An enormous **growth** in their number has taken place in the postwar period with the advent of more **leisure** time, advances in transportation **technology** (particularly jet aircraft) and a wider variety of wholesaler's products. Agents have developed to sell an increasing number of tours, often termed **package tours**, organised by wholesalers. In so doing, they are able to pass on the bulk buying economies generated by the wholesaler to their retail customers. In many cases, the wholesaler owns retail travel agents and thus controls both the production of the **package tour** and retail distribution through a process termed **vertical integration**.

The travel agency business is becoming more competitive, and ownership patterns have altered. Independent agencies owned and managed by an individual businessman have come under increasing competitive pressures from large companies owning multiple travel agency branches and franchised branches of major touristic companies. Independent agents have reacted by cooperating through the pooling of **resources** in various types of **marketing** consortia. Another challenge to be faced is for travel agents to redefine their role and business in the face of bookings of air tickets and **hotel** bookings, among others, by potential tourists directly through the **Internet**, thus bypassing their services.

Further reading

Brendon, P. (1991) *Thomas Cook: 150 Years of Popular Tourism*, London: Secker and Warburg. (Traces the history of Thomas Cook over 150 years.)

NIGEL G. EVANS, UK

Travel and Tourism Analyst

Travel and Tourism Analyst contains studies by authors with expertise in various facets of tourism operation. Each issue contains some five reviews on the **transportation** sector, **outbound** market, **market segmentation**, **accommodation** and **leisure** markets, and occasional studies. Each

underdevelopment

Within **development** theory, the concept of underdevelopment suggests that wealthy capitalist countries have held back the development of so-called **Third World** countries. Tourism in Third World destinations is controlled for the economic benefit of foreign owners, reinforces dependency, lacks involvement of local decision makers, leads to negative sociocultural impacts, and results in the promotion of staged attractions to capture an international tourism market.

PETER MASON, NEW ZEALAND

UNESCO

The main objective of UNESCO is to advance **peace** and collaboration among nations through **education**, science, culture and communications. With 185 member states, UNESCO promotes the advancement, transfer and sharing of knowledge and the exchange of specialised information. UNESCO's World Heritage Centre assists in implementing an international agreement to preserve outstanding cultural and natural **heritage** sites around the world. It is headquartered in Paris.

JONI E. BAKER, USA

unfamiliarity

The process whereby the effects of strangeness are minimised is the management of unfamiliarity. Since tourists are distinguished according to the amount of familiarity or **novelty** they desire, it is important to reduce the foreignness of situations for those recreational and diversionary types seeking reminders of home. This technique is frequently employed by **travel writers** through the use of simile, denigration of locals and establishing **expatriate** connections.

GRAHAM M.S. DANN, UK

uniform system of accounting

The idea of a uniform system of accounts was first developed in the 1920s. One of its first advocates was the German economist and **accounting** theorist Eugen Schmalenbach. The idea was further developed and applied in various other countries, including France, Sweden and the United States, along two rather different lines. Uniform systems of accounts may by either macro- or micro-orientated: that is, they may be designed with a view to national income accounting, economic **planning**, taxation and other governmental concerns; or they may seek to serve the accounting needs of individual business firms. They may also include a bookkeeping structure for recording accounting data (input) as well as a uniform set of financial statements (output), or may be confined to the latter. They may be tailored to the needs of individual industries, or be more general. While in principle uniform systems of accounts (especially those with an industry orienta-

the production of guidebooks for merchants and travellers. These contained information on distances, **accommodation**, transport, costs and hazards. This factual assistance developed even further in the seventeenth and eighteenth century with the **growth** of the **Grand Tour**. Highly detailed guides were produced, and these were further developed in the nineteenth century by specialised guide publishers such as Murray and Baedeker.

Another form of this literature, also with an extensive history, is the account of a journey. In its content it may at times resemble the guidebook, and travel accounts have often served this dual function. From the ancient world, such important accounts include Herodotus's *History of the Persian Wars* (*c*. fifth century BC), which drew on his extensive journeys, and the travels of Chang Ch'ien in Asia (*c*. 130 BC). Accounts of shorter and less eventful journeys also survive, such as Cicero's visits to his holiday villas. For a later period, the movements of Hsuan-Tsang from China in the seventh century AD, Marco Polo in the late thirteenth century and the Arab Ibn Batuta in the fourteenth century must rank among the great travel accounts of all time.

The sixteenth century growth in travel and trade within and beyond Europe resulted in a massive surge in travel accounts. Reading about travel became a staple **diet** for the educated person by the later seventeenth and eighteenth centuries. During the latter century, the pervasive influence of these accounts in literary culture can be seen in its role in the development of the novel where plot, structure, character and style all reveal links with travel literature. This influence can be traced in the works of Swift, Defoe, Fielding, Smollett, Johnson and Sterne, among others. Furthermore, information gathered from travel and recorded in detail had an influence on science and thought of the period.

The great sea voyages, from the fifteenth century onwards, of figures such as Magellan, da Gama, Dampier and Anson appeared in numerous collections. Early accounts were detailed in Hakluyt's *Principall Navigations... of the British Nation* (1589) and De Bry's *Voyages* (1590–1634); late ones appeared in productions like Churchill's *A Collection of Voyages and Travels* (1704). As with the guidebook, the travel account continued as a popular literary genre through the nineteenth century when it was used as a device by writers such as Mark Twain, Henry James and Robert Louis Stevenson. The stress on subjective impressions marked the increasing divide between this form of travel literature and the objective factual detail of the nineteenth-century guidebook. Versions of traveller's tales have maintained their popularity right through to the present, with recent exponents including Eric Newby and Paul Theroux (see **travel writer**).

Travel literature can be considered not only by content but also by its form, existing in both published and unpublished state. This latter form can include the simple notes made during a journey, recording information and impressions. A particularly important form, however, has long been the letter. During the eighteenth century, the letter format was a very popular device for recording travel experiences, for instance Tobias Smollett's *Travels in France and Italy* (1766) and his novel *Humphry Clinker* (1771). In addition to the letter, the literature has often been produced in the form of a diary or journal, again existing in published and unpublished state. The veracity of letters, diaries and journals has always raised problems for research. Even unpublished accounts may be altered to protect the writer or the sensibilities of the audience. In published form, not only can events be altered, but also the popularity of the genre, especially in the eighteenth century, can give rise to totally spurious accounts. The practice of keeping a diary or journal during travel probably reached its peak in the eighteenth century, after which the habit seems to have declined. In the twentieth century, keeping a travel journal is relatively unusual for the general **tourist**.

The literature can also occur in other forms. It may arise in biographies and autobiographies and in poetic form. Indeed, this literature has been produced by a wide variety of people travelling for many purposes, including traders, missionaries, scientists and diplomats, as well as the traveller for pleasure. From the eighteenth century onwards there has also been the professional writer who has utilised its appeal as a vehicle for literary output, satisfying both active visitor and sedentary armchair traveller.

Further reading

Adams, P.G. (1983) *Travel Literature and the Evolution of the Novel*, Lexington, KY: University of Kentucky.

Batten, C.L. (1978) *Pleasurable Instruction: Form and Convention in Eighteenth Century Travel Literature*, Berkeley, CA: University of California. (A detailed critique of the genre in the eighteenth century with useful discussions on authenticity.)

Casson, L. (1974) *Travel in the Ancient World*, London: Allen and Unwin. (An introduction to travel writing and literature in ancient Greece and Rome.)

Fussell, P. (1980) *Abroad: British Literary Travelling Between the Wars*, Oxford: Oxford University Press. (Discusses a range of travel literature in the early twentieth century.)

JOHN TOWNER, UK

travel writer

Imaginative literature has always incidentally featured destinations or forged links between writers and places. In the last hundred years there has been a massive growth in travel writing as a specialist, non-fiction genre in its own right whose prime subject is touristic. Both kinds of writing exert a powerful, but under-researched, influence on tourism.

A.V. SEATON, UK

treaty

Treaties promoting trade in tourism have political and economic ramifications and serve as integral parts of the world's **political economy**, with bilateral agreements as the most common form. Usual objectives of treaties and agreements include increasing two-way tourism, supporting efforts of national tourism organisations and their promotion offices, improving **facilitation**, encouraging reciprocity in industry **investment**, promoting **standardisation** and sharing of **statistics**, suggesting cooperation on international **policy** issues, and fostering consultation and a focus on **education**, **training** and mutual understanding.

A significant agreement with worldwide trade **impact** on tourism is the General Agreement on Tariffs and Trade (GATT). Signed in 1994 by 125 nations, its sub-agreement, the General Agreement on Trade and Services (GATS), sets forth global trading **rules** for service industries providing procedures for liberalisation of trade through intergovernmental negotiations. The member countries agreed to submit commitment schedules specifically outlining national trading rules for each service sector. Over 100 countries have submitted schedules in tourism, more than in any other service sector. GATS is legally binding, with signatory countries agreeing to specific rights and obligations which may be disputed, settled or enforced through established procedures. Its rules guide member countries towards progressive liberalisation of commercial policies including gradual accord of **market** access, national treatment and most-favoured nation status to all members.

The North American Free Trade Agreement (NAFTA) is a regional trilateral treaty arrangement using GATT principles. NAFTA principles apply universally to all sectors including tourism, and exclude only those singled out in the annexes. Under NAFTA, the United States, Canada and Mexico accord national treatment to tourism services. This signifies treatment as favourable as that accorded by a province or state to those residing within that domain. The effect is liberalisation of intra-border trade in tourism services and reduction of barriers. NAFTA requires annual consultations to monitor implementation of tourism provisions which ensures regular review of tourism matters, and encourages more positive regional relations.

Tourism treaties and agreements aim to reduce existing worldwide barriers and deter formation of new ones. They are critical components in free trade in tourism, increasing business opportunities, raising receipts, increasing job opportunities and promoting more peaceful progress toward national and international economic and social goals.

Further reading

Edgell, D.L. (1995) 'A barrier-free future for tourism?', *Tourism Management* 16(2): 107–10. (Interprets implementation of the General Agreement on Trade in Services for tourism.)

Smith, G. (1994) 'The North American Free Trade Agreement (NAFTA) and the implications for the U.S. tourism industry,' *Tourism Management*, October: 323–6. (Identifies aspects of trade in tourism services under NAFTA and highlights implications for the US industry.)

GINGER SMITH, USA
DAVID L. EDGELL, USA

trekking

Walking on trails is increasingly popular among tourists with better **accessibility** to areas of the world such as **Nepal** and the hill country of **Thailand**. This practise is also known as tramping. Overuse on certain routes, such as the Mt Everest base camp, has caused severe negative environmental impacts including deforestation and trailside litter. Conversely, utilising locals as guides can successfully reduce **activities** such as poaching in **protected areas**.

CHARLES G. JOHNSTON, NEW ZEALAND

trend

Forecasting trends consists of trying to foresee how tourism will evolve in the **future** so that public authorities can plan and design appropriate policies and so that private companies can prepare and respond to projected changes. Many different trends are of interest and importance in tourism, but those dealing with **demand** and spending receive special attention. For these, variables used include the number of tourists, spending per tourist and the sector's share of the gross national product.

In order to calculate trends, different methods are used. These may be of a quantitative or qualitative nature. Quantitative methods of forecasting can be divided between those which use techniques based on economic **theory** and those which are purely statistical. Among those most commonly used methods are time series analysis, models of gravity, models of action/auditing/intervention and the Box–Jenkins I method, along with market participation and simulations. All these methods use a single or dual variate and are applied to just one dependent variable, either the number of tourists or tourism spending.

Econometric models include price levels and fictitious variables for unusual occurrences in demand of tourist flows. Often the reason for these analyses is to offer explanations rather than to make predictions, although the results are good enough to forecast trends in **international tourism**. Sectoral analysis can include input–output models or multiple interactive econometric models (like the Cambridge model or the Australian ORANI model). These models contemplate not only tourist demand, but also **investment**, income, taxes and other variables which contribute to tourism's share of the economy. Amongst qualitative methods used, mentions can be made of the brain-storming technique, the Delphi method and the executive-consensus method, all of which have been used by commercial organisations to predict their own demand.

See also: competitive advantage; market analysis; marketing research

ANTONI SASTRE ALBERTI, SPAIN

Trinet

Trinet is the acronym for Tourist Research Information Network, an electronic bulletin board designed for researchers and educators in the field of tourism. Approximately seven hundred tourism researchers in over ninety countries are connected through Trinet. Founded by Jafar Jafari and Pauline Sheldon, communications on this network are varied and include announcements of conferences and other tourism research related events, tables of contents of tourism journals, discourses on tourism research related issues, grant and employment opportunities, requests for information and details of upcoming publications. Discourses have included topics such as **ecotourism**, definitions of tourism, tourism expenditure methodologies, **virtual reality** and other applications of technology, education issues and tourism in island economies. Many subscribers who are educators find these discussions to be good material for class discussions. Each communication is transmitted to the electronic mailboxes of all subscribers. As a result, many one-to-one communications and productive professional relationships ensue from the original communications. Many report significant benefit from these contacts. Trinet is an important supplementary research tool due to the timeliness of the responses. Subscribers sending out requests for information receive responses in minutes, and many researchers have identified important sources and citations quickly through Trinet.

The main server for Trinet is the University of Hawaii's UNIX mainframe computer, which runs the Electronic Bulletin Board (EBB) software called LISTPROC. Trinet is administered by staff at the School of Travel Industry Management, University of Hawaii, who do the subscribing, updating and production of an annual directory of subscribers. This is a closed EBB, which means that those interested in subscribing must first be approved to use the system. This option was chosen to ensure quality message traffic, which is hard to maintain with an open EBB unless each message is moderated (a very time-consuming task). All tourism researchers are invited to subscribe to Trinet. Interested researchers can send an email message to trinet@uhmtravel.tim.hawaii.edu and request to be added to the list. A short biosketch will then be requested to ensure that all subscribers are bone-fide tourism researchers, whether they are in academia, **industry** or **government**. This biosketch is also forwarded to subscribers as an introduction to the new subscriber, and is used for the subscribers' information in the directory. Once subscribed, researchers can freely post messages to Trinet and will automatically receive all messages posted to the EBB by other subscribers. A directory of all subscribers which include names, contact information and research interests is updated annually and sent electronically to all subscribers.

PAULINE J. SHELDON, USA

Tunisia

The Tunisian government launched a **mass tourism development** strategy in 1942 and has since undertaken a major **hotel** construction programme to capitalise on the country's 800-mile Mediterranean coastline. During the 1970s and 1980s, the number of lodging facilities more than doubled and the bed capacity more than tripled, making Tunisia one of the fastest growing tourism economies in the world. Industry officials estimate that more than 200,000 beds should be in place by the end of the century. Although the majority of facilities are in the capital of Tunis, most **growth** has been seen in the coastal areas of Nabeul-Hammamet, the Sahel or eastern littoral, and Djerba-Gades. With the Right Development Plan (1992–6) the government began shifting some tourism development **resources** to the interior zones in Kebili, Tozeur and Gafsa in Tunisia's exotic southwest.

The growth of the industry has been rapid and the capital **investment** has been intense, particularly since the 1970s, and shows no sign of levelling off as entries have maintained a steady rise excepting the year of the Gulf War (1991). Entries into Tunisia show that tourism is very Eurocentric. Although there is a large influx of North African tourists, Europeans stay much longer and spend more money. Due to Tunisia's historical ties with France, most have been French tourists. However, government strategies to promote Tunisia's winter 'fun in the sun' appeal have resulted in more German and Scandinavian tourists. Efforts by the Ministry of Tourism to attract more North American tourists have also begun to show success. Once a small market share, the US and Canadian entries into Tunisia have steadily increased since 1989. The ratification of the Bilateral Investment Treaty between the United States and Tunisia will undoubtedly accelerate the growth of that market.

The economic **benefits** of tourism to Tunisia have been excellent. Trade **statistics** clearly show tourism's primary position and its role in reducing the trade deficit. In addition to its value as a foreign exchange source, this dynamic aspect of Tunisia's economic profile has alleviated the country's high unemployment rate. Although tourism does create many low-skill jobs, the government has countered by establishing successful hotel/restaurant management **training** programmes. To counter the concerns that tourism can adversely **impact** the economy, the government also has launched major environmental protection plans for improving beach and water quality. Tunisia's overall commitment to developing the tourism industry will likely pay large dividends into the twenty-first century.

ROBERT A. POIRER, USA

Turismo em Analise

Turismo em Analise is a Latin American journal published in Portuguese. It favours original multidisciplinary studies which analyse **development** trends of interest to the academic community and tourism professionals. Its goal is to provide a communication channel between the university and public or private organisations and markets. *Turismo em Analise* accepts manuscripts written in Portuguese and Spanish, others submissions are translated into Portuguese. First appearing in 1990, it is published twice yearly by Escola de Comunicaçoes e Artes da Universidade de São Paulo (ISSN 0103–5541).

RENE BARETJE, FRANCE

Turizam

Turizam is regarded as a forum for practitioners and academicians committed to tourism. As one of the oldest journals in the field, its original intent was to inform the tourism industry. Treating tourism as an academic field of investigation today broadens its scope. It publishes full-length articles as well as research notes and reviews. Contents of *Turizam*, a refereed journal (double blind), appear in Croatian and English. First appearing in 1953, it is published quarterly by the Croatian National Tourism Board (ISSN 0494–2639).

RENE BARETJE, FRANCE

Turkey

Turkey forms a bridge between Asia and Europe and has been impacted by the cultures of these two continents. In addition to many historical and cultural assets left by earlier civilisations, Turkey features natural touristic **resources** including beaches, high mountains, rivers and lakes, forests and caves, and fauna and flora. It has a variety of climates ranging from subtropical in the south to very rainy along the Black Sea coastline and a multitude of variation in between.

Turkey is relatively a newcomer to **international tourism**, although the importance of the industry as a source of much needed **foreign exchange** has always been recognised. The First Five-Year Plan of 1963–7 set the objectives of tourism **planning** and stressed the importance of investments in its **infrastructure** and facilities. The subsequent plans gave increasing attention to tourism **development**. Two pieces of legislation, the Foreign Investments Encouragement Law and the Tourism Encouragement Law, provided a multitude of incentives for investors, including easier credit terms, reduced taxes and land allocation on long term leases. These incentives and increased awareness of tourism as a development tool finally paid off. Decreasing government dominance in economic life and increasing private sector involvement began to show results, and in 1990 the number of foreign arrivals reached 5.4 million, or almost twice the volume registered in 1987. Tourism revenues also increased from $1.7 billion in 1987 to $3.3 billion in 1990. The Gulf War was a tremendous blow to flourishing Turkish tourism in 1991, but the impact was temporary due to the shortness of the war. By 1997 the number of tourists was over 9.7 million and the revenues close to $10 billion. It is expected that these figures will double in the early years of the twenty-first century. Currently, tourism receipts approach 40 per cent of the total export earnings of the country. The main issues that need to be addressed relate to environmental protection, fair distribution of the **benefits** derived from tourism, seasonality, and fragmented responsibilities for tourism **policy** determination, implementation and control.

Further reading

State Institute of Statistics, Tourism Statistics, Government of Turkey, web page, http:// www.die.gov.tr. (Contains up-to-date statistical information including charts and graphs.)

TURGUT VAR, USA

turnover

The tourism industry now finds itself critically short of employees. One reason could be related to the baby boomers and their shift in the society. Another factor could be that many thought this would be 'a temporary situation'. Regardless how this problem is viewed, this workforce shortage is changing the face of the industry. According to the **National Restaurant Association**, for instance, in 1995 this sector alone needed another 800,000 employees to fill the jobs already created.

While the **hospitality** industry is not the only one interested in stemming the turnover tide, the situation is not nearly so grim elsewhere. For instance, in the electronics industry (well- known for its high turnover) the rate of turnover is only about 27 per cent. Even in nursing, the rate is only about 40 per cent per year. The average for turnover in all industries in the United States is about 12 per cent annually, down from 17 per cent annually in 1995. Turnover in hospitality averages somewhere between 50–400 per cent employees and 25–200 per cent for managers annually. But it must also be recognised that a large percentage of jobs in tourism are seasonal in nature, an important factor that partly explains the situation.

Turnover costs, on average, are from $3,000–$10,000 per hourly employee. According to the National Restaurant Association, turnover costs for this sector averages about $5,000 per employee. Turnover costs for managers can average $50,000 or more. Typically, many companies associate the cost of losing one trained manager with approximately one year's annual salary, because that is how long it takes for a new manager to become fully productive.

Quality of supervision has been cited by both managers and employees as the number one cause of turnover in this field. In other words, more employees leave their employer because they are unhappy with the quality of supervision they receive than for any other reason. Ineffective communications is the second most frequently cited cause. Other major causes include poor communications between co-workers, poor **training** programmes and lack of **career** ladders.

ROBERT H. WOODS, USA

typology, tourist

Tourist typologies reflect the diversity of individual motivations, styles, interests and values, and the subsequent differences often correlate with specific disciplinary research interests. The historical literature (Towner 1996) ascribes tourism primarily to wealth, or special status as in **pilgrimage** or **war**. As the scientification of tourism progressed, subsequent to the Second World War, typologies have increased in number and specificity. Plog (1964) identified a bell-shaped curve linking tourist **motivation** with **destination**, and described three travel personality types (see **allocentric**).

Typologies based on age and economy dominated during the 1970s, led by Cohen (1972) whose initial typology established two non-institutionalised roles as drifter and explorer, and two institutionalised types, organised mass tourist and **individual mass** tourist. Smith (1977) described the demographic aspects of tourism, in seven levels as numbers increased from explorers to mass and charter tourists, and their heightened impacts upon the host culture and **local** perceptions of tourism. Further, she defined five destination interests and motivations: ethnic, cultural, historical, environmental and recreational. This decade was also marked by the initial polemic between advocates of tourism as a phenomenon of pleasure-seeking tourists and those who search for **authenticity** (MacCannell 1973). Cohen (1979) summarised this diversity as five modes of touristic experience: recreational, diversionary, experiential, experimental and existential.

The decade of the 1980s extended typologies to include historic types such as the **Grand Tour**, north–south tourism, and long-term youth and budget travel, some of which is self-testing (Riley

1988). Graburn (1983) differentiated two types of contemporary tourism, as the annual vacation or holiday break and the **rites of passage** tourism associated with major changes in status such as adulthood or **career** changes. Environmental concerns generated numerous new tourist types related to 'appropriate' or **alternative tourism**, such as ecotourists or green tourists (Smith and Eadington 1992). Postmodernism has dominated the 1990s with renewed interest in levels of reality (Urry 1990), concerns with levels of carrying capacity and sustainability, and types of tourist **lifestyle** and **behaviour** experiences (Mazanec *et al.* 1998). Typologies also serve the industry, describing **market** niches as the basis for promotion and advertising according to the trip **purpose**, group character, **transportation** activities and interests

References

Cohen, E. (1972) 'Toward a sociology of international tourism', *Social Research* 39:164–82.

—— (1979) 'A phenomenology of tourist experiences', *Sociology* 13: 179–202.

Graburn, N. (1983) 'The anthropology of tourism', *Annals of Tourism Research* 10: 9–33.

MacCannell, D. (1976) *The Tourist: A New Theory of the Leisure Class*, New York: Shocken Books.

Mazanec, J., Zins, A. and Dolnicar, S. (1998) 'Analysing tourist behaviour with lifestyle and vacation style typologies', in W. Theobald (ed.), *Global Tourism*, Oxford: Butterworth-Heinemann, 278–96.

Plog, S. (1974) 'Why destinations rise and fall in popularity', *Cornell Hotel and Restaurant Administration Quarterly*, February.

Riley, P. (1988) 'Road culture of international long-term budget travellers', *Annals of Tourism Research* 15: 313–38.

Smith, V. (1977) *Hosts and Guests: The Anthropology of Tourism*, Philadelphia: University of Pennsylvania Press.

Smith, V. and Eadington, W. (1992) *Tourism Alternatives: Potentials and Problems in the Development of Tourism*, Philadelphia: University of Pennsylvania Press.

Towner, J. (1996) *An Historical Geography of Recreation and Tourism in the Western World 1540–1940*, London: Wiley.

Urry, J. (1990) *The Tourist Gaze*, London: Sage.

VALENE L. SMITH, USA

underdevelopment

Within **development** theory, the concept of underdevelopment suggests that wealthy capitalist countries have held back the development of so-called **Third World** countries. Tourism in Third World destinations is controlled for the economic benefit of foreign owners, reinforces dependency, lacks involvement of local decision makers, leads to negative sociocultural impacts, and results in the promotion of staged attractions to capture an international tourism market.

PETER MASON, NEW ZEALAND

UNESCO

The main objective of UNESCO is to advance **peace** and collaboration among nations through **education**, science, culture and communications. With 185 member states, UNESCO promotes the advancement, transfer and sharing of knowledge and the exchange of specialised information. UNESCO's World Heritage Centre assists in implementing an international agreement to preserve outstanding cultural and natural **heritage** sites around the world. It is headquartered in Paris.

JONI E. BAKER, USA

unfamiliarity

The process whereby the effects of strangeness are minimised is the management of unfamiliarity. Since tourists are distinguished according to the amount of familiarity or **novelty** they desire, it is important to reduce the foreignness of situations for those recreational and diversionary types seeking reminders of home. This technique is frequently employed by **travel writers** through the use of simile, denigration of locals and establishing **expatriate** connections.

GRAHAM M.S. DANN, UK

uniform system of accounting

The idea of a uniform system of accounts was first developed in the 1920s. One of its first advocates was the German economist and **accounting** theorist Eugen Schmalenbach. The idea was further developed and applied in various other countries, including France, Sweden and the United States, along two rather different lines. Uniform systems of accounts may by either macro- or micro-orientated: that is, they may be designed with a view to national income accounting, economic **planning**, taxation and other governmental concerns; or they may seek to serve the accounting needs of individual business firms. They may also include a bookkeeping structure for recording accounting data (input) as well as a uniform set of financial statements (output), or may be confined to the latter. They may be tailored to the needs of individual industries, or be more general. While in principle uniform systems of accounts (especially those with an industry orientation) might be international in scope, in practice this is unusual.

The French approach focused on a national uniform plan of accounts for all business enterprises, termed *Le Plan Comptable Général* (the General Accounting Plan). This was a set of model financial statements (balance sheet and income statement) and a bookkeeping structure designed to produce the information required to prepare these financial statements. The key element of this bookkeeping structure concerned the general ledger. This (also known as nominal ledger) is the central 'book' of account, in which information is recorded and accumulated in various accounts so that financial statements can be prepared. Each account has a title, and in modern systems a number, especially as in modern systems the 'book' is in fact a computer file. Accounting data are captured by being coded according to the accounts in which the data are to be recorded; this coding takes place in another set of 'books' known as journals or daybooks, following which the general ledger is updated. Under the General Accounting Plan, not merely is the structure of the general ledger (the set of account titles and numbers known as the *chart of accounts*) laid down for all business enterprises in the country, but the rules for coding accounting data are also laid down.

The General Accounting Plan in France has historically been a concern of **government** and oriented towards financial accounting, taxation matters and gathering data for the national income accounts (in other words, a macro-orientation), rather than with **management accounting** (a micro-orientation). While recent versions have paid more attention to the requirements of management accounting, this is limited by the fact that, under the French approach, only certain industries have by virtue of their specificities their own version of the national plan. An alternative and more micro-oriented approach used in countries such as Germany and Sweden focuses on establishing standard charts of accounts at the level of particular industries rather than the national level. This approach is better able to attend to the requirements of management accounting. However, the focus is still national rather than international.

However, for one industry in particular, the **hotel** industry, there exists the Uniform System of Accounts for Hotels, which is in use at the international level. This system was first developed and published in 1926 by the Hotel Association of New York City. It has since undergone a number of revisions, the eighth of which was in 1986. Given the important part played historically by US corporations in the international **development** of the hotel industry and the highly international nature which characterises the industry, it is easy to understand both the benefits of the Uniform System of Accounts for Hotels and its widespread international use. In particular, the system provides a uniform basis for management accounting in hotels which greatly facilitates the financial controls of these **international tourism** groups. As such, it is both strongly micro-oriented and suitable for international use within the industry. The focus of this hotel system is on a set of standardised financial statements or reports, rather than on standard account titles and numbers or rules for coding accounting data.

SIMON ARCHER, UK

United Kingdom

The United Kingdom is made up of England, Scotland, Wales and Northern Ireland. This country is a significant **destination** for both international and **domestic tourism**, as well as a major generator of **international tourism**. The historical **development** of tourism in the United Kingdom has been influential in shaping trends across the world. The first true **resorts** were developed in this country from the seventeenth century onwards, based on the resources of mineral springs and seawater. In the nineteenth century, Thomas Cook enabled pent-up **demand** for travel to be realised by running **tours** within the country and overseas. In the twentieth century, demand for travel overseas led in the 1960s to a supply-side response by the packaging together of **accommodation**, flights and transfers in the form of the inclusive tour. In the second half of the century, both demand and **supply** grew as a result of social and economic change, technological **innovation** and the emergence of tourism enterprises.

The UK supply of tourism is based on the rich variety of **landscape**, **heritage** and culture. Valued landscapes are recognised as **national parks** and heritage coasts, rivers and lakes form the basis for tourism activities, the built heritage represents some of the world's most important historic buildings, and special events attract international tourists. The varied accommodation base is focused on the cities and coasts. There is an issue of quality in terms of UK accommodation stock, an issue that is exacerbated by the lack of compulsory registration. Transport gateways such as Heathrow, Gatwick and Dover are well known, but other innovative and successful gateways are at Manchester, Glasgow and London City airports. Other ports include Poole and Portsmouth, and the Channel Tunnel. Internal transport is dominantly by road, although tourism transport also takes the form of **rail**, cycle and the waterways.

Demand for **domestic tourism** has been in decline since 1950, with a substantial reduction in nights spent in the country. Short, additional holidays form the growth sector for domestic tourism, with a shift away from traditional destinations (such as seaside resorts) towards rural areas, cities and activity-based holidays. In contrast, outbound international tourism has shown substantial growth since 1945, fuelled by competitive pricing of inclusive tours and the realisation of strong consumer preference for holidays overseas enabled by rises in discretionary income. A structural shift in demand is seeing strong growth for independent travel and for long-haul destinations, although European destinations still dominate. **Inbound international tourism** is based on heritage, culture, **countryside** and ethnic reasons. The pattern of visits is dependent upon currency exchange rates, the **health** of the economy, special events and promotional activities. The origin of overseas tourists is diverse and changing, but is dominated by North America and Western Europe.

Whilst the public sector organisation of tourism *per se* in the United Kingdom occurred late, related **legislation** had been in place earlier (notably the 1949 National Parks and Access to the Countryside Act, and the establishment of a tourist board in Northern Ireland in 1948). Until 1969, tourism was overseen by trade organisations with government responsibility residing in the Department of Trade. In 1969, the Development of Tourism Act created the structure of country tourism boards (for England, Scotland and Wales) and the British Tourist Authority which was charged with the overseas promotion of the country as a whole. The Act also stimulated the creation of a regional tourism board structure, resulting in regional boards for England, Scotland and Wales. At **local** level, substantial resources and efforts are devoted to tourism by those authorities which benefit from it. In the late 1990s this structure is still in place, although the exposure of industry to political scrutiny in both the mid-1980s and the 1990s has shifted political responsibility for it initially into the Department of Employment and then to the Department of National Heritage.

There is no published **policy** for tourism in the United Kingdom; rather, policy has to be inferred from sporadic political action. Throughout most of the twentieth century, the basis for government involvement in it has been the attraction of foreign currency and also regional development. In the mid-1980s emphasis shifted towards tourism as an employment generator, while in the 1990s the Department of Natural Heritage began assembling a national tourism policy. An important new influence in the mid-1990s was the introduction of the Heritage Lottery Fund to finance tourism development (amongst other things). At regional level, there are published strategies for each of the regional boards, while local-level policy is shaped by individual authorities and their industry.

Economically, tourism contributes through spending, employment and regional development. This industry accounts for 1.7 million jobs in the United Kingdom (although the quality of these jobs is often questioned) and makes a contribution to the **balance of payments**. Tourism is used as an **economic development** tool in Scotland, rural areas and the inner cities in Wales, and it contributes to the Exchequer through the air passenger duty introduced in the mid-1990s. Its environmental impact upon the country has received considerable attention since the mid-1980s with many official reports, good practice guides and media attention. What has emerged is recognition of the severe **impact** of tourism at many honey pot sites based upon both the natural and the built heritage. Nonetheless, awareness of these issues, the emergence of **green tourism** initiatives and the adoption of good practice have ameliorated some of the worst impacts. Social consequences have been little researched. In areas of heavy tourist demand, there has been a strong local reaction, as in rural areas where demand for second homes has acted to displace local people and led to the decline in rural services.

Further reading

Boniface, B. and Cooper, C. (1994) *The Geography of Travel and Tourism*, Oxford: Butterworth Heinemann. (Provides detailed treatment of the geography of UK tourism.)

Publications by the country tourist boards and the British Tourist Authority provide the most comprehensive set of reports and statistics relating to tourism in the United Kingdom.

CHRIS COOPER, AUSTRALIA

United States

The thirty-fifth anniversary of the United States Government's formal legislated involvement in international travel and tourism occurred in 1996. Three entities responsible for serving as the federal government's national tourism office have come and gone during this period. The US government agency for tourism emerged from a 1958 study, which concluded that tourism was a unique instrument to promote economic advancement and to contribute to friendly, peaceful relations among nations. A small Office of International Travel was set up in the Department of Commerce to serve as the locus for tourism and as a government–industry liaison. As the diplomatic and monetary value of tourism increased, Congress passed the International Travel Act of 1961, which established the United States Travel Service within the Department of Commerce as the national tourism office mandated to promote international tourism to this country. In 1970, with international arrivals totalling 12.4 million and receipts totalling $2.7 billion, a National Tourism Policy Study was established by the Senate Commerce Committee. Completed in 1978, the four-year study recommended the establishment of a national tourism policy.

In 1980, congress passed the National Tourism Policy Act, which was signed into law in 1981. This act created the United States Travel and Tourism Administration, to replace the previous one. The new office was to implement broad tourism policy initiatives, promote travel to this country as a stimulus to economic stability, support the growth of the tourism industry, reduce the nation's tourism deficit, and foster friendly understanding and appreciation of the United States abroad. The act specifically mandated the establishment of two important tourism policy bodies. An interagency coordinating committee, the Tourism Policy Council, was developed to ensure that the national tourism interest would be fully considered in federal decision making. The industry-based Travel and Tourism Advisory Board advised the Secretary of Commerce on implementation of the National Tourism Policy Act and directed the Assistant Secretary for Tourism Marketing regarding such activity plans for the US Travel and Tourism Administration.

In 1992, the Tourism Policy and Export Promotion Act's passage set new objectives for US government tourism policy. The first ever White House Conference on Travel and Tourism in October 1995 identified key policies and recommended creation of a new national tourism office. The US Travel and Tourism Administration was abolished in April 1996. A month later, Congress transferred its principal activities of tourism policy, international marketing and research to an interim office known as Tourism Industries, located in the International Trade Administration, Department of Commerce. The United States Tourism Organisation Act of 1996 proposed a privately managed, federally chartered United States National Tourism Organisation and nine-member National Tourism Board. Until this organisation is formed, Tourism Industries will provide both government and industry with access to important tourism policy information and research. Key policy issues facing the present office and the proposed organisation include

investment in highway, aviation and public lands **infrastructure**; implementation of a permanent visa waiver programme; achievement of bipartisan party platform support for tourism; elimination of unrelated taxes on tourism; and establishment of national tourism grassroots networks. Foremost is the challenge for this government and industry to work together to forge a coordinated national tourism policy within the proposed organisation capable of competing at the cutting edge of complex and interdependent global affairs in the twenty-first century.

Tourism numbers over time tell a consistent and successful story for this US industry. From 1986 to 1995, domestic and international tourism increased by 40 per cent and 72 per cent, respectively. This country is the first international choice **destination** and top average earner. The power of this industry is in its job creation, export promotion and **return on investment** to help shape tourism policy worldwide. In 1994, **inbound** international tourists numbered over 45 million and generated $78 billion in revenue. Domestic and international tourists spent more than $400 billion in the United States in 1995, making tourism the nation's third largest retail industry behind auto and health services. These expenditures generated 6.3 million jobs directly and $110 billion in payroll, while supporting another 8 million jobs indirectly. Over the past decade, tourism has created jobs at a faster rate than the rest of the economy. It generates $58 billion in **tax** revenue for federal, state and **local** governments. International tourism is a top US export, the fifth overall and first in services. Most international tourists (excluding Mexico and Canada) are from Europe; the second largest group is from Asia and the Middle East, and the third is from South America. International tourists spend $210 million a day, creating a $19.6 billion trade surplus.

In the early 1990s, US passenger air travel has shown substantial increases; operating profits nearly doubled to $2.7 billion, airline **employment** rose 1 per cent to 543,325, and the average salary increased 4 per cent to $43,230. Domestic air traffic has risen 2.7 per cent and international traffic 4.1 per cent. Global alliances and partnerships among airlines' computerised **reservation** systems and international carriers are leading to increased **code sharing** among competitors and the advent of the 'seamless' trip for domestic and international air travellers. Major problems include reciprocity in 'open skies' access to US airports, promotion of multilateral versus bilateral aviation agreements, and an end to competition-thwarting subsidies for national carriers.

America's hotels and motels have experienced increases after recovering from the losses of the 1980s of nearly $14 billion. This decade's overbuilding resulted in excess **supply** that was not halted until the passage of the Tax Reform Act of 1986. Improvements began in 1992, when hotels enjoyed a 2.9 per cent increase in room nights sold. Like aviation, US lodging is globalising through implementation of worldwide reservation system technologies, promotion of international initiatives for sustainable 'green' development, and support for **hotel** consumer awareness campaigns increasing tourist **safety** and **security**. The hotel industry enjoyed its best year ever in 1995 with profits about 3.7 times higher (adjusted for inflation) than in 1979, the previous record year. Casino revenues also increased from $14 billion in 1994 to $16.1 billion in 1995. While cruise line and rental car industry profits remained largely flat in 1995, auto travel increased by 2.2 per cent over 1993. **National park** attendance climbed 2 per cent in 1995, and travel agent sales of airline tickets rose 6 per cent to a new high of $61.2 billion.

From both a cultural and attraction perspective, the United States is a virtual cornucopia. The abundant destinations have resulted in the advantages and problems the United States has experienced as a destination. Any type of tourist 'draw' can be found in the Untied States, from big cities with well-developed cultural entertainment to pristine natural environments. New York City and its Great White Way have long been associated with the theatre, musicals, plays and **shopping**. Conversely, the United States National Park system has been copied worldwide. The **theme park** business is also a well-developed area in tourism, with Bush Gardens, Six Flags, Opryland and the Disney Parks, which have been copied worldwide. The range of its cultural attractions is extremely diverse. Because of the ongoing policy of welcoming immigrants to the United States, various distinct cultural areas have evolved throughout the country. Some of the most prominent of these are in the South, with its image of the plantation **lifestyle**, in the West with its **image** of cowboys and frontier life, and in the Northeast with its image of hard-working, independent people. Ethnic cultures are also evident in the cities and some regions of the country. The regions of the United States have become known for their unique forms of culture and cultural attractions. Due to immigration policies, large numbers of individuals with similar ethnic backgrounds tended to settle in distinct communities, and develop tourism attractions and events which are similar to their homelands, such as Scottish games, Bask and Obon festivals, and ethnic areas in communities such as Chinatown and Little Italy.

The United States has become one of the preeminent destinations in the world due to its size, differing perspectives and diversity of attractions and culture. Because of this success and the economic impact it provides, it is important to understand the various aspects of tourism policy development in this country, and the difficulty in adequately representing all of the interested parties.

Further reading

Smith, G. and Edgell, D.L. (1993) 'Tourism milestones for the millennium: projections and implications of international tourism for the United States through the year 2000,' *Journal of Travel Research*, Summer: 42–7. (Reviews tourism's economic impact and policy issues and trends.)

Travel and Tourism Government Affairs Council (1996) 'The Travel and Tourism Government Affairs Council 1996 Midyear Report', letter to Council members from Chairman Roger H. Ballou. (Offers an overview of US tourism industry's top priority concerns.)

GINGER SMITH, USA
DAVID L. EDGELL, USA
TAYLOR ELLIS, USA

urban recreation

The concept of urban **recreation** covers recreational activity that takes place in an urban **environment** in contrast to a rural setting (see **rural recreation**). Participants in such activities are either urban residents themselves, day **visitors** from rural areas, or **tourists** (see **urban tourism**). The major activities are shopping, visits to **heritage** sites, **museums**, movie theatres, operas, sport and **music** events, and indoor sports activities.

MARTIN OPPERMANN, AUSTRALIA

urban tourism

For a long time, tourism was associated with centrifugal flows of urban residents going to the **countryside** or the seaside for a holiday or an **excursion**. The concept of urban tourism only entered the research agenda in the 1980s, when it became obvious that many cities were developing into important destinations. **Business travel** and city trips have always existed, but the **leisure** motives have become more important and the numbers of urban tourists have increased considerably.

The delay of interest in urban tourism can be explained by a neglect in urban studies to assess the importance of leisure, **recreation** and tourism in an urban **environment**. Further, there was a lack of understanding of the urban tourism system. The difficulty of separating this from non-tourism functions in the wide range of urban activities and the tendency to explore the issue of urban tourism in case studies rather than by conceptual studies have contributed to the slow progress of urban tourism research. Contrary to other destinations, where the **product** (the supply side) as well as the range of activities could be well described, in the multifunctional urban system the identification of the tourism function and the multipurpose character of many visits is far more complicated. Understanding urban tourism and the product **life cycle** of urban destinations implies an integrated approach in the analysis of forms and functions. Research into new methods to identify the role of tourism and tourists in the urban environment is now in full progress and is the objective of many comparative studies.

The recent popularity of publications on urban tourism covers aspects such as trends in **demand**, creating urban attractions and clusters, urban **planning** and **policy** issues, **impact** studies, product–place marketing and resource and **visitor management**. The concepts which were introduced to understand tourism as a system needed to be adapted to the urban context. Cities can be seen as a spatially concentrated spectrum of opportunities (the Tourist Opportunity Spectrum, also referred to as **recreation opportunity spectrum**) in which one can distinguish core elements and secondary elements. The first group refers to the mix of attractions which are unique and interesting and thus capable of attracting tourists to the place, whereas the second group includes the range of urban facilities which support the touristic experience, without being a first motive for the visit. The group of primary elements of the urban product includes both the setting of the place (urban morphology, built **heritage**, green spaces, waterfronts) and the offer of facilities which allow for different activities, such as the cultural **resources** (**museums**, theatres, exhibition halls and so on), sport facilities, the amusement sector (such as casinos and **theme parks**) and the agenda of festivals and events. These core attractions are supported by facilities in the **hospitality** sector (hotels, **restaurants**, pubs) and in the retail trade, including shopping facilities and street markets. The latter group can be considered to be the added value to the urban tourist **experience**. In some cities, the shopping opportunities are becoming so attractive to the extent that they can be considered as a core product for the market of **shopping** tourism (the Mall of America in Minneapolis, USA, is quickly assuming this core position). In most cities the core products belong to the public domain, whereas the supporting facilities result from initiatives in the private sector. This interdependency partners is typical for the supply side of the urban tourism product.

Historic cities, in particular, hold many opportunities to develop tourism products based on cultural heritage resources. The trend towards tourismification of cultural heritage responds to a growing market for **cultural tourism**. The markets for city trips and for cultural tourism and shopping tourism are growing and are strongly interrelated. Several surveys indicate in the ranking of motives of urban tourists, the predominant role of visiting a unique and interesting place. Visiting museums, discovering interesting architecture, learning about the **history** of a place and, above all, seeing the well-known landmarks have become important aspects in the experience. This type of behaviour is typical for city trips and short breaks and also penetrates in the market of business travel. However, distinguishing the market segments in urban tourism remains a difficult exercise because of the complexity of motives and behaviour patterns.

Seeing the high potentials of urban tourism and the wide range of facilities which benefit from such earnings, many cities are now exploring the possibilities of developing tourism as a lever to diversify and stimulate for the urban economy. This has become a key instrument in many urban revitalisation projects, in urban waterfront development plans, in the upgrading of cultural activities (such as festivals and events), in the conservation of historical heritage and even an incentive to redesign urban shopping areas. The success of the urban product mix has even led to a stage of saturation in the product life cycle of several historic cities. The issue of **carrying capacity** and **sustainable development** has become a major concern in urban management policies. In addition, this market of urban destinations, particularly in Europe, is becoming highly competitive. The traditional top destinations such as London and Paris are now competing with numerous 'new' urban choices, such as Berlin, Barcelona, Munich, Prague and Dublin.

Gradually the focus of tourism research is moving from the place-marketing issues towards a discussion on resource and management strategies. The competitive advantage and, as such, the chances of the sustainability of the urban forms and function now lie in developing cultural tourism products with a strong local **identity** (sense of place capacity) and with the image of uniqueness and **authenticity**, despite the strong **globalisation** trends in the tourism market.

The combination of sustainable development in which the physical and social impacts of tourism can be monitored and the economic benefits optimised requires a new approach in urban **planning**, a strategic **marketing** and management policy and a better understanding of the touristic experiences in the urban environment. Understanding the synergy between tourism activities and other urban functions is necessary in order to develop and sustain urban destinations of a high quality both for these temporary and permanent populations.

Further reading

Ashworth, G.J. and Dietvorst, A. (eds) (1995) *Tourism and Spatial Transformations*, Wallingford: CAB International.

Cazes, G. and Potier, F. (1996) *Le Tourisme Urbain*, Paris: Presses universitaires de France.

Getz, D. (1991) *Festivals, Special Events and Tourism*, New York: Van Nostrand Reinhold.

Hinch, T.D. (1996) 'Urban tourism : perspectives on sustainability', *The Journal of Sustainable Tourism* 4(2): 95–110.

Jansen-Verbeke, M. and van Rekom, J. (1996) 'Scanning museum visitors: urban tourism marketing', *Annals of Tourism Research* 23(2): 364–75.

Page, S. (1995) *Urban Tourism*, London: Routledge.

MYRIAM JANSEN-VERBEKE, BELGIUM

US Travel Data Center

The US Travel Data Center (the research department of the **Travel Industry Association of America**), seeks to meet the needs of its members and the industry in general by gathering, conducting, analysing, publishing and disseminating economic and **marketing research** which articulates the economic significance of tourism at national, state and **local** levels; defines the size, characteristics and **growth** of existing and emerging markets; and provides qualitative trend analysis and quantitative forecasts of **future** tourism activity and **impact**. Its headquarters office is located in Washington, DC.

TURGUT VAR, USA

vacation

A vacation is travel for reasons of **recreation** and **leisure**. Vacations in a modern sense developed in the nineteenth century with industrialisation. Economic systems emerged where work patterns gave rise to formal '**holiday** ' periods, people with sufficient income to meet a desire for vacation during these periods, transport systems to permit more people to travel within the constrained time, and infrastructures of attractions and **accommodation** at destinations serviced by the new transport. A spatial and social diffusion has occurred in which more places are explored by increasing numbers of people.

A vacation can offer different experiences, which include the recreational, **escape** from daily stress or routine, **quest** for new experiences and places, search for alternative **lifestyles**, and **pilgrimage** in that the tourist seeks a spiritual rebirth. Based upon the types of experiences being sought, tourists have been segmented into various categories. For example, tourists may be described as ecotourists, adventure seekers, sex tourists, jet setters or mass package holidaymakers based upon their motivations and the **behaviour** in which they indulge. The provision of these different types of vacations lies within the domain of operators. Examples include action vacations with white water rafting, ecoholidays like guided trekking tours through areas of environmental interest, or tours offering opportunities to visit operas and museums. Vacations are thus multivalent, compete with other forms of leisure, and are complex social and economic phenomena.

See also: cognitive dissonance; paid vacation; satisfaction; tours; typology, tourist

CHRIS RYAN, NEW ZEALAND

vacation hinterland

Tourists from urban places move elsewhere to satisfy their **vacation** requirements and come to dominate land uses and economic activity at convenient distances surrounding the city, reflecting amenity values and costs in terms of both travel time and money. These areas of urban dominance are the city's hinterland. Occasionally, the term is used in reverse to refer to the area surrounding a **destination** from which the majority of tourists originate.

See also: distance decay

Further reading

Greer, T. and Wall, G. (1986) 'Recreational hinterlands: a theoretical and empirical analysis', in G. Wall (ed.), *Recreational Land Use in Southern Ontario*, Department of Geography Publication Series No. 14, Waterloo: University of Waterloo, 227–45.

GEOFFREY WALL, CANADA

vacationscape

The term 'vacationscape' was coined by Gunn (1988) to describe the art and practice of integrated design and **development** for tourism. He emphasises structural, physical and cultural/ aesthetic functionalism as broad guiding principles underpinning vacationscape. The creation of satisfying environments for tourism calls for greater awareness and integration of **planning** and design fundamentals.

Further reading

Gunn, C. (1988) *Vacationscape*, New York: Von Nostrand Reinhold.

JOHN J. PIGRAM, AUSTRALIA

valuation *see* contingent valuation; environmental valuation

value added

Value added is the difference between the value of the outputs of a business and the value of its purchased inputs (materials and services). It is in principle a measure of the value added by any business, including tourism, to its inputs in transforming them into outputs. For this purpose, outputs are normally measured at selling prices and inputs at purchase prices.

See also: tax

SIMON ARCHER, UK

values

Values are a set of strongly held beliefs according to which people organise their behavior. They are abstract, learned concepts induced by family, cultural, societal and peer experiences, stabilising during adulthood. Values function as rules or as a set of general strategies that help form and guide attitudes, expectations and behaviour. They can be means, or ends in themselves impacting on tourism decision-making processes and preferences.

Values are often used synonymously with needs and motives. All are closely related as they are organising antecedents of behaviour. Whereas needs and motives are generic terms, values indicate an evaluation of behaviour, persons and/or situations (objects). While values can be biogenic, (including the need for food or shelter), they can also be based on psychogenic needs. 'Self-actualisation', for example, is the process of lifting the real self to the level of the ideal one. Thus people travel for a combination of escape and search needs (Iso-Ahola 1990). Escape refers to perceived obstacles to one's well-being. Search expresses itself in either a tourist's desire for challenges leading to self-fulfillment, or in the search for acknowledgement leading to self-completion, healing, or recreation. As such, these needs are abstract. The actual satisfaction of them is practical and learnt through behaviour (including trial and error). Tourists thus learn how to travel and satisfy their needs over time (such as choice of destinations and activities).

The transformation of tourism experiences into adaptable, learned behaviour can be interpreted as being guided by a person's (a group's or a society's) values. It involves the abstraction of information from previously encountered, similar objects. From an information processing view, experiences generating values and attitudes are learned and stratified clusters of representations. These are the basis for conceptual learning and, when related to each other, form the basis for rule-acquisitions. In turn, their use in problem-solving tasks generates cognitive strategies that constitute learned behaviour and become independent of the actual contents of codes and symbols. In other words, rules are abstracted from specific situations. Destination choice and touristic activities are an expression of learned values perceived (believed) to lead to satisfactory outcomes. In terms of adaptation theory, values are learned strategies to either adapt one's environment or, adapt oneself to a given environment, to meet one's needs.

See also: culture; tradition

References

Iso-Ahola, S.E. (1990) 'Motivation for leisure', in E.L. Jackson and T.L. Burton (eds), *Understanding Leisure and Recreation: Mapping the Past, Chartering the Future*, State College, PA: Venture Publishing, 247–9.

Further reading

Kahle, L.R. (1983) *Social Values and Social Change: Adaptation to Life in America*, New York: Praeger.

JUERGEN GNOTH, NEW ZEALAND

Venezuela

Located in the north of South America, Venezuela features a wide geocultural and biodiversity **landscape**. The political system, legal base optimisation, cultural consolidation, space ordinance, **image** consolidation, area organisation and **investment** promotions support **planning** and **development** of tourism in this country. In 1995, 596,670 tourists spent $1.2 billion in Venezuela.

RAFAEL RODRIGUEZ, VENEZUELA

vertical integration

Vertical integration in tourism takes place when an organisation at one level of the chain of distribution unites with another one in a different sector. This integration can be forward in the direction of the chain, or backward against the direction of the chain. A typical example of vertical integration is when **hotel**, **airline** and **tour** companies merge. Similar to **horizontal integration**, organisations can achieve significant **economies of scale** by expanding vertically.

INMACULADA BENITO, SPAIN
FRANCISCO SASTRE, SPAIN

video

Tourism videos are used as entertainment to enhance touristic **experience**, as an educational device in order to reduce negative **impacts**, and as information about a specific **product** or **destination** to attract tourists. In the latter case, research has shown that such promotional material often merely projects the imagery of the video producer's culture rather than that of the destination.

MONICA HANEFORS, SWEDEN

vineyard

Places where grapevines are planted and cultivated for growing wine grapes are called vineyards. They often have a winery on site where the winemaking process takes place. Grape harvesting starts in September in the Northern Hemisphere and in April in the Southern Hemisphere. In major wine-producing countries, such as the United States, France, Italy, Spain, Germany, Portugal, Australia, Chile, South Africa and Argentina, vineyards are also major tourism attractions where tourists can see grapes harvested and made into wine. At the end of the tour, they can sample wines made from a variety of grapes and from different years known as vintages. Some operations also have attached restaurants where tourists can dine and order their popular labels and/or store where bottles are available for purchase. Vineyard and winery tours are gaining popularity worldwide.

PETER D'SOUZA, USA

virtual reality

By engaging a world model generated by current virtual reality **technology**, an ideal virtual reality would create an **environment** which encompasses all or most of the user's senses and abilities. This medium can create an illusion of some artificial world, or of being present in a remote **location** of the physical **destination**. Virtual reality can produce a powerful sense of conviction that such a world actually exists as imagined. The user, immersed in this artificial world, is able to navigate within it and manipulate it by means of computer-assisted devices. This provides realism of simulations, electronic **representation** with which people interact, artificiality, sensory immersion in a virtual environment, telepresence (to be present somewhere remotely, rather than physically), full body immersion and networked communications.

For tourism, virtual reality becomes a **metaphor** that describes the endeavour of industrialised cultures to develop technological simulation of the real for commercial or social advantage. Destinations such as Walt Disney World develop **pseudo-events** and spectacles which are uniform and predictable, in order to draw as constant a stream of visitors as possible. Such tourism sites are commercialised; they replace natural environments, social structures and cultural events with a simulacrum. The question of **authenticity** arises when the tourist interacts with this technology, the hyperreal (**theme parks**, guided tours, interpretative sites), without being able or willing to exercise any real power or direct influence upon the natural environment. The tourist is provided with an experience of interaction with the **other**, and engagement in the exotic that is prefabricated and illusory, but engaging and encompassing. This decontextualisation ensures an anticipated experience, while maintaining freedom from the confines, expectations and responsibilities of the real world.

See also: authenticity; hyperreality

Further reading

Fjellman, S.M. (1992) *Vinyl Leaves: Walt Disney World and America*, San Francisco: Westview Press. (A critique of Walt Disney World as surreal, as simulacrum.)

Heim, M.N. (1993) *The Metaphysics of Virtual Reality*, NY: Oxford University Press. (Describes the essence of virtual reality, particularly in Chapter 8.)

Lippit, A.M. (1994) 'Virtual annihilation, optics, virtual reality and the discourse of subjectivity', *Criticism – A Quarterly for Literature and the Arts* 36(4): 595–612. (Discusses changing meanings of reality, and virtual reality as a form of liberation.)

WILLIAM CANNON HUNTER, USA

Visions in Leisure and Business

Visions in Leisure and Business is an interdisciplinary forum which seeks to improve delivery systems in this field. The purpose of the journal is isolation and integration of those business processes that relate to the **leisure service** industry, in order to broaden current **theory** and applied methods through stimulation of ideas among traditional and non-traditional aspects of the leisure and business institution. First published in 1982, the journal appears quarterly, published by Appalachian Associates (ISSN 0277–5204).

RENE BARETJE, FRANCE

visiting friends and relatives

The visiting friends and relatives (VFR) market has often been ignored by researchers and marketers who assume that it cannot easily be managed and that its economic importance is minimal. However, research has shown that these tourists are more numerous and have greater economic impacts than was previously thought (Morrison and O'Leary 1995). Further, studies have revealed that this is not a homogeneous group: visiting friends can be quite different from trips to relatives. VFR as a trip purpose should be separated from visiting as a trip activity when segmenting the market (see also **market segmentation**).

The VFR market is related to historical links between regions, and particularly **migration** patterns. It might be expected that attractive destinations generate more such traffic, but family and social benefits actually predominate. Special events can be developed to encourage this sector by appealing to the '**homecoming**' instinct. While it is generally true that the VFR segments are often older, stay longer, use less commercial **accommodation**, spend less per day and are not necessarily lured by promotions, their impacts can nevertheless be substantial. Hosts often take their guests to attractions around the home base; **shopping** and **gift** giving can stimulate retailing; dining out and attending local entertainment and events are common. Research has shown that members of the long-haul VFR market are likely to be more influenced by word-of-mouth information, that many are repeat tourists and that their lifetime value to a **destination** must be considered, especially as many use commercial accommodation for all or part of their trips.

References

Morrison, A. and O'Leary, J. (1995) 'The VFR market: desperately seeking respect', *The Journal of Tourism Studies* 6(1): 2–5.

Further reading

Paci, E. (1994) 'Market segments: the major international VFR markets', *EIU Travel and Tourism Analyst* 6: 36–50.

The Visiting Friends and Relatives Market (1995) special issue of *The Journal of Tourism Studies* 6(1).

DONALD GETZ, CANADA

visitor

A visitor is someone who has left their residence to spend time in a **destination**. Distance travelled, **length of stay** or other criteria may be used to define the term operationally. The **World Tourism Organization** considers 'visitor' to be the basic unit for collecting tourism **statistics**. Visitors consist of **tourists** (overnight visitors) and excursionists (same-day visitors). Some researchers use the visitor and tourist terms interchangeably.

STEPHEN SMITH, CANADA

visitor bureau *see* convention and visitor bureau

voluntary sector

The voluntary sector in tourism comprises all the activities and **services** provided on the basis of volunteering, which in turn refers to unpaid work accomplished within the context of a formal business (public or private) or an association. There are significant difficulties in measuring the impact of the voluntary sector in tourism, as the local authorities often tend to exaggerate the involvement of residents in their community. Further, organisations involved in this **service** are often not coordinated enough to provide reliable data for this purpose.

Reasons for emergence of the volunteering phenomenon, in and beyond tourism, can be related to the positive **attitude** that there is a higher degree of life satisfaction to be gained in helping people or serving the society. For governments, the voluntary sector is sometimes viewed as a substitute for the welfare state; for some people, volunteering is a satisfactory substitute for work (Cohen-Mansfield 1989).

In the tourism industry, volunteering can be offered in relation to sport tourism, cultural tourism and peace/religion tourism, among others. In general, to express support for their community, city or neighbourhood, people tend to do voluntary tasks by helping visitors, especially during local celebrations and special events, as well as national or international events such as sports competitions or exhibitions or Olympic Games. Ecological or cleaning aims is a big part of volunteering motivations, including the conservancy of parks and gardens. For instance, over 1,200 neighbours from New York spent 35,000 hours painting benches and picking up rubbish at Central Park in 1996 (El País 1997). Another is educational and cultural activities for foundations, museums, schools or research centres, many of which are often financed on the base of private donations.

A voluntary service is usually accomplished or coordinated within a **non-profit organisation**. These or other associations can be found in many tourism areas. In fact, the industry has a long tradition of providing food and shelter during times of crisis (earthquakes and other natural disasters) and, more ordinarily, in big cities or urban areas. The industry has also been committed to educational efforts and fund-raising, to helping poor and homeless people, both in the **hospitality** and **restaurant** sectors. In fact, many restaurant chains regularly donate their unused food to some private or public agencies for redistribution to the needy. Hotels also donate food, linen and furniture to shelters and organisations and other appropriate entities.

References

Andregnette, P. (1995) 'Chaîne volontaire et nouvelles technologies: les actions du groupe Hélan en matière de communication et de réservation', *Cahier Espaces* 44(December): 93–7.

Centre d'Etude du Commerce et de la Distribution, Service Hotellerie-Tourisme (1979) *Les chaînes hotellières volontaires: examen critique et bilan*, Paris: CECOD.

DeFranco, A.L. and Kripner, O.M. (1997) 'Hospitality with a heart: a choice for success', *Journal of Hospitality & Tourism Education* 9(1): 5–9.

MARTINA GONZALES GALLARZA, SPAIN

walking tour

Walking tours maximise **experience** of a **landscape** with all senses. A major component of **ecotourism**, walking amongst other things allows the inner person to rediscover **nature**, such as a jungle trail; provides access not possible by motorised **transport**, like mountain treks; delivers a sense of excitement not feasible by vehicle, including walking through a game park; and enables some physical exercise. Urban walking tours facilitate detailed **sightseeing**.

TREVOR SOFIELD, AUSTRALIA

wanderlust

In common usage, a term used to describe a constant urge to travel. More specifically, it describes a type of tourism characterised by a desire to **experience** new places or cultures and a willingness to adapt to **local** conditions, forsaking the familiar for the unusual. It also implies continual, explorer-type travel as opposed to single-resort **vacation** tourism.

RICHARD SHARPLEY, UK

war

Wars have a decidedly negative **impact** on tourism. Media coverage and global attention to politically motivated conflicts can have a severe and instant effect on international and **domestic tourism**. In 1994, for example, threatened and actual attacks of **terrorism** against tourists had a disastrous effect on Egypt's tourism, its leading source of **foreign exchange**. Major wars, such as the Second World War, have brought tourism to a virtual standstill. Tourists visit 'safe' areas where they feel secure. Governments often impose **travel advisories** and strong **safety** and **security** measures, particularly at airports, which also discourage travel.

War and its aftermath underscore the extent to which military conflict – and the preparations for it – threaten the tourism environments, both natural and built. The 1991 Persian Gulf war brought widespread damage to the region's **environment** and people. The largest oil spill in history seeped throughout the slow-to-replenish waters of the Persian Gulf damaging beaches and birds.

War is the antithesis of the image required for tourism to thrive and impedes its **marketing** and **development** efforts. Without over 100 wars and innumerable violent internal conflicts since 1945, tourism's growth presumably would have been much greater. Just in the last decade, for example, tourism declined as a result of violent conflicts in Sri Lanka, the Philippines and Peru, and in 1999 in countries close to the Yugoslav war. Postwar effects of tourism can be positive, with veterans revisiting war areas to show their families the places or cultures they experienced. Political and economic constraints of war excluded Vietnam from much of the 1960–70s tourism bonanza. Its postwar economic policy for 'openness' and establishment of US diplomatic relations now fuel Vietnam's rapid tourism **development**.

In war-torn areas, mere prospects for **peace** can increase tourism. In Northern Ireland, during the seventeen-month ceasefire ending February 1996, tourism rose by 70 per cent and travel inquiries by 40 per cent. Northern Ireland wisely maximised opportunities while they lasted by waging a vigorous campaign to reassert its tourism **product** and present a positive **media** image. In essence, tourism can foster even short-term opportunities for peace which wars and political instability can rapidly undo.

Further reading

Pizam, A. and Mansfeld, Y. (eds) (1995) *Tourism, Crime, and International Security Issues*, New York and London: Wiley. (Discusses the relationship of crime and international security issues for the tourism industry.)

Ryan, C. (1993) 'Crime, violence, terrorism, and tourism: an accidental or intrinsic relationship,' *Tourism Management*, June: 173–83. (Shows the relationship between crime and recreation in tourism locations.)

GINGER SMITH, USA
DAVID L. EDGELL, USA

waste management

Wastes emanating from a tourism operation can be solid, liquid or gaseous, and can include by-products, contaminated, reject, spilt and dated materials, packaging and used containers, kitchen and garden waste, and obsolete equipment. Waste management is concerned with waste avoidance and reduction; reuse, **recycling** and waste treatment; and waste disposal. For tourism undertakings, a further important aspect is the dissemination and adoption of procedures for best practice waste management. The challenge is to reduce to a minimum the materials used in the first place, to recycle and reuse waste materials when practical, and to dispose of residuals safely.

As with **pollution management** generally, self-regulation is a necessary complement to monitoring by regulatory authorities (see also **regulation, self**). A first management step is identification of waste streams from the tourism operations and the processes which generate them. Whereas it is important to identify and act upon major contributors to waste, guests may be more concerned about visible waste items which could affect the quality of their experience. Once waste streams are identified, measures should be taken to reduce the consumption of materials which lead to this situation. These could include ordering in bulk, reducing packaging, using efficient appliances, minimising food wastage and using refillable containers.

Tourism operations should also be assessed for the potential for reuse, **recycling** and treatment of wastes for environmental and economic benefits. Reuse minimises consumption and reduces waste streams. Recycling involves recovering materials meant for disposal and reprocessing them into useful products. Waste separation assists in recycling, as does an efficient collection **service** and recycling **infrastructure**. Small tourism operators can overcome constraints on **recycling** by joining forces. A desirable outcome of recycling is for tourism establishments to use a wider range of recycled products.

Ultimately, a safe, environmentally acceptable means of disposing of wastes should be adopted. In many cases, this will be governed by **legislation**. Hazardous and trade wastes can cause environmental damage if not disposed of properly. It is usually not acceptable for such wastes to be discharged directly or indirectly to surface waters, to sewers or to stormwater drains. Collection and treatment off-site may be required before disposal. In tourism undertakings, care will also be needed with emissions to the air.

Further reading

Commonwealth Department of Tourism (1995) *Best Practice Ecotourism. A Guide to Energy and Waste Minimisation*, Canberra: Commonwealth Department of Tourism.

JOHN J. PIGRAM, AUSTRALIA

wayfinding

The issue of orientation or finding one's way in an unfamiliar **environment** is frequently a concern for the independent traveller. Much of the early research on wayfinding for tourists has been conducted in large indoor settings such as museums and art galleries. If much mental effort is spent negotiating the labyrinths of the building, little capacity remains for attending to the exhibits and the whole experience may be diminished. Visitors who cannot successfully use the cues, signs, **maps** and diagrams of the setting may miss exhibits they want to see, spend time in parts of the building where the exhibits are of lesser interest, and fail to locate essential services such as food and toilets. The issue of wayfinding extends beyond buildings to the use of subways and airports, and to open air settings such as theme parks and touring by automobile.

Much of the research on wayfinding has been directed at identifying the value of good maps and orientation aids (signs, route markers). Specific experimental studies have begun to identify the features of maps which make them most informative in public settings. One broad hypothesis integrating these specific findings is the notion of cognitive steps, where this term refers to a mental transformation of information from the language of the map to the reality the viewer can be expected to encounter. The core-integrating hypothesis is that maps with fewer cognitive steps are preferred and promote better wayfinding. For example, any map drawn in two dimensions is a cognitive step away from the three-dimensional world the individual has to encounter. In general three-dimensional maps are likely to be preferred to two-dimensional abstract maps. Other examples of cognitive steps, with the more difficult element being mentioned first, include black and white maps as opposed to colour, rotated as opposed to fixed ones oriented in the same direction as the viewer and an indexed system as opposed to names on the map. The evidence is consistent with the view that such cognitive steps take longer to process and information obtained from these types is harder to remember.

In addition to maps and signposting devices, good architectural design of tourism facilities can promote easy wayfinding. Both within buildings and in open air environments, clearly visible routes and prominent landmarks aid **visitor** orientation. In novel settings such as unfamiliar cities, there is some evidence that tourists build up a list of landmarks first and then connect them with routes and paths. There may be some **gender** differences in wayfinding, with males showing some greater use of paths and female tourists relying more on landmarks.

PHILIP L. PEARCE, AUSTRALIA

weekend

The weekend forms one of the basic components of people's opportunity for **leisure** time and for short touristic trips, including **visiting friends and relatives**. It generally consists of a one and one-half or two-day period at the end of a five-day working week. In its modern form, it evolved in the United Kingdom during the 1870s; the earliest use of the term, according to the Oxford English Dictionary, was in 1879. The custom of a weekend break spread to Europe and North America largely after the First World War. Today, the weekend has become not simply the end of the week but has gained its own distinct significance in people's leisure lives.

It was industrial societies which developed the clearly defined weekend. Pressure for regulated work schedules eroded previous customs of free time, leaving Sunday as the only frequent period free from labour. But in the United Kingdom from the mid-nineteenth century, a movement grew to make Saturday a half-holiday. The Factory Act of 1850 ended work at 2 pm on Saturday. By the 1890s, a one and a half day holiday was the norm. This distinctive UK habit spread slowly to Europe and North America from the 1920s and 1930s. In 1935, Italy adopted a one and one-half day weekend, and Germany did so after 1945. In France, *le weekend* of two days grew in the 1950s and 1960s.

However, the weekend has never been universally adopted. Japan has been slow to accept this Western custom, and a Friday and Saturday weekend was not common in Israel until the

1980s. At an earlier time, the French Revolution abolished the week and made every tenth day a holiday, while Stalin attempted a cycle of four days work and one day holiday. In Western consumer societies today, flexible working arrangements, agreed or imposed, are perhaps beginning to blur the distinctive character of the traditional weekend as it has evolved in the last hundred years.

See also: paid vacation

Further reading

Bailey, P. (1978) *Leisure and Class in Victorian England*, London: Methuen. (Discussing the weekend through a general study of Victorian leisure.)
Cunningham, H. (1990) 'Leisure and culture', in F.M.L. Thompson (ed.), *The Cambridge Social History of Britain 1750–1950*, vol. 2, Cambridge: Cambridge University Press, 279–339. (Traces the weekend through an analysis of the working week.)
Rybczynski, W. (1991) *Waiting for the Weekend*, New York: Viking. (Presents a wide-ranging study of the subject.)

JOHN TOWNER, UK

wilderness

According to the United States Wilderness Act of 1964, wilderness is an area where the earth and its community of life are untrammelled by human beings, where humans are temporary visitors who do not remain. Such areas are to be forever free of 'permanent improvements' such as roads and built structures. True wilderness areas must be large in size, have low intensities of use and be free of the trappings of civilisation.

Wilderness may be sought by some tourists, particularly ecotourists, for the solitude, the opportunity to commune with **nature**, and the high-quality **hiking** experiences which can be obtained there but, paradoxically, tourism is also a threat to wilderness as such areas have low **carrying capacities** and developments encroach onto hitherto pristine areas. Many motorised recreational activities, such as snowmobiling and power boating, may be banned from wilderness areas, so that they are towards the end of the **recreation opportunity spectrum** providing low-intensity use in relatively natural settings with minimal signs of management. The value ascribed to wilderness has increased over time as its **supply** has been reduced and more people have come to live in cities.

See also: parks

Further reading

Huth, H. (1957) *Nature and the American: Three Centuries of Changing Attitudes*, Berkeley, CA: University of California Press. (Chronicles the growth of the conservation movement and the protection of wild lands.)
Lime, D.R. (ed.) (1990) *Managing America's Enduring Wilderness Resource*, St Paul, MN: Minnesota Extension Service, University of Minnesota. (Addresses many aspects of wilderness preservation, use and management.)
Nash, R. (1967) *Wilderness and the American Mind*, New Haven, CN: Yale University Press. (Examines changing attitudes of Americans towards wilderness.)

GEOFFREY WALL, CANADA

wildland recreation

This encompasses any recreational activity in keeping with the principles of **wilderness** in which humans are to be unobtrusive visitors who draw little attention to their presence, and leave little or no trace of their passing. These **activities** may include primitive **camping**, **hiking** and climbing, but would not involve any motorised vehicles, permanent structures or alteration of the **landscape**.

TOM BROXON, USA

winery *see vineyard*

work ethic

The concept of work ethic is concerned with an accepted standard of occupational behaviour. From an individual perspective, work ethic is reflective of an acquired personal value system whereby underlying personal beliefs and attitudes help determine the individual's work ethic. Overall, work ethic is an accumulation of family influences, situation factors, morals, experiences and peer influences, and thus related to **professionalism** and **code of ethics** in tourism.

RANDALL UPCHURCH, USA

World Bank

Created in 1946, the mission of the World Bank is to alleviate poverty and improve living standards by promoting sustainable growth and investments in people (see also **sustainable development**). To achieve this, it comprises five institutions: the International Bank for Reconstruction and Development, International Development Association, International Finance Company, Multilateral Investment Guarantee Agency and International Centre for the Settlement of Investment Disputes. The World Bank funds major tourism projects in developing countries.

LIONEL BECHEREL, UK

World Leisure and Recreation

As the official journal of the **World Leisure and Recreation Association**, *World Leisure and Recreation* is oriented towards **leisure** across cultures and countries, as well as an international leisure constituency of educators, managers, academics, policymakers, legislators and leisure providers within private and public sectors. The journal is organised on a theme issue basis, but occasionally it publishes other contributions. First appearing in 1965, it is published quarterly by World Leisure and Recreation Association (ISSN 0441–9054).

RENE BARETJE, FRANCE

World Recreation and Leisure Association

The World Recreation and Leisure Association (WRRA) is a worldwide, non-governmental membership organisation, dedicated to discovering and fostering those conditions best permitting **leisure** to serve as a force for human growth, development and well-being. It aims to explore the nature and consequences of leisure experiences for individuals, for groups and for communities; to enhance provisions for leisure education and services; and to advocate for social and environmental development through sustainable and equitable leisure programming. The association provides forums including biennial world congresses, regional conferences, seminars and workshops, and print and electronic media. It delivers programmes and recognises success through various awards such as the Excellence in Leadership Award, Community Excellence Award and Honorary Life Membership.

WLRA has consultative status with the Economic and Social Council and its several commissions, UNESCO and UNICEF regional associations, including the European Leisure and Recreation Association (ELRA), the Latin America Leisure and Recreation Association (ALATIR), and the Australian and New Zealand Association for Leisure Studies (ANZALS).WLRA disseminates and stimulates discussion through *World Leisure and Recreation* and *World Leisure Newsletters*, reports and conference proceedings, world charters and position papers and conventions, and WLRA on the Web. Members are from all parts of the world and from every walk of life, including leisure professionals from public, private and voluntary sectors, researchers and scholars, managers and **policy** makers, leisure educators, concerned members of the community with diverse interests in tourism, **parks** and **recreation** services, the arts and culture, sport and exercise, and others. The headquarters of WLRA are in Okanaga Falls, Canada.

LESLIE M. REID, USA
TURGUT VAR, USA

World Tourism Organization

The World Tourism Organization (WTO) is the leading global organisation in the field of tourism. It serves as a framework of expertise in tourism **policy** and a practical source of tourism know-how in specific issues. Its membership includes 138 countries and territories and more than 350 members of the WTO Business Council, representing local **government**, tourism associations, educational institutions and private sector companies, including airlines, **hotel** groups and **tour operators**. WTO is an inter-governmental body entrusted by the United Nations with the promotion and **development** of tourism. Through tourism, it aims to stimulate economic growth and job creation, provide incentives for protecting the **environment** and heritage of destinations, and promote **peace** and understanding among all the nations of the world.

The World Tourism Organization had its beginnings as the International Union of Official Tourist Publicity Organisations, set up in 1925 in The Hague. It was renamed the International Union for Official Tourism Organisations (IUOTO) after the Second World War, and moved to Geneva. In 1967, IUOTO members approved a resolution transforming it into an inter-governmental organisation empowered to deal on a worldwide basis with all matters concerning tourism and to cooperate with other competent organisations, particularly those of the United Nations system. A recommendation to the same effect was passed in December 1969 by the UN General Assembly, which recognised the 'decisive and central role' the transformed IUOTO should play in the field of world tourism 'in cooperation with the existing machinery within the UN'. IUOTO was renamed the World Tourism Organization, and its first General Assembly was held in Madrid in May 1975. The Secretariat was installed in Madrid early the following year at the invitation of the Spanish government.

The General Assembly, as the supreme organ, is composed of voting delegates representing full members and associate members. Affiliate members and representatives of other international organisations participate as observers. It meets every two years to approve the budget and programme of work, and to debate topics of vital importance to tourism worldwide. Every four years it elects a Secretary-General. The Executive Council is the WTO's governing board. It meets twice a year and is composed of twenty-six members elected by the General Assembly in a ratio of one for every five full members. Associate members and members participate in Executive Council meetings as observers.

The WTO has six regional commissions, in Africa, the Americas, East Asia and the Pacific, Europe, the Middle East and South Asia. The commissions meet at least once a year, and are composed of all the full members and associate members from that region. Business council members from the region participate as observers. Specialised committees of WTO members advise on management and programme content. These include the Programme Committee, the Budget and Finance Committee, the Statistics Steering Committee, the Environment Committee, the Quality Support Committee, the Education Centres Network and the Communications Advisory Council.

The Secretariat is led by the Secretary-General, who supervises the staff at the headquarters. These officials are responsible for implementing the WTO's programme of work and serving the needs of members. The Secretariat also includes a regional support office for Asia-Pacific in Osaka, Japan. The WTO Business Council is supervised by a full time Chief Executive Officer at the Madrid headquarters. A Leadership Forum of the Business Council meets once a year to make programme recommendations to the Secretariat. The WTO exists to help all countries to maximise the positive impacts of tourism, such as job creation, new **infrastructure** and **foreign exchange** earnings, while at the same time minimising negative environmental or sociocultural impacts. Its programmes are grouped into six broad areas.

The Human Resource Development department provides education and **training** that matches the needs of future tourism professionals and employers. In cooperation with its network of nineteen WTO Education and Training Centres throughout the world, this department sets global standards for tourism education. Its TEDQUAL

(Tourism Education Quality) methodology helps member governments and tourism enterprises assess their needs and create programmes to train the burgeoning workforce needed to keep pace with the growth of the industry. The newly launched General Tourism Achievement Test (GTAT) is another tool to encourage standardisation of curricula and make degrees in tourism more internationally comparable. The department offers regional 'educating the educators', 'tourism policy', 'tourism environmental quality' and other seminars, distance learning and *in situ* courses for officials and professionals in tourism, and *practicum* courses at the headquarters for selected tourism officials from member countries. It produces books and other educational materials and administers a limited number of scholarships for postgraduate studies.

The Statistics, Economic Analysis and Market Research department provides the world's most comprehensive tourism statistics – for which WTO is best known – used by member governments, private companies, consulting firms, universities and the media. The WTO sets international standards for tourism measurement and reporting. Its recommendations on tourism statistics were adopted by the United Nations in 1993 and are being implemented by countries around the world, creating a common language of statistics that allows destinations to compare their success with that of their competitors.

The Environment, Planning and Finance department works closely with members and other international organisations to ensure that new tourism developments are properly planned and managed, and thus to protect the natural and cultural environments. Its message of encouraging low-impact sustainable tourism development rather than uncontrolled mass tourism has been embraced in recent years by member countries. They understand that governments, in partnership with the private sector, have a responsibility to keep the environment in good condition for future generations and for the future success of the industry itself.

The Quality of Tourism Development department aims to help member destinations improve quality in order to become more competitive and, at the same time, to ensure **sustainable development**. As global competition in tourism becomes more intense, quality is the factor which can make the difference between success and failure. Basic components of quality include competitivity through trade liberalisation; access, with emphasis on tourists with disabilities; **safety** and **security**, including **health**; and technical standards.

The Cooperation for Development department helps with the transfer of tourism know-how to developing countries, which is one of the fundamental tasks of this intergovernmental organisation. As an executing agency of the United Nations Development Programme, WTO contributes tourism experience to the sustainable development goals of countries throughout the world. Acting on requests from member governments, it secures financing, locates experts and carries out all types of tourism development projects, large and small. In the operating period 1996–7, the organisation carried out development activities in forty-two countries.

The Communications and Documentation department increases awareness about the importance of the industry and the relevance of the organisation. It also acts as a publishing house, producing 20–30 titles each year. The WTO operates a permanent documentation centre for use by members and non-members, with the goal of making the organisation a true clearinghouse for international tourism information.

In this era of raising consumers' expectations, segmentation of **demand**, environmental and cultural concerns, technological change, and increased competitive pressures, the World Tourism Organization aims to play a needed role as the authority setting and managing global frameworks of expertise in tourism. This goal is often accomplished in partnership with national, regional and **local** governments, other inter-governmental organisations and the private sector, as well as concerned non-governmental organisations.

Further reading

World Tourism Organization (1994a) *WTO Basic Documents*, Madrid: WTO.

—— (1994) *Recommendations on Tourism Statistics*, Madrid: WTO.

—— (1995) *World Tourism Directory*, Madrid: WTO.

—— (1996) *Educating the Educators in Tourism*, Madrid: WTO.

EDUARDO FAYOS-SOLÀ, SPAIN
DEBORAH LUHRMAN, SPAIN

World Travel and Tourism Council

The World Travel and Tourism Council (WTTC) is made up of chief executive from all sectors of the tourism industry, including **accommodation**, **catering**, cruises, entertainment, **recreation**, **transportation** and travel-related services. Its central goal is to work with governments to realise the full economic impact of tourism. Its millennium vision is to make tourism a strategic economic and **employment** priority, to move towards open and competitive markets, to pursue **sustainable development**, and to eliminate barriers to growth. Its headquarters are located in Brussels, Belgium.

TURGUT VAR, USA

World Wide Web

The World Wide Web (WWW) is one of the most important information services on the **Internet**. Based on the concept of hyperlinks, the WWW integrates information that is stored on different computers all over the world. Its pages can combine text, graphics, sounds, movie clips, database-queries, forms and many other multimedia components.

The basic element of the WWW is simple text files that describe the content of the page in hypertext markup language (HTML). These webpages are stored on a computer on the Internet that runs a Web server, a programme that allows users to access the stored information. When a Web client, another computer programme running on an Internet-connected computer which might be thousands of miles away, requests this page, the Web client loads it from disk and sends it via the network to the requesting programme. It is the responsibility of the client programme to interpret the embedded HTML tags and to display the page accordingly. Because of this, the designer of the page does not have full control over the appearance of the page. HTML is a standard that is under development and is evolving rapidly.

The WWW derives much of its appeal and flexibility from the Uniform Resource Locator (URL), a general scheme for addressing resources on the Internet. The URL defines the type of the resource, its name, the path, and the computer where it can be found. By specifying its URL, a webpage can refer in a hyperlink to any other publicly available resource on the Internet.

The capabilities of the World Wide Web are dramatically expanded by use of Common Gateway Interface (CGI) scripts and Java applets. They add dynamics to the structure of webpages. Through CGI, a user can start programmes on the server computer and thus see different pages depending on several factors such as the content of a database or the time of the day. Java applets are small programmes that are delivered by the server, but run on the computer of the user. Through this WWW system, much trade and research-oriented tourism data and publications are now readily available.

Further reading

December, J. and Randall, N. (1994) *The World Wide Web Unleashed*, Indianapolis, IN: Sams Publishing.

Flanagan, D. (1996) *Java in a Nutshell*, Sebastopol: O'Reilly & Associates.

Graham, I S. (1996) *HTML Sourcebook*, 2nd edn, New York: Wiley.

Magid, J., Matthews, R.D. and Jones, P. (1995) *The Web Server Book: Tools & Techniques for Building your own Internet Information Site*, Chapel Hill, NC: Ventana Press.

GUNTHER MAIER, AUSTRIA

xenophobia

Xenophobia is an irrational fear or contempt of **strangers** or foreigners. This ancient cultural and political phenomenon is also present in contemporary tourism, mainly manifesting itself in the hostile attitudes of residents towards tourists. Tourism research has not shown a specific interest in this problem, except in a broader context. Sociological studies on the stranger were an important starting point. The tourist can in fact be considered – like the stranger – in transit in a foreign community. The strangerhood perspective was developed to emphasise the cultural distance between the stranger and the integrated community. Xenophobia in tourism can thus be studied by the social sciences from different viewpoints, focusing on aspects such as interpersonal relationships (psychosociological or anthropological host–guest relationships); intercultural communication with stress on groups; and socioeconomic forms of neocolonialism or **imperialism**. Xenophobia should in any case be considered in domestic and international tourism in terms of economic, social and cultural distance, which is accentuated by the type and number of tourists and the rate of tourism development. The **demonstration effect**, stereotyping, **social impacts** and sociocultural change are all related concepts (see also **change, sociocultural**).

Attempts have been made to measure residents' attitudes through an irritation index within the framework of the resort cycle. Xenophobia can occur when the **carrying capacity** is exceeded and the tourist is seen as responsible for all the evils caused by social change. At the beginning of the 1970s, displays of xenophobia were not unusual, ranging from graffiti telling tourists to go home to assaults on foreign cars, hotels and other tourism settlements. The outbursts of xenophobia took place especially where economic, social and cultural differences between the **host and guest** populations were at their greatest.

Generally speaking, over the last two decades tourism has been increasingly considered as an unavoidable component of a world process of **globalisation**, and even interpreted as a vital force for **peace** and understanding. There have also been efforts through **policy**, **impact assessments** and other measures to mitigate the negative consequences of tourism on host societies. Several case studies on the attitudes of residents suggest that the economic benefits of tourism outweigh its social costs. On the other hand, new forms of terrorist xenophobia (see **terrorism**) specifically aimed at tourists are appearing in different parts of the world.

Further reading

Mathieson, A. and Wall, G. (1982) *Tourism: Economic, Physical and Social Impacts*, London: Longman. (Provides some important insights in Chapter 5, dealing with social impacts of tourism.)

GIULI LIEBMAN PARRINELLO, ITALY

Y

yield management

The concept of maximising the revenue by raising or lowering prices in respect to **demand** is known as yield management. The necessary conditions for a successful application of yield management include a fairly fixed capacity, high fixed costs, low variable costs, fluctuations in demand and similarity of **inventory** capacity. Yield management was popularised with the deregulation of the US airline industry and it is extensively used by this and other tourism sectors.

ANTONI SERRA, SPAIN

yield percentage

To demonstrate the variable effect of both average price rates (see **pricing**) and **occupancy rates**, tourism managers develop a comparison focusing upon the yield rates during a given period of time. Managers can increase the yield rate (see **yield management**) by raising rates when **demand** is high.

DAVID G.T. SHORT, NEW ZEALAND

youth pilgrimage

Youth **pilgrimage** is common in all religions, particularly Christianity. The Children's Crusade in 1212 was the first such event. Every two years, pilgrims assemble in Taise, an ecumenical pilgrimage centre, to take part in the World Youth Days celebration together with Pope John Paul II; this event was established by the Pope in 1984–5. There are also European Youth Meetings. Some centres (such as Jasna Gora) specialise in high school graduation and first communion pilgrimages.

ANTONI JACKOWSKI, POLAND

youth tourism

There is nothing regular about youth. Different cultures have defined it, if at all, in many different ways in terms of age, socioeconomic status and personal rights. For most, youth is a **life cycle** stage that precedes full incorporation into the adult world. Modern industrial societies tend to locate it between the beginning of puberty and completion of high school and or college, that is, the time span between 13 and 23–25 years of age, that precedes full legal capacity and entry into the workforce. Usually a distinction is made between early youth or pre-juvenile status (13–17 years) and true youth (18–23 years).

Of relevance to the study of tourism is whether there is anything specific about the touristic **behaviour** of this group in modern industrial countries. There is not much general theoretical or research activity on this field to answer this question. Data are few and far between. Educated guesses and opinions usually surround some hard facts. These extend to three main domains: socioeconomic significance, differences with mainstream tourism in terms of cultural exchanges with the guests, and an alternative or complementary road to tourism **development**.

What is known about international youth tourism flows comes mainly from **World Tourism Organization** sources. From these, an argument can be developed along the following lines: that youth tourism encompasses a distinct group of travellers between 15–24 years of age; that international arrivals in this group have outpaced arrivals in general by a factor of 50 per cent from 1980 to 1990; that regionally, it is decreasing in Europe (although this continent still makes up over two-thirds of all youth arrivals) and increasing fast in North America and in East Asia-Pacific; and that there seems to be a fair potential for growth in this group, as originating countries are affluent societies that will be growing steadily in the future. In order to get a better picture, it should be pointed out that these data are incomplete inasmuch as they do not include **domestic tourism**, which is very important in countries such as the United States.

There are also some educated guesses about the economic relevance of youth tourism. First, even though there are no records of expenditure by young tourists, it is suggested that youngsters have limited travel budgets, as witnessed by the success of touristic guides to 'travel on a shoestring'. Second, this effect is offset by longer periods of stay in their destinations. Third, youth tourism has its own distribution channels (specialised travel agencies, bucket shops and institutional retailers, such as universities, churches and so on) and **accommodation** networks (youth hostels, youth campsites, family homes and so on). Fourth, it is more open to non-programmed and active **behaviour** than mainstream tourism.

It is also said that young tourists are more likely to engage in close encounters with their hosts. In this way, youth tourism would be a key factor in cross-cultural exchanges. Youngsters are more likely to respect the values of their hosts (language, customs, rituals) and oft-times they travel specifically to participate in their activities (festivals, harvests, archaeological sites). There is also some literature on the educational value of these experiences, and this corpus suggests that international students broaden their minds, attitudes and values as a consequence of being exposed to other cultures. From the **Grand Tour** to pilgrimages in Kathmandu, travel has influenced many youngsters.

On the other hand, not everything is positive in these exchanges and enthusiasm should be toned down. Exchanges are often skin deep, and prone to reinforce stereotypes of hosts and guests. Casual attitudes about clothing, sex or drugs on the part of the hosts may lead to annoying situations, and the guests' ignorance of their hosts' cultural realities may develop into misunderstandings. Eventually, a good previous acquaintance with the host culture, even if it only comes from guides and textbooks, may smooth these rough edges.

Finally, these specific features of youth tourism have nourished some expectations that it might be an alternative way to the touristic development of some societies that do not want to be engulfed in the type of growth that mainstream tourism expects. Both in economic terms and in cultural respect, it is said that youth tourism is better. Even though youngsters may not have much to spend, their budgets are impressive in some parts of the planet; they will accept traditional types of accommodation, so that there will be no need for big investments in **infrastructure** or **resorts**; and they are glad to consort with the locals and partake in their activities. However, multiplication effects of their tourism dollars are limited; they will not offer the same opportunities for local employment, and they are definitely unable to stop the flow of migrants to the cities. Even though only the host society should have full authority in deciding which type of tourism development strategy it wants to pursue, exaggerated expectations as to the potential of youth tourism might lead to disappointing experiences.

JULIO R. ARAMBERRI, USA

Z

zoning

Zoning, as it is popularly conceived, is part of the toolkit of command and control **development** planning. Command and control regulatory frameworks such as exclusionary zoning often seem the most convenient methods for development **planning** and protection of the **environment**, largely because both business and government have had more experience with them. In this context, zoning seeks to regulate land uses by separating them based on incompatibility, or allowing like/**compatible** uses to co-exist.

A basic principle of tourism zoning is the conservation of specific environmental features such as wetlands, archaeological and historic sites, important stands of vegetation and unusual geological features. Related to this is the maintenance of visual diversity. Also important is the achievement of successful functional groupings of **resort** facilities and activities, such as **accommodation**, commercial and cultural facilities, and recreation facilities in suitable areas. Buffer zones containing mixtures of tourism facilities and less fragile environmental **preservation** requirements may also be designated.

With appropriate consideration, zoning may also be used to achieve area-wide management. Variations of the basic principle can include performance measures relating to allowable land uses (design, density, servicing standards), mandatory clustering of facilities/attractions, scenic road overlays, controlled circulation networks and access restrictions, mandatory **interpretation** of sites and facilities, and relocation of undesirable buildings to alternative areas. An example of the use of zoning is that of Borobudur National Archaeological Park in Java, Indonesia. In the 1980s, five zones for various types and intensities of land use were established around this important ancient monument and **attraction**. Zone one protects the immediate environment with no development allowed except for landscaping; zone two includes development of facilities for tourist use, park operation and archaeological conservation activities; zone three is designated for access road and smaller monuments, within which land uses are strictly controlled to be compatible with the park; zone four maintains historical scenery; and zone five includes archaeological surveys and the protection of unexcavated archaeological sites. Planning for this park also includes determination of maximum visitor capacities, facility needs and conservation requirements, and an important part of the implementation programme was the relocation of some residents further away from the monument in order to implement the zoning plan.

See also: planning, environmental; protected areas; self-regulation

Further reading

Gunn, C.A. (1994) *Tourism Planning*, 3rd edn, Washington: Taylor & Francis.

World Tourism Organization (1993) *Sustainable Tourism Development: Guide for Local Planners*, Madrid: WTO.

MALCOLM COOPER, AUSTRALIA

Index

Note: Page references in **bold** type indicate articles; those in *italic* indicate authorship of articles in this volume.

Abacus 254
Abernethy, Ted *386–7*
Aborigine **1–2**, 40, 41
  art 30–1
Abu-Alam, Wesal *581–2*
Abu Simbel (Egypt) 187
Acapulco (Mexico) 389
accessibility **2**, 145, 157
  Finland 231
  Greece 262
  of archaeological sites 28
  of attractions 36
  of rainforest canopy 342
accommodation **2–4**, 367, 462
  and carrying capacity 72
  bed and breakfast **50**
  benchmarking **51**
  classification 83
  cruise lines 122
  forecasting 234
  grading system **259**
  in Argentina 30
  in Austria 43
  in Dominican Republic 159
  in Egypt 187
  in France 239
  in Greece 262
  in holiday camps **280–1**
  in mountain regions 397
  in India 302
  in Italy 333–4
  in Malaysia 367
  in Mexico 390
  in The Netherlands 412
  in Peru 435
  in the Philippines 436
  in the Seychelles 532
  in Spain 550
  in Tunisia 606
  in the USA 614
  room night **513**
  see also resort development, integrated; timeshare
ACCOR 239
accounting **4–6**
  and expenses estimation 215
  cash flow **73–4**
  control systems **109–10**
  feasibility study **223–5**
  financial control **229**
  food and beverage cost analysis **232–3**
  gross profit **265**
  profit **466**
  profit centres **466**
  property management system **469**
  ratio analysis **487–8**
  satellite account **519**, 565
  uniform system **610–11**
  see also auditing; management accounting
accounting rate of return models 331
Accreditation Commission for Programs in Hospitality Administration 118
acculturation **6–7**, 77, 91, 124, 520
  and adaptation 10
  and national character 404
  of strangers 559
  see also demonstration effect
ACED-I *see* Association of Conference and Events Directors-International

achievement **7–8**
acquisition *see* merger
Acta Turistica **8**
action research **8–9**, 216
activity **9–10**
  and agrotourism 14
  participation in 137
  recreation education 184
activity space **10**
Adams, G.D. 126
adaptation **10–11**, 563
ad hoc multipliers 399
Adler, Alfred 7, 359
Adler, J. 533
adventure tourism **11**, 237, 389
  exploration **217–18**
  see also bushwalking; quest
advertisements
  binary structure **54**
  clichés 83
  envy in **11–12**
  format of **12**
advertising **12–14**, 377
  and content analysis 108
  bans 210
  European Travel Commission **208**
  exoticism 214
  in communication mix 92
  maps **371–2**
  media planning **385–6**
  of attractions 36
  on radio 487
  on television 575
  product positioning **464**
  significant omissions **535**
  truth in 106
  see also brochure
Aeroperu 435
aesthetics
  environmental **194–5**
  of handicrafts 270
  of nature 409
  of outdoor recreation 420
  role of museums 401
African Travel Association **14**
Africa Travel News 14
Agenda 21 **14**, 54, 353, 506
agglomeration theory 162
agriculture, and economic development 161
agrotourism **14–15**, 222, 455
Aguilo, Eugini *242*
AHMA *see* American Hotel and Motel Association
AID *see* automatic interaction detection
AIDS **15**, 154, 273, 577–8
  and travel restrictions 220
AIEST *see* Association Internationale d'Experts Scientifique de Tourisme
Air Canada 599
Air India 303
air service agreement 319
  see also Bermuda 1
air transport 596
  catering **74**
  charter flights **77**
  demand **137–8**
  Federal Aviation Administration **225**
  see also airline; international aviation, bilateral; international aviation organisations
Air Transport Association of America **15**
air travel demand *see* demand, air travel
aircraft, supersonic **15–16**
  see also Concorde
airline **17**, 368, 596
  class of service 82
  code sharing 86
  demand 137
  deregulation **143–4** , 317, 375, 597
  economies of scale 164
  frequent flyer programme **241**, 501
  mergers **387–8**
  regulation 81
  route system 514
  seamless service **521**, 599
  see also air transport; international aviation organisations
airline distribution systems **16**
Airline Reporting Corporation **16**
airport **17–18**, 246
airport hotels *see* hotel, airport
alarde (Basque celebration) 91
Alaska 29
Alberti, Antoni Sastre *605*
alcohol **18–19**, 159, 539
  and deviant behaviour 150
  licensing **357**
Aleman, Miguel 389
Alemtejo (Portugal) 455
Algarve (Portugal) 455

alienation **19**, 23
  and religion 497
allocentrism **19**, 520
Allahabad (India) 498
Almagor, U. 484
Alps 43, 238, 383, 396–7, 568
alternative tourism **20–1**, 24, 27, 148
  and the environment 194, 507
  Cuba 122
  see also ecotourism
Altman, J.C. 1
Amboseli (Kenya) 343
'ambush' marketing 210
amenity
  at destinations 144–5
  resource-based **505**
  user-oriented **21**
American Express 157
American Hotel and Motel Association **21**, 285
American Marketing Association 375
American Museum of Natural History (New York) 401
American National Restaurant Association 508
American Recreation Society 407
American Resort and Residential Development Association 582
American Society of Travel Agents **21–2**, 453
AMFORHT *see* Association Mondiale pour la Formation Hôtelière et Touristique
Amsterdam (The Netherlands) 412, 470
Amtrak 487
amusement park *see* theme park
Anatolia (journal) **22**
Anderson, B. 408
Anderson, Donald *54–5*, *110–11*, *146*, *146–7*, *213–14*
Anderson, Walter 285
Andes 29, 31, 77, 396
Andrews, Hazel *533–4*
Andronikou, Antonios *132*
animation **22**, 216
  and cruise lines 121
Annals of Tourism Research **22–3**, 24, 172, 526
anomie **23**, 393, 477
Antarctic tourism **23**, 29, 445
Antarctic Treaty 445
anthropology **23–6**
  acculturation **6–7**
  Fourth World 237
  impact of tourism 140
  liminality **359–60**
  link with geography 249
  rite of passage **510–11**
  ritual 365, **511–12**
  see also ethnography; host and guest
anticipation **26**, 34, 36, 275
antiquities, smuggling 539
anti-tourism **27**
APEX system 213
Apollo/Galileo 254
Apostolopoulos, Yorghos *27*, *384*
Appadurai, A. 91. 92
Appleton, J. 347
appropriate tourism **27**
appropriation **27–8**
Aramberri, Julio R. *56–7*, *454–6*, *503*, *633–4*
arbitration in disputes 89
ARC *see* Airline Reporting Corporation
archaeological sites 25
  Kenya 343
  Mexico 389
  Morocco 391
  Peru 435
archaeology **28–9**
  Holy Land 282
Archer, Brian *351*
Archer, Simon *4–6*, *32*, *39–40*, *48–50*, *73–4*, *131*, *229*, *235–7*, *300–1*, *418*, *572–3*, *610–11*, *619*
Arctic tourism **29**, 445, 506
Argentina **29–30**
ARIMA forecasting method 234, 235
Armstrong Gamradt *336*
Arnstein, S. 475
arrival/departure card **30**, 327
art **30–1**
  and craft 271, 548
  authenticity 44
  commercialisation 91
  intellectual property 313
  postmodern 457
articulation, programme **31–2**
ASA *see* air service agreement; Bermuda 1
Asher, D. 172
Ashmolean Museum (Oxford) 400
Ashworth, G.J. *277–8*, *429*, *439*, *454*, *470*, *535–6*
Asia Pacific Journal of Tourism Research **32**
Asia Pacific Tourism Association **32**
ASK *see* available seat kilometres

asset management **32**
  cash flow **73–4**
  control systems **109–10**
association executive marketing organisations 156
Association Internationale d'Experts Scientifiques du Tourisme **33**
Association Mondiale pour la Formation Hôtelière et Touristique **33**
Association of American Geographers 250, 251
Association of Conference and Events Directors-International **33**
Association of Tourism Teachers and Trainers 588
Association of Travel Marketing Executives **33**
ASTA *see* American Society of Tra vel Agents
Aswan (Egypt) 187
ATAA *see* Air Transport Association of America
Atacama desert 77
Atlas mountains 391
attention **33–4**
attitude **34–5**
  manipulation 348
  social representation 472
attraction **35–7**
  environmental quality 481
  marker 373
  politically based 453
  religious **37–8**
attractivity **38**, 144
attribute 571
attribution theory **38–9**
auditing 6, **39–40**
  environmental **195–6**, 454, 496
  marketing 376
Australia **40–2**, 243, 342, 423
  Aborigines **1–2**, 30–1, 40, 41
Australian Tourist Commission 421
Austria **42–3**, 228, 396
Austrian Central Statistical Office 43
authenticity 20, 24, 25, **43–5**, 48, 484, 545
  and commercialism 91
  and cultural conservation 124
  and escape 201
  and experience 216
  and historical tourism 278
  and representation 501
  and virtual reality 621
  ethnic tourism 204
  in art 31
  of festivals 226–7
  of souvenirs 548
  see also heritage; staged authenticity
automated search engines 177
automatic interaction detection **45–6**
automation **46–7**
Automobile Association 259
available seat kilometries 137
aviation, *see* international aviation, bilateral
aviation liberalisation *see* international aviation liberalisation
aviation rights *see* international aviation rights
Avis 69
axis rotation methods 221
Ayers Rock *see* Uluru

Bachmann, Walter 284
back office **48**
backpacking 20
back-stage **48**
Bacon, Sir Francis 333
Baedeker, Karl 268, 269, 568
Baffin-Pangnirtung (Canada) 66
Bahamas **48**
Bahia 56
Baja California (Mexico) 389, 557
Bajio (Mexico) 389
Baker, Joni E. *62*, *602*, *610*
balance of payments **48–50**
  and purchasing power 236
  Brazil 57
  currency controls **131**
  import substitution 300
  Italy 333, 334
Balaton, Lake (Hungary) 291
Balearic Islands 324, 325, 550
Bali 129, 214, 304
Bali Tourist Development Corporation 305
Banff (Canada) 66
Banff National Park (Canada) 427
Bangkok (Thailand) 577
banquet hall catering 74
bar **50**
Baran, Paul 149
Barbados **50**
Barcelona (Spain) 616
Bardolet, Esteban *323–5*
Baretje, René *8*, *22*, *22–3*, *32*, *65*, *109*, *112*, *20*, *23*, *227*, *231* , *321–3*, *338–9*, *339–41*, *356*,

371, 423, 430, 543 , 575, 586, 588, 589, 591, 601–2, 607, 621, 628
Barnett, Lynn A. *442–4*
Barrows, Clayton W. *84–5*
Barth, Joachim E. *361–2, 536, 592*
Barthes, Roland 102, 402, 437, 563
Basho 590
Basque country 238
Bath (UK) 549
Batle Larente, Francesc J. *151*
Baud-Bovey, M. 384
Baum, Tom *119, 169–70, 288–90, 371*
Beaman, Jay 154, *563*
Becherel, Lionel *213, 238–40, 240–1, 301, 420, 628*
bed and breakfast **50**
Bedard, François *66–8*
behaviour **50**
  activity 9
  and activity space 10
  and attitudes 35
  and brochures 59
  and community 93
  and environmental education 181
  and product positioning 464
  change 77
  controlling 348
  demonstration effect **140–1**
  deviance **150–1**
  ecoethics **160**
  educating 183
  in entrepreneurial education 179
  influenced by interpretation 328
  modified for ethnic tourism 205
  motive manipulation **396**
  performance appraisal 432
  political 257
  professional 466
  recreation **50–1**
  related to roles 513
  religious tourists 498
  responsible tourism 507
  ritualistic 511
  social interaction **540**
  time budget studies 581
  tourists in Greece 263
  see also motivation; play; psychology; rules; values
Belize 372
benchmarking **51**, 300–1, 479
benefit–cost analysis **51–3**, 549
benefit segmentation 374–5
benefits **53**
Benito, Immaculada *284, 421, 620*
Berelson, B. 106
Berger, P. 435
Berlin (Germany) 487, 616
Bermuda 1 **53–4**
Bermuda Agreement 317
Berne, Eric 594
Bertalanffy, L. von 570
best practice, in environmental management **198**
Best Western hotels 104
beverage *see* food and beverage cost analysis
Bialowieza National Park (Poland) 444
Biarritz (France) 238
biblical sites **537**
Bieszczady mountains 444
binary structure of advertisements **54**
biological diversity **54**, 427, 492
biosphere reserves 405, 406
Black Sea resorts 516
Blazey, Michael A. *527, 541*
Blue Ridge Parkway (USA) 521
board **54–5**
Bob Evans 113
Boberg, Kevin B. *17–18, 137–8, 143–4, 241, 316–17, 375*
Bodewes, T. 173
Boeing 747 16
Boissevain, Jeremy *383*, 512
Bombay (India) 228
Borneo 304
Boorstin, D.J. 544
borders **55** , 246
Boston (USA) 487
Boulding, Kenneth 295
boundaries *see* borders
Bourdieu, Pierre 102
bowel infections 154
Brah, Nev *198–9, 313*
branding 12, **55–6**, 364, 377
Brazil **56–7**, 342
break **57**
break-even point analysis **58**, 467
Breiter, Deborah *100–1, 488, 513*
Bretton Woods Agreement (1944) 236
Brewer, J.D. 557
British Airways 17, 480

British Hospitality Association 285
British Tourist Authority 285, 612
brochure 13, **58–9**
 and content analysis 107
broker *see* culture broker
Brotherton, Bob *388*
Brown, Frances *307*
Broxon, Tom *65–6*, *277*, *410*, *420–1*, *517*, *627*
Brundtland Report 274, 473, 474, 567
Brunei 423
Bruner, Edward M. 44, *126–7*, *129*, *254*, *409*, *437*, *465*, *552*
Buckley, P.J. 575
Budapest (Hungary) 291
Budget (car rentals) 69
budget hotel **59**
budgetary control **59–61**
budgeting 6, 59, **61**, 110
 and benefit–cost analysis 51
Buenos Aires (Argentina) 29–30
Buhalis, Dimitrios *188*, *262–4*, *307*, *390*, *423–4*
Bureau International de Tourisme Social **62**, 542
Bureau International pour le Tourisme et les Echanges de la Jeunesse **62**
burnout **62–3**
bus systems 595, 596
Bush Gardens (USA) 614
Bushell, Robyn *272–4*, 392, *473–4*, *519–20*
bushwalking **63**, 237
business
 legal aspects 352
 see also management; small business management
Business Council for Sustainable Development 197
business format **63**
 of hotel chains 76
Business Monitor 285
business plans 180
business tourist *see* business travel; convention business; market segmentation; tourist
business travel **63**, 257
 and demand for accommodation 3
 demand 137
 group market **265–6**
 incentive **301**
 see also car rental; meetings, business
Butler, Richard 95, 108, *132*, *266*, *358*, *360*, 372, *521–3*, *563*
buying decisions **64**
Byrne Swain, Margaret *23–6*, *77*, *125*, *542–3*

cafeteria **65**
Cahier Espaces **65**
Cairo (Egypt) 186
Calantone, Roger *380–3*
Callan, Roger J. *104–5*, *259*
Campaign to End Child Prostitution in the Third World 531
camping **65–6**, 237
Canada 29, **66–8**, 243, 423
Canadian Association of Geographers 250, 251
Canadian Museum of Civilisation (Ottawa) 401
Canadian Tourism Commission 67, 68
Canary Islands 325, 383, 550
Cancún (Mexico) 253, 389, 505
Cannes (France) 324, 505
Cannes Film Festival 228–9
Cape d'Agde (France) 238
capital account 48, 49
capital budgets 61
capital expenditure 5
capitalism 148–9, 254
car rental **68–9**, 240, 368, 596
caravan park **69**
career **69–71**
 executive development **213–14**
 paths 174
Caribbean 383, 389, 503
 economic factors in tourism 255, 449
 see also Jamaica
Caribbean Tourism Organisation **71–2**
Carlson Company 113–14, 157
Carmichael, Barbara A. *101–2*
carnival *see* event
Carpathian mountains 444
carrying capacity **72–3**, 249, 298, 383, 506
 and life cycle 95
 and recreation planning 442, 492
 parks 427
 protected areas 470
 recreational **73**
 urban areas 616
 see also crowding
Cartagena de Indias (Colombia) 89
Casablanca (Morocco) 392
Casa de Campo resort 191
cash, asset management 32
cash budgets 1

cash flow 5–6, **73–4**, 331
casino 243–4
see also gambling
Castro, Fidel 122
catering **74**, 91
catering, airline **74**
outsourcing 421
see also food service, contract
causal model **74–5**, 382
causal research 382
Cawley, Mary *332*
Cazes, Georges H. *20–1*, *542*
celebration 226
central place theory 362
central reservation **75**
centrally planned economy **75–6**
China 79
centre–periphery **76**, 250
Centro Oberhausen (Germany) 532
CETUR (Ecuadorian Tourism Corporation) 166
chain hotels **76–7**
Chan Chan citadel (Peru) 435
change **77**
and commoditisation 91
and impact 296
cultural 124, 140
economic 160
in recreational preferences 442
in the workplace 290
profit sensitivity analysis **466–7**
sociocultural **77**
tourism as catalyst 448
see also limit of acceptable change

Channel Tunnel 80
charter, air **77**, 324
Chesser, Jerald *74*, *232*
Chicago Convention on International Civil Aviation (1944) 53, 315, 316–17, 320
Chick, G. *356*
Child, J. 119
child prostitution 530, 531
Chile **77–8**
China **78–9**, 423
and Hong Kong 283
art 31
China National Tourism Administration 79
Ching-Yick Tse, Eliza *112–14*, *114–15*, *192–3*, *398*, *561–2*
Cho, Bae-Haeng *54*, *144–5*
choice set **79–80**, 145–6
Chon, K.S. (Kaye) *31–2*, *155–6*, *323*, *339*
CHRIE *see* Council on Hotel Restaurant and Institutional Education
CHRIE Communique 117–18
Christaller, W. 358, 362
Christmas shopping trips 532
Chulcchi 445
Chunnel **80**
circuit tourism **80**, 302
circulation of money 399
city office **80–1**
civil aviation authority **81**
civil law 352
Clark, Adrian *588*
class **81–2**
and musical taste 402
and political economy 450
conspicuous consumption 105
leisure pursuits 102
class of service **82**
air travel 137
classification **82–3**
and cluster analysis 85
of accommodation 259
of air travellers 138
of corporate culture 128
of recreation 489
of recreation experience 491
of restaurants 508
cleaning schedules 220
clean-up programmes 199
Clefs d'Or 100
Clements, Christine J. *53*
CLIA *see* Cruise Lines International Association
cliché **83**
climate **83–4**, 249
and recreational behaviour 51
greenhouse effect **264**
club **84–5**, 503
Club Managers Association of America 84
Club Mediterranée 191, 239, 253, 281, 426, 505, 582
cluster analysis **85–6**
CMAA *see* Club Managers Association of America
CNTA *see* China National Tourism Administration
coach travel *see* motor coach tourism
code of ethics **86**, 106, 110, 430, 507

environmental **86–7**
see also ecoethics
code sharing **86**
coexistence, and conservation 103
coffee houses 284
cognition **87**, 295, 471
and training effectiveness 178
classification of attractions 37
cognitive dissonance **87–8**
cognitive mapping *see* map; perceptual mapping; wayfinding
Cohen, Erik 19, 44, 91, 92, 109, 126, 157, *213*, *215–16*, 217, *226–7*, *344*, *359–60*, 365, *426*, *438*, *456*, 484, 543, *544–8*, *558–60*, *580*, *591–2*, 608
Coimbra (Portugal) 455
collaborative education **88**
collective bargaining **88–9**
Collison, Fredrick M. *17*, *74*, *81*, *82*, *319–20*, *513–14*, *596–8*
Colombia **89–90**
colonies de vacances 281
colonisation **90**, 452
see also neo-colonialism
Colvin, Jean *166*
commercialisation **90–1**
of festivals 226
of religious tourism 499
commissary 74, **91**
Commission for Sustainable Development 360
commoditisation **91–2**, 383, 449
and host and guest 287
effect on Greece 263
of handicrafts 270–1
of heritage 276
of 'paradise' 426
common law 352
communication
and the environment 194
and history of tourism 279
content analysis **106–8**
discourse **153**
language of tourism **348–9**
of transport failures 483
radio **487**
satellite 310
telemarketing **575**
through interpretation 328
Trinet **605–6**
see also global distribution system; media; public relations; semiotics; sociolinguistics
communication mix **92–3**
communitas concept 359
community **93**
and ethnic groups 203
attitude to tourists 106
see also host and guest
residential recreation **502**
voluntary sector **622–3**
community approach **93–4**
community attitude **94–6**
community development **96**
community planning **96**
community recreation **96**
comparative advantage **96–7**, 147, 162
and import substitution 300
comparative study **97**
compatible **97–8**
compensation administration **98**
competition, and feasibility studies 224
competition analysis *see* marketing audit; marketing plan; marketing research
competitive advantage **98–9**, 162, 375
and entrepreneurship 192
of attractions 36
competitiveness **99–100**
airlines 143
and economies of scale 164
and efficiency 186
European Union 209
complaint
ombudsmen 106
see also cognitive dissonance; loyalty
compliance **100**
computer applications
golf attaché 256
in education 176–7, 182
see also distance education
computer-based recreation 386
computer reservation system 16, 75, 100, 156, 308, 376, 502
and car rental 69
use by consortia 104
computer technology *see* information technology
concentration of activities 162
concentration ratio **100**
concession **100**
concierge 4, **100–1**

Concorde 15–16, 596
conference accommodation 2, 3
  consortium approach 104
  group business market **265–6**
  see also convention business
conflict **101**, 249, 563
  and conservation 103
  in recreational development 421
Congo 31, 342
conjoint analysis **101–2**, 464
connotation **102–3**
conservation 28, 54, 86, **103–4**
  and future of tourism 242
  best practice 198
  cultural **123–5**
  in Greece 263
  in Kenya 344
  of energy 192
  role of museums 401
  see also national park; preservation
Consort hotels 104
consortium **104–5**
conspicuous consumption **105**
constitutional law 257
constructivism 425
consumer choice
  conjoint analysis **101–2**
  market research 382
consumer protection, WTO measures 219
consumer surplus 52
consumerism **105–6**
  and escape 201
contact levels 529–30
content analysis **106–8**
contestable markets **375**
contestation **108**
contingent valuation **108–9**
  environmental factors 199
continuum model **108–9**, 393
Contours **109**, 166
contract 352
  see also management contract
contract food service **233**
contribution margin 467
control system **109–10**
convention and visitor bureau **110–11**, 146, 315–16, 361
convention business **111**, 361, 368
  Finland 231
  Germany 252
  see also meetings, business
convention planners 156
Cooper, Chris 10, *80*, *171*, *302*, *361*, *384*, *413*
Cooper, Malcolm *199–200*, *360*, *440–1*, *458–9*, *635*
Cooper, R. 543
cooperation, in air transport 15, 17
Copacabana (Brazil) 56
copyright 313
core commodities 461
core–periphery 76
  see also dependency theory; pleasure periphery
Cornell Hotel and Restaurant Administration Quarterly **112**
corporate culture *see* culture, corporate
corporate finance **112**
corporate strategy **112–14**
corporate structure **114–15**
corporate taxes 573
corporate travel offices 156
correspondence analysis **115–16**
cosmopolitanism **116**
cost **116**
  advertising 13
  competitive advantage 98–9
  efficiency reductions 186
  of recreational activities 51
  of recycling 494
  operating **417**
  profit sensitivity analysis 467
  travel cost method 602
cost accounting 370
cost–benefit analysis *see* benefit–cost analysis
cost volume profit analysis 467
Costa, J. *55–6*
Costa Brava 324
Costa Rica **116–17**, 342, 506
COTAL *see* Latin American Confederation of Tourist Organisations
Côte d'Azur (France) 505
Coubertin, Baron Pierre de 553
Council on Hotel Restaurant and Institutional Education **117–18**, 170, 284
country house hotel **118**
countryside **118**
courier *see* tour guide
Coutt's Heritage Print and Design 372
Cracow (Poland) 444

craft *see* handicrafts
Craig-Smith, Stephen J. *356*
creative tourism **118**
Crick, M. 326, 512
crime **118**, 523, 524, 575
critical theory 425
critics **118**, 448, 449
  of specific events 209
  see also anti-tourism
Crompton, J.L. 489
cross-cultural education **119**
cross-cultural interaction 472
cross-cultural management **119–20**
cross-cultural study **120–1**
cross-training **121**
Crotts, John C. *108*
Crouch, Geoffrey I. *146*, *351*, *560–1*
crowding **121**, 472
Crowley, Daniel J. *30–1*
CRS *see* computer reservation system
cruise line **121–2**, 255, 368, 597
  Antarctic 445
  class of service 82
  emergency management 188
  Nile 187
Cruise Lines International Association **122**
Csikszentmihalyi, Mihalyi 232
CTC *see* Canadian Tourism Commission
CTO *see* Caribbean Tourism Organisation
Cuba **122–3**, 255
cuisine **123**
  diversity 285
Culler, J. 44
cultural conservation **123–5**
cultural survival **125**
Cultural Survival 124
cultural tourism 92, **125–6**
  and ethnic groups 203
  Italy 334–5
  museums 401
  musical events 402
  Thailand 577
  urban areas 616
  see also Grand Tour
culture
  Aboriginal 1
  adaptation 10
  and demonstration effect 140
  and environmental education 182
  and ethnic tourism 204
  and theme parks 579
  archaeology 28
  commoditisation **91–2**
  consumption of 24
  corporate 113, **127–9**
    and cross-cultural management **119–20**
  cross-cultural education **119**
  diversity in USA 615
  effect on entrepreneurial education 180
  environmental awareness 194
  ethnic groups 203
  festivals **226–7**
  impact of tourism 297
  intellectual property 313
  interpretation 328
  invention of **129**
  organisational **129**
  postmodern 457
  regional 494
  resource evaluation 506
  symbolism 569
  tourism **129–31**
  see also acculturation; anthropology; art; cross-cultural study; cultural tourism; ethnography; heritage; sociology
culture broker **126–7**, 374, 465
  and Grand Tour 260
  ethnic tourism 204, 205
culture change *see* sociocultural change
culture shock **127**, 558, 560
  see also homesickness
curative tourism **131**
currency
  control **131**
  in balance of payments 49
  see also foreign exchange
currency union 235
current account 48, 49
curriculum design **131–2**
Currie, Russell R. *15*, *21–2*, *316*
customs conventions 219
Customs Cooperation Council 219
Cuzco (Peru) 435
CVB *see* convention and visitor bureau
cycle **132**
cycle time maintenance 220
Cyprus 12, **132**, 371–2
Czech Republic **132–3**

Czestochowa (Poland) 444, 498

Dalder, Hans 299
D'Amore, Louis J. *430*
Danish Centre of Tourism Development 142
Danish Tourist Board 142
Dann, Graham M.S. *11–12*, *23*, *48*, *54*, *83*, 102, 120, *222*, *237*, *242*, *265*, *275*, *293*, *348–9*, *366*, *371–2*, *384–5*, *393–5*, *416*, *469*, *477*, *488*, *495*, *535*, *540*, *543–4*, *576*, *591*, *610*
Darien jungle 89
data collection 345, 377, 382
  demographic 139
  marketing research 381
  interviews **327–8**
  on recreational participation 428
data processing
  factor analysis **220–1**
  see also neurocomputing
data sources 382
database marketing **134**, 152
Dateline 407
Daugherty, Carolyn M. *245*, *491*, *502*, *538*
Davidson, R. *265–6*
DBV (German Spa Association) 253
Deal, T.E. 129
death 273
  see also thanatourism
De Beer, E.S. 268
de Burlo, Charles R. *303–4*, *313–14.*
decision making **134**
  and cognition 87
  and entrepreneurial education 180
  and environmental perception 431
  and power 458
  and technology 310
  choice set **79–80**
  destination choice 145
  game theory **245**
  public participation **474–5**
  push–pull factors **477**
  see also investment decision
decision support system **134–5** , 379
decomposition methods 234–5
Decrop, Alain *121–2*
definition **135**
  ethnic self-definition 206
  of domestic tourism 158
  of hospitality 284
  of inbound tourism **301**
  of leisure 355–6
  of management contract 370
  of national park 405
  of play 365
  of product 461, 463
  of quality 479
  of recreation 489
  of tourism 585
  of tourist 590
Defoe, Daniel 589–90
degradation of archaeological sites 29
de Grazia, S. 355
Delphi technique **135–6**, 234, 442, 508, 605
demand **136–7**
  air travel **137–8**
  and demography 139
  and economics of scale 164
  changes *see* input–output analysis
  database marketing **134**
  domestic tourism 158
  effect of terrorism 576
  for accommodation 3
  for non-tourism products 462
  globalisation 255
  in feasibility studies 224
  recreational **138–9**
  see also forecasting
Deming, W.E . 479
democratisation **139**
demography **139–40**
  and domestic tourism 158
  and labour supply 189
demonstration effect **140–1**, 449, 453
Denmark **141–2**
denotation **142**
  of utopian destinations 202
Department of Conservation (New Zealand) 414
Department of National Heritage (UK) 612
Department of Tourism (India) 302
dependency theory 94, **142–3**, 162, 451
  and appropriation 27
deregulation, airline 17, **143–4**, 317–19
  and airport congestion 17–18
  quality factors 482
descriptive models 447
descriptive research 382
design and layout **144**
  aesthetics 195

research 217
design of organisations **419**
de Sola Pool, I. 107
destination **144–5**
and activity space 10
attractivity **38**
awareness through film 228
brochures **58–9**
cycle **132**
economic priorities 163
gravity model forecasts 261
image 295, **296**
internal marketing **314**
International Tourism Reports **323**
intervening opportunities **329**
length of stay **356**
life cycle model 108, 279, **358**
opportunity set **418**
utopian 202
use of transaction tables 311
destination choice **145–6**
destination information system *see* decision support system; geographical information system; marketing, destination
destination life cycle *see* life cycle, destination
destination management **146**
education 168–9
destination management company **146**
destination management organisation **146–7**
destination marketing **378–9**
detective controls 60
developing country **147–8**
development (of tourism) **148–50**
and acculturation 7
and demonstration effect 140
and environmental engineering 196–7
and environmental planning 440–1
and infrastructure 309–10
and product life cycle 463
and protected areas 470
community attitude 94
dependency theory 142
effect of wars 624
entrepreneurial education **179–80**
environmental code of ethics **86–7**
future evolution **242**
government policies 257
see also policy
hard tourism **272**
impact assessment 298
in Australia 40
in Canada 66
in Chile 77–8
in China 78–9
in Costa Rica 117
in Cuba 122
in Cyprus 132
in Czech Republic 133
in Denmark 141
in Finland 230
in France 238
in Germany 252
in Greece 262
in Hong Kong 283
in Hungary 291
in India 302
in Indonesia 305
in Israel 332
in Italy 333
in Japan 336
in Kenya 343
in Malaysia 366
in Mexico 388–9, 390, 505
in Nepal 411
in New Zealand 413
in the Philippines 436
in Poland 444
in Portugal 455
in Russia 515
in Singapore 535
in South Africa 547
in South Korea 345
in Spain 549–50
in Switzerland 568
in Thailand 577
in Tunisia 606
in Turkey 607
in the UK 611–12
in the USA 614
influence of life cycle 357
master plan **384**
regional economics 162, 494
scale **520**
Turismo em Analise **607**
use of demography 139
WTO role 630
zoning **635**

see also destination, life cycle; economic development; political development
development economics 162
development era **150**
deviance **150–1**
Dewar, Keith *125–6*
Dewey, J. 593
diagonal integration 151
diet **151**
differentiation **151**, 162
Ding, Peiyi *195–6, 298–9*
diploma courses 167, 170
Direccion Nacional de Turismo (Peru) 435
direct marketing 13, **152**, 156, 376
directional policy matrix 376
Directory of Hotel and Motel Companies 21
disasters *see* emergency management
disciplinary action **152–3**
discount pricing *see* pricing
discourse **153**
Discovery TV channel 341
discretionary income 414
discretionary travel **153**
discriminant analysis **153–4**
discrimination in employment 190
disease **154**
skin cancer 275
see also illness
disembarkation card *see* arrival/departure card
Disney attractions 92, 128, 193, 281, 457, 505, 578, 579, 614, 621
displacement **154**, 563
disposable income 414
distance decay **155**, 250
distance education **155–6**, 177, 182
executive development 213
in Finland 231
distribution channel **156–7**, 376, 380, 601
economies of scale 164
diversification 113
diversionary tourist **157**
Djerba-Gades (Tunisia) 606
DMO *see* destination management organisation
domestic tourism **158–9**, 448, 451
as import substitution 162
Germany 252
Greece 263
Spain 550
UK 612
Dominican Republic **159**, 191, 255
double-entry book-keeping 5
Douglas, Ngaire *423*
Dowling, Ross K. *86–7, 160, 165, 173–5, 272, 459*
Doxey, G.V. 94
drinking **159**
see also alcohol
drug trafficking and abuse 219
DRV (German Travel Agencies Association) 253
D'Souza, Peter *223, 620*
DTV (German Travel Association) 25
Dublin (Irish Republic) 616
Dufour, R. 402, 403
Dumazdier, J. 354–5
Durkheim, Emile 23, 393
duty-free shopping 532

Eade, John *437*
Eadington, William R. *243–4, 338*, 609
Earth Summit (1992) 14, 54, 197
Eccles, G. *460*
Echtner, Charlotte M. *172–3, 176–7*
ecoethics **160**
ecological economics *see* economics, ecological
Ecological Economics 163
ecologically sustainable tourism **160**, 274, 507
and health 272
in Finland 230
in Malaysia 367
in Nepal 411
resource evaluation 506
uses of interpretation 328
see also sustainable tourism
ecology **160**
and public health 473
Eco-Management and Audit Scheme 197
econometrics, demand forecasting 234, 605
economic development **160–1**, 162
and city offices 81
and conservation 197
and demand for accommodation 2–3
and enclave tourism 191
and gambling 244
and host–guest relationship 95
Brazil 56
Egypt 187
Italy 333, 334
Kenya 343
Portugal 455

role of tourism 250, 452
see also developing country
economic impact **298**
economic leakage *see* leakage
economic multiplier *see* multiplier effect
economics **161–3**
and community attitude 94
balance of payments **48–50**
centrally planned economy **75–6**
comparative advantage **96–7**
competitive advantage **98–9**
competitiveness **99–100**
demand **136–7**
dependency theory **142–3**
ecological **163–4**
foreign exchange **235–7**
globalisation **254–6**
industry **305–6**
input–output analysis **310–12**
investment **330–1**
market **373–4**
multiplier effect 255, 298, 310–11, **398–400**
of conservation 103
of tourism 24, 67, 367, 383
domestic 158
international 324
productivity **464–5**
resources **506–7**
satellite account 519
supply **564–5**
Tourism Economics 586
uniform system of accounting **610–11**
see also political economy
economies of scale **164–5**
economy
informal **307**
leakages **351**
of Greece 263
of Hungary 291
of Mexico 389
performance indicators 433
role of handicrafts 271
role of tourism 529
see also political economy
ecoresort **165**
ecosystem *see* environment; planning, environmental
ecotourism 20, 29, 54, 103, **165–6**, 181, 237, 409, 506
and exploration 217
in Australia 41
in Chile 77
in Costa Rica 117
in Denmark 142
in Israel 332
in national parks 407
in Portugal 455
see also bushwalking; green tourism
Ecotourism Association of Australia 86
Ecotourism Society 86, **166**
ECTWT *see* Ecumenical Coalition on Third World Tourism
Ecuador **166**
Ecumenical Coalition on Third World Tourism **166**, 430, 452, 507, 531
Edgell, David L. Sr *604–5*, *613–15*, *624–5*
education **166–9**, 629
careers of educators 169
collaborative **88**
computer-assisted **169–70**, 182
Council on Hotel Restaurant and Institutional Education **117–18**
cross-cultural **119**
curriculum design **131–2**
distance **155–6**
Ecotourism Society **166**
effectiveness **170–2**
entrepreneurial **172–3**
environmental **173–5**
executive development **213–14**
Finnish University Network for Tourism Studies **231**
graduate distance courses in tourism 155
Institute of Certified Travel Agents **312**
International Society of Travel and Tourism Educators **323**
modular programme **391**
multidisciplinary **179–82**
National Restaurant Association 408
of teachers **573–4**
problems and shortcomings 169
programme articulation 31–2
recreation **184–5**
role of museums 401
systems theory **570–1**
see also career; human resource development
education/industry relationship **175–6**
education level **176–7**

education media **182**
education method **178–9**
education policy **182–4**
effectiveness 465
  efficiency indicators 433
  of PR campaigns 476
efficiency **185–**6, 465
  of energy use 192
  performance indicators 433
  profit margin **466**
effluent control 454
Efteling theme park 412
ego-enhancement *see* motivation; prestige
Egypt **186–8**
  mass tourism results 383
Egyptian General Authority for the Promotion of Tourism 187
EIA *see* environmental impact assessment
Eiffel Tower (Paris) 402–3
Eilat (Israel) 332
Einsiedeln (Switzerland) 498
electronic promotion source **188**
Elgin Marbles 91
Ellis, Taylor *613–15*
Elson, M.J . 10
EMAS *see* Eco-Management and Audit Scheme
emergency management **188–9**
Emizet, Kisangani **539–40**
employment **189**
  and education effectiveness 177
  compensation **98**
  conditions 88–9
  economic factors 162
  gender factors 247
  in developing countries 147
  in European Union 208
  payroll cost analysis **430**
  performance appraisal **432**
  turnover **608**
  see also human resource development; job design analysis; labour relations; recruitment
employment law **189–90**
EMS *see* enviromental management system
enclave tourism **190–1**, 451, 503–4, 505
  see also exclave
encounter **191**
energy **191–2**
English Channel 80
English Tourist Board 259
Enloe, Cynthia *453*
entrepreneurial education **179–80**
entrepreneurship **192–3**, 539
environment **193–5**
  adaptation to 10
  alternative tourism 21
  and attention 33–4
  and carrying capacity **72–3**
  and climate 84
  and compliance 100
  and design and layout 144
  and economic factors 163
  code of ethics **86–7**
  consumer attitudes 106
  ecoethics **160**
  in developing countries 147
  in Greece 263
  integrated environmental management **313**
  interface with human activity 249
  outdoor recreation 185
  limits of acceptable change **360**
  pollution management **454**
  precautionary principle **458–9**
  see also conservation; ecologically sustainable tourism; ecology
environmental aesthetics **195**
environmental auditing **195–6**, 200, 454, 496
environmental codes of ethics *see* code of ethics, environmental
environmental compatibility **196**
environmental damage 52, 496, 507
environmental education **180–2**
environmental engineering **197**
environmental hazards 510
environmental impact *see* impact, environmental
environmental impact assessment **299**, 353, 439
environmental legislation **353–4**
environmental management, best practice **198**
Environmental Management for Hotels 198
environmental management system 195, **198**
environmental perception **431**
environmental planning **440–1**
environmental psychology 472
environmental quality **480–2**
environmental rehabilitation **199–200**
environmental standards *see* environmental auditing
environmental stewardship *see* environmental management, best practice

environmental valuation **200**
  contingent valuation **108**
envy, in advertising **11–12**
Epcot Centre 25
Ernesto, Carlos *29–30*
eroticism 201, **214**
Erstand, Margaret *479–80*
escape **201–2**, 393, 410, 619
  and alientation 19
  and homelessness 282
  diversionary tourism **157**
  see also inversion
escorted tour *see* guided tour
Espaces 202
Estienne, Charles 268
Estudios Turisticos **203**
Estudios y Perspectivas en Turismo **202–3**
ETC *see* European Travel Commission
ethics 465
  code of **86**
ethnic group **203–4**
  and ethnicity 206
  see also minorities
ethnic tourism 92, **204–6**
  and exoticism 214
ethnicity **206–7**
ethnocentrism **207**
ethnography **207**, 478
ethnology 24, **207–8**
ethno-tourism 20
ethos 359
euro 237
Europcar/Budget 69
European Action Plan on Tourism 209
European Commission 208
European Council 208
European Court of Audit 208
European Court of Justice 208
European Monetary System 237
European Monetary Union 237
European Parliament 208
European Travel Commission **208**
European Union **208–9**, 455
  aviation liberalisation 318, 597, 599
  environmental management systems 197
  grading systems 259
  observer status to OAS 418
  tourism policy 430
  travel freedom 220
European Year if Tourism (1990) 209
Eurostar 80
Evans, Nigel G. *584–5*, *601*
event **209–11**
  and significance of food 232
  feasts **225**
  gay and lesbian 283
  golfing 256
  sponsored **552–3**
  see also pseudo-event
event management **211**
Event Management 212
event marketing **211–2**
evolution **212**
Ewing, Gordon *122–3*
exchange rates 235
Excite 177
exclave **217**
excursion **213**
excursionist **213**
executive development **213–14**
exhibition *see* event
exit survey 566
exoticism 204
  see also fantasy
expatriate **214–15**
expectation **215**
  and ideology 294
  see also anticipation
expenses estimation **215**
expenses variances 467–8
experience **215–16**
  and anticipation 26
  learning curve **351**
  recreation **491**
  word-of-mouth accounts 385
  see also knowledge acquisition; psychology
experiential learning 182
experimental research **216–17**
  marketing 382
exploration **217–18**, 484
Exploratorium (San Francisco) 401
exploratory research 382
Expo 234
exponential smoothing models 234
export base theory 399
exposition *see* trade show
externalities **218**

FAA *see* Federal Aviation Administration
facilitation **219–20**
facilities management **220**
factor analysis **220–1**, 432
  and principal components analysis 461
factory tourism **221**, 554
  see also industrial tourism
Fagence, Michael *494*
fairs **221**, 226
  see also events
fairy tales **222**
Falassi, A. 226
familiarity *see* motivation; novelty
fantasy **222**
  and escape 202
  inversion 330
  see also exoticism
farm tourism **222–3**, 455, 514
  see also agrotourism
Farrell, Tracy *409–10*
fast food 74, **223**, 407
  history 285
Fatima shrine (Portugal) 455, 498
Fayos-Solà, Eduardo *164–5*, *186*, *208–9*, *479–80*, *549–52*, *629–31*
feasibility study 180, **223–5**
feasible tourism 97
feast **225**
Federal Aviation Administration 81, **225**
Federal Mediation and Conciliation Service (USA) 346
Federation of International Youth Travel Organisation **225**
feminism **225**
ferry **226**
Fesenmaier, Daniel R. *58–9*, *74–5*, *85–6*, *153–4*, *220–1*, *396*, *397–8*, *418*
festival **226–7**, 333
  Cannes Film Festival 228–9
  religious **227–8**
  San Francisco International Film Festival 229
  see also event
Festival Management and Event Tourism **227**
Fez (morocco) 392
Figuerola Polomo, D. Manuel *312–13*
film **228–9**
finance
  feasibility study **223–5**
  see also accounting; corporate finance; management
financial control **229**, 369
  cash flow **73–4**
financial objectives **229–30**
financial proposals 180
financial statements 6
  see also accounting
Finland **230–1**
Finlayson, J. 1
Finnish University Network for Tourism Studies **231**
First Floor Software 177
FIU Hospitality Review **231**
fixed assets 5
  asset management 32
  audit 40
FIYTO *see* Federation of International Youth Travel Organisations
Fletcher, John *75–6*, *142–3*, *160–1*, *161–3*, *298*, *310–12*, *330–1*, *398–400*, *414* , *417*
Florence (Italy) 260
flow **232**, 580
focus groups 478
FONATUR (Fondo Nacional de Fomento al Turismo) (Mexico) 389
food **232**
  fast food **223**
  feast **225**
  gastronomy **245**
  home delivery **282**
  hygiene regulations 220
  see also catering; cuisine; diet
food and beverage cost analysis **232–3**
food-borne illness 188, **233**, 273, 473, 509, 518
food service 368, 462
  commissary **91**
  concession **100**
  contract **233**
Ford, Richard 268
forecasting **233–5**, 376, 379
  Delphi technique **135–6**
  expenses estimation 215
  gravity model **261**
  recreational demand 442
  revenue 59, **508**
  trends **605**
  use of demography 139
foreign exchange 49, 162, **235–7**, 449

and currency smuggling 539–40
and developing countries 147
Italy 333
loosening of controls 255
foreign independent tours **237**
foreign investment **237**
forest recreation **237**
formulae **237**
Fourth World 204, **237–8**
see also indigenous people
France **238–40**, 268, 325, 542
Plan Comptable Général 611
rail system 487
franchising 3, **240–1**
as distribution channel 156
car rental 69
feasibility study 224
game theory 245
Frank André Gunder 149
Frechtling, Douglas C. *556–7*, *565–6*
freedom **241**, 534
perceived **241**
Freedoms of the Air 317, 319–20
Freeloader 177
frequent flyer programme **241**, 501, 600
Freudenberger, H.J. 62
Fried, Bernard *109–10*
Friel, Martin *50*, *304*, *345*, *374*
Friedman, M. 349
friends and relatives *see* visiting friends and relatives
front stage *see* back stage
Fuenmayor, Maria *71–2*, *349*, *408*
fun **242**
see also tourist as child
function catering 74
FUNTS *see* Finnish University Network for Tourism Studies
future **242**, 580
political factors 257

Gafsa (Tunisia) 606
Galapagos Islands 166, 406
Gallarza, Martina Gonzalez *428–9*, 520, *622–3*
gambling 55, **243–4**
Journal of Gambling Studies **338**
game park reserve **245**
game theory **245**
Gamradt, J. Armstrong *336*
Gandhi, Mahatma 452
Gartner, William C. *134*, *295–6*, *309–10*, *344–5*, *501*, *564–5*, *601*
gastronomy **245**
gateway **246**
GATS *see* General Agreement on Trade and Services
GATT *see* General Agreement on Tariffs and Trade
gaze 25, 92, 130, **246**
see also photography
Gdansk (Poland) 444
GDS *see* global distribution system
Gellner, E. 408
gender **246–8**
and host–guest relationship 287–8
see also feminism
General Agreement on Tariffs and Trade 604
General Agreement on Trade and Services 219–20, 604
General Mills 115
General Tourism Achievement Test 630
general tourism management model 175
Gennep, Arnold van 359, 511
geographical information system **248**
geography **248–51**
and activities 9
and community 93
of business activity 362
recreational **251**
region **494**
tourism impacts 24
use of remote sensing 500
Geomex 372
Gerbner, G. 107
Germany **251–3**
rail system 487
Getz, Donald *96*, *209–11*, *211–12*, *221*, *225*, *226–7*, *387*, *439*, *468–9*, *491*, *502*, *522–3*, *621–2*
Ghawdex 367
ghetto **253–4**
see also enclave; resort
gift **254**
Gil Saura, Irene *555–6*
Giroux, Sharon *269*
GIS *see* geographical information system
GIT *see* group inclusive tour
Global Classroom 552

global distribution system 75, 100, 188, **254**, 308, 376
global village *see* globalisation
global warming 84
globalisation 24, 25, 149–50, **254–6**, 449
  and cross-cultural management 119–20
  and developing economies 148
  of social tourism 542
  of transportation **598–9**
  see also internationalisation
Gnoth, Jurgen *619–20*
Go, Frank M. *77*, *170–2*, *304*
Goeldner, Charles R. *358*, 602
Goeltom, D.R. 170
Gold Coast (Australia) 41, 243
golf tourism **256**, 455, 503
Gomez Viveros, R.C. *78*
Goss-Turner, Steven *2–4*, *288*, *582*
government **256–9**
  incentives for human resource development 290
  role in Spanish tourism 550
  see also policy; tax
Grabowski, Peter *15*, *154*
Graburn, Nelson H.H. 81, *91–2*, *120–1*, *214*, 271, 365, *400–2*, *415–16*, 580, 609
grading system **259**
Graham, Anne *68–9*, *226*, *573*
Grand Tour 27, 81, 250, 255, **259–61**, 268, 269, 326, 333, 400, 583, 584, 590
Grant, Marcus *121*
gratuities *see* tipping
gravity model 250, *261*
  distance decay 155
  forecasting 234
Great Barrier Reef (Australia) 40, 41, 243, 427, 506
Great Wall of China 569
Greece 164, **262–4**, 512
Greek National Tourism Organisation 262, 263
Green Key (Denmark) 142
green marketing **264**
green tourism **264**, 506, 613
  see also jungle tourism
Greenacre, M.J. 116
greenhouse effect **264–5**
Greenland 29
greenspeak **265**
Greenwood, D.J. 91
Grieve, Deborah *118*, *237*, *266*
Gross, W. 128
gross profit **265**
group business market **265–6**
group inclusive tour **266**
group tour *see* guided tour
group travel market **266**
growth **266**
GTAT *see* General Tourism Achievement Test
GTMM *see* general tourism management model
Gu, Zheng 237, *510*
Guadalajara (Mexico) 389
Guadeloupe 498
Guanajuato (Mexico) 389
Guatemala 372
Guba, E. 425
guest **266–7**
  survey **566–7**
guide *see* culture broker; tour guide
guidebook **267–9**, 533, 602
  gay 283
Guide to College Programs in Hospitality and Tourism 117
guided tour **269**
Guild of Guide Lecturers 583
Gulbenkian Museum (Lisbon) 401
Gunn, Clare A. *131–2*, 589, 619

The Hague Declaration on Tourism (1989) 219
Haiti 31
hajj (pilgrimage to Mecca) 130, **270**, 343, 499
Hall, C. Michael *29*, *283*, *302–3*, *304*, *341–2*, *361*, *445–8*, *449–50*, *458*, 465, *579*, *580–1*
hallmark event *see* event; festival
Hamburg 470, 487
Hampden-Turner 127, 128
handicapped **270**, 407
  access for 28
  facilitation measures 219
handicraft **270–2**
  see also creative tourism
Handy, C. 128
Hanefors, Monica *620*
hard tourism **272**
Hardy, Dennis *280–1*
Harrison, David *90*, *147–8*, *148–50*, *530–1*
Harrison, R. 128
Hartshorne, Duncan *100*, *353–4*, *495–6*
Hashimoto, Atsuko *10–11*, *283*
Hassoun, Souad *391–2*
Haukeland, Jan Vidar *415*
Havana (Cuba) 122

Havana Convention (on international aviation) (1929) 316
Havlová, Hana *132–3*
Hawaii 325, 383
Hawkins, Donald E. *169–70*
Haywood, Michael *80–1*, *105–6*, *110–11*
hazard analysis 518–19
hazardous substance control 454
health **272–4**
  WTO standards 219
  see also disease; illness; public health
health hazards 518
Heap, J. 465
Heckscher-Ohlin model 96
hedonic pricing 200
hedonism **275**
Heidelberg (Germany) 252
heliocentrism **275**
Helleiner, Frederick M. *321*
Hely, Thriné *600*
Hendee, J.C. 50
heritage **275–7**
  agreements 496
  and nationalism 408
  legislation 353
  maps 372
  Mexico 389
  museums 401
  patrimony 429
  rural 118
  see also historical tourism
heritage sites 25
  guides 583
heritage tourism 82
  and alienation 19
  Greece 262
  see also nostalgia
Hertz car rentals 69
Hideaways International 13
hiking **277**
Hill, Richard *251–3*
hill stations **277**, 302
Hilton hotels 128, 239
Hilton Head (USA) 505
Himalayas 383, 396, 411
hinterland *see* vacation hinterland
hippie **277**
Hispaniola *see* Dominican Republic; Haiti
historical tourism **277–8**
history **278–80**
  Egypt 186
  of hospitality 284–5
Hitrec, Tomislav *281–2*
Hjalager, Anne-Mette *310*
Hobsbawm, E. 408
Hofstede, G. 119
HOGA (German Hotel and Restaurant Association) 253
Hogg, David *420*
Holden, Peter *166*, *507*
holiday *see* paid vacation; vacation
holiday camp **280–1**
holiday home *see* second home; summer cottage
holiday with pay *see* paid vacation
holistic approach **281**
Hollinshead, Keith 1, *1–2*, *90–1*, *108*, *123–5*, *151*, *153*, *191*, *207*, *232*, *292*, *411*, *420*, *425–6*, *435–6*, *471*, *501*, *509*, 543
Holloway, J. Christopher *583*, *584*
Holy Land **281–2**, 438, 537, 602
Homan, Willem J. 319, *387–8*, *598–9*
home delivery **282**
homecoming **282**
homelands 37
homelessness **282–3**
homesickness **283**
homosexuality **283**
Hong Kong **283**, 423
Hopi 31
horizontal integration 113, 255, **283–4**
hospitality **284–6**
  breaks 58
  economies of scale 164
  education 168, 174
  entrepreneurship 192
  management accounting **369–70**
  rituals 267
Hospitality 284
Hospitality and Tourism Educator 117
Hospitality Financial and Technology Professionals 370
hospitality information system **286**
Hospitality Partnership 170
Hospitality Sales and Marketing Association International **286**
host and guest 24, 95, **286–8**, 367
  and employee training 170
  and enclave tourism 191, 253–4

and marginality 372
and sex tourism 530, 531
community approach **93–4**
continuum model 109
cultural translocation 130
effect of tourism 325, 383
environmental factors 194–5
ethnic tourism 205
local tourism awareness 314
racial factors 486
semiotics 526
social relations **541**
social rituals 267
spatial interaction 552
see also guest
host culture
adaptation to tourism 11
and demonstration effect 140
generational effects 140
HOSTEUR Magazine 117
hotel **288**
airport **288**
association with railways 285
back office **48**
budget **61**
chains **76–7**, 240
classification 83
consortia 104
country house **118**
feasibility study 224
grading system **259**
Hotel and Catering International Management Association 284
Hotel Association of New York City 611
Hotel Training Foundation (UK) 285
hotel representatives 156
hotel schools 167, 174–5
housekeeping 4
property management system 469
HRD *see* human resource development
HSMAI *see* Hospitality Sales and Marketing Association
Huff model 536
Huizinga, Johan 365
human resource development **288–90**, 629
Canada 594
Malaysia 367
see also education; manpower development; training
humour **291**, 469
see also fun
Hungarian Tourism Corporation 291–2
Hungary **291–2**
Hunt, J.D. 171
Hunter, William Cannon *620–1*
Hurghada (Egypt) 194
Husbands, W.C. 10
Husserl, Edmund 435
hyperreality **292**

IAAPA *see* International Association of Amusement Parks and Attractions
IACVB *see* International Association of Convention and Visitor Bureaus
IATA *see* International Air Transport Association
ICOMOS *see* International Council on Monuments and Sites
iconic signs 526
iconography 102, **293**, 570
national character 404
Icoz, Orhan *33*, *208*, *315*
ICTA *see* Institute of Certified Travel Agents
idealism 579
identity **293–4**, 450
and race 486
ideology **294–5**
IET *see* Instituto de Estudios Turísticos
IFTTA *see* International Forum of Travel and Tourism Advocates
IGU *see* International Geographical Union
Iguaçu waterfalls (Brazil) 56
Iguaza Falls (Argentina) 29
IH&RA *see* International Hotel & Restaurant Association
illness 477
food-borne 188, **233**
in India 303
motion sickness **392**
therapeutic recreation 185
tourism-related 272–3
see also disease
image **295–6**
and hyperreality 292
and markers 373
and nationalism 408
and perception 431
iconography **293**, 526

marketing 376
metaphor **388**
of destination 145, **296**
place promotion **468–9**
postcards 456
see also advertising; symbolism
imitation *see* demonstration effect
impact 73, **296–8**, 544
and anti-tourism 27
and codes of ethics 86
and destination marketing 379
and rehabilitation 199
and resort morphology **504–4**
and social control 540
and sociocultural change 77
compatible tourism **97**
concentration ratio **100**
economic **298**
environmental **298–9**
integrated management **312–13**
factors 194
multiplier effect 310
of wars 624
on environment of Greece 263
on identities 293
precautionary principle **458–9**
qualitative research 478
Spain 550
impact assessment
and infrastructure investment 310
environmental **299**
imperialism 24, 90, **299–300**, 449, 513
import substitution **300–1**
inbound **301**, 323
incentive 157, **302–3**
see also reward system
incentive meeting planner 387
incentive travel firm 156
income 4
non-discretionary **414**
income elasticity of demand 136
index, trip **302**
indexical signs 526
India 31, **302–3**
India Tourism Development Corporation 302
Indian Airlines 303
indigenous people **303–4**, 409
Aborigines **1–2**, 30, 40 41
intellectual property 313
indirect tourism **304**
individual mass **304**
individualism **304**
Indonesia 31, **304–5**, 342, 423
induced agents 296
industrial designs 313
industrial economics 162
industrial recreation **305**
industrial tourism **305**
see also factory tourism
industrialised service 306
industry **305–7**
industry attractiveness analysis 376
inevitability **307**
inflation 49, 236
informal economy **307**
information
and attitude change 35
automated systems 46
guidebooks **267–9**
in distribution channels 157
job analysis 337
knowledge acquisition **344–5**
WTO as clearinghouse 630
see also interpretation; marketing information system; media
information centre **307**
see also visitor centre
information processing 472
see also data processing
information source **307**
information technology **308–9**
decision support system 134
diagonal integration 151
promotional use 188
property management system **469**
rapid evolution 177
telemarketing **575**
Information Technology and Tourism **309**
infrastructure **309–10**
and location 362
investment 330
Mexico 389–90
Netherlands 412
Portugal 455
rail development 487
Ingram, Bill 285
Ingram, H. *460*
Inguanez, Joe *367*

injury 273
innovation 179, 192, **310**
input–output analysis **310–12**, 399
Institute of Certified Travel Agents **312**
Institute of Management Accountants 370
Instituto de Estudios Turísticos **312–13**
insurance, WTO measures 219
integrated environmental management **313**
intellectual property 304, **313–14**
interactive learning 181, 183
interactive television 188, 244
Inter-American Travel Congress 418
interest rate parity 236
Interlaken (Switzerland) 568
internal marketing **314**
international rate of return model 331
International Academy for the Study of Tourism **315**, 444
International Air Transport Association 53, **315**, 317, 319
International Association of Amusement Parks and Attractions **315**
International Association of Antarctic Tour Operators 445
International Association of Convention and Visitors Bureaus 111, 168, **316**
international aviation
  Bermuda 1 agreement **53**
  bilateral **316–17**, 598
  liberalisation **317–19**
international aviation organisations **319**
  IATA **315** , 317, 319
international aviation rights **319–20**
International Civil Aviation Organisation 53, 315, 319, **320–1**
International Council on Monuments and Sites 406
international economics 162
International Festivals and Events Association 212
International Fisher Effect 236
International Forum of Travel and Tourism Advocates **321**
International Geographical Union 250, 251, **321**
International Geophysical Year (1957–8) 445
International Hotel and Restaurant Association **321**, 322
International Hotels Enviromental Initiative 198
International Institute for Peace Through Tourism 430
International Journal of Contemporary Hospitality Management **321–2**
International Journal of Heritage 578
International Journal of Hospitality Management **322**
International Journal of Tourism Research **322**
International Labour Organisation **322**
international marketing **380**
International Marketing Review 380
International Monetary Fund 452
international organisation **322**
International Peace Garden 55
international relations 451
International Society of Travel and Tourism Educators **323**
International Sociological Association **323**, 425
International Standards Organisation 197, 479
international tourism **323–5**, 448
  and political stability 453
  Italy's position 334
International Tourism Reports **323**
International Tourism Year (1967) 219
international understanding **325–6**
International Union for Official Tourism Organisations 629
International Union for the Conservation of Nature 181, 405, 406
internationalisation 129, **326**
Internet 46, 309, 310, **327**, 374, 385
  airline reservations 16, 601
  direct selling 156
  distance education 155, 177
  effect on business travel 138
  gambling 244
  in business suites 3
  in Canadian tourism 68
  promotional use 188, 422
  travelogue 229
  see also World Wide Web
interpretation **327–8**, 534
  and ethnography 207
  historical tourism 277–8
  of heritage 275, 506
  through museums 401
Interval International 581, 582
intervening opportunity **329**
interview **329**, 382
  in job design analysis 337
intrinsic motivation **395–6**
Inuit 31, 140, 445

inventory 5, **329**
  and entrepreneurial education 180
  audit 40
  food and beverage cost analysis **232–3**
  of attractions 36–7
  of resources 505
inventory control 32
inversion **329–30**
investment **330–1**
  and location 362
  benefit–cost analysis **51–3**
  foreign 258, 449
  in developing countries 147, 440
  in French tourism 239
  in India 303
  joint venture **338**
  return on 230
investment decision **331–2**
'invisible' current account 49
Ipanema 56
Iran **332**
Ireland **332**
Irish Tourist Board 332
Irridex model 94–5
Irritation Index 541, 632
ISO 14001 197
isochrones 2
Israel 37, 282, **332–3**, 403
ISTTE *see* International Society of Travel and Tourism Educators
Italian State Tourist Office 333, 335
Italy 268, 279, 325, **333–5**
itinerary 80, **335**, 533
  and culture brokers 126
  group inclusive tour **266**
  guidebooks 267, 268
  maps 371
ITT 114
IUCN *see* International Union for the Conservation of Nature
IUOTO *see* International Union for Official Tourism Organisations
Ixtapa (Mexico) 389

Jackowski, Antoni *438, 633*
Jackson, S.E. 62
Jackson County (Missouri) 37
Jacobsen, Jens Kristian Steen *415*
Jafari, Jafar 172, 173, 365, *585–6*, 589, 605
Jamaica **336**
Jandala, Csilla *291–2*
Jansen-Verbeke, Myriam *357–8, 532, 615–17*
Japan **336–7**, 403, 423
  art 31
  maps 371
  rail system 487
Japan airlines 17
Java (Indonesia) 129, 304
Java (software) 177
Jefferson, Thomas 452
Jeng, J.-M. *85–6, 153–4, 220–1, 396, 397–8, 418*
Jenkins, C.L. 171
Jenkins, John M. *103–4, 196–7, 470*
Jensen, Jens Friis *141–2*
Jerusalem 282, 451, 498, 512, 537
job design analysis **387–8**, 493
job description 493
job evaluation 98
job rotation 338
job specification 493
Johns, Nick *464–5, 476*
Johnson, Keith *61, 288, 392*
Johnston, Barbara Rose *486*
Johnston, C.G. *605*
Johnston, Charles S. *68, 305, 410, 417, 503–4*
joint venture **338**
Jones, Peter 39, *65, 91, 100, 216–17, 233, 282, 329,* 465, *496, 507–8, 549*
Jones, Tom *220*
Journal of Applied Recreation Research 338
Journal of Gambling Studies **338**
Journal of Hospitality and Leisure Marketing **338–9**
Journal of Hospitality and Tourism Research 117, 284, **339**
Journal of International Hospitality, Leisure and Tourism Management **339**
Journal of Leisure Research **339**, 407
Journal of Park and Recreation Administration **339**
Journal of Restaurant and Foodservice Marketing **339–40**
Journal of Sustainable Tourism **340**
Journal of Tourism Studies B340
Journal of Travel & Tourism Marketing **340**
Journal of Travel Research **340**, 602
Journal of Vacation Marketing **340–1**
journalism **341–2**
Jules-Rosette, B. 92, 526
jungle tourism **341–2**

Kaabah **343**
Kahn, Alfred 143
Kakadu national park (Australia) 41, 228
Kamp, Kathryn A. *28–9*
Kandy (Sri Lanka) 512
Kant, Immanuel 465
Kasavana, Michael L. *46–7*
Kathmandu (Nepal) 406, 411
Kattara, Hanan *131*
Kebili (Tunisia) 606
Keller, Peter *568–9*
Kemmuna 367
Kennedy, A.A. 129
Kenya 11, **343–4**
Kenya Tourist Board 344
Kenya Tourist Development Corporation 344
Kenya Utalii College 344
keying **344**, 349
Keynes, J.M. 398
Keynesian multiplier models 399
Khan, Mahmoud *117–18*, *408*
Khan, Maryam *21*, *321*
Kim, Kyung-Hwan *419*
Kim, Y.J. Edward *127*, *512–13*, *514*
King, Brian *322,* 503
Kinnaird, Vivian *225*, *246–8*
Kirk, David *48*, *518–19*, *529–30*
Kirtland (Ohio) 37
Kleiber, D.A. 355
Kleinwort Benson Securities 4
Klemm, M. 575
Kluckhohn, C. 123
Kneasfsey, M. 126
knowledge acquisition **344–5**
Korea **345**, 423, 531
Kostiainen, Auvo A. *203–4*
Kostroma (Russia) 315
Kraas-Schneider, F. 203
Kreul, Lee M. *265*, *466*
Kroeber, A.L. 123
Kuala Lumpur (Malaysia) 367
Kuhn, Thomas 425
Kurtzman, Joseph *256*, *282*, *553*
Kurz, Helmut *385–6*, *475–6*

labour relations **346–7**
Ladkins, A. 71
Lafargue, Paul 354
Lake District Declaration (1987) 406
land tenure **347**
landscape 249, **347**, 450
historical research 280
national parks 406
landscape evaluation **347–8**
Lanfant, Marie-Franhoise 124
Langer, E.J. 472
language of tourism **348–9**
Lapland 230
La Romana (Dominican Republic) 191
Lascaux (France) 29
Lash, S. 201
Las Vegas (USA) 243, 244, 457, 554, 555, 565
Latin American Confederation of Tourist Organisations **349**
Laurent, A. 119
Lavender, R. 226
law **349–50**
international 448, 452
regulatory agencies **496**
see also legislation
Lawson, F. 384
Lawton, M.P. 355
layout *see* design and layout
leadership **350–1**
executive development **213–14**
leakage **351**, 399
and globalisation 255
learning curve **351**
Leblon (Brazil) 56
Lebruto, Stephen M. *57–8*, *59–61*, *61*, *112*, *215*, *223–5*, *229–30*, *232–3*, *286*, *331–2*, *369–70*, *430*, *466–8*, *469*, *508*
Lee, Bang Sik *32*
Lee, Bong-Koo *166*, *322*, *419*
Lee, Choong-Ki *423*
Lee, Christine L.H. *338*, *491–2*
legal aspects **352–3**, 369
Legal Issues in Recreation Administration 407
legislation 352, **353**
constitutional law 257, 353
environmental **353–4**, 495–6
environmental rehabilitation 198, 199
European Union 208
intellectual property 313
labour relations 346
licensing **357**
occupational safety **417**

paid holidays 281, 425
Russian 516
tourism-related, in UK 612
Legoland (Denmark) 142
Leiper, Neil *172–4*, *305–7*, 490, *570–1*, *589*, *589–91*
leisure **354–6**
and recreation 489, 490
and religion 497
intrinsic motivation **395–6**
Journal of Leisure Research **339**, 407
Managing Leisure **371**
recreational media **386**
Visions in Leisure and Business **621**
World Recreation and Leisure Association **628**
leisure parks 281
Leisure Sciences **356**
Leisure Studies **356**
leisure tourist **356**
length of stay **356**
Leontief, Wassily W. 311
Leshan 498
Leslie, David *281*
less developed countries 147
see also Third World; Fourth World
Lett, James 365
Lew, Alan A. *35–7*, *404–5*, *414–15*, *500*
Lhasa (Tibet) 498
Li, Sarah *275–7*, *569–70*
liabilities 5
Liansheng, Zhang *78–9*
liberal arts programmes 167, 175
liberalisation *see* international aviation, liberalisation
licensing **357**, 496
life cycle 95, **357–8**
and marginality 372
and prestige 460
destination 108, 279, **358**
hospitality industry 398
product **357**, 460, **463**
lifeseeing **358**
lifestyle 151, **359**
and future of tourism 242
and life cycle 357
and public health 273
environmental implications 19
inversion 330
market segmentation 375
liminality 150, **359–60**, 443, 510
limit of acceptable change 298, **360**, 427
protected areas 470
Lincoln, J.R. 119
Lisbon (Portugal) 455
literary tourism **360**
Little, J. 134
Littrel, Mary A. *270–2*
loans 5
repayment 74
local **360–1**
local authorities 258
local organisation **361**
localisation curve **361**
Locarno-Ascona (Switzerland) 568
location **361–2**
of attractions 36
nearest neighbour analysis **410**
used in films 228
see also site analysis
location quotient **362**
location theory 162
locational analysis **362–3**
Lockwood, Andrew *284–6*, *527–9*, *530*
lodging 284
Lodging 21
London 616
London Film Commission 228
Long, Veronica *222–3*, *390–1*, *579–80*
long range **364**
longitudinal study **363–4**
Lord of Sipan (Peru) 435
loss (accounting) 5
LOT (Polish airline) 444
lotteries 243, 244
Lourdes (France) 10, 498, 565
Louvre (Paris) 401
Low, Linda *535*
loyalty **364**, 376
and personal selling 434
database marketing **134**
incentive travel 301
programmes 104
related to quality 480
see also relationship marketing
Lucas, R.C. 10
Luckman, T.L. 435
ludic behaviour 330, **365**, 443
and liminality 359

Lufthansa 599
Luhrman, Deborah *629–31*
Lumbina 498
Lumiere, Louis and Auguste 228
Lundgren, J. 2
Luxor (Egypt) 187
Luz Rufilanchas, Maria *209*
Lycos 177
Lynch, K. 10
Lyon (France) 487
Lytham and St Anne's Golf Club 256

Maasai Mara (Kenya) 343, 427
Maastricht Treaty (1992) 208, 237
Macao 244
MacCannell, Dean 19, 36, 44, 48, 201, *373*, *457–8*, 484, 511, *525–7*, 544–5, *553–5*
macroeconomics 161
Madurodam (The Hague) 412
magazine **366**
magic **366**
Maier, Gunther *631*
maintenance of facilities 220
Maitland, Robert *116*
Makkah *see* Mecca
malaria 154
Malaysia **366–7**, 423
Malaysian Tourism Promotion Board 366
Mall of America 532, 616
Malta **367**, 512
Man and the Biosphere programme 406
management **367–9**
  and control systems 109–10
  and corporate structure **114–15**
  and future of tourism 242
  and organisation design 419
  corporate strategy **112–14**
  decision support system **134–5**
  destination **146**
  disciplinary action **152–3**
  education 167, 168, 174
  entrepreneurial education **179–80**
  executive development **213–14**
  facilities **220**
  financial control **229**
  general tourism management model 175
  innovation 310
  integrated environmental **313**
  Journal of International Hospitality, Leisure and Tourism Management **339**
  Journal of Park and Recreation Administration **339**
  leadership **350–1**
  of emergencies **188**
  of events **209–10**
  of heritage 276
  of parks 427
  of pollution **454**
  of quality 479
  of services 528
  of transportation services 482
  performance indicators **432–3**
  profit variance analysis **467–8**
  property management system **469**
  ratio analysis **487–8**
  roles 350
  sales force **517**
  small business **538–9**
  strategic planning **561–2**
  Tourism Management **586**
  use of demography 139
  see also budgeting; SWOT analysis
management accounting 6, **369–70**
management contract **370–1**
Managing Leisure **371**
Mandalay (Burma) 498
Manila (Philippines) 436
Manila Declaration on World Tourism (1980) 219
Mannell, Roger C. *241*, *354–6*, *395–6*, *418*
Manning, F. 226
manpower, and location 362
  see also employment
manpower development **371**
Manrai, Ajay K. *431–2*
Manrai, Lalita A. *431–2*
Mansfeld, Yoel *332–3*, *509*, *517*, *523–5*
map **371–2**, 570
  orienteering **420**
  perceptual **431–2**
  scenic resources 347
  see also geographical information system; way-finding
marginality 150, **372–3**
marker 36, **373**
  exoticism 214
market **373–4**
market analysis **374**
market development 113

market research *see* marketing; marketing research
market segmentation **374–5**, 376
  adventure tourism 11
  advertising 13
  airlines 138
  alternative tourism **20–1**
  and activities 9
  and anti-tourism 27
  and differential pricing 460
  automatic interaction detection **45–6**
  branding **55–6**
  cluster analysis 85
  conjoint analysis 101
  direct marketing 152
  psychographics **471**
  see also target marketing
market systems, economics 161
marketing 369, **357–8**
  and accessibility 2
  Austrian tourism 43
  and consumerism 106
  and product life cycle 463
  city offices 80–1
  communication mix 92
  consortia 104
  customer satisfaction **520**
  database **134**, 152
  destination **378–9**
  destination management 146
  direct **152**
  events 210
  green **264**
  hotel chains 76
  human resource implications 289
  in Germany 253
  internal **314**
  international **380**
  Journal of Hospitality and Leisure Marketing **388–9**
  Journal of Restaurant and Foodservice Marketing **339–40**
  Journal of Travel Tourism Marketing **340**
  Journal of Vacation Marketing **340–1**
  literary connotations 360
  media coverage 341
  media planning **385–6**
  of airlines 17
  of attractions 36
  personal selling **434–5**
  product positioning **464**
  regional organisations **494–5**
  relationship **496**
  role of images 145
  sales promotion **517–18**
  strategic **560–1**
  target **572**
  telemarketing **575**
  trade show **592**
  use of demography 139
  use of discriminant analysis 154
  use of regional divisions 494
  see also franchising; overseas office; tourism office; promotion, place
marketing audit 376, **378**
  of events 209
marketing communication *see* communication mix
marketing information system 46, **379–80**
marketing mix 376, **380–1**
marketing objective 376, 380, **381**, 525
  use of direct marketing 152
marketing plan 81, 376, **381**
  role of national character 404
marketing research 13–14, **381–3**
  multidimensional scaling 397
marketing strategy 431
markets, contestable **375**
Marrakech (Morocco) 392
Marseilles (France) 487
Marx, Karl 19, 23, 81, 146, 452
Maslach, C. 62
Maaslow, A.H. 216, 520, 525
Mason, Peter *8–9*, *610*
mass tourism 81, 257, **383**
  and dependency theory 143
  and package tours 424
  and responsible tourism 507
  and souvenirs 548
  charter flights 77
  control through language 348
  critics **118**
  Cuba 122
  future of 242
  history 324, 333
  organised **384**
  placelessness **439**
  Portugal 455
  pseudo-events **471**
  sun, sand, sea and sex **564**
  see also democratisation; group inclusive tour;

hard
tourism; tour
master plan **384**
Masuria lakes (Poland) 444
Mataatua Declaration on Cultural and Intellectual Property Rights of Indigenous Peoples (1993) 314
Mateos, Lopez 389
Matthews, Harry *256–9*, *448–9*
mature market *see* senior market
Mazanec, Josef A. *374–5*, *375–8*, *387*, *412–13*, *464*, *525*
Mazatlan (Mexico) 389
McDonald's 113, 120, 128, 223, 508
McDonnell, Ian *30*
McGuckin, Eric *237–8*
McGuirk, P. *223*
McIntosh, Robert W. *88*, 183
McKercher, Bob *158–9*, *301–2*
McLellan, Gina K. *396*
MDS *see* multidimensional scaling
Mecca 498, 500
Kaabah **343**
see also hajj
media **384–5**, 428
and public relations 476
recreational **386**
media planning **385–6**
mediation in disputes 89
medical resources 188
Medina (Saudi Arabia) 498
Mediterranean 383
Medjugorje (Croatia) 498, 565
meeting planner 156, **386–7**
meetings, business **387**
Meetings Professionals International 387
mega-event *see* event; festival
Meis, Scott M. *66–8*
Meknes (Morocco) 392
Melbourne (Australia) 41
mementoes 547, 548
see also souvenir
Menia (Egypt) 187
Mercer, David *50–1*, *138–9*, *410–11*, *428*
merchandising **387**
merger **387–8**
Mesalles, Lluis *89–90*, *159*, *435*
metaphor **388**, 547–8
methodology **388**, 544
experimental research 217
Metropolitan Museum (New York) 401
Mexico **388–90**, 505, 557
Mexico City 389
Meyer-Arendt, Klaus J. *190–1*, *503–4*, *504–5*
Michelin 259, 268
microeconomics 162
middleman **390**
Middleton, Victor T.C. *378–9*, 424
migration 139, **390**
of ethnic groups 203
People on the Move **430**
Mihalič, Tanja *241*, *500–1*, *538*
Mill, John Stuart 354
Mill, R.C. 585
Milne, Simon *122–3*
mindfulness/mindlessness 472
Ministry of Culture and Islamic Guidance (Iran) 332
Ministry of Culture, Arts and Tourism (Malaysia) 366
Ministry of Economic Affairs (Australia) 43
Ministry of Tourism (Egypt) 187
Ministry of Tourism (Greece) 262
Ministry of Tourism and Aviation (Nepal) 411
minorities **390–1**
ethnic group 203
Mintzberg. H. 350
Miossec, I.M. 362
MIS *see* marketing information system
mission statement 562
Mitchell, Lisle S. *1*, *251*, *505*
mobile unit catering 74
model **391**
forecasting 234
game theory 245
modernisation theory 149
modernity **391**, 415, 544
modular programme **391**
Mohamad, Sulong *441–2*
Mombasa (Kenya) 343
Momsen, Janet Henshall *514–15*
Monaco 238
Mondor, Philip E. *593–4*
monitoring impact of tourism 297
Monte Carlo 244
Monterrey (Mexico) 389
Montreal (Canada) 66
Montreal, Declaration of (1996) 542

Montreux (Switzerland) 568
Moore, Barrington 149
Moore, K. 490
Moore, Roland S. *81–2*, *207–8*, *557–8*
Moreno, Josephine M. *270–2*
Morocco **391–2**
morphology, resort **504–5**
Morrison, Alastair M. *11*, *361*, *421–2*, *463–4*, *494–5*, 585
Morrison, Alison J. *50*, *118*
Moscardo, Gianna *9–10*, *33–4*, *87*, 95, *327–8*, *461*, *505–6*
Moscow 515
motel **392**
motion sickness **392**
motivation 24, 105, **393–5**, 544
  and achievement 7–8
  and activity 9
  and anomie 23
  and demography 139
  and escape 201
  and ethnic tourism 205
  extrinsic 395
  in entrepreneurial education 179
  intrinsic **395–6**
  of play 442
  of recreation 489–90
  push–pull factors **477**
  self-actualisation 525
  theories 433–4, 471–2
  see also incentive; values
motive manipulation **396**
motor coach tourism **396**
Mount Tremblant (Canada) 66
mountaineering **396–7**, 411
Moutinho, Luiz *64*, *364*, *378*
movements, mapping 249, 250
Mozambique 347
Mugler, Joseph *358–9*
Muller, Hansruedi *314*
multidimensional scaling **397–8**, 432
multidisciplinary education **172–4**
multinational firm 147, 324, **398**
  and globalisation 255
  and import substitution 301
  and internationalisation 326
  enclave development 191
  expatriate staff **214**
multiplier effect 255, 298, 310–11, **398–400**
Munich (Germany) 252, 616
Murphy, Laurie *145–6*, *540*, *541*
Murphy P.E. 95
Murray, John 268, 269
Murthy, Bvsan *127–9*
Museo Nacional de Anthropologia (Mexico City) 401
museum 28, 222, **400–2**
  Amsterdam 412
  and authenticity 44
  historical theme parks 578
  manufacturing industry 554
Museum of Modern Art (New York) 401
Museum of Modern Art (San Francisco) 401
music **402**, 435
  ethnic 92
myth 215, **402–3**, 534
  see also symbolism

Nabeul-Hammamet (Tunisia) 606
Nafziger, E.W. 161
Nahanni River (Canada) 66
Naim, Ahmed *391–2*
Nairobi 343
Naisbitt, John 240
Nakuru, Lake (Kenya) 343
Nansen, Fridtjof 538
Naples (Italy) 260
Nash, Dennison *6–7*, *207*, *299–300*
Nassau (Bahamas) 48
national character **404–5**
National Ecotourism Association (Australia) 41
National Labour Relations Board (USA) 346
National Liaison Group 170
National Lottery (UK) 244
national organisations, in UK 612
national park 118, **405–7**, 414, 506, 547
  conservation 103
  historical research 279–80
  USA 614
  World Heritage Site 276
National Recreation and Park Association (USA) **407**
National Restaurant Association 285, **407–8**, 608
National Tour Foundation **408**
national tourism administration 408
  see also city office; destination management organisation; information centre; local organisation; marketing, destination; overseas

office; regional organisation; tourism organisations
National Tourism Association (Russia) 516
National Tourism Committee (Hungary) 291
National Tourism Policy Study (USA) 613
national vocational qualifications 189
nationalism **408–9**
and ethnicity 206
native Americans 557–8
natives **409**
ethnic tourism 204
exoticism 214
professional **465**
Natural History Museum (London) 401
nature **409–10**
valuation 506
see also national park; park; rebirth
nature preservation 103
nature tourism 165, 237, 409, **410**, 414
Chile 77
Ecuador 166
Finland 230
in national parks 407
see also green tourism
nature trail **410**
Navajo 31
Nazca Lines (Peru) 435
nearest neighbour analysis **410**
Nebel, E.C. 350
need 520
recreational **410–11**
see also values
Negev desert 332
neo-colonialism 57, 105, **411**, 449
and appropriation 27
and authenticity 45
and enclave tourism 191
and multinational firms 301
see also colonisation
Nepal 403, **411–12**, 605
adventure tourism 11, 397
Nepal Tourism Board 411
net present value models 331
Netherlands **412**
Neulinger, J. 355
Neumyer, M.H. and E.S. 489
neurocomputing 134, 377, **412–13**
new product development **413**
New Zealand **413–14**, 423
adventure tourism 11, 414
Recreation Survey 428
New Zealand Tourism Association 414
New Zealand Tourism Board 414
New Zealand Tourist Industry Federation 86
newspaper travel sections 341, 366, 385
see also media
NGO *see* non-governmental organisation
Niagara Falls 55
Nice (France) 238, 324, 333, 505
Nile 186
NNW *see* neurocomputing
Nolan, Mary Lee *65–6*, *420–1*
non-discretionary income **414**
non-governmental organisation **414–15**
non-profit organisation **415**, 622
North America Free Trade Area 220, 604
Northern Ireland 625
see also Ireland
Northern Ireland Tourist Board 259
Norway **415**, 538
nostalgia 201, 395, 403, **415–16**, 457, 477
and alienation 19
and heritage 276, 415
see also regression
nouvelle cuisine 123
Nouvelles Frontières 239
novelty **416**
Nowliss, Michael *123*, *245*
NRA *see* National Restaurant Association
NRPA *see* National Recreation and Park Association
Nuryanti, Wiendu *304–5*
Nusa Dua (Bali) 305
Nyberg, Lars *568*

Oakes, Tim *204–6*, *206–7*, *293–4*
OAS *see* Organisation of American States
Oaxaca (Mexico) 389
occupancy rate 59, **417**
occupational safety **417**
Odgers, P. 476
OECD *see* Organisation for Economic Cooperation and Development
off-premise catering 74
off-road vehicle **417**
Office of National Tourism (Australia) 42
Office of Tourism and Sport (New Zealand) 414
O'Halloran, Robert M. *100–11*, *592*

Okechuku, C. 119
Okrant, Mark J. *139–40*
Olsen, Michael D. *560*, *562*
Olympic Airways 262
Olympic Games 211, 234, 428, 250, 553
on-the-job training 289, 593
Ong Lei Tin, Jackie *504*
operating budgets 61
operating cost **417**
opinion *see* attitude
Oppermann, Martin *264*, *277*, *396–7*, *490*, *514*, *615*
opportunity cost 161, **418**
opportunity set **418**
Opryland 614
optimal arousal **418**
Orbis Travel Agency (Poland) 444
Ordnance Survey (UK) 372
O'Reilly, Ainsley *48*
organic agents 296
organic bunching 564
organisation *see* tourism organisations; trade associations
organisation charts 109–10
organisation culture *see* culture, corporate; culture, organisational
organisation design **418**
Organisation of American States **419**
Organisation for Economic Cooperation and Development **419**
  state of the environment reports 481
Orient Express 82, 285, 597
orientalism **419–20**, 513
orienteering **420**
origin–linkage–destination models 144
Osaka (Japan) 336, 487
other **420** , 558
Ottawa Charter (1986) 473
outback 41
outbound tourism 323, **420**
outdoor recreation 185, **420–1**, 445, 492, 514
  travel cost valuation 199
Outing 341
outsourcing **421**
overseas office **421–2**
Owen, Robert 354
Ozment, John *80*, *426–7*, *482–3*, *487*, *595–6*

Paajanen, Marja *51–3*, *136–7*, *218*, *473*
Pacific Ocean 383, 389
Pacific Asia Travel Association 273, 322, 411, **423**, 495, 586
Pacific Rim **423**, 503
Pacific Tourism Review **423**
package tour 255, **423–4**, 582–3, 584, 601
  and escape 201
  choice set **79**
  Japan 336
Page, Stephen *329*
paid vacation **424–5**
  legislation 281
Palio (Italian horse race) 91, 130
pampas 29
Pan-American Highway 89, 435
Papa John's International 113
Papua New Guinea 31
parade *see* event
paradigm **425–6**
paradise **46**, 534
  and escape 202
  fantasy 222
paratransit **426–7**
Paris 260, 470, 487, 616
Paris Convention (on international aviation) (1919) 53, 316
park **427–8**, 622
  see also national park; theme park
Parrinello, G.L. *118*, 394, *632*
participation
  recreation **428**
  see also public participation
partnership **428–9**
  event management 210
passports 219, 220
pastoral care **429**
PATA *see* Pacific Asia Travel Association
Patagonia 77
patents 313
path analysis 75
patrimony **429**
Pattaya (Thailand) 577
Pau (France) 238
Pawanteh, Latiffah *386*
payback models 331
Payne, Robert J. *427–8*, *492*, *506–7*
payroll cost analysis **430**
peace 326, **430**, 624

Pearce, D. 10
Pearce, D.G. **586–8**
Pearce, Philip L. *7–8*, *19*, *26*, *34–5*, *38–9*, *50*, 95, 217, *269*, 393, 394, *408*, *459–60*, *471–3*, *525*, *578–9*, *626*
Pêcheux, M. 1
Pechlaner, H. *22*
Pedro, Aurora *164–5*, *212*
Peirce, Charles Sanders 526
Peltonen, Arvo *231*
Pender, Lesley *16*, *77*
People on the Move **430**
Pepsi Cola 115
perception *see* advertising; anticipation; freedom, perceived; image; mapping, perceptual; marketing; risk, perceived; satisfaction
perception, environmental **431**
perceptual mapping **431–2**
performance appraisal **432**
  see also performance standard
performance indicators 229, **432–3**
performance standard **433–4**
  see also performance appraisal
Perrot. S. *33*
personal documents, in historical research 279
personal selling **434–5**
Peru **435**, 624
Pfaffenberg, Carl *58*
phenomenology **435–6**
Philippines 423, **436–7**, 531, 624
photography 245, 342, **437**
  in historical research 279
  prohibition 214
Phuket (Thailand) 577
Phyloxenia project 209
physical geography 249
Pieper, J. 226
piety **437**
Pigram, John J. *40–2*, *51*, *144*, *191–2*, *193–5*, *197*, *198*, *431*, *432–3*, *454*, *480–2*, *493–4*, *619*, *625*
Pihlström, Bengt *230–1*
pilgrim 19, 24, 27, 227, 255, **438**
  and domestic tourism 159
  and fun 140
  and Grand Tour 260
  and liminality 359
  early guidebooks 268
  in India 302
  in Israel 332
  in Portugal 455
  rest houses 284
  to Lourdes 10
  see also piety
pilgrimage **438**, 498, 517, 537, 602
  youth **633**
  see also *hajj*; shrine
pilgrimage route **438**
pilgrimage site **438–9**, 498, 499, 500
Pine, Ray *214–15*, *574–5*
pitch-and-putt 256
Pitt-Rivers, J. 287
Pizam, Abraham *62–3*, *199–20*, 325, *367–9*
Pizza Hut 508
placelessness **439**
Plan Comptable Général 611
planning **439–40**, 451
  community 94, **96**, 439
  environmental 196, **440–1**
  event management **209–10**
  financial controls 229
  long range **364**
  master plan **384**
  precautionary principle **458–9**
  product **463–4**
  public participation 475
  recreation 410, **441–2**
  regional 494
  resource evaluation **505–6**
  Thailand 577
  use of remote sensing 500
  WTO involvement 630
  zoning **635**
  see also forecasting
play 151, 242, 365, **442–4**
Playa Tambor (Costa Rica) 117
playgrounds 502
playing fields 502
pleasure periphery **444**, 457
pleasure tourist **444**
Plog, S. 109, 358, 471
Pointcast 177
Poland **444–5**, 512
polar regions **445**
policy **445–8**, 450
  and place promotion **468–9**
  environmental 440
  European Union 208, 430
  France 239

Germany 253
Greece 262
Hungary 291
India 302
influencing impact of tourism 297
international aviation 316
Iran 332
Italy 335
Kenya 343–4
Malaysia 366
Mexico 389
New Zealand 414
on tourism education 574
Peru 435
Russia 516
social tourism 542
Spain 550–1
Switzerland 568–9
Tourism Management **586**
Turkey 607
United Kingdom 612
USA 613
see also World Tourism Organisation
political development **448–9**
political economy **449–50**, 458
political science **450–2**
political socialisation **452–3**
political stability 451, **453**, 509, 601
pollution 194
legislation 354
licensing 496
pricing 459
pollution management **454**
population *see* demography
pornography **454**
porters 4
portfolio model 376
Porto (Portugal) 455
Portugal **454–6**
positioning *see* product positioning
postcard **456–7**
postgraduate education 168, 171, 289
post-industrial **457**
postmodernism 24, 415, **457–8**
and authenticity 45
and differentiation 151
and escape 201
postpositivism 425
Potter, Robert B. *326*, *388*
power 450, **458**
and political socialisation 452
political economy 449
Poznan (Poland) 444
Prague (Czech Republic) 616
precautionary principle **458–9**
preservation **459**
of archaeological sites 28–9
see also conservation
prestige **459–60**, 484
and professionalism 466
pricing 377, 380, 459, **460**
and demand for air travel 137
predatory 600
transportation **600**
Prince Edward Island (Canada) 13
Prince of Wales Business Leaders Forum 198
principal components analysis 115, 220, **461**
product 380, **461–3**
buying decisions **64**
definition 463
demand **136–7**
development 113
five-layer model 462
innovation 310
life cycle **357**, 460, **463**
new product development **413**
role of destination 145
supply and demand 564
product planning 381, **463–4**
product positioning 374, 376, **464**
production
economies of scale **164–5**
efficiency **185–6**
process 462
productivity **464–5**
Professional Convention Management Association 387
professional native **465**
professionalism **465**
profit 5, 162, **466**
and cash flow 73
and the business market 3
break-even point analysis **57–8**
financial objectives 229–30
gross **265**
return on investment 112, 230, 460, **508**
see also opportunity cost
profit centre 6, **466**

profit margin **466**
profit sensitivity analysis **466–7**
profit variance analysis **467–8**
profitability, and quality 480
project-based learning 182
promotion 380, 381
  place 367, **468–9**
    and expectation 215
    brochures **58–9**
    cuisine 123
    greenspeak **263**
    Kenya 344
    Netherlands 412
    Peru 435
    role of national character 404
    through overseas offices 421–2
  puns in **469**
  see also sales promotions
promotion mix *see* marketing; marketing mix; sales promotion Promperu 435
property management system 32, **469**
prostitution 55, 148, 150, 383, 437, 341, **470**, 507, 530, 577
protected area **470**
protected landcapes/seascapes 406
prototypical tourism form **470–1**
Providencia Island (Colombia) 89
Provincial Tourist Offices (Italy) 333
Przeclawski, Krzysztof *359*, *444–5*
pseudo-event **471**, 501, 621
psychographics **471**
psychology **471–3**
  cognition **87**
  connotation **102–3**
  cognitive dissonance **87–8**
  motivation 393
  of leisure 355
  of satisfaction 520
  optimal arousal **418**
  perceived freedom **241**
  social interaction 540
  transactional analysis **594–5**
Public Authority for Conference Centre (Egypt) 187
public goods **473**
public health 273, **473–4**
public participation **474–5**
public policy 445–6
  see also policy
public–private partnership 429
public recreation *see* park; recreation
public relations 377, **475–6**
  and community attitude 94
public sector economics 162–3
publicity *see* public relations
Puerto Plata (Dominican Republic) 159
Puerto Rico 503
Puerto Vallarta (Mexico) 389
puns in promotion **469**
purchasing **476**
purchasing power parity 236
purpose **477**
push–pull factors 393, **477**
Pyrennees 238, 514

qualitative research 233–4, **478–9**, 508
quality **479–80**
  and education issues 184
  and efficiency 186
  and package tours 424
  attitudinal training 170
  classification 82–3
  competitive advantage 98–9
  environmental **480–2**
  hallmark event 211
  of theme parks 579
  service **530**
  transportation **482–3**
  WTO involvement 630
  see also standardisation
quantitative method 233, **483–4**, 508, 605
Quesnay, François 311
quest **484–5**
  see also Grand Tour
queue behaviour 217
quick service *see* fast food

Rabat (Morocco) 392
race **486**
radio **487**
Radisson Hotels 157
rail **487**, 595, 596, 597
  and hotel development 285
  class of service 82
rainforest 342
Ramada hotels 240
Ranger, T. 408

ratio analysis 112, **487–8**
rebirth **488**
reception 4, **488**
records **30**
recreation 462, **488–90**
  and carrying capacity **72–3**
  benefit–cost analysis 51–2
  community facilities 96
  demand **138–9**
  industrial **305**
  integrated resort facilities 503
  Journal of Applied Recreation Research **338**
  outdoor **420–1**
  participation **428**
  planning **441–2**
  residential **502**
  rural **514**
  social **541**
  urban **615**
  wildland **627**
  World Recreation and Leisure Association **628**
Recreation and Parks Law Reporter 407
recreation behaviour **50–1**
recreation business district **490**, 504
recreation centre **491**
recreation education **184–5**
recreation experience **491**
recreation manager **491–2**
recreation opportunity spectrum 9, 427, 442, **492**
  protected areas 470
  urban areas 616
recreation participation *see* participation, recreation
recreation planning *see* planning, recreation
recreational carrying capacity *see* carrying capacity, recreational
recreational demand *see* demand, recreational
recreational geography **251**
recreational hazards 273
recreational media **386**
recreational need **410–11**
recreational sport **563**
recreational tourist **591**
recruitment **493**
recycling 198, **493–4**
region **494**
  UK tourist boards 612
  see also trading area
regional economics 162
regional organisation **494–5**
regional synthesis 249
register 348, **495**
Register 21
regression **495**
regulation 597
  self **495–6**
  see also legal aspects
regulatory agency **496**
rehabilitation, environmental **198–9**
Reichheld, F.F. 480
Reid, Laurel J. *152*
Reid, Leslie M. *628*
relationship marketing **496**
  see also loyalty
relatives *see* visiting friends and relatives
relics **497**
  smuggling 539
religion **497–500**
  biblical sites **537**
  pastoral care **429**
  relics **497**
  sacred sites **536**
  see also piety; pilgrim
religious attraction *see* attraction, religious
religious centre **500**
religious festival **227–8**
religious schism, and ethnic groups 203
religious ritual 511
religious sites 37–8, 332, **537**
religious souvenir *see* souvenir, religious
remote sensing **500**
renewable energy sources 191–2
rent **500–1**
repeat tourist **501**, 512
representation 450, **501**
  see also virtual reality
research and development **502**
  action research **8–9**
  Annals of Tourism Research **22–3**, 24
  Asia Pacific Journal of Tourism Research **32**
  Association Internationale d'Experts Scientifiques du Tourisme 33
  automatic interaction detection **45–6**
  Cahier d'Espace **65**
  carrying capacity 72
  causal modelling **74–5**
  cluster analysis **85–6**
  comparative study **97**

content analysis **106–8**
contingent valuation **108**
cross-cultural studies **120–1**
demonstration effect **140–1**
destination choice 145
discriminant analysis **1153–4**
economic effects of tourism 161–2
ecotourism 41
Ecotourism Society **166**
education for 171
Estudios Turistocos **203**
Estudios y Perspectivas en Turisme **202–3**
ethnography 207
evolution of tourism 278–9
experimental **216–17**
factor analysis **220–1**
gender in tourism 247
guest survey **566–7**
Information Technology and Tourism **309**
Instituto de Estudios Turísticos **312–13**
International Academy for the Study of Tourism **315**
International Geographical Union **321**
International Journal of Tourism Research **322**
interview **327–8**
Journal of Applied Recreation Research **338**
Journal of Gambling Studies **338**
Journal of Hospitality and Tourism Research **339**
Journal of Leisure Research **339**, 407
Journal of Restaurant and Foodservice Marketing **339–40**
Journal of Tourism Studies **340**
Journal of Travel Research **340**, 602
knowledge acquisition 345
Leisure Sciences **356**
longitudinal studies **363–4**
methodology **338**
multidimensional scaling 397
on behaviour 50
on events 212
on guided tours 269
on handicrafts and tourism 271
on health and tourism 272
on ideology in tourism 294
on image 295–6
on impact 297
on recreation 488
on recreation participation 428
on recreational geography **251**
on resort morphology **504–5**
Pacific Tourism Review **423**
qualitative **478–9**
quantitative **483–4**
Society and Leisure/Loisir et Société **543**
sociolinguistics **543–4**
survey design 471, 565–6
systems theory **570–1**
Teoros **575**
time budget studies 581
tourism 585
Tourism Analysis **586**
tourism and conservation 104
Tourism Management **586**
tourism policy 446–7
Tourism Recreation Research **588**
translating into education 172
Travel and Tourism Research Association 340, **602**
Trinet **605–6**
US Travel Data Center 617
see also marketing research; theory
research vacations 25
reservation 4, **502**
central **75**
code sharing 86
Internet 327
property management system 469
see also computer reservation system
residential recreation **502**
resort **503**
Canada 66
class factors 81
curative 131
enclave tourism **190–1**
reducing local contacts 141
see also ghetto
resort club **503**
Resort Condominiums International 581
resort development, integrated **503–4**
resort enclave **504**
resort hotel **504**
resort morphology **504–5**
resource-based amenity **505**
resource evaluation **505–6**
resources **506–7**
responsible tourism **507**

restaurant **507–8**
  classification 83
  franchising 240
  history 284–5
  National Restaurant Association 285, **407–8**, 608
  space allocation **549**
  see also cafeteria
restaurant cars (rail) 285
Restaurant USA 408
retailing *see* shopping
return on investment 112, 230, **508**
  and pricing 460
  and seasonality 521
reunions 282
revenue expenditure 5
revenue forecasting 59, **508**
revenue variances 467
reward system 433–4
  for ethical behaviour 466
  see also incentive
rhetoric **509**, 576
Richter, Linda K. 117, *411–12*, *436–7*, *450–3*, *576–8*
Riera, Antoni *200*
Riesman, D. 354
Rif mountains 391
Riley, M. 71
Ring of Fire Museum (Osaka) 401
Rinschede, Gisbert *37–8*, *500*
Rio de Janeiro (Brazil) 56, 57
Rio Declaration (1992) 54
risk **509**
  adventure tourism 11, 397
  and achievement 7, 8
  and corporate culture 128
  climatic 84
  entrepreneurial 179, 192
  in Colombia 89
  perceived **510**, 524
  political factors 451, 509, 510
  precautionary principle **458–9**
  tourism-related 273
risk analysis **509–10**
Ritchie, J.R. Brent *166–9*, 171, 173, *177–8*, *182–4*, *266*, *391*, *573–4*
Ritchie, Robin J.B. *63*, *111*
rite of passage **510–11**
ritual 365, **511–12**, 554
ritual inversion 81
Robinson, Mike *221*, *305*
Robledo, Marco Antonio *134*, *496*, *575*
Rocky Mountains 37, 396, 514
Rojas, Mariano *388–90*
roles **512–13**
romanticism **513**
Rome 260, 268, 333, 354, 438, 498, 499, 500, 537
room night **513**
Roper, Angela *62*, *76–7*, *370–1*
Ross, Glenn F. *69–71*, *93–4*, *94–6*, *350–1*
Rousseau, J.-J. 44
route system 17–18, **513–14**
Royal and Ancient Golf Club 256
Royal Geographical Society 217
royalties 313
rules **514**
  and discipline 153
  and roles 513
  see also treaty
rural recreation **514**
rural tourism **514–15**
Russia 29, 230, **515–16**
Rutas Caleteras (Colombia) 89
Rwanda 3432
Ryan, Chris *140–1*, *150–1*, *283*, *335*, *391*, *541*, *591*, *618*

Sabbath 354
Sabre 254
sacred journey **517**
sacred site **536**
safari 437, 516, **517**
safety 509, **517**, 523, 601
  and adventure tourism 11
  occupational **417**
  standards 106
safety training 188
Sahel (Tunisia) 606
Said, Edward 420
St Moritz (Switzerland) 568
St Petersburg (Russia) 515
St Tropez (France) 505
sales force management **517**
sales promotions 377, **517–18**
salmonella 233
Salzburg (Austria) 228
Samburu (Kenya) 343

Samoyed 445
sampling 382
sampling frame 566
San Andres Island (Colombia) 89
San Damiano (Italy) 439
San Francisco International Film Festival 229
San Miguel Allende (Mexico) 389
San Sebastian de Garabandal (Spain) 439
Sancho-Perez, Amparo **150**
Sandals holiday complexes 253
sanitation 219, 509, **518–19**
Santiago (Chile) 78
Santiago de Compostella (Spain) 438, 537
Sasser, W.E. Jr 480
Sastre, Francisco *284*, *421*, *620*
satellite account **519**, 565
satellite communications 310
satisfaction 492, **519–20**
  and attribution theory 38–9
  and motivation 393, 394
  aspect of consumerism 106
  customer **520**
  monitoring 377
  quality factors 479
Say's Law 565
scale of development **520**
Scandinavia 29
Scandinavian Airlines System 599
scenery *see* landscape
scenic drive **521**
Schein, E.H. 70, 71, 127, 128
Schichman, S. 128
Schmalenbach, Eugen 610
Schmidgall, Raymond J. *487–8*
Schmidt, Marcus *115–16*, *380*
Schutz, Alfred 394
Schwartz, Zvi *245*
Science Museum (London) 401
Sea World 578
seamless service **521**, 599
seasonality 51, 83, 250, **251–2**
  and employment law 190
  and incentive travel 301
  and trained workforce 169
  environmental impact 194
  extending season 564
  Grand Tour 260
  of convention business 111
  religious tourism 498
Seaton, A.V. *27*, *106–8* , *294–5*, *341*, *366*, *391*, *487*, *575*, *578*, *604*
second home 65, 501, 515, **522–3**
  and dependency theory 143
  and domestic tourism 158
  in France 239
  increasing prices 141
secular tourism **588**
security 424, 509, **523–5**, 558–9
  in Israel 332
  in Mexico 390
  in Thailand 578
  WTO documents 219
seeing *see* gaze
Seers, Dudley 149
segmentation
  a posteriori 374, 375, **525**
  a priori 374, **525**
  see also market segmentation
Selanniemi, Tom *159*, *215*, *365*, *510–11*
self-actualisation **525**
self-challenges 553
self-discovery **525**
self-regulation **495–6**
Selim, Abdal-Rahman *219–20*
Selwyn, Tom 44, *102–3*, *116*, *142*, *201–2*, *283*, *286–8*, *402–3*, *408–9*, *444*, *457*, *511–12*, *513*, *534–5*, *562–3*, *592–3*
semiotics of tourism 36, 102, 348, 373, **525–7**, 545–6
senior tourism **527**
Serengeti National Park 427
Serra, Antoni *633*
service **527–9**
  professionalism 465
service charge 582
service delivery system **529–30**
  purchasing **476**
service quality **530**
services 305–6
  buying decisions **64**
  classification 82–3
  distribution channels **156–7**
Sessa, Alberto *335–5*
sex tourism 15, 24, 148, 222, 436, **530–1**, 577
Seychelles **531–2**
Shangri La myth 202, 403
Sharpley, Richard *19*, *27–8*, *82–3*, *108–9*, *139*,

157, *253–4*, *277*, *329–30*, *393–4*, *415*, *477*, *547*, *624*
Sheldon, Pauline J. *75*, *100*, *254*, *308–9*, *605–6*
Shengela, Natela *515–16*
Sheppard, Anthony G. *15–16*, *86*, *225*
Sheraton 114, 119 128, 240
Shinkansen 82, 487
ships 597
  see also cruise line; ferry
Shivers, J.S. 489
Shock, Patti J. *592*
shopping **532**, 616
  cross-border 55
  for handicrafts 271
  factory tourism 221
Short, David G.T. *417*, *502*, *633*
shrine **533**
  see also pilgrimage
sight **533**
sightseeing **533–4**, 624
  and markers 373
  coach tours 396
  guided tours **269**
sign **534–5**, 570
  see also semiotics of tourism; wayfinding
significant omission **535**
Simmel, G. 23, 287, 559
Simmons, David G. *63*, *73*, *237*, *405–7*, *413–14*, *488–90*
simulation, of impact of tourism 297
Singapore 366, 423, **535**
Singapore Airline 128
Singapore Tourism Promotion Board 535
Singh, Shalini *182–4*
Singh, Tej Vir *182–4*
single-entry book-keeping 5
site **535–6**
  archaeological 28
  biblical **537**
  connotation 102
  sacred **536**
site analysis **536**
Six Flags theme parks 578, 614
skiing 230, **538**, 568
skill-training programmes 168, 170
  recreational activities 185
skin cancer 275
Sklair, L. 265
slogans 313
Slovenia **538**
small business management **538–9**
Smeral, Egon *96–7*
Smith, Adam 590
Smith, Ginger *604–5*, *613–15*, *624–5*
Smith, Russell Arthur *97–8*, *366–7*, *364*
Smith, Stephen L.J. 21, *38*, *100*, *135*, *135–6*, *261*, *354–6*, *362–3*, *410*, *461–3*, *519*, *602*, *622*
Smith, Valene L. 121, 140, 217, *222–3*, 287, 372, *390–1*, *445*, *579–80*, *608–9*
smuggling 451, **539–40**
snowboarding 538
Soccer World Cup (1998) 239
Sochi (Russia) 516
social control 76, **540**
social costs 52
social interaction 93, **540**
social psychology 472
social recreation **541**
social relations 457, 458, **541**
social representation 34, 35
Social Science Research Council (USA) 6
social situation **541**
  and action research 8–9
social tourism 20, **542**
  see also democratisation
socialisation **542–3**
  political **452–3**
Society and Leisure/Loisir et Société **543**
sociocultural change *see* change, sociocultural
sociolinguistics **543–4**
sociology **544–7**
  and motivation 393–4
  International Sociological Association **323**
  of leisure 354–5
  political development **448–9**
  stranger **559–60**
  see also paradigm
Sofield, Trevor *53–4*, *275–7*, *347*, *521*, *569–70*, *624*
soft tourism **547**
solar heating 192
Sontag, Susan 437
South Africa 243, **547**
South African National Parks 547
South Pacific 31
souvenir 30, 105, 484, 532, **547–8**
  nationalistic 408
  postcards 456

relics **497**
religious **548–9**
ritual aspects 512
spa **549**
spa holidays 43, 250, 272
historical research 279
space allocation **549**
Spain 325, **549–52**
environmental degradation 383
gambling 243
spatial analysis 249
spatial interaction **552**, 563
special interest group **552**
Spice Islands 305
spill-over effects 218
sponsored event 210, **552–3**
sport, recreational **553**
sports event *see* event
sports tourism **553**
India 303
Sports Tourism International Council 553
Sri Lanka 140, 624
staged authenticity 216, 276, 545, **553–5**, 592
standardisation **555–6**
in shopping 532
see also quality
Stanger, C. 35
Stansfield, C. 358
STAR alliance 599
star rating system 83
state occasion *see* event
state of the environment reports 481
state park *see* park
statistics 279, **556–7**
discriminant analysis **153–4**
principal components analysis 115, 220, **461**
quantitative analysis **483–4**
sources 285
visitor 622
WTO 630
Statistics Canada 139
stereotype **557–8**
of national character 404
stock control 32
stocks *see* inventory
Stonequist, E.V. 472
strangeness **558–9**
stranger **559–60**
hospitality rituals 267
see also xenophobia
strategic business unit 113, **560**
strategic marketing **560–1**
strategic planning **561–2**
and conservation 103
see also SWOT analysis
strategy formulation **562**
stress
burnout **62–3**
culture shock **127**, 558, 560
see also impact
structuralism **562–3**
study tours 552
substitution **563**
Sulawesi 304
Sumatra 304
summer camps 281
summer cottage **563**
see also second home
Sun International 243
sun, sand, sea and sex **564**
Sung, Heidi H. *11*
sunlust 275, 365, **564**
Superbowl 226
Superfund programme (USA) 199
supply **564–5**
recreational opportunities 138
survey 375, **565–6**
and product planning 463
design 471
destination image 296
guest 377, **566–7**
in media planning 385
interview **327–8**, 382
of purpose of tourism **477**
of recreational participation 428
sustainable development 24, 27, 163, 474, **567–8**
Agenda 21 **14**, 54, 353, 506
and alternative tourism 21
and environmental planning 441
CTO conferences 72
energy 191
environmental best practice 198
environmental engineering 196
environmental management systems 197
greenspeak **265**
in Australia 42
in urban areas 616, 617
sustainable tourism 97, 453

and conservation 103
and environmental education 181, 182
Charter (1995) 54
environmental audits 196
ecotourism **165–6**
Journal of Sustainable Tourism **340**
Kenya 344
management 492
resource evaluation 506
see also ecologically sustainable tourism
Suzdal (Russia) 515
Sweden **568**
Switzerland 269, 324, 542, **568–9**
SWOT analysis 112, 300, 376, 463, 562, **569**
Sydney (Australia) 41
symbiosis, and conservation 103
symbolism 511–12, 526, **569–70**
of heritage 275
see also myth
System One/Amadeus 254
systems theory **570–1**, 589
and ecological economics 163
and multidisciplinary education 173–4

Tahiti 202
Taipei (Taiwan) 228
Taise 633
Taiwan 423, **572**
talk *see* discourse
Tankovič, Semso *270*
Tanzania 11, 31
target marketing 375, **572**
and image 295
Tatra mountains 444
tax **572–3**
taxi **573**, 596
Tayeb, M. 119
teacher education **573–5**
Teare, Richard *55–6, 460*
technology **574–5**
aid to exploration 217
airline distribution systems 16
and airport congestion 17
and quality of service 556
and asset management 32
and Canadian tourism 68
and demand for accommodation 3
automation **46–7**
human interface 217
virtual reality **620**
see also information technology
TEDQUAL 629–30
Tel Aviv (Israel) 332
telemarketing 152, **575**
telemedicine 273
television **575**
Telfer, David J. *153, 217–18, 325–6, 372–3, 484–5*
Teoros **575**
terrorism 118, 220, 436, 453, 509, 523, **575–6**, 624
tertiary industries 306
testimony **576**
TGI Fridays 508
TGV (train à grande vitesse) 487
Thai International Airlines 577, 599
Thailand 31, 342, 423, 531, **576–8**, 605
thanatourism **578**
theme park 25, 222, 368–9, **578–9**
and film 228
economies of scale 164
USA 614
theory **579**
therapeutic recreation 420, 549
third-party effects 218
Third World **579–80**, 610
and adaptation 11
and responsible tourism 507
anthropology 25
criticisms of tourism 449
Ecumenical Coalition on Third World Tourism **166**, 430, 452, 507, 531
infrastructure 310, 324
modelling tourism structure 144
Thomas Cook 157, 255, 280, 400, 424, 583, 584, 601, 611
Thurot, J.M. and G. 81
TIAA *see* Travel Industry Association of America
TIAC *see* Tourist Industry Association of Canada
Tijuana (Mexico) 389
time **580**
longitudinal studies **363–4**
nostalgia **415**
perception, and achievement 8
time budget **580–1**
time series data 363
time-series forecasting techniques 234
timeshare 523, **581–2**
Timeshare Council 582
Timmermans, H. 101

Timothy, Dallen J. *55, 184–5, 212, 246*
tipping **582**
Tivoli (Denmark) 142
Tocqueville, Alexis de 452
Tokyo 487
Torajans (Indonesia) 214
Toronto (Canada) 66
total quality management 186
touchscreen terminals 47
tour **582–3**
tour guide 269, 576, **583–4**
  as culture broker 126, 181
tour leader *see* guided tour; tour guide
tour operator 368. 374, 424, **584–5**
tour wholesaler **585**
tourism **585–6**
  alternative **20–1**, 24, 27, 148
  and acculturation 7
  and conservation 103
  and culture 129–31
  and liminality 359
  attitudes of world religions 498–9
  benefit–cost analysis **51–3**
  benefits **53**
  boards **54–5**
  cultural 192, **125–6**
  domestic **158–9**, 448, 451
  economic importance 49, 161
  effect of borders 55
  effects on race 486
  evolution **212**
  global warming effects 84
  government interaction 257, 258
  in centrally planned economies 75–6
  in European Union 208
  indirect **304**
  prototypical **470–1**
  rural **514–15**
  secular **588**
  social 20, **542**
  urban **615–17**
  see also carrying capacity; international tourism
Tourism 588
Tourism Academy (Egypt) 188
Tourism Analysis **586**
tourism bureaux 361
  as culture brokers 126
  destination management 146
  in China 79
  in Greece 262
  regional 495
  see also city office; convention and visitor bureau; overseas office
Tourism Code (1980) 219
Tourism Concern 124, 531
Tourism Consultants Group 588
Tourism Development Authority (Egypt) 187, 188
Tourism Economics **586**
Tourism Education Quality (TEDQUAL) 629–30
tourism/hospitality education 168
Tourism Industries (USA) 613
Tourism Management **586**
tourism organisations **586–8**
  see also trade associations
Tourism Organisation of Thailand 577
Tourism Recreation Research **588**
Tourism Society **588**
tourism system **589**
Tourisme **589**
tourist **589–91**
  activity space 10
  classification 83
  educating 183
  recreational **591**
  repeat 501**, 512**
  self-organised travel 237
  typology 10, 217, 484, **608–9**
tourist as child 222, 275, 348, **591**
tourist culture *see* culture, tourism
Tourist Industry Association of Canada 67
Tourist Research Information Network *see* Trinet
Tourist Review 33, **591**
tourist space **591–2**
tourist trap **592**
TOURISTinfo 321
Towner, John *259–61, 267–9, 278–80, 263–4, 424–5, 602–4 , 626–7*
Tozeur (Tunisia) 606
TQM *see* total quality management
trade associations
  African Travel Association **14**
  Air Transport Association of America **15**
  Airline Reporting Corporation **16**
  American Hotel and Motel Association **21**, 285
  American National Restaurant Association 508
  American Resort and Residential Development Association 582
  American Society of Travel Agents **21–2**, 453

Asia Pacific Tourism Association **32**
Association Mondiale pour la Formation Hôtelière et Touristique **33**
Association of Travel Marketing Executives **33**
Bureau International de Tourisme Social **62**
Bureau International pour le Tourisme et les Echanges de la Jeunesse **62**
Caribbean Tourism Organisation **71–2**
Clefs d'Or 100
Cruise Lines International Association **122**
European Travel Commission **208**
Federation of International Youth Travel Organisation **225**
Hospitality Sales and Marketing Association **286**
in Canada 67
in Germany 253
International Association of Amusement Parks and Attractions **316**
International Association of Convention and Visitor Bureaus 111, 168, **315–16**
International Hotel and Restaurant Association **321**, 322
National Restaurant Association (USA) 285, **407–8**, 608
New Zealand Tourism Industry Association 414
Pacific Asia Travel Association 273, 322, 411, **423**, 495, 586
Timeshare Council 582
see also tourism organisations
trade fairs 221
see also events
trade show **592**
trade unions *see* collective bargaining; labour relations
trademarks 313
trading area **592**
tradition **592–3**
conservation 124
see also culture
train *see* rail
training **593–4**
and adventure tourism 11
by hotel chains 76
cross-training **121**
curriculum design **131–2**
in-house programmes 170
national vocational qualifications 189
professional 466
tour guides 269, 584
see also career; education; human resource development
transaction processing facility 308
transactional analysis **594–5**
transactions table 311
transit **595–6**
transportation 368, 462, **596–8**
and accessibility 2
and Australian tourism 40
and the environment 194
availability 482–3
class of service **82**
globalisation **598–9**
paratransit **426–7**
quality **482–3**
route system **513–14**
speed factor 482
see also airline; cruise line; rail; taxi; transit
transport cost theory 362
transportation pricing **600**
transportation service quality *see* quality, transportation
Trans-Siberian Railway 285, 516
travel **600–1**
and accessibility 2
travel advisory **601**, 624
travel agency 368, 424, **601**
airline distribution systems 16
hotel reservations via 4
Travel and Tourism Analyst **601–2**
Travel and Tourism Research Association 340, **602**
travel cost method **602**
travel cost valuation 199
Travel Industry Association of America **602**, 617
travel literature **602–4**
Travel Marketing Decisions Magazine 33
travel narratives 268, 385, 603
travel-related hazards 273
see also hazard analysis
travel restrictions 219
see also facilitation
travel writer **604**
Traveller 341
treaty **604–5**
trekking 411, **605**
trend **605**
trend curve analysis 234
Trinet **605–6**
trip index **302**

tronc 582
TTRA *see* Travel and Tourism Research Association
Tunbridge Wells (UK) 549
Tungus 445
Tunisia **606–7**
Turismo em Analise **607**
Turizam **607**
Turkey **697–8**
Turner, Victor 359, 580
turnover **608**
TWA 128
Tyler, Duncan *470–1*
typology, tourist 10, 217, 484, **608–9**
  continuum models 109
  diversionary **157**

Ueno Park (Tokyo) 401
Uluru (Australia) 30, 40, 41
uncertainty, precautionary principle **458–9**
underdevelopment 149, **610**
undergraduate programmes 167, 171, 289
  multidisciplinary 173
  policy frameworks 174
UNESCO 455, **610**
unfamiliarity **610**
uniform system of accounting **610–11**
Uniform System of Accounts for Hotels 611
Uniglobe Travel 240
United Airlines 599
United Kingdom 325, **611–13**
  gambling 243
  nationalistic souvenirs 408–9
United Nations 452, 629
  see also World Tourism Organisation
United Nations Conference on International Travel and Tourism (1963) 219
United States 325, 423, **613–15**
  gambling 243, 244
  rail development 487
universities, tourism sub-departments 173
Upchurch, Randall S. *86*, *465–6*, *628*
urban recreation **615**
urban tourism **615–17**
  historical 277
  to factories *221*
Urry, John 201, *246*, 459, 544
US Travel Data Center 139, **617**
Utah (USA) 37
Uysal, Muzaffer S. *87–8*, *156–7*

vacation **618**
vacation hinterland **618**
vacationscape **619**
Valdes Peninsula (Argentina) 29
valuation
  of environmental factors 199
  see also contingent valuation; environmental valuation
value added **619**
value added tax 573
values **619–20**
  professionalism **465–6**
  work ethic **628**
  see also behaviour
Vancouver (Canada) 66
van den Bergh, J.C.J.M. 163
Van Hoof, Hubert B. *533*
Var, Turgut *14*, *16*, *33*, *62*, *122*, *225*, *286*, *312*, *315*, *316*, *320–1*, *323*, *407*, *419*, *607–8*, *617*, *628*, *31*
Varadero (Cuba) 122
Varanasi 498
variable labour expenses 468
variances 467
Varig 599
Vavrik, Ursula A.L. *45–6*
Vayudoot 303
Veblen, T. 102, 105, 354, 359
Vellas, François *238–40*
Vellenga, David B. *319*, *387–8*, *598–9*
Veluwe (Netherlands) 412
Venezuela **60**
Venice 260
vertical integration 157, 255, **620**
VFR *see* visiting friends and relatives
Victoria and Albert Museum (London) 401
Victoria Falls (Zimbabwe/Zambia) 55
video **620**
video checkout 469
video images 177
Vietnam 624
viewdata 16
vineyard **620**
virtual reality 309, 545, **620–1**
visa 219, 220

vision quest 484–5
Visions in Leisure and Business **621**
visiting friends and relatives **621–2**
visitor **622**
visitor bureau *see* convention and visitor bureau; tourism bureau
visitor centres 328
  and quality of service 556
Vladimir (Russia) 515
Vogt, Christine *434–5, 517–18*
voluntary sector **622–3**
volunteers 28
von Thünen, J.H. 362
Vukonič, Boris *277–8, 429, 438–9, 497–500, 517, 533, 536–7, 548–9, 588*

Wachtel, Jeffrey M. *594–5*
Wadworth, James 268
Wahab, Salah *14, 186–8, 189–90, 349–50, 352–3*
Wai Man, V.Y. 119
Walker, John R. *18–19, 50, 357*
walking tour **624**
Wall, Geoffrey *2, 10, 14–15, 72–3, 76, 79–80, 83–4, 93, 96, 97, 101,155, 165–6, 248, 248–51, 264–5, 291, 296–8, 307, 329, 347, 347–8, 356, 390, 439–90, 474–5, 521, 549, 567–8, 660–1, 618, 627*
Walle, Alf H. *478–9, 483–4*
wanderlust **624**
Wang, Ning *43–5*
Wanhill, Stephen R.C. *373–4*
war 523, **624–5**
Warsaw (Poland) 444
Washington, DC 487
Washington Weekly Newsletter 408
waste management **625**
  see also recycling
Waugh, William L. Jr *188–9, 509–10*
wayfinding **626**
wealth 4
weather 83
  see also climate
Weber, Max 81, 148, 359, 393
Webster, Kathryn *151, 233*
weekend 402, **626–7**
Weiermaier, K. *42–3, 98–9, 99–100*
Weiler, Betty *552*
Werthner, Hannes *309*
West Edmonton Mall (Canada) 532
Westlake, John 171, *189*
Whippy 582
Whistler (Canada) 66
Wieliczka (Poland) 444
wilderness **627**
  activity space 10
  Canada 66
  carrying capacity 72
  conservation 103, 124
  exoticism 214
  vision quest 484–5, 558
wilderness hiking 65
wildland recreation **627**
wildlife viewing 342, 343, 390, 410, 437, 445, 547
Wilkinson, Paul F. *254–6*
Williams, Raymond 123, 125
Wilson, David *105, 118, 520, 531–2*
winery tours 620
winter sports 538, 568
Witt, Stephen F. *233–5*
Witzel, Ineke *412*
Wöber, K. *134–5, 379–80*
Wolfenden, John A.J. *163–4*
Wong, Robert A.G. *569*
Woods, Robert H. *88–9, 98, 152–3, 337–8, 346–7, 417, 432, 433–4, 493, 608*
Woodside, Arch G. *12–14, 92–3*
work displays 554
work ethic 354, **628**
work placement in education 171
working capital 32
World Bank 159, 452, **628**
World Commission on Environment and Development 274, 473, 474, 567
World Council of Indigenous Peoples 303
World Health Organisation 272, 273, 473
World Heritage Convention (1978) 406
World Heritage Convention (1982) 275, 276
world heritage sites 276, 305, 405, 406–7
World Leisure and Recreation **628**
World Recreation and Leisure Association **628**
World Tourism Organisation 20, 86, 135, 139, 273, 301, 315, 322, 325, 360, 450, 452, 455, 462, 507, 557, 567, 586, **629–31**, 634
  facilitation instruments 219
  hotel guidelines 285
World Travel and Tourism Council **631**
World Wide Web 177, 310, 327, 345, 422, **631**

see also Internet
World's Fairs 211, 221, 226, 450
Worldspan 254
worldview *see* paradigm
WRLA *see* World Leisure and Recreation Association
WTO *see* World Tourism Organisation
WTCC *see* World Travel and Tourism Council

xenophobia **632**
Xi, Yu Ming *121*

Yamashita, Tetsuro *336–7*
Yaroslavl (Russia) 515
Yee, Jordan *228–9*
Yellowstone National Park (USA) 405
yield management 138, 245, 460, **633**
yield percentage **633**
Yosemite National Park (USA) 427
youth pilgrimage **633**
youth tourism 62, **633–4**
and music 402
Federation of International Youth Travel Organisation **225**
Youth Travel International 225
Yucatan (Mexico) 389, 505
Yugoslav conflict 624

Zagreb University 8
Zaragosa, Isabel *388–90*
Zargham, Hamid *332*
Zauhar, John *256*, *282*, *553*
zero-base budgeting (ZBB) 59–60
Zimbabwe 342
Zins, Andreas *566–7*, *572*
zoning **635**

Milton Keynes UK
Ingram Content Group UK Ltd.
UKHW051900071024
449327UK00025B/2039